T0215048

Lecture Notes in Computer Science 9224

Commenced Publication in 1973
Founding and Former Series Editors:
Gerhard Goos, Juris Hartmanis, and Jan van Leeuwen

Advanced Research in Computing and Software Science
Subline of Lecture Notes in Computer Science

More information about this series at http://www.springer.com/series/7407

Ernst W. Mayr (Ed.)

Graph-Theoretic Concepts in Computer Science

41st International Workshop, WG 2015
Garching, Germany, June 17–19, 2015
Revised Papers

 Springer

Editor
Ernst W. Mayr
TU München Institut für Informatik
Garching
Germany

ISSN 0302-9743   ISSN 1611-3349   (electronic)
Lecture Notes in Computer Science
ISBN 978-3-662-53173-0   ISBN 978-3-662-53174-7   (eBook)
DOI 10.1007/978-3-662-53174-7

Library of Congress Control Number: 2016947500

LNCS Sublibrary: SL1 – Theoretical Computer Science and General Issues

Printed on acid-free paper

This Springer imprint is published by Springer Nature
The registered company is Springer-Verlag GmbH Berlin Heidelberg

Preface

The 41st International Workshop on Graph-Theoretic Concepts in Computer Science (WG 2015) was held in Garching near Munich in Germany, during June 17–19, 2015. The WG conference series has a long tradition. Since 1975, it has taken place 23 times in Germany, four times in The Netherlands, three times in France, twice in Austria and in the Czech Republic, as well as once in each of Italy, Slovakia, Switzerland, Norway, the UK, Greece, and Israel. The WG conferences aim to connect theory and practice by demonstrating how graph-theoretic concepts can be applied to various areas of computer science and by extracting new graph problems from applications. Their goal is to present new research results and to identify and explore directions of future research. WG 2015 had 79 submissions. Each submission was carefully reviewed by three members of the Program Committee. The Program Committee then accepted 32 papers for presentation at WG 2015.

The WG 2015 Best Paper Award, sponsored by Springer, was awarded to Konstantinos Stavropoulos and his co-authors Martin Grohe, Stephan Kreutzer, Roman Rabinovich, and Sebastian Siebertz, for their paper on "Colouring and Covering Nowhere Dense Graphs." The program also included three inspiring invited talks: Daniel Paulusma (Durham University, UK) gave a talk on "Open Problems on Graph Coloring for Special Graph Classes," Shmuel Zaks (Technion, Haifa, Israel) spoke "On the Complexity of Approximation and On-line Scheduling Problems with Applications to Optical Networks," and Rolf Niedermeier (TU Berlin, Germany) presented "Parameterized Algorithmics for Graph Modification Problems: On Interactions with Heuristics."

We would like to thank all the authors of the papers submitted to WG 2015, the speakers of the 32 contributed and the three invited talks, the members of the Program Committee, and all the 130 external reviewers. Special thanks also go to the Leibniz Supercomputing Centre (LRZ) of the Bavarian Academy of Sciences and Humanities for providing space and support for the sessions and the coffee breaks, and to the members of the local Organizing Committee and the members of the Chair for Efficient Algorithms of the Technical University of Munich (TUM), whose effort made the conference run smoothly and led to such a successful event. Finally, we want to express our thanks for the financial support we received from Springer for the best paper award and from Deutsche Forschungsgemeinschaft (DFG) for (most of) the conference participants.

May 2016   Ernst W. Mayr

Organization

Program Committee

Hajo Broersma	Twente, The Netherlands
L. Sunil Chandran	Bangalore, India
Jianer Chen	College Station, USA
Victor Chepoi	Marseille, France
Pinar Heggernes	Bergen, Norway
Juraj Hromkovič	Zurich, Switzerland
Klaus Jansen	Kiel, Germany
Michael Kaufmann	Tübingen, Germany
Jan Kratochvíl	Prague, Czech Republic
Dieter Kratsch	Metz, France
Van Bang Le	Rostock, Germany
Ernst W. Mayr	München (Chair), Germany
Ross McConnell	Fort Collins, USA
Bojan Mohar	Burnaby, Canada
Haiko Müller	Leeds, UK
Christophe Paul	Montpellier, France
Dieter Rautenbach	Ulm, Germany
Dimitrios Thilikos	Montpellier/Athens, France/Greece
Oren Weimann	Haifa, Israel

Organizing Committee

Ernst Bayer	TUM, München, Germany
Christine Lissner	TUM, München, Germany
Ernst W. Mayr	TUM, München (Chair), Germany
Helga Tyroller	LRZ, Garching, Germany

The conference received ample support from the Department of Informatics of the Technical University of Munich (TUM) and also (and in particular) the Leibniz Supercomputing Centre (LRZ) of the Bavarian Academy of Sciences and Humanities.

Additional Reviewers

Patrizio Angelini	Michael Bekos
Jasine Babu	Maria Paola Bianchi
Kfir Barhum	Markus Blaeser
Julien Baste	Hans L. Bodlaender

Hans-Joachim Boeckenhauer
Edouard Bonnet
Paul Bonsma
Magnus Bordewich
Marin Bougeret
Till Bruckdorfer
Roberto Bruni
Jeremie Chalopin
Dimitris Chatzidimitriou
Rajesh Chitnis
Basile Couëtoux
Bruno Courcelle
Konrad Kazimierz Dabrowski
Guillaume Ducoffe
Vida Dujmović
Martin Dyer
Thomas Erlebach
Bertrand Estellon
Josep Fàbrega
Stefan Felsner
Qilong Feng
Jiri Fiala
Valentin Garnero
Michael Gentner
Archontia Giannopoulou
Petr Golovach
Martin Golumbic
Laurent Gourves
Sathish Govindarajan
Alexander Grigoriev
Jiong Guo
Gregory Gutin
Frederic Havet
Danny Hermelin
Bart M.P. Jansen
Vít Jelínek
Felix Joos
Tomas Kaiser
Christos Kaklamanis
Frank Kammer
Mamadou Moustapha Kanté
Telikepalli Kavitha
Ralf Klasing
Kolja Knauer
Ekkehard Köhler
Dennis Komm

Miklós Krész
Murali Krishnan
Robert Krug
Sacha Krug
Piyush Kurur
O-Joung Kwon
Arnaud Labourel
Benjamin Leveque
Bernard Lidicky
Mathieu Liedloff
Andrzej Lingas
Giuseppe Liotta
Antoine Lobstein
Anna Lubiw
Christian Lwenstein
Jens Maberg
Frederic Maffray
Spyridon Maniatis
Martin Mares
Rogers Mathew
Klaus Meer
Daniel Meister
George Mertzios
Neeldhara Misra
Pranabendu Misra
Dieter Mitsche
Lalla Mouatadid
Shay Mozes
Guyslain Naves
Jaroslav Nesetril
Jan Obdrzalek
Pascal Ochem
Daniel Paulusma
Chris Pinkau
Alexandre Pinlou
Sheung-Hung Poon
Andrzej Proskurowski
Roman Rabinovich
Deepak Rajendraprasad
Michael Rao
Jean-Florent Raymond
Igor Razgon
Clément Requilé
Ignaz Rutter
Katarzyna Rybarczyk
Ignasi Sau

Ingo Schiermeyer
Chintan Shah
Feng Shi
Somnath Sikdar
Nitin Singh
Naveen Sivadasan
R. Sritharan
Björn Steffen
Lorna Stewart
Jayme Szwarcfiter
Jan Arne Telle
Ioan Todinca
Stefan Toman
Hanjo Täubig
Torsten Ueckerdt

Ali Vakilian
Petru Valicov
Erik Jan van Leeuwen
Rob van Stee
Yann Vaxès
Sundar Vishwanathan
Jan Vondrak
Jeremias Weihmann
Samuel Wilson
Steve Wismath
Gerhard J. Woeginger
Bang Ye Wu
Mingyu Xiao
Jie You
Christian Zielke

Sponsors

We gratefully acknowledge generous support for the conference by Deutsche Forschungs-gemeinschaft (DFG), project MA 890/14–1, and by Springer-Verlag for the Best Paper Award.

More details on the conference are available at
http://wwwmayr.in.tum.de/konferenzen/WG2015/

Contents

Design and Analysis

Computational Geometry

Structural Graph Theory

Graph Drawing

Fixed Parameter Tractability

Invited Talks

Parameterized Algorithmics for Graph Modification Problems: On Interactions with Heuristics

Christian Komusiewicz, André Nichterlein, and Rolf Niedermeier[✉]

Institut für Softwaretechnik und Theoretische Informatik,
TU Berlin, Berlin, Germany
{christian.komusiewicz,andre.nichterlein,rolf.niedermeier}@tu-berlin.de

Abstract. In graph modification problems, one is given a graph G and the goal is to apply a minimum number of modification operations (such as edge deletions) to G such that the resulting graph fulfills a certain property. For example, the CLUSTER DELETION problem asks to delete as few edges as possible such that the resulting graph is a disjoint union of cliques. Graph modification problems appear in numerous applications, including the analysis of biological and social networks. Typically, graph modification problems are NP-hard, making them natural candidates for parameterized complexity studies. We discuss several fruitful interactions between the development of fixed-parameter algorithms and the design of heuristics for graph modification problems, featuring quite different aspects of mutual benefits.

1 Introduction

Graph modification problems lie in the intersection of algorithmics, graph theory, and network analysis.[1] Formally, a graph modification problem is given as follows.

GRAPH MODIFICATION
Input: A graph $G = (V, E)$, a graph property Π, and an integer $k \in \mathbb{N}$.
Question: Can G be transformed with at most k modification operations into a graph satisfying Π?

Herein, graph modification operations include edge deletions, insertions, and contractions, and vertex deletions. Classic examples for Π are "being edgeless" (this is known as VERTEX COVER when the allowed modification operation is vertex deletion) and "being a disjoint union of cliques" (this is known as CLUSTER EDITING when the allowed modification operations are edge deletion and insertion).

---

[1] Also refer to the 2014 Dagstuhl Seminar 14071 on "Graph Modification Problems" organized by Hans L. Bodlaender, Pinar Heggernes, and Daniel Lokshtanov [5]. Liu et al. [34] survey kernelization algorithms for graph modification problems.

© Springer-Verlag Berlin Heidelberg 2016
E.W. Mayr (Ed.): WG 2015, LNCS 9224, pp. 3–15, 2016.
DOI: 10.1007/978-3-662-53174-7_1

We will deal with simple and natural graph modification problems that are motivated by real-world applications. In these applications, the common way of solving these problems is via heuristics.

We present four main themes on how the interaction between parameterized algorithmics and heuristics can take place, each time illustrated by some "key" graph modification problems.

In Sect. 2, we consider a graph-based clustering problem that has been defined only implicitly by means of a greedy heuristic [26]. We describe how a natural NP-hard parameterized problem (referred to as HIGHLY CONNECTED DELETION) can be derived from this, and how this leads to further insight into the corresponding clustering approach [28].

In Sect. 3, starting with a practically successful heuristic for anonymizing social networks [33] (the corresponding NP-hard problem is known as DEGREE ANONYMITY), we describe how a closer inspection yields that either the corresponding approach provides optimal solutions in polynomial time or one can derive a polynomial-size problem kernel with respect to the parameter maximum vertex degree of the underlying graph [23]. Moreover, we briefly indicate how this led—in a feedback loop, so to speak—to improvements also for the heuristic approach [21].

In Sect. 4, we study parameterized local search—the parameter is the degree of locality [14]. Local search is a key technique in combinatorial optimization and the design of "improvement heuristics". We address both limitations and prospects of this approach. We discuss, among others, the NP-hard example problems VERTEX COVER and FEEDBACK ARC SET IN TOURNAMENTS.

In Sect. 5, we finally discuss how one may speed up parameterized algorithms by a clever use of heuristics. In particular, we discuss parameterization above lower bounds derived from linear programming relaxations [35] (here the key example is the NP-hard VERTEX COVER problem), and the idea of programming by optimization [22,27] (here the key example is the NP-hard CLUSTER EDITING problem). We draw some final conclusions in Sect. 6.

Preliminaries. We assume familiarity with fundamental concepts of graph theory, algorithms, and complexity.

A *parameterized problem* is a set of instances of the form $(\mathcal{I}, k)$, where $\mathcal{I} \in \Sigma^*$ for a finite alphabet Σ, and $k \in \mathbb{N}$ is the *parameter*. A parameterized problem Q is *fixed-parameter tractable*, shortly FPT, if there exists an algorithm that on input $(\mathcal{I}, k)$ decides whether $(\mathcal{I}, k)$ is a yes-instance of Q in time $f(k) \cdot |\mathcal{I}|^{O(1)}$, where f is a computable function independent of $|\mathcal{I}|$. A parameterized problem Q is *kernelizable* if there exists a polynomial-time algorithm that maps an instance $(\mathcal{I}, k)$ of Q to an instance $(\mathcal{I}', k')$ of Q such that $|\mathcal{I}'| \leq \lambda(k)$ for some computable function λ, $k' \leq \lambda(k)$, and $(\mathcal{I}, k)$ is a yes-instance of Q if and only if $(\mathcal{I}', k')$ is a yes-instance of Q. The instance $(\mathcal{I}', k')$ is called a *kernel* of $(\mathcal{I}, k)$.

A problem that is W[1]-hard does not admit a fixed-parameter algorithm, unless the widely believed conjecture FPT $\neq$ W[1] fails.

2 From Heuristics to Parameterized Problems

In the following, we illustrate how the consideration of heuristic algorithms may lead to the definition of new interesting graph modification problems. Our example concerns the interplay between two standard approaches for graph-based data clustering.

One approach is to formalize desired properties of clusters and then to find a clustering of the graph such that the output clusters fulfill these properties. This clustering can be obtained by modifying the input graph for example by deleting edges so that all remaining edges are only inside clusters. Starting with CLUSTER EDITING [19], there are by now numerous parameterized algorithmics studies on graph modification problems related to clustering, varying on the cluster graph definition [4,11,15,20,32], the modification operation [29], or both [3]. Most of the examples of variants of CLUSTER EDITING evolved primarily from a graph-theoretic interest.

Another approach is to define the clustering algorithmically, that is, to describe an algorithm that outputs a clustering and to analyze the properties of the clusters that are produced by the algorithm. In this section, we discuss how the consideration of a popular and natural clustering algorithm due to Hartuv and Shamir [26] leads to the definition of the graph modification problem HIGHLY CONNECTED DELETION. This is our key example for how to obtain practically motivated parameterized graph modification problems by a closer inspection of known heuristics. The study of this new problem then may yield new challenges for parameterized algorithmics and, furthermore, provide a better understanding of the strengths and weaknesses of the original heuristic algorithms. We will first discuss the original algorithm and then how we obtain the definition of HIGHLY CONNECTED DELETION from this algorithm.

Hartuv and Shamir [26] posed the following connectivity demands on each cluster: the *edge connectivity* $\lambda(G)$ of a graph G is the minimum number of edges whose deletion results in a disconnected graph, and a graph G with n vertices is called *highly connected* if $\lambda(G) > n/2$.[2] The algorithm by Hartuv and Shamir [26] partitions the vertex set of the given graph such that each partition set is highly connected by iteratively deleting the edges of a minimum cut in a connected component that is not yet highly connected. The output clusters of the algorithm are the connected components of the remaining graph which are then highly connected. The definition of being highly connected ensures several useful cluster properties, for example that at least half of the possible edges are present within each cluster and that each cluster has diameter at most two [26].

While Hartuv and Shamir's algorithm guarantees to output a partitioning into *highly connected* subgraphs, it iteratively uses a greedy step to find small edge sets to delete. As a consequence, it is not ensured that the partitioning comes along with a *minimum number of edge deletions* making the resulting graphs consist of highly connected components. This naturally leads to the edge

---

[2] An equivalent characterization is that a graph is highly connected if each vertex has degree greater than $n/2$ [9].

deletion problem HIGHLY CONNECTED DELETION where the goal is to minimize the number of edge deletions; this optimization goal is addressed only implicitly by Hartuv and Shamir's algorithm.

HIGHLY CONNECTED DELETION
Input: An undirected graph $G = (V, E)$ and an integer $k \in \mathbb{N}$.
Question: Is there an edge set $E' \subseteq E$ of size at most k such that in
$G' = (V, E \setminus E')$ all connected components are highly connected?

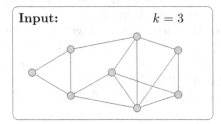
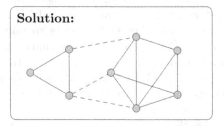

Interestingly, in the worst case the algorithm by Hartuv and Shamir [26] does not give a good approximation for the optimization version of HIGHLY CONNECTED DELETION. Consider two cliques with vertex sets $u_1, \ldots, u_n$ and $v_1, \ldots, v_n$, respectively, and the additional edges $\{u_i, v_i\}$ for $2 \leq i \leq n$. Then these additional edges form a solution set of size $n - 1$; however, Hartuv and Shamir's algorithm will (with unlucky choices of minimum cuts) transform one of the two cliques into an independent set by repeatedly cutting off one vertex, thereby deleting $n(n + 1)/2 - 1$ edges.

The following theoretical results are known for HIGHLY CONNECTED DELETION [28]. It is NP-hard even on 4-regular graphs and, provided the Exponential Time Hypothesis (ETH) [30] is correct, cannot be solved in subexponential time. On the positive side, there is a kernelization that can in polynomial time reduce an instance to one containing at most $10 \cdot k^{1.5}$ vertices, and an FPT algorithm that solves HIGHLY CONNECTED DELETION in $O(3^{4k} \cdot k^2 + n^{O(1)})$ time.

As to the relevance of parameterized algorithmics for HIGHLY CONNECTED DELETION, one has to note that the mentioned FPT algorithm is impractical. In terms of exact solutions, an integer linear programming formulation combined with data reduction rules (partially coming from the kernelization results), however, performs reasonably well [28]. Even when relaxing the goal to find exact solutions for HIGHLY CONNECTED DELETION, data reduction turned out to be beneficial in combination with heuristics (improving running time and solution quality) [28]. In a nutshell, the most practical contribution of parameterized algorithmics in this example is the development of efficient and effective data reduction rules, also helping to improve inexact solutions based on heuristics. A further benefit of considering a formally defined edge modification problem HIGHLY CONNECTED DELETION is that the objective is now independent of a heuristic method used to find it. Thus, it becomes possible to evaluate the biological quality of the objective [28].

As to potential for future research with respect to HIGHLY CONNECTED DELETION, so far other modification operations combined with the used cluster graph model are unexplored. Improvements on the known kernelization for HIGHLY CONNECTED DELETION may have direct practical impact. Moreover, a first step to make the FPT algorithm more practical could be to devise a faster FPT algorithm that relies only on branching (the current algorithm uses dynamic programming in a subroutine). Finally, besides striving for improvements with respect to the standard parameter "number of edge deletions", the investigation of other parameterizations may be interesting as well.

From a more general perspective, however, it remains to "remodel" further heuristic algorithms into natural parameterized problems.

3   Interpreting Heuristics with FPT Methods

While in the previous section we derived a natural parameterized problem (HIGHLY CONNECTED DELETION) from a simple and effective greedy heuristic, in this section we demonstrate that the tools of parameterized complexity analysis and, in particular, kernelization, may be beneficial in understanding and improving a known heuristic on the one side, and in providing a rigorous mathematical analysis on the other side. Here, we have examples in the context of graph completion problems, our key example here being the DEGREE ANONYMITY problem arising in the context of anonymizing social networks.

For many scientific disciplines, including the understanding of the spread of diseases in a globalized world or power consumption habits with impacts on energy efficiency, the availability of social network data becomes more and more important. To respect privacy issues, there is a strong demand to anonymize the associated data in a preprocessing phase [18]. If a graph contains only few vertices with some distinguished feature, then this might allow the identification (and violation of privacy) of the underlying real-world entities with that particular feature. Hence, in order to ensure pretty good privacy and anonymity behavior, every vertex should share its feature with many other vertices. In a landmark paper, Liu and Terzi [33] (also see Clarkson et al. [10] for an extended version) considered the vertex degrees as feature; see Wu et al. [41] for other features considered in the literature. Correspondingly, a graph is called ℓ-anonymous if for each vertex there are at least $\ell - 1$ other vertices of the same degree. Therein, different values of ℓ reflect different privacy demands and the natural computational task arises to perform few changes to a graph in order to make it ℓ-anonymous.

DEGREE ANONYMITY
Input: An undirected graph $G = (V, E)$ and two integers $k, \ell \in \mathbb{N}$.
Question: Is there an edge set $S \subseteq \binom{V}{2} \setminus E$ of size at most k such that $G + S$ is ℓ-anonymous?

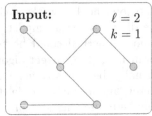 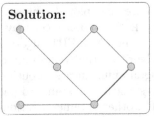

The central parameterized complexity result for DEGREE ANONYMITY is that it has a polynomial-size problem kernel when parameterized by the maximum vertex degree Δ of the input graph [23]. In other words, there is a polynomial-time algorithm that transforms any input instance into an equivalent instance with $O(\Delta^7)$ vertices. Indeed, one encounters a "win-win" situation when proving this result: Liu and Terzi's heuristic strategy [33] finds an optimal solution when the size k of a minimum solution is larger than $2\Delta^4$. Hence, either one can solve the problem in polynomial time or the solution size is "small". As a consequence, one can bound k in $O(\Delta^4)$ and, hence, a polynomial kernel for the combined parameter (Δ, k) actually is also a polynomial kernel only for Δ. While this kernelization directly implies fixed-parameter tractability for DEGREE ANONYMITY parameterized by Δ, there is also an FPT algorithm running in $O(\Delta^{O(\Delta^4)} + (k\ell + \Delta)\Delta kn)$ time.

The ideas behind the "win-win" situation generalize to further graph completion problems where the task is to insert edges so that the degree sequence of the resulting graph fulfills some prescribed property Π [17]. Furthermore, an experimental evaluation of the usefulness of the theoretical results on the "win-win" situation delivered encouraging results even beyond the theoretical guarantees, that is, when $k < 2\Delta^4$ [21,40]. This led to an enhancement of the heuristic due to Liu and Terzi [33] which substantially improves on the previously known theoretical and empirical running times. As for HIGHLY CONNECTED DELETION, previously known heuristic solutions could be substantially improved in terms of solution quality.

Finally, we mention in passing that making a graph ℓ-anonymous was studied from a parameterized point of view using also several other graph modification operations [2,7,25]. All these studies are of purely theoretical nature and there are only little positive algorithmic results; links with heuristic algorithm design are missing.

From a general perspective, the quest arising from the findings for DEGREE ANONYMITY is to provide further examples where parameterized complexity analysis sheds new light on known heuristics, both theoretically and practically. A good starting point might be the heuristic of Lu et al. [36] which clusters the vertices and then anonymizes each cluster. Here, the question is whether such a practical link between anonymization and clustering could be complemented with theoretical results. Obviously, these studies should not be limited to problems arising in anonymization but to graph modification problems from different application areas.

4   Improving Heuristic Solutions with FPT Algorithms

Local search is a generic algorithmic paradigm that yields good heuristics for many optimization problems. The idea is to start with any feasible solution and then search for a better one in the local neighborhood of this solution. This search is continued until a locally optimal solution is found. For graph modification problems, a feasible solution S is any set of modification operations that transforms the input graph into one that satisfies the graph property Π. The local neighborhood of S is usually defined as the sets of modification operations that can be obtained by adding and removing at most k vertices from S. This type of neighborhood is called k-exchange neighborhood.

An obvious approach to obtain more powerful local search algorithms is to reduce the running time needed for searching the local neighborhood. This could enable a local search algorithm to examine larger neighborhoods and reduce the likelihood to remain in a locally optimal but globally suboptimal solution. Usually, the size of the k-exchange neighborhood in an n-vertex graph is upper-bounded by $n^{f(k)}$ for some function f. In parameterized algorithmics, a natural question is whether it is necessary to consider all elements of this neighborhood or whether the neighborhood can be searched faster, that is, in $f(k) \cdot n^{O(1)}$ time.

For many vertex deletion problems this is not the case [14]. For example, in the local search variant of VERTEX COVER, one is given a *vertex cover S*, that is, a vertex set S such that deleting S from a graph G results in an independent set. The task is to find a smaller vertex cover S' by adding and removing at most k vertices from S.

LOCAL SEARCH VERTEX COVER
Input: An undirected graph $G = (V, E)$, a vertex cover S of G, and an integer $k \in \mathbb{N}$.
Question: Is there a vertex cover $S' \subseteq V$ such that $|S'| < |S|$ and $|(S \setminus S') \cup (S' \setminus S)| \leq k$?

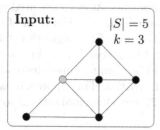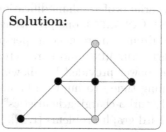

Unfortunately, unless W[1] = FPT, there is no FPT algorithm for LOCAL SEARCH VERTEX COVER parameterized by k [14]. Positive results were obtained for special cases. For example, LOCAL SEARCH VERTEX COVER and many other local search variants of vertex deletion problems are fixed-parameter tractable on planar graphs [14]. These results, however, are based on the technique of locally bounded treewidth. As a consequence, the resulting algorithms might not be useful in practice.

Positive results were obtained for FEEDBACK ARC SET IN TOURNAMENTS which is the problem of transforming a tournament, that is, a directed graph in which every pair of vertices is connected by exactly one of the two possible arcs, into an acyclic graph by a minimum number of arc deletions. Here, the local search problem is fixed-parameter tractable. More precisely, given a set S of arc deletions that makes a given tournament acyclic, it can be decided in $2^{O(\sqrt{k}\log k)} \cdot n^{O(1)}$ time whether there is a set S' that can be obtained from S by adding and removing at most k arcs [16].

This positive result seems to be rooted in the combinatorially restricted nature of tournaments and *not* in the fact that FEEDBACK ARC SET IN TOURNAMENTS is an arc modification problem: The local search variant of the similarly simple CLUSTER EDITING problem is not fixed-parameter tractable unless W[1] = FPT [13].

Summarizing, the natural idea of parameterized local search faces two major obstacles. The first obstacle is that, as discussed above, many local search problems are probably not fixed-parameter tractable. The second obstacle is that, so far, none of the parameterized local search algorithms for graph modification problems have been shown to be useful in practice. One encouraging result was obtained for INCREMENTAL LIST COLORING [24]. Here, the input is a graph with a list-coloring that colors all graph vertices except one. The task is to obtain a list-coloring that also colors v and disagrees with the old list-coloring on at most c vertices. Thus, the new solution is searched within the neighborhood of the old solution. This problem can be solved in $k^c \cdot n^{O(1)}$ time where k is the maximum size of any color list in the input. The crucial observation is that this local search-like approach can be embedded in a coloring heuristic that outperforms the standard greedy coloring algorithm in terms of the coloring number. Since INCREMENTAL LIST COLORING is W[1]-hard with respect to the parameter c, the key to success seems to be the consideration of the combined parameter (k, c).

A goal for future research should thus be to obtain similar success stories for local search variants of graph modification problems. As demonstrated by INCREMENTAL LIST COLORING, one promising route is the consideration of combined parameters. From a more general perspective, the FPT algorithm for INCREMENTAL LIST COLORING and parameterized local search have in common that they use the power provided by allowing FPT running time—instead of polynomial running time—to improve known heuristics. This approach, which has been coined "turbo-charging heuristics" [12], has close connections to dynamic versions of hard graph problems [1,12].

5 Heuristic Tuning of Parameterized Algorithms

Heuristics are often used to boost the performance of exact algorithms in practice. A prominent example here is the branch-and-bound concept where heuristic lower and upper bounds restrict the search space for search-tree algorithms [38].

Better heuristic bounds give a smaller search space and thus faster exact algorithms. When analyzed in the classic complexity framework, the theoretical running time improvements due to the heuristic bounds are (if at all) marginal compared to the speed-ups observed in practice. Here, parameterized algorithmics can be used to give some theoretical explanation for experimental observations by using the above-guarantee parameterization [37]. As the name suggests, the parameter is the difference between the size of an optimal solution and a given lower bound. Fixed-parameter tractability with respect to the above-guarantee parameter then shows that the problem of finding a solution close to the lower bound is "easy". Thus, if the corresponding lower bound is close to the optimum, then the corresponding algorithm using this lower bound is fast—in practice *and* in theory.

An example for above-lower bound parameterization is VERTEX COVER. One lower bound on the size of a VERTEX COVER is the value ℓ of a linear programming (LP) relaxation. The well-known LP relaxation is as follows:

$$\text{Minimize} \qquad \sum_{v \in V} x_v$$

$$\text{subject to} \qquad x_u + x_v \geq 1, \qquad \forall \{u, v\} \in E$$
$$x_v \geq 0, \qquad \forall v \in V.$$

It is known that in an optimal solution for the LP relaxation each variable has value 0, 1/2, or 1 [39].

VERTEX COVER ABOVE LP

Input: An undirected graph $G = (V, E)$, an integer $k \in \mathbb{N}$, and a rational number $\ell \in \mathbb{Q}$ denoting the value of the LP relaxation.

Question: Is there a vertex subset $S \subseteq V$ of size at most k such that $G[V \setminus S]$ is edgeless?

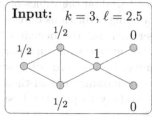

Input: $k = 3$, $\ell = 2.5$

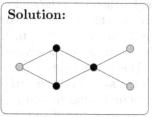

Solution:

Lokshtanov et al. [35] presented an algorithm solving VERTEX COVER ABOVE LP in $2.32^{k-\ell} \cdot n^{O(1)}$ time. On a high level, this algorithm starts with the lower bound and uses, after some preprocessing, a standard search-tree algorithm. Thus, a good lower bound allows not only in practice, but also in theory for an efficient algorithm solving VERTEX COVER. Moreover, the fixed-parameter tractability result now may help explaining why heuristics can successfully exploit the lower bound provided by the LP relaxation.

Another example for heuristic tuning of algorithms is programming by optimization [27]. This is a helpful and powerful tool for developing fast implementations. Here, the basic idea is that the implementation leaves open several design choices for different parts of the algorithm—these are settled later when train-ing the algorithm with real-world instances. Then, for the final configuration of the implementation, let a program choose from the alternatives in such a way that the performance is optimized on a representative set of instances. Here, the automated optimizer can give an answer to the following questions:

- Given several alternative implementations for one subproblem (for example different sorting algorithms or different lower bounds), which one should be chosen?
- Should a certain data reduction rule be applied?
- What are the "best" values for certain "magic" or "hidden" constants? For example, should a data reduction rule be applied in every second level of the search tree or every fourth level?

The programming by optimization approach has led to a state-of-the-art solver for CLUSTER EDITING [22]. This solver combines one sophisticated data reduc-tion rule and a branch-and-bound algorithm. The solver outperforms previ-ous algorithms which are based on integer linear programming (ILP) and pure branch-and-bound. Thus, with the help of programming by optimization, imple-mentations of parameterized algorithms may successfully compete with ILP-based algorithms.

On a high level, programming by optimization can be seen as a heuristic counterpart to parameterized algorithmics: Parameterized algorithmics provides theoretical bounds on the running time of algorithms and the effectiveness of data reduction rules. These bounds depend on the parameter. Thus, to solve a problem for a specific type of data, one should measure different parameters and choose, based on this measurement, the most promising data reduction rules and algorithms. With programming by optimization, this choice is made automati-cally, based on the performance of the algorithm on a given representative set of test instances. Furthermore, the choice is not based on the values of parameters but directly on the efficiency of the corresponding algorithms on the test data.

A goal for future research is to further increase the benefit obtained by combining the strengths of programming by optimization and parameterized algorithmics. This could be done, for example, by first providing several FPT algorithms for the same problem with different parameters and then using pro-gramming by optimization to find a good strategy to pick the best algorithm depending on the structure of an input instance.

6   Conclusions

As Karp [31] pointed out, one of the most pressing challenges in theoretical com-puter science is to contribute to a better understanding why many heuristics work so well in practice. In particular, a formal footing of the construction of heuristic algorithms is considered highly desirable. This task is also closely connected to (hidden) structure detection in real-world input instances. We discussed several routes to a beneficial interaction between heuristic and parameterized algorithm design

To date, a clear majority of research results in parameterized algorithmics is of purely theoretical nature. A natural way to increase the practical impact of parameterized algorithmics is to seek fruitful interactions with the field of heuristic algorithm design. We believe that particularly graph (modification) problems may be a forerunner in offering numerous fruitful research opportunities in this direction.

So far the strongest impact achieved by parameterized algorithmics on practical computing and heuristics is due to kernelization, and polynomial-time data reduction techniques in general. Notably, often data reduction rules seemingly not strong enough to provide kernelization results may still have strong practical impact. Moreover, a general route for future research is to develop heuristic algorithms in parallel with performing a parameterized complexity analysis (particularly, in terms of kernelization). As results for graph modification problems in this direction demonstrate, there are good prospects to win something in both worlds.

Finally, in this paper we focused on NP-hard graph modification problems for illustrative examples. It goes without saying that our general remarks and observations are not limited to graph modification problems only but clearly extend to further graph problems and fields beyond, e.g. string algorithms [8] or computational social choice [6].

Acknowledgment. We are grateful to Till Fluschnik and Vincent Froese for feedback to our manuscript.

References

1. Abu-Khzam, F.N., Egan, J., Fellows, M.R., Rosamond, F.A., Shaw, P.: On the parameterized complexity of dynamic problems. Theoret. Comput. Sci. **607**, 426–434 (2015)
2. Bazgan, C., Bredereck, R., Hartung, S., Nichterlein, A., Woeginger, G.J.: Finding large degree-anonymous subgraphs is hard. Theoret. Comput. Sci. **622**, 90–110 (2016)
3. van Bevern, R., Moser, H., Niedermeier, R.: Approximation and tidying - a problem kernel for s-plex cluster vertex deletion. Algorithmica **62**(3–4), 930–950 (2012)
4. Bodlaender, H.L., Fellows, M.R., Heggernes, P., Mancini, F., Papadopoulos, C., Rosamond, F.A.: Clustering with partial information. Theoret. Comput. Sci. **411**(7–9), 1202–1211 (2010)
5. Bodlaender, H.L., Heggernes, P., Lokshtanov, D.: Graph modification problems (Dagstuhl seminar 14071). Dagstuhl Rep. **4**(2), 38–59 (2014)
6. Bredereck, R., Chen, J., Faliszewski, P., Guo, J., Niedermeier, R., Woeginger, G.J.: Parameterized algorithmics for computational social choice: nine research challenges. Tsinghua Sci. Technol. **19**(4), 358–373 (2014)
7. Bredereck, R., Froese, V., Hartung, S., Nichterlein, A., Niedermeier, R., Talmon, N.: The complexity of degree anonymization by vertex addition. Theoret. Comput. Sci. **607**, 16–34 (2015)

8. Bulteau, L., Hüffner, F., Komusiewicz, C., Niedermeier, R.: Multivariate algorithmics for NP-hard string problems. Bull. EATCS **114**, 31–73 (2014)
9. Chartrand, G.: A graph-theoretic approach to a communications problem. SIAM J. Appl. Math. **14**(4), 778–781 (1966)
10. Clarkson, K.L., Liu, K., Terzi, E.: Towards identity anonymization in social networks. In: Yu, P.S., Han, J., Faloutsos, C. (eds.) Link Mining: Models, Algorithms, and Applications, pp. 359–385. Springer, New York (2010)
11. Damaschke, P., Mogren, O.: Editing simple graphs. J. Graph Algorithms Appl. **18**(4), 557–576 (2014)
12. Downey, R.G., Egan, J., Fellows, M.R., Rosamond, F.A., Shaw, P.: Dynamic dominating set and turbo-charging greedy heuristics. Tsinghua Sci. Technol. **19**(4), 329–337 (2014)
13. Dörnfelder, M., Guo, J., Komusiewicz, C., Weller, M.: On the parameterized complexity of consensus clustering. Theoret. Comput. Sci. **542**, 71–82 (2014)
14. Fellows, M.R., Fomin, F.V., Lokshtanov, D., Rosamond, F.A., Saurabh, S., Villanger, Y.: Local search: is brute-force avoidable? J. Comput. Syst. Sci. **78**(3), 707–719 (2012)
15. Fellows, M.R., Guo, J., Komusiewicz, C., Niedermeier, R., Uhlmann, J.: Graph-based data clustering with overlaps. Discrete Optim. **8**(1), 2–17 (2011)
16. Fomin, F.V., Lokshtanov, D., Raman, V., Saurabh, S.: Fast local search algorithm for weighted feedback arc set in tournaments. In: Proceedings of the Twenty-Fourth AAAI Conference on Artificial Intelligence, (AAAI 2010), pp. 65–70 (2010)
17. Froese, V., Nichterlein, A., Niedermeier, R.: Win-win kernelization for degree sequence completion problems. J. Comput. Syst. Sci. **82**(6), 1100–1111 (2016)
18. Fung, B.C.M., Wang, K., Chen, R., Yu, P.S.: Privacy-preserving data publishing: a survey of recent developments. ACM Comput. Surv. **42**(4), 14:1–14:53 (2010)
19. Gramm, J., Guo, J., Hüffner, F., Niedermeier, R.: Graph-modeled data clustering: exact algorithms for clique generation. Theor. Comput. Syst. **38**(4), 373–392 (2005)
20. Guo, J., Komusiewicz, C., Niedermeier, R., Uhlmann, J.: A more relaxed model for graph-based data clustering: s-plex cluster editing. SIAM J. Discrete Math. **24**(4), 1662–1683 (2010)
21. Hartung, S., Hoffmann, C., Nichterlein, A.: Improved upper and lower bound heuristics for degree anonymization in social networks. In: Gudmundsson, J., Katajainen, J. (eds.) SEA 2014. LNCS, vol. 8504, pp. 376–387. Springer, Heidelberg (2014)
22. Hartung, S., Hoos, H.H.: Programming by optimisation meets parameterised algorithmics: a case study for cluster editing. In: Jourdan, L., Dhaenens, C., Marmion, M.-E. (eds.) LION 9 2015. LNCS, vol. 8994, pp. 43–58. Springer, Heidelberg (2015)
23. Hartung, S., Nichterlein, A., Niedermeier, R., Suchý, O.: A refined complexity analysis of degree anonymization in graphs. Inf. Comput. **243**, 249–262 (2015)
24. Hartung, S., Niedermeier, R.: Incremental list coloring of graphs, parameterized by conservation. Theoret. Comput. Sci. **494**, 86–98 (2013)
25. Hartung, S., Talmon, N.: The complexity of degree anonymization by graph contractions. In: Jain, R., Jain, S., Stephan, F. (eds.) TAMC 2015. LNCS, vol. 9076, pp. 260–271. Springer, Heidelberg (2015)
26. Hartuv, E., Shamir, R.: A clustering algorithm based on graph connectivity. Inf. Process. Lett. **76**(4–6), 175–181 (2000)
27. Hoos, H.H.: Programming by optimization. Commun. ACM **55**(2), 70–80 (2012)
28. Hüffner, F., Komusiewicz, C., Liebtrau, A., Niedermeier, R.: Partitioning biological networks into highly connected clusters with maximum edge coverage. IEEE/ACM Trans. Comput. Biol. Bioinf. **11**(3), 455–467 (2014)

29. Hüffner, F., Komusiewicz, C., Moser, H., Niedermeier, R.: Fixed-parameter algorithms for cluster vertex deletion. Theor. Comput. Syst. **47**(1), 196–217 (2010)
30. Impagliazzo, R., Paturi, R., Zane, F.: Which problems have strongly exponential complexity? J. Comput. Syst. Sci. **63**(4), 512–530 (2001)
31. Karp, R.M.: Heuristic algorithms in computational molecular biology. J. Comput. Syst. Sci. **77**(1), 122–128 (2011)
32. Liu, H., Zhang, P., Zhu, D.: On editing graphs into 2-club clusters. In: Snoeyink, J., Lu, P., Su, K., Wang, L. (eds.) AAIM 2012 and FAW 2012. LNCS, vol. 7285, pp. 235–246. Springer, Heidelberg (2012)
33. Liu, K., Terzi, E.: Towards identity anonymization on graphs. In: Proceedings of the ACM SIGMOD International Conference on Management of Data (SIGMOD 2008), pp. 93–106 (2008)
34. Liu, Y., Wang, J., Guo, J.: An overview of kernelization algorithms for graph modification problems. Tsinghua Sci. Technol. **19**(4), 346–357 (2014)
35. Lokshtanov, D., Narayanaswamy, N.S., Raman, V., Ramanujan, M.S., Saurabh, S.: Faster parameterized algorithms using linear programming. ACM Trans. Algorithms **11**(2), 15:1–15:31 (2014)
36. Lu, X., Song, Y., Bressan, S.: Fast identity anonymization on graphs. In: Liddle, S.W., Schewe, K.-D., Tjoa, A.M., Zhou, X. (eds.) DEXA 2012, Part I. LNCS, vol. 7446, pp. 281–295. Springer, Heidelberg (2012)
37. Mahajan, M., Raman, V.: Parameterizing above guaranteed values: MaxSat and MaxCut. J. Algorithms **31**(2), 335–354 (1999)
38. Mehlhorn, K., Sanders, P.: Algorithms and Data Structures: The Basic Toolbox. Springer, Heidelberg (2008)
39. Nemhauser, G.L., Trotter, L.E.: Properties of vertex packing and independence system polyhedra. Math. Program. **6**(1), 48–61 (1974)
40. Nichterlein, A.: Degree-Constrained Editing of Small-Degree Graphs. Ph.D. thesis, TU Berlin (2015)
41. Wu, X., Ying, X., Liu, K., Chen, L.: A survey of privacy-preservation of graphs and social networks. In: Aggarwal, C.C., Wang, H. (eds.) Managing and Mining Graph Data. Advances in Database Systems, vol. 40, pp. 421–453. Springer, Heidelberg (2010)

Open Problems on Graph Coloring
for Special Graph Classes

Daniël Paulusma[(✉)]

School of Engineering and Computing Sciences, Durham University
Science Laboratories, South Road, Durham DH1 3LE, UK
daniel.paulusma@durham.ac.uk

Abstract. For a given graph G and integer k, the COLORING problem is that of testing whether G has a k-coloring, that is, whether there exists a vertex mapping $c : V \rightarrow \{1, 2, \ldots\}$ such that $c(u) \neq c(v)$ for every edge $uv \in E$. We survey known results on the computational complexity of COLORING for graph classes that are hereditary or for which some graph parameter is bounded. We also consider coloring variants, such as precoloring extensions and list colorings and give some open problems in the area of on-line coloring.

1   Introduction

Graph coloring is a central topic in Computer Science due to a high number of theoretical and practical applications. Within both structural and algorithmic graph theory, many graph coloring variants and generalizations have been studied. Besides the well-known text-book of Toft and Jensen [65], several survey papers appeared over the years. For instance, the survey of Tuza [70] considered the graph coloring problem and variants of it, in which local restrictions are imposed on the coloring (e.g. precoloring extensions and list colorings) whereas the survey of Randerath and Schiermeyer [61] considered structural and complexity aspects of graph colorings for hereditary graph classes.

As graph coloring and many of its variants are computationally hard on general graphs, it is natural to restrict the input graph to some special graph class. This topic has been extensively studied in the literature. A recent survey of Golovach et al. [27] updated several parts of the two aforementioned survey papers [61,70] and the survey of Chudnovsky [13] by primarily focussing on computational complexity aspects of graph coloring for graph classes characterized by one or two forbidden induced subgraphs. As noted by Golovach et al. [27], the task to collect complexity results for graph coloring restricted to other graph classes might be beyond the scope of a single paper. Our aim is therefore to discuss a number of such results and open problems not mentioned in [27]. In particular we will identify a number of gaps in existing complexity results.

The survey is organized as follows. In Sect. 2 we state some terminology. Then, in Sect. 3, we consider the (classical) complexity of graph coloring, precoloring extension and list colorings for a number of graph classes characterized

The author was supported by EPSRC (EP/K025090/1).

E.W. Mayr (Ed.): WG 2015, LNCS 9224, pp. 16–30, 2016.
DOI: 10.1007/978-3-662-53174-7_2

by more than two forbidden induced subgraphs and also for some graph classes for which some graph parameter is bounded. We discuss parameterized coloring problems in Sect. 4 and on-line coloring problems in Sect. 5. We briefly consider graph homomorphisms in Sect. 6.

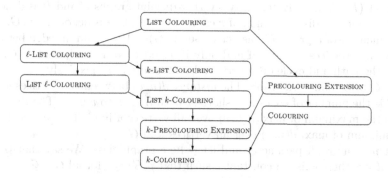

Fig. 1. Relationships between COLORING and its variants as shown in [27]. An arrow from one problem to another indicates that the latter is a special case of the former; k and ℓ are any two integers for which $\ell \geq k$.

2 Preliminaries

A *coloring* of a graph $G = (V, E)$ is a vertex mapping $c : V \to \{1, 2, \ldots\}$ with the additional condition that $c(u) \neq c(v)$ whenever $uv \in E$. We call $c(u)$ the *color* of u. If $1 \leq c(u) \leq k$ for all $u \in V$ then c is also called a *k-coloring* of G. We say that G is *k-colorable* if a k-coloring of G exists. The *chromatic number* $\chi(G)$ of G is the smallest integer k for which G is k-colorable. The COLORING problem is that of deciding whether a graph G is k-colorable for some given integer k. If k is fixed (that is, not part of the input) we obtain the k-COLORING problem.

A *k-precoloring* of a graph $G = (V, E)$ is a mapping $c_W : W \to \{1, 2, \ldots k\}$ for some subset $W \subseteq V$. We say that a k-coloring c of G is an extension of a k-precoloring c_W of G if $c(v) = c_W(v)$ for each $v \in W$. For a given graph G, a positive integer k and a k-precoloring c_W of G, the PRECOLORING EXTENSION problem asks whether c_W can be extended to a k-coloring of G. If k is fixed we denote this problem as the k-PRECOLORING EXTENSION problem.

A *list assignment* of a graph $G = (V, E)$ is a function L with domain V such that for each vertex $u \in V$, $L(u)$ is a subset of $\{1, 2, \ldots\}$. This set is called the *list* of *admissible* colors for u. If $L(u) \subseteq \{1, \ldots, k\}$ for each $u \in V$ then L is also called a *k-list assignment*. The *size* of a list assignment L is the maximum list size $|L(u)|$ over all vertices $u \in V$. A coloring c *respects* L if $c(u) \in L(u)$ for all $u \in V$. This leads to the following three problems. Given a graph G with a list assignment L, the LIST COLORING problem is that of testing whether G has a coloring that respects L. If inputs are restricted to pairs (G, L) where L has size at most ℓ then we obtain the ℓ-LIST COLORING problem, and if each L is a k-list assignment, we obtain the LIST k-COLORING problem. See Fig. 1 for a display of the relationships between the seven problems defined above.

Let G be a graph and $\{H_1, \ldots, H_p\}$ be a set of graphs. Then G is said to be $(H_1, \ldots, H_p)$-*free* if G has no *induced* subgraph isomorphic to a graph in $\{H_1, \ldots, H_p\}$. If $p = 1$, we may write that G is H_1-free. The disjoint union $(V(G) \cup V(H), E(G) \cup E(H))$ of two vertex-disjoint graphs G and H is denoted by $G + H$ and the disjoint union of r copies of a graph G is denoted by rG. The complement $\overline{G}$ of a graph G has vertex set $V(\overline{G}) = V(G)$ and an edge between two distinct vertices u and v if and only if u and v are not adjacent in G. We denote the path and cycle on n vertices by P_n and C_n, respectively.

Let G be a connected graph. The *distance* $d(u, v)$ between two vertices u and v in G is the number of edges in a shortest path from u to v in G. The *diameter* of G is the maximum of $\max_v d(u, v)$ over all vertices u in G. The *radius* of G is the minimum of $\max_v d(u, v)$ over all vertices u in G.

Let p be a graph parameter and let $\mathcal{G}$ be a graph class. We say that $\mathcal{G}$ has *bounded* p if there exists a constant c such that $p(G) \leq c$ for all $G \in \mathcal{G}$.

3   Classical Complexity

The problems k-COLORING, k-PRECOLORING EXTENSION, LIST k-COLORING and k-LIST COLORING are polynomial-time solvable for general graphs if $k \leq 2$ and NP-complete if $k \geq 3$ [52,67]. As mentioned, we refer to the recent survey [27] for an overview of known results for these problems when restricted to H-free graphs and (H_1, H_2)-free graphs. In this section we consider a number of other graph classes.

Cycle-free Graphs. A *hole* is a cycle of on at least four vertices. An *antihole* is the complement of a hole. A cycle, hole or antihole is *even* if it contains an even number of vertices; otherwise it is *odd*. An (anti)hole is *long* if it has at least five vertices. A graph is *odd-hole-free* or *odd-antihole-free* if it contains no induced odd holes or no induced odd antiholes, respectively. In a similar way we define (even-)hole-free, (even-)antihole-free, long-hole-free, long-antihole-free, odd-cycle-free and (odd-)anticycle-free graphs.

Grötschel et al. [30] proved that COLORING is polynomial-time solvable for perfect graphs, or equivalently (due to the Strong Perfect Graph Theorem [14]) for graphs that are odd-hole-free and odd-antihole-free. Note that hole-free graphs and antihole-free graphs are perfect. Hence COLORING is also polynomial-time solvable for hole-free graphs and antihole-free graphs, and thus for cycle-free graphs (forests) and anticycle-free graphs (coforests).

Král' et al. [49] proved that COLORING is NP-complete for $(2P_2, C_5)$-free graphs and also for (C_3, C_4, C_5)-free graphs. Consequently, COLORING is NP-complete for long-hole-free graphs and long-antihole-free graphs, and thus for odd-hole-free graphs and odd-antihole free graphs. In contrast, COLORING is polynomial-time solvable for odd-cycle-free graphs (bipartite graphs) and odd-anticycle-free graphs (cobipartite graphs). If we change the parity of the forbidden cycles from odd to even, then we obtain two long-standing open problems;

we refer to the survey of Vušković [68] for more on even-cycle-free graphs (even-hole-free graphs).

Open Problem 1. *Determine the complexity of* COLORING *for even-cycle-free graphs and even-anticycle-free graphs.*

Table 1. The complexity of k-PRECOLORING EXTENSION and LIST k-COLORING on P_t-free bipartite graphs for fixed k and t.

r	k-PRECOLORING EXT.				LIST k-COLORING			
	$k = 3$	$k = 4$	$k = 5$	$k \geq 6$	$k = 3$	$k = 4$	$k = 5$	$k \geq 6$
$t \leq 6$	P	P	P	P	P	P	P	P
$t = 7$	?	?	?	?	?	?	?	?
$t = 8$	?	?	?	?	?	NP-c	NP-c	NP-c
$t = 9$	?	?	?	?	?	NP-c	NP-c	NP-c
$t \geq 10$	?	NP-c	NP-c	NP-c	?	NP-c	NP-c	NP-c

Bipartite and Chordal Bipartite Graphs. A graph is *chordal bipartite* if it is bipartite and every induced cycle has exactly four vertices. Hujter and Tuza [41] proved that PRECOLORING EXTENSION is linear-time solvable on P_5-free bipartite graphs (which are chordal bipartite) and NP-complete for P_6-free chordal bipartite graphs. Kratochvíl [50] answered two of their open problems [40] by proving that 3-PRECOLORING EXTENSION is NP-complete for planar bipartite graphs and that 5-PRECOLORING EXTENSION is NP-complete for P_{14}-free bipartite graphs. The latter result was strengthened by Huang et al. [39], who proved that, for all $k \geq 4$, k-PRECOLORING EXTENSION is NP-complete for P_{10}-free chordal bipartite graphs. The same authors [39] also proved that LIST 4-COLORING is NP-complete for P_8-free chordal bipartite graphs.

Brandstädt et al. [5] proved that the class of (C_3, P_6)-free graphs has bounded clique-width. By combining their result with results of Kobler and Rotics [48] and Oum and Seymour [62] we find that, for all $k \geq 1$, LIST k-COLORING is polynomial-time solvable on (C_3, P_6)-free graphs (see also e.g. [39]). Table 1 summarizes the above results for P_t-free bipartite graphs.

Open Problem 2. *Determine the complexity of the problems* k-PRECOLORING EXTENSION *and* LIST k-COLORING *for the missing cases in Table 1.*

Huang et al. [39] posed the following two open problems.

Open Problem 3. *Determine the complexity of the problems* LIST 3-COLORING *and* 3-PRECOLORING EXTENSION *for the class of chordal bipartite graphs.*

Planar Graphs and Graphs of Bounded Vertex Degree. To recall two classic results, Garey et al. [26] proved that 3-COLORING is NP-complete even for planar graphs of maximum degree 4, whereas every planar graph is 4-colorable by the Four Color Theorem [2]. Chlebík and Chlebíková [12] strengthened the aforementioned result of Kratochvíl [50] for planar bipartite graphs by proving that 3-PRECOLORING EXTENSION is NP-complete even for planar bipartite graphs of maximum degree 4. The same authors [12] also showed that LIST 3-COLORING is NP-complete for 3-regular planar bipartite graphs but that PRECOLORING EXTENSION is polynomial-time solvable for arbitrary graphs of maximum degree at most 3.

Demange and de Werra [17] proved that 3-PRECOLORING EXTENSION is NP-complete for subgrids (which are induced subgraphs of grids and hence have maximum degree at most 4) and that LIST 3-COLORING is NP-complete even for subgrids of maximum degree at most 3, whereas Kratochvíl and Tuza [51] showed that LIST COLORING is polynomial-time solvable for graphs of maximum degree 2. Demange and de Werra [17] also proved that LIST 4-COLORING is NP-complete for grids, and they posed the following open problem.

Open Problem 4. *Determine the complexity of* PRECOLORING EXTENSION *for grids.*

When consider graphs of bounded degree that are not necessarily planar a full complexity classification is known (see [16]) for COLORING, LIST COLORING and PRECOLORING EXTENSION and also for k-LIST COLORING, LIST k-COLORING and k-PRECOLORING EXTENSION. However, no dichotomy is known for k-COLORING restricted to graphs of maximum degree at most d, but some partial results have been obtained. Molloy and Reed [60] classified the complexity for all pairs (k, d) for sufficiently large d, whereas Emden-Weinert et al. [19] showed that k-COLORING is NP-complete for graphs of maximum degree at most $k + \lceil \sqrt{k} \rceil - 1$. By combining the latter result with Brooks' Theorem [8], we find that the smallest open case is the following problem.

Open Problem 5. *Determine the complexity of* 5-COLORING *for graphs of maximum degree* 6.

Graphs of Bounded Diameter. By using a reduction from 3-COLORING via adding dominating vertices one can easily show that k-COLORING is NP-complete for graphs of diameter d for all pairs (k, d) with $k \geq 3$ and $d \geq 2$ except for two notorious cases, namely $(k, d) \in \{(3, 2), (3, 3)\}$. Mertzios and Spirakis [57] settled the case $(k, d) = (3, 3)$. They proved that, for every $0 \leq \epsilon < 1$, 3-COLORING is NP-complete even for classes of triangle-free graphs $G = (V, E)$ of diameter 3, radius 2 and minimum degree $\delta = \Theta(|V|^\epsilon)$.

We note that, for every $k \geq 1$ and $p \geq 1$, the problems k-COLORING and k-PRECOLORING EXTENSION are polynomially equivalent on the class of graphs of diameter at most p. This can be seen as follows. Firstly, k-COLORING is a special case of k-PRECOLORING EXTENSION. Secondly, if we are given a graph G of diameter at most p with a k-precoloring c_W for some $W \subseteq V(G)$, then we identify any two vertices of W that are colored alike. Afterwards all precolored vertices have a distinct color and we add an edge between any two of them that are not adjacent already. This results in a graph G' of diameter p and a k-precoloring $c'_{W'}$ defined on some subset $W' \subseteq V(G')$, such that $c'_{W'}$ can be extended to a k-coloring of G' if and only if c_W can be extended to a k-precoloring of G. Moreover, the set W' forms a clique of size at most k in G' meaning that we may just as well uncolor these vertices, that is, $c'_{W'}$ can be extended to a k-coloring of G' if and only if G' is k-colorable. Hence we only need to consider:

Open Problem 6. *Determine the complexity of the problems* 3-COLORING *and* LIST 3-COLORING *for graphs of diameter* 2.

Graphs of Bounded Asteroidal Number. An *asteroidal triple* in a graph is a set of three mutually non-adjacent vertices such that each two of them are joined by a path that avoids the neighborhood of the third. An *asteroidal set* in a graph G is an independent set $S \subseteq V(G)$, such that every set of three vertices of S forms an asteroidal triple. The *asteroidal number* is the size of a largest asteroidal set in G. Note that graphs with asteroidal number at most 2 have no asteroidal triple. These graphs are also known as *AT-free* graphs. Stacho [63] proved that 3-COLORING is polynomial-time solvable on AT-free graphs. Later, Kratsch and Müller [53] extended this result by showing that even LISTk-COLORING is polynomial-time solvable on these graphs for every fixed integer $k \geq 1$. Marx [56] proved that PRECOLORING EXTENSION is NP-complete for proper interval graphs, which form a subclass of AT-free graphs. It follows from a result of Jansen [42] (see [28]) that ℓ-LIST COLORING ($\ell \geq 3$) is NP-complete for $3P_1$-free graphs, and thus for AT-free graphs. However, the following problem, posed by Broersma et al. [7] in 1999, is still open.

Open Problem 7. *Determine the complexity of* COLORING *for AT-free graphs.*

Král' et al. [49] proved that COLORING is NP-complete for $4P_1$-free graphs and thus for graphs with asteroidal number at most 3.

Open Problem 8. *Determine, for every* $k \geq 3$ *and* $p \geq 3$, *the complexity of* k-COLORING, k-PRECOLORING EXTENSION *and* LIST k-COLORING *on graphs with asteroidal number at most* p.

4   Parameterized Coloring Problems

A problem is called *fixed-parameter tractable* (FPT) if every instance (I, p) of it can be solved in time $f(p)|I|^{O(1)}$ where f is a computable function that only depends on p. If k-COLORING is polynomial-time solvable for some graph class $\mathcal{G}$ for every integer k (and COLORING is not known to be polynomial-time solvable for $\mathcal{G}$) then k is a natural parameter to consider. We refer to [27] for a survey on parameterized complexity results (and open problems) with this parameter for classes of graphs characterized by one or two forbidden induced subgraphs. Here, we only mention the following open problem of Kratsch and Müller [53].

Open Problem 9. *Is* COLORING *fixed-parameter tractable for AT-free graphs when parameterized by* k*?*

Because 3-COLORING is NP-complete in general, other parameters have been considered. For instance, Marx [55] proved that PRECOLORING EXTENSION parameterized by the number of precolored vertices is W[1]-hard for interval graphs. We survey a number of other results below (see also Table 2).

Arnborg and Proskurowski [3] proved that COLORING is FPT when parameterized by the treewidth of the input graph. Fellows et al. [21] showed that PRECOLORING EXTENSION and LIST COLORING are W[1]-hard with this parameter. On the positive side, LIST COLORING is polynomial-time solvable for any graph class of bounded treewidth, as shown by Jansen and Scheffler [44].

It is known [15] that the clique-width of a graph G is at most $2^{\text{tw}(G)-1}$, where $\text{tw}(G)$ denotes the treewidth of G. Moreover, by combining results of Kobler and Rotics [48] and Oum and Seymour [62], one finds that COLORING is polynomial-time solvable for any graph class of bounded clique-width. Hence, it is natural to research whether one can improve the FPT result for COLORING from treewidth to clique-width. However, Fomin et al. [24] showed that COLORING is W[1]-hard when parameterized by the clique-width of the input graph. We also note that PRECOLORING EXTENSION is NP-complete for distance-hereditary graphs [4], which have clique-width at most 3 [29].

The *vertex cover number* of a graph G is the size of a smallest subset $U \subseteq V(G)$, such that $G - U$ is edgeless. Fiala et al. [23] proved that, with this parameter, PRECOLORING EXTENSION is FPT and LIST COLORING is W[1]-hard even for split graphs. It can be observed [23] that the treewidth of a graph is at most its vertex cover number. Hence, the aforementioned result of Jansen and Scheffler [44] implies that LIST COLORING is polynomial-time solvable for any graph class of bounded vertex cover number.

The *twin cover number* of a graph G is the size of a smallest subset $U \subseteq V(G)$, such that every two adjacent vertices in $G - U$ have the same closed neighborhood in G; note that $G - U$ is a disjoint union of cliques. Ganian [25] proved that PRECOLORING EXTENSION is FPT when parameterized by the twin cover number of the input graph. As the twin cover number of a graph is at most its vertex cover number (by definition), this result strengthens the aforementioned result of Fiala, Golovach and Kratochvíl [23]. For the same reason, LIST COLORING is W[1]-hard when parameterized by the twin cover number.

The *cluster vertex deletion number* of a graph G is the size of a smallest subset $U \subseteq V(G)$, such that $G - U$ is a disjoint union of cliques. Note that the cluster vertex deletion number of a graph G is at most its twin cover number. However, in contrast to the aforementioned FPT result of Ganian [25], Doucha and Kratochvíl [18] proved that PRECOLORING EXTENSION is W[1]-hard when parameterized by the cluster vertex deletion number of the input graph. The same authors [18] also showed that COLORING is FPT with this parameter. It is easily seen that LIST COLORING is polynomial-time solvable for any graph class $\mathcal{G}$ of bounded cluster vertex deletion number. We guess a coloring of the set U of a graph $G \in \mathcal{G}$ that respects the given list of each vertex of U. We then remove all vertices of U from G after adjusting the lists of the vertices in $G - U$ accordingly. As $G - U$ is a disjoint union of cliques we can solve LIST COLORING in polynomial time (see e.g. [9]). Since $|U|$ is bounded, the maximum number of colorings of U that we need to guess is polynomial. Hence, the result follows.

For an integer $c \geq 1$, the *c-bounded cluster vertex deletion number* of a graph G is the size of a smallest subset $U \subseteq V(G)$, such that $G - U$ is a disjoint union of cliques of size at most c. Doucha and Kratochvíl [18] proved that, for every fixed integer $c \geq 1$, PRECOLORING EXTENSION is FPT when parameterized by the c-bounded cluster vertex deletion number of the input graph.

Table 2. The complexity of COLORING, PRECOLORING EXTENSION and LIST COLOR-ING for various graph parameters. All problems are in XP for each parameter except when para-NP-complete. The relationships, as given in [18], between rows 1–6 are: $1 \leq 2 \leq 4 \leq 6$ and $1 \leq 3 \leq 5 \leq 6$ and $3 \leq 4$, where $x \leq y$ means that parameter x is bounded by (some function of) parameter y. Hence, membership in FPT or XP for a problem with parameter x carries over to y, and W[1]-hardness for y carries over to x.

	COLORING	PRECOLORING EXT.	LIST COL.
clique-width	W[1]-hard	para-NP-c	para-NP-c
treewidth	FPT	W[1]-hard	W[1]-hard
cluster vertex deletion number	FPT	W[1]-hard	W[1]-hard
c-bounded cluster vertex deletion number	FPT	FPT	W[1]-hard
twin cover number	FPT	FPT	W[1]-hard
vertex cover number	FPT	FPT	W[1]-hard

Recently, Aboulker et al. [1] considered the number p_k of vertices of degree at least $k+1$ of the graph G in an instance (G, k) of COLORING. This parameter is motivated by Brooks' theorem [8]: if $p_k(G) = 0$ then G is k-colorable unless G is a complete graph or an odd cycle. They showed that COLORING is FPT when parameterized by p_k.

We now discuss some graph classes that fall under the "distance from triviality" framework, introduced by Guo et al. [31]. For a graph class $\mathcal{F}$ and an integer p we define four classes of "almost $\mathcal{F}$" graphs, i.e. graphs that are, in some sense, "distance" p apart from $\mathcal{F}$, namely the classes $\mathcal{F} + pe$, $\mathcal{F} - pe$, $\mathcal{F} + pv$ and $\mathcal{F} - pv$, which consist of all graphs that can be modified into a graph of $\mathcal{F}$ by deleting at most p edges, adding at most p edges, deleting at most p vertices and adding at most p vertices, respectively. As Grötschel et al. [30] proved that COLORING is polynomial-time solvable on perfect graphs, COLORING was studied from a parameterized point of view for various subclasses $\mathcal{F}$ of perfect graphs. We survey a number of these results below. For every result mentioned, p is the chosen parameter (see also Table 3 for an overview). In this context a *modulator* of a graph is a set of at most p edges or vertices whose removal or addition makes the graph a member of $\mathcal{F}$.

Table 3. The complexity of COLORING, when parameterized by p, for classes close to some subclass $\mathcal{F}$ of perfect graphs. The polynomial and FPT cases hold even if a modulator is not part of the input.

	$+pe$	$-pe$	$+pv$	$-pv$
Bipartite	para-NP-c	P	para-NP-c	P
Chordal	FPT	FPT	W[1]-hard	P
Interval	FPT	FPT	W[1]-hard	P
Split	FPT	FPT	W[1]-hard	P
Comparability	para-NP-c	?	?	P
Complete	P	FPT	FPT	P

Cai [9] proved that COLORING is FPT on split$+pe$ graphs and W[1]-hard for split$+pv$ graphs. The same author [9] also proved that whenever COLORING is polynomial-time solvable on a graph class $\mathcal{F}$ that is closed under edge contraction and a modulator is given, then COLORING is FPT on $\mathcal{F}-pe$. As a result, COLORING is FPT for split$-pe$ graphs, and also for interval$-pe$ graphs[1] and chordal$-pe$ graphs. Note that we obtain polynomial-time solvability for $\mathcal{F} - pv$ whenever $\mathcal{F}$ is a class of perfect graphs closed under vertex deletion(as in that case it holds that $\mathcal{F} - pv = \mathcal{F}$). Cai [9] also proved that COLORING is NP-complete for bipartite$+2v$ graphs and for bipartite$+3e$ graphs but linear-time solvable onbipartite$+1v$ and bipartite$+2e$ graphs. Marx [55] showed that COLORING is FPT[2] on interval$+pe$ graphs and also on chordal$+pe$ graphs but W[1]-hard for interval$+pv$ graphs and for chordal$+pv$ graphs.

An (undirected) graph is a *comparability graph* if there exists an assignment of exactly one direction to each of its edges such that (u, w) is a directed edge whenever (u, v) and (v, w) are directed edges. Takenaga and Higashide [64] proved that COLORING, restricted to comparability$+pe$ graphs, is polynomial-time solvable for $p = 1$ and NP-complete for $p \geq 2$. They also proved that COLORING is polynomial-time solvable on comparability$-1e$ graphs.

Open Problem 10. *Determine the complexity of COLORING for the classes of comparability$-pe$ graphs and comparability$+pv$ graphs when parameterized by p.*

We now consider the class of complete graphs. It is known that LIST COLORING is FPT for complete$-pe$ graphs [28]. We observe that LIST COLORING is polynomial-time solvable on complete$+pe$ graphs and complete$-pv$ graphs (because such graphs are complete and LIST COLORING is polynomial-time solvable even on block graphs [4]). In contrast to the aforementioned W[1]-hardness

---

[1] Villanger et al. [66] proved afterward that a modulator can be computed in FPT time.

[2] For the two FPT results it was proven later, namely by Cao [10] and Marx [54], respectively, that a modulator does not have to be part of the input (but can be computed in FPT time as well).

results of COLORING for split+pv, interval+pv and chordal+pv graphs, it holds that COLORING is FPT for complete+pv graphs, as shown by Cai [9]. Golovach et al. [28] posed the following open problem.

Open Problem 11. *Determine the complexity of* LIST COLORING *and* PRE-COLORING EXTENSION *for complete+pv graphs when parameterized by p.*

Jansen and Kratsch [43] and de Weijer [69] considered the k-COLORING problem for various graph classes $\mathcal{F} + pv$ in order to obtain polynomial kernels (also some negative results are shown, for instance, 3-COLORING on path+pv has no polynomial kernel unless NP$\subseteq$ coNP/poly [43]).

5   On-Line Coloring

In this section we focus on the on-line setting of graph coloring. On-line coloring algorithms were introduced by Gyárfás and Lehel [35] to model a rectangle packing problem related to dynamical storage allocation. In this setting the graph is presented vertex by vertex according to some externally determined ordering. An *on-line coloring* algorithm irrevocably colors the vertices when they come in by using a strategy that depends only on the subgraph induced by the revealed vertices and their colors. A well-known example of an on-line coloring algorithm is `First-Fit` which assigns, starting from the empty graph, each new vertex the least color from $\{1, 2, \ldots\}$ that does not appear in its neighborhood. We refer to the survey of Kierstead [45] for more details.

Non-surprisingly, the number of colors used by an on-line coloring algorithm for an arbitrary graph G can be much larger than the chromatic number of G. Below we define three measures for the performance of an on-line algorithm on graphs of some specified class $\mathcal{G}$ after first giving some additional terminology.

Let $AOL(G)$ be the (finite) set of all on-line coloring algorithms for a graph G. Let $\Pi(G)$ be the set of all permutations of the vertices of G. For $A \in AOL(G)$ and $\pi \in \Pi(G)$, let $\chi_A(G, \pi)$ denote the number of colors used by A when the vertices of G are presented to A according to π. The A-*chromatic number* $\chi_A(G)$ of G is the largest number of colors used by A to color G, that is,

$$\chi_A(G) = \max_{\pi \in \Pi(G)} \chi_A(G, \pi).$$

An algorithm A is an on-line coloring algorithm for some graph class $\mathcal{G}$ if $A \in AOL(G)$ for every $G \in \mathcal{G}$. We let $AOL(\mathcal{G})$ be the set of on-line coloring algorithms for $\mathcal{G}$.

A natural performance measure for an on-line coloring algorithm, introduced by Gyárfás and Lehel [35], is to determine whether the number of colors it uses on any graph $G \in \mathcal{G}$ is bounded from above by a function that only depends on the chromatic number of G. Formally, for a graph class $\mathcal{G}$, we say that an algorithm $A \in AOL(\mathcal{G})$ is *competitive* if there exists a χ-*bounding* function f, that is, a function f such that

$$\chi_A(G) \le f(\chi(G)) \text{ for every } G \in \mathcal{G}.$$

In that case $\mathcal{G}$ is said to be *on-line χ- bounded*. For example, the class of P_4-free graphs is on-line χ-bounded, because `First-Fit` colors every P_4-free graph G with $\chi(G)$ colors [35]. It is also known that `First-Fit` is competitive for interval graphs with a linear χ-bounding function [46]. Consequently, the class of interval graphs is on-line χ-bounded as well. Every class of graphs with bounded independence number [11] is also on-line χ-bounded, just as the class of P_5-free graphs [33]. In fact, Gyárfás and Lehel [33] proved a stronger statement, namely that the class of P_5-free graphs is *on-lineω-bounded*, that is, there exists an on-line coloring algorithm A and a function g, called an ω-*bounding* function, such that $\chi_A(G) \le g(\omega(G))$ for every P_5-free graph G. This result has been extended by Kierstead et al. [47] who proved that, for every tree T of radius at most 2, the class of T-free graphs is on-line ω-bounded (with a superexponential ω-bounding function). As a special case of their result we find that the class of cocomparability graphs is on-line ω-bounded. More recently, Felsner et al. [22] gave a cubic ω-bounding function for a subclass of cocomparability graphs, namely for the class of intersection graphs of convex sets spanned between two lines (their algorithm uses the intersection representation as input).

Despite all the above results there exist many graph classes, such as the class of trees [35], for which no competitive on-line coloring algorithm exists. These negative results lead to a natural definition of a weaker form of competitiveness, namely on-line competitiveness, which is defined as follows. The *on-line chromatic number* $\chi_{OL}(G)$ of G is the smallest number of colors used by any on-line coloring algorithm for G, that is,

$$\chi_{OL}(G) = \min_{A \in AOL(G)} \chi_A(G).$$

Then, for a graph class $\mathcal{G}$, an algorithm $A \in AOL(\mathcal{G})$ is said to be *on-line competitive* if there exists a function h such that

$$\chi_A(G) \le h(\chi_{OL}(G)) \text{ for every } G \in \mathcal{G}.$$

In that case $\mathcal{G}$ is said to be *on-line χ_{OL}-bounded*. This performance measure was coined by Gyárfás et al. [32], who proved that the class of graphs with girth at least 5 is on-line χ_{OL}-bounded, but results of this type have been obtained before the term was formally introduced. For instance, Gyárfás and Lehel [34] proved that, for any tree T, `First-Fit` uses $\chi_{OL}(T)$ colors.

By combining known and new results, Broersma et al. [6] proved that, for all bipartite graphs H on at most five vertices, the class of H-free bipartite graphs is on-line χ_{OL}-bounded. If H has six or more vertices the situation is not clear. For instance it is not known whether the class of C_6-free bipartite graphs is on-line χ_{OL}-bounded. In fact this is not even known for its subclass of chordal bipartite graphs.

Open Problem 12. *Is the class of chordal bipartite graphs on-line χ_{OL}-bounded?*

Now consider P_t-free bipartite graphs. Broersma et al. [6] proved that the class of P_7-free bipartite graphs is on-line χ_{OL}-bounded. The algorithm behind their result is based on a certain way of coloring the vertices of a complete bipartite graph with classes $\{u_1, \ldots, u_m\}$ and $\{v_1, \ldots, v_m\}$ minus a perfect matching $\{u_1 v_1, u_2 v_2 \ldots, u_m v_m\}$. If the ordering is $u_1, v_1, u_2, v_2, \ldots, u_m, v_m$, then First-Fit assigns colors $1, 1, 2, 2, \ldots, m, m$, so m colors in total. However, assigning colors 1, 1, 2, 3 to the first four vertices u_1, v_1, u_2, v_2 in this ordering requires only three colors in total. The algorithm for P_7-free bipartite graphs expands on this approach and uses two disjoint lists of colors for the bipartition classes of each connected component in the subgraph revealed so far. Then, whenever two connected components are glued together by an incoming vertex, it tries to prevent the "mixing" of these color lists as much as possible. Recently, Micek and Wiechert refined and extended this approach. In this way they could prove that the classes of P_8-free bipartite graphs [58] and even P_9-free bipartite graphs [59] are on-line χ_{OL}-bounded. This leads to the following open problem.

Open Problem 13. *Is the class of P_k-free bipartite graphs on-line χ_{OL}-bounded for every $k \geq 1$?*

As P_5-free graphs and P_6-free bipartite graphs are on-line ω-bounded and on-line χ_{OL}-bounded, respectively, the following problem from [6] is of interest as well.

Open Problem 14. *Is the class of (C_3, P_6)-free graphs on-line χ_{OL}-bounded?*

Solving Open Problems 12–14 may lead to new on-line χ_{OL}-bounded classes of graphs. Unlike the competitive variant, no negative results are known.

Open Problem 15. *Is the class of all graphs on-line χ_{OL}-bounded?*

6   Conclusions

We surveyed a number of results and open problems for COLORING for restricted graph classes. We did so both in an off-line and on-line setting and also considered the more general variants PRECOLORING EXTENSION and LIST COLORING. Another way of generalizing the concept of graph coloring is to consider graph homomorphisms. A *graph homomorphism* from a graph G to a graph H is a mapping $f : V(G) \to V(H)$ such that $f(u)f(v) \in E_H$ whenever $uv \in E_G$. For a fixed graph H, the problem H-HOMOMORPHISM tests whether a given graph G allows a homomorphism to H. If we choose H to be the complete graph on k vertices, then this problem is equivalent to k-COLORING. We refer to the survey of Hell and Nešetřil [37] for more on graph homomorphisms.

The classical result in the area of graph homomorphisms is the Hell-Nešetřil dichotomy theorem [38] which states that H-HOMOMORPHISM is solvable in polynomial time if H is bipartite, and NP-complete otherwise. It is a natural question whether tractability results can be obtained for non-bipartite graphs H when the input is restricted to some graph class. Not so many results are known in this direction for hereditary graph classes, but to give an example, Enright et al. [20] proved that, for every fixed graph H, the list version of H-HOMOMORPHISM is polynomial-time solvable for a superclass of graphs that contains the classes of permutation graphs and interval graphs.

References

1. Aboulker, P., Brettell, N., Havet, F., Marx, D., Trotignon, N.: Colouring graphs with constraints on connectivity, Manuscript. arXiv:1505.01616
2. Appel, K., Haken, W.: Every planar map is four colorable. In: Contemporary Mathematics, vol. 89. AMS Bookstore (1989)
3. Arnborg, S., Proskurowski, A.: Linear time algorithms for NP-hard problems restricted to partial k-trees. Discrete Appl. Math. **23**, 11–24 (1989)
4. Bonomo, F., Durán, G., Marenco, J.: Exploring the complexity boundary between coloring and list-coloring. Ann. Oper. Res. **169**, 3–16 (2009)
5. Brandstädt, A., Klembt, T., Mahfud, S.: P_6- and triangle-free graphs revisited: structure and bounded clique-width. Discrete Math. Theor. Comput. Sci. **8**, 173–188 (2006)
6. Broersma, H.J., Capponi, A., Paulusma, D.: A new algorithm for on-line coloring bipartite graphs. SIAM J. Discrete Math. **22**, 72–91 (2008)
7. Broersma, H.J., Kloks, T., Kratsch, D., Müller, H.: Independent sets in asteroidal triple-free graphs. SIAM J. Discrete Math. **12**, 276–287 (1999)
8. Brooks, R.L.: On colouring the nodes of a network. Math. Proc. Cambridge Philos. Soc. **37**, 194–197 (1941)
9. Cai, L.: Parameterized complexity of vertex coloring. Discrete Appl. Math. **127**, 415–429 (2003)
10. Cao, Y.: Linear recognition of almost (unit) interval graphs, Manuscript. arXiv:1403.1515
11. Cieślik, I., Kozik, M., Micek, P.: On-line coloring of I_s-free graphs and co-planar graphs. Discrete Math. Theor. Comput. Sci. Proc. **AF**, 61–68 (2006)
12. Chlebík, M., Chlebíková, J.: Hard coloring problems in low degree planar bipartite graphs. Discrete Appl. Math. **154**, 1960–1965 (2006)
13. Chudnovsky, M.: Coloring graphs with forbidden induced subgraphs. Proc. ICM **IV**, 291–302 (2014)
14. Chudnovsky, M., Robertson, N., Seymour, P.D., Thomas, R.: The strong perfect graph theorem. Ann. Math. **164**, 51–229 (2006)
15. Corneil, D.G., Rotics, U.: On the relationship between clique-width and treewidth. SIAM J. Comput. **34**, 825–847 (2005)
16. Dabrowski, K.K., Dross, F., Johnson, M., Paulusma, D.: Filling the complexity gaps for colouring planar, bounded degree graphs, Manuscript. arXiv:1506.06564
17. Demange, M., de Werra, D.: On some coloring problems in grids. Theoret. Comput. Sci. **472**, 9–27 (2013)
18. Doucha, M., Kratochvíl, J.: Cluster vertex deletion: a parameterization between vertex cover and clique-width. In: Rovan, B., Sassone, V., Widmayer, P. (eds.) MFCS 2012. LNCS, vol. 7464, pp. 348–359. Springer, Heidelberg (2012)
19. Emden-Weinert, T., Hougardy, S., Kreuter, B.: Uniquely colourable graphs and the hardness of colouring graphs of large girth. Comb. Probab. Comput. **7**, 375–386 (1998)
20. Enright, J., Stewart, L., Tardos, G.: On list coloring and list homomorphism of permutation and interval graphs. SIAM J. Discrete Math. **28**, 1675–1685 (2014)
21. Fellows, M.R., Fomin, F.V., Lokshtanov, D., Rosamond, F., Saurabh, S., Szeider, S., Thomassen, C.: On the complexity of some colorful problems parameterized by treewidth. Inf. Comput. **209**, 143–153 (2011)
22. Felsner, S., Micek, P., Ueckerdt, T.: On-line coloring between two lines. In: Proceedings SoCG 2015, LIPIcs, vol. 34, pp. 630–641 (2015)

23. Fiala, J., Golovach, P.A., Kratochvíl, J.: Parameterized complexity of coloring problems: treewidth versus vertex cover. Theoret. Comput. Sci. **412**, 2514–2523 (2011)
24. Fomin, F.V., Golovach, P.A., Lokshtanov, D., Saurabh, S.: Clique-width: on the price of generality. In: Proceedings of SODA 2009, pp. 825–834 (2009)
25. Ganian, R.: Twin-cover: beyond vertex cover in parameterized algorithmics. In: Marx, D., Rossmanith, P. (eds.) IPEC 2011. LNCS, vol. 7112, pp. 259–271. Springer, Heidelberg (2012)
26. Garey, M.R., Johnson, D.S., Stockmeyer, L.J.: Some simplified NP-complete graph problems. In: Proceedings of STOC, pp. 47–63 (1974)
27. Golovach, P.A., Johnson, M., Paulusma, D., Song, J.: A survey on the computational complexity of coloring graphs with forbidden subgraphs, Manuscript. arXiv:1407.1482
28. Golovach, P.A., Paulusma, D., Song, J.: Closing complexity gaps for coloring problems on H-free graphs. Inf. Comput. **237**, 20–21 (2014)
29. Golumbic, M.C., Rotics, U.: On the clique-width of some perfect graph classes. Int. J. Found. Computer Sci. **11**, 423–443 (2000)
30. Grötschel, M., Lovász, L., Schrijver, A.: Polynomial algorithms for perfect graphs. Ann. Discret. Math. **21**, 325–356 (1984)
31. Guo, J., Hüffner, F., Niedermeier, R.: A structural view on parameterizing problems: distance from triviality. IWPEC 2004. LNCS, vol. 3162, pp. 162–173. Springer, Heidelberg (2004)
32. Gyárfás, A., Király, Z., Lehel, J.: On-line competitive coloring algorithms. Technical report TR-9703-1 (1997)
33. Gyárfás, A., Lehel, J.: Effective on-line coloring of P_5-free graphs. Combinatorica **11**, 181–184 (1991)
34. Gyárfás, A., Lehel, J.: First fit and on-line chromatic number of families of graphs. Ars Combinatorica **29C**, 168–176 (1990)
35. Gyárfás, A., Lehel, J.: On-line and first-fit colorings of graphs. J. Graph Theory **12**, 217–227 (1988)
36. Golovach, P.A., Paulusma, D.: List coloring in the absence of two subgraphs. Discrete Appl. Math. **166**, 123–130 (2014)
37. Hell, P., Nešetřil, J.: Graphs and Homomorphisms. Oxford Univ, Press (2004)
38. Hell, P., Nešetřil, J.: On the complexity of H-coloring. J. Comb. Theory Ser. B **48**, 92–110 (1990)
39. Huang, S., Johnson, M., Paulusma, D.: Narrowing the complexity gap for coloring (C_s, P_t)-Free Graphs. Comput. J. (to appear)
40. Hujter, M., Tuza, Z.: Precoloring extension. II. Graph classes related to bipartite graphs. Acta Math. Univ. Comenianae **LXII**, 1–11 (1993)
41. Hujter, M., Tuza, Z.: Precoloring extension. III. Classes of perfect graphs. Comb. Probab. Comput. **5**, 35–56 (1996)
42. Jansen, K.: Complexity results for the optimum cost chromatic partition problem. Universität Trier, Mathematik/Informatik, Forschungsbericht, pp. 96–41 (1996)
43. Jansen, B.M.P., Kratsch, S.: Data reduction for graph coloring problems. FCT 2011. LNCS, vol. 6914, pp. 90–101. Springer, Heidelberg (2011)
44. Jansen, K., Scheffler, P.: Generalized coloring for tree-like graphs. Discrete Appl. Math. **75**, 135–155 (1997)
45. Kierstead, H.A.: Coloring graphs on-line. In: Fiat, A. (ed.) Online Algorithms 1996. LNCS, vol. 1442, pp. 281–305. Springer, Heidelberg (1998)
46. Kierstead, H.A.: The linearity of first-fit coloring of interval graphs. SIAM J. Discrete Math. **1**, 526–530 (1988)

47. Kierstead, H.A., Penrice, S.G., Trotter, W.T.: On-line coloring and recursive graph theory. SIAM J. Discrete Math. **7**, 72–89 (1994)
48. Kobler, D., Rotics, U.: Edge dominating set and colorings on graphs with fixed clique-width. Discrete Appl. Math. **126**, 197–221 (2003)
49. Král', D., Kratochvíl, J., Tuza, Z., Woeginger, G.J.: Complexity of coloring graphs without forbidden induced subgraphs. WG 2001. LNCS, vol. 2204, p. 254. Springer, Heidelberg (2001)
50. Kratochvíl, J.: Precoloring extension with fixed color bound. Acta Mathematica Universitatis Comenianae **62**, 139–153 (1993)
51. Kratochvíl, J., Tsuza, Z.: Algorithmic complexity of list colorings. Discrete Appl. Math. **50**, 297–302 (1994)
52. Lovász, L.: Coverings and coloring of hypergraphs. In: Proceedings of 4th Southeastern Conference on Combinatorics, Graph Theory, and Computing, Utilitas Math, pp. 3–12 (1973)
53. Kratsch, D., Müller, H.: Colouring AT-free graphs. ESA 2012. LNCS, vol. 7501, pp. 707–718. Springer, Heidelberg (2012)
54. Marx, D.: Chordal deletion is fixed-parameter tractable. Algorithmica **57**, 747–768 (2010)
55. Marx, D.: Parameterized coloring problems on chordal graphs. Theoret. Comput. Sci. **351**, 407–424 (2006)
56. Marx, D.: Precoloring extension on unit interval graphs. Discrete Appl. Math. **154**, 995–1002 (2006)
57. Mertzios, G.B., Spirakis, P.G.: Algorithms and almost tight results for 3-colorability of small diameter graphs, Algorithmica (to appear)
58. Micek, P., Wiechert, V.: An on-line competitive algorithm for coloring P8-free bipartite graphs. In: Ahn, H.-K., Shin, C.-S. (eds.) ISAAC 2014. LNCS, vol. 8889, pp. 516–527. Springer, Heidelberg (2014)
59. Micek, P., Wiechert, V.: An on-line competitive algorithm for coloring bipartite graphs without long induced paths, Manuscript. arXiv:1502.00859
60. Molloy, M., Reed, B.: Colouring graphs when the number of colours is almost the maximum degree. J. Comb. Theory, Ser. B **109**, 134–195 (2014)
61. Randerath, B., Schiermeyer, I.: Vertex coloring and forbidden subgraphs - a survey. Graphs Comb. **20**, 1–40 (2004)
62. Oum, S.-L., Seymour, P.D.: Approximating clique-width and branch-width. J. Comb. Theory Ser. B **96**, 514–528 (2006)
63. Stacho, J.: 3-coloring AT-free graphs in polynomial time. Algorithmica **64**, 384–399 (2012)
64. Takenaga, Y., Higashide, K.: Vertex coloring of comparability+ke and $-ke$ graphs. In: Fomin, F.V. (ed.) WG 2006. LNCS, vol. 4271, pp. 102–112. Springer, Heidelberg (2006)
65. Toft, B., Jensen, J.R.: Graph Coloring Problems. Wiley, New York (1995)
66. Villanger, Y., Heggernes, P., Paul, C., Telle, J.A.: Interval completion is fixed parameter tractable. SIAM J. Comput. **38**, 2007–2020 (2009)
67. Vizing, V.G.: Coloring the vertices of a graph in prescribed colors. In: Diskret. Analiz., no. 29, Metody Diskret. Anal. v. Teorii Kodov i Shem, vol. 101, pp. 3–10 (1976)
68. Vušković, K.: Even-hole-free graphs: a survey. Appl. Anal. Discrete Math. **4**, 219–240 (2010)
69. de Weijer, P.: Kernelization upper bounds for parameterized graph coloring problems. MSc Thesis, Utrecht University (2013)
70. Tuza, Z.: Graph colorings with local restrictions - a survey. Discussiones Mathematicae Graph Theory **17**, 161–228 (1997)

On the Complexity of Approximation and Online Scheduling Problems with Applications to Optical Networks

Shmuel Zaks$^{(\boxtimes)}$

Department of Computer Science, Technion, Haifa, Israel
zaks@cs.technion.ac.il

Abstract. We present scheduling problems that stem from optical networks, and discuss their complexity. We present lower bounds and inapproximability results for several optimization problems. They include offline and online scenarios, and concern problems that optimize the use of components in the optical networks, specifically Add-Drop Multiplexers (ADMs) and regenerators.

1 Introduction - Optical Networks

1.1 Background and Problem Definition

Background: Optical wavelength-division multiplexing (WDM) is the most promising technology today that enables us to deal with the enormous growth of traffic in communication networks, like the Internet. Optical fibers using WDM technology can carry around 80 wavelengths (colors) in real networks and up to few hundreds in testbeds. As satisfactory solutions have been found for various coloring problems, the focus of studies shifts from the number of colors to the hardware cost. These new measures provide better understanding for designing and routing in optical networks.

A communication between a pair of nodes is done via a *lightpath*, which is assigned a certain wavelength. In graph-theoretic terms, a lightpath is a simple path in the network, with a color assigned to it. We concentrate on the hardware cost, in terms of ADMs and regenerators.

ADMs: Each lightpath uses two Add-Drop Multiplexers (ADMs), one at each endpoint. If two adjacent lightpaths, i.e. lightpaths sharing a common endpoint, are assigned the same wavelength, then they can use the same ADM, provided their concatenation is a simple path. An ADM may be shared by at most two lightpaths. The total cost considered is the total number of ADMs. For a detailed technical explanation see [14].

Stated in graph-theoretic terms, we are given a set of paths $\mathcal{P}$, and need to assign them colors, such that two edge-intersecting paths must get different colors. (The issue of vertex-intersecting paths will not be discussed here.) Such a color assignment is termed a *legal coloring*. Path that share an endpoint can

E.W. Mayr (Ed.): WG 2015, LNCS 9224, pp. 31–46, 2016.
DOI: 10.1007/978-3-662-53174-7_3

get the same color. The cost is measured in terms of the total number of ADMs. As each path is using two ADMs, one at each endpoint, the total number of ADMs is $2|\mathcal{P}|$. When two paths that share an endpoint get the same color, we save one ADM, so the cost is $2|\mathcal{P}|$ minus the number of these savings. The goal is to minimize the total number of ADMs. We thus the following minimization problem:

ADM MINIMIZATION (MINADM)

Input: A graph $G = (V, E)$, a set $\mathcal{P}$ of simple paths in G.
Output: A legal coloring of $\mathcal{P}$.
Objective: Minimize the number of ADMs.

Regenerators: The energy of the signal along a lightpath decreases and thus amplifiers are used every fixed distance. Yet, as the amplifiers introduce noise into the signal there is a need to place a regenerator every at most d hops.

The *length* of a lightpath is the number of edges it contains. The *internal vertices* (resp. *edges*) of a lightpath or a path ℓ are the vertices (resp. edges) in ℓ different from the first and the last one. Given an integer d, a lightpath ℓ is d-*satisfied* if there are no d consecutive internal vertices in ℓ without a regenerator. A set of lightpaths is d-*satisfied* if each of its lightpaths is d-satisfied. Given p sets of lightpaths $L_1, \ldots, L_p$, with $L_i = \{\ell_{i,j} \mid 1 \leq j \leq x_i\}$ (that is, x_i is the number of lightpaths in the set L_i), we consider the union of all lightpaths in the p sets $\cup L_i = \{\ell_{i,j} \mid 1 \leq i \leq p, 1 \leq j \leq x_i\}$. An *assignment* of regenerators is a function $reg : V \times \cup L_i \to \{0, 1\}$, where $reg(v, \ell) = 1$ if and only if a regenerator is used at vertex v by lightpath ℓ. When one set of lightpaths is given, the number of regenerators put at node v is $reg(v)$. We present two problems.

The first problem considers a scenario where we are given a finite set of p possible traffic patterns (each given by a set of lightpaths), and our objective is to place the minimum number of regenerators at the nodes so that each of the traffic patterns is satisfied. Thus, given $p \geq 1$ sets of lightpaths, and a distance $d \geq 1$, we need to determine the smallest number of regenerators that d-satisfy each of the p sets. Formally, for two fixed integers $d, p \geq 1$, the optimization problem we study is defined as follows.

(d, p)-TOTAL REGENERATORS ((d, p)-TR)

Input: An undirected graph $G = (V, E)$ and p sets of lightpaths $\mathcal{L} = \{L_1, \ldots, L_p\}$.

Output: A function $reg : V \times \cup L_i \to \{0, 1\}$ s.t. each lightpath in $\cup L_i$ is d-satisfied.

Objective: Minimize $\sum_{v \in V} reg(v)$ $(reg(v) = \max_{1 \leq i \leq p} \sum_{\ell \in L_i} reg(v, \ell))$.

When $p = 1$ (that is, when there is a single set of requests) the problem is trivially solvable in polynomial time, as the regenerators can be placed for each lightpath independently. The case $d = 1$ is not interesting either, as for each internal vertex $v \in V$ and each $\ell \in \cup L_i$, $reg(v, \ell) = 1$, so there is only one feasible solution, which is optimal.

The second problem deals with the case where there is a limit imposed by the technology on the number of regenerators that can be placed in a network node [5,11]. We denote this limit by k and refer to the case where this limit is not likely to be reached by any regenerator placement as $k = \infty$. When $k = \infty$ we consider the *regenerator location problem* (RLP) where the objective is to minimize the number of nodes that are assigned regenerators. When k is bounded there are inputs for which there is no feasible regenerator placement that satisfies both conditions. For example, consider the case $d = 2$ and $k = 1$, and three identical lightpaths $u - v - w - x$. Each of these lightpaths must have a regenerator either at v or w, and this is clearly impossible. In this case we consider the *Path Maximization Problem* (MAxPATH) that seeks for regenerator placements that serve as many lightpaths as possible, as follows:

PATH MAXIMIZATION (MAxPATH)

Input: An undirected graph $G = (V, E)$, a set $\mathcal{P}$ of paths in G, $d, k \geq 1$.
Output: A regenerator assignment reg for which $reg(v) \leq k$ for every node $v \in V$.
Objective: Maximize the number of paths of $\mathcal{P}$ that are d-satisfied.

Approximation Algorithms: Given an NP-hard minimization problem Π, we say that a polynomial-time algorithm $\mathcal{A}$ is an α-approximation algorithm for Π, with $\alpha \geq 1$, if for any instance of Π, algorithm $\mathcal{A}$ finds a feasible solution with cost at most α times the cost of an optimal solution. In complexity theory, the class APX (Approximable) contains all NP-hard optimization problems that can be approximated within a constant factor. The subclass PTAS (Polynomial Time Approximation Scheme) contains the problems that can be approximated in polynomial time within a ratio $1 + \varepsilon$ for *any* fixed $\varepsilon > 0$. In some sense, these problems can be considered to be *easy* NP-hard problems. Since, assuming P $\neq$ NP, there is a strict inclusion of PTAS in APX (for instance, MINIMUM VERTEX COVER $\in$ APX $\setminus$ PTAS), an APX-hardness result for a problem implies the non-existence of a PTAS unless P = NP.

Online Algorithms: An online minimization algorithm is said to be c-competitive if for any input, it produces a solution that is at most c times that used by an optimal offline algorithm (see [4]).

The motivation for the online scenario in the context of optical networks stems from the need to utilize the cost of use of the optical network. We assume that the switching equipment (ADMs or regenerators) is already installed in the network. Once a lightpath arrives, we need to assign it a color, and our target is to optimize the objective function.

1.2 Previous Work

We list below some previous works regarding optimization problems for ADMs and regenerators.

ADMs: Minimizing the number of ADMs in optical networks is a main research topic in recent studies. The problem was introduced in [14] for the ring topology. An approximation algorithm for the ring topology with approximation ratio of $\frac{3}{2}$ was presented in [7], and was improved in [9, 21] to $\frac{10}{7} + \epsilon$ and $\frac{10}{7}$, respectively.

For general topology [8] described an algorithm with approximation ratio of $\frac{8}{5}$. The same problem was studied in [6] and an algorithm with an approximation ratio of $\frac{3}{2} + \epsilon$ was presented. This algorithm is further analyzed in [13].

The problem of online path coloring is studied in earlier works, such as [16]. The problem studied in these works has a different objective function, namely the number of colors.

Regenerators: Placement of regenerators in optical networks has become an active area in recent years. Most of the researches have focused on the technological aspects of the problems. Moreover, heuristics and simulations have been performed in order to reduce the number of regenerators are performed in (e.g., [5, 10, 15, 19, 22, 24, 25]). The regenerator location problem (RLP) was shown to be NP-complete in [5], followed by heuristics and simulations. In [11] theoretical results for the offline version of RLP are presented. The authors study four variants of the problem, depending on whether the number k of regenerators per node is bounded, and whether the routings of the requests are given. Regarding the complexity of the problem, they present polynomial-time algorithms and NP-completeness results for a variety of special cases.

We note that while considering the path topology, RLP has implications for the following scheduling problem: Assume a company has n cars and that car i needs to be serviced within every at most d days between day a_i and b_i. Furthermore, assume that the garage can serve at most k cars per day and charges a certain cost each time the garage is used. The objective is to service the cars in the fewest number of days and hence minimizing the number of times the garage is used.

Other objective functions have also been considered in the context of regenerator placement. e.g., in [17] the problem of minimizing the total number of regenerators is studied under other settings.

2   In This Paper

We present scheduling problems that stem from optical networks, and discuss their complexity. We present lower bounds and inapproximability results for several optimization problems. They include offline and online scenarios, and concern problems that optimize the use of components in the optical networks, specifically ADMs and regenerators.

We start with the (d, p)-Total Regenerators ((d, p)-TR) problem. In scheduling this corresponds to the case where we have p sets of jobs, each needs a service within d time units (d−satisfied), and we need to place a smallest number of machines such that for each set A, each of the jobs in A is d-satisfied. In [17] we provided hardness results and approximation algorithms for the (d, p)-TR problem. We proved that for any fixed integers $d, p \geq 2$, (d, p)-TR does not admit a PTAS unless P = NP, even if the underlying graph G has maximum degree at most 3, and the lightpaths have length at most $2d$. We complemented this hardness result with a constant-factor approximation algorithm with ratio $\min\{p, H_{d \cdot p} - 1/2\}$, where $H_n = \sum_{i=1}^{n} \frac{1}{i}$ is the n-th harmonic number. We proved that (d, p)-TR is polynomial-time solvable in paths when all the lightpaths share the first (or the last) edge, as well as when the maximum number of lightpaths sharing an edge is bounded. In this paper we show that (d, p)-TR does not admit a PTAS; this result is presented in Sect. 3.

We then study the online version of the ADM minimization (MINADM) problem. This corresponds to scheduling jobs to machines, where the cost is associated with opening a machine, closing a machine, or moving a machine from one job to another. In [20] we showed a competitive ratio of $\frac{7}{4}$ for any network topology, including rings of size at least four, $\frac{5}{3}$ for a triangle network, and $\frac{3}{2}$ for a path topology, and showed that these results are best possible. In this paper we present the lower bound of $\frac{3}{2}$ for a path topology; this result is presented in Sect. 4.

We continue by discussing the online version of the Path Maximization (MAXPATH) problem, following [18]. When there is a bound on the number of regenerators in a single node, there is not necessarily a solution for a given input. We distinguish between feasible inputs and infeasible ones. For the latter case our objective is to satisfy the maximum number of lightpaths. For a path topology we consider the case where $d = 2$, and show a lower bound of $\sqrt{l}/2$ for the competitive ratio (where l is the number of internal nodes of the longest lightpath) on infeasible inputs, and a tight bound of 3 for the competitive ratio on feasible inputs.In scheduling this corresponds to the case where we are given a set of predetermined routes in a network whose topology is a graph G, with servers at the nodes (a server can serve only one user), and a positive integer d. Each route represents a user, who is satisfied when it can use a server every at most d consecutive edges.In this paper we present the above-mentioned lower bounds of $\sqrt{l}/2$ for the competitive ratio for general instances which may be infeasible, and the tight bound of 3 for the competitive ratio of deterministic online algorithms for feasible instances; these results are presented in Sect. 5.

Last we consider the problem of minimizing the number of regenerators, in the case where one regenerator can serve up to g lightpaths. This corresponds to the notion of *grooming* in optical networks. In scheduling this corresponds to the case in which up to g jobs can be processed simultaneously by a single machine. The goal is to assign the jobs to machines so that the total busy time is minimized. The problem is known to be NP-hard already for $g = 2$. Following [12], we present an algorithm whose competittive ratio is between 3 and 4; this is shown in Sect. 6.

3 (d, p)-Total Regenerators

In this section we prove that, unless P = NP, (d, p)-TR does not admit a PTAS for any $d, p \geq 2$, even if the underlying graph G has maximum degree at most 3 and the lightpaths have length $\mathcal{O}(d)$. Before this, we need two technical results to be used in the reductions.

MINIMUM VERTEX COVER is known to be APX-hard in cubic graphs [2]. By a simple reduction, we prove in the following lemma that MINIMUM VERTEX COVER is also APX-hard in a class of graphs with degree at most 3 and high girth, which will be used in the sequel.

Lemma 1. MINIMUM VERTEX COVER *is* APX-*hard in the class of graphs* $\mathcal{H}$ *obtained from cubic graphs by subdividing each edge twice.*

Proof. Given a cubic graph G, let H the graph obtained from G by subdividing each each twice. That is, each edge $\{u, v\}$ gets replaced by 3 edges $\{u, u_e\}$, $\{u_e, v_e\}$, and $\{v_e, v\}$, where u_e, v_e are two new vertices. We now claim that

$$\text{OPT}_{\text{VC}}(H) = \text{OPT}_{\text{VC}}(G) + |E(G)| , \tag{1}$$

where OPT_{VC} indicates the size of a minimum vertex cover. Indeed, let $S_G \subseteq V(G)$ be a vertex cover of G. We proceed to build a vertex cover S_H of H of size $|S_G| + |E(G)|$. First, include in S_H all the vertices in S_G. Then, for each 3 edges $\{u, u_e\}$, $\{u_e, v_e\}$, and $\{v_e, v\}$ of H corresponding to edge $\{u, v\} \in E(G)$, the edge $\{u_e, v_e\}$ is *not* covered by S_G, and at least one of $\{u, u_e\}$ and $\{v_e, v\}$ is covered by S_G. Therefore, adding either u_e or v_e to S_H covers the three edges $\{u, u_e\}$, $\{u_e, v_e\}$, and $\{v_e, v\}$. This procedure defines a vertex cover of H of size $|S_G| + |E(G)|$. Conversely, let $S_H \subseteq V(H)$ be a vertex cover of H, and let us construct a vertex cover S_G of G of size at most $|S_H| - |E(G)|$. We shall see that we can construct S_G from S_H by decreasing the cardinality of S_H by at least one for each edge of G. Indeed, consider the three edges $\{u, u_e\}$, $\{u_e, v_e\}$, and $\{v_e, v\}$ of H corresponding to an edge $e = \{u, v\} \in E(G)$. Note that at least one of u_e and v_e belongs to S_H. If both $u_e, v_e \in S_H$, add either u or v to S_G if none of u, v was already in S_G. Otherwise, if exactly one of u_e and v_e (say, u_e) belongs to S_H, then at least one of u and v must also belong also to S_H, and do not add any new vertex to S_G.

Note that as G is cubic, each vertex in a solution S_G covers exactly 3 edges, so $|E(G)| \leq 3 \cdot \text{OPT}_{\text{VC}}(G)$.

In order to prove the lemma, assume for contradiction that there exists a PTAS for MINIMUM VERTEX COVER in $\mathcal{H}$. That is, for any $\varepsilon > 0$, we can find in polynomial time a solution $S_H \subseteq V(H)$ such that $|S_H| \leq (1+\varepsilon) \cdot \text{OPT}_{\text{VC}}(H)$. By the above discussion, we can find a solution $S_G \subseteq V(G)$ such that

$$
\begin{aligned}
|S_G| &\leq |S_H| - |E(G)| \\
&\leq (1+\varepsilon) \cdot \text{OPT}_{\text{VC}}(H) - |E(G)| \\
&= (1+\varepsilon) \cdot (\text{OPT}_{\text{VC}}(G) + |E(G)|) - |E(G)| \\
&= (1+\varepsilon) \cdot \text{OPT}_{\text{VC}}(G) + \varepsilon \cdot |E(G)| \\
&\leq (1+\varepsilon) \cdot \text{OPT}_{\text{VC}}(G) + 3\varepsilon \cdot \text{OPT}_{\text{VC}}(G) \\
&= (1+4\varepsilon) \cdot \text{OPT}_{\text{VC}}(G) ,
\end{aligned}
$$

(a)   (b)

Fig. 1. (a) A two-coloring of the edges of the Petersen graph (grey and black) such that each monochromatic component is a path of length at most 5. (b) Construction of the lightpaths from a path of length 2 for several values of d, in the proof of Theorem 2. Full dots correspond to vertices of the VERTEX COVER instance (called *black* in the proof).

where we have used Eq. (1) and the fact that $|E(G)| \leq 3 \cdot \text{OPT}_{\text{VC}}(G)$. That is, the existence of a PTAS for MINIMUM VERTEX COVER in the class of graphs $\mathcal{H}$ would imply the existence of a PTAS in the class of cubic graphs, which is a contradiction by [2] unless P = NP.   □

It is known that the edges of any cubic graph can be two-colored such that each monochromatic connected component is a path (of any length) [1]. In fact, solving a conjecture of Bermond *et al.* [3], Thomassen proved [23] a stronger result: the edges of any cubic graph can be two-colored such that each monochromatic connected component is a path of length at most 5 (see Fig. 1(a) for an example). In addition, the aforementioned colorings can be found in polynomial time [1,23]. Note that in such a coloring of a cubic graph, each vertex appears exactly once as an endpoint of a path, and exactly once as an internal vertex of another path. We next show that these results can be easily strengthened for the family of graphs $\mathcal{H}$ defined in Lemma 1.

Lemma 2. *Let $\mathcal{H}$ be the class of graphs obtained from cubic graphs by subdividing each edge twice. The edges of any graph in $\mathcal{H}$ can be two-colored in polynomial time such that each monochromatic connected component is a path of length at most 2.*

Proof. Let $H \in \mathcal{H}$ be a graph obtained from a cubic graph G by subdividing each each twice. That is, edge $\{u, v\}$ of G gets replaced by 3 edges $\{u, u_e\}$, $\{u_e, v_e\}$, and $\{v_e, v\}$ in H. Find a two-coloring of the edges of G such that each monochromatic connected component is a path, using [1] or [23]. To color the edges of H, do the following for each edge $\{u, v\}$ of G: color $\{u, u_e\}$ and $\{v_e, v\}$ with the same color as $\{u, v\}$, and color $\{u_e, v_e\}$ with the other color. It is then easy to check that each monochromatic connected component of the obtained two-coloring of H is a path of length at most 2.   □

We now present the main results of this section. For the sake of presentation, we first present in Theorem 1 the result for the case $d = p = 2$, and then we show in Theorem 2 how to extend the reduction to any fixed $d, p \geq 2$.

Theorem 1. *$(2,2)$-TR does not admit a PTAS unless P = NP, even if G has maximum degree at most 3 and the lightpaths have length at most 4.*

Proof. The reduction is from MINIMUM VERTEX COVER (VC for short) in the class of graphs $\mathcal{H}$ obtained from cubic graphs by subdividing each edge twice, which does not admit a PTAS by Lemma 1 unless P = NP. Note that by construction any graph in $\mathcal{H}$ has girth at least 9. Given a graph $H \in \mathcal{H}$ as instance of VC, we proceed to build an instance of (2, 2)-TR. We set $G = H$, so G has maximum degree at most 3.

To define the two sets of lightpaths L_1 and L_2, let $\{E_1, E_2\}$ be the partition of $E(H)$ given by the two-coloring of Lemma 2. Therefore, each connected component of $H[E_1]$ and $H[E_2]$ is a path of length at most 2. Each such path in $H[E_1]$ (resp. $H[E_2]$) will correspond to a lightpath in L_1 (resp. L_2), which we proceed to define. A key observation is that, as the paths of the two-coloring have length at most 2, if any endpoint v of such a path P had one neighbor in $V(P)$, it would create a triangle, a contradiction to the fact that the girth of H is at least 9. Therefore, as the vertices of H have degree 2 or 3, any endpoint v of a path P has at least one neighbor in $V(H) \setminus V(P)$.

We are now ready to define the lightpaths. Let P be a path with endpoints u, v, and let u' (resp. v') be a neighbor of u (resp. v) in $V(H) \setminus V(P)$, such that $u' \neq v'$ (such distinct vertices u', v' exist because P has length at most 2 and H has girth at least 9; in fact we only need H to have girth at least 5). The lightpath associated with P consists of the concatenation of $\{u', u\}$, P, and $\{v, v'\}$. Therefore, the length of each lightpath is at most 4. This completes the construction of the instance of (2, 2)-TR. Observe that since we assume that $d = 2$, regenerators must be placed in such a way that all the internal edges of a lightpath (that is, all the edges except the first and the last one) have a regenerator in at least one of their endpoints. We can assume without loss of generality that no regenerator serves at the endpoints of a lightpath, as the removal of such regenerators does not alter the feasibility of a solution. Note that in our construction, each vertex of G appears as an internal vertex in at most two lightpaths, one (possibly) in L_1 and the other one (possibly) in L_2, so we can assume that $reg(v) \leq 1$ for any $v \in V(G)$.

We now claim that $\mathrm{OPT_{VC}}(H) = \mathrm{OPT}_{(2,2)-\mathrm{TR}}(G, \{L_1, L_2\})$.

Indeed, let first $S \subseteq V(H)$ be a vertex cover of H. Placing one regenerator at each vertex belonging to S defines a feasible solution to (2, 2)-TR in G with cost $|S|$, as at least one endpoint of each internal edge of each lightpath contains a regenerator. Therefore, $\mathrm{OPT_{VC}}(H) \geq \mathrm{OPT}_{(2,2)-\mathrm{TR}}(G, \{L_1, L_2\})$.

Conversely, suppose we are given a solution to (2,2)-TR in G using r regenerators. Since E_1 and E_2 are a partition of $E(G) = E(H)$ and the set of internal edges of the lightpaths in L_1 (resp. L_2) is exactly E_1 (resp. E_2), the regenerators placed at the endpoints of the internal edges of the lightpaths constitute a vertex cover of H of size at most r. Therefore, $\mathrm{OPT_{VC}}(H) \leq \mathrm{OPT}_{(2,2)-\mathrm{TR}}(G, \{L_1, L_2\})$.

Summarizing, since $\mathrm{OPT_{VC}}(H) = \mathrm{OPT}_{(2,2)-\mathrm{TR}}(G, \{L_1, L_2\})$ and any feasible solution to $\mathrm{OPT}_{(2,2)-\mathrm{TR}}(G, \{L_1, L_2\})$ using r regenerators defines a vertex cover of H of size at most r, the existence of a PTAS for (2, 2)-TR would imply the existence of a PTAS for VERTEX COVER in the class of graphs $\mathcal{H}$, which is a contradiction by Lemma 1, unless P = NP.   □

The result can be extended to general values of d and p (for details see [17]):

Theorem 2. (d, p)-TR *does not admit a PTAS for any $d \geq 2$ and any $p \geq 2$ unless $P = NP$, even if the underlying graph G satisfies $\Delta(G) \leq 3$ and the lightpaths have length at most $2d$.*

4   Online ADMs Minimization

Theorem 3. *For any $\epsilon > 0$, there is no $(\frac{3}{2} - \epsilon)$-competitive deterministic algorithm for path topology.*

Proof. Let G be a path with $2k$ nodes $u_1, v_1, u_2, v_2, ..., u_k, v_k$, $u_1 < v_1 < u_2 < v_2 < ... < u_k < v_k$. Let ALG be any deterministic algorithm. The value of k will be determined later.

The adversary works in two phases. In the first phase the input is $a_1, a_2, ..., a_k$ where $\forall i, a_i = (u_i, v_i)$. In the second phase the input depends on the decisions made by ALG during the first phase. Let $w(a_i)$ be the color assigned by the algorithm to path a_i. For every $1 \leq i < k$, if $w(a_i) = w(a_{i+1})$ then the input contains two paths $b_i = (u_1, u_{i+1})$ and $b'_i = (v_i, v_k)$, otherwise the input contains one path $c_i = (v_i, u_{i+1})$.

Let $0 \leq x \leq k - 1$ be the number of times $w(a_i) = w(a_{i+1})$ is satisfied. Then $w(a_i) \neq w(a_{i+1})$ is satisfied $k - 1 - x$ times.

During the first phase the algorithm uses $2k$ ADMs, one for each node.

For the paths b_i and b'_i, let $\lambda = w(a_i)(= w(a_{i+1}))$. λ is not feasible neither for b_i nor for b'_i. Then the algorithm assigns other colors to b_i and b'_i, and it uses 4 ADMs, for a total of $4x$ ADMs.

For the path c_i, let $\lambda = w(a_i)$ and $\lambda' = w(a_{i+1})(\neq \lambda)$, coloring c_i with one of these colors ALG uses one ADM, otherwise it uses 2 ADMs. Therefore for the paths c_i, ALG uses at least $k - 1 - x$ ADMs.

Summing up, we get that ALG uses at least $2k + 4x + (k - 1 - x) = 3(k + x) - 1$ ADMs.

On the other hand the following solution is possible. For any consecutive paths $c_i, c_{i+1}, ..., c_{i+j}$ color such that $w(b_{i-1}) = w(a_i) = w(c_i) = w(a_{i+1}) = w(c_{i+1}) = ... = w(c_{i+j}) = w(a_{i+j+1}) = w(b'_{i+j+1})$. This solutions use $2k + 2x$ ADMs, one ADM at each u_i, v_i, x additional ADMs at u_1, and x additional ADMs at v_k.

Therefore the competitive ratio of ALG is at least $\frac{3(k+x)-1}{2(k+x)} = \frac{3}{2} - \frac{1}{2(k+x)} \geq \frac{3}{2} - \frac{1}{2k}$. For any $\epsilon > 0$ we can choose $k > \frac{1}{2\epsilon}$, so that the competitive ratio of ALG is bigger then $\frac{3}{2} - \epsilon$. $\qquad\square$

5   Online Path Maximization

We consider the simple instance of the MaxPATH problem, i.e. the case where the network is a path, $d = 2$, and at most one regenerator can be place in one location.

We say that an instance is *feasible*, if there is a regenerator assignment that d-satisfies all the paths in $\mathcal{P}$, and *infeasible* otherwise. We show that if the input instance is infeasible, no online algorithm (for MAXPATH) has a small competitive ratio; precisely, we show that no online algorithm is better than $\sqrt{l}/2$-competitive, where l is the length of the longest path in the input. We then focus on feasible instances.

For infeasible instances we have the following result:

Lemma 3. *Consider the path topology. For $k = 1$ and $d = 2$, any deterministic online algorithm for MAXPATH has a competitive ratio at least $\sqrt{l}/2$, where l is the number of internal vertices of the longest path.*

Proof. The adversary first releases a path of length $l+1$ with l internal vertices. The online algorithm has to satisfy this path, otherwise, the competitive ratio is unbounded. Then the adversary releases $\sqrt{l}$ paths along the first path each with $\sqrt{l}$ (disjoint) internal vertices. If the online algorithm does not satisfy any of these paths, the competitive ratio is at least $\sqrt{l}$ and we are done. Suppose x of these paths are satisfied. In order to make the first path and these x paths 2-satisfied, there is one regenerator placed in each node along these x paths. For each of these x paths P, the adversary releases $\sqrt{l}/2$ paths along P each with two (disjoint) internal vertices. The online algorithm is not able to satisfy any of these short paths and the total number of 2-satisfied paths is $x + 1$. On the other hand, the optimal offline algorithm satisfies all the paths except the first path of length l, i.e., $\sqrt{l} + x\sqrt{l}/2$ paths. As a result, the competitive ratio of the online algorithm is $\frac{(x+2)\sqrt{l}}{2(x+1)} > \sqrt{l}/2$. $\square$

We now consider feasible instances, that is, instances, where there exists a placement of regenerators such that all paths are satisfied. We prove that, for feasible instances, there is a tight bound of 3 for the competitive ratio. Here we show a lower bound of 3 for the competitive ratio of every deterministic online algorithm for feasible instances. We also show (see [18]) an online algorithm which achieves the competitive ratio of 3.

Note that a regenerator assignment 2-satisfies a path P if and only if it constitutes a vertex cover of the edges of P, except its first and last edges. Therefore, in this section, for simplicity we assume that the leftmost and rightmost edges of the paths have been removed and a regenerator assignment is a vertex cover of the edges of the paths.

Theorem 4. *Any deterministic online algorithm for MAXPATH has a competitive ratio at least 3 even when the instance is restricted to feasible ones on path topologies and $k = 1, d = 2$.*

Proof. We will prove that, for every $\varepsilon > 0$, there exist infinitely many inputs such that every algorithm has competitive ratio at least $3 - \varepsilon$. Choose an integer n, such that $\frac{2}{n+1} < \varepsilon$. The adversary provides initially a path P_0 with $13n - 2$ edges. The algorithm must satisfy the path P_0, since otherwise the adversary stops and the competitive ratio is infinite. We divide P_0 into n subpaths P_i, $i = 1, 2, \ldots, n$, with 11 edges each, where between two consecutive subpaths there exist two edges.

Consider any such subpath P_i, $i = 1, 2, \ldots, n$. Suppose that there exist two edges ab and cd of P_i, where $\{a, b\} \cap \{c, d\} = \emptyset$, such that $reg(a, P_0) = reg(b, P_0) = reg(c, P_0) = reg(d, P_0) = 1$. Then the adversary provides next the paths $P_{i,1} = (a, b)$ and $P_{i,2} = (c, d)$. These two paths $P_{i,1}$ and $P_{i,2}$ cannot be satisfied, since each of the vertices a, b, c, d has a regenerator for path P_0. So the competitive ratio of the algorithm is at least 3.

We thus can assume that there do not exist such edges ab and cd for any of the P_i's. That is, there exist at most three consecutive vertices u_1, u_2, u_3 of P_i, such that $reg(u_1, P_0) = reg(u_2, P_0) = reg(u_3, P_0) = 1$, while for every other edge uu' of P_i, there exists a regenerator for P_0 either on vertex u or on vertex u'. Then, there exist five consecutive vertices $v_1^i, v_2^i, v_3^i, v_4^i, v_5^i$ of P_i, such that $reg(v_1^i, P_0) = reg(v_3^i, P_0) = reg(v_5^i, P_0) = 1$ and $reg(v_2^i, P_0) = reg(v_4^i, P_0) = 0$.

The adversary now provides the path $P_i' = (v_2^i, v_3^i, v_4^i)$. Thus, since $reg(v_3^i, P_0) = 1$ and $reg(v_2^i, P_0) = reg(v_4^i, P_0) = 0$, the only way that the algorithm can satisfy P_i' is to place regenerators for P_i' at the vertices v_2^i and v_4^i (that is, $reg(v_2^i, P_i') = reg(v_4^i, P_i') = 1$).

The adversary proceeds as follows. In the case where the algorithm chooses not to satisfy the path P_i', the adversary does not provide any other path that shares edges with P_i. Otherwise, if the algorithm satisfies P_i', then the adversary provides the paths $P_i'' = (v_1^i, v_2^i)$ and $P_i''' = (v_4^i, v_5^i)$ (see Fig. 2). In this case, $reg(v_2^i, P_i') = reg(v_4^i, P_i') = 1$ and $reg(v_1^i, P_0) = reg(v_5^i, P_0) = 1$, and thus the paths P_i'' and P_i''' remain unsatisfied by the algorithm.

We now show that this instance is indeed feasible. Actually, we show that even the instance that includes P_0, all the paths $P_i' = (v_2^i, v_3^i, v_4^i)$, all the paths $P_i'' = (v_1^i, v_2^i)$ and all the paths $P_i''' = (v_4^i, v_5^i)$ is feasible. To see this, we put regenerators at the nodes v_1^i, v_3^i, v_5^i, that will satisfy $P_i'' = (v_1^i, v_2^i)$, $P_i' = (v_2^i, v_3^i, v_4^i)$ $P_i''' = (v_4^i, v_5^i)$, respectively. We then put regenerators at all other nodes (including the nodes v_2^i, v_4^i), which clearly satisfies P_0.

Denote by h the number of subpaths P_i, for which the adversary adds the path P_i', P_i'' and P_i'''. Thus the number of subpaths P_i, for which the adversary adds P_i', but not P_i'' or P_i''', is $n - h$. The total number of paths that the adversary provided is thus $1 + 3h + (n - h) = 1 + n + 2h$. The number of paths satisfied by the algorithm is $1 + h$. That is, the competitive ratio of the algorithm is $\frac{1+n+2h}{1+h} = 3 + \frac{n-2-h}{1+h}$. Therefore, since $h \leq n$, it follows that the competitive

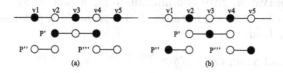

(a)   (b)

Fig. 2. Adversary for Lemma 4. (a) The online assignment where P'' and P''' cannot be satisfied. (b) The optimal assignment where all paths are satisfied.

ratio of the algorithm is at least $3 - \frac{2}{1+n} > 3 - \varepsilon$. Since this holds for every $\varepsilon > 0$, it follows that any deterministic online algorithm has competitive ratio at least 3. This completes the proof of the lemma.   □

6   Minimization of Regenerators with Grooming (Parallel Scheduling)

6.1   Preliminaries

We consider a scheduling problem in which a bounded number of jobs can be processed simultaneously by a single machine. The input is a set of n jobs $\mathcal{J} = \{J_1, \ldots, J_n\}$. Each job, J_j, is associated with an interval $[s_j, c_j]$ along which it should be processed. Also given is the parallelism parameter $g \geq 1$, which is the maximal number of jobs that can be processed simultaneously by a single machine. Each machine operates along a contiguous time interval, called its *busy interval*, which contains all the intervals corresponding to the jobs it processes. The goal is to assign the jobs to machines so that the total busy time is minimized.

The problem is known to be NP-hard already for $g = 2$. We present an algorithm whose competitive ratio is between 3 and 4.

Unless specified otherwise, we use lowercase letters for indices and upper case letters for jobs, time intervals and machines; also, calligraphic characters are used for sets of jobs, sets of intervals and sets of machines.

Definition 1. *Given a time interval $I = [s, c]$, the length of I is $len(I) = c - s$. This extends to a set $\mathcal{I}$ of intervals; namely, the length of $\mathcal{I}$ is $len(\mathcal{I}) = \sum_{I \in \mathcal{I}} len(I)$.*

Definition 2. *For a set $\mathcal{I}$ of intervals we define the span of $\mathcal{I}$ as $span(\mathcal{I}) = len(\cup \mathcal{I})$.*

Note that $span(\mathcal{I}) \leq len(\mathcal{I})$ and equality holds if and only if $\mathcal{I}$ is a set of pairwise disjoint intervals. Because jobs are given by time intervals, we use the above definitions also for jobs and sets of jobs, respectively.

Definition 3. *For any instance $\mathcal{J}$ and parallelism parameter $g \geq 1$, $OPT(\mathcal{J})$ denotes the cost of an optimal solution, that is, a solution in which the total busy time of the machines is minimized.*

The next observation gives two immediate lower bounds for the cost of any solution. *Observation* For any instance $\mathcal{J}$ and parallelism parameter $g \geq 1$, the following bounds hold:

- The parallelism bound: $OPT(\mathcal{J}) \geq \frac{len(\mathcal{J})}{g}$.
- The span bound: $OPT(\mathcal{J}) \geq span(\mathcal{J})$.

The parallelism bound holds since g is the maximum parallelism that can be achieved in any solution. The span bound holds since at any time $t \in \cup \mathcal{J}$ at least one machine is working.

W.l.o.g., we assume that the interval graph induced by the jobs is connected; otherwise, the problem can be solved by considering each connected component separately. Clearly, in any optimal solution, no machine is busy during intervals with no active jobs. w.l.o.g. we assume that each machine is busy along a contiguous interval. Given any solution we denote by $\mathcal{J}_i$ the set of jobs assigned to machine M_i by the solution. The cost of M_i is the length of its busy interval, i.e. $busy_i = span(\mathcal{J}_i)$.

6.2 Algorithm FirstFit

In this section we present an algorithm for general instances and show that its approximation ratio is between 3 and 4.

Algorithm FirstFit schedules the jobs greedily by considering them one after the other, from longest to shortest. Each job is scheduled to the first machine it can fit.

---

1. Sort the jobs in non-increasing order of length, i.e., $len(J_1) \geq len(J_2) \geq \ldots \geq len(J_n)$.
2. Consider the jobs by the above order: assign the next job, J_j, to the first machine that can process it, i.e., find the minimum value of $i \geq 1$ such that, at any time $t \in J_j$, M_i is processing at most $g - 1$ jobs. If no such machine exists, then open a new machine for J_j.

We show that algorithm FirstFit is a 4-approximation algorithm. Formally,

Theorem 5. *For any instance $\mathcal{J}$, $FirstFit(\mathcal{J}) \leq 4 \cdot OPT(\mathcal{J})$.*

Proof. The proof is based on the following observation (see [12]):

Observation: Let J be a job assigned to machine M_i by FirstFit, for some $i \geq 2$. For any machine $M_k, (k < i)$, there is at least one time $t_{i,k}(J) \in J$ and a set $s_{i,k}(J)$ of g jobs assigned to M_k such that, for every $J' \in s_{i,k}(J)$, it holds that (a) $t_{i,k}(J) \in J'$, and (b) $len(J') \geq len(J)$.

The following key lemma (for a proof see [12]) is needed in the proof of Theorem 5:

Lemma 4. *For any $i \geq 1$, $len(\mathcal{J}_i) \geq \frac{g}{3} span(\mathcal{J}_{i+1})$.*

Proof. Combining the span bound and Lemma 4 we can now complete the analysis of the algorithm. By definition, all the jobs in $\mathcal{J}_{i+1}$ are assigned to one machine, i.e. M_{i+1}. For such a set the cost of the assignment is exactly its span. Thus, $FirstFit(\mathcal{J}_{i+1}) = busy_{i+1} = span(\mathcal{J}_{i+1}) \leq \frac{3}{g} len(\mathcal{J}_i)$. Let $m \geq 1$ be the number of machines used by FirstFit. Then

$$\sum_{i=2}^{m} FirstFit(\mathcal{J}_i) = \sum_{i=1}^{m-1} FirstFit(\mathcal{J}_{i+1})$$

$$\leq \frac{3}{g} \sum_{i=1}^{m-1} len(\mathcal{J}_i)$$

$$< \frac{3}{g} \sum_{i=1}^{m} len(\mathcal{J}_i) = \frac{3}{g} len(\mathcal{J}) \leq 3 \cdot OPT(\mathcal{J})$$

where the last inequality follows from the parallelism bound.

Now, using the span bound, we have that $FirstFit(\mathcal{J}_1) = busy_1 = span(\mathcal{J}_1)$ $\leq span(\mathcal{J}) \leq OPT(\mathcal{J})$. Therefore, $FirstFit(\mathcal{J}) \leq 4 \cdot OPT(\mathcal{J})$.   □

Theorem 6. *For any $\varepsilon > 0$, there are infinitely many instances $\mathcal{J}$ having infinitely many input sizes, such that $FirstFit(\mathcal{J}) > (3 - \varepsilon) \cdot OPT(\mathcal{J})$.*

Proof. Consider the instance $\mathcal{J}$ depicted in Fig. 3.

For this instance OPT uses one machine in the interval $[0, 1]$, one machine in the interval $[2 - 2\varepsilon', 3 - 2\varepsilon']$, and $g - 1$ machines in the interval $[1 - \varepsilon', 2 - \varepsilon']$, for a total cost of $OPT(\mathcal{J}) = g + 1$. In contrast, FirstFit may use g machines in the interval $[0, 3 - 2\varepsilon']$ for a total cost of $FirstFit(\mathcal{J}) = (3 - 2\varepsilon')g = (3 - 2\varepsilon')\frac{g}{g+1}OPT(\mathcal{J})$. Choosing g and ε' appropriately (for example, $\varepsilon' = \varepsilon/4$ and $g \geq 6/\varepsilon - 1$) we get that $FirstFit(\mathcal{J}) > (3 - \varepsilon)OPT(\mathcal{J})$.   □

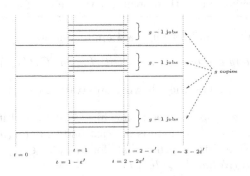

Fig. 3. Lower bound scenario

Combining Theorems 5 and 6, we get:

Theorem 7. *The approximation ratio of FirstFit is between 3 and 4.*

References

1. Akiyama, J., Chvátal, V.: A short proof of the linear arboricity for cubic graphs. Bull. Liberal Arts Sci. Nippon Med. Sch. **2**, 1–3 (1981)
2. Alimonti, P., Kann, V.: Some APX-completeness results for cubic graphs. Theoret. Comput. Sci. **237**(1–2), 123–134 (2000)
3. Bermond, J.-C., Fouquet, J.L., Habib, M., Péroche, B.: On linear k-arboricity. Discrete Math. **52**, 123–132 (1984)
4. Borodin, A., El-Yaniv, R.: Online Computation and Competitive Analysis. Cambridge University Press, Cambridge (1998)
5. Chen, S., Ljubic, I., Raghavan, S.: The regenerator location problem. Networks, **55**(3), 205–220 (2010)
6. Călinescu, G., Frieder, O., Wan, P.-J.: Minimizing electronic line terminals for automatic ring protection in general wdm optical networks. IEEE J. Sel. Area Commun. **20**(1), 183–189 (2002)
7. Călinescu, G., Wan, P.-J.: Traffic partition in wdm/sonet rings to minimize sonet ADMs. J. Comb. Optim. **6**(4), 425–453 (2002)
8. Eilam, T., Moran, S., Zaks, S.: Lightpath arrangement in survivable rings to minimize the switching cost. IEEE J. Sel. Area Commun. **20**(1), 172–182 (2002)
9. Epstein, L., Levin, A.: Better bounds for minimizing SONET ADMs. In: 2nd Workshop on Approximation and Online Algorithms, Bergen, Norway, September 2004
10. Fedrizzi, R., Galimberti, G.M., Gerstel, O., Martinelli, G., Salvadori, E., Saradhi, C.V., Tanzi, A., Zanardi, A.: Traffic independent heuristics for regenerator site selection for providing any-to-any optical connectivity. In: Proceedings of IEEE/OSA Conference on Optical Fiber Communications (OFC) (2010)
11. Flammini, M., Marchetti-Spaccamela, A., Monaco, G., Moscardelli, L., Zaks, S.: On the complexity of the regenerator placement problem in optical networks. IEEE/ACM Trans. Networking **19**(2), 498–511 (2011)
12. Flammini, M., Monaco, G., Moscardelli, L., Shachnai, H., Shalom, M., Tamir, T., Zaks, S.: Minimizing total busy time in parallel scheduling with application to optical networks. Theor. Comput. Sci. **411**(40–42), 3553–3562 (2010)
13. Flammini, M., Shalom, M., Zaks, S.: On minimizing the number of ADMs in a general topology optical network. In: Dolev, S. (ed.) DISC 2006. LNCS, vol. 4167, pp. 459–473. Springer, Heidelberg (2006)
14. Gerstel, O., Lin, P., Sasaki, G.: Wavelength assignment in a WDM ring to minimize cost of embedded SONET rings. In: INFOCOM 1998, Seventeenth Annual Joint Conference of the IEEE Computer and Communications Societies (1998)
15. Kim, S.W., Seo, S.W.: Regenerator placement algorithms for connection establishment in all-optical networks. IEEE Proc. Commun. **148**(1), 25–30 (2001)
16. Leonardi, S., Vitaletti, A.: Randomized lower bounds for online path coloring. In: Rolim, J.D.P., Serna, M., Luby, M. (eds.) RANDOM 1998. LNCS, vol. 1518, pp. 232–247. Springer, Heidelberg (1998)
17. Mertzios, G.B., Sau, I., Shalom, M., Zaks, S.: Placing regenerators in optical networks to satisfy multiple sets of requests. IEEE Trans. Networking **20**(6), 1870–1879 (2012)
18. Mertzios, G.B., Shalom, M., Wong, P.W.H., Zaks, S.: Online regenerator placement. In: Fernàndez Anta, A., Lipari, G., Roy, M. (eds.) OPODIS 2011. LNCS, vol. 7109, pp. 4–17. Springer, Heidelberg (2011)

19. Pachnicke, S., Paschenda, T., Krummrich, P.M.: Physical impairment based regenerator placement and routing in translucent optical networks. In: Optical Fiber Communication Conference and Exposition and The National Fiber Optic Engineers Conference, p. OWA2. Optical Society of America (2008)

20. Shalom, M., Wong, P.W., Zaks, S.: Optimal on-line colorings for minimizing the number of adms in optical networks. J. Discrete Algorithms **8**(2), 174–188 (2010)

21. Shalom, M., Zaks, S.: A $10/7 + \epsilon$ approximation scheme for minimizing the number of ADMs in SONET rings. In: First Annual International Conference on Broadband Networks, San-José, California, USA, pp. 254–262, October 2004

22. Sriram, K., Griffith, D., Su, R., Golmie, N.: Static vs. Dynamic Regenerator Assignment in Optical Switches: models and Cost Trade-offs. Workshop on High Performance Switching and Routing (HPSR), pp. 151–155 (2004)

23. Thomassen, C.: Two-coloring the edges of a cubic graph such that each monochromatic component is a path of length at most 5. J. Comb. Theor. Ser. B **75**(1), 100–109 (1999)

24. Yang, X., Ramamurthy, B.: Dynamic routing in translucent WDM optical networks. In: Proceedings of the IEEE International Conference on Communications (ICC), pp. 955–971 (2002)

25. Yang, X., Ramamurthy, B.: Sparse regeneration in translucent wavelength-routed optical networks: Architecture, network design and wavelength routing. Photonic Netw. Commun. **10**(1), 39–53 (2005)

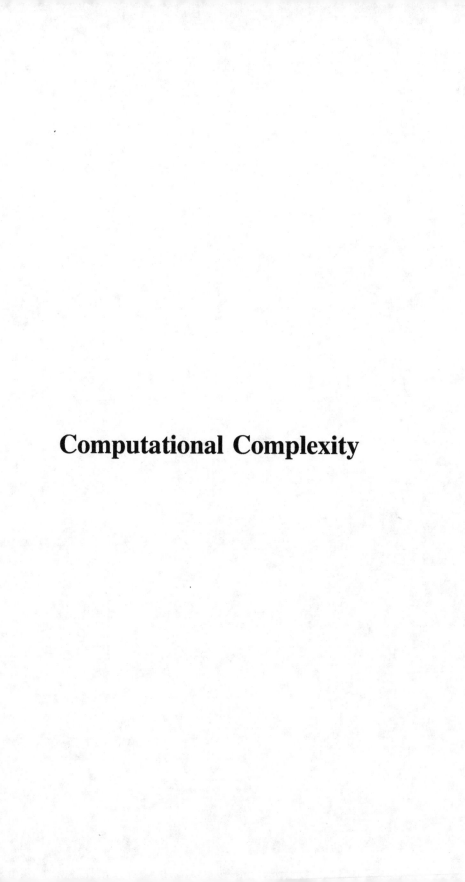

Computational Complexity

The Stable Fixtures Problem with Payments

Péter Biró[1,2], Walter Kern[3], Daniël Paulusma[4(✉)], and Péter Wojuteczky[1]

[1] Institute of Economics, Hungarian Academy of Sciences,
H-1112 Budaörsi út 45, Budapest, Hungary
peter.biro@krtk.mta.hu, peter.wojuteczky@gmail.com
[2] Department of Operations Research and Actuarial Sciences,
Corvinus University of Budapest, Budapest, Hungary
[3] Faculty of Electrical Engineering, Mathematics and Computer Science,
University of Twente, P.O. Box 217, 7500 AE Enschede, Netherlands
w.kern@math.utwente.nl
[4] School of Engineering and Computing Sciences, Durham University,
Science Laboratories, South Road, Durham DH1 3LE, UK
daniel.paulusma@durham.ac.uk

Abstract. We generalize two well-known game-theoretic models by introducing multiple partners matching games, defined by a graph $G = (N, E)$, with an integer vertex capacity function b and an edge weighting w. The set N consists of a number of players that are to form a set $M \subseteq E$ of 2-player coalitions ij with value $w(ij)$, such that each player i is in at most $b(i)$ coalitions. A payoff is a mapping $p : N \times N \to \mathbb{R}$ with $p(i, j) + p(j, i) = w(ij)$ if $ij \in M$ and $p(i, j) = p(j, i) = 0$ if $ij \notin M$. The pair (M, p) is called a solution. A pair of players i, j with $ij \in E \setminus M$ blocks a solution (M, p) if i, j can form, possibly only after withdrawing from one of their existing 2-player coalitions, a new 2-player coalition in which they are mutually better off. A solution is stable if it has no blocking pairs. We give a polynomial-time algorithm that either finds that no stable solution exists, or obtains a stable solution. Previously this result was only known for multiple partners assignment games, which correspond to the case where G is bipartite (Sotomayor 1992) and for the case where $b \equiv 1$ (Biro et al. 2012). We also characterize the set of stable solutions of a multiple partners matching game in two different ways and initiate a study on the core of the corresponding cooperative game, where coalitions of any size may be formed.

1   Introduction

Consider a group of soccer teams participating in a series of friendly games with each other off-season. Suppose each team has some specific target number

P. Biró—Supported by the Hungarian Academy of Sciences under Momentum Programme LD-004/2010, by the Hungarian Scientific Research Fund - OTKA (no. K108673), and by János Bolyai Research Scholarship of the Hungarian Academy of Sciences.

D. Paulusma—Supported by EPSRC Grant EP/K025090/1.

E.W. Mayr (Ed.): WG 2015, LNCS 9224, pp. 49–63, 2016.
DOI: 10.1007/978-3-662-53174-7_4

of games it wants to play. For logistic reasons, not every two teams can play against each other. Each game brings in some revenue, which is to be shared by the two teams involved. The revenue of a game may depend on several factors, such as the popularity of the two teams involved or the soccer stadium in which the game is played. In particular, at the time when the schedule for these games is prepared, the expected gain may well depend on future outcomes in the current season (which are in general difficult to predict [15]). In this paper, we assume for simplicity that the revenues are known. Is it possible to construct a *stable* fixture of games, that is, a schedule such that there exist no two unmatched teams that are better off by playing against each other? Note that this could require them to first cancel one of their others games in order not to exceed their targets.

The above example describes the problem introduced in this paper. We model it as follows. A *multiple partners matching game* is a triple (G, b, w), where $G = (N, E)$ is a finite undirected graph on n vertices and m edges with no loops and no multiple edges, $b : N \to \mathbb{Z}_+$ is a nonnegative integer function called a *vertex capacity function*, and $w : E \to \mathbb{R}_+$ is a nonnegative edge weighting. The set N is called the *player set*. There exists an edge $ij \in E$ if and only if players i, j can form a 2-player coalition. A set $M \subseteq E$ is a *b-matching* if every player i is incident to at most $b(i)$ edges of M. So, a b-matching is a set of 2-player coalitions, in which no player is involved in more 2-player coalitions than described by her capacity. If $ij \in M$ then i and j are *matched* by M. With M we associate a binary vector $x^M : E \to \{0, 1\}$ called the *characteristic function* of M, which is defined by $x^M(ij) = 1$ for all $ij \in M$ and $x^M(ij) = 0$ for all $ij \in E \setminus M$. Then we can write $\sum_{j:ij \in E} x^M(ij) \le b(i)$ for all $i \in N$ as an alternative way to state the capacity condition. The *value* of a 2-player coalition i, j with $ij \in E$ is given by $w(ij)$.

A nonnegative function $p : N \times N \to \mathbb{R}_+$ is a *payoff* with respect to M if $p(i, j) + p(j, i) = w(ij)$ for $ij \in M$ and $p(i, j) = p(j, i) = 0$ for $ij \notin M$; we also say that M and p are *compatible*. Note that p prescribes how the value $w(ij)$ of a 2-player coalition $\{i, j\}$ is distributed amongst i and j, ensuring that non-coalitions between two players yield a zero payoff. A pair (M, p), where M is a b-matching and p is a payoff compatible with M, is a *solution* for (G, b, w). If M is a 1-matching (i.e., a *matching*) then, for each $i \in N$, we have $p(i, j) > 0$ for at most one player $j \neq i$, which must be matched to i. Hence, if $b \equiv 1$, we assume, with slight abuse of notation, that p is a nonnegative function defined on N.

Let (M, p) be a solution. Two players i, j with $ij \in E \setminus M$ may decide to form a new 2-player coalition if they are "better off", even if one or both of them must first leave an existing 2-player coalition in M (in order not to exceed their individual capacity). To describe this formally we define a *utility function* $u_p : N \to \mathbb{R}_+$, related to payoff p. If i is *saturated* by M, that is, if i is incident with $b(i)$ edges in M, then we let $u_p(i) = \min\{p(i, j) \mid x^M(ij) = 1\}$ be the worst payoff $p(i, j)$ of any 2-player coalition i is involved in. Otherwise, i is *unsaturated* by M and we define $u_p(i) = 0$. A pair i, j with $ij \in E \setminus M$ *blocks* (M, p) if $u_p(i) + u_p(j) < w(ij)$. We say that (M, p) is *stable* if it has no blocking pairs, or equivalently, if every edge $ij \in E \setminus M$ satisfies the *stability condition* $u_p(i) + u_p(j) \geq w(ij)$. We can now define our problem formally:

STABLE FIXTURE WITH PAYMENTS (SFP)
Instance: A multiple partners matching game (G, b, w)
Question: Does (G, b, w) have a stable solution?

So far, we modelled only situations in which 2-player coalitions can be formed. Allowing coalitions of any size is a natural and well-studied setting in the area of Cooperative Game Theory. Moreover, as we will discuss, there exist close relationships between stable solutions and their counterpart in the second setting, the so-called core allocations, which we define below.

A *cooperative game with transferable utilities* (TU-game) is a pair (N, v), where N is a set of n *players* and a *value function* $v : 2^N \to \mathbb{R}_+$ with $v(\emptyset) = 0$. It is usually assumed that the *grand coalition* N is formed. Then the central problem is how to allocate the *total value* $v(N)$ to the individual players in N. In this context, a *payoff* (or *allocation*) is a vector $p \in \mathbb{R}^N$ with $p(N) = v(N)$, where we write $p(S) = \sum_{i \in S} p(i)$ for $S \subseteq N$. The *core* of a TU-game consists of all allocations $p \in \mathbb{R}^N$ satisfying

$$\begin{aligned} p(S) &\geq v(S), \quad \emptyset \neq S \subseteq N \\ p(N) &= v(N) \end{aligned} \tag{1}$$

A core allocation is seen as reasonable, because it offers no incentive for a subset of players to leave the grand coalition and form a coalition on their own. However, a TU-game may have an empty core. Hence, the most interesting computational complexity questions (given an input game) are:

1. Is the core nonempty?
2. Can we exhibit a vector in the core – provided there is any?
3. Does a given vector $p \in \mathbb{R}^N$ belong to the core?

In the literature both polynomial-time and NP-hardness results are known for each of these three questions (see e.g. [7]). An efficient algorithm for answering question 3 implies that questions 1–2 can be solved in polynomial time as well. This follows from the work of Grötschel et al. [11,12] who proved, by refining the ellipsoid method of Khachiyan [16], that an efficient algorithm for solving the separation problem for a polyhedron P implies a polynomial-time algorithm that either finds that P is empty, or obtains a vector of P.

We define the TU-game (N, v) that corresponds with a multiple partners matching game (G, b, w) by setting, for every $S \subseteq N$,

$$v(S) = w(M_S) = \sum_{e \in M_S} w(e),$$

where M_S is a maximum weight b-matching in the subgraph of G induced by S (we define $v(S) = 0$ if S induces an edgeless graph). We say that (N, v) is *defined* on (G, b, w) but, unless confusion is possible, we may also call (N, v) a multiple partners matching game. If we say that the payoff vector p of a stable solution is a core allocation, we mean that the *total payoff vector* $p^t \in \mathbb{R}^N$ defined by $p^t(i) = \sum_{ij \in E} p(i, j)$ for all $i \in N$ is a core allocation.

Example. Let G be the 4-vertex cycle $v_1 v_2 v_3 v_4 v_1$. We define a vertex capacity function b by $b(v_1) = b(v_2) = 1$ and $b(v_3) = b(v_4) = 2$, and an edge weighting w by $w(v_1 v_2) = 3$ and $w(v_2 v_3) = w(v_3 v_4) = w(v_4 v_1) = 1$. The pair (M, p) with $M = \{v_1 v_2, v_3 v_4\}$ and $p(v_1, v_2) = p(v_2, v_1) = \frac{3}{2}$, $p(v_3, v_4) = p(v_4, v_3) = \frac{1}{2}$ and $p(v_2, v_3) = p(v_3, v_2) = p(v_4, v_1) = p(v_1, v_4) = 0$ is a solution for the multiple partners matching game (G, b, w). Note that $u_p(v_1) = u_p(v_2) = \frac{3}{2}$ and $u_p(v_3) = u_p(v_4) = 0$. We find that (M, p) is even a stable solution, because $u_p(v_2) + u_p(v_3) = \frac{3}{2} \geq 1 = w(v_2 v_3)$ and $u_p(v_4) + u_p(v_1) = \frac{3}{2} \geq 1 = w(v_4 v_1)$ (note that we only need to verify the stability condition for edges outside the matching). Moreover, the total payoff vector p^t given by $p^t(v_1) = p^t(v_2) = \frac{3}{2}$ and $p^t(v_3) = p^t(v_4) = \frac{1}{2}$ is readily seen to be a core allocation of the corresponding TU-game. In Sect. 3 we will give an example of a multiple partners matching game with no stable solutions for which the corresponding TU-game has a nonempty core.

Before stating our results for multiple partners matching games in both settings we first discuss some existing work. As we will see, our model in both settings generalizes (or relaxes) several well-known models.

Known Results. The first model that we discuss is related to the famous *stable marriage problem* (SM), defined by Gale and Shapley [9] as follows. Given sets I and J of men and women, respectively, let each player have a strict preference ordering over the opposite set of players. A set of marriages is a matching in the underlying bipartite graph with partition classes I and J. Such a matching is stable if there is no unmarried pair, who would prefer to marry each other instead of a possible other partner. Gale and Shapley [9] proved that every instance of this problem has a stable matching and gave a linear-time algorithm that finds one. The main assumptions in this model are

   (i) monogamy: each player is matched to at most one other player (1-matching);
  (ii) opposite-sex: every match is between players from I and J (bipartiteness);
 (iii) no dowry: only cardinal preferences are considered (no payments).

Dropping one or more of these three conditions leads to seven new models, one of which corresponds to our model of multiple partner matching games, namely the one, in which *none* of the three conditions (i)–(iii) is imposed. Below we briefly survey the other six models (see also Table 1).

In the first three models that we discuss, payments are not allowed. Hence, the notion of a (core) allocation is meaningless for these three models.

Not (i). If we allow bigamy, that is, if we allow general b-matchings instead of only 1-matchings, we obtain the *many-to-many stable matching problem*, which generalizes the stable marriage problem. The problem variant, in which we demand that $b(i) = 1$ for each player $i \in I$, is called the *college admission problem* [9],

Table 1. The eight different models; "mp" stands for "multiple partners", and the highlighted case is the new model introduced and considered in this paper.

		Opposite-sex	Different-sex allowed
Monogamy	No dowry	SM (stable marriage)	SR (stable roommates)
	Dowry	SMP & assignment game	SRP & matching game
Bigamy allowed	No dowry	many-to-many stable matching	SF (stable fixtures)
	Dowry	MPA & mp assignment game	**SFP & mp matching game**

which is also known as the *many-to-one stable matching problem* [22] and as the *hospital/residents problem* [20]. Gale and Shapley [9] proved that every instance of the college admission problem has a stable matching and gave a linear-time algorithm that finds one. Baïou and Balinski [1] proved these two results for the (more general) many-to-many stable matching problem.

Not (ii). If we allow same-sex marriages, so the underlying graph may be non-bipartite, then we get the *stable roommates problem* (SR), also defined by Gale and Shapley [9]. They proved that, unlike the previously discussed models, in this model a stable matching does not always exist. Irving [13] gave a linear-time algorithm for finding a stable matching (if there exists one).

Not (i) & (ii). Allowing bigamy *and* same-sex marriages leads to the *stable fixtures problem* (SF), which generalizes the stable roommates problem. Hence, a stable matching does not always exist. Irving and Scott [14] gave a linear-time algorithm for finding a stable matching (if there exists one). Cechlárová and Fleiner [5] defined the more general *multiple activities problem*, in which the underlying graph may have multiple edges. They proved that even in this setting a stable matching can be found in polynomial time (if there exists one). Moreover, they also showed that this problem can be reduced to SR by a polynomial size graph construction.

In the remaining three models we allow payments to individual players.

Not (iii). If we allow dowry then we obtain an *assignment game*, which is a multiple partners matching game (G, b, w) where G is bipartite and $b \equiv 1$. In this case the SFP problem is known as the *stable marriage problem with payments* (SMP). Koopmans and Beckmann [18] proved that every instance of SMP has a stable solution. Shapley and Shubik [23] proved that every core allocation is a payoff vector in a stable solution and vice versa. Consequently, every assignment game has a nonempty core. It is possible to obtain a stable solution in polynomial time and also to give affirmative answers to questions 1–3 about the core of an assignment game; in the next paragraph we explain that this holds even if we allow same-sex marriages.

Not (ii) & (iii). If we allow dowry and same-sex marriages then we obtain a *matching game*, which is a multiple partners matching game (G, b, w) where $b \equiv 1$. In this case the SFP problem is called the *stable roommates problem with payments* (SRP). The following two observations are well-known [4,8] and easy to verify. First, a payoff p is a core allocation of a matching game if and only if there exists a matching M, such that (M,p) is a stable solution. Second, for matching games, the coalitions in the system of inequalities (1) may be restricted to 2-player coalitions. The latter means that question 3, on core membership, can be answered in linear time. We also obtain polynomial-time algorithms for solving questions 1–2, about core nonemptiness, and finding a core allocation, and thus finding a stable solution; the restriction to 2-player coalitions even allows one to use the ellipsoid method of Khachiyan [16] directly. In a previous paper [4], we circumvented the ellipsoid method and presented an $O(nm + n^2 \log n)$-time algorithm that either finds that the core is empty, or obtains a core allocation.

Not (i) & (iii). If we allow dowry and bigamy then we obtain a *multiple partners assignment game*, which is a multiple partners matching game (G, b, w) where G is bipartite. In this case the SFP problem is called the *multiple partners assignment problem* (MPA). Just as matching games, multiple partners assignment games generalize assignment games. Sotomayor proved the following, which answers questions 1–2 positively (the answer to question 3 is still open).

Theorem 1 ([24]). *Every multiple partners assignment game has at least one stable solution, which can be found in polynomial time. Moreover, for every stable solution (M,p) it holds that M has maximum weight, p is a core allocation and every other maximum weight b-matching is compatible with p.*

Our Results. In Sect. 2 we will prove that SFP is polynomial-time solvable. This generalizes the aforementioned corresponding results for SRP and MPA, respectively. Our proof technique is based on a reduction to MPA. Moreover, we characterize the set of stable solutions for a given instance of SFP via a reduction to SRP. We do this via linear programming techniques that show a close relationship between optimal solutions in the dual LP for SFP and stable solutions in the reduced instance of SRP.

In Sect. 3 we first prove that also for multiple partners matching games, the payoff vectors in stable solutions are always core allocations. We then prove that core membership, which corresponds to question 3, is polynomial-time solvable for multiple partner matching games defined on a triple (G, b, w) with $b \leq 2$, that is, with $b(i) \leq 2$ for all $i \in N$. Due to the aforementioned result of Grötschel et al. [11,12] we also obtain efficient answers to questions 1–2 (for $b \leq 2$). In our proof, we make a connection to the tramp steamer problem [17].

Finally, in Sect. 4, we give some directions for future work.

2   Stable Fixtures with Payments

In order to prove our results we will use two known results as lemmas.

Lemma 1 ([3]). *If (M,p) is a stable solution of an instance $(G, 1, w)$ of SRP then M has maximum weight, and every maximum weight matching M' of G is compatible with p.*

Lemma 2 ([19]). *Let G be a graph with vertex capacity function b and edge weighting w. Then it is possible to find a maximum-weight b-matching of G in $O(n^2 m \log(n^2/m))$ time.*

2.1 Characterizing Stable Solutions of SFP

In this subsection we show a correspondence of stable solutions of an instance (G, b, w) of SFP with stable solutions of a corresponding instance $(G', 1, w')$ of SRP, where G' is a graph of size $O(n^3)$, and with integral optimal solutions of an LP relaxation.

We first explain how to construct $(G', 1, w')$; see also Fig. 1. Our construction is based on a well-known construction, which was introduced by Tutte [25] for nonbipartite graphs with no edge weights. For each player $i \in N$ with capacity $b(i)$ we create $b(i)$ copies, $i^1, i^2, \ldots i^{b(i)}$ in N'. For each edge $ij \in E$ we create four players, $\overline{i}_j, i_j, j_i, \overline{j}_i$, with edges $i^s \overline{i}_j$ for $s = 1 \ldots b(i)$, $\overline{i}_j i_j$, $i_j j_i$, $j_i \overline{j}_i$ and $\overline{j}_i j^t$ for $t = 1 \ldots b(j)$, each with weight $w(ij)$. This completes our construction. We write $G' = (N', E')$. Note that G' is bipartite if and only if G is bipartite. Hence, our construction also reduces an instance of MPA to an instance of SMP.

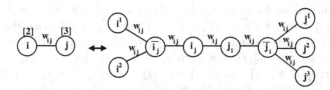

Fig. 1. The construction of an SPR instance $(G', 1, w')$ from a SFP instance (G, b, w).

Given an instance (G, b, w) of SFP, we formulate the corresponding primal LP, denoted by Primal-(G, b, w), as follows.

$$\max \sum_{ij \in E} w(ij) x(ij) \tag{P-obj}$$

$$\text{subject to} \quad \sum_{j: ij \in E} x(ij) \leq b(i) \text{ for each } i \in N \tag{a}$$

$$0 \leq x(ij) \leq 1 \text{ for each } ij \in E. \tag{b}$$

The integral solutions of this LP are the b-matchings of instance (G, b, w). We now formulate the dual LP, denoted by Dual-(G, b, w).

$$\min \sum_{i \in N} b(i) y(i) + \sum_{ij \in E} d(ij) \tag{D-obj}$$

$$\text{subject to}\quad y(i) + y(j) + d(ij) \geq w(ij) \text{ for each } ij \in E, \tag{a'}$$
$$\text{where } 0 \leq y(i) \text{ for all } i \in N, \text{ and } 0 \leq d(ij) \text{ for all } ij \in E.$$

Note that for an optimal dual solution (y, d), it holds that $d(ij) = [w(ij) - y(i) - y(j)]_+$ (where the latter notation means $max\{w(ij) - y(i) - y(j), 0\}$). We are now ready to prove our main result.

Theorem 2. *Let (G, b, w) be an instance of* SFP. *The following statements are equivalent.*

1. *(G, b, w) has a stable solution.*
2. *$(G', 1, w')$ has a stable solution.*
3. *Primal-(G, b, w) has an integral optimal solution.*

Proof. We prove in three separate statements that 1 implies 2, 2 implies 3 and 3 implies 1.

1⇒2. Given a stable solution (M, p) of (G, b, w), we define (M', p') of $(G', 1, w')$ as follows. Recall that we define p' as a function on N, as $b \equiv 1$. The payments of the copies will be the same as the minimum payments of the original players, that is, for each $i \in N$, let $p'(i^s) = u_p(i)$ for every $s = 1 \ldots b(i)$. For each $ij \in M$, if j is i's s-th partner for some $s \in \{1 \ldots b(i)\}$ and i is j's t-th partner for some $t \in \{1 \ldots b(j)\}$ then let $i^s \overline{i_j} \in M'$, $i_j j_i \in M'$ and $\overline{j_i} j^t \in M'$ with the following payments: $p'(i_j) = p(i, j)$ and $p'(\overline{i_j}) = w(ij) - u_p(i)$, and similarly $p'(j_i) = p(j, i)$ and $p'(\overline{j_i}) = w(ij) - u_p(j)$. For each $ij \in E \setminus M$, let $\overline{i_j} i_j$, $j_i \overline{j_i} \in M'$ with $p'(i_j) = min\{u_p(i), w(ij)\}$ and $p'(\overline{i_j}) = w(ij) - min\{u_p(i), w(ij)\}$, and similarly $p'(\overline{j_i}) = w(ij) - min\{u_p(j), w(ij)\}$ and $p'(j_i) = min\{u_p(j), w(ij)\}$. We observe that (M', p') is a solution for $(G', 1, w')$. In order to prove that (M', p') is a stable solution we have to check the stability condition for each edge not in M'. As $u_{p'}(i) = p'(i)$ for any $i \in N'$, this comes down to checking whether $p'(i) + p'(j) \geq w(ij)$ for all $i, j \in N'$. For each edge $i^s \overline{i_j}$ not in M' there are two cases. In the first case, when $p'(\overline{i_j}) = w(ij) - u_p(i)$, the condition is satisfied by equality, because $p'(i^s) = u_p(i)$. In the other case, we have $p'(\overline{i_j}) = w(ij) - min\{u_p(i), w(ij)\}$, so $p'(i^s) + p'(\overline{i_j}) = u_p(i) + w(ij) - min\{u_p(i), w(ij)\} \geq w(ij)$. For each edge $\overline{i_j} i_j$ not in M', $p'(\overline{i_j}) = w(ij) - u_p(i)$ and $p'(i_j) = p(i, j)$, so $p'(\overline{i_j}) + p'(i_j) \geq w(ij)$ as $u_p(i) \leq p(i, j)$. Finally, if $i_j j_i$ is not in M' then in the first case, when any payoff of the middle players is $w(ij)$, the requirement is trivially true, and in the second case when the payoffs of both players differ from $w(ij)$ the stability of (M, p) implies that $p'(i_j) + p'(j_i) = u_p(i) + u_p(j) \geq w(ij)$.

2⇒3. Suppose that (M', p') be a stable solution for $(G', 1, w')$. We will first prove that (M', p') can be transformed into a stable solution (M'', p') of $(G', 1, w')$, where M'' is a matching that we may obtain by the above reduction from (G, b, w) to $(G', 1, w')$. From Lemma 1 we know that M' is a maximum weight matching in $(G', 1, w')$. For a player set X let $E(X)$ denote the set of edges incident to any player in X. Let $E_{ij} = E(\{\overline{i_j}, i_j, j_i, \overline{j_i}\})$. Considering the edge set E_{ij}, either three or two edges should be contained in a maximum weight matching M'. If $|E_{ij} \cap M'| = 3$ then M' should contain $i_j j_i$, $i^s \overline{i_j}$ for some

$s \in \{1, \ldots, b(i)\}$ and $\overline{j_i} j^t$ for some $t \in \{1, \ldots, b(j)\}$. Let M'' contain the same edges in this case. If $|E_{ij} \cap M'| = 2$ then let $\{\overline{i_j} i_j, j_i \overline{j_i}\} \subseteq M''$. Therefore M'' is a maximum weight matching, and from Lemma 1, (M'', p') is a stable solution in $(G', 1, w')$, which can be obtained from a reduction from (G, b, w) to $(G', 1, w')$, as described in the first part of our proof. Now we note that $p'(i^s) = p'(i^t)$ for any indices $s, t \in \{1 \ldots b(i)\}$. This is because $p'(i^s) < p'(i^t)$ would imply that $p'(i^s) > 0$, so i^s must be covered by M', say, $i^s \overline{i_j} \in M'$, where $p'(i^s) + p'(\overline{i_j}) = w(ij)$, and hence $i^r \overline{i_j}$ would be blocking, a contradiction. Therefore we can set $y(i) = p'(i^s)$ for every copy i^s of $i \in N$, as this is well-defined. Together with $d(ij) = [w(ij) - y(i) - y(j)]_+$ for any $ij \in E$ we get a feasible solution (y, d) of Dual-(G, b, w). Now we define an integral feasible solution x of Primal-(G, b, w) as follows. Let $x(ij) = 1$ if $i_j j_i \in M''$ and $x(ij) = 0$ otherwise. We prove that (y, d) and x are both optimal solutions as they satisfy the complementary slackness conditions.

- If $\sum_{j:ij \in E} x(ij) < b(i)$ then some copy of $i \in N$, say i^s, is unmatched in M'', therefore $y(i) = p'(i^s) = 0$.
- If $x(ij) < 1$ for some $ij \in E$ then $i_j j_i \notin M''$ and therefore $y(i) + y(j) \geq p'(i_j) + p'(j_i) \geq w(ij)$, where the first inequality is implied by the stability condition for $i^s \overline{i_j} \notin M''$ and $j^t \overline{j_i} \notin M''$, and the second inequality is implied by the stability condition for $i_j j_i \notin M''$. As a consequence $d(ij) = [w(ij) - y(i) - y(j)]_+ = 0$ should hold.
- If $x(ij) > 0$ then $i_j j_i \in M''$ and therefore $y(i) + y(j) \leq p'(i_j) + p'(j_i) = w(ij)$, where the inequality is implied by the stability condition for $i_j \overline{i_j} \notin M''$ and $j_i \overline{j_i} \notin M''$. As a consequence $d(ij) = w(ij) - y(i) - y(j)$ and the dual condition (a') is binding.

Hence x is an optimal integral solution for Primal-(G, b, w), as required.

3⇒1. From an optimal solution x of Primal-(G, b, w) and optimal dual solution (y, d) of Dual-(G, b, w) we create a stable solution (M, p) for (G, b, w) as follows. Let M be the b-matching defined by the characteristic function x. For each $ij \in M$ we choose $\xi(i, j) \geq 0$ and $\xi(j, i) \geq 0$ with $\xi(i, j) + \xi(j, i) = d(ij)$ and define $p(i, j) = y(i) + \xi(i, j)$, and otherwise we define $p(i, j) = 0$. These are valid payments, as $x(ij) = 1$ implies $p(i, j) + p(j, i) = y(i) + \xi(i, j) + y(j) + \xi(j, i) = y(i) + y(j) + d(ij) = w(ij)$, where the last equality is coming from the fact that condition (a') is binding. Now we show that $u_p(i) \geq y(i)$ for every $i \in N$. If i is unsaturated by M then $u_p(i) = y(i) = 0$ again by the complementary slackness condition for $y(i)$. If i is saturated then for every $ij \in M$ we have $p(i, j) \geq y(i)$ by our setting of $p(i, j)$, and therefore $u_p(i) \geq y(i)$ by the definition of $u_p(i)$. Finally, if $ij \notin M$ then $x(ij) < 1$ implies $d(ij) = 0$. Consequently, $u_p(i) + u_p(j) \geq y(i) + y(j) \geq w(ij)$, where the last equality is due to the fact that (a') is tight. This completes our proof. $\qquad \Box$

Remark 1. From the proof of Theorem 2 we note that there is a one-to-one correspondence between the stable payments in an SFP instance (G, b, w) and the utilities of some of the players in its reduced SPR instance $(G', 1, w')$, namely a pair (M, p) is a stable solution for (G, b, w) if and only if there exists a stable solution (M', p') for $(G', 1, w')$ with $p'(i_j) = p(i, j)$ for every $ij \in M$.

Remark 2. Assume (G, b, w) is a solvable instance of SFP. Then there is a one-to-one correspondence between the dual variables y of an optimal solution (y, d) of Dual-(G, b, w) and the stable payoffs of the players' copies in the reduction G'.

2.2   Solving SFP Efficiently

In order to solve SFP on an instance (G, b, w), we construct the instance $(G', 1, w')$ of SRP. This takes $O(n^3)$ time as we may assume without loss of generality that $b(i) \leq n$ for all $i \in N$ and thus $|V(G')| = \sum_{i \in N} b(i) + 4m = O(n^2)$ and $|E(G')| \leq \sum_{i \in N} b(i)n + 3m \leq n^3 + 3m = O(n^3)$. We then use the aforementioned algorithm of Biró et al. [4] to compute in $O(n'm' + n'^2 \log n') = O(n^5)$ time a stable solution for $(G', 1, w')$ or else conclude that $(G', 1, w')$ has no stable solution. In the first case we can modify the stable solution into a stable solution of (G, b, w) in $O(n^3)$ time, as described in the proof of Theorem 2. In the second case, Theorem 2 tells us that (G, b, w) has no stable solution. The total running time is $O(n^5)$. Below we present an algorithm that solves SFP in $O(n^2 m \log(n^2/m))$ time.

A *half-b-matching* in a graph $G = (N, E)$ with an integer vertex capacity function b and an edge weighting w is an edge mapping f that maps each edge e to a value in $\{0, \frac{1}{2}, 1\}$, such that $\sum_{e:v \in e} f(e) \leq b(v)$ for each $v \in N$. The weight of f is $w(f) = \sum_{e \in E} w(e) f(e)$.

Let (G, b, w) be an instance of SFP. We define its *duplicated* instance $(\hat{G}, \hat{b}, \hat{w})$ of MPA as follows. We replace each player i of G by two players i' and i'' in $\hat{G}$ with the same capacities, that is, we set $\hat{b}(i') = \hat{b}(i'') = b(i)$. Moreover, we replace each edge ij by two edges $i'j''$ and $i''j'$ with half-weights, that is, we set $\hat{w}(i'j'') = \hat{w}(i''j') = \frac{1}{2} w(ij)$.

In a previous work [4], three of us proved the following statement for instances of SRP only. In our next theorem we generalize this result for instances of SFP by using similar arguments (proof omitted due to space restrictions).

Theorem 3. *An instance (G, b, w) of SFP admits a stable solution if and only if the maximum weight of a b-matching in G is equal to the maximum weight of a half-b-matching in G.*

We note that the maximum weight of a b-matching can be computed in $O(n^2 m \log(n^2/m))$ time, as described in Lemma 2. The maximum weight of a half-b-matching can be computed in the same running time, since the maximum weight of a half-b-matching in (G, b, w) is the same as the maximum weight of a b-matching in a duplicated bipartite graph $(\hat{G}, \hat{b}, \hat{w})$, as explained in the proof of Theorem 3. This leads to the following result.

Theorem 4. *SFP can be solved in $O(n^2 m \log(n^2/m))$ time.*

3   Core Properties

We first present the following result, which is in line with corresponding results for the other models and thus shows that the notion of stability is well defined with respect to the core definition. We omit the proof due to page restrictions.

Proposition 1. *The payoff vector of every stable solution of a multiple partners matching game is a core allocation.*

In contrast to our results in Sect. 2, the analysis of the core cannot be reduced to the the case in which we have unit vertex capacities. We illustrate this by giving the following example that shows that the core of a multiple partners matching game can be nonempty whilst there exist no stable solutions.

Example 1. Take a *diamond*, that is, a cycle on three vertices s_1, s_2, s_3 to which we add a fourth vertex u with edges $s_2 u$ and $s_3 u$. We set $b(s_i) = 2$ for $i = 1, 2, 3$ and $b(u) = 1$, and $w \equiv 1$. Then Theorem 3 tells us that a stable solution does not exist, since the maximum weight of a b-matching is 3, whilst the maximum weight of a half-b-matching is $3\frac{1}{2}$ (for the latter, take $f(s_1 s_2) = f(s_1 s_3) = 1$ and $f(s_2 s_3) = f(s_2 u) = f(s_3 u) = \frac{1}{2}$). However, the total payoff vector p^t defined by $p^t(s_i) = 1$ for $i = 1, \ldots, 3$ and $p^t(u) = 0$, which corresponds to, say, the b-matching $M = \{s_1 s_2, s_1 s_3, s_2 s_3\}$ with payments $p(s_1, s_2) = 1$, $p(s_2, s_3) = 1$ and $p(s_3, s_1) = 1$ and zero payments for the other edges is in the core.

Moreover, the core of a multiple partners matching game may be empty. This is clear for $b \equiv 1$ (for instance, take a triangle on three vertices and let $w \equiv 1$) but may also be the case for $b \neq 1$. We illustrate the latter by presenting the following example, which shows that the core of a b-matching game may be empty even if $b \equiv 2$.

Example 2. Take a *net*, that is, a cycle on three vertices s_1, s_2 and s_3 to which we add a pendant (degree 1) vertex t_i to s_i for $i = 1, 2, 3$. We set $b \equiv 2$ and $w \equiv 1$. Then $v(N) = 4$. We may assume without loss of generality that $p(t_i) = 0$ for $i = 1, 2, 3$. Then, by symmetry, $p(s_i) = \frac{4}{3}$ for $i = 1, 2, 3$. Take the coalition $S = \{s_1, s_2, t_1, t_2\}$. It holds that $p(S) = 2 \times \frac{4}{3} < 3 = v(S)$ and thus the core of this game is empty.

In what follows, we analyze the case $b(i) \leq 2$ for $i = 1, \ldots, n$. Note that a player i with $b(i) = 0$ necessarily gets 0 payoff in any core allocation, so the problem reduces trivially to $G \backslash \{i\}$. For this reason we assume $b(i) \geq 1$ for all $i \in N$ in the following.

Our main result is that answering question 3, on testing core membership, can be done in polynomial time, hereby answering question 3 positively. Recall that this implies the existence of a polynomial-time algorithm that either finds that the core is empty, or else obtains a core allocation. As mentioned in Sect. 1, our algorithm uses an algorithm that solves the tramp steamer problem, which we formally define below, as a subproblem.

Let $G = (N, E)$ be a graph with an edge weighting $p : E \rightarrow \mathbb{R}_+$ called the *profit function* and an edge weighting $w : E \rightarrow \mathbb{R}_+$ called the *cost function*. Let $C = (N_C, E_C)$ be a (simple) cycle of G. The *profit-to-cost ratio* for a cycle C is $\frac{p(C)}{w(C)}$ where we write $p(C) = p(E_C)$ and $w(C) = w(E_C)$. The tramp steamer problem is that of finding a cycle C with maximum profit-to-cost ratio. This problem is well-known to be polynomial-time solvable both for directed and undirected graphs (see [17], or [21] for a treatment of more general "fractional optimization" problems).

Lemma 3. *The tramp steamer problem can be solved in polynomial time.*

We are now ready to prove the main result of this section.

Theorem 5. *It is possible to test in polynomial time if an allocation is in the core of a multiple partners matching game defined on a triple (G, b, w) with $b \leq 2$.*

Proof. Let (N, v) be a multiple partners matching game defined on a triple (G, b, w), where $b(i) \leq 2$ for all $i \in N$. Given $S \subseteq N$, a maximum weight b-matching in $G[S]$ is composed of cycles and paths. Hence the core can be alternatively described by the following (slightly smaller) set of constraints:

$$
\begin{aligned}
p(C) &\geq w(C), \text{ for all cycles } C \in \mathcal{C} \\
p(P) &\geq w(P), \text{ for all paths } \ P \in \mathcal{P} \qquad (2) \\
p(N) &= v(N).
\end{aligned}
$$

Here, $\mathcal{C}$ stands for the set of simple cycles $C \subseteq E$ in G with $b(i) = 2$ for all $i \in V(C)$. Similarly, $\mathcal{P}$ stands for the set of simple paths P with $b(i) = 2$ for all inner points on P.

Given $p \in \mathbb{R}^N$, we can of course easily check whether $p(N) = v(N)$ holds by computing a maximum weight b-matching in G, which can be done in polynomial time by Lemma 2. Thus we are left with the inequalities for cycles and paths in (2).

We deal with the cycles first. Let $N_2 := \{i \in N \mid b(i) = 2\}$ and $G_2 := G[N_2]$. In the induced graph $G_2 = (N_2, E_2)$, we "discharge" the given allocations $p(i)$ to the edges by setting $p(i, j) := (p(i) + p(j))/2$ for every edge ij in G_2. This defines an edge weighting $p : E_2 \rightarrow \mathbb{R}$ such that, obviously, the core constraints for cycles are equivalent to

$$
\max_{C \in \mathcal{C}} \frac{w(C)}{p(C)} \leq 1,
$$

where the maximum is taken over all cycles in G_2. Hence we obtained an instance of the tramp steamer problem, which is polynomial-time solvable by Lemma 3. Note that by solving the above minimization problem we either find that all cycle constraints in (2) are satisfied or we end up with a particular cycle C corresponding to a violated core inequality. (The latter is of particular importance if we intend to use the "membership oracle" as a subroutine for the ellipsoid method.)

In what follows, we thus assume that all cycle constraints in (2) are satisfied by the given vector $p \in \mathbb{R}^N$ and turn to the path constraints. We process these separately for all possible endpoints $i_0, j_0 \in N$ (with $i_0 \neq j_0$) and all possible lengths $k = 1, ..., n - 1$. Let $\mathcal{P}_k(i_0, j_0) \subseteq \mathcal{P}$ denote the set of (simple) $i_0 - j_0$-paths of length k in G. We construct a corresponding auxiliary graph $G_k(i_0, j_0)$, a subgraph of $G \times P_{k+1}$, the product of G with a path of length k. To this end, let $N_2^{(1)}, ..., N_2^{(k-1)}$ be $k - 1$ copies of N_2. The vertex set of $G_k(i_0, j_0)$ is then $\{i_0, j_0\} \cup N_1^{(1)} \cup ... \cup N_2^{(k-1)}$. We denote the copy of $i \in N_2$ in $N_2^{(r)}$ by $i^{(r)}$. The edges of $G_k(i_0, j_0)$ and their weights $\bar{w}$ can then be defined as

$$(i_0, j^{(1)}) \quad \text{for } i_0 j \in E \quad \text{with weight } \bar{w}(i_0 j) := p(i_0) + p(j)/2 - w(i_0 j)$$
$$(i^{(r-1)}, j^{(r)}) \text{ for } ij \in E \quad \text{with weight } \bar{w}(ij) := (p(i) + p(j))/2 - w(ij)$$
$$(i^{(k-1)}, j_0) \quad \text{for } ij_0 \in E \quad \text{with weight } \bar{w}(ij_0) := p(i)/2 + p(j_0) - w(ij_0).$$

We claim that $p(P) \geq w(P)$ holds for all $P \in \mathcal{P}$ if and only if the (w.r.t. $\bar{w}$) shortest $i_0 - j_0$-path in $G_k(i_0, j_0)$ has weight ≥ 0 for all $i_0 \neq j_0$ and $k = 1, .., n-1$. Then what is left to do, in order to verify whether $p(P) \geq w(P)$ holds for all $P \in \mathcal{P}$, is to solve $O(n^3)$ instances of the shortest path problem, each of which have size $O(n^2)$ by using the well-known Bellman-Ford algorithm [2].

First suppose some $P \in \mathcal{P}$ has $p(P) < w(P)$. Let i_0 and j_0 denote its endpoints and let k denote its length. Then $P \in \mathcal{P}_k(i_0, j_0)$ corresponds to an $i_0 - j_0$-path $\bar{P}$ in $G_k(i_0, j_0)$ of weight $\bar{w}(\bar{P}) < 0$. Now we will show that $p(P) \geq w(P)$ for all $P \in \mathcal{P}$ implies $\bar{w}(\bar{P}) \geq 0$ for all $i_0 - j_0$-paths $\bar{P}$ in any $G_k(i_0, j_0)$. Indeed, an $i_0 - j_0$-path $\bar{P}$ visiting players $i_0, i_1^{(1)}, ..., i_{k-1}^{(k-1)}, j_0$ corresponds to a simple $i_0 - j_0$ path $P \subseteq E$ in G plus possibly a number of cycles $C_1, ..., C_s \subseteq E$. Furthermore, by definition of $\bar{w}$, we have $\bar{w}(\bar{P}) = p(P) - w(P) + \sum_i p(C_i) - w(C_i) \geq p(P) - w(P)$, as we assume that $p(C) \geq w(C)$ holds for all cycles. Hence, indeed, $\bar{w}(\bar{P}) \geq 0$, as claimed. □

4 Future Work

We finish our paper with some directions for future research. Our reduction from SFP to SRP might be used to generalize more known results from SRP to SFP. For instance, can we generalize the path to stability result of Biró et al. [3] for SRP to be valid for SFP as well?

We do not know the answers for questions 1–3 on the core of multiple partners matching games (G, b, w) when $b(i) \geq 3$ for some player i. In particular, we tend to believe that testing core membership is NP-hard (for example, the case $b \equiv 3$ looks close to the maximum 3-regular subgraph problem, which is NP-complete, see e.g. [10]). However, we did not succeed in finding a proof. We recall that testing core membership is also open for multiple partners assignment games (except when $b \leq 2$).

Chalkiadakis et al. [6] defined cooperative games with overlapping coalitions, where players can be involved in coalitions with different intensities, leading to three alternative core definitions. It would be interesting to study the problem of finding a stable solutions and questions 1–3 in these settings. To illustrate this, suppose that the set of soccer teams from the example in Sect. 1 consists of international teams. Then it seems realistic that the home team needs to spend fewer days for playing the game than the visiting team, which must travel from another country. Hence, every team now has a number of days for playing friendly games instead of an upper limit (target) on the number of such games.

References

1. Baïou, M., Balinski, M.: Many-to-many matching: stable polyandrous polygamy (or polygamous polyandry). Discrete Appl. Math. **101**, 1–12 (2000)
2. Bellman, R.: On a routing problem. Q. Appl. Math. **16**, 87–90 (1958)
3. Biró, P., Bomhoff, M., Golovach, P.A., Kern, W., Paulusma, D.: Solutions for the stable roommates problem with payments. Theoret. Comput. Sci. **540–541**, 53–61 (2014)
4. Biró, P., Kern, W., Paulusma, D.: Computing solutions for matching games. Int. J. Game Theory **41**, 75–90 (2012)
5. Cechlárová, K., Fleiner, T.: On a generalization of the stable roommates problem. ACM Trans. Algorithms **1**, 143–156 (2005)
6. Chalkiadakis, G., Elkind, E., Markakis, E., Polurkov, M., Jennings, N.R.: Cooperative games with overlapping coalitions. J. Artif. Intell. Res. **39**, 179–216 (2010)
7. Deng, X., Ibaraki, T., Nagamochi, H.: Algorithmic aspects of the core of combinatorial optimization games. Math. Oper. Res. **24**, 751–766 (1999)
8. Eriksson, K., Karlander, J.: Stable outcomes of the roommate game with transferable utility. Int. J. Game Theory **29**, 555–569 (2001)
9. Gale, D., Shapley, L.S.: College admissions and the stability of marriage. Am. Math. Mon. **69**, 9–15 (1962)
10. Garey, M., Johnson, D.: Computers and Intractability. A Guide to the Theory of NP-Completeness. Freeman, San Francisco (1979)
11. Grötschel, M., Lovász, L., Schrijver, A.: The ellipsoid method, its consequences in combinatorial optimization. Combinatorica **1**, 169–197 (1981). [corrigendum: Combinatorica **4**, 291–295 (1984)]
12. Grötschel, M., Lovász, L., Schrijver, A.: Geometric Algorithms and Combinatorial Optimization, 2nd edn. Springer, Berlin (1993)
13. Irving, R.W.: An efficient algorithm for the stable roommates problem. J. Algorithms **6**, 577–595 (1985)
14. Irving, R.W., Scott, S.: The stable fixtures problem - a many-to-many extension of stable roommates. Discrete Appl. Math. **155**, 2118–2129 (2007)
15. Kern, W., Paulusma, D.: The new FIFA rules are hard: complexity aspects of sport competitions. Discrete Appl. Math. **108**, 317–323 (2001)
16. Khachiyan, L.G.: A polynomial algorithm in linear programming. Soviet Math. Dokl. **20**, 191–194 (1979)
17. Lawler, E.: Combinatorial Optimization. Courier Dover Publ., New York (1976)
18. Koopmans, T.C., Beckmann, M.: Assignment problems and the location of economic activities. Econometrica **25**, 53–76 (1957)
19. Letchford, A.N., Reinelt, G., Theis, D.O.: Odd minimum cut sets and b-matchings revisited. SIAM J. Discrete Math. **22**, 1480–1487 (2008)
20. Manlove, D.: Algorithmics of Matching Under Preferences, Series on Theoretical Computer Science, vol. 2. World Scientific (2013)

21. Megiddo, N.: Combinatorial optimization with rational objective functions. Math. OR **4**, 414–424 (1979)
22. Roth, A.E., Sotomayor, M.: Two-Sided Matching: A Study in Game-Theoretic Modeling and Analysis. Cambridge University Press, Cambridge (1990)
23. Shapley, L.S., Shubik, M.: The assignment game I: the core. Int. J. Game Theory **1**, 111–130 (1972)
24. Sotomayor, M.: The multiple partners game. In: Equilibrium and Dynamics: Essays in Honor of David Gale. Macmillan Press Ltd, New York (1992)
25. Tutte, W.T.: A short proof of the factor factor theorem for finite graphs. Can. J. Math. **6**, 347–352 (1954)

Complexity of Secure Sets

Bernhard Bliem[(✉)] and Stefan Woltran

Institute of Information Systems 184/2, TU Wien,
Favoritenstrasse 9–11, 1040 Vienna, Austria
{bliem,woltran}@dbai.tuwien.ac.at

Abstract. A secure set S in a graph is defined as a set of vertices such that for any $X \subseteq S$ the majority of vertices in the neighborhood of X belongs to S. It is known that deciding whether a set S is secure in a graph is co-NP-complete. However, it is still open how this result contributes to the actual complexity of deciding whether, for a given graph G and integer k, a non-empty secure set for G of size at most k exists. While membership in the class Σ_2^P is rather easy to see for this existence problem, showing Σ_2^P-hardness is quite involved. In this paper, we provide such a hardness result, hence classifying the secure set existence problem as Σ_2^P-complete. We do so by first showing hardness for a variant of the problem, which we then reduce step-by-step to secure set existence. In total, we obtain eight new completeness results for different variants of the secure set existence problem.

Keywords: Computational complexity · Complexity analysis · Secure sets

1 Introduction

Secure sets in graphs were introduced by Brigham et al. [3] as a strengthening of defensive alliances. Secure sets can be applied to, for instance, opinion research for analyzing group behaviour and identifying insusceptible peer groups, or to strategic settings where entities occupy spots on a map such that they can defend themselves against attacks from neighbors. A secure set of a graph $G = (V, E)$ is a set $S \subseteq V$ such that for each subset $X \subseteq S$, the number of vertices in $N[X] \cap S$ is not less than the number of vertices in $N[X] \setminus S$, where $N[X]$ denotes the closed neighborhood of X in G, i.e., X together with all vertices adjacent to X. The SECURE SET problem, which we address in this paper, can be specified as follows: Given a graph $G = (V, E)$ and an integer k, does there exists a non-empty secure set S of G such that $|S| \leq k$?

It is known that deciding whether a given set S is secure in a graph is co-NP-complete [6], indicating that the problem of finding (non-trivial) secure sets is very hard. Unfortunately, the exact complexity of this problem has so far remained unresolved. This is an unsatisfactory state of affairs because it leaves the possibility open that existing approaches for solving the problem (e.g., [1]) are

© Springer-Verlag Berlin Heidelberg 2016
E.W. Mayr (Ed.): WG 2015, LNCS 9224, pp. 64–77, 2016.
DOI: 10.1007/978-3-662-53174-7_5

suboptimal in that they employ unnecessarily powerful programming techniques. Hence we require a precise complexity-theoretic classification of the problem.

The main contribution of our paper is to show that SECURE SET is Σ_2^P-complete. Unlike the co-NP-hardness proof for secure set verification of [6], which uses a reduction from DOMINATING SET, we base our proof on a reduction from a problem in the area of logic. To be specific, we first show that the canonical Σ_2^P-complete problem QSAT$_2$ can be reduced to a variant of SECURE SET where (i) vertices can be forced to be in or out of every solution, (ii) pairs of vertices can be specified to indicate that every solution must contain exactly one element of each such pair, and (iii) each solution contains exactly a given number of elements. In order to prove the desired complexity result, we then successively reduce this variant to the standard SECURE SET problem.

Membership in the class Σ_2^P is obvious, as checking if a guessed candidate is indeed a solution can be done with a call to an oracle for co-NP [6]; in fact, [1] presents a poly-time reduction to Answer Set Programming [2] and thus implicitly shows this result. Together with our corresponding hardness result, it thus follows that SECURE SET is Σ_2^P-complete, and it turns out that all the problem variants we consider in this paper are Σ_2^P-complete.

We thus complete the picture of the precise complexity of the SECURE SET problem. Our results underline that SECURE SET is among the few rather natural problems in graph theory that are complete for the second layer of the polynomial hierarchy(like, e.g., CLIQUE COLORING [7] or 2-COLORING EXTENSION [8]). Therefore, our results allow us to conclude that approaches as suggested in [1] for solving the problem are indeed adequate from a complexity-theoretic point of view.

2   Background

All graphs in this paper are simple. Given a graph $G = (V, E)$, the *open neighborhood* of a vertex $v \in V$, denoted by $N_G(v)$, is the set of all vertices adjacent to v, and $N_G[v] = N_G(v) \cup \{v\}$ is called the *closed neighborhood* of v. Let $S \subseteq V$ be a set of vertices. We abuse notation by writing $N_G(S)$ and $N_G[S]$ to denote $\bigcup_{v \in S} N_G(v)$ and $\bigcup_{v \in S} N_G[v]$, respectively. If it is clear from the context which graph is meant, we write $N(\cdot)$ and $N[\cdot]$ instead of $N_G(\cdot)$ and $N_G[\cdot]$, respectively.

Definition 1. Given a graph $G = (V, E)$, a set $S \subseteq V$ is *secure in* G if for each $X \subseteq S$ it holds that $|N[X] \cap S| \geq |N[X] \setminus S|$.

We often write "S is secure" instead of "S is secure in G" if it is clear from the context which graph is meant. By definition, the empty set is secure in any graph. Thus, in the following decision problems we ask for secure sets of size at least 1.

The following is our main problem:

SECURE SET

Input: A graph $G = (V, E)$ and an integer k with $1 \leq k \leq |V|$

Question: Is there a set $S \subseteq V$ with $1 \leq |S| \leq k$ that is secure?

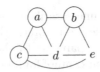

Fig. 1. A graph with a minimum non-empty secure set indicated by circled nodes

Figure 1 shows a graph together with a minimum non-empty secure set $S = \{a, b, c\}$. Observe that for any $X \subseteq S$ the condition $|N[X] \cap S| \geq |N[X] \setminus S|$ is satisfied.

We now define three variants of the SECURE SET problem that we require in our proofs. SECURE SET^F generalizes the SECURE SET problem by designating some "forbidden" vertices that may never be in any solution. This variant can be formalized as follows:

SECURE SET^F

    Input: A graph $G = (V, E)$, an integer k with $1 \leq k \leq |V|$ and a set $V_\square \subseteq V$

Question: Is there a set $S \subseteq V \setminus V_\square$ with $1 \leq |S| \leq k$ that is secure?

SECURE SET^{FN} is a further generalization that, in addition, allows "necessary" vertices to be specified that must occur in every solution.

SECURE SET^{FN}

    Input: A graph $G = (V, E)$, an integer k with $1 \leq k \leq |V|$, a set $V_\square \subseteq V$ and a set $V_\triangle \subseteq V$

Question: Is there a set $S \subseteq V \setminus V_\square$ with $V_\triangle \subseteq S$ and $1 \leq |S| \leq k$ that is secure?

Finally, we introduce the generalization SECURE SET^{FNC}. Here we may state pairs of "complementary" vertices where any solution must contain exactly one element of such a pair.

SECURE SET^{FNC}

    Input: A graph $G = (V, E)$, an integer k with $1 \leq k \leq |V|$, a set $V_\square \subseteq V$, a set $V_\triangle \subseteq V$ and a set $C \subseteq V^2$

Question: Is there a set $S \subseteq V \setminus V_\square$ with $V_\triangle \subseteq S$ and $1 \leq |S| \leq k$ that is secure and, for each pair $(a, b) \in C$, contains either a or b?

While the SECURE SET problem asks for secure sets of size *at most* k, we also consider the EXACT SECURE SET problem that concerns secure sets of size *exactly* k. Note that a secure set may become insecure by adding or removing elements, so this is a non-trivial problem variant. Analogously, we can also define exact versions of the three generalizations of SECURE SET presented above.

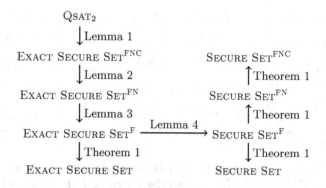

Fig. 2. Reduction strategy for proving Σ_2^P-hardness of SECURE SET

We often use the terms *attackers* and *defenders* of a subset X of a secure set candidate S. By these we mean elements of $N[X] \setminus S$ and $N[X] \cap S$, respectively. To show that a subset X of a secure set candidate S is *not* a witness to S being insecure, we sometimes employ the notion of *matchings* of the attackers of X with dedicated defenders in $N[X] \cap S$: If we are able to find an injective mapping $\mu : N[X] \setminus S \to N[X] \cap S$, then obviously $|N[X] \setminus S| \leq |N[X] \cap S|$. Given such a matching μ, we say that a defender d *repels* an attack on X by an attacker a whenever $\mu(a) = d$. When we say that a set of defenders D *can repel* attacks on X from a set of attackers A, we mean that there is a matching that assigns each element of A a dedicated defender in D.

3   Complexity Results

We prove Σ_2^P-hardness of the SECURE SET problem and several variants by a chain of polynomial reductions from QSAT₂. Our proof strategy is illustrated in Fig. 2. In Theorem 1, we finally obtain Σ_2^P-completeness for eight problems.

Lemma 1. EXACT SECURE SET^{FNC} *is* Σ_2^P-*hard.*

Proof. We reduce from QSAT₂ to EXACT SECURE SET^{FNC}. We are given a quantified Boolean formula $\varphi = \exists x_1 \ldots \exists x_{n_x} \forall y_1 \ldots \forall y_{n_y} \psi$, where ψ is in 3-DNF and contains n_t terms. We assume that no term contains both a variable and its complement (since such a term can never be satisfied) and that each term contains at least one universally quantified variable (since φ is trivially true otherwise).

We construct an instance $(G, k, V_\triangle, V_\square, C)$ of EXACT SECURE SET^{FNC}, where the set of vertices of $G = (V, E)$ is the union of the following sets:

$$X = \{x_1, \ldots, x_{n_x}\} \qquad \overline{X} = \{\overline{x_1}, \ldots, \overline{x_{n_x}}\} \qquad T = \{t_1, \ldots t_{n_t}\}$$

$$Y = \{y_1, \ldots, y_{n_y}\} \qquad \overline{Y} = \{\overline{y_1}, \ldots, \overline{y_{n_y}}\} \qquad \overline{T} = \{\overline{t_1}, \ldots \overline{t_{n_t}}\}$$

$$Y_\triangle = \{y_{i,j}^\triangle, \overline{y_{i,j}}^\triangle \mid 1 \le i \le n_y,\ 1 \le j \le n_t\} \qquad Y_\triangle' = \{y_j^\triangle \mid 1 \le j \le n_t - 1\}$$

$$Y_\square = \{y_{i,j}^\square \mid 1 \le i \le n_y,\ 1 \le j \le n_t + 1\} \qquad H = \{d_1^\square, d_2^\square, \overline{t}^\square\}$$

$$\overline{T}_\square = \{\overline{t_1}^\square, \ldots \overline{t_{n_t}}^\square\} \quad \overline{T}_\triangle = \{\overline{t_1}^\triangle, \ldots \overline{t_{n_t}}^\triangle\} \qquad T' = \{t_1', \ldots, t_{n_t}'\}$$

$$\overline{T'} = \{\overline{t_1'}, \ldots \overline{t_{n_t}'}\} \qquad T_\square' = \{t_1'^\square, \ldots, t_{n_t}'^\square\} \qquad \overline{T'}_\square = \{\overline{t_1'}^\square, \ldots, \overline{t_{n_t}'}^\square\}$$

Next we define the set of edges. In the following, whenever we speak of a literal in the context of the graph G, we mean the vertex corresponding to that literal (i.e., some x_i, $\overline{x_i}$, y_i or $\overline{y_i}$), and we proceed similarly for terms. Furthermore, when we are dealing with a literal l, then $\overline{l}$ shall denote the complement of l. For any term t_i, let $L_X(t_i)$ and $L_Y(t_i)$ denote the set of existentially and universally quantified literals, respectively, in t_i.

$$
\begin{aligned}
E = \ & \Big\{ (\overline{t_i}, \overline{t}^\square), (\overline{t_i}, \overline{t_i}^\triangle), (t_i', t_i'^\square), (\overline{t_i'}, \overline{t_i'}^\square) \mid t_i \in T \Big\} \cup \big(T' \times (Y \cup \overline{Y}) \big) \\
& \cup \Big\{ (l, \overline{t_i}^\square), (l, \overline{t_i}) \mid t_i \in T,\ l \in L_X(t_i) \Big\} \cup \Big\{ (l, \overline{t_i'}) \mid t_i \in T,\ l \in L_Y(t_i) \Big\} \\
& \cup \Big\{ (d_1^\square, \overline{t_i}) \mid t_i \in T,\ |L_X(t_i)| \le 1 \Big\} \cup \Big\{ (d_2^\square, \overline{t_i}) \mid t_i \in T,\ L_X(t_i) = \emptyset \Big\} \\
& \cup \Big\{ (y_i, y_{i,j}^\triangle), (\overline{y_i}, \overline{y_{i,j}}^\triangle) \mid 1 \le i \le n_y,\ 1 \le j \le n_t \Big\} \\
& \cup \Big\{ (y_i, y_{i,j}^\square), (\overline{y_i}, y_{i,j}^\square) \mid y_{i,j}^\square \in Y_\square \Big\} \cup \big(Y_\triangle' \times (Y \cup \overline{Y}) \big)
\end{aligned}
$$

Finally, we define $V_\triangle = Y \cup \overline{Y} \cup Y_\triangle \cup Y_\triangle' \cup \overline{T}_\triangle$, $V_\square = Y_\square \cup \overline{T}_\square \cup T_\square' \cup \overline{T'}_\square \cup H$, $C = \{(x_i, \overline{x_i}) \mid 1 \le i \le n_x\} \cup \{(t_i, \overline{t_i}), (\overline{t_i}, t_i'), (t_i', \overline{t_i'}) \mid 1 \le i \le n_t\}$, and $k = |V_\triangle| + n_x + 2n_t$. For an illustration, see Fig. 3.

The following observations are crucial: Elements of $X \cup \overline{X}$ are only adjacent to vertices from $\overline{T}_\square$ and $\overline{T}$. For any i, each element of $X \cup \overline{X}$ is adjacent to $\overline{t_i}^\square \in \overline{T}_\square$ if and only if it is adjacent to $\overline{t_i} \in \overline{T}$. Furthermore, for any i, j, if x_i or $\overline{x_i}$ is adjacent to $\overline{t_j}$, then the term t_j is falsified by setting the variable x_i to true or false, respectively. Finally, for any i, j, if y_i or $\overline{y_i}$ is adjacent to $\overline{t_j'}$, then the term t_j is falsified by setting the variable y_i to true or false, respectively.

We claim that φ is true if and only if $(G, k, V_\triangle, V_\square, C)$ is a positive instance of EXACT SECURE SET^{FNC}. Since, for each φ, the corresponding instance $(G, k, V_\triangle, V_\square, C)$ can be constructed in time polynomial in the size of φ, Σ_2^P-hardness follows.

"Only if" direction. If φ is true, then there is an assignment I to $x_1, \ldots, x_{n_x}$ such that, for all assignments extending I to $y_1, \ldots, y_{n_y}$, some term in ψ is satisfied.

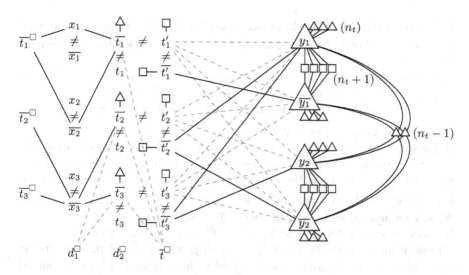

Fig. 3. Graph corresponding to the QSAT$_2$ formula $\exists x_1 \exists x_2 \exists x_3 \ \forall y_1 \forall y_2 \ \big((\neg x_1 \wedge x_2 \wedge y_1) \vee (x_3 \wedge \neg y_1 \wedge y_2) \vee (x_3 \wedge \neg y_1 \wedge \neg y_2) \big)$. "$\neq$" combines complementary vertices from C. To avoid clutter, we omit labels for vertices $Y_\triangle$, $Y'_\triangle$, $Y_\square$, $\overline{T}_\triangle$, $T'_\square$ and $\overline{T}'_\square$, and we draw some edges in a dashed style.

We define a set

$$S = V_\triangle \cup \{x_i \in X \mid I(x_i) = \text{true}\} \cup \{\overline{x_i} \in \overline{X} \mid I(x_i) = \text{false}\}$$
$$\cup \{\overline{t_i} \in \overline{T}, \ \overline{t_i'} \in \overline{T'} \mid \text{there is some } l \in L_X(t_i) \text{ such that } I \not\models l\}$$
$$\cup \{t_i \in T, \ t_i' \in T' \mid \text{for all } l \in L_X(t_i) \text{ it holds that } I \models l\}.$$

We observe that $|S| = k$, $V_\square \cap S = \emptyset$, $V_\triangle \subseteq S$, and that for any $(a, b) \in C$ it holds that $a \in S$ if and only if $b \notin S$. By construction, whenever some element of $X \cup \overline{X}$ is in S, then all its neighbors in $\overline{T}$ are in S; and whenever some $\overline{t_i}$ is in S, then some neighbor of $\overline{t_i}$ in $X \cup \overline{X}$ is in S.

We claim that S is a secure set in G. Let R be an arbitrary subset of S. We show that R has at least as many defenders as attackers by matching each attacker of R with a dedicated defender in $N[R] \cap S$. We distinguish cases regarding the origins of the attacks on R.

- We match each attacker $\overline{t_i}^{\,\square} \in \overline{T}_\square$ with $\overline{t_i}$. Since $\overline{t_i}^{\,\square}$ attacks R, R must contain some element of $X \cup \overline{X}$ that is adjacent to $\overline{t_i}^{\,\square}$ and thus also to $\overline{t_i}$, so $\overline{t_i} \in N[R] \cap S$.
- Each attacker from $X \cup \overline{X} \cup \{d_1^\square, d_2^\square\}$ is adjacent to some $\overline{t_i} \in \overline{T} \cap R$. We match that attacker with $\overline{t_i}^{\,\triangle}$, which is adjacent to $\overline{t_i}$. Note that it cannot be the case that $\overline{t_i}$ is attacked by more than one vertex in $X \cup \overline{X} \cup \{d_1^\square, d_2^\square\}$ because $\overline{t_i}$ has exactly two neighbors from that set and would not be in S if neither of these neighbors was in S.

- If $\overline{t}^{\square}$ attacks R, then it attacks at least one element of $\overline{T} \cap R$, which is adjacent to some element of $X \cup \overline{X}$ that is also in S. We match $\overline{t}^{\square}$ with any such element of $X \cup \overline{X}$.
- Any attack from some $\overline{t}_i \in \overline{T}$ on R must be on $\overline{t}_i^{\triangle}$. Since $\overline{t}_i \notin S$, $\overline{t}_i^{\triangle}$ is not consumed for repelling an attack on $\overline{t}_i$, so we match $\overline{t}_i$ with $\overline{t}_i^{\triangle}$.
- If some $t_i'^{\square} \in T'_{\square}$ attacks R (by attacking t_i'), we match $t_i'^{\square}$ with t_i'.
- Analogously, we match each attacker $\overline{t_i'}^{\square} \in \overline{T'}_{\square}$ with $\overline{t_i'}$.
- If, for some i with $1 \leq i \leq n_y$, the vertices $y_{i,j}^{\square}$ for $1 \leq j \leq n_t + 1$ attack R, then we distinguish the following cases: If y_i is in R, then the adjacent vertices $y_{i,j}^{\triangle}$ for $1 \leq j \leq n_t$ are in the neighborhood of R, too. We then match each $y_{i,j}^{\square}$ with $y_{i,j}^{\triangle}$ for $1 \leq j \leq n_t$, and we match $y_{i,n_t+1}^{\square}$ with y_i. Otherwise, $\overline{y_i}$ is in R, and we proceed symmetrically using $\overline{y_{i,j}}^{\triangle}$ and $\overline{y_i}$ as matches.
- In order to account for attacks from $T' \cup \overline{T'}$ on R, we distinguish two cases.

First, if for some i with $1 \leq i \leq n_y$, both y_i and $\overline{y_i}$ are in R, then, in the step before, we have matched each $y_{i,j}^{\square}$ with the respective $y_{i,j}^{\triangle}$ or y_i, but all $\overline{y_{i,j}}^{\triangle}$ are still free. These vertices can repel all attacks from $T' \cup \overline{T'}$, as there are at most n_t such attacks.

Otherwise, we show that there are at most $n_t - 1$ attacks from $T' \cup \overline{T'}$, and they can be repelled using $Y_{\triangle}'$. Consider the (partial) assignment J that assigns the same values to the variables $x_1, \ldots, x_{n_x}$ as the assignment I above, and, for any variable y_i, sets y_i to true or false if R contains the vertex y_i or $\overline{y_i}$, respectively. By assumption we know that our assignment to $x_1, \ldots, x_{n_x}$ is such that for all assignments to $y_1, \ldots, y_{n_y}$ some term t_i in ψ is true. In particular, it must therefore hold that J falsifies no existentially quantified literal in t_i. Then, by construction of S, the vertex $\overline{t_i'}$ is not in S. We also know that J falsifies no universally quantified literal in t_i. But then the vertices from $Y \cup \overline{Y}$ adjacent to the vertex $\overline{t_i'}$ are not in R due to our construction of J, so $\overline{t_i'}$ does not attack any vertex in R. From this it follows that there are at most $n_t - 1$ attacks from $T' \cup \overline{T'}$ on R. We can repel all these attacks using the vertices $y_1^{\triangle}, \ldots, y_{n_t-1}^{\triangle}$.

This allows us to conclude $|N[R] \cap S| \geq |N[R] \setminus S|$. Therefore S is secure.

"If" direction. Suppose S is a secure set in G honoring the conditions regarding forbidden, necessary and complementary vertices. If S contains some $l \in X \cup \overline{X}$, then $N(l) \cap \overline{T} \subseteq S$, as the number of neighbors l has in $\overline{T}$ is equal to the number of its other neighbors, which are all in $\overline{T}_{\square}$ and attack l since $\overline{T}_{\square} \subseteq V_{\square}$. If S contains some $\overline{t}_i \in \overline{T}$, then $\overline{t}_i$ must be adjacent to some element of $X \cup \overline{X}$ that is also in S. Otherwise $\overline{t}_i$ would have three attackers ($\overline{t}^{\square}$ and two vertices from $X \cup \overline{X} \cup \{d_1^{\square}, d_2^{\square}\}$) but only two defenders ($\overline{t}_i^{\triangle}$ and itself).

We construct an interpretation I on the variables $x_1, \ldots, x_{n_x}$ that sets exactly those x_i to true where the corresponding vertex x_i is in S, and we claim that for each extension of I to the universally quantified variables there is a satisfied term in ψ. To see this, suppose to the contrary that some assignment J to all variables extends I but falsifies all terms in ψ. Then we define a set R consisting of all vertices y_i such that $J(y_i) = \text{true}$, all vertices $\overline{y_i}$ such that $J(y_i) = \text{false}$, and all vertices in $(T' \cup \overline{T'}) \cap S$ that are adjacent to these vertices y_i or $\overline{y_i}$. Clearly, R is a subset of S. R has $|R|$ defenders due to itself, $n_t - 1$ defenders due to $Y'_\triangle$, and $n_y \cdot n_t$ defenders due to $N(R) \cap Y_\triangle$. This amounts to $|N[R] \cap S| = |R| + n_t - 1 + n_y \cdot n_t$.

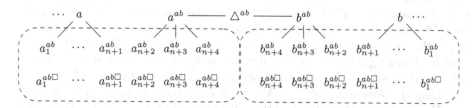

Fig. 4. Gadget resulting from reducing an EXACT SECURE SET^{FNC} instance with $(a, b) \in C$ to an equivalent EXACT SECURE SET^{FN} instance. Dashed rounded rectangles designate cliques.

On the other hand, there are n_t attacks on R from $T' \cup \overline{T'}$. This is because for any term t_i in ψ one of the following cases applies: (a) The term t_i is falsified already by I. Then $\overline{t'_i} \in S$ and thus $t'_i \notin S$. The vertex t'_i, however, is adjacent to any element of $Y \cup \overline{Y}$, so it attacks R. (b) The term t_i is not falsified by I but by J. Then $\overline{t'_i} \notin S$, and $L_Y(t_i)$ contains some literal l with $\overline{l} \in N(\overline{t'_i})$ and $J \models \overline{l}$, so $\overline{l}$ is in R and attacked by $\overline{t'_i}$.

In addition to these n_t attackers, R has $|R \cap (T' \cup \overline{T'})|$ attackers in $N(R) \cap (T'_\square \cup \overline{T'}_\square)$, as well as $n_y \cdot (n_t + 1)$ attackers in $Y_\square$. Since $|R| = n_y + |R \cap (T' \cup \overline{T'})|$, we obtain in total

$$|N[R] \setminus S| = n_t + |R \cap (T' \cup \overline{T'})| + n_y \cdot (n_t + 1) = |R| + n_t + n_y \cdot n_t > |N[R] \cap S|.$$

This contradicts S being secure, so for each extension of I to the universally quantified vertices, ψ is true; hence φ is true. $\square$

Lemma 2. EXACT SECURE SET^{FN} *is Σ_2^P-hard.*

Proof. We give a polynomial reduction from EXACT SECURE SET^{FNC}.

Let an instance of EXACT SECURE SET^{FNC} be given by $G = (V, E)$, $k > 0$, $V_\square \subseteq V$, $V_\triangle \subseteq V$ and $C \subseteq V^2$, and let $n = |V|$. For each $(a, b) \in C$, we introduce the set of new vertices $A^{ab} = \{a^{ab}, b^{ab}, \triangle^{ab}\}$ as well as, for any $x \in \{a, b\}$, sets of new vertices $C^{ab}_{x\bigcirc} = \{x^{ab}_1, \ldots, x^{ab}_{n+1}\}$, $D^{ab}_{x\bigcirc} = \{x^{ab}_{n+2}, \ldots, x^{ab}_{n+4}\}$, $C^{ab}_{x\square} = \{x^{ab\square}_1, \ldots, x^{ab\square}_{n+1}\}$ and $D^{ab}_{x\square} = \{x^{ab\square}_{n+2}, \ldots, x^{ab\square}_{n+4}\}$; see also Fig. 4 how these vertices are used. We now construct a graph $G' = (V', E')$ as follows:

$$V' = V \cup \bigcup_{(a,b) \in C} (A^{ab} \cup C^{ab}_{a\bigcirc} \cup C^{ab}_{b\bigcirc} \cup C^{ab}_{a\square} \cup C^{ab}_{b\square} \cup D^{ab}_{a\bigcirc} \cup D^{ab}_{b\bigcirc} \cup D^{ab}_{a\square} \cup D^{ab}_{b\square})$$

$$E' = E \cup \bigcup_{(a,b) \in C} \bigcup_{x \in \{a,b\}} (\{(\triangle^{ab}, x^{ab})\} \cup (\{x\} \times C^{ab}_{x\bigcirc}) \cup (\{x^{ab}\} \times D^{ab}_{x\bigcirc})$$

$$\cup \{(s, t) \in (C^{ab}_{x\bigcirc} \cup C^{ab}_{x\square} \cup D^{ab}_{x\bigcirc} \cup D^{ab}_{x\square})^2 \mid s \neq t\})$$

Finally, we define $k' = k + |C| \cdot (n+6)$, $V'_\square = V_\square \cup \bigcup_{(a,b) \in C} (C^{ab}_{a\square} \cup C^{ab}_{b\square} \cup D^{ab}_{a\square} \cup D^{ab}_{b\square})$ and $V'_\triangle = V_\triangle \cup \bigcup_{(a,b) \in C} \{\triangle^{ab}\})$.

The intention of our construction is, for each $(a,b) \in C$, $x \in \{a,b\}$ and any solution S, that (1) $x \in S$ if and only if $C^{ab}_{x\square} \cap S \neq \emptyset$; (2) $C^{ab}_{x\square} \cup D^{ab}_{x\square} \subseteq S$ if $(C^{ab}_{x\square} \cup D^{ab}_{x\square}) \cap S \neq \emptyset$; (3) $D^{ab}_{x\square} \cap S \neq \emptyset$ if and only if $x^{ab} \in S$; (4) $\triangle^{ab} \in S$ and therefore $a^{ab} \in S$ or $b^{ab} \in S$. Note that we have chosen $|D^{ab}_{x\square}| = 3$ in order to ensure the "if" direction of (3) by making sure x^{ab} has more neighbors from $D^{ab}_{x\square}$ than other neighbors (including itself). Furthermore, these conditions imply that a^{ab} and b^{ab} cannot both be in S, as otherwise $|S| > k'$. We claim that $(G, k, V_\square, V_\triangle, C)$ is a positive instance of EXACT SECURE SET^{FNC} if and only if $(G', k', V'_\square, V'_\triangle)$ is a positive instance of EXACT SECURE SET^{FN}.

"Only if" direction. Let $S \subseteq V \setminus V_\square$ be a secure set in G with $|S| = k$ and $V_\triangle \subseteq S$ such that $|S \cap \{a,b\}| = 1$ whenever $(a,b) \in C$. We construct $S' = S \cup \bigcup_{(a,b) \in C,\, x \in S \cap \{a,b\}} (\{\triangle^{ab}, x^{ab}\} \cup C^{ab}_{x\square} \cup D^{ab}_{x\square})$ and observe that $S' \cap V'_\square = \emptyset$, $V'_\triangle \subseteq S'$, and $|S'| = k'$. Let X' be an arbitrary subset of S'. Since S is secure and $X' \cap V \subseteq S$, there is a matching $\mu : N_G[X' \cap V] \setminus S \to N_G[X' \cap V] \cap S$. We now construct a matching $\mu' : N_{G'}[X'] \setminus S' \to N_{G'}[X'] \cap S'$. For any attacker v of X' in G', we distinguish three cases. (1) If v is some $x^{ab\square}_i \in C^{ab}_{x\square} \cup D^{ab}_{x\square}$ for some $(a,b) \in C$ and $x \in \{a,b\}$, we set $\mu'(v) = x^{ab}_i$. This element is in $N_{G'}[X']$ since all nodes in X' that v can attack are adjacent to it. (2) If v is a^{ab} or b^{ab} for some $(a,b) \in C$, its only neighbor in X' can be $\triangle^{ab}$ (by our construction of S'), and we set $\mu'(v) = \triangle^{ab}$. (3) Otherwise v is in $N_G[X' \cap V] \setminus S$ (by our construction of S'). Since the codomain of μ is a subset of the codomain of μ', we may set $\mu'(v) = \mu(v)$. Since μ' is injective, each attack on X' in G' can be repelled by S'. Thus S' is secure in G'.

"If" direction. Let $S' \subseteq V' \setminus V'_\square$ be a secure set in G' with $|S'| = k'$ and $V'_\triangle \subseteq S'$. First we make the following observations for each $(a,b) \in C$ and each $x \in \{a,b\}$: (a) If $x \in S'$, then $C^{ab}_{x\square} \cap S' \neq \emptyset$, since x only has at most $n-1$ neighbors not in $C^{ab}_{x\square}$. (b) Similarly, if $x^{ab} \in S'$, then $D^{ab}_{x\square} \cap S' \neq \emptyset$, as x^{ab} only has one neighbor not in $D^{ab}_{x\square}$. (c) If some $x^{ab}_i \in C^{ab}_{x\square}$ is in S', then $C^{ab}_{x\square} \cup D^{ab}_{x\square} \cup \{x\} \subseteq S'$, since x^{ab}_i has $n+4$ attackers from $C^{ab}_{x\square} \cup D^{ab}_{x\square}$ and only $n+4$ other neighbors. (d) If some $x^{ab}_i \in D^{ab}_{x\square}$ is in S', then $C^{ab}_{x\square} \cup D^{ab}_{x\square} \cup \{x^{ab}\} \subseteq S'$ for the same reason. So for any $(a,b) \in C$ and $x \in \{a,b\}$, S' contains either all or none of $\{x, x^{ab}\} \cup C^{ab}_{x\square} \cup D^{ab}_{x\square}$.

For each $(a,b) \in C$, S' contains a^{ab} or b^{ab}, since $\triangle^{ab} \in S'$, whose neighbors are a^{ab} and b^{ab}. It follows that $|S'| > |C| \cdot (n+6)$ even if S' contains only one of each $(a,b) \in C$. If, for some $(a,b) \in C$, S' contained both a and b, we could derive a contradiction to $|S'| = k'$ because then $|S'| > (|C|+1) \cdot (n+6) > k'$. So S' contains either a or b for any $(a,b) \in C$.

We construct the set $S = S' \cap V$ and observe that $|S| = k$, $V_\triangle \subseteq S$, $V_\square \cap S = \emptyset$, and $|S \cap \{a,b\}| = 1$ for each $(a,b) \in C$. Let $X \subseteq S$ and $X' = X \cup \bigcup_{(a,b) \in C,\, x \in X \cap \{a,b\}} C^{ab}_{x\square}$. For each $C^{ab}_{x\square}$ in X', there are $|C^{ab}_{x\square} \cup D^{ab}_{x\square}| = n+4$ additional defenders and $|C^{ab}_{x\square} \cup D^{ab}_{x\square}| = n+4$ additional attackers of X' in G' compared to X in G; so $|N_{G'}[X'] \cap S'| - |N_G[X] \cap S| = |N_{G'}[X'] \setminus S'| - |N_G[X] \setminus S|$.

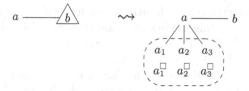

Fig. 5. Transformation of a sample SECURE SET^{FN} instance with $k = 1$ to an equivalent SECURE SET^F instance with $k' = 4$. The dashed rounded rectangle designates a clique.

Clearly $X' \subseteq S'$, so $|N_{G'}[X'] \cap S'| \geq |N_{G'}[X'] \setminus S'|$ as S' is secure in G'. We conclude $|N_G[X] \cap S| \geq |N_G[X] \setminus S|$. Hence S is secure in G.   □

Lemma 3. EXACT SECURE SET^F *is* Σ_2^P*-hard.*

Proof. We give a polynomial reduction from EXACT SECURE SET^{FN}.

Let an instance of EXACT SECURE SET^{FN} be given by $G = (V, E)$, $k > 0$, $V_\square \subseteq V$ and $V_\triangle \subseteq V$. Let $n = |V|$ and $V_\bigcirc = V \setminus (V_\square \cup V_\triangle)$. We assume $k \geq |V_\triangle|$, as the instance is trivially negative otherwise. We define for each $v \in V_\bigcirc$ the set of new vertices $C_v = \{v_1, \ldots, v_{n+1}, v_1^\square, \ldots, v_{n+1}^\square\}$, and we use shorthand notation $C_v^\bigcirc = \{v_1, \ldots, v_{n+1}\}$ and $C_v^\square = \{v_1^\square, \ldots, v_{n+1}^\square\}$. The intention is for each $v_i^\square$ to be forbidden and for each v_i to be in a secure set if and only if v is in it at the same time. We define the graph $G' = (V', E')$ with $V' = V \cup \bigcup_{v \in V_\bigcirc} C_v$ and

$$E' = E \cup \{(v, u) \mid v \in V_\bigcirc, \, u \in C_v^\bigcirc\} \cup \{(s, t) \in C_v^2 \mid v \in V_\bigcirc, \, s \neq t\}.$$

Furthermore, we define $V'_\square = V_\square \cup \bigcup_{v \in V_\bigcirc} C_v^\square$ and $k' = |V_\triangle| + (k - |V_\triangle|) \cdot (n+2)$. An example for this construction is given in Fig. 5. We claim that $(G, k, V_\square, V_\triangle)$ is a positive instance of EXACT SECURE SET^{FN} if and only if $(G', k', V'_\square)$ is a positive instance of EXACT SECURE SET^F.

"Only if" direction. Let $S \subseteq V \setminus V_\square$ be a secure set in G with $|S| = k$ and $V_\triangle \subseteq S$. We construct $S' = S \cup \bigcup_{v \in S \cap V_\bigcirc} C_v^\bigcirc$. It holds that $|S'| = |S| + (|S| - |V_\triangle|) \cdot (n+1) = k + (k - |V_\triangle|) \cdot (n+1) = |V_\triangle| + (k - |V_\triangle|) \cdot (n+2) = k'$. To show that S' is secure in G', let $X' \subseteq S'$. Since S is secure in G and $X' \cap V \subseteq S$, there is a matching $\mu : N_G[X' \cap V] \setminus S \to N_G[X' \cap V] \cap S$. We now construct a matching $\mu' : N_{G'}[X'] \setminus S' \to N_{G'}[X'] \cap S'$. For any attacker a of X' in G', we distinguish two cases: If a is some is some $v_i^\square \in C_v^\square$ for some $v \in V_\bigcirc$, there is some $v_j \in C_v^\bigcirc$ in X'. We set $\mu'(v_i^\square) = v_i$, which is in $N_{G'}[X'] \cap S'$ since $C_v^\bigcirc$ forms a clique. Otherwise a is in $N_G[X' \cap V] \setminus S$ (by our construction of S'). Since the codomain of μ is a subset of the codomain of μ', we may set $\mu'(a) = \mu(a)$. Since μ' is injective, each attack on X' in G' can be repelled by S'. Thus S' is secure in G'.

"If" direction. Let $S' \subseteq V' \setminus V'_\square$ be a secure set in G' with $|S'| = k'$. If S' contains some $v_i \in C_v^\bigcirc$ for some $v \in V_\bigcirc$, then $\{v\} \cup C_v^\bigcirc \subseteq S'$, as v_i has $n + 1$ attackers in $C_v^\square$ and only $n + 1$ neighbors not in $C_v^\square$. It is impossible for S' to contain some $v \in V_\bigcirc$ but none of $C_v^\bigcirc$, as v has at most $n - 1$ neighbors not in $C_v^\bigcirc$

and $|C_v^O| = n+1$. From these considerations we derive $|S'| = |S' \cap V_\triangle| + |S' \cap V_O| \cdot (n+2)$. Since $|V_\triangle| \leq n$, this implies $|S'| \bmod (n+2) = |S' \cap V_\triangle|$. At the same time, $|S'| = k' = |V_\triangle| + (k - |V_\triangle|) \cdot (n+2)$, which implies $|S'| \bmod (n+2) = |V_\triangle|$. We conclude $|S' \cap V_\triangle| = |V_\triangle|$, so $V_\triangle \subseteq S'$.

We construct $S = S' \cap V$ and claim that it is secure in G. It is easy to see that $|S| = k$; otherwise we would obtain a contradiction to $|S'| = k'$ as we have seen that, for any $v \in V_O$, S' contains all elements of $C_v^O \cup \{v\}$ whenever it contains any of them. Let X be an arbitrary subset of S. We construct $X' = X \cup \bigcup_{v \in X \cap V_O} C_v^O$ and observe that each C_v^O we put into X' entails $|C_v^O|$ additional defenders and $|C_v^\square| = |C_v^O|$ additional attackers of X' in G' compared to X in G; so $|N_{G'}[X'] \cap S'| - |N_G[X] \cap S| = |N_{G'}[X'] \setminus S'| - |N_G[X] \setminus S|$. Clearly $X' \subseteq S'$, so $|N_{G'}[X'] \cap S'| \geq |N_{G'}[X'] \setminus S'|$ as S' is secure in G'. Consequently, $|N_G[X] \cap S| \geq |N_G[X] \setminus S|$. Hence S is secure in G. □

Lemma 4. SECURE SET^F is Σ_2^P-hard.

Proof. We give a polynomial reduction from EXACT SECURE SET^F.

Let an instance of EXACT SECURE SET^F be given by $G = (V, E)$, $k > 0$ and $V_\square \subseteq V$. Let n denote $|V|$. We define for each $v \in V$ the sets of new vertices $A_v^O = \{v_0, v_1, \ldots, v_{n+1}\}$, $A_v^\square = \{v_0^\square, v_1^\square, \ldots, v_{n+1}^\square\}$, and denote by A_v^+ the set $A_v^O \setminus \{v_0\}$ and by A_v the set $A_v^O \cup A_v^\square$. We also introduce the new vertices $W = \{w_1, \ldots, w_n\}$ and $F^\square = \{f_1^\square, \ldots, f_k^\square\}$, and we write A_0 to denote $\{v_0 \in A_v \mid v \in V\}$ and B to denote $A_0 \cup W \cup F^\square$. We define the graph $G' = (V', E')$ with $V' = V \cup W \cup F^\square \cup \bigcup_{v \in V} A_v$ and

$$E' = E \cup \{(v, u) \mid v \in V, \ u \in A_v^+\} \cup \{(s, t) \in A_v^2 \mid v \in V, \ s \neq t\}$$
$$\cup \{(s, t) \in B^2 \mid s \neq t\}.$$

Furthermore, we define $V_\square' = V_\square \cup F^\square \cup \bigcup_{v \in V} A_v^\square$ and $k' = k \cdot (n + 3) + n$.

Our construction is illustrated in Fig. 6. For any $v \in V$, the intention for v_i is to be in a secure set if and only if v is in it. The intention behind the clique formed by B is that W shall be part of any secure set and is used for repelling attacks from $F^\square$ and from those elements of A_0 that are not in the secure set. This enforces that all secure sets in G' contain k elements of V. We claim that $(G, k, V_\square)$ is a positive instance of EXACT SECURE SET^F if and only if $(G', k', V_\square')$ is a positive instance of SECURE SET^F.

"Only if" direction. Let $S \subseteq V \setminus V_\square$ be a secure set in G with $|S| = k$. We construct $S' = S \cup W \cup \bigcup_{v \in S} A_v^O$ and observe that $|S'| = k'$. Let X' be an arbitrary subset of S'. Since S is secure and $X' \cap V \subseteq S$, there is a matching $\mu : N_G[X' \cap V] \setminus S \to N_G[X' \cap V] \cap S$. We now construct a matching $\mu' : N_{G'}[X'] \setminus S' \to N_{G'}[X'] \cap S'$ by distinguishing three cases: (1) For any $v \in V$, if any $v_i^\square \in A_v^\square$ attacks X', there is some $v_j \in A_v^O$ in X'. We set $\mu'(v_i^\square) = v_i$, which is in $N_{G'}[X'] \cap S'$ since A_v^O forms a clique. (2) If any of the $2n + k$ elements of B attacks X', $B \cap X' \neq \emptyset$. By construction, $|A_0 \cap S'| = k$ and $W \subseteq S'$, and we know $|W| = n$. So there are exactly n attacks from B. They can all be repelled using the n vertices in W, which are all in $N_{G'}[X'] \cap S'$ as B forms a clique. (3)

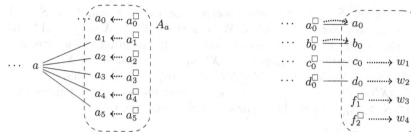

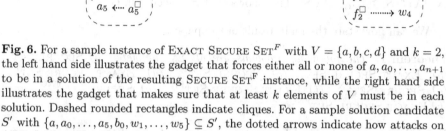

Fig. 6. For a sample instance of EXACT SECURE SET^F with $V = \{a, b, c, d\}$ and $k = 2$, the left hand side illustrates the gadget that forces either all or none of $a, a_0, \ldots, a_{n+1}$ to be in a solution of the resulting SECURE SET^F instance, while the right hand side illustrates the gadget that makes sure that at least k elements of V must be in each solution. Dashed rounded rectangles indicate cliques. For a sample solution candidate S' with $\{a, a_0, \ldots, a_5, b_0, w_1, \ldots, w_5\} \subseteq S'$, the dotted arrows indicate how attacks on $\{a_0, b_0\}$ can be repelled.

Any $v \in V$ attacking X' is in $N_G[X' \cap V] \setminus S$, due to our construction of S'. Since the codomain of μ is a subset of the codomain of μ', we may set $\mu'(v) = \mu(v)$. Clearly, μ' is injective, so S' can repel each attack on X' in G'. Thus S' is secure in G'.

"If" direction. Let $S' \subseteq V' \setminus V'_\square$ be a secure set in G' with $1 \leq |S'| \leq k'$. First we make the following observations: (a) If $v \in S'$ for some $v \in V$, then some element of A_v^+ must be in S' as well, since v has $n+1$ neighbors in A_v^+, while it only has at most $n-1$ neighbors not in A_v^+. (b) If $v_0 \in S'$ for some $v \in V$, then some $v_i \in A_v$ is in S'; otherwise v_0 would have at least $2n + 3 + k$ attackers due to $A_v \setminus \{v_0\}$ and $F^\square$, while v_0 has only $2n - 1$ other neighbors (due to W and $A_0 \setminus \{v_0\}$). (c) If $v_i \in S'$ for some $v \in V$ and $v_i \in A_v^+$, then $A_v^\circ \cup \{v\} \subseteq S'$, since v_i has $n+2$ attackers from $A_v^\square$ and only $n+2$ other neighbors (due to $A_v^\circ \setminus \{v_i\}$ and v). (d) If $w_i \in S'$ for some $w_i \in W$, then $v_0 \in S'$ for some $v \in V$; otherwise w_i would have at least $n + k$ attackers due to $A_0 \cup F^\square$ and at most n defenders (due to W).

It follows that, for any $v \in V$, S' contains either all or none of the elements in $\{v, v_0, v_1, \ldots, v_{n+1}\}$. This implies that S' cannot contain more than k elements of V, since $|S'| \leq k'$. Moreover, the case distinction shows that if *any* vertex is in S', then in particular some $v \in V$ is in S'.

As S' is not empty, we may conclude that $v_0 \in S'$ for some $v \in V$. Since v_0 has $n+2$ attackers from $A_v^\square$, k attackers from $F^\square$ and at least $n - k$ attackers from A_0 (as at most k elements of A_0 can be in S'), v_0 in total has at least $2n + 2$ attackers. At the same time, v_0 has $n+2$ defenders from A_v° (including itself), at most n defenders from W, and no other defenders. So v_0 has at most $2n + 2$ defenders. This shows that S' must contain W and at least k elements of V; otherwise v_0 would have more attackers than defenders. It follows in particular that S' contains *exactly* k elements of V.

We construct $S = S' \cap V$ and observe that $|S| = k$ and $V_{\square} \cap S = \emptyset$. Let X be an arbitrary subset of S. We construct $X' = X \cup \bigcup_{v \in X} A_v^+$ and observe that each A_v^+ we put into X' entails $|A_v^{\circ}|$ additional defenders and $|A_v^{\square}| = |A_v^{\circ}|$ additional attackers of X' in G' compared to X in G; so $|N_{G'}[X'] \cap S'| - |N_G[X] \cap S| = |N_{G'}[X'] \setminus S'| - |N_G[X] \setminus S|$. We have $X' \subseteq S'$ and $|N_{G'}[X'] \cap S'| \geq |N_{G'}[X'] \setminus S'|$ as S' is secure in G'. It follows that $|N_G[X] \cap S| \geq |N_G[X] \setminus S|$. This shows that S is secure in G.

We can now state the main result of the paper.

Theorem 1. *The following problems are all Σ_2^P-complete: (a)* SECURE SET, *(b)* EXACT SECURE SET, *(c)* SECURE SET^F, *(d)* EXACT SECURE SET^F, *(e)* SECURE SET^{FN}, *(f)* EXACT SECURE SET^{FN}, *(g)* SECURE SET^{FNC}, *and (h)* EXACT SECURE SET^{FNC}.

Proof. For membership in Σ_2^P, given an instance of one of the problems under consideration, we guess a subset S of the vertices such that, depending on the problem variant, the size of S is either at most k or exactly k, and S additionally respects the applicable conditions regarding forbidden, necessary and complementary vertices. Checking that S is secure can be done with an NP oracle [6].

We recall our proof strategy illustrated in Fig. 2. Σ_2^P-hardness of (h), (f), (d) and (c) follows from Lemmas 1–4. Σ_2^P-hardness of (e) and (g) follows directly from (c). We now show Σ_2^P-hardness of SECURE SET (a) by a polynomial reduction from SECURE SET^F (c). Let an instance of SECURE SET^F be given by $G = (V, E)$, $k > 0$ and $V_{\square} \subseteq V$. For each $f \in V_{\square}$, we introduce new vertices $f_1, \ldots, f_{2k}$, and we define a graph $G' = (V', E')$ with $V' = V \cup \{f_1, \ldots, f_{2k} \mid f \in V_{\square}\}$ and $E' = E \cup \{(f_i, f_j) \mid f \in V_{\square}, 1 \leq i < j \leq 2k\} \cup \{(f, f_i) \mid f \in V_{\square}, 1 \leq i \leq 2k\}$. The instance $(G, k, V_{\square})$ of SECURE SET^F possesses the same solutions as the instance (G', k) of SECURE SET: Each secure set S in G is also secure in G' because the subgraph of G induced by $N_G[S]$ is isomorphic to the subgraph of G' induced by $N_{G'}[S]$. On the other hand, a secure set S' in G' with $|S'| \leq k$ cannot not contain any $v \in \{f, f_1, \ldots, f_{2k}\}$ for any $f \in V_{\square}$, as $|N[v]| \geq 2k + 1$. Hence S' is also secure in G as the subgraphs induced by the respective neighborhoods are again isomorphic. The same argument also proves Σ_2^P-hardness of EXACT SECURE SET (b) by reduction from EXACT SECURE SET^F (d). $\quad\square$

4   Conclusion

In this work, we have solved a complexity problem in graph theory that, to the best of our knowledge, has remained open since the introduction of secure sets [3] in 2007. We have shown that the problem of deciding whether, for a given graph G and integer k, G possesses a non-empty secure set of size at most k is Σ_2^P-complete. We moreover obtained Σ_2^P-completeness for seven further variants of this problem. In future work, we would like to identify subclasses of graphs that make the problems under consideration easier. Additionally, we plan to contribute results on the parameterized complexity of secure sets beyond the already known fixed-parameter tractability result from [5]. In particular, we believe that recently introduced classes (above NP) due to [4] are of interest for secure sets.

Acknowledgments. This work was supported by the Austrian Science Fund (FWF) projects P25607 and Y698. We would like to thank the reviewers for their valuable comments and Herbert Fleischner for drawing our attention to the problem of secure sets.

References

1. Abseher, M., Bliem, B., Charwat, G., Dusberger, F., Woltran, S.: Computing secure sets in graphs using answer set programming. In: Proceedings of ASPOCP 2014 (2014)
2. Brewka, G., Eiter, T., Truszczyński, M.: Answer set programming at a glance. Commun. ACM **54**(12), 92–103 (2011)
3. Brigham, R.C., Dutton, R.D., Hedetniemi, S.T.: Security in graphs. Discrete Appl. Math. **155**(13), 1708–1714 (2007)
4. de Haan, R., Szeider, S.: The parameterized complexity of reasoning problems beyond NP. In: Proceedings of KR 2014, pp. 82–91 (2014)
5. Enciso, R.I., Dutton, R.D.: Parameterized complexity of secure sets. Congr. Numer. **189**, 161–168 (2008)
6. Ho, Y.Y.: Global secure sets of trees and grid-like graphs. Ph.D. thesis, University of Central Florida, Orlando, USA (2011)
7. Marx, D.: Complexity of clique coloring and related problems. Theor. Comput. Sci. **412**(29), 3487–3500 (2011)
8. Szeider, S.: Generalizations of matched CNF formulas. Ann. Math. Artif. Intell. **43**(1), 223–238 (2005)

Efficient Domination for Some Subclasses of P_6-free Graphs in Polynomial Time

Andreas Brandstädt[1], Elaine M. Eschen[2], and Erik Friese[3(✉)]

[1] Institut für Informatik, Universität Rostock, 18051 Rostock, Germany
ab@informatik.uni-rostock.de
[2] West Virginia University, Morgantown, WV, USA
Elaine.Eschen@mail.wvu.edu
[3] Institut für Mathematik, Universität Rostock,
Ulmenstr. 69, 18057 Rostock, Germany
erik.friese@uni-rostock.de

Abstract. Let G be a finite undirected graph. A vertex *dominates* itself and all its neighbors in G. A vertex set D is an *efficient dominating set* (*e.d.* for short) of G if every vertex of G is dominated by exactly one vertex of D. The *Efficient Domination* (ED) problem, which asks for the existence of an e.d. in G, is known to be $\mathbb{NP}$-complete even for very restricted graph classes such as P_7-free chordal graphs. The ED problem on a graph G can be reduced to the Maximum Weight Independent Set (MWIS) problem on the square of G. The complexity of the ED problem is an open question for P_6-free graphs and was open even for the subclass of P_6-free chordal graphs. In this paper, we show that squares of P_6-free chordal graphs that have an e.d. are chordal; this even holds for the larger class of (P_6, house, hole, domino)-free graphs. This implies that ED/WeightedED is solvable in polynomial time for (P_6, house, hole, domino)-free graphs; in particular, for P_6-free chordal graphs. Moreover, based on our result that squares of P_6-free graphs that have an e.d. are hole-free and some properties concerning odd antiholes, we show that squares of (P_6, house)-free graphs ((P_6, bull)-free graphs, respectively) that have an e.d. are perfect. This implies that ED/WeightedED is solvable in polynomial time for (P_6, house)-free graphs and for (P_6, bull)-free graphs (the time bound for (P_6, house, hole, domino)-free graphs is better than that for (P_6, house)-free graphs). The complexity of the ED problem for P_6-free graphs remains an open question.

Keywords: Efficient domination · Chordal graphs · Hole-free graphs · (house, hole, domino)-free graphs · P_6-free graphs · Polynomial-time algorithm

1 Introduction

Let $G = (V, E)$ be a finite undirected graph. A vertex $v \in V$ *dominates* itself and its neighbors. A vertex subset $D \subseteq V$ is an *efficient dominating set* (*e.d.* for

© Springer-Verlag Berlin Heidelberg 2016
E.W. Mayr (Ed.): WG 2015, LNCS 9224, pp. 78–89, 2016.
DOI: 10.1007/978-3-662-53174-7_6

short) of G if every vertex of G is dominated by exactly one vertex in D. Note that not every graph has an e.d.; the EFFICIENT DOMINATING SET (ED) problem asks for the existence of an e.d. in a given graph G. If a vertex weight function $\omega : V \to \mathbb{N}$ is given, the WEIGHTED EFFICIENT DOMINATING SET (WED) problem asks for a minimum weight e.d. in G, if there is one, or for determining that G has no e.d. The importance of the ED problem mostly results from the fact that the ED problem for a graph G is a special case of the EXACT COVER problem for hypergraphs (problem [SP2] of [13]); ED is the Exact Cover problem for the closed neighborhood hypergraph of G.

For a graph F, a graph G is called F-*free* if G contains no induced subgraph isomorphic to F.

We denote by $G + H$ the disjoint union of graphs G and H. Let P_k denote a chordless path with k vertices, and let $2P_k$ denote $P_k + P_k$, and correspondingly for kP_2. The *claw* is the 4-vertex tree with three vertices of degree 1.

Many papers have studied the complexity of ED on special graph classes - see e.g. [7,16] for references. In particular, a standard reduction from the Exact Cover problem shows that ED remains NP-complete for $2P_3$-free chordal graphs and for bipartite graphs. Moreover, it is known to be NP-complete for line graphs and thus, for claw-free graphs.

A *linear forest* is a graph whose components are paths; equivalently, it is a graph that is cycle-free and claw-free. The NP-completeness of ED on chordal graphs, on bipartite graphs and on claw-free graphs implies: If F is not a linear forest, then ED is NP-complete on F-free graphs. This motivates the analysis of ED/WED on F-free graphs for linear forests F. For F-free graphs, where F is a linear forest, the only remaining open case is the complexity of ED on P_6-free graphs (see [3]).

The main results of this paper are the following:

- If G is (P_6, HHD)-free and has an e.d., then G^2 is chordal. Then using a subsequently described reduction of ED/WED on G to the Maximum Weight Independent Set (MWIS) problem on G^2, we obtain a polynomial time solution for ED/WED on this class of graphs, since MWIS is solvable in polynomial time on chordal graphs. This also gives a dichotomy result for P_k-free chordal graphs, since ED is NP-complete for P_7-free chordal graphs.
- If G is P_6-free and has an e.d., then G^2 is hole-free. This does not yet imply that ED for P_6-free graphs is solvable in polynomial time, since the MWIS problem for hole-free graphs is an open question but it leads to further results on ED for subclasses of P_6-free graphs.
- If G is P_6-free and has an e.d., then odd antiholes in G^2 have very special structure. Analyzing the structure of C_4 realizations in G^2, we obtain a polynomial time solution of ED/WED for (P_6, house)-free graphs and for (P_6, bull)-free graphs, since in this case, G^2 is perfect if G has an e.d.

2   Basic Notions and Results

2.1   Some Basic Notions

All graphs considered in this paper are finite, undirected and simple (i.e., without loops and multiple edges). For a graph G, let $V(G)$ or simply V denote its vertex set and $E(G)$ or simply E its edge set; throughout this paper, let $|V| = n$ and $|E| = m$. We can assume that G is connected (otherwise, ED can be solved separately for its components); thus, $m \geq n - 1$. For $U \subseteq V$, let $G[U]$ denote the subgraph of G induced by U.

For a vertex $v \in V$, $N(v) = \{u \in V \mid uv \in E\}$ denotes its (*open*) *neighborhood*, and $N[v] = \{v\} \cup N(v)$ denotes its *closed neighborhood*. A vertex v *sees* the vertices in $N(v)$ and *misses* all the others. Let $d_G(v, w)$ denote the distance between v and w in G.

Let P_k denote a chordless path with k vertices, and let C_k denote a chordless cycle with k vertices. Chordless cycles C_k with $k \geq 5$ are called *holes*. The complement graph $\overline{P_5}$ is also called *house*. *Domino* has six vertices and can be obtained by adding a vertex y to a P_5 $x_1, \ldots, x_5$ with edges $x_i x_{i+1}$, $1 \leq i \leq 4$, such that $yx_1 \in E$, $yx_3 \in E$, and $yx_5 \in E$; see Fig. 1. A graph is *chordal* if it is C_k-free for every $k \geq 4$. A graph is (*house, hole, domino*)-free (*HHD-free* for short) if it has no induced subgraph isomorphic to a house, hole or domino. Obviously, chordal graphs are HHD-free, and G is (P_6, HHD)-free if and only if G is (P_6, C_5, C_6, house, domino)-free. The importance of HHD-free graphs as a natural generalization of chordal graphs is illustrated by various characterizations of them such as: G is HHD-free if and only if G is $(5, 2)$-chordal (see e.g. [5]).

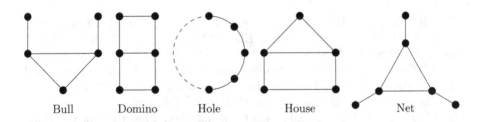

Fig. 1. Some special graphs.

2.2   Reducing the ED Problem on a Graph
to the MWIS Problem on its Square

The *square* of a graph $G = (V, E)$ is the graph $G^2 = (V, E^2)$ such that $uv \in E^2$ if and only if $d_G(u, v) \in \{1, 2\}$. In [6,14,16], the following relationship between the ED problem on a graph G and the maximum weight independent set (MWIS) problem on G^2 is used:

Lemma 1. *Let $G = (V, E)$ be a graph and $\omega(v) := |N[v]|$ a vertex weight function for G. Then the following are equivalent for any subset $D \subseteq V$:*

(i) D is an efficient dominating set in G.
(ii) D is a minimum weight dominating set in G with $\omega(D) = |V|$.
(iii) D is a maximum weight independent set in G^2 with $\omega(D) = |V|$.

Thus, the ED problem on a graph class $\mathcal{C}$ can be reduced to the MWIS problem on the squares of graphs in $\mathcal{C}$. In [2], this is extended to the vertex-weighted version WED of the ED problem.

3   Squares of (P_6, HHD)-Free Graphs that have an e.d. are Chordal

Obviously, the square of a chordal graph can contain a C_4 as for example, the complete 4-sun shows. If we additionally require that the graph is P_6-free and has an e.d., the situation is different: The main result of this section is Theorem 1, which shows that for any graph G that is (P_6, HHD)-free and has an e.d., its square G^2 is chordal, i.e., C_k-free for every $k \geq 4$. Theorem 2 in Sect. 4.1 shows that G^2 is C_k-free for every $k \geq 5$ for the larger class of P_6-free graphs, but its proof is long and technically involved. For the special case of (P_6, HHD)-free graphs, we give a direct proof here since it is much shorter than the proof of Theorem 2.

Theorem 1. *If G is a (P_6, HHD)-free graph that has an e.d., then G^2 is chordal.*

For the proof of Theorem 1, we first prove several lemmas. Let G be a (P_6, HHD)-free graph with an e.d. D. Suppose that G^2 contains a chordless cycle C_k C with vertices $v_1, \ldots, v_k$, $k \geq 4$; we call these the *real vertices* of C and denote them by $R(C) = \{v_1, \ldots, v_k\}$. For $d_G(v_i, v_{i+1}) = 2$ (index arithmetic is modulo k throughout this section), let x_i be a common neighbor of v_i and v_{i+1}; we call these x_i vertices the *auxiliary vertices* of C and denote the set of these vertices by $A(C)$. Let $V(C) = R(C) \cup A(C)$ denote the set of vertices (real and auxiliary) in G *realizing* a C_k C in G^2; we call $V(C)$ a *cycle embedding*.

Observation 1. *For every $i \in \{1, \ldots, k\}$, $d_G(v_i, v_{i+1}) \leq 2$. Also, $d_G(v_i, v_j) > 2$ if v_i and v_j are not consecutive in the C_k C in G^2. In particular, if $d_G(v_i, v_{i+1}) = 1$ then $d_G(v_{i+1}, v_{i+2}) = 2$ and $d_G(v_{i-1}, v_i) = 2$. Clearly, auxiliary vertices are pairwise distinct, and for every x_i, $v_j x_i \notin E$ for all $j \notin \{i, i+1\}$.*

We claim that there are k distinct auxiliary vertices $x_1, \ldots, x_k$ in $V(C)$:

Lemma 2. *For all $i \in \{1, \ldots, k\}$, $d_G(v_i, v_{i+1}) = 2$.*

Proof. Without loss of generality, suppose $v_1 v_2 \in E$. Then $d_G(v_2, v_3) = 2$ and $d_G(v_k, v_1) = 2$.

Case $k = 4$: If further $v_3 v_4 \in E$, then $V(C)$ induces either a C_6 or a domino in G which is a contradiction. Thus, $d_G(v_3, v_4) = 2$ and there is a vertex x_3.

Since $\{v_1, v_2, x_2, v_3, x_3, v_4\}$ does not induce a P_6, we have $x_2x_3 \in E$.
Since $\{v_2, v_1, x_4, v_4, x_3, v_3\}$ does not induce a P_6, we have $x_3x_4 \in E$.
Since $\{v_3, x_2, v_2, v_1, x_4, v_4\}$ does not induce a P_6, we have $x_2x_4 \in E$.

Now, $\{v_1, v_2, x_2, x_3, x_4\}$ induces a house which is a contradiction. Thus, we have $d_G(v_i, v_{i+1}) = 2$ for all $i \in \{1, \ldots, 4\}$ and Lemma 2 holds for $k = 4$.

For $k \geq 5$, since $\{v_k, x_k, v_1, v_2, x_2, v_3\}$ does not induce a P_6, we have $x_kx_2 \in E$. If x_k and x_2 have a common neighbor x_i, $2 < i < k$, then $\{v_1, v_2, x_2, x_k, x_i\}$ induces a house; thus for all i with $2 < i < k$, we have:

$$(*) \quad x_2x_i \notin E \text{ or } x_kx_i \notin E.$$

Case $k = 5$: Without loss of generality, suppose $d_G(v_3, v_4) = 2$.
Since $\{v_1, v_2, x_2, v_3, x_3, v_4\}$ does not induce a P_6, we have $x_2x_3 \in E$; thus, $x_3x_5 \notin E$ by $(*)$.
Since $\{v_1, v_2, x_2, x_3, v_4, v_5\}$ does not induce a P_6, we have $d_G(v_4, v_5) = 2$ and thus, there is a vertex x_4.
Since $\{v_2, v_1, x_5, v_5, x_4, v_4\}$ does not induce a P_6, we have $x_4x_5 \in E$, which implies $x_2x_4 \notin E$ by $(*)$.
Since $\{x_2, x_3, v_4, x_4, x_5\}$ does not induce a C_5, we have $x_3x_4 \in E$; but now, $G[\{x_2, x_3, v_4, x_4, x_5\}]$ is a house, which is a contradiction. Thus, Lemma 2 holds for $k = 5$.

Case $k > 5$: Since $\{v_4, v_3, x_2, x_k, v_k, v_{k-1}\}$ does not induce a P_6, we have either $d_G(v_3, v_4) = 2$ or $d_G(v_k, v_{k-1}) = 2$; without loss of generality, let $d_G(v_3, v_4) = 2$ and there is a vertex x_3.
Since $\{v_4, x_3, v_3, x_2, v_2, v_1\}$ does not induce a P_6, we have $x_2x_3 \in E$, which implies $x_3x_k \notin E$ by $(*)$.
Since $\{v_4, x_3, x_2, x_k, v_k, v_{k-1}\}$ does not induce a P_6, we have $d_G(v_k, v_{k-1}) = 2$ and thus, there is a vertex x_{k-1}.
Since $\{v_{k-1}, x_{k-1}, v_k, x_k, v_1, v_2\}$ does not induce a P_6, we have $x_{k-1}x_k \in E$, which implies $x_2x_{k-1} \notin E$ by $(*)$.
Since $\{v_4, x_3, x_2, x_k, x_{k-1}, v_{k-1}\}$ does not induce a C_6 or P_6, we have $x_3x_{k-1} \in E$ and now $G[\{v_1, v_2, x_2, x_k, x_3, x_{k-1}\}]$ is a domino, which is a contradiction. Thus, Lemma 2 holds for $k > 5$. $\qquad \square$

Lemma 3. *For all $i \in \{1, \ldots, k\}$, $x_ix_{i+1} \in E$.*

Proof. Without loss of generality, suppose to the contrary that $x_1x_2 \notin E$. Then, since $\{v_1, x_1, v_2, x_2, v_3, x_3\}$ does not induce a P_6, we have $x_1x_3 \in E$ or $x_2x_3 \in E$. If $x_1x_3 \in E$ then $x_2x_3 \in E$, else $\{x_1, v_2, x_2, v_3, x_3\}$ induces a C_5. Thus, $x_2x_3 \in E$. Since $\{v_1, x_1, v_2, x_2, x_3, v_4\}$ does not induce a P_6, we have $x_1x_3 \in E$. But, now $\{x_1, v_2, x_2, v_3, x_3\}$ induces a house, which is a contradiction; thus, $x_1x_2 \in E$ and Lemma 3 is shown. $\qquad \square$

Lemma 4. $D \cap \{v_1, \ldots, v_k, x_1, \ldots, x_k\} = \emptyset$.

Proof. First suppose to the contrary that $D \cap \{v_1, \ldots, v_k\} \neq \emptyset$; without loss of generality, let $v_1 \in D$. Then $v_2 \notin D$ and $v_k \notin D$, but they must be dominated by D-vertices, say $d_2, d_k \in D$ with $d_2 v_2 \in E$ and $d_k v_k \in E$. Since $d_G(v_2, v_k) > 2$, $d_2 \neq d_k$. Also, $d_2 \notin \{x_1, x_2\}$ and $d_k \notin \{x_{k-1}, x_k\}$. Now $G[\{d_2, v_2, x_1, x_k, v_k, d_k\}]$ is a P_6, which is a contradiction. Thus, $D \cap \{v_1, \ldots, v_k\} = \emptyset$.

Now suppose to the contrary that $D \cap \{x_1, \ldots, x_k\} \neq \emptyset$; without loss of generality, let $x_1 \in D$. We know already that $v_3, v_k \notin D$ and thus, there is $d_3 \in D$ with $d_3 v_3 \in E$ such that $d_3 \notin \{x_2, x_3\}$. If $k = 4$ and $d_3 v_4 \in E$, then $G[\{v_3, d_3, v_4, x_4, x_1, v_2\}]$ is a P_6; thus, $d_3 v_4 \notin E$. If $k \geq 5$ then $d_3 v_k \notin E$ since $d_G(v_3, v_k) > 2$. So there must be $d_k \in D$ with $d_k v_k \in E$ such that $d_k \neq d_3$ and $d_k \notin \{x_{k-1}, x_k\}$. If $x_2 x_k \notin E$ then $G[\{d_3, v_3, x_2, x_1, x_k, v_k\}]$ is a P_6, and if $x_2 x_k \in E$ then $G[\{d_3, v_3, x_2, x_k, v_k, d_k\}]$ is a P_6, which is a contradiction. Thus, Lemma 4 is shown. □

For all $i \in \{1, \ldots, k\}$, let $d_i \in D$ be the vertex with $d_i v_i \in E$. We claim that $d_1, \ldots, d_k$ are pairwise distinct:

Corollary 1. *For all $i \in \{1, \ldots, k\}$, d_i has exactly one neighbor in $\{v_1, \ldots, v_k\}$.*

Proof. As in Observation 1, a D-vertex cannot see both v_i and v_j if v_i and v_j are not consecutive in the C_k C in G^2. Suppose without loss of generality $d_1 = d_2$, i.e., $d_1 v_1 \in E$ and $d_1 v_2 \in E$. Then x_1 can be replaced by d_1 in the cycle embedding and thus, d_1 is an auxiliary vertex in D which contradicts Lemma 4. □

Lemma 5. *For all $i \in \{1, \ldots, k\}$, $d_i x_{i-1} \notin E$ or $d_i x_i \notin E$.*

Proof. Without loss of generality, assume that $d_1 x_k \in E$ and $d_1 x_1 \in E$. Then $G[\{d_2, v_2, x_1, x_k, v_k, d_k\}]$ is a P_6, which is a contradiction. □

Lemma 6. *For all $i \in \{1, \ldots, k\}$, if $d_i x_i \notin E$ then $d_i x_{i+1} \notin E$, and if $d_i x_{i-1} \notin E$ then $d_i x_{i-2} \notin E$.*

Proof. Assume that $d_i x_i \notin E$ and $d_i x_{i+1} \in E$. Then $\{d_i, v_i, x_i, x_{i+1}, v_{i+1}\}$ induces a house, which is a contradiction. Thus, if $d_i x_i \notin E$ then $d_i x_{i+1} \notin E$ and similarly, if $d_i x_{i-1} \notin E$ then $d_i x_{i-2} \notin E$. □

Lemma 7. *For all $i \in \{1, \ldots, k\}$, $d_i x_i \notin E$ implies $d_{i+2} x_{i+1} \in E$ and $d_{i+2} x_{i+2} \notin E$.*

Proof. Assume without loss of generality that $d_1 x_1 \notin E$. Then, by Lemma 6, $d_1 x_2 \notin E$ and if $d_3 x_1 \in E$ then $d_3 x_2 \in E$. Since $\{d_1, v_1, x_1, x_2, v_3, d_3\}$ does not induce a P_6, we have $d_3 x_2 \in E$ or $d_3 x_1 \in E$, which implies $d_3 x_2 \in E$. Then, by Lemma 5, $d_3 x_3 \notin E$. □

For an odd hole, repeating the argument of Lemma 7 on $d_3 x_3 \notin E$, and so on, determines all the edges and non-edges between D-vertices and auxiliary vertices. For C_4 or an even hole, repeating the argument determines the edges for every second D-vertex, but then a second round (using the fact that $d_2 x_2 \notin E$) determines the remaining edges and non-edges.

Proof of Theorem 1. First suppose that C is a C_4 in G^2. Then, since G is HHD-free, we have $x_1 x_3 \in E$ or $x_2 x_4 \in E$; without loss of generality say $x_1 x_3 \in E$.

Moreover, $d_1 x_1 \notin E$ or $d_2 x_1 \notin E$; without loss of generality say $d_1 x_1 \notin E$. Then by Lemma 7, $d_3 x_2 \in E$ (and thus, $d_2 x_2 \notin E$) and $d_3 x_3 \notin E$ holds and repeating the same arguments, we get $d_4 x_3 \in E$, $d_4 x_4 \notin E$, and $d_2 x_1 \in E$, but now $G[\{d_1, v_1, x_1, x_3, v_3, d_3\}]$ is a P_6 which is a contradiction.

Now suppose that C is a C_k in G^2 for some $k \geq 5$. Then, since G is HHD-free, there is an edge $x_i x_j \in E$ where $j \notin \{i-1, i+1\}$. Then in the case that $d_i x_j \notin E$ (and thus also $d_j x_j \notin E$), $G[\{d_i, v_i, x_i, x_j, v_j, d_j\}]$ is a P_6. The case when $d_{i+1} x_i \notin E$ is symmetric. Thus, we have a contradiction. This concludes the proof of Theorem 1.   □

Corollary 2. *For (P_6, HHD)-free graphs, the WED problem is solvable in polynomial time.*

Proof. By Lemma 1, the ED problem for G can be reduced to the MWIS problem for G^2. By Theorem 1, G^2 is chordal. By the result of Frank [9], the MWIS problem can be solved in linear time for chordal graphs. Thus, for (P_6, HHD)-free graphs, the ED problem is solvable in polynomial time. By [2], the WED problem can be solved in polynomial time for the same class.   □

4   Some Properties of P_6-Free Graphs that have an e.d.

4.1   Squares of P_6-Free Graphs that have an e.d. are Hole-Free

The main result of this subsection is Theorem 2 which shows that for any P_6-free graph G with an e.d., its square G^2 is hole-free. This result is based on the unpublished thesis [10]. It would imply that ED is solvable in polynomial time for P_6-free graphs if the MWIS problem for hole-free graphs is solvable in polynomial time, but the complexity of the MWIS problem for hole-free graphs is an open question. We will use Theorem 2, however, in subsequent sections for finding a polynomial time solution for (P_6, house)-free graphs $((P_6, \text{bull})$-free graphs, respectively).

Theorem 2. *If G is a P_6-free graph that has an e.d., then G^2 is C_k-free for any $k \geq 5$.*

4.2   Odd Antiholes in Squares of P_6-Free Graphs that have an e.d.

Our main reason for considering odd antiholes in squares of P_6-free graphs with an e.d. is the famous Strong Perfect Graph Theorem [8] saying that a graph is perfect if and only if it is odd-hole-free and odd-antihole-free. If one were able to exclude odd antiholes in the squares of P_6-free graphs with an e.d., it would mean that G^2 is perfect and thus, ED would be solvable in polynomial time for P_6-free graphs. Some partial results in this direction are described subsequently.

Throughout this subsection, let $G = (V, E)$ be a P_6-free graph with an e.d. D, and let $G^2 = (V, E^2)$. Let C be an odd antihole in G^2 with real vertices $R(C)$ and auxiliary vertices $A(C)$ as before. Since by Theorem 2, we know that C_5

is impossible in G^2, we can assume that C is a $\overline{C_{2k+1}}$ for $k \geq 3$. Obviously, $|D \cap R(C)| \leq 2$ since the distance between any two D-vertices is at least 3; D is an independent vertex set in G^2, and the independence number of an odd antihole is 2. The main result of this section, namely Theorem 3, is based on [10] and shows that no real vertex of an odd antihole C is in D:

Theorem 3. *If G is a P_6-free graph that has an e.d. D and C is an odd antihole in G^2, then $|D \cap R(C)| = 0$.*

4.3 C_4 in Squares of P_6-Free Graphs that have an e.d.

Let $G = (V, E)$ be a P_6-free graph with an e.d. D, and let $G^2 = (V, E^2)$ as defined above. By Theorem 2, we know that the square of a P_6-free graph with an e.d. is C_k-free for any $k \geq 5$. For considering the ED problem on some subclasses of P_6-free graphs, it is useful to analyze how a C_4 in G^2 can be realized. In particular, Lemma 13 is helpful in various cases, and Lemmas 11 and 12 are used for solving ED on (P_6,house)-free graphs.

As before, let C be a C_4 in G^2 with real vertices $R(C) = \{v_1, v_2, v_3, v_4\}$ such that $v_i v_{i+1}$ are adjacent in G^2 (index arithmetic is modulo 4), and with auxiliary vertices $A(C)$. Let the auxiliary vertex x_i be a common neighbor of v_i and v_{i+1}; $x_i \in A(C)$ if and only if $v_i v_{i+1} \notin E$.

For this subsection, we assume that $D \cap R(C) = \emptyset$. This assumption is motivated by Theorem 3 which says for a P_6-free graph G with an e.d. D, in an odd antihole C of G^2, no real vertex of C is in D; subsequently, we will consider a C_4 that is an induced subgraph of an odd antihole in G^2 where G is a P_6-free graph with an e.d. Let $d_i \in D$ denote the D-neighbor of v_i. Clearly, v_i and v_{i+2} have distinct D-neighbors for $i = 1$ and $i = 2$. There are the following types:

Type 1. $R(C)$ is dominated by two D-vertices; say, v_1, v_2 are dominated by $d_1 \in D$, and v_3, v_4 are dominated by $d_3 \in D$.

Type 1.1. $v_1 v_2 \notin E$, $v_3 v_4 \notin E$.

Lemma 8. *For any C_4 of type 1.1, $v_2 v_3 \in E$ and $v_1 v_4 \in E$ holds.*

Type 1.2. $v_1 v_2 \in E$, $v_3 v_4 \notin E$.
     Since $v_1 v_2 \in E$, we have $v_2 v_3 \notin E$ and $v_1 v_4 \notin E$.

Lemma 9. *For any C_4 of type 1.2, $d_3 x_2 \in E$ and $d_3 x_4 \in E$ holds.*

Type 1.3. $v_1 v_2 \in E$, $v_3 v_4 \in E$.
     Since $v_1 v_2 \in E$ and $v_3 v_4 \in E$, we have $v_2 v_3 \notin E$ and $v_1 v_4 \notin E$.

Lemma 10. *For any C_4 of type 1.3, we have: If neither d_1 nor d_3 dominates x_2 then $x_2 x_4 \in E$ and either $d_1 x_4 \in E$ or $d_3 x_4 \in E$.*

Type 2. $R(C)$ is dominated by three distinct D-vertices; say, v_1, v_2 are dominated by $d_1 \in D$, v_3 is dominated by $d_3 \in D$, and v_4 is dominated by $d_4 \in D$, $d_3 \neq d_4$.

Type 2.1. $v_1 v_2 \notin E$.

Lemma 11. *For any C_4 C of type 2.1, the following conditions hold:*

(i) $v_3v_4 \notin E$, $v_2v_3 \notin E$, and $v_1v_4 \notin E$. The auxiliary vertices x_2, x_3, x_4 are pairwise adjacent in G.

(ii) $d_1x_3 \in E$. Moreover, $d_1x_2 \notin E$ or $d_1x_4 \notin E$, and $d_1x_2 \notin E$ implies $d_3x_2 \in E$, while $d_1x_4 \notin E$ implies $d_4x_4 \in E$.

Note that $\{d_1, x_2, x_3, v_2, v_3\}$ induces a house in G if $d_1x_2 \notin E$, and thus, in any case of type 2.1, G contains a house.

Type 2.2. $v_1v_2 \in E$. Then by the distance properties, $v_2v_3 \notin E$ and $v_1v_4 \notin E$.

Lemma 12. *For any C_4 C of type 2.2, the following conditions hold:*

(i) *If* $v_3v_4 \in E$ *then* $d_4x_2 \in E$, $d_3x_4 \in E$ *and* $x_2x_4 \in E$.

(ii) *If* $v_3v_4 \notin E$ *then* x_2, x_3, x_4 *are pairwise adjacent in* G, $d_3x_4 \in E$ *or* $d_3x_3 \in E$, *and* $d_4x_2 \in E$ *or* $d_4x_3 \in E$. *Moreover,* $d_1x_2 \notin E$ *or* $d_1x_4 \notin E$, $d_3x_2 \notin E$ *or* $d_3x_3 \notin E$, *and* $d_4x_4 \notin E$ *or* $d_4x_3 \notin E$. *At most one of* x_2, x_4 *is dominated by a vertex* $d \in D$, $d \neq d_1, d_3, d_4$.

Note that in any case of type 2.2, G contains a house.

Type 3. $R(C)$ is dominated by four pairwise distinct D-vertices d_1, d_2, d_3, d_4. This type is excluded by the following:

Lemma 13. *For at least one pair* $i, j \in \{1, 2, 3, 4\}$, $i \neq j$, $d_i = d_j$ *holds.*

Corollary 3. *If G is $(P_6,$ house)-free graph that has an e.d. D and C is a C_4 in G^2 such that none of its real vertices is in D, then $R(C)$ is dominated by exactly two D-vertices.*

5   ED for (P_6, House)-Free Graphs and (P_6, Bull)-Free Graphs in Polynomial Time

Throughout this section, let G be a P_6-free graph that has an e.d. D. The aim of this section is to show that for (P_6, house)-free graphs (for (P_6, bull)-free graphs, respectively), the ED problem is solvable in polynomial time. Independently, for (P_6, bull)-free graphs, ED was solved in polynomial time by Karthick [12] using a different approach.

Theorem 4. *If G is a (P_6, house)-free graph that has an e.d., then G^2 is odd-antihole-free.*

Proof. Let G be a (P_6, house)-free graph with an e.d. D. Suppose to the contrary that G^2 contains an odd antihole H with real vertices $v_1, v_2, \ldots, v_{2k+1}$, $k \geq 3$, that are consecutively *co-adjacent* (i.e., nonadjacent in G^2). By Theorem 3, $D \cap \{v_1, v_2, \ldots, v_{2k+1}\} = \emptyset$ holds. Clearly, the neighborhood of any vertex $d \in D$ in H is a clique in G^2, and the clique cover number of H in G^2 is 3. Thus, the number of D-vertices dominating H is at least 3. Without loss of generality, let d_1 dominate v_1 and let d_2 dominate v_2. Since v_1 and v_2 are co-adjacent in H, $d_1 \neq d_2$ holds. If there is a vertex $d \in D$, $d \neq d_1, d \neq d_2$ dominating a vertex in $v_4, v_5, \ldots, v_{2k}$, then there is a C_4 in H that is dominated by at least three D-vertices, which contradicts Corollary 3. Thus, assume that d_1 and d_2 dominate all of $v_4, v_5, \ldots, v_{2k}$, and without loss of generality, let d dominate v_3. Then consider the C_4 C induced by $\{v_2, v_3, v_5, v_6\}$. By assumption, d_1 and d_2 dominate v_5 and v_6 and since v_5 and v_6 are co-adjacent in C, the D-vertices dominating v_5 and v_6 are distinct. Thus, C has three distinct D-vertices which contradicts Corollary 3. This shows Theorem 4.   □

Theorem 5. *If G is a $(P_6, bull)$-free graph that has an e.d., then G^2 is odd-antihole-free.*

Proof. Let G be a $(P_6, bull)$-free graph with an e.d. D, and suppose to the contrary that G^2 contains an odd antihole H with real vertices $v_1, v_2, \ldots, v_{2k+1}$, $k \geq 3$, that are consecutively co-adjacent. As before, $D \cap \{v_1, v_2, \ldots, v_{2k+1}\} = \emptyset$ holds, by Theorem 3. We first show:

Claim 1. *For every C_4 in H with real vertices u_1, u_2, u_3, u_4, for exactly two values of i, $1 \leq i \leq 4$, $u_i u_{i+1} \in E$ holds.*

Proof of Claim 1. Let u_1, u_2, u_3, u_4 be a C_4 in G^2 with $d_G(u_i, u_{i+1}) \leq 2$ (as before, let x_i be a common neighbor of u_i and u_{i+1} if $u_i u_{i+1} \notin E$) and suppose that for at most one i, $u_i u_{i+1} \in E$ holds.

First suppose that there is exactly one edge $u_i u_{i+1}$, say $u_1 u_2 \in E$. Then $u_2 u_3 \notin E$, $u_3 u_4 \notin E$, and $u_4 u_1 \notin E$. Since $\{u_1, u_2, x_2, u_3, x_3, u_4\}$ does not induce a P_6, we have $x_2 x_3 \in E$ but, now $\{u_2, x_2, u_3, x_3, u_4\}$ induces a bull, which is a contradiction.

Thus, for all i, $u_i u_{i+1} \notin E$ holds. We have seen already that if $x_i x_{i+1} \in E$ for some i, then $\{u_i, x_i, u_{i+1}, u_{i+2}, x_{i+1}\}$ induces a bull. Thus, for all i, $x_i x_{i+1} \notin E$. Since $\{u_1, x_1, u_2, x_2, u_3, x_3\}$ does not induce a P_6, we have $x_1 x_3 \in E$ and similarly we have $x_2 x_4 \in E$. By Lemmas 8 and 13, we know that the C_4 has exactly three D-vertices d_1, d_3, d_4; say d_1 sees u_1 and u_2, d_3 sees u_3, and d_4 sees u_4. Recall that $d_1 x_2 \notin E$ and $d_1 x_4 \notin E$ since G is assumed to be bull-free.
Since $\{u_1, d_1, u_2, x_2, u_3, d_3\}$ does not induce a P_6, we have $x_2 d_3 \in E$.
Since $\{u_2, d_1, u_1, x_4, u_4, d_4\}$ does not induce a P_6, we have $x_4 d_4 \in E$.
Since $\{u_2, x_2, u_3, x_3, d_3\}$ does not induce a bull, we have $x_3 d_3 \in E$.
Since $\{u_1, x_4, u_4, x_3, d_4\}$ does not induce a bull, we have $x_3 d_4 \in E$, which is a contradiction showing Claim 1. ◇

By Claim 1, every C_4 in the odd antihole H of G^2 has exactly two edges in E. We apply this as follows:

Claim 2. *For all $i, 1 \leq i \leq 2k+1$, we have: If $v_i v_{i+2} \in E$ then $v_{i+1} v_{i+3} \in E$ (index arithmetic is modulo $2k+1$). In particular, if for some i, $v_i v_{i+2} \in E$ then for all i, $1 \leq i \leq 2k+1$, $v_i v_{i+2} \in E$.*

Proof of Claim 2. Let $v_1 v_3 \in E$. Then, by the distance conditions, $v_1 v_4 \notin E$ and $v_3 v_{2k+1} \notin E$. Considering the C_4 in G^2 induced by $\{v_1, v_3, v_4, v_{2k+1}\}$, we have

$v_4v_{2k+1} \in E$, which implies $v_5v_{2k+1} \notin E$. Considering the C_4 in G^2 induced by $\{v_1, v_4, v_5, v_{2k+1}\}$, we have $v_5v_1 \in E$, which implies $v_5v_2 \notin E$. Considering the C_4 in G^2 induced by $\{v_1, v_2, v_4, v_5\}$, we have $v_2v_4 \in E$. Applying this repeatedly along the odd antihole H, we obtain $v_iv_{i+2} \in E$ for all i. ◇

Now first assume that for one i, $v_iv_{i+2} \in E$ holds; say, $v_1v_3 \in E$. Then we consider the C_4s with v_1, v_{2k+1} and the opposite pairs v_i, v_{i+1}, $3 \leq i \leq 2k-2$, and we obtain an alternating sequence of edges and non-edges for v_1, i.e., $v_1v_i \in E$ for all odd i, $3 \leq i \leq 2k - 1$ and $v_1v_i \notin E$ for all even i, $4 \leq i \leq 2k - 2$, and considering the C_4 in G^2 induced by $\{v_1, v_{2k-2}, v_{2k-1}, v_{2k+1}\}$, we obtain $v_{2k+1}v_{2k-1} \notin E$, which contradicts Claim 2.

Thus, suppose that for all $i, 1 \leq i \leq 2k + 1$, $v_iv_{i+2} \notin E$ holds. Since by assumption, every C_4 has exactly two E-edges, we can assume that $v_1v_i \in E$ for some i. Then, by the distance conditions, $v_1v_{i-1} \notin E$, $v_1v_{i+1} \notin E$, $v_iv_2 \notin E$, and $v_iv_{2k+1} \notin E$. By the C_4 argument, $v_2v_{i-1} \in E$ and $v_{2k+1}v_{i+1} \in E$ follows. Repeatedly applying the distance argument and the C_4 argument implies that finally, for some j, $v_jv_{j+2} \in E$, which is a contradiction that concludes the proof of Theorem 5.   □

Corollary 4. *For (P_6, house)-free graphs and (P_6, bull)-free graphs, the WED problem is solvable in polynomial time.*

Proof. First suppose that G is (P_6, house)-free. By Theorem 2, for a P_6-free graph G with an e.d., G^2 is hole-free. By Lemma 4, G^2 is odd-antihole-free. If G^2 is odd-hole-free and odd-antihole-free then, by the Strong Perfect Graph Theorem [8], G^2 is perfect. By [11], MWIS is solvable in polynomial time for perfect graphs. By Lemma 1, the ED problem on G can be transformed into the MWIS problem on G^2. Thus, ED is solvable in polynomial time on (P_6, house)-free graphs. By [2], the WED problem can be solved in polynomial time for the same class.

Now suppose that G is (P_6, bull)-free. By Lemma 5, G^2 is odd-antihole-free, and thus, G^2 is perfect. Hence, WED is solvable in polynomial time on (P_6, bull)-free graphs by the same arguments as above.   □

6 Conclusion

The main results of this paper are Theorems 1, 2, 3, 4, and 5 solving ED in polynomial time for P_6-free chordal graphs, (P_6, HHD)-free graphs, (P_6, house)-free graphs, (P_6, bull)-free graphs, respectively.

For some other subclasses of P_6-free graphs, ED has been solved in polynomial time, such as for $(P_6, S_{1,2,2})$-free graphs [7] and for $(P_6, S_{1,1,3})$-free graphs [12]; see also [4] where the time bound for ED on (P_6, bull)-free graphs is improved. Meanwhile, Karthick (private communication) also showed that ED can be solved in polynomial time on (P_6, net)-free graphs, which extends the result for (P_6, bull)-free graphs.

Conjecture [10]. *If G is a P_6-free graph that has an e.d., then G^2 is perfect.*

Remark in Press. Very recently, Lokshtanov et al. [15] showed that the ED problem on P_6-free graphs can be solved in polynomial time, using a completely different approach.

Acknowledgments. The first and second authors gratefully acknowledge support from the West Virginia University NSF ADVANCE Sponsorship Program, and the first author thanks Van Bang Le for discussions about the Efficient Domination problem.

References

1. Brandstädt, A., Eschen, E.M., Friese, E.: Efficient domination for some subclasses of P_6-free graphs in polynomial time, arXiv:1503.00091v1 (2015)
2. Brandstädt, A., Fičur, P., Leitert, A., Milanič, M.: Polynomial-time algorithms for weighted efficient domination problems in AT-free graphs and dually chordal graphs. Inf. Process. Lett. **115**, 256–262 (2015)
3. Brandstädt, A., Giakoumakis, V.: Weighted efficient domination for $(P_5 + kP_2)$-free graphs in polynomial time, arXiv:1407.4593v1 (2014)
4. Brandstädt, A., Karthick, T., Weighted efficient domination in classes of P_6-free graphs, arXiv:1503.06025v1 (2015)
5. Brandstädt, A., Le, V.B., Spinrad, J.P.: Graph classes: a survey. In: SIAM Monographs on Discrete Mathematics Application, vol. 3. SIAM, Philadelphia (1999)
6. Brandstädt, A., Leitert, A., Rautenbach, D.: Efficient dominating and edge dominating sets for graphs and hypergraphs. In: Chao, K.-M., Hsu, T., Lee, D.-T. (eds.) ISAAC 2012. LNCS, vol. 7676, pp. 267–277. Springer, Heidelberg (2012)
7. Brandstädt, A., Milanič, M., Nevries, R.: New polynomial cases of the weighted efficient domination problem. In: Chatterjee, K., Sgall, J. (eds.) MFCS 2013. LNCS, vol. 8087, pp. 195–206. Springer, Heidelberg (2013)
8. Chudnovsky, M., Robertson, N., Seymour, P., Thomas, R.: The strong perfect graph theorem. Ann. Math. **164**, 51–229 (2006)
9. Frank, A.: Some polynomial algorithms for certain graphs and hypergraphs. In: Proceedings of 5th British Combinatorial Conference 1976, Aberdeen, Congressus Numerantium No. XV, pp. 211–226 (1975)
10. Friese, E.: Das efficient-domination-problem auf P_6-freien graphen, Master Thesis. University of Rostock, Germany (2013) (in German)
11. Grötschel, M., Lovász, L., Schrijver, A.: The ellipsoid method and its consequences in combinatorial optimization. Combinatorica **1**, 169–197 (1981)
12. Karthick, T.: Weighted Efficient Domination for Certain Classes of P_6-free Graphs, Manuscript (2015)
13. Garey, M.R., Johnson, D.S.: Computers and Intractability - A Guide to the Theory of NP-completeness. Freeman, San Francisco (1979)
14. Leitert, A.: Das dominating induced matching problem für azyklische hypergraphen, Diploma Thesis. University of Rostock, Germany (2012) (in German)
15. Lokshtanov, D., Pilipczuk, M., van Leeuwen, E.J.: Independence and efficient domination on P6-free graphs, arXiv:1507.02163v1 (2015)
16. Milanič, M.: Hereditary efficiently dominatable graphs. J. Graph Theor. **73**, 400–424 (2013)

On the Tree Search Problem
with Non-uniform Costs

Ferdinando Cicalese[1(✉)], Balázs Keszegh[2], Bernard Lidický[3],
Dömötör Pálvölgyi[4], and Tomáš Valla[5]

[1] Department of Computer Science, University of Verona, Verona, Italy
cicalese@dia.unisa.it
[2] Rényi Institute, Budapest, Hungary
keszegh.balazs@renyi.mta.hu
[3] Department of Mathematics, Iowa State University, Ames, USA
lidicky@iastate.edu
[4] Eötvös University, Budapest, Hungary
dom@cs.elte.hu
[5] Faculty of Information Technology, Czech Technical University,
Prague, Czech Republic
tomas.valla@fit.cvut.cz

Abstract. Searching in partially ordered structures has been considered
in the context of information retrieval and efficient tree-like indices, as
well as in hierarchy based knowledge representation. In this paper we
focus on tree-like partial orders and consider the problem of identifying
an initially unknown vertex in a tree by asking edge queries: an edge
query e returns the component of $T - e$ containing the vertex sought for,
while incurring some known cost $c(e)$.

The Tree Search Problem with Non-Uniform Cost is the following:
given a tree T on n vertices, each edge having an associated cost, construct a strategy that minimizes the total cost of the identification in the
worst case.

Finding the strategy guaranteeing the minimum possible cost is an
NP-complete problem already for input trees of degree 3 or diameter 6.
The best known approximation guarantee was an $O(\log n / \log \log \log n)$-
approximation algorithm of [Cicalese et al. TCS 2012].

B. Keszegh—Research supported by Hungarian National Science Fund (OTKA),
under grant PD 108406 and under grant NN 102029 (EUROGIGA project GraDR
10-EuroGIGA-OP-003) and the János Bolyai Research Scholarship of the Hungarian
Academy of Sciences.
B. Lidický—Research is partially supported by NSF grants DMS-1266016 and DMS-
1600390.
D. Pálvölgyi—Research supported by Hungarian National Science Fund (OTKA),
under grant PD 104386 and under grant NN 102029 (EUROGIGA project GraDR
10-EuroGIGA-OP-003) and the János Bolyai Research Scholarship of the Hungarian
Academy of Sciences.
T. Valla—Supported by the Centre of Excellence – Inst. for Theor. Comp. Sci.
(project P202/12/G061 of GA ČR).

© Springer-Verlag Berlin Heidelberg 2016
E.W. Mayr (Ed.): WG 2015, LNCS 9224, pp. 90–102, 2016.
DOI: 10.1007/978-3-662-53174-7_7

We improve upon the above results both from the algorithmic and the computational complexity point of view: We provide a novel algorithm that provides an $O(\frac{\log n}{\log \log n})$-approximation of the cost of the optimal strategy. In addition, we show that finding an optimal strategy is NP-hard even when the input tree is a spider of diameter 6, i.e., at most one vertex has degree larger than 2.

1   Introduction

The design of efficient procedures for searching in a discrete structure is a fundamental problem in discrete mathematics [1,2] and computer science [10]. Searching is a basic primitive for building and managing operations of an information system as ordering, updating, and retrieval. The typical example of a search procedure is binary search which allows to retrieve an element in a sorted list of size n by only looking at $O(\log n)$ elements of the list. If no order can be assumed on the list, then it is known that any procedure will have to look at the complete list in the worst case. Besides these two well characterized extremes, extensive work has also been devoted to the case where the underlying structure of the search space is a partial order. Partial orders can be used to model lack of information on the totally ordered elements of the search space [12] or can naturally arise from the relationship among the elements of the search space, like in hierarchies used to model knowledge representation [15], or in tree-like indices for information retrieval of large databases [3]. For more about applications of tree search see the end of this section.

In this paper, we focus on the case where the underlying search space is a tree-like partially ordered set and tests have nonuniform costs. We investigate the following problem.

THE TREE SEARCH PROBLEM WITH NON-UNIFORM COSTS

Input: A tree $T = (V, E)$, $|V| = n$, with non-negative rational costs assigned to the edges defined by a $c : e \in E \mapsto c(e) \in \mathbb{Q}$.

Output: A strategy that minimizes (in the worst case) the cost spent to identify an initially unknown vertex x of T by using *edge queries*. An *edge query* $e = \{u, v\} \in E$ asks for the subtree T_u or T_v which contains x, where T_u and T_v are the (maximal connected) components of $T - e$, including the vertex u and v respectively. The cost of the query e is $c(e)$. The cost of identifying a vertex x is the sum of the costs of the queries asked.

More formally, a strategy for the Tree Search Problem with nonuniform costs over the tree T is a *decision tree* D which is a rooted binary tree with $|V|$ leaves where every leaf ℓ is associated with one vertex $v \in V$ and every internal node[1] $\nu \in V(D)$ is associated with one test $e = \{u, v\} \in E$. The outgoing edges from ν are associated with the possible outcomes of the query, namely, to the case

---

[1] For the sake of avoiding confusion between the input tree and the decision tree, we will reserve the term vertex for the elements of V and the term *node* for the vertices of the decision tree D.

where the vertex to identify lies in T_u or T_v respectively. Every vertex has at least one associated leaf. The actual identification process can be obtained from D starting with the query associated to the root and moving towards the leaves based on the answers received. When a leaf ℓ is reached, the associated vertex is output (see also Fig. 1 in the appendix for an example).

Given a decision tree D, for each vertex $v \in V(T)$, let $cost^D(v)$ be the sum of costs of the edges associated to nodes on the path from the root of D to the leaf identifying v. This is the total cost of the queries performed when the strategy D is used and v is the vertex to be identified.

In addition, let the cost of D be defined by

$$cost(D) = \max_{v \in V(T)} cost^D(v).$$

This is the worst-case cost of identifying a vertex of T by the decision tree D. The optimal cost of a decision tree for the instance represented by the tree T and the cost assignment c is given by

$$OPT(T, c) = \min_D cost(D),$$

where the min is over all decision trees D for the instance (T, c).

Previous Results and Related Work. The Tree Search Problem has been first studied under the name of tree edge ranking [5,7,9,11,13], motivated by multi-part product assembly. In [11] it was shown that in the case where the tests have uniform cost, an optimal strategy can be found in linear time. A linear algorithm for searching in a tree with uniform cost was also provided in [14]. Independently of the above articles, the first paper where the problem is considered in terms of searching in a tree is [3], where the more general problem of searching in a poset was also addressed.

The variant considered here in which the costs of the tests are non-uniform was first studied by Dereniowski [6] in the context of edge ranking. In this paper, the problem was proved NP-complete for trees of diameter at most 10. Dereniowski also provided an $O(\log n)$ approximation algorithm. In [4] Cicalese et al. showed that the tree search problem with non-uniform costs is strongly NP-complete already for input trees of diameter 6, or maximum degree 3, moreover, these results are tight. In fact, in [4], a polynomial time algorithm computing the optimal solution is also provided for diameter 5 instances and an $O(n^2)$ algorithm for the case where the input tree is a path. For arbitrary trees, Cicalese et al. provided an $O(\frac{\log n}{\log \log \log n})$-approximation algorithm.

Our Result. Our contribution is both on the algorithmic and on the complexity side. On the one hand, we provide a new approximation algorithm for the tree search problem with non-uniform costs which improves upon the best known guarantee given in [4]. In Sect. 3 we will prove the following result.

Theorem 1. *There is an $O(\log n / \log \log n)$-approximation algorithm for the Weighted Tree Search Problem that runs in polynomial time in n.*

A *spider graph* (henceforth simply referred to as a *spider*) is a tree with at most one vertex of degree larger than 2.

In this paper, we also show that the tree search problem with non-uniform costs is NP-hard already when the input tree is a *spider* of diameter 6.

More About Applications. We discuss some scenarios in which the problem of searching in trees with non-uniform costs naturally arises.

Consider the problem of locating a buggy module in a program in which the dependencies between different modules can be represented by a tree. For each module we can verify the correct behavior independently. Such a verification may consist in checking, for instance, whether all branches and statements in a given module work properly. For different modules, the cost of using the checking procedure can be different (here the cost might refer to the time to complete the check). In such a situation, it is important to device a debugging strategy that minimizes the cost incurred in order to locate the buggy module in the worst case.

Checking for consistency in different sites keeping distributed copies of tree-like data structures (e.g., file systems) can be performed by maintaining at each node some check sum information about the subtree rooted at that node. Tree search can be used to identify the presence of "buggy nodes", and efficiently identifying the inconsistent part in the structure, rather than retransmitting or exhaustively checking the whole data structure. In [3], an application of this model in the area of information retrieval is also described.

Another example comes from a class of problems which is in some sense dual to the previous ones: deciding the assembly schedule of a multi-part device. Assume that the set of pairs of parts that must be assembled together can be represented by a tree. Each assembly operation requires some (given) amount of time to be performed and while assembling two pieces, the same pieces cannot be involved in any other assembly operations. At any time different pairs of parts can be assembled in parallel. The problem is to define the schedule of assembly operations which minimize the total time spent to completely assembly the device. The schedule is an edge ranking of the tree defined by the assembly operations. By reversing the order of the assembly operation in the schedule we obtain a decision tree for the problem of searching in the tree of the assembly operation where each edge cost is equal to the cost of the corresponding assembly.

2   Basic Lower and Upper Bounds

In this section we provide some preliminary results which will be useful in the analysis of our algorithm presented in the next section. We introduce some lower bounds on the cost of the optimal decision tree for a given instance of the problem. We also recall two exact algorithms for constructing optimal decision trees which were given in [4]. The first is an exponential time dynamic programming algorithm which works for any input tree. The second is a quadratic time algorithm for instances where the input tree is a path. Finally, we show a construction of 2-approximation decision trees for spider graphs.

Let T denote the input tree and c the cost function. It is not hard to see that, given a decision tree D for T, we can extract from it a decision tree for the instance of the problem defined on a subtree T' of T and the restriction of c to the vertices in T'. For this, we can repeatedly apply the following operation: if in D there is a node ν associated with an edge $e = \{u, v\}$, such that T_u (resp. T_v) is included $T - T'$, then remove the node ν together with the subtree rooted at the child of ν corresponding to the case where the vertex to identify is in T_u (resp. T_v). Let D' be the resulting decision tree when the above step cannot be performed any more. Then, clearly $cost(D', c) \leq cost(D, c)$. We have shown the following (also observed in [4]).

Lemma 1. *Let T' be a subtree of T. Then, $OPT(T, c) \geq OPT(T', c)$.*

Another immediate observation is that for a given input tree T, the value $OPT(T, c)$ is monotonically non-decreasing with respect to the cost of any edge. This is recorded in the following.

Lemma 2. *Let c and c' be cost assignments on a tree T such that $c'(e) \leq c(e)$ for every $e \in E(T)$. Then, $OPT(T, c) \geq OPT(T, c')$.*

The next proposition shows that subdividing an edge cannot decrease the cost of the optimal decision tree.

Proposition 1. *Let c be a cost assignment on a tree T. Let $v \in V(T)$ have exactly two neighbors $u_1, u_2 \in V(T)$. If T' is obtained from $T - v$ by adding the edge $\{u_1, u_2\}$ and c' is obtained from c by setting $c'(u_1 u_2) = \min\{c(u_1 v), c(u_2 v)\}$, then $OPT(T, c) \geq OPT(T', c')$.*

Proof. Let D be an optimal decision tree for the instance (T, c). Let us assume without loss of generality that in D the node ν_1 associated with $e_1 = \{u_1, v\}$ is an ancestor of the node ν_2 associated with $e_2 = \{u_2, v\}$. Notice that one of the children of ν_2 is a leaf associated with the vertex v. Let $\tilde{D}$ be the subtree of D rooted at the non-leaf child of ν_2.

Let D' be the decision tree obtained from D by associating the node ν_1 to the edge $e = \{u_1, u_2\}$ and replacing the subtree rooted at ν_2 with the subtree $\tilde{D}$.

It is not hard to see that D' is a proper decision tree for T'. We also have that for any vertex z of T' which is associated to a leaf in $\tilde{D}$ it holds that $cost^{D'}(z) = cost^D(z) - c(e_1) - c(e_2) + c'(u_1 u_2)$, and for any other vertex z of T' we have $cost^{D'}(z) = cost^D(z) - c(e_1) + c'(u_1 u_2)$ or $cost^{D'}(z) \leq cost^D(z)$. It follows that $OPT(T', c') \leq cost(D') \leq cost(D) = OPT(T, c)$. $\square$

The following two results from [4] provide exact algorithms for the construction of optimal strategies. More precisely, Proposition 2 provides an exponential dynamic programming based algorithm for general trees. Theorem 2 gives an $O(n^2)$ time algorithm for the special case where the input tree is a path and will be useful in the analysis of our main algorithm and also in Lemma 3 regarding the spider tree.

Proposition 2 [4]. *Let T be an edge-weighted tree on n vertices. Then an optimal decision tree for T can be constructed in $O(2^n n)$ time.*

The following theorem was proved by Cicalese et al. in [4] and will be useful later in the analysis of our algorithm and also in the following lemma regarding the spider tree.

Theorem 2 [4]. *There is an $O(n^2)$ time algorithm that constructs an optimal decision tree D for a given weighted path on n vertices.*

Note that for a star T any decision tree D has the same cost, since all the edges have to be asked in the worst case. Hence, for a tree T such that there is only one node with degree greater than 1 we have $OPT(T, c) = \sum_{e \in E(T)} c(e)$, for any cost function c.

Definition 1. *A tree T is a spider if there is at most one vertex in T of degree greater than two. We refer to this vertex as the head (or center) of the spider. In the special case when all vertices have degree at most 2 then an arbitrarily chosen vertex of degree 2 is designated to be the head of the spider. Moreover, each path from the head of the spider to one of the leaves will be referred to as a leg of the spider.*

Lemma 3. *Let T be a spider. Then there is an algorithm which computes a 2-approximate decision tree D for T and runs in time $O(n^2)$.*

Proof. If T is a path, then by Theorem 2 there exists an algorithm computing the optimal decision tree in $O(n^2)$ time. Assume T is not a path. Then T contains exactly one vertex v of degree at least three. Let S_v be the star induced by v and the vertices adjacent to v. Let us denote by $w_1, \ldots, w_k$ the vertices adjacent to v, where $k = \deg(v)$. By Theorem 2, for every $i \in \{1, \ldots, k\}$ we construct the optimal decision tree D_i for the path component C_i of $T - v$ containing w_i in time $O(|C_i|^2)$. Note that the total running time for construction of $D_1, \ldots, D_k$ is $O(n^2)$. Finally, for S_v we compute the optimal decision tree D_v (in $O(n)$ time). The decision tree D for T is obtained from D_v by replacing the node corresponding to w_i by the root of D_i for every $i \in \{1, \ldots, k\}$. Clearly, the algorithm runs in $O(n^2)$ time and $cost(D) \leq OPT(S_v, c) + \max_{1 \leq i \leq k}\{OPT(C_i, c)\} \leq 2OPT(T, c)$. The last inequality follows because by Lemma 1 both $OPT(S_v, c)$ and $\max_{1 \leq i \leq k}\{OPT(C_i, c)\}$ are lower bounds on $OPT(T, c)$. □

3   The Algorithm

In this section we present Algorithm TS for the tree search problem—we also provide a pseudocode of the algorithm in the appendix.

Let n be the size of the input tree and $t = 2^{\lfloor \log \log n \rfloor + 2}$ be a parameter fixed for the whole run of the algorithm. It holds that $2 \log n \leq t \leq 4 \log n$.

The basic idea of our algorithm is to construct a subtree S of the input tree T such that: (i) we can construct a decision tree for S whose cost is at most a constant times the cost of an optimal decision tree for S; (ii) each component of $T - S$ has size not larger than $|T|/t$.

This will allow us to build a decision tree for T by assembling the decision tree for S with the decision trees recursively constructed for the components of $T - S$. The constant approximation guarantee on S and the fact that, due to the size of the subtrees on which we recur, we need at most $O(\frac{\log n}{\log \log n})$ levels of recursion to show that our algorithm gives an $O(\frac{\log n}{\log \log n})$ approximation.

The Subtree S. We iteratively build subtrees $S_0 \subset S_1 \subset \cdots \subset S_t \subseteq T$. Starting with the empty tree S_0, in every iteration $i \in \{1, \dots, t\}$ we pick a centroid[2] x_i of the largest component of the forest $T - S_{i-1}$. The subtree S_i is set to be the minimal subtree containing x_i and S_{i-1}. If for some i we have that $S_i = T$, then we set $S = S_i = T$ and we stop the iterations. If all t iterations are completed, then we set $S = S_t$.

We have the following lemma—which establishes (ii) above.

Lemma 4. *If H is a component of $T - S$, then $|H| \leq |T|/\log|T|$.*

Proof. We prove by induction on k that after 2^k iterations all components of $T - S_{2^k}$ have size at most $|T|/2^{k-1}$. Let $k = 0$. We observe that by the definition of centroid, after $1 = 2^k$ iteration all components of $T - S_1$ have size at most $|T|/2 \leq 2|T| = |T|/2^{k-1}$. This establishes the basis of our induction.

Now fix some $k > 0$ and assume (induction hypothesis) that after 2^{k-1} iterations all components of $T - S_{2^{k-1}}$ have size at most $|T|/2^{k-2}$. Among these there are $p \leq 2^{k-1}$ components that have size at least $|T|/2^{k-1}$. In the next p iterations we will choose a centroid in each of these components, one by one. Choosing a centroid in a component H splits H into parts that have size at most half of H, thus after $2^k = 2^{k-1} + p$ steps all components of $T - S_{2^k}$ have size at most $|T|/2^{k-1}$.

Thus, if the process of constructing S is stopped after $t = 2^{\lfloor \log \log n \rfloor + 2}$ iterations all components have size at most $|T|/2^{\lfloor \log \log n \rfloor + 1} \leq |T|/\log n$. On the other hand, if the process of constructing S is stopped at some iteration $i < t$, then it means that $S = T$ and trivially $|H| = 1$. □

The Decision Tree for S. Let X contain all x_i for $i \in \{1, \dots, t\}$ and vertices of degree at least three in S. Note that $|X| \leq 2t - 2$ for $t \geq 2$ and $|X| = 1 = 2t - 1$ for $t = 1$. Indeed, by induction on k, this is true for $k = 1, 2$. Adding a new x_i, S_i is a tree which is the union of S_{i-1} and a path reaching to x_i, thus S_i has at most one more vertex of degree at least three than S_{i-1}. Together with x_i, $|X|$ increases by at most two in a step. Let $P_{u,v}$ be the path of T whose endpoints are vertices u and v.

We define an auxiliary tree Y on the vertex set X in which the paths of T between the vertices of X are replaced by 'shortcut' edges. Vertices $u, v \in X$ form an edge of Y if u and v are the only vertices of X of the path $P_{u,v}$ in

---

[2] Recall that a *centroid* of a tree T is a vertex v such that any component of $T - v$ has size at most $|T|/2$.

T with endpoints u and v. Let $e_{uv} = \arg\min_{e \in P_{u,v}} c(e)$ (the edge of $P_{u,v}$ with minimal cost) and $c_Y(uv) = c(e_{uv})$. Let $Z = \bigcup_{uv \in E(Y)} e_{uv}$. By Proposition 2, we can compute an optimal decision tree D_Y for Y in time $O(2^{2t}t)$ which is polynomial in n (Fig. 2).

Let D_X be obtained from D_Y by changing the label of every internal node from uv to e_{uv}, for each $uv \in E(Y)$. The tree D_X is not a decision tree for S, however, leaves of D_X correspond to components of $S - Z$. Notice that $cost(D_X) = cost(D_Y) = OPT(Y, c_Y)$.

Since every component C of $S - Z$ contains at most one vertex of degree at least three, every such component is a spider. By Lemma 3, a decision tree D_C for each such component $C \in S - Z$ can be computed in $O(n^2)$ time with approximation ratio 2.

We can now obtain the decision tree D_S for S by replacing each leaf in D_X with the decision tree for the corresponding component in $S - Z$. We have

$$
\begin{aligned}
\frac{cost(D_S)}{OPT(S, c)} &\leq \frac{cost(D_X) + \max_{C \in S-Z} cost(D_C)}{OPT(S, c)} \\
&\leq \frac{cost(D_X)}{OPT(Y, c_Y)} + \max_{C \in S-Z} \frac{cost(D_C)}{OPT(C, c)} \leq 3,
\end{aligned}
\tag{1}
$$

where the second inequality holds because a repeated application of Proposition 1 implies $OPT(Y, c_Y) \leq OPT(S, c)$ and Lemma 1 implies $OPT(C, c) \leq OPT(S, c)$.

Assembling the Pieces in the Decision Tree for T. Let v be a vertex in S with a neighbor not in S, let S_v be the star induced by v and its neighbors outside $V(S)$.

Let D_v be a decision tree for S_v (notice that they all have the same cost). For every neighbor $w \notin V(S)$ of v we compute recursively the decision tree D_w for the component H_w of $T - S$ containing w and replace the leaf node of D_v associated to w with the root of D_w. The result is a decision tree D'_v for the subtree of T including S_v and all the components of $T - S$ including some neighbor w of v.

In order to obtain a decision tree D_T for T we now modify D_S as follows: for each vertex v in S with a neighbor not in S, replace the leaf in D_S associated with v with the decision tree D'_v computed above.

The Approximation Guarantee for D_T. Let $APP(T) = \frac{cost(D_T)}{OPT(T,c)}$ denote the approximation ratio obtained by Algorithm TS on the instance (T, c). Let $APP(k) = \max_{|T| \leq k} APP(T)$.

Lemma 5. For any tree T on n vertices and any cost assignment c, we have $APP(T) \leq 4 \log n / \log \log n$.

Proof. For every $1 \leq k \leq n$ let $f(k) = \max\{1, 4 \log k / \log \log n\}$. We shall prove by induction on k that $APP(k) \leq f(k)$, which implies the statement of the lemma.

If $|T| \leq t$, then our algorithm builds an optimal decision tree, thus $APP(k) = 1 \leq f(k)$ for $k \leq t$. This establishes the induction base.

Choose a tree T as in the statement of the lemma such that $APP(T) = APP(n)$. Let S and Y be the substructures of T built by the algorithm as described above. Let $\tilde{V}$ be the set of vertices of S with some neighbor not in S. For each $w \notin V(S)$ let H_w be the component of $T - S$ containing w. Let $\mathcal{H}$ be the set of components of $T - S$. Then, by construction, we have

$$APP(T) = \frac{ALG(T)}{OPT(T)} \tag{2}$$

$$\leq \frac{cost(D_S) + \max_{v \in \tilde{V}} cost(D_v) + \max_{w \notin V(S)} cost(D_w)}{OPT(T,c)} \tag{3}$$

$$\leq \frac{cost(D_S)}{OPT(S,c)} + \max_{v \in \tilde{V}} \frac{cost(D_v)}{OPT(S_v,c)} + \max_{w \notin V(S)} \frac{cost(D_w)}{OPT(H_w,c)} \tag{4}$$

$$\leq 4 + \max_{H \in \mathcal{H}} \frac{ALG(H)}{OPT(H,c)} = 4 + \max_{H \in \mathcal{H}} \{APP(H)\} \tag{5}$$

$$\leq 4 + \max_{H \in \mathcal{H}} f(|H|) \leq 4 + f(|T|/\log n) \tag{6}$$

$$= 4 + f(n/\log n) = 4 + \frac{4 \log \frac{n}{\log n}}{\log \log n} = \frac{4 \log n}{\log \log n}, \tag{7}$$

where

- (4) follows from (3) because of $OPT(S,c), OPT(S_v,c), OPT(H_w,c) \leq OPT(T,c)$ (Lemma 1)
- (5) follows from (4) because of (1) we have $\frac{cost(D_S)}{OPT(S,c)} \leq 3$ and because any decision tree for a star S_v has the same cost, hence also equal to $OPT(S_v,c)$
- in (6) the first inequality follows by induction and the second inequality by Lemma 4
- (7) follows from (6) because of $|T| = n$ and the definition of $f(\cdot)$.   □

Lemma 6. *For a tree T on n vertices, the Algorithm TS builds the decision tree D_T in time polynomial in n.*

Proof. If $|T| \leq t$, then the algorithm builds an optimal decision tree for T in time $O(2^t \cdot t) = O(n^4)$ using the construction from Proposition 2. Otherwise, every iteration needed to build the subtree S (lines 7–11 of the algorithm) introduces one new vertex x_i and at most one other vertex of degree at least three, thus $|X| \leq 2t - 1$. Proposition 2 then implies that an optimal decision tree D_Y for Y can be computed in time $O(2^{2t} \cdot 2t)$ which is polynomial in n. By Lemma 3, the 2-approximation decision tree D_H for H can be computed in $O(n^2)$ time. Building the decision tree D_v for the stars S_v takes $O(|S_v|)$ time (line 30). The rest of the algorithm, not counting the recursion on line 34, needs time $O(n^2)$. As the recursion is for a graph whose size is at most half of the original, the overall algorithm running time is polynomial in n.   □

Lemmas 5 and 6 now imply Theorem 1.

4   Tree Search with Non-uniform Costs is NP-hard on Spider Graphs

In this section we provide a new hardness result which contributes to refining the separation between hard and polynomial instances of the tree search problem with non-uniform costs. We show that the problem of finding a minimum cost decision tree is hard even for instances where the input graph is a spider and the length of every leg is three.

Our reduction is from the NP-complete Balanced Partition problem [8, A3.2], a special case of the Partition problem. The input of the Partition problem is given by a set of numbers, $\{a_i \mid i \in [m]\}$, and our goal is to find an index set I such that $\sum_{i \in I} a_i = \sum_{i \notin I} a_i$. In the Balanced Partition problem it is further required that $|I| = m/2$, i.e., there are the same number of numbers in both parts of the partition. (This implies that for a non-trivial input m has to be even.) Because of this, we can also suppose that all the numbers have roughly the same size as adding a contant to each will not affect the set of solutions. This implies that we can suppose that $a_i < 2a_j$ and $\sum_{i \in I} a_i < \sum_{i \notin I} a_i$ for any $|I| < m/2$.

From a set of numbers $\{a_i \mid i \in [m]\}$ with the above properties, we construct an instance (S, c) for the tree search problem with non-uniform costs, where S is a spider. Each leg will correspond to a number. Therefore, we will speak of the ith leg as the leg corresponding to the ith number. For each $i \in [m]$, the ith leg will consist of three edges: the one closest to the head will be called *femur* (and referred to as f_i), the middle edge will be called *tibia* (and referred to as t_i), the end will be called the *tarsus* (and referred to as s_i). The cost function is defined as follows: For each $i \in [m]$, we set $c(f_i) = 2a_i$; $c(t_i) = a_i$ and $c(s_i) = N$, with $N = \sum_{i \in [m]} a_i$.

It is easy to see that in an optimal strategy, for each $i \in [m]$ the tarsus is always queried last among the edges on the ith leg. Given a decision tree D, we denote by F the set of indices of the legs for which, in D, the node associated with the query to the tibia is an ancestor[3] of the node associated with the query to the femur. Then, we have the following proposition, whose proof is omitted from this extended abstract.

Proposition 3. *There is an optimal decision tree D with $F \neq \emptyset$ and such that:*
   (i) for any $i \in F$ and $j \in [m] \setminus F$ the node of D associated with the jth femur is an ancestor of the node associated with the ith tibia.
   (ii) for any $i, j \in F$ the node of D associated with the ith tibia is an ancestor of the node associated with the jth femur.

By this proposition, we can assume that in the optimal decision tree D for at least one leg of the spider the first edge queried is a tibia. In addition, in D, there is a root to leaf path where first all femora not in F are queried, then all tibiae in F, and finally all femora in F. Then, the cost of such a decision tree is given

---

[3] A node ν is an ancestor of another node ν' if ν lies on the path connecting ν' to the head.

by the maximum of the cost of the above mentioned path ($\sum_{i\notin I} 2a_i + \sum_{i\in I} 3a_i$) and the costs of the paths to the leaves on the legs, which either start with a femur with index not in F (cost $\leq \sum_{i\notin I} 2a_i + \max_{i\notin I} a_i + N$) or with a tibia with index in F (cost $\leq \sum_{i\notin I} 2a_i + \sum_{i\in I} a_i + N$). This last value is indeed attained for the leaf of the last leg starting with a tibia. Using the fact that for all i, j we have $a_i < 2a_j$, this last value, $\sum_{i\notin I} 2a_i + \sum_{i\in I} a_i + N$, is always greater than the cost for leaves on legs starting with femora, $\sum_{i\notin I} 2a_i + \max_{i\notin I} a_i + N$ if $m \geq 4$. Therefore, the cost of the optimal solution is given by the following expression

$$OPT(S, c) = \min_{\emptyset \subset I \subseteq [m]} \max \left\{ \sum_{i\notin I} 2a_i + \sum_{i\in I} a_i + N; \sum_{i\notin I} 2a_i + \sum_{i\in I} 3a_i \right\}.$$

Using $N = \sum_{i\in[m]} a_i$, the two sums are equal if and only if $\sum_{i\notin I} a_i = \sum_{i\in I} a_i$. Therefore $OPT(S, c) \geq \frac{5}{2} \sum_{i\in[m]} a_i$ and equality holds if and only if $\sum_{i\notin I} a_i = \sum_{i\in I} a_i$, which is only possible if $|I| = m/2$ (recall that we could suppose $\sum_{i\in I} a_i < \sum_{i\notin I} a_i$ for any $|I| < m/2$). This is equivalent to having a Balanced Partition of the numbers, so we have finished the reduction.

Acknowledgment. We are very grateful to Balázs Patkós for organizing 5th Emléktábla Workshop where we collaborated on this paper.

Appendix

Figures

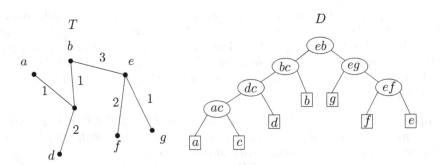

Fig. 1. An example of the tree search problem, T is the input tree and D is a decision tree with $cost(D) = 7 = cost^D(a) = cost^D(c)$. If the vertices of the tree T represent the parts of a device to assemble, the decision tree corresponds to the assembly procedure that at time 0 joins e with b; then at time 3 joins b with c and e with g. At time 4 the joining of d with c and e with f is started. Finally, at time 6 part a is joined with part c and the procedure ends by time 7.

Algorithm TS. Tree Search Algorithm

1: **function** MAIN(tree T, cost c)
2:     $t \leftarrow 2^{\lfloor \log\log |T| \rfloor + 2}$
3:     **Output** $D \leftarrow$ TREESEARCH(T, c, t)
4: **end function**
5: **function** TREESEARCH(tree T, costs c, t)
6:     **if** $|T| \leq t$ **then return** optimal decision tree D_X for T computed by Proposition 2
7:     $S_0 \leftarrow \emptyset$
8:     **for all** $i = 1, \ldots, t$ **do**
9:         $x_i \leftarrow$ centroid of a maximum size component of $T - S_{i-1}$
10:        $S_i \leftarrow$ smallest subtree containing x_i and S_{i-1}
11:    **end for**
12:    $S \leftarrow S_t$
13:    $X \leftarrow \{x_i |\ i = 1, \ldots, t\} \cup \{v \in V(S) |\ \deg_S(v) \geq 3\}$
14:    $Y \leftarrow$ tree on vertex set X, $uv \in E(Y)$ iff $X \cap P_{u,v} = \{u, v\}$
15:    **for all** $uv \in E(Y)$ **do**
16:        $c_Y(uv) \leftarrow \min_{e \in P_{u,v}} c(e)$
17:        $e_{uv} \leftarrow$ edge of $P_{u,v}$ with minimum cost
18:    **end for**
19:    $Z \leftarrow \bigcup_{uv \in E(Y)} e_{uv}$
20:    Compute optimal decision tree D_Y for (Y, c_Y) by Proposition 2
21:    **for all** $uv \in E(Y)$ **do**
22:        Replace label of uv in D_Y by e_{uv}
23:    **end for**
24:    **for all** components H of $Y - Z$ **do**
25:        ▷ H contains at most one vertex of degree 3 or more, i.e., H is a spider
26:        Compute 2-approximate decision tree D_H for H by Lemma 3
27:        replace the leaf $k \in D_Y$ corresponding to H by the root of D_H
28:    **end for**
29:    **for all** $v \in V(S)$ with a neighbor not in S **do**
30:        $S_v \leftarrow$ star induced by v and its neighbors outside of $V(S)$
31:        Construct decision tree D_v for (S_v, c)
32:        **for all** $w \in S_v \setminus \{v\}$ **do**
33:            $U \leftarrow$ component of $T - S$ containing w
34:            $D_w \leftarrow$ TREESEARCH(U, c, t)
35:            leaf of D_v corresponding to $w \leftarrow$ root of D_w
36:        **end for**
37:        replace the leaf of D_Y associated to v by the root of D_v
38:    **end for**
39:    **return** D_Y
40: **end function**

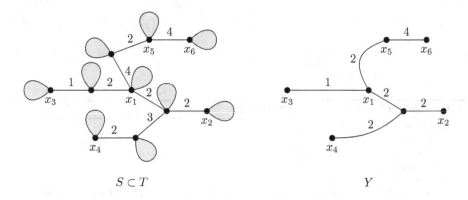

$S \subset T$ Y

Fig. 2. An example of the tree S, the important set of vertices X and the auxiliary tree Y in the construction of Sect. 3

References

1. Ahlswede, R., Wegener, I.: Search Problems. Wiley, Chichester-New York (1987)
2. Aigner, M.: Combinatorial Search. Wiley-Teubner, New York-Stuttgart (1988)
3. Ben-Asher, Y., Farchi, E., Newman, I.: Optimal search in trees. SIAM J. Comput. **28**(6), 2090–2102 (1999)
4. Cicalese, F., Jacobs, T., Laber, E., Valentim, C.: The binary identification problem for weighted trees. Theor. Comput. Sci. **459**, 100–112 (2012)
5. de la Torre, P., Greenlaw, R., Schäffer, A.: Optimal edge ranking of trees in polynomial time. Algorithmica **13**(6), 592–618 (1995)
6. Dereniowski, D.: Edge ranking of weighted trees. Discrete Appl. Math. **154**, 1198–1209 (2006)
7. Dereniowski, D.: Edge ranking and searching in partial orders. Discrete Appl. Math. **156**(13), 2493–2500 (2008)
8. Garey, M.R., Johnson, D.S.: Computer and Intractability. W.H. Freeman & Co., New York (1979)
9. Iyer, A.V., Ratliff, H.D., Vijayan, G.: On an edge ranking problem of trees and graphs. Discrete Appl. Math. **30**(1), 43–52 (1991)
10. Knuth, D.: Searching and Sorting. The Art of Computer Programming, vol. 3. Addison-Wesley, Reading (1998)
11. Lam, T.W., Yue, F.L.: Optimal edge ranking of trees in linear time. In: Proceedings of the Ninth Annual ACM-SIAM Symposium on Discrete Algorithms, SODA 1998, pp. 436–445, Philadelphia, PA, USA, Society for Industrial and Applied Mathematics (1998)
12. Linial, N., Saks, M.: Searching order structures. J. Algorithms **6**, 86–103 (1985)
13. Makino, K., Uno, Y., Ibaraki, T.: On minimum edge ranking spanning trees. J. Algorithms **38**, 411–437 (2001)
14. Mozes, S., Onak, K., Weimann, O.: Finding an optimal tree searching strategy in linear time. In: Proceedings of the 19th Annual ACM-SIAM Symposium on Discrete Algorithms (SODA 2008), pp. 1096–1105 (2008)
15. Wermelinger, M.: Searching Efficiently in Posets. New University of Lisbon, Topics in Programming Technology (1993)

An $\mathcal{O}(n^2)$ Time Algorithm for the Minimal Permutation Completion Problem

Christophe Crespelle[1]($\boxtimes$), Anthony Perez[2], and Ioan Todinca[2]

[1] Université Claude Bernard Lyon 1 and CNRS, DANTE/INRIA,
LIP UMR CNRS 5668, ENS de Lyon, Université de Lyon and Institute
of Mathematics, Vietnam Academy of Science and Technology,
18 Hoang Quoc Viet, Hanoi, Vietnam
`christophe.crespelle@inria.fr`
[2] University of Orléans, INSA Centre Val de Loire,
LIFO EA 4022, 45067 Orléans, France
`{anthony.perez,ioan.todinca}@univ-orleans.fr`

Abstract. We provide an $O(n^2)$ time algorithm computing a minimal permutation completion of an arbitrary graph $G = (V, E)$, i.e., a permutation graph $H = (V, F)$ on the same vertex set, such that $E \subseteq F$ and F is inclusion-minimal among all possibilities.

1 Introduction

In graph modification problems, we are given an arbitrary input graph G and the goal is to transform it, using a small number of "modifications", into a graph satisfying some property Π. Typically, modifications consist in adding and/or removing edges and/or vertices. Here we consider the case where we are only allowed to add edges to the input graph $G = (V, E)$, transforming it into a super-graph $H = (V, F)$, such that H belongs to some required class of graphs.

Probably the most famous problem of this kind is MINIMUM FILL-IN, where the goal is to add as few edges as possible in order to obtain a *chordal* graph H. A graph is chordal if it has no induced cycles with four or more vertices. The problem being NP-hard [15], it triggered the attention to a simpler one, where we are only required to compute an inclusion-minimal chordal supergraph H of G. Such a graph is called a *minimal triangulation* of G, and the MINIMAL TRIANGULATION problem has been known to be polynomial since 1976 [12,14]. The problem can be solved in time $O(nm)$ [14], and when the graph is dense the current best algorithm is the one of Heggernes et al. [7]. A detailed survey on this problem is provided in [5].

Minimal completions into other graph classes have been intensively studied. There are polynomial algorithms computing minimal completions into *interval graphs* [4,11], *proper interval graphs* [13], *split graphs* [8], *cographs* [9] and *comparability graphs* [6]. The *minimum* versions of these problems are NP-complete (see, e.g., the thesis of Mancini [10] for further discussion and references).

In this paper we consider the problem of minimal completions into *permutation* graphs. A graph is a permutation graph if we can assign to each vertex a

© Springer-Verlag Berlin Heidelberg 2016
E.W. Mayr (Ed.): WG 2015, LNCS 9224, pp. 103–115, 2016.
DOI: 10.1007/978-3-662-53174-7_8

segment, all segments having an endpoint on a "top" line and the other one on a parallel "bottom" line, such that two vertices are adjacent if and only if the two corresponding segments intersect. Such a representation is called a *permutation model* of the graph. We give an $O(n^2)$ algorithm computing a minimal permutation completion of an arbitrary graph. This is, to the best of our knowledge, the first polynomial algorithm for the problem. Let us point out that computing a permutation completion with a minimum number of edges is NP-hard [2].

Our result is based on a vertex-incremental approach, also used for other types of completions. More specifically, we take the vertices of the input graph one by one, in an arbitrary order, and at each step we add the new vertex x_i to the previously computed minimal permutation completion H_{i-1}. The new minimal permutation completion H_i is obtained by only adding edges between x_i and the rest of the graph. Very informally, if we are given a permutation model of H_{i-1}, we need to insert the segment of x_i in a minimal way. In Sect. 3 we consider the case when H_{i-1} has a unique permutation model and we provide an efficient computation, in $O(n)$ time, of such an insertion position. Somehow surprisingly, even this task is non-trivial (the similar algorithm is very simple in the case of minimal interval completions). Then we need to take into account (Sect. 4) the fact that H_{i-1} may have many different models, fortunately they are all encoded in its *modular decomposition*. Eventually, we compute the modular decomposition of the new completion H_i. This is done thanks to the algorithm of Crespelle and Paul [3], which incrementally maintains the modular decomposition of permutation graphs.

2   Preliminaries

Every graph $G = (V, E)$ considered here will be finite, undirected, loopless and simple. We denote $V(G)$ the set of vertices of G and $n = |V(G)|$. The edge between vertices x and y will arbitrarily be denoted either xy or yx. For a subset $S \subseteq V$ of vertices, we denote by $G[S]$ the subgraph of G induced by S.

A graph $G = (V, E)$ is a *permutation graph* if and only if it admits a *permutation model* (π_1, π_2), i.e. two one-to-one mappings $\pi_1, \pi_2 : V \to \{1, \ldots, n\}$ such that two vertices x and y are adjacent in G if and only if $(\pi_1(x) - \pi_1(y))(\pi_2(x) - \pi_2(y)) < 0$. An equivalent geometric definition of a permutation model for G is to associate to each vertex x a segment, each segment having one endpoint on a *top line* and the other on a parallel *bottom line* and all endpoints being pairwise distinct, such that two vertices x and y are adjacent if and only if the corresponding segments intersect. The correspondence between these two definitions is that the order in which appear the endpoints on the top (resp. bottom) line of the permutation model is the order defined by mapping π_1 (resp. π_2). We equally use these two visions in the rest of the article, depending on which one is more convenient for our purpose.

Note that a permutation model can be encoded by storing mappings π_1 and π_2 as arrays, which uses $O(n)$ space under the usual assumption that integers smaller then n are stored in constant space. Then, the condition for x and y to be adjacent can be tested in $O(1)$ time by two comparisons of integers.

Let now $G = (V, E)$ be an arbitrary graph. A *permutation completion* of G is a permutation graph $H = (V, F)$, on the same vertex set, such that $E \subseteq F$. If, moreover, set F is inclusion-minimal under these constraints, we say that H is a *minimal permutation completion* of G. In this paper we deal only with permutation completions, thus we will sometimes simply refer to these completions as (minimal) completions, omitting the term "permutation".

Note that every graph G has a permutation completion: one can simply add all the missing edges to G and observe that the complete graph is a permutation graph. Permutation graphs are also *hereditary*, i.e., an induced subgraph of a permutation graph is also a permutation graph (it is sufficient to restrict an permutation model of the original graph to the vertices of the induced subgraph).

Our approach for computing a minimal completion of an arbitrary graph G is incremental, in the sense that we take the vertices of G one by one, in an arbitrary order $(x_1, \ldots, x_n)$, and at step i we compute a minimal permutation completion H_i of $G_i = G[\{x_1, \ldots, x_i\}]$ from a minimal permutation completion H_{i-1} of G_{i-1}, by adding only edges incident to x_i. This is possible thanks to the following observation that is general to all hereditary graph classes that are also stable by addition of a universal vertex, in particular for permutation graphs.

Lemma 1 (see, e.g., [11]). *Let G be an arbitrary graph and let H be a minimal permutation completion of G. Consider a new graph $G' = G + x$, obtained by adding to G a new vertex x adjacent to an arbitrary set $N(x)$ of vertices of G. There is a minimal permutation completion H' of G' such that $H' - x = H$.*

For any subset $W \subseteq V(G)$ of vertices, we say that we *fill* W in H' if we make all the vertices of $W \setminus N(x)$ adjacent to x in the completion H' of $G + x$.

The New Problem. From now on, we consider the following problem, with slightly modified notations.

Let G be a permutation graph, and let $G + x$ be the graph obtained by adding to G a new vertex x adjacent to some set $N(x)$ of vertices of G. Our goal is to compute a minimal permutation completion H of $G + x$ such that $H - x = G$.

We will solve this problem in $O(n)$ time, where n is the number of vertices of G. As we shall see, graph G will be given together with its *modular decomposition*, which is known to encode all its possible permutation models. Our algorithm will compute the set $N'(x) \supseteq N(x)$ of neighbours of x in graph H, and will update this data structure for the completion H, thanks to the algorithm of [3].

3   Minimal Completion Respecting a Permutation Model

Definition 1 (Minimal completion respecting a permutation model).
Given a permutation model (π_1, π_2) of a permutation graph G and a vertex x to insert in G, a permutation completion H of $G + x$ respects (π_1, π_2) if there exists a permutation model of H such that removing x from it results in (π_1, π_2). Moreover, if H is inclusion minimal among such completions, then H is called a minimal permutation completion respecting (π_1, π_2).

The goal of this section is an $O(n)$ algorithm computing a minimal permutation completion respecting a given permutation model of a graph G.

Theorem 1. *Given a permutation model (π_1, π_2) of a permutation graph G and a vertex x to insert in G together with its neighbourhood $N(x)$, a minimal permutation completion H of $G + x$ respecting (π_1, π_2) can be computed in $\mathcal{O}(n)$ time. The output is a permutation model of H.*

This does not provide in general a minimal completion of G as a permutation graph may admit many different permutation models. Nevertheless, such a completion is minimal when graph G is prime (in which case it admits a unique permutation model, up to symmetries, and this is the only case where we will use this approach in our global algorithm. However, the algorithm we design in this section is not specific to prime graphs and we present it in the general setting.

In order to obtain a permutation model of H respecting (π_1, π_2), we extend the definition of functions π_1 and π_2, initially defined from $V(G)$ to $\{1, \ldots, n\}$, to vertex x as well, in the following way.

Definition 2 (Insertion position). *An insertion position of x in the permutation model (π_1, π_2) is a couple $(\pi_1(x), \pi_2(x)) = (i + 0.5, j + 0.5)$ for some integers $i, j \in \{0, \ldots, n\}$.*

Observe that for insertion position $(\pi_1(x), \pi_2(x))$ a necessary and sufficient condition to give a permutation completion of $G + x$ is that $\{z \in N(x) \mid \pi_1(z) < \pi_1(x)\} = \{t \in N(x) \mid \pi_2(t) > \pi_2(x)\}$. We now characterise all such insertion positions of x by a family of *slots* (I_j, J_j), i.e., couples of intervals, such that a position $Pos = (p_1, p_2)$ can be assigned to x in order to obtain a completion of $G + x$ if and only if (p_1, p_2) is contained in one of the slots (I_j, J_j). Moreover, the intervals I_j (resp. J_j) will be pairwise disjoint. To this purpose, we adopt the following notations, see their illustration on Fig. 1.

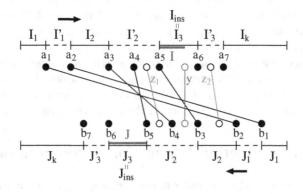

Fig. 1. Some definitions and notations.

Let $l = |N(x)|$, $a_1, a_2, \ldots, a_l$ denote the neighbours of x in **increasing** order in π_1, i.e. $\pi_1(a_1) < \pi_1(a_2) < \cdots < \pi_1(a_l)$, and $b_1, b_2, \ldots, b_l$ denote the neighbours of x in **decreasing** order in π_2, i.e. $\pi_2(b_l) < \cdots < \pi_2(b_2) < \pi_2(b_1)$. For $1 \leq i \leq l-1$, we also denote $A_i = \{a_1, \ldots, a_i\}$ and $B_i = \{b_1, \ldots, b_i\}$. We define $Valid = \{i \in \{1, \ldots, l-1\} \mid A_i = B_i\}$. Let $k = |Valid| + 2$ and we denote the elements of $Valid$ by $v_2, v_3, \ldots, v_{k-1}$ in increasing order. Then, for any $j \in \{2, \ldots, k-1\}$, let $I_j = [\pi_1(a_{v_j}), \pi_1(a_{v_j+1})]$ and $J_j = [\pi_2(b_{v_j+1}), \pi_2(b_{v_j})]$ (cf. Fig. 1). We also denote $I_1 = [0, \pi_1(a_1)]$, $J_1 = [\pi_2(b_1), n+1]$, $I_k = [\pi_1(a_l), n+1]$ and $J_k = [0, \pi_2(b_l)]$. Note that the intervals I_j, $1 \leq j \leq k$, are pairwise disjoint and numbered from left to right in π_1, while the intervals J_j are pairwise disjoint and numbered from right to left in π_2.

With these notations, the insertion positions $(\pi_1(x), \pi_2(x))$ that define a completion H of $G + x$ (i.e., such that x is at least adjacent in H to all the vertices of $N(x)$) are exactly the insertion positions such that $\pi_1(x) \in I_i$ and $\pi_2(x) \in J_i$, for some $i \in \{1, \ldots, k\}$. Our next goal is to choose, among all slots (I_i, J_i), an insertion slot (I_{ins}, J_{ins}) that will provide an optimal insertion position (Definition 3). Then, through a sequence of lemmas, we prove that this slot contains indeed a position providing a minimal permutation completion compatible with our permutation model (Lemma 3).

Given an interval I, let $l(I)$ (resp. $r(I)$) denote the left endpoint (resp. right endpoint) of I. For any $i \in \{1, \ldots, k-1\}$, we define $I'_i = [r(I_i), l(I_{i+1})]$ (cf. Fig. 1), and $I'_k = \varnothing$. Symmetrically, in π_2, we denote $J'_i = [r(J_{i+1}), l(J_i)]$, and $J'_k = \varnothing$. Finally, we denote $\hat{I}_i = I_i \cup I'_i$ and $\hat{J}_i = J_i \cup J'_i$.

Definition 3 (Forced, forwarding, insertion slot and vertex y). *A vertex z is forced if $z \in N(x)$ or there exists $i \in \{1, \ldots, k\}$ such that $\pi_1(z) \in I'_i$ and $\pi_2(z) \in J'_i$.*

We say that $i \in \{1, \ldots, k\}$ is forwarding if all vertices y such that $\pi_1(y) \in \bigcup_{1 \leq j \leq i} \hat{I}_j$ and $\pi_2(y) \in \bigcup_{1 \leq j \leq i} \hat{J}_j$ are forced. The insertion slot (I_{ins}, J_{ins}) is defined by $ins = \min\{i \in \{1, \ldots, k\} \mid i$ is not forwarding$\}$ and we denote by y a non-forced vertex that makes ins not forwarding.

First, note that forced vertices are adjacent to x in any completion of $G + x$ respecting (π_1, π_2). Of course, it may happen that all i are forwarding. In this case, x is adjacent to all the vertices of G in any completion respecting (π_1, π_2). Consequently, the unique minimal such completion is easily obtained by inserting x in any arbitrary slot (I_i, J_i), with $i \in \{i, \ldots, k\}$. In the following, we do not consider this case anymore and we assume that there exists a slot i which is not forwarding. Then, the insertion slot ins and the vertex y are well defined. Observe that, by minimality of ins, we either have $\pi_1(y) \in \hat{I}_{ins}$ and $\pi_2(y) \in \bigcup_{1 \leq j \leq ins} \hat{J}_j$, or $\pi_2(y) \in \hat{J}_{ins}$ and $\pi_1(y) \in \bigcup_{1 \leq j \leq ins} \hat{I}_j$.

We make some crucial observations about the slot (I_{ins}, J_{ins}), that will allow to prove that one can find a minimal completion by inserting x in this slot.

Lemma 2. *There is a completion obtained by inserting x in the slot (I_{ins}, J_{ins}) such that x is not adjacent to y in this completion.*

Moreover, this holds exactly for insertion positions in a sub-slot $(I, J) \subseteq (I_{ins}, J_{ins})$ defined as follows. Denote $I_y^- = [0, \pi_1(y)]$, $I_y^+ = [\pi_1(y), n+1]$, $J_y^- = [0, \pi_2(y)]$ and $J_y^+ = [\pi_2(y), n+1]$. We distinguish two cases:

1. *if $\pi_1(y) \in I_{ins}$ or $\pi_2(y) \in J_{ins}$ but not both, then there exists a unique $\alpha \in \{-, +\}$ s.t. $I_{ins} \cap I_y^\alpha \neq \varnothing$ and $J_{ins} \cap J_y^\alpha \neq \varnothing$, and we set $(I, J) = (I_{ins} \cap I_y^\alpha, J_{ins} \cap J_y^\alpha)$;*
2. *otherwise, $(\pi_1(y), \pi_2(y)) \in (I_{ins}, J_{ins})$ and then there are two suitable slots $(I, J) = (I_{ins} \cap I_y^-, J_{ins} \cap J_y^-)$ and $(I, J) = (I_{ins} \cap I_y^+, J_{ins} \cap J_y^+)$.*

Lemma 3. *Let C be a completion obtained by inserting x in the slot (I_{ins}, J_{ins}) such that x is not adjacent to y in the completion, and C is minimal among such completions. Then, C is a minimal completion of $G + x$ respecting (π_1, π_2).*

Sketch of Proof. Such a completion C exists by Lemma 2. Remind that all the completions respecting (π_1, π_2) are obtained by inserting x in some slot (I_i, J_i), with $1 \leq i \leq k$. Moreover, if $i > ins$ it is straightforward to see that the completions obtained by inserting x in (I_i, J_i) make x adjacent to y and are therefore not subgraphs of completion C. Now, let D be a completion obtained by inserting x in (I_i, J_i) with $i < ins$. Consider the completion C' obtained by inserting x at position $(l(I_{ins}) + 0.5, r(J_{ins}) - 0.5)$ in (I_{ins}, J_{ins}). The fact that $ins - 1$ is forwarding implies that C' is contained in D. If x is adjacent to y in C', clearly D is not a subgraph of C. If x is not adjacent to y in C', then C' is not a strict subgraph of C (by minimality of C), thus D is not a strict subgraph of C. $\square$

All insertion positions in (I_{ins}, J_{ins}) avoiding to make x adjacent to y are characterised by Lemma 2 as the insertion positions in a slot (I, J) of subintervals of respectively I_{ins} and J_{ins} (actually two couples in the second case of the lemma). Our next goal is to compute an optimal insertion position contained in a given slot (I, J), Lemma 4 below. We need the following definition.

Definition 4 ((Rightmost) left-stable insertion position). *Let (I, J) be two intervals with endpoints in $\{0, \ldots, n + 1\}$. An insertion position $Pos = (p_1, p_2)$ for x in (I, J) is left-stable if all vertices z such that $l(I) < \pi_1(z) < p_1$ (resp. $l(J) < \pi_2(z) < p_2$) satisfy $\pi_2(z) < p_2$ (resp. $\pi_1(z) < p_1$).*

Moreover, there is a unique left-stable insertion position Pos, called the rightmost one, such that there is no left-stable insertion position $Pos' = (p_1', p_2')$ strictly on the right of Pos in (I, J), i.e. different from Pos and having $p_1' \geq p_1$ and $p_2' \geq p_2$.

The uniqueness of the rightmost left-stable insertion position comes from the fact that two left-stable insertion positions (p_1, p_2) and (p_1', p_2') that cross each other define another left-stable insertion position $(\max\{p_1, p_1'\}, \max\{p_2, p_2'\})$ which is strictly on the right of both of them.

Lemma 4. *Let $Pos = (p_1, p_2)$ be the rightmost left-stable insertion position in (I, J). Then, the completion obtained by inserting x at position Pos is minimal among completions obtained by insertion of x in (I, J).*

Sketch of Proof. Let Pos' be another insertion position in (I, J) and let C' (resp. C) be the completion obtained from Pos' (resp. Pos). We distinguish three cases to show that C' is not strictly contained in C. Firstly, if the segment of Pos' crosses the one of Pos, then, because of the left-stability of Pos, in the order where the endpoint of Pos is on the right side of the one of Pos', there is one vertex z whose segment crosses Pos' and not Pos. Then, C' is not contained in C. Now, if Pos' is strictly on the left of Pos, the left-stability of Pos implies that there is no vertex with one endpoint on the right of Pos and one between Pos and Pos'. Then, C is contained in C' and the conclusion follows. Finally, if Pos' is strictly on the right of Pos, then, by definition of Pos, Pos' is not left-stable. And since Pos is left-stable, there exists one vertex with one endpoint between Pos and Pos' and the other one on the right side of Pos', which implies that C' is not contained in C.   □

The Algorithm. Our algorithm is in two steps. The first step determines the insertion slot (I_{ins}, J_{ins}). To that purpose we first compute the slots (I_i, J_i) defined above, in $O(n)$ time as shown in [3]. Then, by increasing i, we scan the vertices having one endpoint in the slot (I_i, J_i) until we find one non-forced vertex y, which can be tested in $O(1)$ time. We then have $ins = i$ and a suitable vertex y as in Definition 3. This takes $O(n)$ time.

The second step inserts x in (I_{ins}, J_{ins}). completion is minimal among those respecting (π_1, π_2). By Lemma 2, either there is a unique slot (I, J) that contains the optimal insertion position, or there are two possible slots (I_r, J_r) and (I_l, J_l) such that the first one is on the right. In the first case, we find (by an algorithm described a little below) the rightmost left-stable insertion position in (I, J), and it is optimal by Lemma 4. In the second case, we compute the rightmost left-stable insertion positions Pos_r and Pos_l in each of these slots (I_r, J_r) and (I_l, J_l). If the completion C_r obtained from Pos_r is included in the one C_l obtained from Pos_l, then Lemma 4 ensures that C_r is minimal among the completions respecting (π_1, π_2) and the algorithm returns Pos_r. Otherwise (when C_r is not contained in C_l), the algorithm returns position Pos_l. The correctness of this choice comes from the fact that if there exists a completion C' obtained by insertion of x in (I_r, J_r) that is contained in C_l, one can show that C_r is also necessarily contained in C_l. Note that inclusion of C_r into C_l can be easily tested in $O(n)$ time by scanning one order of the permutation model.

It remains to show how to compute the rightmost left-stable insertion position in a slot (I, J) in $\mathcal{O}(n)$ time. We start from the left-stable position $(p_1, p_2) = (l(I) + 0.5, l(J) + 0.5)$ and while it preserves left-stability, we increment p_1 (resp. p_2) by one. When it is no longer possible, the vertices $nextR(p_1)$ and $nextR(p_2)$ that are immediately on the right of p_1 and p_2 respectively satisfy $\pi_2(nextR(p_1)) > p_2$ and $\pi_1(nextR(p_2)) > p_1$. We then start a search for the leftmost left-stable position which is strictly on the right of (p_1, p_2) in I, if it exists. We set $b_2 \leftarrow \pi_2(nextR(p_1))$ to that purpose, and also $b_1 \leftarrow \pi_1(nextR(p_2))$, and we increment both p_1 and p_2. We then continue to scan π_1 (resp. π_2) by iteratively incrementing p_1 (resp. p_2), without going beyond b_1 (resp. b_2), and for every vertex z encountered in π_1 (resp. π_2) we set $b_2 \leftarrow max\{b_2, \pi_2(z)\}$ (resp. $b_1 \leftarrow max\{b_1, \pi_1(z)\}$). We proceed in this way simultaneously for p_1 and p_2 in an asynchronous manner until we get both $p_1 = b_1 - 0.5$ and $p_2 = b_2 - 0.5$, when no further increment is possible. At this stage, if (p_1, p_2) is no longer in (I, J) then the algorithm returns the previous left-stable position, say (p_1^{prev}, p_2^{prev}), and stops. Otherwise, (p_1, p_2) is the leftmost left-stable position on the right of (p_1^{prev}, p_2^{prev}) that we were looking for. Then, the algorithm continues to look for other left-stable positions on the right of current (p_1, p_2) by starting again at the beginning of this paragraph. When this process ends, we get the rightmost left-stable position in (I, J). As every treatment during the scan of π_1 and π_2 takes constant time, the overall complexity is $O(n)$. This achieves the proof of Theorem 1.

4   General Minimal Completion of $G + x$

For our general algorithm computing a minimal permutation completion of $G + x$, we consider the modular decomposition of G. For each node u of the modular decomposition tree T, we denote by $V[u]$ the subset of vertices of G appearing in the subtree rooted in u, and by $G[u]$ the subgraph of G induced by these vertices. We will denote by $G[u] + x$ the graph obtained from $G[u]$ by adding x with neighbourhood $N(x) \cap V[u]$.

A subset $W \subseteq V(G)$ of vertices is said to be *hit* in $G + x$ (resp. in a completion H of $G + x$) if it intersects the neighbourhood $N(x)$ of x in $G + x$ (resp. the neighbourhood $N'(x)$ of x in H). If W is contained in $N(x)$ (resp. $N'(x)$), we say that W is *full* in $G + x$ (resp. H). If W is hit but not full, we say it is *mixed*. We also say that a node u of the modular decomposition tree T of G is hit, full or mixed according to the status of its associated set of vertices $V[u]$. When we omit to precise it, the graph referred to in these notions is $G + x$. Observe that the set of hit nodes of T can be computed in $\mathcal{O}(n)$ time by a bottom-up parsing of T.

For each node u of T, $P_u = (V_u, E_u)$ denotes the corresponding quotient graph. If u is a *prime* node, P_u is prime, and it is stored together with a permutation model. If u is a *series* or *parallel* node, then graph P_u is a complete graph, respectively an independent set.

In this section, it is more convenient to work with the geometric version of a permutation model where each vertex is represented by a segment, all segments have an extremity on the *top line* of the model and another one on the *bottom* line. By [1], any permutation model of $G[u]$ is obtained from a model of P_u by *expanding each segment corresponding to a vertex v_i of P_u into a model of $G[v_i]$* as follows. The segment of v_i is enlarged into a parallelogram by enlarging its top (resp. bottom) endpoint into a top (resp. bottom) edge, lying on the top (resp. bottom) line of the permutation model. These expansions are such that the top edges (resp. bottom edges) of the parallelograms are pairwise disjoint, and they appear on the top (resp. bottom) line in the same order as the endpoints of the segments. Therefore, for any pair of vertices v_i and v_j of P_u, the corresponding parallelograms intersect if and only if $v_i v_j$ is an edge of P_u. Now, for each v_i (recall that v_i is a child of u in the modular decomposition tree), we insert a model of $G[v_i]$ into the parallelogram of v_i.

Definition 5 (Contracted graph $P_u + x$). *The contracted graph $P_u + x$ is the graph obtained from P_u by adding the vertex x with neighbourhood $N_{P_u+x}(x) = \{v_i \in V_u \mid v_i \text{ is hit}\}$.*

Let H_u be a minimal permutation completion of $P_u + x$. In order to obtain a permutation completion of $G[u] + x$, we could consider each node v_i adjacent to x in H_u and fill the set $V[v_i]$ (i.e., make all vertices of $V[v_i]$ adjacent to x). It is not hard to see that this construction yields a permutation completion of $G[u] + x$. Indeed, take a permutation model H_u^{mod} of H_u, and expand, for each node v_j of P_u, the segment of v_j into a parallelogram (the segment of x remains unchanged). By inserting a model of $G[v_j]$ into the parallelogram of v_j, we obtain a model for the completion of $G[u] + x$.

Such a completion is not necessarily minimal, because in the construction above we can sometimes "shift" the top and/or bottom endpoint of segment x inside a parallelogram and save some edges in the completion, by avoiding to have the segment of x cross some of the segments inside this parallelogram. As we shall see, there are at most two hit children v_i of u for which we do not fill $V[v_i]$. Depending on the cases, for such a child v_i, either we can use any minimal permutation completion of $G[v_i]$, or we have to use one which is minimal among those satisfying an additional constraint, which we call an *external-minimal permutation completion* (see Definition 6 below). This is the reason why our algorithm actually computes not just one but two permutation completions of $G[u] + x$ for each node u of the modular decomposition tree: one minimal permutation completion and one external-minimal permutation completion.

Definition 6 (External-minimal permutation completion). *A permutation completion H of $G + x$ is said to be an external permutation completion if H has a permutation model such that the bottom endpoint of segment x is the leftmost among all bottom endpoints of the segments of the model.*

If, moreover, H is minimal among all external completions, then H is called an external-minimal permutation completion.

Since any model can be reversed from left to right or upside-down, in the definition above we could replace "leftmost" by "rightmost" and "bottom" by "top".

The notion of *representation* defined below is central in the rest of the paper. We use it to construct a minimal completion of $G[u] + x$ from a minimal completion of $P_u + x$ and minimal and external-minimal completions of $G[v_i] + x$.

Definition 7 (Representation). *Let H_u be a permutation completion of $P_u + x$. A representation of H_u is a triplet $(H_u^{mod}, v_{top}, v_{bot})$ or a couple (H_u^{mod}, v_{top}) or a single element H_u^{mod} such that:*

- *H_u^{mod} is a permutation model of H_u, and*
- *v_{top} (resp. v_{bot}) is a vertex of P_u that is adjacent to x in H_u and such that the top (resp. bottom) endpoint of segment v_{top} (resp. v_{bot}) is next to the top (resp. bottom) endpoint of segment x in H_u^{mod}.*

Observe that we might have the situation that both v_{top} and v_{bot} exist and are equal. The following definition shows how to use a representation of H_u in order to get a permutation completion of $G[u] + x$.

Definition 8 (Permutation completion resulting from a representation). *Let H_u be a permutation completion of $P_u + x$. Consider a representation $\mathcal{R}$ of H_u, with $\mathcal{R} = (H_u, v_{top}, v_{bot})$ or $\mathcal{R} = (H_u, v_{top})$ or $\mathcal{R} = H_u$, and assume that we are given, for each node $v_\alpha \in \{v_{top}, v_{bot}\}$, the neighbourhood N'_α (resp. N_α^{ext}) of x in a minimal (resp. an external-minimal) permutation completion of $G[v_\alpha] + x$.*

We construct a vertex set N' (initially empty) as follows. For each $v_i \in V_u$ adjacent to x in H_u and distinct from v_{top} and v_{bot}, we add $V[v_i]$ to N'. Then, if at least one of v_{top} and v_{bot} exists, we distinguish two cases:

1. *if both v_{top} and v_{bot} exist and $v_{top} = v_{bot}$, we add N'_{top} to N',*
2. *otherwise, i.e. if v_{bot} does not exist or if both v_{top} and v_{bot} exist and are not equal, for each $v_\alpha \in \{v_{top}, v_{bot}\}$, we add N_α^{ext} to N'.*

The permutation completion of $G[u] + x$ resulting from the representation $\mathcal{R}$ of H_u is the one obtained from $G[u] + x$ by filling N'.

The above definition is correct as filling N' indeed results in a permutation completion of $G + x$.

Finally, we refine the notion of representation for that of an *optimal representation*, based on a minimal permutation completion of $P_u + x$. The supplementary conditions required for a representation to be optimal ensure that the resulting permutation completion is minimal (Theorem 2 below).

Definition 9 (Optimal representation). *Let H_u be a minimal permutation completion of $P_u + x$. For each node v_i of P_u, let N'_i be the neighbourhood of x in a minimal completion of $G[v_i] + x$, and let N_i^{ext} be the neighbourhood of x in an external-minimal completion of $G[v_i] + x$.*

1. *If there is a representation $\mathcal{R}_1 = (H_u^{mod}, v_{top}, v_{bot})$ of H_u such that v_{top} and v_{bot} are different and $N_{top}^{ext} \subsetneq V[v_{top}]$ and $N_{bot}^{ext} \subsetneq V[v_{bot}]$, we say that H_u is of the first type and $\mathcal{R}_1$ is an optimal representation of H_u.*
2. *If H_u is not of the first type, but there exists a representation $\mathcal{R}_2 = (H_u^{mod}, v_{top}, v_{bot})$ of H_u such that $v_{top} = v_{bot}$ and $N'_{top} \subsetneq V[v_{top}]$, we say that H_u is of the second type and $\mathcal{R}_2$ is an optimal representation of H_u.*
3. *If H_u is not of the first or second type, but there exists a representation $\mathcal{R}_3 = (H_u^{mod}, v_{top})$ such that $N_{top}^{ext} \subsetneq V[v_{top}]$, we say that H_u is of the third type and $\mathcal{R}_3$ is an optimal representation of H_u.*
4. *If none of the above holds, we say that H_u is of the fourth type, and the fourth-type optimal representation is simply H_u^{mod}, where H_u^{mod} is any model of H_u.*

We have a similar definition for producing an external-minimal completion of $G[u] + x$. We now prove the combinatorial theorem which constitutes the base of our algorithm for minimal permutation completion.

Theorem 2. *Let H_u be a minimal (resp. external-minimal) permutation completion of $P_u + x$ and let $\mathcal{R}$ be an optimal (resp. external-optimal) representation of H_u. Then, the permutation completion H of $G[u] + x$ resulting from $\mathcal{R}$ is minimal (resp. external-minimal).*

Sketch of Proof. For lack of space, we only give a flavour of the proof in the case where H_u is a minimal completion of $P_u + x$ and H_u is of the first type. Assume for contradiction that there is some permutation completion H' of $G[u] + x$ such that H' is a strict subgraph of H. Apply the following transformation to a permutation model of H': keep the segment of x and for each node v_i in P_u keep only one vertex of $V[v_i]$ and choose it adjacent to x in H' if there exists one such vertex in $V[v_i]$. The model obtained is the one of a permutation completion H'_u of $P_u + x$. Applying the same process to a permutation model of H results in a model H_u^{mod} of H_u defining an optimal representation of H_u of the first type. Moreover, one can show that because H' is a subgraph of H, then H'_u is also a subgraph of H_u. Therefore, $H'_u = H_u$ by minimality of H_u. As a consequence, for every node v_i such that $V[v_i]$ is hit in H (hence v_i is adjacent to x in H_u) the set $V[v_i]$ is also hit in H' (because v_i is adjacent to x in H'_u).

Let now y be a vertex that is adjacent to x in H but not in H'. Let v_m be the node of P_u such that $y \in V[v_m]$, necessarily v_m is mixed in H'. Observe that, by construction (see Definition 8), since H_u is of the first type, then $V[v_{top}]$ and $V[v_{bot}]$ are mixed in H, and so in H' from what precedes. Also observe that in any completion of $G[u] + x$, the only children v_i of u such that $V[v_i]$ is mixed are those whose parallelogram contains at least one endpoint of x. Then, in any permutation model $H^{mod'}$ of H', the parallelogram of v_m contains one endpoint of x and only one, because the parallelograms of both v_{top} and v_{bot} also contain one endpoint of x and at least one of v_{top} and v_{bot} is different from v_m (v_{top} and v_{bot} are distinct since H_u is of the first type). Consequently, restricting $H^{mod'}$ to $V[v_m] \cup \{x\}$ shows that $H'[V[v_m] \cup \{x\}]$ is an external completion of $G[v_m] + x$. Moreover, by construction (see Definition 8), $H[V[v_m] \cup \{x\}]$ is an external-minimal completion of $G[v_m] + x$, contradicting the fact that $H'[V[v_m] \cup \{x\}]$ is a strict subgraph of $H[V[v_m] \cup \{x\}]$. This proves the theorem for the case where H_u is of the first type. □

The Algorithm. Our $\mathcal{O}(n)$-time algorithm computing a minimal permutation completion of $G + x$ proceeds by a bottom-up computation of two functions, $MinIns(u)$ and $MinInsExt(u)$, for each node u. Function $MinIns(u)$ computes a minimal permutation completion of $G[u] + x$ (with the corresponding neighbourhood $N(x) \cap V[u]$). It returns the neighbourhood N'_u of x in this completion. Function $MinInsExt(u)$ computes a minimal-external permutation completion of $G[u] + x$, i.e., it returns the neighbourhood N_u^{ext} of x in such a completion. These functions will only be called on hit nodes, from bottom to top. Therefore we can assume that, when we treat a node u, we already have computed the sets N'_v and N_v^{ext} for each hit child v of u in the modular decomposition tree T. The case when u is a (hit) leaf is trivial: both functions return as neighbourhood of x the vertex of G that labels this leaf. Now consider an internal node u, we show how to determine the type of H_u and an optimal (resp. external-optimal) representation.

For lack of space, we do not describe the (easier) cases where node u is *parallel* or *series*, but we sketch the case where u is *prime*. In this case, P_u admits a unique permutation model (π_1, π_2) (up to symmetries) and therefore all completions H_u of P_u+x respect (π_1, π_2). Then, we can get a minimal completion H_u by using the algorithm of Theorem 1, in $O(k)$ time, where k is the number of children of node u in T. We have to pay attention to the fact that there may have other insertion positions in (π_1, π_2) defining the same completion H_u, some of them possibly resulting in a smaller type for H_u, in which case we must use one of smallest type in order to get an optimal representation for H_u. Fortunately, Theorem 4 of [3] implies that there exists at most one other insertion position and that it can be obtained from the first one in $O(k)$ time. Consequently, our algorithm simply computes the two possible insertion positions. For each of them, it checks the two conditions $N_\alpha^{ext} \subsetneq V[v_\alpha]$ and $N_\alpha' \subsetneq V[v_\alpha]$ for the vertices $v_\alpha \in V_u$ that have one endpoint next to one endpoint of x in π_1 or in π_2. Then, it decides which one of the two insertion positions gives the smaller type and uses it in an optimal representation $\mathcal{R}$ of H_u. Finally, we compute the set N_u' of neighbours of x in the completion resulting from $\mathcal{R}$ as in Definition 8. We do not detail the computation of the set N_u^{ext}, which is similar and much simpler (in particular it does not need the algorithm of Theorem 1).

With a careful implementation of the algorithm, we compute in $O(n)$ time the neighbourhood N' of x in a minimal completion H of $G + x$. Using the algorithm of [3], we transform the modular decomposition of G into the one of H in time $O(n)$. By incrementally applying this algorithm we conclude:

Theorem 3. *There is an $O(n^2)$ time algorithm computing a minimal permutation completion of an arbitrary input graph.*

References

1. Bergeron, A., Chauve, C., de Montgolfier, F., Raffinot, M.: Computing common intervals of K permutations, with applications to modular decomposition of graphs. In: Brodal, G.S., Leonardi, S. (eds.) ESA 2005. LNCS, vol. 3669, pp. 779–790. Springer, Heidelberg (2005)
2. Burzyn, P., Bonomo, F., Durán, G.: NP-completeness results for edge modification problems. Discrete Appl. Math. **154**(13), 1824–1844 (2006)
3. Crespelle, C., Paul, C.: Fully dynamic algorithm for recognition and modular decomposition of permutation graphs. Algorithmica **58**(2), 405–432 (2010)
4. Crespelle, C., Todinca, I.: An $O(n^2)$-time algorithm for the minimal interval completion problem. Theor. Comput. Sci. **494**, 75–85 (2013)
5. Heggernes, P.: Minimal triangulations of graphs: a survey. Discrete Math. **306**(3), 297–317 (2006)
6. Heggernes, P., Mancini, F., Papadopoulos, C.: Minimal comparability completions of arbitrary graphs. Discrete Appl. Math. **156**(5), 705–718 (2008)
7. Heggernes, P., Telle, J.A., Villanger, Y.: Computing minimal triangulations in time $O(n^{\alpha \log n}) = o(n^{2.376})$. SIAM J. Discrete Math. **19**(4), 900–913 (2005)

8. Heggernes, P., Mancini, F.: Minimal split completions. Discrete Appl. Math. **157**(12), 2659–2669 (2009)
9. Lokshtanov, D., Mancini, F., Papadopoulos, C.: Characterizing and computing minimal cograph completions. Discrete Appl. Math. **158**(7), 755–764 (2010)
10. Mancini, F.: Graph Modification Problems Related to Graph Classes. Ph.D. thesis, University of Bergen, Norway (2008)
11. Ohtsuki, T., Mori, H., Kashiwabara, T., Fujisawa, T.: On minimal augmentation of a graph to obtain an interval graph. J. Comput. Syst. Sci. **22**(1), 60–97 (1981)
12. Ohtsuki, T.: A fast algorithm for finding an optimal ordering for vertex elimination on a graph. SIAM J. Comput. **5**(1), 133–145 (1976)
13. Rapaport, I., Suchan, K., Todinca, I.: Minimal proper interval completions. Inf. Process. Lett. **5**, 195–202 (2008)
14. Rose, D.J., Tarjan, R.E., Lueker, G.S.: Algorithmic aspects of vertex elimination on graphs. SIAM J. Comput. **5**(2), 266–283 (1976)
15. Yannakakis, M.: Computing the minimum fill-in is NP-complete. SIAM. J. Algebraic Discrete Methods **2**(1), 77–79 (1981)

On the Number of Minimal Separators
in Graphs

Serge Gaspers[1,2(✉)] and Simon Mackenzie[1,2]

[1] The University of New South Wales, Sydney, Australia
{sergeg,simonwm}@cse.unsw.edu.au
[2] NICTA, Sydney, Australia

Abstract. We consider the largest number of minimal separators a graph on n vertices can have.

- We give a new proof that this number is in $O\left(\left(\frac{1+\sqrt{5}}{2}\right)^n \cdot n\right)$.
- We prove that this number is in $\omega\left(1.4457^n\right)$, improving on the previous best lower bound of $\Omega(3^{n/3}) \subseteq \omega(1.4422^n)$.

This gives also an improved lower bound on the number of potential maximal cliques in a graph. We would like to emphasize that our proofs are short, simple, and elementary.

1 Introduction

For a graph $G = (V, E)$, and two vertices $a, b \in V$, a vertex subset $S \subseteq V \setminus \{a, b\}$ is an (a, b)-*separator* if a and b are in different connected components of $G - S$, the graph obtained from G by removing the vertices in S. An (a, b)-separator is *minimal* if it does not contain another (a, b)-separator as a subset. A vertex subset $S \subset V$ is a *minimal separator* in G if it is a minimal (a, b)-separator for some pair of distinct vertices $a, b \in V$.

By $\mathsf{sep}(G)$, we denote the number of minimal separators in the graph G. By $\mathsf{sep}(n)$, we denote the maximum number of minimal separators, taken over all graphs on n vertices.

Potential maximal cliques are closely related to minimal separators, especially in the context of chordal graphs. A graph is *chordal* if every induced cycle has length 3. A *triangulation* of a graph G is a chordal supergraph of G obtained by adding edges. A graph H is a *minimal triangulation* of G if it is a triangulation of G and G has no other triangulation that is a subgraph of H. A vertex set is a *potential maximal clique* in G if it is a maximal clique in at least one minimal triangulation of G.

By $\mathsf{pmc}(G)$, we denote the number of potential maximal cliques in the graph G. By $\mathsf{pmc}(n)$, we denote the maximum number of potential maximal cliques, taken over all graphs on n vertices.

Minimal separators and potential maximal cliques have been studied extensively [1–3,10,13,16,17,20–22]. Upper bounds on $\mathsf{sep}(n)$ are used to upper bound the running time of algorithms for enumerating all minimal separators [1,17,21]. Bounds on both $\mathsf{sep}(n)$ and $\mathsf{pmc}(n)$ are used in analyses of algorithmic running

© Springer-Verlag Berlin Heidelberg 2016
E.W. Mayr (Ed.): WG 2015, LNCS 9224, pp. 116–121, 2016.
DOI: 10.1007/978-3-662-53174-7_9

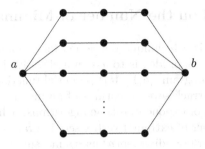

Fig. 1. Melon graphs have $\Omega(3^{n/3})$ minimal separators.

times for computing the treewidth and minimum fill-in of a graph [3,10,13], and for computing a maximum induced subgraph isomorphic to a graph from a family of bounded treewidth graphs [11].

Our Results. Fomin et al. [10] proved that $\mathsf{sep}(n) \in O\,(1.7087^n)$. Fomin and Villanger [13] improved the upper bound and showed that $\mathsf{sep}(n) \in O\,(\rho^n \cdot n)$, where $\rho = \frac{1+\sqrt{5}}{2} = 1.6180\ldots^1$. We prove the same upper bound with simpler arguments.

As for lower bounds, it is known [10] that $\mathsf{sep}(n) \in \Omega(3^{n/3})$; see Fig. 1. We improve on this lower bound by giving an infinite family of graphs with $\omega\,(1.4457^n)$ minimal separators. This answers an open question raised numerous times (see, e.g., [9,10]), for example by Fomin and Kratsch [9, p. 100], who state

> It is an open question, whether the number of minimal separators in every n-vertex graph is $O^*(3^{n/3})$.

Here, the O^*-notation is similar to the O-notation, but hides polynomial factors.

As a corollary, we have that there is an infinite family of graphs, all with $\omega(1.4457^n)$ potential maximal cliques. This answers another open question on lower bounds for the number of potential maximal cliques in graphs. For example, Fomin and Villanger [12] state

> There are graphs with roughly $3^{n/3} \approx 1.442^n$ potential maximal cliques [10]. Let us remind that by the classical result of Moon and Moser [19] (see also Miller and Muller [18]) that the number of maximal cliques in a graph on n vertices is at most $3^{n/3}$. Can it be that the right upper bound on the number of potential maximal cliques is also roughly $3^{n/3}$? By Theorem 3.2, this would yield a dramatic improvement for many moderate exponential algorithms.

Preliminaries. We use standard graph notation from [4]. For an edge uv in a graph G, we denote by G/uv the graph obtained from G by contracting the edge uv, i.e., making u adjacent to $N_G(\{u,v\})$ and removing v.

---

[1] The bound stated in [13] is $O(1.6181^n)$, but this stronger bound can be derived from their proof.

2   Upper Bound on the Number of Minimal Separators

Measure and Conquer is a technique developed for the analysis of exponential time algorithms [7]. Its main idea is to assign a cleverly chosen (sometimes, by solving mathematical programs [5,14,15]) potential function to the instance – a so-called *measure* – to track several features of an instance in solving it. While developed in the realm of exponential-time algorithms, it has also been used to upper bound the number of extremal vertex sets in graphs (see, e.g., [6,8]).

Our new proof upper bounding $\mathsf{sep}(n)$ uses a measure that takes into account the number of vertices of the graph and the difference in size between the separated components of the graph. This simple trick allows us to avoid several complications from [13], including the use of the firefighter lemma (Lemma 3.1 in [13]), fixing the size of the separators, the discussion of "full components", and distinguishing between separators of size at most $n/3$ and at least $n/3$.

Theorem 1. $\mathsf{sep}(n) \in O(\rho^n \cdot n)$, where $\rho = \frac{1+\sqrt{5}}{2} = 1.6180...$ is the golden ratio.

Proof. Let $G = (V, E)$ be any graph on n vertices containing vertex $a \in V$. For $d \leq |V|$, an $[a, d]$-separation is a partition (A, S, B) of V such that

- $a \in A$,
- $G[A]$ is connected,
- S is a minimal (a, b)-separator for some vertex $b \in B$, and
- $|A| \leq |B| - d$.

Let $\mathsf{sep}_a(G, d)$ denote the number of $[a, d]$-separations in G. By symmetry, $\mathsf{sep}_a(G, 0)$ upper bounds the number of minimal separators in G up to a factor $O(n)$. To upper bound $\mathsf{sep}_a(G, d)$, we will use the measure

$$\mu(G, d) = |V(G)| - d.$$

The theorem will follow from the claim that $\mathsf{sep}_a(G, d) \leq \rho^{\mu(G,d)}$ for $0 \leq d \leq |V|$.

If $\mu(G, d) = 0$, then $d = |V|$ and $\mathsf{sep}_a(G, d) = 0$ since there is no $A \subseteq V$ with $|A| \leq 0$ and $a \in A$. If $d_G(a) = 0$, then there is at most one $[a, d]$-separation, which is $(\{a\}, \emptyset, V \setminus \{a\})$. Therefore, assume $\mu(G, d) \geq 1$, a has at least one neighbor, and assume the claim holds for smaller measures. Consider a vertex $u \in N(a)$. For every $[a, d]$-separation (A, S, B), either $u \in S$ or $u \in A$. Therefore, we can upper bound the $[a, d]$-separations (A, S, B) counted in $\mathsf{sep}_a(G, d)$ with $u \in S$ by $\rho^{\mu(G-\{u\},d)} = \rho^{\mu(G,d)-1}$, and those with $u \notin S$ by $\rho^{\mu(G/au,d+1)} = \rho^{\mu(G,d)-2}$. It remains to observe that $\rho^{\mu(G,d)-1} + \rho^{\mu(G,d)-2} = \rho^{\mu(G,d)}$. ☐

3   Lower Bound on the Maximum Number of Minimal Separators

In the melon graph in Fig. 1, each horizontal layer implies a choice between 3 vertices. Each such choice also 'costs' 3 vertices. The new construction improves the bound by adding vertical choices on top of the horizontal choices. This is achieved by 'sacrificing' horizontal choices and adding 18 vertices for 126 horizontal layers. For most minimal separators, two horizontal choices are sacrificed, and their order matters, giving $126 \cdot 125$ possibilities, at the cost of 24 vertices (18 new vertices and 6 from the horizontal choices). Since $126 \cdot 125 > 3^{24/3}$, this gives more choices than the melon graph on the same number of vertices.

Theorem 2. $\mathrm{sep}(n) \in \omega(1.4457^n)$.

Proof. We prove the theorem by exhibiting a family of graphs $\{G_1, G_2, \dots\}$ and lower bounding their number of minimal separators.

Let $p = 9$ and $q = 4$. Let $I = \{1, \dots, 3\}$, $J = \{1, \dots, \binom{p}{q}\}$, and $K = \{1, \dots, p\}$. The graph G_1 is constructed as follows (see Fig. 2). Construct disjoint vertex sets $V_i = \{v_{i,j} : j \in J\}$ for each $i \in I$, and the vertex sets $U = \{u_i : i \in K\}$ and $W = \{w_i : i \in K\}$. The vertex set of G_1 is $\{a, b\} \cup U \cup W \cup \bigcup_{i \in I} V_i$. The edge set of G_1 is obtained by first adding the paths $(v_{1,j}, v_{2,j}, v_{3,j})$ for all $j \in J$ and the edges $\{au_i : i \in K\}$ and $\{w_i b : i \in K\}$. Then, add edges between U and V_1 such that each vertex in V_1 has q neighbors in U and no two vertices from V_1 have the same neighbors in U. This is possible, since there are $\binom{p}{q}$ distinct subsets of U of size q. Similarly, add edges between V_3 and W such that each vertex in V_3 has q neighbors in W and no two vertices from V_3 have the same neighbors in W. The graph G_ℓ, $\ell \geq 2$, is obtained from ℓ disjoint copies of G_1, merging the copies of a, and merging the copies of b.

Let us now consider the sets $\mathcal{S}_{r,s}$ of minimal (a, b)-separators in G_1 that contain r vertices from U and s vertices from W. The separators in $\mathcal{S}_{0,0}$ contain one vertex among $\{v_{1,j}, v_{2,j}, v_{3,j}\}$ for each $j \in J$, giving $|\mathcal{S}_{0,0}| = 3^{|J|}$. The separators in $\mathcal{S}_{0,s}$ with $s \geq q$ contain s vertices from W, leaving $\binom{p}{q} - \binom{s}{q}$ vertices from V_3 with at least one neighbor in W. For each such vertex $v_{3,j}$, the path to a is interrupted by selecting one vertex among $\{v_{1,j}, v_{2,j}, v_{3,j}\}$. In total, $|\mathcal{S}_{0,s}| \geq \binom{p}{s} \cdot 3^{\binom{p}{q} - \binom{s}{q}}$. Similarly, $|\mathcal{S}_{r,0}| \geq \binom{p}{r} \cdot 3^{\binom{p}{q} - \binom{r}{q}}$. For $r, s \geq j$, separators containing $S_U \subseteq U$ and $S_W \subseteq W$ are not minimal if $S_U \subseteq S_W$ or $S_W \subseteq S_U$. Otherwise, $S_U \cup S_W$ can be extended to a minimal (a, b)-separator by selecting one vertex from $\{v_{1,j}, v_{2,j}, v_{3,j}\}$ for each $j \in J$ such that $v_{1,j}$ has at least

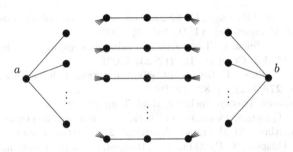

Fig. 2. The graph G_1 has 126 vertices adjacent to a and 9 adjacent to b.

one neighbor in $U \setminus S_U$ and $v_{3,j}$ has at least one neighbor in $W \setminus S_W$. In total, $|\mathcal{S}_{r,s}| \geq \left(\binom{p}{r} \cdot \binom{p}{s} - \binom{p}{\max(r,s)} \cdot \binom{\max(r,s)}{\min(r,s)} \right) \cdot 3^{\binom{p}{q} - \binom{r}{q} - \binom{s}{q}}$. Therefore, the number of minimal separators of G_1 is at least $x = 3^{\binom{p}{q}} + 2 \cdot \sum_{s=q}^{p} \left(\binom{p}{s} \cdot 3^{\binom{p}{q} - \binom{s}{q}} \right) + \sum_{s=q}^{p} \sum_{r=q}^{p} \left(\binom{p}{r} \cdot \binom{p}{s} - \binom{p}{\max(r,s)} \cdot \binom{\max(r,s)}{\min(r,s)} \right) \cdot 3^{\binom{p}{q} - \binom{r}{q} - \binom{s}{q}} > 2.4603 \cdot 10^{63}$.

Minimal (a,b)-separators for G_ℓ are obtained by taking the union of minimal separators for the copies of G_1. Their number is therefore at least $x^\ell = x^{\frac{n-2}{3|J|+2|K|}} \in \omega(1.4457^n)$, where G_ℓ has $n = \ell(3\binom{p}{q} + 2p) + 2$ vertices.   □

Based on results from [2], Bouchitté and Todinca [3] observed that the number of potential maximal cliques in a graph is at least the number of minimal separators divided by the number of vertices n. Therefore, we arrive at the following corollary of Theorem 2.

Corollary 1. $\mathsf{pmc}(n) \in \omega(1.4457^n)$.

4   Conclusion

We have given a simpler proof for the best known asymptotic upper bound on $\mathsf{sep}(n)$, and we have improved the best known lower bound from $\Omega(3^{n/3})$ to $\omega(1.4457^n)$, thereby reducing the gap between the current best lower and upper bound. Before our work, it seemed reasonable to believe that $\mathsf{sep}(n)$ could be asymptotically equal to $3^{n/3}$, up to polynomial factors. We showed that this is not the case, and we believe there is room to further improve the lower bound.

Acknowledgments. We thank Yota Otachi and Hitoshi Iwai for pointing out an issue in an earlier version of the lower bound proof.

NICTA is funded by the Australian Government through the Department of Communications and the Australian Research Council through the ICT Centre of Excellence Program. Serge Gaspers is the recipient of an Australian Research Council Discovery Early Career Researcher Award (project number DE120101761) and a Future Fellowship (project number FT140100048).

References

1. Berry, A., Bordat, J.P., Cogis, O.: Generating all the minimal separators of a graph. Int. J. Found. Comput. Sci. **11**(3), 397–403 (2000)
2. Bouchitté, V., Todinca, I.: Treewidth and minimum fill-in: grouping the minimal separators. SIAM J. Comput. **31**, 212–232 (2001)
3. Bouchitté, V., Todinca, I.: Listing all potential maximal cliques of a graph. Theor. Comput. Sci. **276**(1–2), 17–32 (2002)
4. Diestel, R.: Graph Theory. Springer, Heidelberg (2010)
5. Eppstein, D.: Quasiconvex analysis of multivariate recurrence equations for backtracking algorithms. ACM Trans. Algorithms **2**(4), 492–509 (2006)
6. Fomin, F.V., Gaspers, S., Pyatkin, A.V., Razgon, I.: On the minimum feedback vertex set problem: exact and enumeration algorithms. Algorithmica **52**(2), 293–307 (2008)

7. Fomin, F.V., Grandoni, F., Kratsch, D.: A measure & conquer approach for the analysis of exact algorithms. J. ACM **56**(5), 1–32 (2009)
8. Fomin, F.V., Grandoni, F., Pyatkin, A.V., Stepanov, A.A.: Combinatorial bounds via measure and conquer: bounding minimal dominating sets and applications. ACM Trans. Algorithms **5**(1), 1–17 (2008)
9. Fomin, F.V., Kratsch, D.: Exact Exponential Algorithms. Springer, Heidelberg (2010)
10. Fomin, F.V., Kratsch, D., Todinca, I., Villanger, Y.: Exact algorithms for treewidth and minimum fill-in. SIAM J. Comput. **38**(3), 1058–1079 (2008)
11. Fomin, F.V., Todinca, I., Villanger, Y.: Large induced subgraphs via triangulations and CMSO. SIAM J. Comput. **44**(1), 54–87 (2015)
12. Fomin, F.V., Villanger, Y.: Finding induced subgraphs via minimal triangulations. In: Proceedings of the 27th International Symposium on Theoretical Aspects of Computer Science (STACS 2010). LIPIcs, vol. 5, pp. 383–394. Schloss Dagstuhl - Leibniz-Zentrum fuer Informatik (2010)
13. Fomin, F.V., Villanger, Y.: Treewidth computation and extremal combinatorics. Combinatorica **32**(3), 289–308 (2012)
14. Gaspers, S.: Exponential Time Algorithms: Structures, Measures, and Bounds. VDM Verlag Dr. Mueller e.K. (2010)
15. Gaspers, S., Sorkin, G.B.: A universally fastest algorithm for Max 2-Sat, Max 2-CSP, and everything in between. J. Comput. Syst. Sci. **78**(1), 305–335 (2012)
16. Heggernes, P.: Minimal triangulations of graphs: a survey. Discrete Math. **306**(3), 297–317 (2006)
17. Kloks, T., Kratsch, D.: Listing all minimal separators of a graph. SIAM J. Comput. **27**(3), 605–613 (1998)
18. Miller, R.E., Muller, D.E.: A problem of maximum consistent subsets. IBM Research Report RC-240, J.T. Watson Research Center, Yorktown Heights, NY (1960)
19. Moon, J.W., Moser, L.: On cliques in graphs. Isr. J. Math. **3**, 23–28 (1965)
20. Parra, A., Scheffler, P.: Characterizations and algorithmic applications of chordal graph embeddings. Discrete Appl. Math. **79**(1–3), 171–188 (1997)
21. Shen, H., Liang, W.: Efficient enumeration of all minimal separators in a graph. Theor. Comput. Sci. **180**(1–2), 169–180 (1997)
22. Villanger, Y.: Improved exponential-time algorithms for treewidth and minimum fill-in. In: Correa, J.R., Hevia, A., Kiwi, M. (eds.) LATIN 2006. LNCS, vol. 3887, pp. 800–811. Springer, Heidelberg (2006)

Efficient Farthest-Point Queries
in Two-terminal Series-parallel Networks

Carsten Grimm[1,2](✉)

[1] Fakultät für Informatik, Otto-von-Guericke-Universität Magdeburg,
Magdeburg, Germany
carsten.grimm@ovgu.de
[2] School of Computer Science, Carleton University, Ottawa, ON, Canada

Abstract. Consider the continuum of points along the edges of a network, i.e., a connected, undirected graph with positive edge weights. We measure the distance between these points in terms of the weighted shortest path distance, called the *network distance*. Within this metric space, we study farthest points and farthest distances. We introduce a data structure supporting queries for the farthest distance and the farthest points on two-terminal series-parallel networks. This data structure supports farthest-point queries in $O(k + \log n)$ time after $O(n \log p)$ construction time, where k is the number of farthest points, n is the size of the network, and p parallel operations are required to generate the network.

1 Introduction

Consider a geometric network with positive edge weights. For any two points on this network (i.e., points that may be vertices or in the interior of an edge), their *network distance* is the weight of a weighted shortest-path connecting them. Within this metric space, we study farthest points and farthest distances. We introduce a data structure supporting queries for the farthest distance and the farthest points on two-terminal series-parallel networks.

As a prototype application, imagine the task to find the ideal location for a new hospital within the network formed by the streets of a city. One criterion for this optimization would be the emergency unit response time, i.e., the worst-case time an emergency crew needs to drive from the hospital to the site of an accident. However, a location might be optimal in terms of emergency unit response time, but unacceptable with respect to another criterion such as construction costs. We provide a data structure that would allow a decision maker to quickly compare various locations in terms of emergency unit response time.

We obtain our data structure for two-terminal series-parallel networks by studying simpler networks reflecting parallel structure (parallel-path) and serial structure (bead-chains). Combining these insights, we support queries on flat series-parallel networks (abacus). Finally, we decompose series-parallel networks into a tree of nested abaci and combine their associated data structures.

© Springer-Verlag Berlin Heidelberg 2016
E.W. Mayr (Ed.): WG 2015, LNCS 9224, pp. 122–137, 2016.
DOI: 10.1007/978-3-662-53174-7_10

Our focus on supporting human decision makers with data structures deviates from the common one-shot optimization problems in location analysis, where we assume that only one factor determines suitable locations for some facility in a network. Moreover, we illustrate new ways of exploiting different parallel structures of networks that may be useful for tackling related problems.

1.1 Preliminaries

A *network* is defined as a simple, connected, undirected graph $N = (V, E)$ with positive edge weights. We write w_{uv} to denote the weight of the edge $uv \in E$ that connects the vertices $u, v \in V$. A point p on edge uv subdivides uv into two sub-edges up and pv with $w_{up} = \lambda w_{uv}$ and $w_{pv} = (1 - \lambda)w_{uv}$, for some $\lambda \in [0, 1]$.[1] We write $p \in uv$ when p is on edge uv and $p \in N$ when p is on some edge of N. The *network distance* between $p, q \in N$, denoted by $d_N(p, q)$, is measured as the weighted length of a shortest path from p to q in N. We denote the *farthest distance* from p by $\bar{d}_N(p)$, i.e., $\bar{d}_N(p) = \max_{q \in N} d_N(p, q)$. Accordingly, we say a point $\bar{p}$ on N is farthest from p if and only if $d_N(p, \bar{p}) = \bar{d}_N(p)$.

We develop data structures supporting the following queries in a network N. Given a point p on N, what is the farthest distance from p? What are the farthest points from p in N? We refer to the former as *farthest-distance query* and to the latter as *farthest-point query*. The query point p is represented by the edge uv containing p together with the value $\lambda \in [0, 1]$ such that $w_{up} = \lambda w_{uv}$.

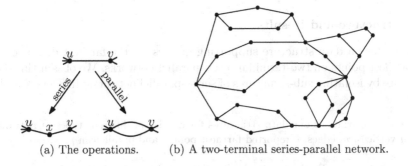

(a) The operations.   (b) A two-terminal series-parallel network.

Fig. 1. The operations (a) that generate two-terminal series-parallel networks (b).

The term *series-parallel network* stems from the following two operations that are illustrated in Fig. 1. The *series operation* splits an existing edge uv into two new edges ux and xv where x is a new vertex. The *parallel operation* creates a copy of an existing edge. A network N is *two-terminal series-parallel* when its underlying graph[2] can be generated from a single edge uv using a sequence of series and parallel operations; the vertices u and v are called *terminals* of N.

---

[1] Observe that $p \notin V$ when $\lambda \in (0, 1)$ in which case none of the sub-edges up and pv are edges in E. When $\lambda = 0$ or $\lambda = 1$, the point p coincides with u and v, respectively.

[2] The final graph is simple even if intermediate graphs have loops and multiple edges.

We refer to the number of parallel operations required to generate N as the *parallelism* of N and to the number of series operations as the *serialism* of N.

A network is called *series-parallel* when every bi-connected component is *two-terminal series-parallel* with respect to any two vertices. In this work, we only consider bi-connected networks; in future work, we shall adapt our treatment of multiple bi-connected components from cacti [5] to series-parallel networks.

1.2 Related Work

Duffin [8] studies series-parallel networks to compute the resistance of circuit boards. He characterizes three equivalent definitions of series-parallel networks and establishes their planarity. Two-terminal series-parallel networks admit linear time solutions for several problems that are NP-hard on general networks [3,16]. Since series-parallel networks have tree-with two [6], this applies to all problems with efficient algorithms on networks with bounded treewidth [1].

A network Voronoi diagram subdivides a network depending on which site is closest [12] or farthest [9,15] among a finite set of sites. Any data structure for farthest-point queries on a network represents a network farthest-point Voronoi diagram where all points on the network are considered sites [4].

A *continuous absolute center* is a point on a network with minimum farthest-distance. Computing a continuous absolute center takes $O(n)$ time on cacti [2] and $O(m^2 \log n)$ time on general networks [13]. As a by-product, we obtain all continuous absolute centers of a series-parallel network in $O(n \log p)$ time.

1.3 Structure and Results

We introduce a data structure supporting queries for the farthest distance and the farthest points on two-terminal series-parallel networks. We obtain this data structure by isolating sub-structures of series-parallel networks: In Sects. 2 and 3,

Table 1. The traits of our data structures for queries in different types of networks, with n vertices, m edges, k reported farthest points, and parallelism p.

Type	Farthest-point query	Size	Construction time	Reference
General	$O(k + \log n)$	$O(m^2)$	$O(m^2 \log n)$	[4]
Tree	$O(k)$	$O(n)$	$O(n)$	[5]
Cycle	$O(\log n)$	$O(n)$	$O(n)$	[5]
Uni-Cyclic	$O(k + \log n)$	$O(n)$	$O(n)$	[5]
Cactus	$O(k + \log n)$	$O(n)$	$O(n)$	[5]
Parallel-Path	$O(k + \log n)$	$O(n)$	$O(n)$	this work
Bead-Chain	$O(k + \log n)$	$O(n)$	$O(n)$	this work
Abacus	$O(k + \log n)$	$O(n)$	$O(n \log p)$	this work
Series-Parallel	$O(k + \log n)$	$O(n)$	$O(n \log p)$	this work

we study networks consisting of parallel paths and networks consisting of a cycle with attached paths (bead-chains), respectively. In Sect. 4, we combine these results into abacus networks, which are series-parallel networks without nested structures. Finally, we combine these intermediate data structures to obtain our main result in Sect. 5. Table 1 summarizes the characteristics of the proposed data structures and compares them to previous results.

Due to space limitations, we provide the proofs for our claims in this work with additional illustrations in an extended version of this paper [10].

2   Parallel-Path Networks

A *parallel-path network* consists of a bundle of edge disjoint paths connecting two vertices u and v, as illustrated in Fig. 2. In terms of series-parallel networks, parallel-path networks are generated from an edge uv using a sequence of parallel operations followed by a sequence of series operations.

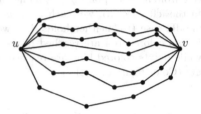

Fig. 2. A parallel-path two-terminal series-parallel network with parallelism $p = 7$.

Let $P_1, P_2, \ldots, P_p$ be the paths of weighted lengths $w_1 \leq w_2 \leq \cdots \leq w_p$ between the terminals u and v in a parallel-path network N. Consider a shortest path tree[3] from a query point $q \in N$. As depicted in Fig. 3, there are three cases: either all shortest paths from q reach v via u (*left case*), or all shortest paths from q reach u via v (*right case*), or neither (*middle case*). We distinguish the three cases using the following notation. Let $\bar{x}_i$ denote the farthest point from $x \in \{u, v\}$ among the points of path P_i, i.e., $\bar{x}_i$ is a point on P_i such that $d(x, \bar{x}_i) = \max_{y \in P_i} d(x, y)$. Together with Fig. 3, the next lemma justifies our choice of the names *left case*, *middle case*, and *right case*.

Lemma 1. *Consider a query q from the i-th path of a parallel-path network.*

(i) We are in the left case when q lies on the sub-path from u to $\bar{v}_i$ with $q \neq \bar{v}_i$,
(ii) we are in the middle case when q lies on the sub-path from $\bar{v}_i$ to $\bar{u}_i$, and
(iii) we are in the right case when q lies on the sub-path from $\bar{u}_i$ to v with $q \neq \bar{u}_i$.

Using Lemma 1, we deal with the three cases as follows.

Left Case and Right Case. In the left case, every shortest path from $q \in P_i$ to any point outside of P_i leaves P_i through u. Hence, the farthest point from q on P_j with $j \neq i$ is the farthest point $\bar{u}_j$ from u on P_j. The distance from

---

[3] More precisely, we consider *extended shortest path trees* [15] which result from splitting each non-tree edge st of a shortest path tree into two sub-edges sx and xt, where all points on sx reach the root through s and all points on xt reach the root through t.

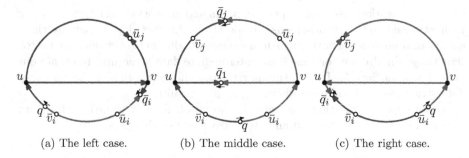

(a) The left case.   (b) The middle case.   (c) The right case.

Fig. 3. The three cases for queries in parallel-path networks. Consider the shortest path tree from a query point $q \in P_i$ along the paths P_1 (center), P_i (bottom), and P_j (top). In the left case (a), we reach v via u. In the right case (c), we reach u via v. In the middle case (b), neither holds, i.e., we *enter* path P_1 from both terminals u and v. Points colored red are reached fastest via a path through u or towards u along uv, while points colored blue are reached fastest via a path through v or towards v along uv. (Color figure online)

q to $\bar{q}_j$ is $d(q, \bar{q}_j) = d(q, u) + d(u, \bar{u}_j) = w_{qu} + \frac{w_1 + w_j}{2}$. On the other hand, the farthest point $\bar{q}_i$ from q on P_i itself moves from $\bar{u}_i$ to v as q moves from u to $\bar{v}_i$ maintaining a distance of $d(q, \bar{q}_i) = \frac{w_1 + w_i}{2}$. Therefore, the farthest distance from q in N is

$$\bar{d}(q) = \max \left[\frac{w_1 + w_i}{2}, \max_{j \neq i} \left(w_{qu} + \frac{w_1 + w_j}{2} \right) \right]$$

$$= \begin{cases} w_{qu} + \dfrac{w_1 + w_p}{2} & \text{if } i \neq p \\ w_{qu} + \dfrac{w_1 + w_{p-1}}{2} & \text{if } i = p \text{ and } \dfrac{w_p - w_{p-1}}{2} \leq w_{qu} \\ \dfrac{w_1 + w_p}{2} & \text{if } i = p \text{ and } \dfrac{w_p - w_{p-1}}{2} \geq w_{qu} \end{cases}$$

The first case means that, for queries from anywhere other than P_p, the farthest points lie on the u-v-paths of maximum length. The second and third case distinguish whether P_p contains a farthest point for queries from P_p itself. Accordingly, we answer a farthest point query from $q \in P_i$ in the left case as follows.

– If $i \neq p$, we report all $\bar{u}_j$ where $w_j = w_p$ and $i \neq j$.
– If $i = p$ and $\frac{w_p - w_{p-1}}{2} \leq w_{qu}$, we report all $\bar{u}_j$ where $w_j = w_{p-1}$ and $j \neq p$.
– If $i = p$ and $\frac{w_p - w_{p-1}}{2} \geq w_{qu}$, we report the farthest point $\bar{q}_p$ from q on P_p using a binary search along the sub-path of P_p from $\bar{u}_p$ to v.

The overlap between the last two cases covers the boundary case when a query from P_p yields a farthest point on P_p itself and farthest points on other u-v-paths.

Swapping u and v above yields the procedure for the right case. Thus, answering farthest-point queries takes $O(k + \log n)$ time in the left and right cases.

Middle Case. In the middle case, there are no farthest points from q on P_i itself and every path P_j with $j \neq i$ contains points that we reach from q via u as well as points that we reach from q via v. Let $\bar{q}_j$ be the farthest point from q along the cycle formed by P_j and P_i. Since the distance from q to $\bar{q}_j$ is $d(q, \bar{q}_j) = \frac{w_i + w_j}{2}$, the farthest distance $\bar{d}(q)$ from q in N is

$$\bar{d}(q) = \max_{j \neq i} \left(\frac{w_i + w_j}{2} \right) = \begin{cases} \dfrac{w_i + w_p}{2} & \text{if } i \neq p \\ \dfrac{w_p + w_{p-1}}{2} & \text{if } i = p \end{cases}.$$

The first case applies for queries from anywhere other than P_p who have their farthest points on the longest u-v-paths, i.e., on the paths P_j with $w_j = w_p$. The second case applies for queries from P_p who have their farthest points on the second longest u-v-paths, i.e., on the paths P_j with $w_j = w_{p-1}$. Using binary search, we can answer a farthest point query from $q \in P_i$ in the middle case by reporting the points $\bar{q}_j$ on those k paths P_j that contain farthest points from q. To improve the resulting query time of $O(k \log n)$, we take a closer look at the position of $\bar{q}_j$ relative to $\bar{u}_j$ and $\bar{v}_j$. As illustrated in Fig. 4, the farthest point $\bar{q}_j$ from $q \in P_i$ along P_j moves from $\bar{u}_j$ to $\bar{v}_j$ as q moves from $\bar{v}_i$ to $\bar{u}_i$.

Lemma 2. *Let N be a parallel-path network with terminals u and v. For any point $x \in N$, let $\bar{x}_i$ denote the farthest point from x along the u-v-path P_i.*

(i) The sub-path from $\bar{v}_i$ to $\bar{u}_i$ has length $d(u,v)$.

(ii) For every point q along the sub-path from $\bar{v}_i$ to $\bar{u}_i$, the sub-path from $\bar{v}_i$ to q has the same length as the sub-path from $\bar{u}_j$ to $\bar{q}_j$ for any $j \neq i$.

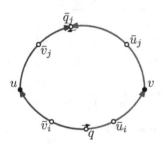

Fig. 4. The positions of the points along the cycle $P_i \cup P_j$ in Lemma 2.

Using Lemma 2, we interpret the searches for $\bar{q}_j$ on the sub-path from $\bar{v}_j$ to $\bar{u}_j$, as a single search *with a common key $\bar{q}$* in *multiple lists* (the $\bar{v}_i$-$\bar{u}_i$-sub-paths) of *comparable search keys* (the vertices along these sub-paths). Using $O(n)$ time, we construct a fractional cascading data structure [7] supporting predecessor queries on the sub-paths from $\bar{v}_j$ to $\bar{u}_j$ for those paths P_j where $w_j = w_{p-1}$.[4]

We answer a farthest-point query from $q \in P_i$ as follows. If $i \neq p$, we locate and report $\bar{q}_p$ along P_p in $O(\log n)$ time. If $i = p$ or $w_p = w_{p-1}$, the remaining farthest points from q are the $\bar{q}_j$ where $j \neq i$ and $w_j = w_{p-1}$; we report them in $O(k + \log n)$ time using the fractional cascading data structure. This query might report a point on P_i, which would be $\bar{q}_i$ for queries from outside P_i. For queries from within P_i, we omit this artifact.

Theorem 1. *Let N be a parallel-path network with n vertices. There is a data structure with $O(n)$ size and $O(n)$ construction time supporting $O(k + \log n)$-time farthest-point queries on N, where k is the number of farthest points.*

---

[4] We consider paths of length w_{p-1} instead of w_p, because we treat P_p separately.

3 Bead-Chain Networks

A *bead-chain network* consists of a main cycle
with attached arcs so that each arc returns
to the cycle before the next one begins. An
example is depicted in Fig. 5. Bead-chains are
series-parallel networks where we first subdivide a cycle using series operations, then we
apply at most one parallel operation to each
edge of this cycle followed by series operations
that further subdivide the arcs and cycle.

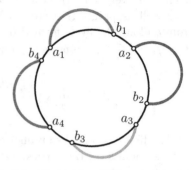

Consider a bead-chain network N with
main cycle C and arcs $\alpha_1, \ldots, \alpha_s$. Let a_i and
b_i be the vertices connecting C with the i-th
arc. Without loss of generality, the path β_i
from a_i to b_i along C is at most as long as α_i.
Otherwise, we swap the roles of α_i and β_i.

Fig. 5. A bead-chain with four arcs
(colored) around its cycle (black).
(Color figure online)

We first study the shape of the function $\hat{d}_i(x)$ that describes the farthest
distance from points along the main cycle to any point on the i-th arc, i.e.,
$\hat{d}_i(x) = \max_{y \in \alpha_i} d(x, y)$. When considering only the i-th arc, we have a parallel-
path network with three paths. Let $\bar{x}$ denote the farthest point from $x \in C$ on
C itself and let $\hat{x}_i$ denote the farthest point from x on arc α_i. From the analysis
in the previous section, we know that $\bar{d}_i(x)$ has the shape depicted in Fig. 6.

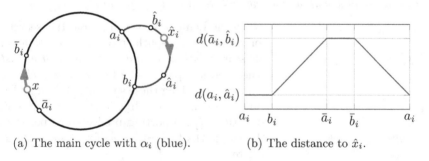

(a) The main cycle with α_i (blue).   (b) The distance to $\hat{x}_i$.

Fig. 6. The shape of the function $\hat{d}_i(x)$ describing the distance from $x \in C$ to the
farthest point $\hat{x}_i$ from x among the points on the i-th arc α_i.

When walking along the main cycle, we encounter a_i, b_i, $\bar{a}_i$, and $\bar{b}_i$ in this
order or its reverse. From a_i to b_i, the point $\hat{x}_i$ moves from $\hat{a}_i$ to $\hat{b}_i$ maintaining
a constant distance. From b_i to $\bar{a}_i$, the point $\hat{x}_i$ stays at $\hat{b}_i$ increasing in distance.
From $\bar{a}_i$ to $\bar{b}_i$, the point $\hat{x}_i$ moves from $\hat{b}_i$ back to $\hat{a}_i$, again, at a constant distance.
Finally, $\hat{x}_i$ stays at $\hat{a}_i$ with decreasing distance when x moves from $\bar{b}_i$ to a_i.

Since the farthest distance changes at the same rate when we move towards
or away from the current farthest-point, the increasing and decreasing segments
of any two functions $\hat{d}_i$ and $\hat{d}_j$ have the same slope except for their sign.

The height of the upper envelope $\hat{D}$ of the functions $\hat{d}_1, \ldots, \hat{d}_s$ at $q \in C$ indicates the farthest distance from q to any point on the arcs and the i-th arc contains a farthest point from q when $\hat{d}_i$ coincides with $\hat{D}$ at q. We construct $\hat{D}$ in linear time using the shape of the functions $\hat{d}_1, \ldots, \hat{d}_s$ described above.

We need to consider an arc separately from the other arcs when it is *too long*. We call an arc α_i *overlong* when the path β_i is longer then remainder γ_i of the main cycle. Figure 7 illustrates an overlong arc α_i with its function $\hat{d}_i$.

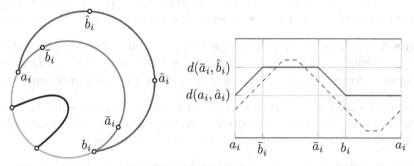

(a) An overlong arc (blue).   (b) The farthest distance to an overlong arc.

Fig. 7. An overlong arc α_i (blue) in a bead-chain network where β_i (green) is longer then the remaining cycle γ_i (orange). The shape of $\hat{d}_i$ is the same as for non-overlong arcs, but its high plateau may horizontally overlap with the high plateau of other arcs. (Color figure online)

Lemma 3. *Every bead-chain network N has at most one overlong arc and the functions $\hat{d}_1, \ldots, \hat{d}_s$ of the remaining arcs $\alpha_1, \ldots, \alpha_s$ satisfy the following.*

(i) The high plateaus of $\hat{d}_1, \ldots, \hat{d}_s$ appear in the order as their arcs $\alpha_1, \ldots, \alpha_s$ appear along the cycle and no two high plateaus overlap horizontally.
(ii) The low plateaus of $\hat{d}_1, \ldots, \hat{d}_s$ appear in the order as their arcs $\alpha_1, \ldots, \alpha_s$ appear along the cycle and no two low plateaus overlap horizontally.

As suggested by Lemma 3, we incrementally construct the upper envelope of the functions $\hat{d}_i$ corresponding to non-overlong arcs and treat a potential overlong arc separately. When performing a farthest-point query from the cycle, we first determine the farthest distance to the overlong arc and the farthest distance to all other arcs. Depending on the answer we report farthest points accordingly.

Lemma 4. *Let $\alpha_1, \ldots, \alpha_s$ be the arcs of a bead-chain that has no overlong arc. Computing the upper envelope $\hat{D}$ of $\hat{d}_1, \ldots, \hat{d}_s$ takes $O(s)$ time.*

To answer a farthest-point query from $q \in C$, we need to find its farthest arcs, i.e., the arcs containing farthest points from q. Suppose each point along the main cycle C has exactly one farthest arc. Then we could subdivide C depending on which arc is farthest and answer a farthest-arc query by identifying the function among $\hat{d}_1, \ldots, \hat{d}_s$ that defines the upper envelope $\hat{D}$ on the sub-edge containing q. On the other hand, there could be multiple farthest arcs when several functions among $\hat{d}_1, \ldots, \hat{d}_s$ have overlapping increasing or decreasing segments. In this case, we could store the at most two farthest arcs from plateaus directly with the corresponding segments of $\hat{D}$. However, storing the farthest arcs from increasing and decreasing segments directly would lead to a quadratic construction time. Instead, we rely on the following observation. An arc α is considered *relevant* when there exists some point $x \in C$ such that α is a farthest arc for x and α is considered *irrelevant* when there is no such point on the main cycle.

Lemma 5. *Let α_i, α_j, and α_k be arcs that appear in this order in a bead-chain without overlong arc. The arc α_j is irrelevant when α_i and α_k are farthest arcs from some query point q such that $\hat{d}_i$ and $\hat{d}_k$ are both decreasing/increasing at q.*

Corollary 1. *Let q be a point on the cycle of a bead-chain with no overlong arc. The farthest arcs from q that correspond to decreasing/increasing segments of $\hat{D}$ form one consecutive sub-list of the circular list of relevant arcs.*

Using Corollary 1, we answer farthest-arc queries from the main cycle of a bead-chain network without overlong arc as follows. When $\hat{D}$ has a plateau at the query point q, we report the at most two farthest arcs stored with this plateau. When $\hat{D}$ has an increasing/decreasing segment at q, we first report the farthest arc α that is stored directly with this segment. We report the remaining farthest arcs by cycling through the circular list of relevant arcs starting from α in both directions until we reach a relevant arc that is no longer farthest from q.

Theorem 2. *Let N be a bead-chain network with n vertices. There is a data structure with $O(n)$ size and $O(n)$ construction time supporting $O(k + \log n)$-time farthest-point queries on N, where k is the number of farthest points.*

4   Abacus Networks

An *abacus* is a network A consisting of a parallel-path network N with arcs attached to its parallel paths, as illustrated in Fig. 8. Let $P_1, \ldots, P_p$ be the parallel paths of N and let B_i be the i-th parallel path with attached arcs.

We split farthest-point queries in an abacus into an *inward query* and an *outward query*: an inward query considers farthest points on the bead-chain containing the query point; an outward query considers farthest points on the remaining bead-chains. We first perform the farthest distance version of inward and outward queries before reporting farthest points where appropriate. Figure 9 illustrates how we treat inward and outward queries in the following.

For an inward query from q on B_i, we construct the bead-chain network B_i' consisting of B_i with an additional edge from u to v of weight $d(u, v)$, as illustrated in Fig. 9a. Since B_i' preserves distances from A, the farthest points from $q \in B_i'$ are the farthest points from q among the points on B_i in A.

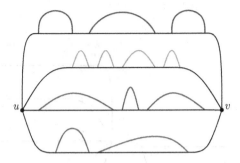

Fig. 8. An abacus with the arcs (colored) attached to its parallel-path network (black). (Color figure online)

For outward queries in an abacus, we distinguish the same three cases as for parallel-path networks: we are in the left case when every shortest path tree reaches u before v, we are in the right case when every shortest path tree reaches v before u, and we are in the middle case otherwise. Analogously to Lemma 1, the left case applies when we are within distance $d(u, \bar{v}_i)$ from u and the right case applies when we are within distance $d(v, \bar{u}_i)$ from v.

For an outward query from $q \in B_i$ in the left case, q has the same farthest points as u outside of B_i. During the construction of the networks $B'_1, \ldots, B'_p$ for inward queries, we determine a list L_j of the farthest points from u in B'_j. Similarly to our treatment of the left case for parallel-path networks, we only keep the list achieving the highest farthest distance and the lists achieving the second highest farthest distance. With this preparation, answering the query for q amounts to reporting the entries of the appropriate lists L_j with $j \neq i$.

For middle case outward queries, we proceed along the following four steps: First, we translate every outward query from an arc of B_i to an outward query from the path P_i, i.e., we argue that it suffices to consider outward queries from the parallel paths (Fig. 9d). Second, we translate outward queries from P_i to outward queries from a virtual edge $\tilde{e}$ connecting the terminals (9e). Third, we speed up queries from the virtual edge by superimposing the data structures for the bead-chains $B_1 \cup \tilde{e}, \ldots, B_p \cup \tilde{e}$, i.e., by conceptually collapsing the parallel chains (Fig. 9f). Finally, we recover the correct answer to the original outward query from the answer obtained with an outward query from the virtual edge.

Lemma 6. *Let α be an arc in an abacus and let β be the other path connecting the endpoints of α. For every point $q \in \alpha$ in the middle case, there is a point $q' \in \beta$ such that q' has the same outward farthest points as q.*

We introduce a virtual edge $\tilde{e}$ from u to v of length w_p, i.e., the length of the longest u-v-path P_p in the underlying parallel-path network, as illustrated in Fig. 9e. Let $\bar{u}$ be the farthest point from u on $\tilde{e}$ and let $\bar{v}$ be the farthest point from v on $\tilde{e}$. From Lemma 2, we know that the sub-edge $\bar{u}\bar{v}$ of $\tilde{e}$ has length $d(u, v)$ and, thus, the same length as the sub-path from $\bar{u}_i$ to $\bar{v}_i$ on each parallel

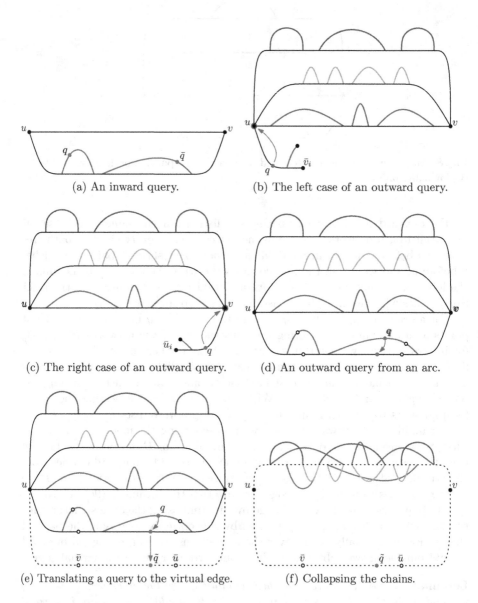

(a) An inward query.

(b) The left case of an outward query.

(c) The right case of an outward query.

(d) An outward query from an arc.

(e) Translating a query to the virtual edge.

(f) Collapsing the chains.

Fig. 9. Inward (a) and outward (b–f) queries for the abacus network from Fig. 8. Inward queries are answered in the bead-chain containing the query (a). Outward queries in the side case are answered with queries form the terminals (b, c). Outward queries in the middle case from arcs are translated to queries from the path (d) and then to queries from a virtual edge (e). From the perspective of the virtual edge, we conceptually collapse all bead-chains of the abacus to support virtual queries (f).

path P_i. We translate an outward query from $q \in P_i$ to a query from the unique point $\tilde{q}$ on $\tilde{e}$ such that $\tilde{q}$ has the same distance to $\bar{u}$ and to $\bar{v}$ as q to $\bar{u}_i$ and to $\bar{v}_i$.

Lemma 7. *For $q \in P_i$ in the middle case, the farthest points from q in $P_i \cup B_j$ are the farthest points from $\tilde{q}$ in $\tilde{e} \cup B_j$ for every $j \neq i$.*

It would be too inefficient to inspect each bead-chain network $B_j \cup \tilde{e}$ with $j \neq i$ to answer an outward query from $q \in P_i$. Instead, we first determine the upper envelopes of the farthest-arc distances $\hat{D}_1, \ldots, \hat{D}_p$ along $\tilde{e}$ in each $B_1 \cup \tilde{e}, \ldots, B_p \cup \tilde{e}$ and then compute their upper envelope U_1 as well as their second level U_2, i.e., the upper envelope of what remains when we remove the segments of the upper envelope. Computing the upper envelope and the second level takes $O(n \log p)$ time, e.g., using plane sweep. Using fractional cascading, we support constant time jumps between corresponding segments of U_1 and U_2. The resulting structure occupies $O(n)$ space, since each of the $O(n)$ arcs along any bead-chain contributes at most four bending points to U_1 and U_2.

We answer an outward query from $q \in P_i$ in the middle case by translating q to $\tilde{q}$. When the segment defining U_1 at $\tilde{q}$ is from some arc α of B_j with $j \neq i$, then α contains an outward farthest point from q. When the segment defining U_1 at $\tilde{q}$ corresponds to an arc of B_i, then we jump down to U_2, which leads us to an arc containing an outward farthest point from q. We report the remaining arcs with outward farthest points by walking $\tilde{q}$ along U_1 and U_2. In order to skip long sequences of segments from B_i, we introduce pointers along U_1 to the next segment from another chain in either direction. Answering outward queries in the middle case takes $O(k + \log n)$ time after $O(n \log p)$ construction time.

Theorem 3. *Let N be an abacus with n vertices and p chains. There is a data structure of size $O(n)$ with $O(n \log p)$ construction time supporting farthest-point queries on N in $O(k + \log n)$ time, where k is the number of farthest points.*

5   Two-Terminal Series-Parallel Networks

Consider a two-terminal series-parallel network N. By undoing all possible series operations and all possible parallel operations in alternating rounds, we reduce N to an edge connecting its terminals and decompose N into paths that reflect its creation history. The colors in Fig. 10 illustrate this decomposition.

Lemma 8. *Let N be a series-parallel network with parallelism p and serialism s. Identifying the terminals of N and reconstructing its creation takes $O(s+p)$ time.*

Once we know the terminals u and v of N, we compute the shortest path distances from u and from v in $O(n \log p)$ time.[5] Consulting the creation history, we determine a maximal parallel-path sub-network P of N with terminals u and v. As illustrated in Figs. 11 and 12, every bi-connected component X of N that is attached to some path of P between vertices a and b is again a two-terminal

---

[5] The priority queue in Dijkstra's algorithm manages never more than p entries.

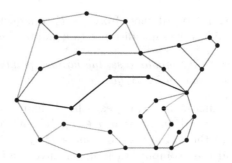

Fig. 10. A two-terminal series-parallel network with colors indicating the parallel operations in a potential creation history: starting with a single red edge, we create a parallel yellow and a parallel blue edge. Then we subdivide the blue edge using series operations until we create a parallel purple edge for one of the blue edges and so forth. (Color figure online)

series-parallel network with terminals a and b. We recurse on these bi-connected components. When this recursion returns, we know a longest a-b-path in X and attach an arc from a to b of this length to P. The resulting network is an abacus A. The abaci created during the recursion form a tree T with root A. Alongside with this decomposition we also create our data structures for the nested abaci.

The size of the resulting data structure remains $O(n)$, since the data structure for each nested abaci consumes space linear in the number of its vertices and each vertex in any nested abaci can be charged to one of the series or parallel operations required to generate the original network.

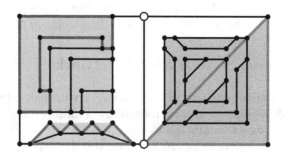

Fig. 11. A two terminal series-parallel network with terminals marked as empty circles. Replacing the three nested structures (blue, red, green) with arcs yields the abacus network A shown at the root of the corresponding abacus tree in Fig. 12. Each colored cycle consists of the shortest path and the longest path connecting the terminals of each to-be-replaced structure; their weighted length determines the weight of the arcs in A. (Color figure online)

We translate a query q to a query in the abacus A; queries from a bi-connected component X attached to P in N will be placed on the corresponding arc of A.

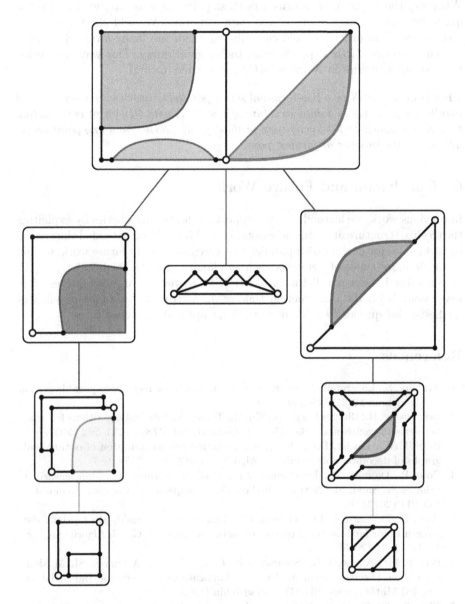

Fig. 12. The tree of nested abaci for the two-terminal series-parallel network from Fig. 11. The inner nodes of this tree correspond to two-terminal networks with nested structures that are indicated with colors; the leaves correspond to abacus networks without nested structures. A query would start at the root abacus and cascade into nested structures when necessary. For instance, when a query at the root abacus yields a farthest point on the blue arc and a farthest point on the red arc, we would perform subsequent queries in the abaci stored in the left and middle child of the root. (Color figure online)

Whenever the query in A returns a farthest point on some arc, we cascade the query into the corresponding nested data structure. We add shortcuts to the abacus tree $\mathcal{T}$ in order to avoid cascading through too many levels of $\mathcal{T}$ without encountering farthest-points from the original query. This way, answering farthest-point queries in N takes $O(k + \log n)$ time in total.

Theorem 4. *Let N be a two-terminal series-parallel network with n vertices and parallelism p. There is a data structure of size $O(n)$ with $O(n \log p)$ construction time that supports $O(k + \log n)$-time farthest-point queries from any point on N, where k is the number of farthest points.*

6   Conclusion and Future Work

In previous work, we learned how to support farthest-point queries by exploiting the treelike structure of cactus networks [5]. In this work, we extended the arsenal by techniques for dealing with parallel structures, as well. In future work, we aim to tackle more types of networks such as planar networks, k-almost trees [11], or generalized series-parallel networks [14]. Moreover, we are also interested in lower bounds on the construction time of data structures supporting efficient farthest-point queries to guide our search for optimal data structures.

References

1. Arnborg, S., Lagergren, J., Seese, D.: Easy problems for tree-decomposable graphs. J. Algorithms **12**(2), 308–340 (1991)
2. Ben-Moshe, B., Bhattacharya, B., Shi, Q., Tamir, A.: Efficient algorithms for center problems in cactus networks. Theoret. Comput. Sci. **378**(3), 237–252 (2007)
3. Bern, M.W., Lawler, E.L., Wong, A.L.: Linear-time computation of optimal subgraphs of decomposable graphs. J. Algorithms **8**(2), 216–235 (1987)
4. Bose, P., Dannies, K., De Carufel, J.L., Doell, C., Grimm, C., Maheshwari, A., Schirra, S., Smid, M.: Network farthest-point diagrams. J. Comput. Geom. **4**(1), 182–211 (2013)
5. Bose, P., De Carufel, J.L., Grimm, C., Maheshwari, A., Smid, M.: Optimal data structures for farthest-point queries in cactus networks. J. Graph Algorithms Appl. **19**(1), 11–41 (2015)
6. Brandstädt, A., Le, V.B., Spinrad, J.P.: Graph Classes: A Survey. SIAM Monographs on Discrete Mathematics and Applications. Society for Industrial and Applied Mathematics (SIAM), Philadelphia (1999)
7. Chazelle, B., Guibas, L.J.: Fractional cascading: I. A data structuring technique. Algorithmica **1**(2), 133–162 (1986)
8. Duffin, R.J.: Topology of series-parallel networks. J. Math. Anal. Appl. **10**(2), 303–318 (1965)
9. Erwig, M.: The graph Voronoi diagram with applications. Networks **36**(3), 156–163 (2000)
10. Grimm, C.: Efficient farthest-point queries in two-terminal series-parallel networks. CoRR abs/1503.01706 (2015). http://arxiv.org/abs/1503.01706

11. Gurevich, Y., Stockmeyer, L.J., Vishkin, U.: Solving NP-hard problems on graphs that are almost trees and an application to facility location problems. J. ACM **31**(3), 459–473 (1984)

12. Hakimi, S.L., Labbé, M., Schmeichel, E.: The Voronoi partition of a network and its implications in location theory. ORSA J. Comput. **4**(4), 412–417 (1992)

13. Hansen, P., Labbé, M., Nicolas, B.: The continuous center set of a network. Discrete Appl. Math. **30**(2–3), 181–195 (1991)

14. Korneyenko, N.M.: Combinatorial algorithms on a class of graphs. Discrete Appl. Math. **54**(2–3), 215–217 (1994)

15. Okabe, A., Satoh, T., Furuta, T., Suzuki, A., Okano, K.: Generalized network Voronoi diagrams: concepts, computational methods, and applications. Int. J. Geogr. Inf. Sci. **22**(9), 965–994 (2008)

16. Takamizawa, K., Nishizeki, T., Saito, N.: Linear-time computability of combinatorial problems on series-parallel graphs. J. ACM **29**(3), 623–641 (1982)

A Polynomial Delay Algorithm for Enumerating Minimal Dominating Sets in Chordal Graphs

Mamadou Moustapha Kanté[1]([✉]), Vincent Limouzy[1], Arnaud Mary[2],
Lhouari Nourine[1], and Takeaki Uno[3]

[1] Clermont-Université, Université Blaise Pascal, LIMOS, CNRS,
Clermont-Ferrand, France
kante@isima.fr
[2] Université Claude Bernard Lyon 1, LBBE, CNRS, Villeurbanne, France
[3] National Institute of Informatics, Tokyo, Japan

Abstract. An output-polynomial algorithm for the listing of minimal dominating sets in graphs is a challenging open problem and is known to be equivalent to the well-known Transversal problem which asks for an output-polynomial algorithm for listing the set of minimal transversals in hypergraphs. We give a polynomial delay algorithm to list the set of minimal dominating sets in chordal graphs, an important and well-studied graph class where such an algorithm was not known. The algorithm uses a new decomposition method of chordal graphs based on clique trees.

1 Introduction

A *hypergraph* $\mathcal{H}$ is a pair $(V, \mathcal{E})$ where V is a finite set and $\mathcal{E} \subseteq 2^V$ is called the set of *hyperedges*. Hypergraphs generalize graphs where each hyperedge has size at most 2. Given a hypergraph $\mathcal{H} := (V, \mathcal{E})$ and $\mathcal{C} \subseteq 2^V$, an *output-polynomial algorithm* for $\mathcal{C}$ is an enumeration algorithm for $\mathcal{C}$ whose running time is bounded by a polynomial depending on the sum of the sizes of $\mathcal{H}$ and $\mathcal{C}$. One of the central problem in the area of enumeration algorithm is the existence of an output-polynomial algorithm for the set of *minimal transversals* in hypergraphs, and is known as the *Transversal problem* or *Hypergraph dualization*. A *minimal transversal (or hitting set)* in a hypergraph $(V, \mathcal{E})$ is an inclusion-wise minimal subset T of V that intersects with every hyperedge in $\mathcal{E}$. The transversal problem has several applications in artificial intelligence [7,8], game theory [13,19], databases [1–3], integer linear programming [2,3], to cite few. Despite the interest in Transversal problem the best known algorithm is the quasi-polynomial time algorithm by Fredman and Khachiyan [9] which runs in time $O(N^{\log(N)})$ where N is the cumulated size of the given hypergraph and its set of minimal transversals. However, there exist several classes of hypergraphs where an output-polynomial algorithm is known (see for instance [7,8,15] for some examples). Moreover, several particular subsets of vertices in graphs are special cases of transversals in hypergraphs and for some of them an output-polynomial algorithm is known, *e.g.*, maximal independent sets, minimal vertex-covers, maximal (perfect) matchings, spanning trees, etc.

© Springer-Verlag Berlin Heidelberg 2016
E.W. Mayr (Ed.): WG 2015, LNCS 9224, pp. 138–153, 2016.
DOI: 10.1007/978-3-662-53174-7_11

In this paper we are interested in the particular case of the Transversal problem, namely the enumeration of *minimal dominating sets* in graphs (DOM-ENUM problem). A *minimal dominating set* in a graph is an inclusion-wise subset D of the vertex set such that every vertex is either in D or has a neighbor in D. In other words D is a minimal dominating set of G if it is a minimal transversal of the *closed neighborhoods* of G, where the *closed neighborhood of a vertex* x is the set containing x and its neighbors. Since in important graph classes an output-polynomial algorithm for the DOM-ENUM problem is a direct consequence of already tractable cases for the Transversal problem, *e.g.*, minor-closed classes of graphs, graphs of bounded degree, it is natural to ask whether an output-polynomial algorithm exists for the DOM-ENUM problem. However, it is proved in [15] that there exists an output-polynomial algorithm for the DOM-ENUM problem if and only if there exists one for the Transversal problem, and this remains true even if we restrict the DOM-ENUM problem to the co-bipartite graphs. This is surprising, but has the advantage of bringing tools from structural graph theory to this difficult problem and is particularly true for the DOM-ENUM problem since in several graph classes output-polynomial algorithms were obtained using the structure of the graphs: graphs of bounded clique-width [4], split graphs [15], interval and permutation graphs [16], line graphs [14,17], etc.

Since the DOM-ENUM problem in co-bipartite graphs is as difficult as the Transversal problem and co-bipartite graphs are a subclass of hole-free graphs, *i.e.*, graphs with no cycles of length greater than or equal to 5, one can ask whether by restricting ourselves to graphs without cycles of length greater than 4, which are exactly *chordal* graphs [6], one cannot expect an output-polynomial algorithm. In fact for several subclasses of chordal graphs an output-polynomial algorithm is already known, *e.g.*, split graphs, chordal P_6-free graphs [15], undi-rected path graphs [14]. Furthermore, chordal graphs have a nice structure, namely the well-known *clique tree* which has been used to solve several algo-rithmic questions in chordal graphs. We prove the following.

Theorem 1. *There exists a polynomial delay algorithm for the* DOM-ENUM *problem in chordal graphs which uses polynomial space.*

An enumeration algorithm is *polynomial delay* if the maximum computation time between two outputs is polynomial in the input size, thus polynomial delay algorithm is output polynomial time. Notice that there exist problems where an output-polynomial algorithm is known and no polynomial delay algorithm exists unless $P = NP$ [20].

Chordal graphs admit several linear structures (*e.g.*, *perfect elimination ordering*) and tree structures (*e.g.*, *clique trees*). The existence of these struc-tures makes many problems polynomially solvable in chordal graphs. For exam-ple, using a clique tree we can split a chordal graph into several subgraphs by removing a clique. This decomposition leads to a dynamic programming algo-rithm for maximum independent set problem by considering the cases that each vertex of the clique is included in the independent set, since any independent set can include at most one vertex of the clique. However, dominating set may include several vertices in a clique, thus this approach is not applicable directly. To the best of our knowledge, there is no good way to deal with this difficulty, and this can explain why minimum dominating set problem is NP-complete. In this paper, we propose to use an "anti-chain" of cliques to decompose chordal graphs. The anti-chain decomposes a graph into several subgraphs, thus the solutions with respect to the anti-chain are obtained by the combination of the solutions of the subgraphs. Since the number of such subgraphs is limited, dynamic programming approach does work. This approach is more powerful than usual decomposition with cliques, in the sense that we can overcome the above difficulty when dealing with the minimal dominating set enumeration problem, thus, gives a new method for designing algorithms for chordal graphs.

2   Preliminaries

An algorithm is said to be *output-polynomial* if the running time is bounded by a polynomial in the input and output sizes. The delay is the maximum computation time between two outputs, pre-processing, and post-processing. If the delay is polynomial in the input size, the algorithm is called *polynomial delay*.

We refer to [5] for our graph terminology. We deal only with finite simple loopless undirected graphs. The vertex set of a graph G is denoted by V_G and its edge set by E_G. An edge between two vertices x and y is denoted by xy (yx respectively). Let G be a graph. The subgraph of G induced by $X \subseteq V_G$, denoted by $G[X]$, is the graph $(X, (X \times X) \cap E_G)$. For a vertex x of G we denote by $N_G(x)$ the set of neighbors of x, *i.e.*, the set $\{y \in V_G \mid xy \in E_G\}$, and we let $N_G[x]$, the *closed neighborhood of* x, be $N_G(x) \cup \{x\}$. For $S \subseteq V_G$, let $N_G[S]$ denote $\bigcup_{x \in S} N_G[x]$. (We will remove the subscript when the graph is clear from the context and this will be the case for all sub or superscripts in the paper.) We say that a vertex x is *dominated* by a vertex y if $x \in N_G[y]$. A *dominating set* of G is a subset D of V_G such that every vertex of G is dominated by a vertex in D. A dominating set is *minimal* if it includes no other dominating set. For $D \subseteq V_G$, a vertex y is a *private neighbor* of $x \in D$ if $N_G[y] \cap D = \{x\}$; the set of private neighbors of a vertex $x \in D$ is denoted by $P(D, x)$. $D \subseteq V_G$ is an *irredundant set* of G if $P(D, x) \neq \varnothing$ for all $x \in D$. $D \subseteq V_G$ is a minimal dominating set of G if and only if D is a dominating set of G and D is an irredundant set.

A *clique* of G is a subset C of G that induces a complete graph, and a *maximal clique* is a clique C of G such that $C \cup \{x\}$ is not a clique for all $x \in V_G \backslash C$. We denote by $\mathcal{C}_G$ the set of maximal cliques of G.

For a *rooted tree* T, let us denote by $\preceq_T$ the relation where $u \preceq_T v$ if v is on the unique path from the root to u; if $u \preceq_T v$ then v is called an *ancestor* of u and u a *descendant* of v. Two nodes u and v of a rooted tree T are *incomparable* if $u \npreceq_T v$ and $v \npreceq_T u$. Given a node u of a rooted T the subtree of T rooted at u is the tree $T[\{v \in V_T \mid v \preceq_T u\}]$ which is rooted at u.

A graph G is called *chordal* if it does not contain chordless cycles of length greater than or equal to 4. From [11] with every chordal graph G, one can associate a tree that we denote by $\mathcal{T}_G$, called *clique tree*, whose nodes are the maximal cliques of G and such that for every vertex $x \in V_G$ the set $\mathcal{T}_G(x) :=$ $\{C \in V(\mathcal{T}_G) \mid$ the maximal clique C contains $x\}$ is a subtree of $\mathcal{T}_G$. Moreover, for every chordal graph G one can compute a clique tree in linear time (see for instance [10]). In the rest of the paper all clique trees are considered rooted.

Let $\mathcal{T}_G$ be a clique tree of a chordal graph G and let us denote its root by C_r. For each $C \in \mathcal{C}_G$, let us denote by $Pa(C)$ its parent and let $f(C) := C \backslash Pa(C)$, *i.e.*, the set of vertices in C that are not in any maximal clique C' ancestor of C. Notice that $\{f(C) \mid C \in \mathcal{T}_G\}$ is a partition of V_G. For each vertex $x \in V_G$, we denote by $C(x)$ the maximal clique C satisfying $x \in f(C)$. Notice that $C(x)$ is uniquely defined since exactly one maximal clique C satisfies $x \in f(C)$. For $C \in \mathcal{C}_G$, the subtree rooted at C is denoted by $\mathcal{T}_G(C)$, and the set of vertices $\bigcup_{C' \in \mathcal{T}_G(C)} f(C')$ is denoted by $V(C)$.

Property 1. Any clique tree $\mathcal{T}_G$ of a chordal graph G satisfies the following.
1. For each $C \in \mathcal{C}_G$, and each $x \in V_G \backslash V(C)$ either $(\{x\} \times f(C)) \subseteq E_G$ or $(\{x\} \times f(C)) \cap E_G = \varnothing$.
2. For any two incomparable C and C' in $\mathcal{C}_G$, we have $(f(C) \times f(C')) \cap E_G = \varnothing$.

For $S \subseteq V_G$, let $\mathcal{C}(S)$ denote the set $\{C(x) \mid x \in S\}$, $Up(S)$ the set of vertices x in V_G such that $C(x)$ is a proper ancestor of a clique $C \in \mathcal{C}(S)$ and $Uncov(S)$ be the vertex set $Up(S) \backslash N_G[S]$, *i.e.*, the set of vertices in $Up(S)$ not dominated by S. For a vertex x, $Up(x)$ denotes $Up(\{x\})$.

A subset $A \subseteq V_G$ is an *antichain* if (1) for any two vertices x and y in A we have $x \notin Up(y)$ and $y \notin Up(x)$, (2) for each vertex $z \in V_G \backslash Up(A)$, $A \cap (C(z) \cup Up(z)) \neq \varnothing$. Intuitively, A is an antichain if $\mathcal{C}(A)$ is a maximal set of pairwise incomparable maximal cliques. Given $S \subseteq V_G$, the *top-set* $A(S)$ is defined as the set of vertices of S included in the upmost cliques in $\mathcal{C}(S)$ that are not descendants of any other in $\mathcal{C}(S)$, i.e., $A(S) := \{x \in S \mid C(x)$ is in $\max_{\preceq_\mathcal{T}}\{\mathcal{C}(S)\}\}$.

If $S \neq \varnothing$, let $\mathcal{L}(S)$ be the set of maximal cliques C satisfying (1) no descendant of C is in $\mathcal{C}(S)$, (2) some descendant of $Pa(C)$ is in $\mathcal{C}(S)$. In other words, $\mathcal{L}(S)$ is the set of upmost maximal cliques no descendant of which intersects with $\mathcal{C}(S)$, i.e., $\mathcal{L}(S) := \max_{\preceq_\mathcal{T}}\{C \in \mathcal{C}_G \mid C$ has no descendant in $\mathcal{C}(S)\}$. If $S = \varnothing$, let $\mathcal{L}(S)$ be $\{C_r\}$. We denote by $\mathcal{L}'(S)$ the set $\max_{\preceq_\mathcal{T}}\{C' \in \mathcal{T}(C) \mid C \in \mathcal{L}(S)$ and $C' \cap S = \varnothing\}$.

We suppose that any clique tree $\mathcal{T}$ is numbered by a pre-order of the visit of a depth-first search. In this numbering, the numbers of the nodes in any subtree forms an interval of the numbers. It is worth noticing that this ordering is a linear extension of the descendant-ancestor relation. We say that a clique is smaller than another clique when its number in the ordering is smaller than the other's. We also extend this numbering to the vertices of the corresponding graph such that the number of a vertex x is smaller than that of a vertex y whenever $C(x)$ is smaller than $C(y)$. We also say that a vertex is smaller than another vertex if its number is smaller than the other's. For a vertex set S, $tail(S)$ denotes the largest vertex in S. A *prefix* of a vertex set S is a subset

$S' \subseteq S$ such that no vertex in $S \backslash S'$ is smaller than $tail(S')$. A *partial antichain* is a prefix of an antichain. We allow the $\varnothing$ to be a partial antichain.

Following this ordering of the vertices of a chordal graph G, a minimal dominating set D is said to be *greedily obtained* if we initially let $D := V_G$ and recursively apply the following rule: if D is not minimal, find the smallest vertex x in D such that $D \backslash \{x\}$ is a dominating set and set $D := D \backslash \{x\}$. Notice that given a graph G there is one greedily obtained minimal dominating set.

3   When Simplicity Means NP-Hardness

A typical way for the enumeration of combinatorial objects is the *backtracking* technique. We start from the emptyset, and in each iteration, we choose an element x, and partition the problem into two subproblems: the enumeration of those including x, and the enumeration of those not including x, and recursively solve these enumeration problems. If we can check the so called EXTENSION PROBLEM in polynomial time, then the algorithm is polynomial delay and uses only polynomial space. The EXTENSION PROBLEM is to answer the existence of an object including S and that does not intersect with X, where S is the set (partial solution) that we have already chosen in the ancestor iterations, and that includes all elements we decided to put in the output solution, and X is the set that we decided not to include in the output solution.

It is known that the EXTENSION PROBLEM for minimal dominating set enumeration is NP-complete [18], and one can even prove that it is still NP-complete in split graphs, which is a proper subclass of chordal graphs.

Proposition 1. *The* EXTENSION PROBLEM *is* NP-*complete in split graphs.*

Proof. It is proved in [18] that the following problem is NP-complete: Given G and $A \subset V_G$ decide whether there exists a minimal dominating set of G containing A. We reduce it to the EXTENSION PROBLEM in split graphs. Let G be a graph, and let $V'_G := \{x' \mid x \in V_G\}$ a disjoint copy of V_G. We let $Split(G)$ be the split graph with vertex set $V_G \cup V'_G$ where V_G and V'_G are respectively the clique and the independent set in $Split(G)$; now xy' is an edge if $x \in N_G[y]$. Now it is easy to check that asking whether there exists a minimal dominating set of G that contains $A \subset V_G$ is equivalent to asking whether there exists a minimal dominating set of $Split(G)$ that contains A and does not intersect with $V'_G \backslash A'$ where $A' := N_{Split(G)}[A] \cap V'_G$. □

Nevertheless, split graphs have a good structure and in the paper [15], it is proved that if $S \cup X$ induces a clique the EXTENSION PROBLEM in split graphs can be solved in polynomial time and this combined with the structure of minimal dominating sets in split graphs lead to a polynomial delay algorithm for the DOM-ENUM problem in split graphs. Chordal graphs also have a good tree structure induced by clique trees. Thus, by following this tree structure, the EXTENSION PROBLEM seems to be solvable. In precise, we consider the case in which a path $\mathcal{P}$, from the root, of the clique tree satisfies that both

$V(C) \cap (S \cup X) \neq \varnothing$ and $V(C) \nsubseteq (S \cup X)$ holds only for cliques C included in $\mathcal{P}$. In other words, the condition is that for any clique $C \notin \mathcal{P}$ whose parent is in $\mathcal{P}$, either $V(C) \cap (S \cup X) = \varnothing$ (totally not determined) or $V(C) \subseteq (S \cup X)$ (totally determined) holds. The solutions are partially determined on the path $\mathcal{P}$, and thus the EXTENSION PROBLEM seems to be polynomial. However, Theorem 2 states that the problem is actually NP-complete.

Theorem 2. *The* EXTENSION PROBLEM *is* NP-*complete in chordal graphs even if a path* $\mathcal{P}$, *from the root, of the clique tree satisfies that any child* C *of a clique in* $\mathcal{P}$ *satisfies either* $V(C) \cap (S \cup X) = \varnothing$ *or* $V(C) \subseteq (S \cup X)$.

Proof. We reduce SAT to our problem. Let φ be an instance of SAT with $x_1, \ldots, x_n$ the variables and $c_1, \ldots, c_m$ the clauses of φ. We construct a chordal graph as follows. The vertex set of the graph is

$$\{x_1, \ldots, x_n, c_1, \ldots, c_m, p_1, \ldots, p_n, \bar{p}_1, \ldots, \bar{p}_n, l_1, \ldots, l_n\} \bigcup$$
$$\{\bar{l}_1, \ldots, \bar{l}_n, y_1, \ldots, y_n, z_1, \ldots, z_n, q_1, \ldots, q_n, \bar{q}_1, \ldots, \bar{q}_n\},$$

where l_i and $\bar{l}_i$ are literals representing respectively x_i and $\bar{x}_i$ (notice that if one literal does not appear, the corresponding vertex is not created). Since with every clique tree one can associate a unique chordal graph, we will construct the clique tree of the chordal graph. For each $1 \leqslant i \leqslant n$, we let $C(l_i)$ and $C(\bar{l}_i)$ be the set of clauses containing the literal l_i and $\bar{l}_i$ respectively. We let its root be $C_r := \{c_1, \ldots, c_m, p_1, \ldots, p_n, \bar{p}_1, \ldots, \bar{p}_n\}$. The other maximal cliques are defined as follows. For each $1 \leqslant i \leqslant n$, we let $C_{x_i} = \{x_i, p_i, \bar{p}_i\}$, $C_{y_i} = \{y_i, x_i\}$, $C_{z_i} = \{y_i, z_i\}$, $C_{q_i} = \{q_i, l_i\}$, $C_{\bar{q}_i} = \{\bar{q}_i, \bar{l}_i\}$, $C_{l_i} = \{l_i, p_i\} \cup C(l_i)$, and $C_{\bar{l}_i} = \{\bar{l}_i, \bar{p}_i\} \cup C(\bar{l}_i)$ with the following parent-child relation: C_{x_i}, C_{l_i} and $C_{\bar{l}_i}$ are the children of C_r, C_{y_i} is the only child of C_{x_i} and C_{z_i} is the only child of C_{y_i}, C_{q_i} and $C_{\bar{q}_i}$ are the only children of C_{l_i} and $C_{\bar{l}_i}$ respectively. It is easy to check that the constructed tree is indeed a clique tree. See Fig. 1 for an illustration.

We set $S := \{x_1, \ldots, x_n, y_1, \ldots, y_n\}$ and $X := \{z_1, \ldots, z_n, c_1, \ldots, c_m\} \cup \{p_1, \ldots, p_n, \bar{p}_1, \ldots, \bar{p}_n\}$ and $\mathcal{P} := \{C_r\}$. For each $1 \leqslant i \leqslant n$, we have by construction $V(C_{x_i}) \subseteq S \cup X$, and $(V(C_{l_i}) \cup V(C_{\bar{l}_i})) \cap (S \cup X) = \varnothing$. Therefore, for any maximal clique C child of C_r, either $V(C) \cap (S \cup X) = \varnothing$, or $V(C) \subseteq (S \cup X)$ holds, thus the condition of the statement holds.

One can easily check that any satisfiable assignment of φ leads to a minimal dominating set containing S and that does not intersect X. Let us prove the converse direction. We observe when we choose both l_i and $\bar{l}_i$ in the dominating set, x_i loses its private neighbors. Thus, any minimal dominating set can include at most one of them. On the other hand, exactly one of l_i and q_i (resp., $\bar{l}_i$ and $\bar{q}_i$) must be included in any minimal dominating set, so that it dominates l_i and q_i (resp., $\bar{l}_i$ and $\bar{q}_i$), and both must be private neighbors of the chosen one. Moreover, to dominate each clause c_j, at least one literal of c_j has to be included in any minimal dominating set. Hence, for any minimal dominating set D including S and not intersect with X, the set of literals included in D corresponds to a satisfiable assignment. Therefore, the answer of the EXTENSION PROBLEM is yes if and only if φ has a satisfiable assignment. $\quad\square$

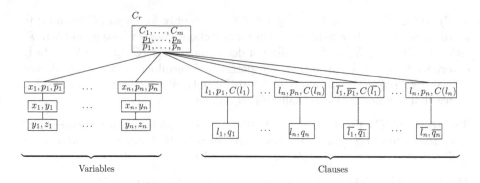

Fig. 1. An illustration of the construction of Theorem 2.

To overcome these difficulties, we will follow another approach. In fact the NP-hardness comes from the fact that the root clique can have both undominated vertices and private neighbors of several vertices of S. In the following, we will introduce a new strategy for the enumeration, that repeatedly enumerates antichains in levelwise manner. Indeed for any minimal dominating set D of a chordal graph G, one can easily check that the set $A(D)$ is an antichain that moreover dominates $Up(A(D))$. Our strategy consists in enumerating such antichains and for each such antichain A enumerates the minimal dominating sets D such that $A(D) = A$. Let's be more precise in the forthcoming sections.

4 (K_1, K_2)-Extensions

From now on we consider a fixed chordal graph G and clique tree $\mathcal{T}$ of G with root C_r so that we do not need to recall them in the statements. Let $K_1, K_2 \subseteq C_r$ be given disjoint sets that are decided to be included in the solution. Intuitively, we are considering the subgraph induced by a subtree of the clique tree rooted at C_r, and K_1 and K_2 are vertices that we already decided to include in the solution, such that vertices in K_2 have private neighbors outside the subgraph, and vertices of K_1 do not. Without confusion we denote $K_1 \cup K_2$ by K. A (K_1, K_2)-*extension* of a partial antichain A is a vertex set D such that $(A \cup K) \subseteq D$ and $D \backslash (A \cup K) \subseteq \bigcup_{C \in \mathcal{L}(A \cup K)} V(C)$. Observe that if D is a (K_1, K_2)-extension of A, then A is a prefix of $A(D)$. When the partial antichain is not specified, (K_1, K_2)-extension is that for the empty partial antichain. A (K_1, K_2)-extension D is *feasible* if it is a dominating set and $P(D, x) \neq \varnothing$ for all $x \in D \backslash K_2$. A partial antichain A is (K_1, K_2)-*extendable* if it has a feasible (K_1, K_2)-extension.

Let us briefly explain the ideas of the algorithm and why we introduce (K_1, K_2)-extensions. We first observe that for any minimal dominating set D of G, its top-set is an $(\varnothing, \varnothing)$-extendable antichain. Moreover, $D \backslash A(D)$ is composed of vertices below $A(D)$, *i.e.*, any vertex in $D \backslash A(D)$ is included in $V(C) \backslash C$ for some $C \in \mathcal{C}(D)$. Using this, we can partition the minimal dominating sets according to their top-sets. Since these top-sets are $(\varnothing, \varnothing)$-extendable, we enumerate all $(\varnothing, \varnothing)$-extendable antichains, and for each $(\varnothing, \varnothing)$-extendable antichain A, enumerate all minimal dominating sets whose top-set is A. As by definition of (K_1, K_2)-extendable for some disjoint $K_1, K_2 \subseteq C_r$, for each $(\varnothing, \varnothing)$-antichain A there is at least one minimal dominating set whose top-set is A. Therefore, each output $(\varnothing, \varnothing)$-antichain will give rise to a solution. This is one of the key to polynomial delay.

Now for a minimal dominating set D and a clique $C \in \mathcal{C}(A(D))$, each vertex x in $D \cap (V(C) \cup C)$ cannot have a private neighbor in another $G[V(C') \cup C']$ for some other $C' \in \mathcal{C}(A(D))$. Therefore, we can treat each $G[V(C) \cup C]$ independently. However, for each $C \in \mathcal{C}(A(D))$ the set $D \cap (V(C) \cup C)$ is not necessarily a minimal dominating set of $G[V(C) \cup C]$ since $D \cap C$ may be equal to a singleton $\{x\}$ with x having a private neighbor in $Up(A(D))$. In such cases we are looking in $G[V(C) \cup C]$ a dominating set D' of $G[V(C) \cup C]$ containing x where x does not necessarily have a private neighbor, but all the other vertices in D' do, i.e., D' is a feasible $(\varnothing, \{x\})$-extension in $G[V(C) \cup C]$ with clique tree $\mathcal{T}(C)$. This situation is what exactly motivated the notion of (K_1, K_2)-extensions.

Assume now we are given a pair (K_1, K_2) of disjoint sets in C_r and a (K_1, K_2)-extendable antichain A. Now contrary to $(\varnothing, \varnothing)$-antichains we can have a vertex x in $K := K_1 \cup K_2$ that belongs to several cliques in A. So we cannot independently make recursive calls in $G[V(C) \cup C]$ for each $C \in \mathcal{C}(A)$. But, for each feasible (K_1, K_2)-extension of A and each $C \in \mathcal{C}(A)$ the set $D \cap (V(C) \cup C)$ is a feasible (K_C^1, K_C^2)-extension of $G[V(C) \cup C]$ for some disjoint K_C^1 and K_C^2 in $(A \cup K) \cap C$. Now the whole task is to define for each $C \in \mathcal{C}(A)$ the sets K_C^1 and K_C^2 in $(A \cup K) \cap C$ in such a way that by combining all these feasible (K_C^1, K_C^2)-extensions we obtain a feasible (K_1, K_2)-extension of A, and also any feasible (K_1, K_2)-extension can be obtained in that way. Actually, the way of setting K_C^1 and K_C^2 is the key, and is described in the next section.

Let us prove some technical lemmas about (K_1, K_2)-extensions needed for proving the correctness of our algorithm. For $C \in \mathcal{C}_G$ and $x \in C$, let $\mathcal{F}(C, x) := \{C' \preceq_T C \text{ and } C' \in \mathcal{L}'(x)\}$, and let $D_C(x)$ denotes a vertex set composed of

1. $Z \subseteq V(C) \cap \left(\bigcup\limits_{C' \in \mathcal{F}(C,x)} C' \right)$ such that $|Z \cap C'| = |Z \cap f(C')| = 1$ for all $C' \in \mathcal{F}(C, x)$,

2. a greedily obtained minimal dominating set of $G[(V(C) \backslash N_G[x]) \backslash N_G[Z]]$.

If $x \notin C$, then we let $D_C(x)$ be a greedily obtained minimal dominating set of $G[V(C)]$.

Property 2. Let $C \in \mathcal{C}_G$ and let $x \in V_G$. Then $D_C(x)$ is an irredundant set in $G[V(C)]$ and every vertex in $V(C) \backslash N_G[x]$ is dominated by $D_C(x)$.

Proof. We first prove that $D_C(x)$ is irredundant. Since each minimal dominating set is also an irredundant set, we can assume that $x \in C$. By definition of Z we have that $\{x\} \times Z \cap E_G = \varnothing$. Moreover, by Property 1 (2) no two vertices of Z are adjacent. Since by construction of $D_C(x)\backslash Z$ no vertex in $D_C(x)\backslash Z$ is adjacent to a vertex of Z, we can conclude that for each $z \in Z$ we have $z \in P(D_C(x), z)$. Moreover, since $(D_C(x)\backslash Z) \cap N_G[Z] = \varnothing$ and $D_C(x)\backslash Z$ is a minimal dominating set of $G[(V(C)\backslash N_G[x]) \setminus N_G[Z]]$, we can conclude that $P(D_C(x), y) \neq \varnothing$ for all $y \in (D_C(x)\backslash N_G[x]) \setminus N_G[Z]$.

Let us now prove that $V(C)\backslash N_G[x]$ is dominated by $D_C(x)$. If $x \notin C$, then $D_C(x)$ is a minimal dominating set of $G[V(C)]$ and then we are done. So, assume that $x \in C$ and let $y \in V(C)\backslash N_G[x]$. Then $C(y)$ is necessarily a descendant of a clique $C' \in \mathcal{L}'(x)$ and such that $C' \preceq_T C$. So, either $y \in N_G[Z]$ or $y \notin N_G[Z]$. In both cases, it is dominated by $D_C(x)$. $\qquad\square$

Given disjoint sets $K_1, K_2 \subseteq C_r$, $D \subseteq V_G\backslash K$ and $x \in D \cup K_1$, a vertex $y \in P(D \cup K, x)$ is said *safe* if either $x = y$, or the following two conditions are satisfied

(S1) $N_G(y) \cap V(C) \subseteq N_G[D_C(y)]$ for all $C \in \mathcal{L}'(D \cup K)$ with $y \in C$ and,
(S2) for each $z \in N_G[y] \cap Uncov(D \cup K)$, there is $C \in \mathcal{L}'(D \cup K)$ such that $z \in N_G[D_C(y)]$.

A vertex $x \in D$ is said *safe* if one of its private neighbors is safe.

Property 3. Let $x \in D \cup K_1$ and let $y \in P(D \cup K, x)$ be safe for x. Then $V(C)\backslash\{y\} \subseteq N_G[D_C(y)]$ for all $C \in \mathcal{L}'(D \cup K)$ with $y \in C$.

Proof. By Property 2 $V(C)\backslash N_G[y]$ is dominated by $D_C(y)$. By definition of safety $N_G(y)$ is dominated by $D_C(y)$. Therefore $V(C)\backslash\{y\}$ is dominated by $D_C(y)$ for all $C \in \mathcal{L}'(D \cup K)$ with $y \in C$. $\qquad\square$

Lemma 1. *Let A be a partial antichain and let $x \in A \cup K_1$. For $y \in P(A \cup K, x)$ that is non-safe, no (K_1, K_2)-extension D of A that is a dominating set satisfies that $y \in P(D, x)$.*

Proof. Since y is not safe, we have $x \neq y$, and therefore y violates one of the two conditions (S1) or (S2) to be safe. Suppose that (S1) is not satisfied, *i.e.*, there is a clique $C \in \mathcal{L}'(A \cup K), y \in C$ such that there is a vertex z in $(N_G(y) \cap V(C))\backslash N_G[D_C(y)]$. Thus, any (K_1, K_2)-extension D of A that is a dominating set includes some vertices in $N_G[y]$ other than x, thus y is not a private neighbor of x.

Suppose now that (S2) is not satisfied, *i.e.*, there is a vertex $z \in N_G[y] \cap Uncov(A \cup K)$ such that no clique $C \in \mathcal{L}'(A \cup K)$ satisfies $z \in N_G[D_C(y)]$. It implies from the definition of $D_C(y)$ that no vertex in $V(C)\backslash N_G[y]$ is adjacent to z in all cliques $C \in \mathcal{L}'(A \cup K)$. Thus, as in the previous case, in any (K_1, K_2)-extension D of A, y is not a private neighbor of x unless D is not a dominating set. $\qquad\square$

Lemma 2. *Let A be a partial antichain and let $x \in A \cup K_1$ be safe. Then there is $y \in P(A \cup K, x)$ that is safe and such that $y \in V(C(x))$.*

Proof. The statement holds if $x \in P(A \cup K, x)$. If not, $C(x)$ includes another vertex in $A \cup K$, and it is adjacent to any vertex in $N_G[x] \backslash V(C(x))$ by Property 1. Thus all its safe private neighbors are always in $V(C(x))$. □

Lemma 3. *A partial antichain A is (K_1, K_2)-extendable if and only if the following two conditions are satisfied*

1. *any vertex in $Uncov(A \cup K)$ is included in a clique of $\mathcal{L}'(A \cup K)$,*
2. *all vertices in $A \cup K_1$ are safe.*

Proof. Let A be a (K_1, K_2)-extendable partial antichain. If (1) is not satisfied, there is a vertex $z \in Uncov(A \cup K)$ that is not included in any clique of $\mathcal{L}'(A \cup K)$, and by definition of (K_1, K_2)-extension no (K_1, K_2)-extension of A can dominate it. So (1) is always satisfied. Now, if (2) is not satisfied, there is a non-safe vertex x in $A \cup K_1$, thus all $y \in P(A \cup K, x)$ are non-safe. By Lemma 1 it follows that $P(D, x) = \varnothing$ for each (K_1, K_2)-extension D of A that is a dominating set, and then (2) is always satisfied.

Suppose now that the two conditions hold. For each $x \in A \cup K_1$ let us choose one safe private neighbor and let us denote the set of all these safe private neighbors by S. We consider a (K_1, K_2)-extension D generated from $A \cup K$ as follows. First of all notice that from the definition of private neighbor and safety for each $C \in \mathcal{L}'(A \cup K)$, $|C \cap S| \leqslant 1$. So, let $\mathcal{L}_1 := \{C \in \mathcal{L}'(A \cup K) \mid |C \cap S| = 1\}$ and $\mathcal{L}_0 := \{C \in \mathcal{L}'(A \cup K) \mid |C \cap S| = 0\}$. It is clear that $\{\mathcal{L}_0, \mathcal{L}_1\}$ is a bipartition of $\mathcal{L}'(A \cup K)$. Let $z \in S$. Now let

$$D := (A \cup K) \cup \left(\bigcup_{C \in \mathcal{L}_1, C \cap S = \{y\}} D_C(y) \right) \cup \left(\bigcup_{C \in \mathcal{L}_0} D_C(z) \right).$$

D is clearly a (K_1, K_2)-extension of A. By definition of $D_C(y)$ for each vertex $x \in D \backslash (A \cup K)$ we have that $P(D, x) \neq \varnothing$. It is moreover easy to check that for each $x \in A \cup K_1$, we have that $S \cap P(A \cup K, x) \in P(D, x)$. Thus, from Property 1, $P(D, x) \neq \varnothing$ for all $x \in D \backslash K_2$. Each vertex in $N_G[A \cup K]$ is dominated. Moreover, since for each $C \in \mathcal{L}_0$ we have $z \notin C$, by definition of $D_C(z)$ we have $V(C)$ is also dominated. Now, let $C \in \mathcal{L}_1$ and let $C \cap S = \{y\}$. We know from Property 2 that $V(C) \backslash N_G[y]$ is dominated by $D_C(y)$ and y is dominated by $A \cup K$ since y is safe for some vertex in $A \cup K_1$. So, it remains to show that $N_G(y) \cap V(C)$ is dominated. By the definition of safety we know that the two conditions (S1) and (S2) are satisfied, *i.e.*, $N_G(y) \cap V(C)$ is dominated. □

As a corollary we have the following.

Lemma 4. *For any partial antichain A one can check in polynomial time whether A is (K_1, K_2)-extendable.*

Proof. By Lemma 3 it is enough to check if (1) all vertices in $A \cup K_1$ are safe and (2) each vertex in $Uncov(A \cup K)$ is included in a clique in $\mathcal{L}'(A \cup K)$. Since (2) can be easily checked in polynomial time from G and a clique tree of G, it remains to show that (1) can be checked in polynomial time. A vertex $x \in A \cup K_1$ is safe if either $x \in P(A \cup K_1, x)$ or there exists a safe $y \in V(C(x)) \cap P(A \cup K_1, x)$ by Lemma 2. But by the definition of safety for each $y \in V(C(x)) \cap P(A \cup K_1, x)$ the conditions (S1) and (S2) are of course checkable in polynomial time from G and a clique tree of G. $\square$

5   The Algorithm

Our enumeration strategy is composed of nested enumerations: enumeration of (K_1, K_2)-extendable antichains, for each (K_1, K_2)-extendable antichain A and each $C \in \mathcal{C}(A)$ define K_C^1 and K_C^2 and enumerate all the feasible (K_C^1, K_C^2)-extensions, and finally the combinations of all these (K_C^1, K_C^2)-extensions. Since any minimal dominating set is a feasible extension of some $(\varnothing, \varnothing)$-extendable antichain, the completeness of the enumeration is trivial. The rest of the section is as follows. We first show how to enumerate (K_1, K_2)-extendable antichains for some fixed (K_1, K_2). Then we show, given a (K_1, K_2)-extendable antichain A, how to define K_C^1 and K_C^2 for each $C \in \mathcal{C}(A)$ and how to combine all the feasible (K_C^1, K_C^2)-extensions in order to obtain all feasible (K_1, K_2)-extensions of A. Before assuming that we can perform both tasks with polynomial delay and use only polynomial space let us show that we can enumerate with polynomial delay and in polynomial space all the feasible (K_1, K_2)-extensions.

Enumeration of (K_1, K_2)-Extensions. The algorithm for enumerating all the feasible (K_1, K_2)-extensions, including the case of the root of the recursion, is composed of the (K_1, K_2)-extendable antichain enumeration and of the enumeration of combinations of the feasible (K_C^1, K_C^2)-extensions for appropriate (K_C^1, K_C^2). It can be described as follows.

Algorithm EnumKExtension$(G, \mathcal{T}, K_1, K_2)$
    G:graph, $\mathcal{T}$:clique tree
    1. **for** each antichain A output by EnumAntichain$(G, \mathcal{T}, K_1, K_2, \varnothing)$ **do**
    2.   **output** each solution of EnumCombination$(G, \mathcal{T}, K_1, K_2, A, A \cup K)$
    3. **end for**

Assume that EnumAntichain$(G, \mathcal{T}, K_1, K_2, \varnothing)$ enumerates all (K_1, K_2)-extendable antichains (Lemma 5) and EnumCombination$(G, \mathcal{T}, K_1, K_2, A, A \cup K)$ enumerates all feasible (K_1, K_2)-extensions of A (Lemma 8), both with polynomial delay and use polynomial space. Then we have the following.

Theorem 3. *The call* EnumKExtension $(G, \mathcal{T}, K_1, K_2)$ *enumerates all feasible (K_1, K_2)-extensions in polynomial delay and uses polynomial space.*

Proof. By definition for every feasible (K_1, K_2)-extension D the top-set $A(D)$ is a (K_1, K_2)-extendable antichain. So by Lemmas 5 and 8 below every feasible (K_1, K_2)-extension is output. From the definition of (K_1, K_2)-extendable antichains every call in Step 1 outputs at least one feasible (K_1, K_2)-extension. Therefore, EnumKExtension $(G, \mathcal{T}, K_1, K_2)$ enumerates all feasible (K_1, K_2)-extensions. Furthermore, since the algorithms EnumAntichain$(G, \mathcal{T}, K_1, K_2, \varnothing)$ and EnumCombination$(G, \mathcal{T}, K_1, K_2, A, A \cup K)$ run in polynomial delay and use polynomial space we can conclude that EnumKExtension $(G, \mathcal{T}, K_1, K_2)$ runs in polynomial delay and uses polynomial space. $\square$

Enumeration of Antichains. Our strategy is to enumerate all (K_1, K_2)-extendable partial antichains by an ordinary backtracking algorithm, that repeatedly appends a vertex x to the current solution S with $x > tail(S)$. In this algorithm, any (K_1, K_2)-extendable partial antichain A is obtained from $A \backslash tail(A)$. Since $A \backslash tail(A)$ is a prefix of A, any (K_1, K_2)-extendable partial antichain is generated from another (K_1, K_2)-extendable partial antichain. This implies that the set of (K_1, K_2)-extendable partial antichains satisfies a kind of monotone property, and thus we can enumerate all (K_1, K_2)-extendable partial antichains with passing through only (K_1, K_2)-extendable partial antichains. The algorithm is described as follows.

Algorithm EnumAntichain$(G, \mathcal{T}, K_1, K_2, A)$
   G:graph, $\mathcal{T}$:clique tree, A:(K_1, K_2)-extendable partial antichain
1. **if** A is an antichain **then output** A;
2. **for** each vertex $z > tail(A)$ **do**
3.   **if** $A \cup \{z\}$ is a (K_1, K_2)-extendable partial antichain **then**
     call EnumAntichain$(G, \mathcal{T}, K_1, K_2, A \cup \{z\})$
4. **end for**

Lemma 5. *The call* EnumAntichain$(G, \mathcal{T}, K_1, K_2, \varnothing)$ *enumerates all* (K_1, K_2)-*extendable antichains in polynomial delay and uses polynomial space.*

Proof. Observe that for any (K_1, K_2)-extendable partial antichain A, $A \backslash tail(A)$ is a (K_1, K_2)-extendable partial antichain. Thus, one can easily prove by induction that the iteration inputting A is recursively called only by the iteration inputting $A \backslash tail(A)$. Therefore, all (K_1, K_2)-extendable partial antichains are generated by this algorithm without repetition. For a (K_1, K_2)-extendable partial antichain A, there is at least one feasible (K_1, K_2)-extension D. By the definition of a feasible (K_1, K_2)-extension, $A(D \backslash K)$ is a (K_1, K_2)-extendable antichain with A as a prefix. This implies that at least one descendant of any iteration outputs an antichain, and every leaf of the recursion tree outputs an antichain. Then, the delay is bounded by the maximum computation time of an iteration multiplied by the depth of the recursion. The depth is at most $|V_G|$, thus the algorithm is polynomial delay since the loop at Step 2 runs at most n times and the (K_1, K_2)-extendability check can be done in polynomial time by

Lemma 4. Since the depth is bounded by $|V_G|$, the algorithm uses obviously a polynomial space.   □

Enumeration of Combinations. We now show, given a (K_1, K_2)-extendable antichain A, how to enumerate with polynomial delay and in polynomial space all feasible (K_1, K_2)-extensions of A by computing for each $C \in \mathcal{C}(A)$ all the (K_C^1, K_C^2)-extensions of $G[V(C) \cup C]$ for appropriate K_C^1 and K_C^2 and combining all of them. Note that the set A is the top-set of any feasible (K_1, K_2)-extension if and only if the (K_1, K_2)-extension is that of A. For pruning redundant partial combinations, we introduce the notion of a partial (K_1, K_2)-extension.

A vertex set $D \supseteq A \cup K$ is called a *partial (K_1, K_2)-extension* of A if there is a feasible (K_1, K_2)-extension D' of A such that $D \setminus (A \cup K)$ is a prefix of $D' \setminus (A \cup K)$, and all the vertices in $V(C(x))$ for $x \in A$ is dominated by D if x is smaller than $tail(D \setminus (A \cup K))$. Our strategy is to enumerate all partial (K_1, K_2)-extensions of A, similar to the antichain enumeration.

For a partial (K_1, K_2)-extension D of A, let $C^*(D)$ be the largest clique C in $\mathcal{C}(A)$ such that $(D \setminus (A \cup K)) \cap V(C) \neq \varnothing$, and $C_*(D)$ be the smallest clique C in $\mathcal{C}(A)$ such that a vertex in $V(C)$ is not dominated by D. Informally $C^*(D)$ is the last clique $C \in \mathcal{C}(A)$ such that $V(C)$ is dominated by D, and $C_*(D)$ the first clique in $\mathcal{C}(A)$ such that $V(C)$ is not dominated by D.

To enumerate all partial (K_1, K_2)-extensions of A and find all (K_1, K_2)-extensions of A, we start from $D = A \cup K$ and repeatedly add a $(K_{C_*(D)}^1, K_{C_*(D)}^2)$-extension of $G[V(C_*(D)) \cup C_*(D)]$ to D for appropriate $(K_{C_*(D)}^1, K_{C_*(D)}^2)$, while keeping extendability. We now characterize the possible pairs $(K_{C_*(D)}^1, K_{C_*(D)}^2)$. Let $Q(C')$ be the vertices x in $K \cup A$ that has no safe private neighbor in $V(C) \cup C, C > C'$, and none of its private neighbor in $P(K \cup A \cup D, x)$ is included in $Up(A) \setminus C'$ or in $V(C), C < C'$. In other words $Q(C')$ is the set of vertices in $K \cup A$ that we must give a private neighbor in $V(C') \cup C'$ for any (K_1, K_2)-extension of A containing D.

Lemma 6. *For a non-empty partial (K_1, K_2)-extension D, $D \cap (V(C^*(D)) \cup C^*(D))$ is a feasible (K_1', K_2')-extension in $G[V(C^*(D)) \cup C^*(D)]$ where $K_1' = Q(C^*(D))$ and $K_2' = ((A \cup K) \cap C^*(D)) \setminus K_1'$.*

Proof. By definitions of partial (K_1, K_2)-extension and of C^*, $D \cap (V(C^*(D)) \cup C^*(D))$ dominates $V(C^*(D))$. Moreover, every vertex x in $Q(C^*(D))$ has a private neighbor only in $V(C^*(D)) \cup C^*(D)$, and moreover $x \in C^*(D)$. Thus, the statement holds.   □

Lemma 7. *Let D be a partial (K_1, K_2)-extension of A and suppose that $C_*(D)$ exists. For any feasible (K_1', K_2')-extension D' in $G[V(C_*(D)) \cup C_*(D)]$ where $K_1' = Q(C_*(D))$, $K_2' = ((A \cup K) \cap C_*(D)) \setminus K_1'$, $D \cup D'$ is a partial (K_1, K_2)-extension of A.*

Proof. As in the proof of Lemma 3, we choose one private neighbor for vertices in $A \cup K$ that have safe private neighbors in $V(C), C > C_*(D)$ and let S be the set of these selected vertices. Then we let $\mathcal{L}_1 := \{C \in \mathcal{L}'(A \cup K) \mid C >$

$C_*(D), |C \cap S| = 1\}$ and $\mathcal{L}_0 := \{C \in \mathcal{L}'(A \cup K) \mid C > C_*(D), |C \cap S| = 0\}$. Let $z \in S$. Now let

$$D^* := (A \cup K \cup D \cup D') \cup \left(\bigcup_{C \in \mathcal{L}_1, C \cap S = \{y\}} D_C(y) \right) \cup \left(\bigcup_{C \in \mathcal{L}_0} D_C(z) \right).$$

According to the proof of Lemma 3, D^* is a feasible (K_1, K_2)-extension of A. □

We can now describe the algorithm.

Algorithm EnumCombination$(G, \mathcal{T}, K_1, K_2, A, D)$
    G:graph, $\mathcal{T}$:clique tree, A:(K_1, K_2)-extendable antichain
    D: a partial (K_1, K_2)-extension of A
1. **if** $C_*(D)$ does not exist **then output** D; **return**
2. $K_1' := Q(C_*(D))$, $K_2' := ((A \cup K) \cap C_*(D)) \backslash K_1'$
3. **for each** D' output by EnumKExtension$(G[V(C_*(D)) \cup C_*(D)], \mathcal{T}(C^*(D)), K_1', K_2')$
4.   **call** EnumCombination$(G, \mathcal{T}, K_1, K_2, A, D \cup D')$
5. **end for**

Lemma 8. *The call* EnumCombination$(G, \mathcal{T}, K_1, K_2, A, A \cup K)$ *enumerates all feasible* (K_1, K_2)-*extensions whose top-set is* A *in polynomial delay and uses polynomial space.*

Proof. From Lemma 6, the iteration of a partial (K_1, K_2)-extension D of A is generated only from the iteration of $D \backslash (V(C^*(D) \backslash C^*(D)))$. This assures that the algorithm enumerates all partial (K_1, K_2)-extensions of A without duplication. From Lemma 7, there is at least one feasible (K_1, K_2)-extension D' of A including the partial (K_1, K_2)-extension D of A that is the input of the iteration. Thus, all the leaf iterations of the recursion of this algorithm always outputs a feasible (K_1, K_2)-extension of A. Now the delay is bounded by the maximum computation time of an iteration multiplied by the depth of the recursion. The depth is at most $|V_G|$, thus the algorithm is polynomial delay since EnumKExtension runs with polynomial delay. Since the depth is at most $|V_G|$, the algorithm is obviously polynomial space. □

Proof (of Theorem 1). By definition every minimal dominating set of G is a feasible $(\varnothing, \varnothing)$-extension. Therefore, the call EnumKExtension $(G, \mathcal{T}, \varnothing, \varnothing)$ enumerates all minimal dominating sets in polynomial delay and polynomial space by Theorem 3. □

6   Conclusion

We have proved that one can list all the minimal dominating sets of a chordal graph with polynomial delay and in polynomial space. The result enlarged the classes in that minimal dominating set enumeration is output-polynomially solvable. However, the problem is still open for several graph classes such as bipartite graphs and unit-disk graphs. In particular, chordal bipartite graph admits an output-polynomial algorithm [12]. Applying our decomposition technique to chordal bipartite graphs is an interesting future research.

References

1. Agrawal, R., Mannila, H., Srikant, R., Toivonen, H., Verkamo, A.I.: Fast discovery of association rules. In: Advances in Knowledge Discovery and Data Mining, pp. 307–328. AAAI/MIT Press (1996)
2. Boros, E., Gurvich, V., Khachiyan, L., Makino, K.: Dual-bounded generating problems: partial and multiple transversals of a hypergraph. SIAM J. Comput. **30**(6), 2036–2050 (2000)
3. Boros, E., Gurvich, V., Khachiyan, L., Makino, K.: Generating weighted transversals of a hypergraph. In: Rutgers University, pp. 13–22 (2000)
4. Courcelle, B.: Linear delay enumeration and monadic second-order logic. Discrete Appl. Math. **157**(12), 2675–2700 (2009)
5. Diestel, R.: Graph Theory (Graduate Texts in Mathematics). Springer, Heidelberg (2005)
6. Dirac, G.A.: On rigid circuit graphs. Abhandlungen Aus Dem Mathematischen Seminare der Universität Hamburg **25**(1–2), 71–76 (1961)
7. Eiter, T., Gottlob, G.: Identifying the minimal transversals of a hypergraph and related problems. SIAM J. Comput. **24**(6), 1278–1304 (1995)
8. Eiter, T., Gottlob, G., Makino, K.: New results on monotone dualization and generating hypergraph transversals. SIAM J. Comput. **32**(2), 514–537 (2003)
9. Fredman, M.L., Khachiyan, L.: On the complexity of dualization of monotone disjunctive normal forms. J. Algorithms **21**(3), 618–628 (1996)
10. Galinier, P., Habib, M., Paul, C.: Chordal graphs and their clique graphs. In: Nagl, M. (ed.) WG 1995. LNCS, vol. 1017, pp. 358–371. Springer, Heidelberg (1995)
11. Gavril, F.: The intersection graphs of subtrees in trees are exactly the chordal graphs. J. Comb. Theor. Ser. B **16**(1), 47–56 (1974)
12. Golovach, P.A., Heggernes, P., Kante, M.M., Kratsch, D., Villanger, Y.: Enumerating minimal dominating sets in chordal bipartite graphs. Discrete Appl. Math. **199**, 30–36 (2015)
13. Gurvich, V.A.: On theory of multistep games. USSR Comput. Math. Math. Phys. **13**(6), 143–161 (1973)
14. Kanté, M.M., Limouzy, V., Mary, A., Nourine, L.: On the neighbourhood helly of some graph classes and applications to the enumeration of minimal dominating sets. In: Chao, K.-M., Hsu, T., Lee, D.-T. (eds.) ISAAC 2012. LNCS, vol. 7676, pp. 289–298. Springer, Heidelberg (2012)
15. Kanté, M.M., Limouzy, V., Mary, A., Nourine, L.: On the enumeration of minimal dominating sets and related notions. SIAM J. Discrete Math. **28**(4), 1916–1929 (2014)
16. Kanté, M.M., Limouzy, V., Mary, A., Nourine, L., Uno, T.: On the enumeration and counting of minimal dominating sets in interval and permutation graphs. In: Cai, L., Cheng, S.-W., Lam, T.-W. (eds.) Algorithms and Computation. LNCS, vol. 8283, pp. 339–349. Springer, Heidelberg (2013)

17. Kanté, M.M., Limouzy, V., Mary, A., Nourine, L., Uno, T.: Polynomial delay algorithm for listing minimal edge dominating sets in graphs. In: Dehne, F., Sack, J.-R., Stege, U. (eds.) WADS 2015. LNCS, vol. 9214, pp. 446–457. Springer, Heidelberg (2015)
18. Mary, A.: Énumeration des Dominants Minimaux d'un graphe. Ph.D. thesis, Université Blaise Pascal (2013)
19. Ramamurthy, K.G.: Coherent Structures and Simple Games. Theory and Decision Library. Game Theory, Mathematical Programming and Operations Research: Series C, vol. 6. Springer, Heidelberg (1990)
20. Strozecki, Y.: Enumeration Complexity and Matroid Decomposition. Ph.D. thesis, Université Paris Diderot - Paris 7 (2010)

Finding Paths in Grids with Forbidden Transitions

Mamadou Moustapha Kanté[1], Fatima Zahra Moataz[2,3], Benjamin Momège[2,3], and Nicolas Nisse[2,3(✉)]

[1] Clermont-Université, Université Blaise Pascal, LIMOS, CNRS, Clermont-Ferrand, France
[2] Université Nice Sophia Antipolis, CNRS, I3S, UMR 7271, Sophia Antipolis, France
[3] Inria, Sophia Antipolis, France
nicolas.nisse@inria.fr

Abstract. A transition in a graph is a pair of adjacent edges. Given a graph $G = (V, E)$, a set of forbidden transitions $\mathcal{F} \subseteq E \times E$ and two vertices $s, t \in V$, we study the problem of finding a path from s to t which uses none of the forbidden transitions of $\mathcal{F}$. This means that it is forbidden for the path to consecutively use two edges forming a pair in $\mathcal{F}$. The study of this problem is motivated by routing in road networks in which forbidden transitions are associated to prohibited turns as well as routing in optical networks with asymmetric nodes, which are nodes where a signal on an ingress port can only reach a subset of egress ports. If the path is not required to be elementary, the problem can be solved in polynomial time. On the other side, if the path has to be elementary, the problem is known to be NP-complete in general graphs [Szeider 2003]. In this paper, we study the problem of finding an elementary path avoiding forbidden transitions in planar graphs. We prove that the problem is NP-complete in planar graphs and particularly in grids. In addition, we show that the problem can be solved in polynomial time in graphs with bounded treewidth. More precisely, we show that there is an algorithm which solves the problem in time $O(k\Delta^2(3k\Delta)^{2k}n)$ in n-node graphs with treewidth at most k and maximum degree Δ.

1   Introduction

Driving in New-York is not easy. Not only because of the rush hours and the taxi drivers, but because of the no-left, no-right and no U-turn signs. Even in a "grid-like" city like New-York, prohibited turns might force a driver to cross several times the same intersection before eventually reaching their destination. In this paper, we give hints explaining why it is difficult to deal with forbidden-turn signs when driving in grid-like road networks.

This work is supported by the associated Inria team AlDyNet, the project ECOS-Sud/CONICYT Chile C12E03, and a grant from the "Conseil régional Provence Alpes-Côte d'Azur".

E.W. Mayr (Ed.): WG 2015, LNCS 9224, pp. 154–168, 2016.
DOI: 10.1007/978-3-662-53174-7_12

Let $G = (V, E)$ be a graph. A transition in G is a pair of two distinct edges incident to a same vertex. Let $\mathcal{F} \subseteq E \times E$ be a set of forbidden transitions in G. We say that a not necessarily elementary path $P = (v_0, \ldots, v_q)$ is $\mathcal{F}$-valid if it contains none of the transitions of $\mathcal{F}$, i.e., $\{\{v_{i-1}, v_i\}, \{v_i, v_{i+1}\}\} \notin \mathcal{F}$ for $i \in \{1, \ldots, q-1\}$. Given G, $\mathcal{F}$ and two vertices s and t, the Path Avoiding Forbidden Transitions (PAFT) problem consists in finding an $\mathcal{F}$-valid s-t-path.

The PAFT problem arises in many contexts. In optical networks, nodes can be highly asymmetric with respect to their switching capabilities as pointed out in [2]. Indeed, an optical node might have some restrictions on its internal connectivity and that, consequently, signal on a certain ingress port can only reach a subset of the egress ports. As explained in [2,4,9], a node can be asymmetrically configured for many reasons such as the limitation on the number of physical ports of optical switch components and the low cost of asymmetric nodes compared to symmetric ones. The existence of asymmetric nodes adds some connectivity constraints in the network. This has motivated some studies to re-investigate, under the assumption of the existence of asymmetric nodes, some classical problems in optical network, such as routing [2,4,11] and protection with node-disjoint paths [9]. These studies do not highlight the computational complexity of the problems they consider. We point out here that the optical nodes configured asymmetrically can be modeled as vertices with forbidden transitions and the routing problem is an application of PAFT. The study of PAFT is also motivated by its relevance to vehicle routing. In road networks, it is possible that some roads are closed due to traffic jams, construction, etc. It is also frequent to encounter no-left, no-right and no U-turn signs at intersections. These prohibited roads and turns can be modeled by forbidden transitions.

When the PAFT problem is studied, a distinction has to be made according to whether the path to find is elementary (cannot repeat vertices) or non-elementary. Indeed, PAFT can be solved in polynomial time [8] for the non-elementary case while finding an elementary path avoiding forbidden transitions has been proved NP-complete in [15]. In this paper, we study the elementary version of the PAFT problem in planar graphs and more particularly in grids. Our interest for planar graphs is motivated by the fact that they are closely related to road networks. They are also an interesting special case to study while trying to capture the difficulty of the problem. Furthermore, to the best of our knowledge, this case has not been addressed before in the literature.

Related work. PAFT is a special case of the problem of finding a path avoiding forbidden paths (PFP) introduced in [18]. Given a graph G, two vertices s and t, and a set $\mathcal{S}$ of forbidden paths, PFP aims at finding an s-t-path which contains no path of $\mathcal{S}$ as a subpath. When the forbidden paths are composed of exactly two edges, PFP is equivalent to PAFT. Many papers address the non-elementary version of PFP, proposing exact polynomial solutions [1,10,18]. The elementary counterpart has been recently studied in [13] where a mathematical formulation is given and two solution approaches are developed and tested. The computational complexity of the elementary PFP can be deduced from the complexity of PAFT which has been established in [15]. Szeider proved in [15] that finding an elementary path avoiding forbidden transitions is NP-complete and gave a complexity classification of the problem according to the types of the forbidden transitions. The NP-completeness proof in [15] does not extend to planar graphs.

PAFT is also a generalization of the problem of finding a properly colored path in an edge-colored graph (PEC). Given an edge-colored graph G^c and two vertices s and t, the PEC problem aims at finding an s-t-path such that any consecutive two edges have different colors. It is easy to see that PEC is equivalent to PAFT when the set of forbidden transitions consists of all pairs of adjacent edges that have the same color. The PEC problem is proved to be NP-complete in directed graphs [7] which directly implies that the PAFT problem is NP-complete in directed graphs[1].

Contribution. Our main contribution is the proof that the PAFT problem is NP-complete in grids. We also prove that the problem can be solved in time $O(k\Delta^2(3k\Delta)^{2k}n)$ in n-node graphs with treewidth at most k and maximum degree Δ. In other words, we prove that the PAFT problem is FPT in $k + \Delta$. Our NP-completeness result strengthens the one of Szeider [15] established in 2003 and extends to the problem of PFP.

The paper is organized as follows. The problem of PAFT is formally stated in Sect. 2. In Sect. 3, the problem is proven NP-complete in grids. A polynomial time algorithm for graphs with bounded treewidth is presented in Sect. 4. Finally, some directions for future work are presented in Sect. 5

2   Problem Statement

Let $G = (V, E)$ be a graph. Given a subgraph H of G, a *transition* in H is a (not ordered) set of two distinct edges of H incident to a same vertex. Namely, $\{e, f\}$ is a transition if $e, f \in E(H)$, $e \neq f$ and $e \cap f \neq \emptyset$. Let $\mathcal{T}$ denote the set of all transitions in G. Let $\mathcal{F} \subseteq \mathcal{T}$ be a set of *forbidden transitions*. A transition in $\mathcal{A} = \mathcal{T} \setminus \mathcal{F}$ is said *allowed*. A *path* is any sequence $(v_0, v_1, \cdots, v_r)$ of vertices such that $v_i \neq v_j$ for any $0 \leq i < j \leq r$ and $e_i = \{v_i, v_{i+1}\} \in E$ for any $0 \leq i < r$. Given two vertices s and t in G, a path $P = (v_0, v_1, \cdots, v_r)$ is called an *s-t-path* if $v_0 = s$ and $v_r = t$. Finally, a path $P = (v_0, v_1, \cdots, v_r)$ is *$\mathcal{F}$-valid* if any transition in P is allowed, i.e., $\{e_i, e_{i+1}\} \notin \mathcal{F}$ for any $0 \leq i < r$.

Problem 1 (Problem of Finding a Path Avoiding Forbidden Transitions, PAFT). Given a graph $G = (V, E)$, a set $\mathcal{F}$ of forbidden transitions and two vertices $s, t \in V$. Is there an $\mathcal{F}$-valid s-t-path in G?

---

[1] Note that, in [7], the authors state that their result can be extended to planar graphs. However, there is a mistake in the proof of the corresponding Corollary 7: to make their graph planar, vertices are added when edges intersect. Unfortunately, this transformation does not preserve the fact that the path is elementary.

3   NP-completeness in Grids

We start by proving that the PAFT problem is NP-complete in grids. We first prove that it is NP-complete in planar graphs with maximum degree at most 8 by a reduction from 3-SAT. Then, we propose simple transformations to reduce the maximum degree and prove that the PAFT problem is NP-complete in planar graphs with degree at most 4. Finally, we prove it is NP-complete in grids.

Lemma 1. *The PAFT problem is NP-complete in planar graphs with maximum degree 8.*

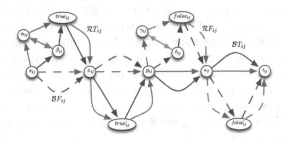

Fig. 1. Example of the gadget-graph G_{ij} for variable v_i, and $j \leq m$. Brown (resp. green) edge is added if v_i appears positively (resp., negatively) in C_j. If $v_i \notin C_j$, none of the green nor brown edge appear. (Color figure online)

Proof. The problem is clearly in NP. We prove the hardness using a reduction from the 3-SAT problem. Let Φ be an instance of 3-SAT, i.e., Φ is a boolean formula with variables $\{v_1, \cdots, v_n\}$ and clauses $\{C_1, \cdots, C_m\}$. We build a grid-like planar graph G where rows correspond to clauses and columns correspond to variables. In what follows, the colors are only used to make the presentation easier. Moreover, we consider undirected graphs but, since the forbidden transitions can simulate orientations, the figures are depicted with directed arcs for ease of presentation. Please note also that we use a multigraph in the reduction for the sake of simplicity. This multigraph can easily be transformed into a simple graph without changing the maximum degree.

Gadget G_{ij}. For any $i \leq n$ and $j \leq m$, we define the gadget G_{ij} depicted in Fig. 1 and that consists of 4 edge-disjoint paths from s_{ij} to t_{ij}: two "blue" paths BT_{ij} and BF_{ij}, and two "red" paths RT_{ij} and RF_{ij} defined as follows.

- $RT_{ij} = (s_{ij}, \alpha_{ij}, true_{ij}, x_{ij}, true'_{ij}, y_{ij}, z_{ij}, t_{ij})$;
- $BT_{ij} = (s_{ij}, \beta_{ij}, true_{ij}, x_{ij}, true'_{ij}, y_{ij}, z_{ij}, t_{ij})$;
- $RF_{ij} = (s_{ij}, x_{ij}, y_{ij}, \gamma_{ij}, false_{ij}, z_{ij}, false'_{ij}, t_{ij})$;
- $BF_{ij} = (s_{ij}, x_{ij}, y_{ij}, \delta_{ij}, false_{ij}, z_{ij}, false'_{ij}, t_{ij})$.

The forbidden transitions $\mathcal{F}_{ij}$ of the gadget G_{ij} are defined in such a way that the only way to go from s_{ij} to t_{ij} is by following one of the paths in $\{\mathcal{B}T_{ij}, \mathcal{B}F_{ij}, \mathcal{R}T_{ij}, \mathcal{R}F_{ij}\}$. It is forbidden to use any transition consisting of two edges from two different paths of the set $\{\mathcal{B}T_{ij}, \mathcal{B}F_{ij}, \mathcal{R}T_{ij}, \mathcal{R}F_{ij}\}$.

Intuitively, assigning the variable v_i to $True$ will be equivalent to choosing one of the paths $\mathcal{B}T_{ij}$ or $\mathcal{R}T_{ij}$ (called *positive* paths) depicted with full lines in Fig. 1. Respectively, assigning v_i to $False$ will correspond to choosing one of the paths $\mathcal{B}F_{ij}$ or $\mathcal{R}F_{ij}$ (called *negative* paths) and depicted by dotted line in Fig. 1.

So far, it is *a priori* not possible to start from s_{ij} by one path and arrive in t_{ij} by another path. In particular, the color by which s_{ij} is left must be the same by which t_{ij} is reached. If Variable v_i appears in Clause C_j, we add one edge to G_{ij} as follows. If v_i appears positively in C_j, we add the *brown* edge $\{\alpha_{ij}, \beta_{ij}\}$ that creates a "bridge" between $\mathcal{B}T_{ij}$ and $\mathcal{R}T_{ij}$. Similarly, if v_i appears negatively in C_j, we add the *green* edge $\{\gamma_{ij}, \delta_{ij}\}$ that creates a "bridge" between $\mathcal{B}F_{ij}$ and $\mathcal{R}F_{ij}$. When the gadget G_{ij} contains a brown (resp. green) edge, all the transitions containing the brown (resp. green) edge are allowed; this makes it possible to switch between the positive (resp. negative) paths $\mathcal{B}T_{ij}$ and $\mathcal{R}T_{ij}$ (resp. $\mathcal{B}F_{ij}$ and $\mathcal{R}F_{ij}$) when going from s_{ij} to t_{ij}. Hence, if v_i appears in C_j, it will be possible to start from s_{ij} with one color and arrive to t_{ij} with a different one. Note that, the type of path (positive or negative) cannot be modified between s_{ij} and t_{ij}.

The following easy claims characterize the $\mathcal{F}_{ij}$-valid s_{ij}-t_{ij}-paths in G_{ij}.

Claim 1. *The $\mathcal{F}_{ij}$-valid s_{ij}-t_{ij}-paths in G_{ij} are $\mathcal{R}T_{ij}, \mathcal{B}T_{ij}, \mathcal{R}F_{ij}, \mathcal{B}F_{ij}$ and*

- *if variable v_i appears positively in Clause C_j:*
  - *the path $\mathcal{R}\mathcal{B}T_{ij}$ that starts with the first edge $\{s_{ij}, \alpha_{ij}\}$ of $\mathcal{R}T_{ij}$, then uses brown edge $\{\alpha_{ij}, \beta_{ij}\}$ and ends with all edges of $\mathcal{B}T_{ij}$ but the first one;*
  - *the path $\mathcal{B}\mathcal{R}T_{ij}$ that starts with the first edge $\{s_{ij}, \beta_{ij}\}$ of $\mathcal{B}T_{ij}$, then uses brown edge $\{\alpha_{ij}, \beta_{ij}\}$ and ends with all edges of $\mathcal{R}T_{ij}$ but the first one;*
- *if variable v_i appears negatively in Clause C_j:*
  - *the path $\mathcal{R}\mathcal{B}F_{ij}$ that starts with the subpath $(s_{ij}, x_{ij}, y_{ij}, \gamma_{ij})$ of $\mathcal{R}F_{ij}$, then uses green edge $\{\gamma_{ij}, \delta_{ij}\}$ and ends with the subpath of $\mathcal{B}F_{ij}$ that starts at δ_{ij} and ends at t_{ij};*
  - *the path $\mathcal{B}\mathcal{R}F_{ij}$ that starts with the subpath $(s_{ij}, x_{ij}, y_{ij}, \delta_{ij})$ of $\mathcal{B}F_{ij}$, then uses green edge $\{\delta_{ij}, \gamma_{ij}\}$ and ends with the subpath of $\mathcal{R}F_{ij}$ that starts at γ_{ij} and ends at t_{ij};*

Claim 2. *Let P be a $\mathcal{F}_{ij}$-valid $s_{ij}t_{ij}$-paths in G_{ij}. Then, either*

- *P passes through $true_{ij}$ and $true'_{ij}$ and does not pass through $false_{ij}$ nor $false'_{ij}$, or*
- *P passes through $false_{ij}$ and $false'_{ij}$ and does not pass through $true_{ij}$ nor $true'_{ij}$.*

Claim 3. *Let P be a $\mathcal{F}_{ij}$-valid $s_{ij}t_{ij}$-paths in G_{ij}. Then the first and last edges of P have different colors if and only if P uses a green or a brown edge, i.e., if $P \in \{\mathcal{R}\mathcal{B}T_{ij}, \mathcal{B}\mathcal{R}T_{ij}, \mathcal{R}\mathcal{B}F_{ij}, \mathcal{B}\mathcal{R}F_{ij}\}$.*

Clause-graph G_j. For any $j \leq m$, the Clause-gadget G_j is built by combining the graphs G_{ij}, $i \leq n$, in a "line" (see Fig. 2). The subgraphs G_{ij} are combined from "left to right" (for $i = 1$ to n) if j is odd and from "right to left" (for $i = n$ to 1) otherwise. In more details, for any $j \leq m$, G_j is obtained from a copy of each gadget G_{ij}, $1 \leq i \leq n$, and two additional vertices s_j and t_j as follows:

- If j is odd, the subgraph G_j starts with a red edge $\{s_j, s_{1j}\}$ and then, for $1 < i \leq n$, the vertices s_{ij} and $t_{i-1,j}$ are identified. Finally, there is a blue edge from t_{nj} to vertex t_j.
- If j is even, the subgraph G_j starts with a blue edge $\{s_j, s_{nj}\}$ and then, for $1 < i \leq n$, the vertices t_{ij} and $s_{i-1,j}$ are identified. Finally, there is a red edge from t_{1j} to vertex t_j.

The forbidden transitions $\mathcal{F}_j$ include, besides all transitions in $\mathcal{F}_{ij}$, $i = 1, \ldots, n$, new transitions which are defined such that, when passing from a gadget G_{ij} to the next one, the same color must be used. This means that if we enter a vertex $t_{ij} = s_{i,j+1}$ by an edge with a given color, the same color must be used to leave this vertex. However, in such vertices, we can change the type (positive or negative) of path.

Note that if we enter a Clause-graph with a red (resp. blue) edge, we can only leave it with a blue (resp. red) edge. This means that a path must change its color inside the Clause-graph, and must hence use a brown or green edge in some gadget-graph. The use of a brown (resp. green) forces a variable that appears positively (resp. negatively) in the clause to be set to true (resp. false) and validates the Clause.

The key property of G_j relates to the structure of $\mathcal{F}_j$-valid paths from s_j to t_j, which we summarize in Claims 4 and 5.

Claim 4. *Any $\mathcal{F}_j$-valid path P from s_j to t_j in G_j consists of the concatenation of:*

Case j odd. *the red edge $\{s_j, s_{1j}\}$, then the concatenation of $\mathcal{F}_{ij}$-valid paths from s_{ij} to t_{ij} in G_{ij}, for $1 \leq i \leq n$ in this order (from $i = 1$ to n), and finally the blue edge $\{t_{nj}, t_j\}$;*

Case j even. *the blue edge $\{s_j, s_{nj}\}$, then the concatenation of $\mathcal{F}_{ij}$-valid paths from s_{ij} to t_{ij} in G_{ij}, for $1 \leq i \leq n$ in the reverse order (from $i = n$ to 1), and finally the red edge $\{t_{1j}, t_j\}$.*

By the previous claim, for any $\mathcal{F}_j$-valid path P from s_j to t_j, the colors of the first and last edges differ. Hence, by Claim 3 and the definition of the allowed transitions between two gadgets:

Claim 5. *Any $\mathcal{F}_j$-valid path P from s_j to t_j must use a green or a brown edge in a gadget G_{ij} for some $1 \leq i \leq n$.*

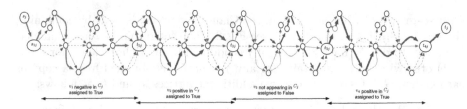

Fig. 2. Case j odd. Clause-graph G_j for a clause $C_j = \bar{v}_1 \vee v_2 \vee v_4$ in a formula with 4 variables. The bold path corresponds to an assignment of v_1, v_2 and v_4 to *True*, and of v_3 to *False* (Color figure online)

Main graph. To conclude, we have to be sure that the assignment of the variables is coherent between the clauses. For this purpose, let us combine the subgraphs G_j, $j \leq m$, as follows (see Fig. 3). First, for any $1 \leq j < m$, let us identify t_j and s_{j+1}. Then, some vertices (depicted in grey in Fig. 3) of G_{ij} are identified with vertices of $G_{i,j+1}$ in such a way that using a positive (resp., negative) path in G_{ij} forces the use of the same type of path in $G_{i,j+1}$. That is, the choice of the path used in G_{ij} is transferred to $G_{i,j+1}$ and therefore it corresponds to a truth assignment for Variable v_i.

Namely, for each $1 \leq j < m$ and for each $1 \leq i \leq n$, we identify the vertices $true_{i,j+1}$ and $false'_{ij}$ on the one hand, and the vertices $true'_{ij}$ and $false_{i,j+1}$ on the other hand to obtain the "grey" vertices. Finally, forbidden transitions $\mathcal{F}$ of G, include, besides all transitions in $\mathcal{F}_j$ for $j = 1, \ldots, m$, new transitions which are defined in order to forbid "crossing" a grey vertex, i.e., it is not possible to go from $G_{i,j}$ to $G_{i,j+1}$ via a grey vertex. The following claims present the key properties of an $\mathcal{F}$-valid path in G.

Claim 6. *Any $\mathcal{F}$-valid path P from s_1 to t_m in G consists of the concatenation of $\mathcal{F}_j$-valid paths from s_j to t_j in G_j from $j = 1$ to m.*

Claim 7. *Let P be an $\mathcal{F}$-valid $s_1 t_m$-path in G. Then, for any $1 \leq i \leq n$, either*

- *for any $1 \leq j \leq m$, the subpath of P between s_{ij} and t_{ij} passes through $true_{ij}$ and $true'_{ij}$ and does not pass through $false_{ij}$ nor $false'_{ij}$, or*

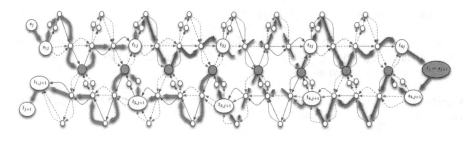

Fig. 3. Combining $C_j = \bar{v}_1 \vee v_2 \vee v_4$ and $C_{j+1} = v_2 \vee \bar{v}_3 \vee \bar{v}_4$ (Case j odd). (Color figure online)

– for any $1 \leq j \leq m$, the subpath of P between s_{ij} and t_{ij} passes through $false_{ij}$ and $false'_{ij}$ and does not pass through $true_{ij}$ nor $true'_{ij}$.

Proof. By Claims 4 and 6, for any $1 \leq i \leq n$ and any $1 \leq j \leq m$, there is a subpath P_{ij} of P that goes from s_{ij} to t_{ij}. Moreover, the paths P_{ij} are pairwise vertex-disjoint.

For $1 \leq i \leq n$, by Claim 2, P_{i1} either passes through $true_{i1}$ and $true'_{i1}$, or through $false_{i1}$ and $false'_{i1}$. Let us assume that we are in the first case (the second case can be handled symmetrically). We prove by induction on $j \leq m$ that P_{ij} passes through $true_{ij}$ and $true'_{ij}$ and does not pass through $false_{ij}$ nor $false'_{ij}$.

Indeed, if P passes through $true_{ij} = false'_{i,j+1}$ and $true'_{ij} = false_{i,j+1}$, then $P_{i,j+1}$ cannot use $false_{i,j+1}$ nor $false'_{i,j+1}$ since P_{ij} and $P_{i,j+1}$ are vertex-disjoint. By Claim 2, $P_{i,j+1}$ passes through $true_{i,j+1}$ and $true'_{i,j+1}$. □

Note that $(G, \mathcal{F})$ can be constructed in polynomial-time. Moreover, G is clearly planar with maximum degree 8. Hence, the next claim allows to prove Lemma 1.

Claim 8. Φ *is satisfiable if and only if there is an* $\mathcal{F}$*-valid* s_1*-*t_m*-path in* G.

Proof. Let φ be a truth assignment which satisfies Φ. We can build an $\mathcal{F}$-valid s_1-t_m-path in G as follows. For each row $1 \leq j \leq m$, we build a path P_j from s_i to t_j by concatenating the paths P_{ij}, $1 \leq j \leq m$, which are built as follows. Among the variables that appear in C_j, let v_k be the variable with the smallest index, which satisfies the clause.

– For $1 \leq i < k$, if $\varphi(v_i) = true$, then $P_{ij} = \mathcal{R}T_{ij}$ if j is odd and $P_{ij} = \mathcal{B}T_{ij}$ if j is even, respectively. If $\varphi(v_i) = false$, then $P_{ij} = \mathcal{R}F_{ij}$ if j is odd, and $P_{ij} = \mathcal{B}F_{ij}$ if j is even.
– If $\varphi(v_k) = true$, then $P_{ij} = \mathcal{R}\mathcal{B}T_{ij}$ if j is odd, and $P_{ij} = \mathcal{B}\mathcal{R}T_{ij}$ if j is even. If $\varphi(v_k) = false$, then $P_{ij} = \mathcal{R}\mathcal{B}F_{ij}$ if j is odd, and $P_{ij} = \mathcal{B}\mathcal{R}F_{ij}$ if j is even.
– For $k < i \leq n$, if $\varphi(v_i) = true$, then $P_{ij} = \mathcal{B}T_{ij}$ if j is odd, and $P_{ij} = \mathcal{R}T_{ij}$ if j is even. If $\varphi(v_i) = false$, then $P_{ij} = \mathcal{B}F_{ij}$ if j is odd, and $P_{ij} = \mathcal{R}F_{ij}$ otherwise.

The path P obtained from the concatenation of paths P_j for $1 \leq j \leq m$ is an $\mathcal{F}$-valid path from s_1 to t_m.

Now let us suppose that there is an $\mathcal{F}$-valid path P from s_1 to t_m. According to Claim 7, for any $1 \leq i \leq n$, for any $1 \leq j \leq m$, P passes through $true_{ij}$ and $true'_{ij}$ or for any $1 \leq j \leq m$, P passes through $false_{ij}$ and $false'_{ij}$. Let us then consider the truth assignment φ of Φ such that for each $1 \leq i \leq n$:

– If P uses $true_{ij}$ and $true'_{ij}$ in all rows $1 \leq j \leq m$, then $\varphi(v_i) = true$.
– If P uses $false_{ij}$ and $false'_{ij}$ in all rows $1 \leq j \leq m$, then $\varphi(v_i) = false$.

Thanks to Claim 7, φ is a valid truth assignment. We need to prove that φ satisfies Φ. According to Claim 6, for each row $1 \leq j \leq m$, P contains an $\mathcal{F}_j$-valid path P_j from s_j to t_j. Each path P_j uses a green or a brown edge as stated by Claim 3. With respect to the possible ways to use a green or a brown edge which are stated in Claim 2, the use of a brown edge in P_j forces P_j (and hence P) to use, for a variable v_i that appears positively in C_j, the vertices $true_{ij}$ and $true'_{ij}$. Similarly, the use of a green edge in P_j forces P_j (and hence P) to use, for a variable v_i that appears negatively in C_j, the vertices $false_{ij}$ and $false'_{ij}$. This means that for each clause C_j, for one of the variables that appear in C_j which we denote v_i, $\varphi(v_i) = true$ (resp. $\varphi(v_i) = false$) if v_i appears positively (resp. negatively) in C_j. Thus, the truth assignment φ satisfies Φ. $\square$

Lemma 2. *The PAFT problem is NP-complete in planar graphs with maximum degree 4.*

Proof. The graph G used in the reduction of the proof of Lemma 1 is planar and each vertex of G has either degree 8, degree 5 or degree at most 4. Let $\mathcal{E}$ be the planar embedding of G that is obtained by embedding the smaller gadgets as in Figs. 1, 2 and 3. We transform G into a planar graph G' with maximum degree 4 and an associated set of forbidden transitions $\mathcal{F}'$ such that finding an $\mathcal{F}$-valid path in G is equivalent to finding an $\mathcal{F}'$-valid path in G'. The transformation goes as follows.

Vertices of degree 5. The vertices of degree 5 in G are the vertices s_{1j} and t_{nj} for j odd, and s_{nj} and t_{1j} for j even. We transform these to vertices of degree 3 as follows. For j odd, we remove the two blue edges incident to s_{1j} and the two red edges incident to t_{nj}. For j is even, we remove the two red edges incident to s_{nj} and the two blue edges incident to t_{1j}. The removed edges do not belong to any allowed transitions around the vertices s_{1j}, t_{nj}, s_{nj} and t_{1j}. Therefore, their removal does not affect the elementary $\mathcal{F}$-valid paths from s_1 to t_m.

Vertices of degree 8. We replace each vertex v of degree 8 by a gadget g_v of maximum degree 4. Gadget g_v is designed such that it can be crossed at most once by a path of G' and only if the edges used to enter and leave g_v correspond to an allowed transition around v. According to its corresponding transitions and to the planar embedding of its adjacent vertices, a vertex v of degree 8 of G is of one of 3 different types. We present in what follows these types as well as the corresponding gadget g_v for each type.

Type 1: The edges incident to v are $E(v) = \{e, e', f, f', g, g', h, h'\}$ and the allowed transitions around v are $A(v) = \{\{e, e'\}, \{f, f'\}, \{g, g'\}, \{h, h'\}\}$. The edges incident to v appear in the planar embedding $\mathcal{E}$ as presented in Fig. 4a. In the graph G, v is one of the vertices x_{ij} or z_{ij}, $1 \leq i \leq n$, $1 \leq j \leq m$.

In this case, g_v is built as follows. For each $\alpha \in E(v)$, a vertex v_α is created. For each $\{\alpha, \beta\} \in A(v)$, vertices v_α and v_β are linked with a path $P_{\alpha\beta}$ of length four. The four paths $P_{\alpha\beta}$, $\{\alpha, \beta\} \in A(v)$ are pairwise intersecting in distinct vertices as illustrated in Fig. 4b. The allowed transitions in g_v are the transitions of the paths $P_{\alpha\beta}$, $\{\alpha, \beta\} \in A(v)$. Now to replace v with g_v

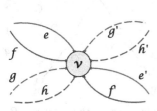

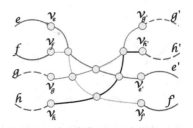

(a) A vertex v of degree 8 and allowed transitions $A(v) = \{\{e,e'\},\{f,f'\}, \{g,g'\},\{h,h'\}\}$ (edges are ordered as in the planar embedding $\mathcal{E}$ of G)

(b) Gadget g_v: the paths $P_{ee'}$, $P_{ff'}$, $P_{gg'}$, and $P_{hh'}$ are respectively the pink, yellow, green and black paths. Transitions around vertices v_α, $\alpha \in E(v)$ and transitions of paths $P_{ee'}$, $P_{ff'}$, $P_{gg'}$, and $P_{hh'}$ are allowed

Fig. 4. Type 1

in G, each edge $\alpha \in E(v)$ of G is linked to vertex v_α of g_v. The gadget g_v is planar, and edges $\alpha \in E(v)$ are connected to it in the same "order" they are connected to v in the planar embedding $\mathcal{E}$ of G as illustrated in Fig. 4. Note the gadget g_v cannot be crossed twice with the same path, i.e., no path has two subpaths in g_v, otherwise the path is not simple. Moreover, g_v can be crossed if and only if the edges used to enter and leave form an allowed transition around v.

Type 2: The edges incident to v are $E(v) = \{e,e',f,f',g,g',h,h'\}$ and the allowed transitions around v are $A(v) = \{\{e,e'\},\{f,f'\},\{g,g'\},\{h,h'\}\}$. The edges incident to v appear in the planar embedding $\mathcal{E}$ as presented in Fig. 5a. In the graph G, v is one of the vertices $true_{ij}$, $true'_{ij}$, $false_{ij}$, $false'_{ij}$, or y_{ij}, $1 \le i \le n$, $1 \le j \le m$.

In this case, g_v is built as follows. For each $\alpha \in E(v)$, a vertex v_α is created. For each $\{\alpha,\beta\} \in A(v)$, vertices v_α and v_β are linked with a path $P_{\alpha\beta}$ of length 7. Each two of the four paths $P_{\alpha\beta}$, $\{\alpha,\beta\} \in A(v)$ intersect in two different vertices as illustrated in Fig. 5b. The allowed transitions in g_v are the transitions of the paths $P_{\alpha\beta}$, $\{\alpha,\beta\} \in A(v)$. Now to replace v with g_v in G, each edge $\alpha \in E(v)$ of G is linked to vertex v_α of g_v. The gadget g_v is planar, and edges $\alpha \in E(v)$ are connected to it in the same "order" they were connected to v in the planar embedding $\mathcal{E}$ of G as illustrated in Fig. 5. Note that the gadget g_v cannot be crossed twice with the same path, i.e., no path has two subpaths in g_v, otherwise the path is not simple. Moreover, g_v can be crossed if and only if the edges used to enter and leave form an allowed transition around v.

Type 3: The edges incident to v are $E(v) = \{e,e',f,f',g,g',h,h'\}$ and the allowed transitions around v are $A(v) = \{\{e,e'\},\{e,g'\},\{f,f'\},\{f,h'\}, \{g,g'\},\{g,e'\},\{h,h'\},\{h,f'\}\}$. The edges incident to v appear in the planar embedding $\mathcal{E}$ as depicted in Fig. 6a. In the graph G, v is one of the vertices s_{ij}, $1 \le i \le n$, $1 \le j \le m$.

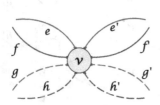
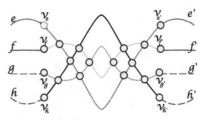

(a) A vertex v of degree 8 and allowed transitions $A(v) = \{\{e, e'\}, \{e, f'\}, \{f, f'\}, \{f, e'\}, \{g, g'\}, \{g, h'\}, \{h, h'\}, \{h, g'\}\}$ (edges are ordered as in the planar embedding $\mathcal{E}$ of G)

(b) Gadget g_v: the paths $P_{ee'}$, $P_{ff'}, P_{gg'}$, and $P_{hh'}$ are respectively the pink, yellow, green and black paths. Transitions around vertices v_α, $\alpha \in E(v)$ and transitions of paths $P_{ee'}$, $P_{ff'}, P_{gg'}$, and $P_{hh'}$ are allowed

Fig. 5. Type 2

In this case, g_v is built as follows. Let $A_1(v) = \{\{e, e'\}, \{f, f'\}, \{g, g'\}, \{h, h'\}\}$. For each $\{\alpha, \beta\} \in A_1(v)$, vertices v_α and v_β are linked with a path $P_{\alpha\beta}$ of length 6. Each two of the paths $P_{\alpha\beta}$ intersect in one or two vertices as illustrated in Fig. 6b. Furthermore, we add two edges linking the paths $P_{ee'}$ and $P_{gg'}$, and $P_{ff'}$ and $P_{hh'}$, respectively, depicted by orange edges in Fig. 6b. Now to replace v with g_v in G, each edge $\alpha \in E(v)$ of G is linked to vertex v_α of g_v. The gadget g_v is planar, and edges $i \in E(v)$ are connected to it in the same "order" they were connected to v in the planar embedding $\mathcal{E}$ of G as illustrated in Fig. 6.

For $\{\alpha, \beta\} \in A(v) \setminus A_1(v)$, let α' and β' be the vertices such that $\{\alpha, \alpha'\}, \{\beta, \beta'\} \in A_1(v)$. We define the path $P_{\alpha\beta}$, as the path which starts

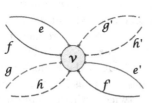
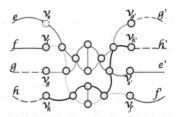

(a) A vertex v of degree 8 and allowed transitions $A(v) = \{\{e, e'\}, \{e, g'\}, \{f, f'\}, \{f, h'\}, \{g, g'\}, \{g, e'\}, \{h, h'\}, \{h, f'\}\}$ (edges are ordered as in the planar embedding $\mathcal{E}$ of G)

(b) Gadget g_v: the paths $P_{ee'}$, $P_{ff'}, P_{gg'}$, and $P_{hh'}$ are respectively the pink, yellow, green and black paths. transitions around vertices v_α, $\alpha \in E(v)$, transitions of the paths $P_{ee'}$, $P_{ff'}, P_{gg'}$, and $P_{hh'}$, and transitions containing the orange edge are allowed

Fig. 6. Type 3 (Color figure online)

at v_α, uses a subpath of $P_{\alpha\alpha'}$, an orange edge and then a subpath of $P_{\beta\beta'}$ and then ends at v_r.

Note that the gadget g_v cannot be crossed twice with the same path, i.e., no path has two subpaths in g_v, otherwise the path is not simple. Moreover, g_v can be crossed if and only if the edges used to enter and leave form an allowed transition around v.

The graph G' obtained from G after applying the transformations described above is planar and has maximum degree 4. The set of forbidden transitions $\mathcal{F}'$ consists of the transitions of the set $\mathcal{F}$ and the forbidden transitions of the gadgets g_v as described above.

Let us now suppose that there is an $\mathcal{F}$-valid path P from s to t in G. Let P' be the s-t-path of G' constructed as follows: P' uses all edges used by P. Furthermore, if P uses a degree 8 vertex with a transition $\{e, e'\}$ then P' uses e, subpath $P_{ee'}$, and e'. The path P' is $\mathcal{F}'$-valid.

Now, let us suppose that there is an $\mathcal{F}'$-valid path P' from s to t in G'. If P' only uses edges from G, then it can be considered as an $\mathcal{F}$-valid path from s to t in G. If P' uses an edge that is not in G, then P' crosses one of the gadgets g_v. Gadgets g_v are designed such that they can be crossed at most once by a path because otherwise the path is not simple, furthermore, the edges used to enter and leave the gadget form an allowed transition in G. This implies that the intersection of each gadget g_v with the path P' is either the empty set or exactly one subpath of P'. If it is a subpath then the edges surrounding it in P' form an allowed transition in G. We can then remove the edges of P' that do not belong to G to obtain an $\mathcal{F}$-valid path P in G.

Theorem 1. *The problem of finding a path avoiding forbidden transitions is NP-complete in grids.*

Proof. To prove the theorem we use the notion of planar grid embedding [16]. A planar grid embedding of a graph G is a mapping Q of G into a grid such that Q maps each vertex of G into a distinct vertex of the grid and each edge e of G into a path of the grid $Q(e)$ whose endpoints are mappings of vertices linked by e. For every pair $\{e, e'\}$ of edges of G, the corresponding paths $Q(e)$ and $Q(e')$ have no points in common, except, possibly, the endpoints. It has been proved in [17], that if $G = (V, E)$ is a planar graph such that $|V| = n$ and $\Delta \leq 4$, then a planar grid embedding of G in a grid of size at most $9n^2$ can be found in polynomial-time. Let us consider an instance of the problem of finding a path avoiding forbidden transitions in a planar graph $G = (V, E)$ of maximum degree at most 4 with a set of allowed transitions $\mathcal{A}$ ($\mathcal{A} = E \times E \setminus \mathcal{F}$). Let Q be a grid planar embedding of G into a grid K of size at most $O(|V|^2)$. Finding a PAFT between two nodes s and t in G with the set $\mathcal{A}$ is equivalent to finding a PAFT between the nodes $Q(s)$ and $Q(t)$ in K with the set of allowed transitions $\mathcal{A}'$ defined such that for each $e \in E$, all the transitions in the path $Q(e)$ are allowed, and for each $\{e, e'\} \in \mathcal{A}$, the pair of edges of $Q(e)$ and $Q(e')$, which share a vertex, is an allowed transition. □

4   Parameterized Complexity

On the positive side, by using dynamic programming on a tree-decomposition of the input graph, we prove that the problem is FPT when the parameter is the sum of the treewidth and the maximum degree.

A *tree-decomposition* of a graph [14] is a way to represent G by a family of subsets of its vertex-set organized in a tree-like manner and satisfying some connectivity property. The *treewidth* of G measures the proximity of G to a tree. More formally, a tree decomposition of $G = (V, E)$ is a pair $(T, \mathcal{X})$ where $\mathcal{X} = \{X_t | t \in V(T)\}$ is a family of subsets, called *bags*, of V, and T is a tree, such that:

1. $\bigcup_{t \in V(T)} X_t = V$, and
2. for any edge $uv \in E$, there is a bag X_t (for some node $t \in V(T)$) containing both u and v, and
3. for any vertex $v \in V$, the set $\{t \in V(T) | v \in X_t\}$ induces a subtree of T.

The *width* of a tree-decomposition $(T, \mathcal{X})$ is $max_{t \in V(T)} |X_t| - 1$ and its *size* is order $|V(T)|$ of T. The treewidth of G, denoted by $tw(G)$, is the minimum width over all possible tree-decompositions of G.

Theorem 2 proves that when the treewidth of the graph is bounded, the PAFT can be solved in polynomial time. Complete proof of the theorem can be found in [12].

Theorem 2. *The PAFT Problem is FPT when parameterized by $k + \Delta$ where k is the treewidth and Δ is the maximum degree. In particular, given a tree-decomposition of width k of the input graph, PAFT can be solved in time $O(k\Delta^2(3k\Delta)^{2k}n)$.*

The Algorithm uses dynamic programming techniques and its key idea is similar to the one used to find a Hamiltonian cycle in graphs with bounded treewidth [3].

In more details, let $G = (V, E)$ be a graph with bounded treewidth k, $\mathcal{F}$ a set of forbidden transitions, and s and t two vertices of V. Let $(T, \mathcal{X})$ be a tree-decomposition of width k of G rooted in an arbitrary node. Let $G[A]$ be the subgraph of G induced by the set of vertices A. For each $u \in V(T)$, we denote by X_u, T_u and V_u the set of vertices of the bag corresponding to u, the subtree of T rooted at u, and the set of vertices of the bags corresponding to the nodes of T_u, respectively.

If there exists an $\mathcal{F}$-valid path P from s to t, then the intersection of this path with $G[V_u]$ for a node $u \in T$ consists of a set of paths avoiding forbidden transitions each having both endpoints in X_u. If $t \in V_u$, then one of the paths has only one endpoint in X_u. With respect to the parts of path P that are in $G[V_u]$, vertices in X_u can be partitioned into three subsets X_u^0, X_u^1, and X_u^2 which are the vertices of degree 0, 1 and 2 in $P \cap G[V_u]$, respectively. Furthermore, a matching M of X_u^1 decides which vertices are endpoints of the same subpath and a set of edges S defines which edges incident to X_u^1 are in P. For each node $u \in T$ and each subproblem $(X_u^0, X_u^1, X_u^2, M, S)$ where (X_u^0, X_u^1, X_u^2) is a partition of X_u, M is a matching of X_u^1 and S is a set of edges incident to the vertices of X_u^1, we need to check if there exists a set of paths avoiding forbidden transitions in V_u such that their endpoints are exactly X_u^1 according to the matching M, they contain the edges of S and the vertices of X_u^2 and they do not contain any vertex of X_u^0. For each node, we will need to solve at most $3^{k+1}k^k\Delta^k$ subproblems; there are at most 3^{k+1} possible partitions of the vertices of X_u into the 3 different sets, k^k possible matchings for a set of k elements and Δ possible edges for each element of X_u^1.

Complexity for planar graphs. The complexity of many dynamic programming algorithms on graphs with bounded treewidth can be improved for planar graphs using their planarity and properties. A speedup can usually be achieved, by using instead of the classical tree-decompositions, new decompositions that have been defined for planar graphs such as the *sphere cut branch decomposition* [6] and the *geometric tree-decomposition* [5]. In particular, these techniques have been used to improve the complexity of the algorithm which solves the Hamiltonian cycle problem in planar graphs from $O((3k)^k)$ to $O(6^k)$, where k is the treewidth. This is primarily due to the fact that the number of matchings to consider in such special tree-decompositions is bounded by 2^{k+1} as proved in [6]. This is roughly due to the fact that, in these special tree-decompositions, the vertices of a bag form a Jordan curve. This curve does not intersect with the curves corresponding to the edges of the planar embedding of the graph. Since in the matching we need to define, two matched vertices correspond to the endpoints of a subpath and the subpaths cannot cross each other, the number of valid matchings is bounded by a Catalan number. Taking this into consideration, our algorithm can run in time $O(k\Delta^2(6\Delta)^{2k}n))$ for planar graphs.

5 Conclusion

We have proven that the problem of finding a path avoiding forbidden transitions is NP-complete even in well-structured graphs as grids. We have also proved that PAFT can be solved in polynomial time when the treewidth is bounded. We believe that the PAFT is actually $W[1]$-hard when parameterized by the treewidth. Future work might focus on proving this conjecture. Another interesting direction in the study of PAFT could be to consider the optimization problem where the objective is to find a path with minimum number of forbidden transitions and to investigate possible approximation solutions.

References

1. Ahmed, M., Lubiw, A.: Shortest paths avoiding forbidden subpaths. Networks **61**(4), 322–334 (2013)
2. Bernstein, G., Lee, Y., Gavler, A., Martensson, J.: Modeling WDM wavelength switching systems for use in GMPLS and automated path computation. IEEE Opt. Commun. Networking **1**(1), 187–195 (2009)

3. Bodlaender, H.L.: Dynamic programming on graphs with bounded treewidth. In: Lepistö, T., Salomaa, A. (eds.) Automata, Languages and Programming. LNCS, vol. 317, pp. 105–118. Springer, Heidelberg (1988)
4. Chen, Y., Hua, N., Wan, X., Zhang, H., Zheng, X.: Dynamic lightpath provisioning in optical WDM mesh networks with asymmetric nodes. Photon Netw. Commun. 25(3), 166–177 (2013)
5. Dorn, F.: Dynamic programming and planarity: improved tree-decomposition based algorithms. Discrete Appl. Math. 158(7), 800–808 (2010). Third Workshop on Graph Classes, Optimization, and Width Parameters Eugene, Oregon, USA, October 2007
6. Dorn, F., Penninkx, E., Bodlaender, H.L., Fomin, F.V.: Efficient exact algorithms on planar graphs: exploiting sphere cut branch decompositions. In: Brodal, G.S., Leonardi, S. (eds.) ESA 2005. LNCS, vol. 3669, pp. 95–106. Springer, Heidelberg (2005)
7. Gourvès, L., Lyra, A., Martinhon, C.A., Monnot, J.: Complexity of trails, paths and circuits in arc-colored digraphs. Discrete Appl. Math. 161(6), 819–828 (2013)
8. Gutiérrez, E., Medaglia, A.L.: Labeling algorithm for the shortest path problem with turn prohibitions with application to large-scale road networks. Ann. Oper. Res. 157(1), 169–182 (2008)
9. Hashiguchi, T., Tajima, K., Takita, Y., Naito, T.: Node-disjoint paths search in WDM networks with asymmetric nodes. In: 2011 15th International Conference on Optical Network Design and Modeling (ONDM), pp. 1–6 (2011)
10. Hsu, C.-C., Chen, D.-R., Ding, H.-Y.: An efficient algorithm for the shortest path problem with forbidden paths. In: Hua, A., Chang, S.-L. (eds.) ICA3PP 2009. LNCS, vol. 5574, pp. 638–650. Springer, Heidelberg (2009)
11. Jaumard, B., Kien, D.: Optimizing ROADM configuration in WDM networks. In: 2014 16th International Telecommunications Network Strategy and Planning Symposium (Networks), pp. 1–7 (2014)
12. Kanté, M.M., Moataz, F.Z., Momège, B., Nisse, N.: Finding Paths in Grids with Forbidden Transitions. Inria Sophia Antipolis, Univeristé Nice Sophia Antipolis, CNRS, Research report (2015)
13. Pugliese, L.D.P., Guerriero, F.: Shortest path problem with forbidden paths: the elementary version. Eur. J. Oper. Res. 227(2), 254–267 (2013)
14. Robertson, N., Seymour, P.D.: Graph minors. II. algorithmic aspects of tree-width. J. Algorithms 7(3), 309–322 (1986)
15. Szeider, S.: Finding paths in graphs avoiding forbidden transitions. Discrete Appl. Math. 126(2–3), 261–273 (2003)
16. Tamassia, R.: On embedding a graph in the grid with the minimum number of bends. SIAM J. Comput. 16(3), 421–444 (1987)
17. Valiant, L.G.: Universality considerations in VLSI circuits. IEEE Trans. Comput. 30(2), 135–140 (1981)
18. Villeneuve, D., Desaulniers, G.: The shortest path problem with forbidden paths. Eur. J. Oper. Res. 165(1), 97–107 (2005)

The Maximum Time of 2-neighbour Bootstrap Percolation in Grid Graphs and Parametrized Results

Thiago Marcilon[✉] and Rudini Sampaio

Dept. Computação, Universidade Federal do Ceará, Fortaleza, Brazil
{thiagomarcilon,rudini}@lia.ufc.br

Abstract. In 2-neighborhood bootstrap percolation on a graph G, an infection spreads according to the following deterministic rule: infected vertices of G remain infected forever and in consecutive rounds healthy vertices with at least two already infected neighbors become infected. Percolation occurs if eventually every vertex is infected. The maximum time $t(G)$ is the maximum number of rounds needed to eventually infect the entire vertex set. In 2013, it was proved by Benevides et al. [10] that $t(G)$ is NP-hard for planar graphs and that deciding whether $t(G) \geq k$ is polynomial time solvable for $k \leq 2$, but is NP-complete for $k \geq 4$. They left two open problems about the complexity for $k = 3$ and for planar bipartite graphs. In 2014, we solved the first problem [24]. In this paper, we solve the second one by proving that $t(G)$ is NP-complete even in grid graphs with maximum degree 3. We also prove that $t(G)$ is polynomial time solvable for solid grid graphs with maximum degree 3. Moreover, we prove that the percolation time problem is fixed parameter tractable with respect to the parameter treewidth $+ k$ and maximum degree $+ k$. Finally, we obtain polynomial time algorithms for several graphs with few P_4's, as cographs and P_4-sparse graphs.

Keywords: 2-Neighbor bootstrap percolation · Maximum percolation time · Grid graph · Fixed parameter tractability · Treewidth

1 Introduction

We consider a problem in which an infection spreads over the vertices of a connected simple graph G following a deterministic spreading rule in such a way that an infected vertex will remain infected forever. Given a set $S \subseteq V(G)$ of initially infected vertices, we build a sequence $S_0, S_1, S_2, \dots$ in which $S_0 = S$ and S_{i+1} is obtained from S_i using such spreading rule.

Under r-neighbor bootstrap percolation on a graph G, the spreading rule is a threshold rule in which S_{i+1} is obtained from S_i by adding to it the vertices of G which have at least r neighbors in S_i. We say that a set S infects a vertex v at time i if $v \in S_i \backslash S_{i-1}$. Let, for any set of vertices S and vertex v of G, $t_r(G, S, v)$ be the minimum t such that v belongs to S_t or, if there is no t such that v

© Springer-Verlag Berlin Heidelberg 2016
E.W. Mayr (Ed.): WG 2015, LNCS 9224, pp. 169–185, 2016.
DOI: 10.1007/978-3-662-53174-7_13

belongs to S_t, then $t_r(G, S, v) = \infty$. Also, we say that a set S_0 infects G, or that S_0 is a percolating set of G, if eventually every vertex of G becomes infected, that is, there exists a t such that $S_t = V(G)$. If S is a percolating set of G, then we define $t_r(G, S)$ as the minimum t such that $S_t = V(G)$. Also, define the *percolation time of* G as $t_r(G) = \max\{t_r(G, S) : S \text{ is a percolating set of } G\}$. In this paper, we shall focus on the case where $r = 2$ and in such case we omit the subscript of the notations $t_r(G, S)$ and $t_r(G)$. Also, from the notation $t(G, S)$, when the parameter G is clear from context, it will be omitted.

Bootstrap percolation was introduced by Chalupa, Leath and Reich [15] as a model for certain interacting particle systems in physics. Since then it has found applications in clustering phenomena, sandpiles [20], and many other areas of statistical physics, as well as in neural networks [1] and computer science [19].

There are two broad classes of questions one can ask about bootstrap percolation. The first, and the most extensively studied, is what happens when the initial configuration S_0 is chosen randomly under some probability distribution? For example, vertices are included in S_0 independently with some fixed probability p. One would like to know how likely percolation is to occur, and if it does occur, how long it takes. The answer to these questions is now well understood for various types of graphs [5,7,8,13,22].

The second broad class of questions is the one of extremal questions. For example, what is the smallest or largest size of a percolating set with a given property? The size of the smallest percolating set in the d-dimensional grid, $[n]^d$, was studied by Pete and a summary can be found in [6]. Morris [25] and Riedl [28] studied the maximum size of minimal percolating sets on the square grid $[n]^2$ and the hypercube $\{0, 1\}^d$, respectively, answering a question posed by Bollobás. However, the problem of finding the smallest percolating set is NP-hard even on subgraphs of the square grid [2] and it is APX-hard even for bipartite graphs with maximum degree four [17]. Moreover, it is hard [16] to approximate within a ratio $O(2^{\log^{1-\varepsilon} n})$, for any $\varepsilon > 0$, unless $NP \subseteq DTIME(n^{polylog(n)})$.

Another type of question is: what is the minimum or maximum time that percolation can take, given that S_0 satisfies certain properties? Recently, Przykucki [27] determined the precise value of the maximum percolation time on the hypercube $2^{[n]}$ as a function of n, and Benevides and Przykucki [11,12] have similar results for the square grid $[n]^2$, also answering a question posed by Bollobás. In particular, they have a polynomial time dynamic programming algorithm to compute the maximum percolation time on rectangular grids [11].

Here, we consider the decision version of the Percolation Time Problem, as stated below.

PERCOLATION TIME
Input: A graph G and an integer k.
Question: Is $t(G) \geq k$?

In 2013, Benevides et al. [10], among other results, proved that the Percolation Time Problem is polynomial time solvable for $k \leq 2$, but is NP-complete for $k \geq 4$ and, when restricted to bipartite graphs, it is NP-complete for $k \geq 7$.

Moreover, it was proved that the Percolation Time Problem is NP-complete for planar graphs. They left three open questions about the complexity for $k = 3$ in general graphs, the complexity for $3 \leq k \leq 6$ in bipartite graphs and the complexity for planar bipartite graphs.

In 2014, the first and the second questions were solved [24]: it was proved that the Percolation Time Problem is $O(mn^5)$-time solvable for $k = 3$ in general graphs and, when restricted to bipartite graphs, it is $O(mn^3)$-time solvable for $k = 3$, it is $O(m^2n^9)$-time solvable for $k = 4$ and it is NP-complete for $k \geq 5$.

In this paper, we solve the third question of [10]. We prove that the Percolation Time Problem is NP-complete for planar bipartite graphs. In fact, we prove a stronger result: the NP-completeness for grid graphs, which are induced subgraphs of grids, with maximum degree 3.

There are NP-hard problems in grid graphs which are polynomial time solvable for solid grid graphs. For example, the Hamiltonian cycle problem is NP-complete for grid graphs [23], but it is polynomial time solvable for solid grid graphs [30]. Motivated by the work of [11] for rectangular grids, we obtain in this paper a polynomial time algorithm for solid grid graphs with maximum degree 3.

Finally, we prove several complexity results for $t(G)$ in graphs with bounded maximum degree and bounded treewidth, some of which implies fixed parameter tractable algorithms for the Percolation Time Problem. Moreover, we obtain polynomial time algorithms for $(q, q - 4)$-graphs, for any fixed q, which are the graphs such that every subset of at most q vertices induces at most $q - 4$ P_4's. Cographs and P_4-sparse graphs are exactly the $(4, 0)$-graphs and the $(5, 1)$-graphs, respectively. These algorithms are fixed parameter tractable on the parameter q.

2 Percolation Time Problem in Grid Graphs with $\Delta = 3$

In this section, we prove that the Percolation Time Problem is NP-complete in grid graphs with maximum degree $\Delta = 3$. We also show that, when the graph is a grid graph with $\Delta = 3$ and $k = O(\log n)$, the Percolation Time Problem can be solved in polynomial time. But, first, let us define a S-infection path and, then, prove two lemmas that will be useful in the proofs.

Let $t(G, S, v)$ be the time where S infects v in G or, if S does not infect v in G, then let $t(G, S, v) = \infty$. Let S be a percolating set. A path $P = v_0, v_1, \ldots, v_n$ is a $\{S, G\}$-infection path if and only if, for all $0 \leq i \leq n - 1$, $t(G, S, v_i) < t(G, S, v_{i+1})$. In both notations, when the parameter G is clear from context, it will be omitted

Note that if $t(G, S, v) = k$ then there is a $\{S, G\}$-infection path $v_1, \ldots, v_k = v$, where each vertex v_i is such that $t(G, S, v_i) = i$. The next lemma, which is valid for every graph with maximum degree 3, is the main technical lemma of this section. Due to space restrictions, its proof will be omitted.

Lemma 1. *Let G be any connected graph with $\Delta = 3$ and k a non-negative integer. Then, $t(G) \geq k$ if and only if G has an induced path P where either all*

vertices in $V(P)$ have degree 3 and $|E(P)| \geq 2k - 2$ or all vertices in $V(P)$ have degree 3, except for one of his extremities, which has degree 2, and $|E(P)| \geq k-1$.

Before proving the NP-completeness result of this section, we use Lemma 1 to show that the Percolation Time Problem is polynomial time solvable for $k = O(\log n)$ when the graph has maximum degree 3.

Theorem 1. *If G is a graph with maximum degree 3, then deciding whether $t(G) \geq k$ can be done in polynomial time for $k = O(\log n)$.*

Proof (sketch of the proof). We can decide whether $t(G) \geq k$ by making use of a modified version of the depth-first search. This version of the depth-first search with maximum search depth l traverses all paths with $l + 1$ vertices starting from some vertex v. For each $v \in V(G)$, we will run this version of the depth-first search starting in v. If $d(v) = 2$, we run the modified depth-first search with maximum search depth $k - 1$. If $d(v) = 3$, we run the modified depth-first search with maximum search depth $2k - 2$. If there is a vertex v such that the depth-first search that starts in v finds a path that is an induced path, reaches the maximum depth and passes only by vertices of degree 3, except maybe for v, then, by Lemma 1, $t(G) \geq k$. Otherwise, $t(G) < k$.

Now, let us show that this algorithm runs in polynomial time. For each vertex v in G, there is at most $3 \cdot 2^{k-2}$ paths of length k in G that starts in v, for any k. In this case, since $k = O(\log n)$, there are at most $3 \cdot 2^{O(\log n)} = 3n^{O(1)}$ paths of length k in G that starts in v, which is a polynomial on n. Therefore, since the depth-first search traverses all paths with length equals to the maximum depth once for each vertex in $V(G)$, then our algorithm runs in time $O(n \cdot 2^k)$, which is polynomial in n since $k = O(\log n)$. ∎

Thus, if $k = O(\log n)$, we can find whether $t(G) \geq k$ in polynomial time for every graph G with $\Delta(G) = 3$. However, the following theorem states that the Percolation Time Problem is NP-complete, even when G is restricted to be a grid graph with $\Delta = 3$.

Theorem 2. *Deciding whether $t(G) \geq k$ is NP-complete when the input G is restricted to be a grid graph with $\Delta(G) \leq 3$.*

Proof (sketch of the proof). Clearly, the problem is in NP. To prove that the problem is also NP-hard, we obtained a reduction from the Longest Path problem with input restricted to be grid graphs with maximum degree 3. The Longest Path problem with input restricted to be grid graphs with maximum degree 3 is a NP-complete problem because the Hamiltonian Path Problem with input restricted to be grid graphs with maximum degree 3 is also NP-complete [26] and there is a trivial reduction from the Hamiltonian Path Problem to the Longest Path problem that does not change the input graph: G has an Hamiltonian Path if and only if G has a path greater or equal to $n - 1$.

Consider the following reduction from the Longest Path Problem's instance (G, k) where G is restricted to be a grid graph with maximum degree 3 to the

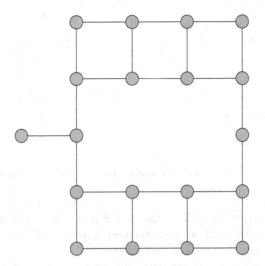

Fig. 1. Grid graph with $\Delta = 3$

Percolation Time Problem's instance $(G', 3k + 2)$ where G' is also a grid graph with maximum degree 3: Multiply the scale of the grid G by three. Each edge in G becomes a path in G' with 4 vertices where the vertices at the extremities are vertices that were originally in G. Let us call an *original vertex* the vertices in G' that were originally in G. After that, for each original vertex v, if $d(v) < 3$, add to G' $3 - d(v)$ vertices in any free position in the grid adjacent to v and link them to v. Thus, after we do that, each original vertex has degree 3 in G'. Henceforth, if a vertex in G' is not an original vertex at this point, then we will call it an *auxiliary vertex*. Note that each auxiliary vertex is adjacent to exactly one original vertex and each original vertex is adjacent to 3 auxiliary vertices.

After that, for each auxiliary vertex v, add a new vertex adjacent to v in the following manner: if the original neighbor of v is located above it, add a vertex adjacent to v at his left position, if there is not one there already, and link it to v. If the original neighbor of v is located below it, add a vertex adjacent to v at his right position, if there is not one there already, and link it to v. If the original neighbor of v is located at his left position, add a vertex adjacent to v at the position below it, if there is not one there already, and link it to v. If the original neighbor of v is located at his right position, add a vertex adjacent to v at the position above it, if there is not one there already, and link it to v. The Fig. 2 show how a 4×4 block will look like in G' before and after we add these vertices.

Then, for each auxiliary vertex v, if $d(v) = 2$, add a new vertex adjacent to v in the following position: if the original neighbor of v is at the left position of v, add a vertex adjacent to v at his right position. If the original neighbor of v is at the right position of v, add a vertex adjacent to v at his left position. If the original neighbor of v is below v, add a vertex adjacent to v above v. If the original neighbor of v is above v, add a vertex adjacent to v below v.

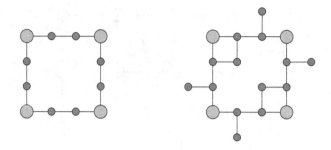

Fig. 2. 4×4 block before and after addition of the auxiliary vertices' neighbors

Thus, the construction of G' is finished. Since G is a grid graph and, every time an original vertex and an auxiliary vertex are in adjacent positions in the grid, they are linked, then G' is a grid graph.

Note that all original and auxiliary vertices have degree 3 and they are the only vertices that have degree 3. Let us call *corner vertex* all the vertices that have degree 2 in G'. Also, note that, for each corner vertex, there is exactly one original vertex at distance 2 of it, and, for each original vertex, there is exactly one corner vertex at distance 2 of it. This happens because each original vertex has degree exactly three. Let f be the bijective function that maps each original vertex to the corner vertex that is at distance 2 of it. The Fig. 3 shows the reduction applied to the grid graph of the Fig. 1. It is worth noting that, in G', a path P that has only original and auxiliary vertices and starts with an original vertex, it has length multiple of 3 if and only if it ends in an original vertex. Also, for each 3 consecutive vertices of this path, two are auxiliary vertices and one is an original vertex.

Now, let us prove that G has a path of length $\geq k$ if and only if $t(G') \geq 3k+2$.

Suppose that G is a grid graph with maximum degree 3 that has a path of length $\geq k$. Let us prove that $t(G') \geq 3k+2$. Since G has a path P of length $\geq k$, we have that G' has an induced path P of length $\geq 3k$ that passes by the same path that P passes, which implies that P passes only by original and auxiliary vertices. Note that, when an auxiliary vertex is in P, his auxiliary neighbor is also in P.

Let v and v' be the extremities of P and $f(v') = q'$. Since v is an original vertex, then let w be any auxiliary neighbor of v that is not in $V(P)$. Note that all neighbors of w, except v, are not in $V(P)$. Let r be the vertex auxiliary neighbor of v' that is in P and let P' be the induced path that we obtain from P by adding w, by removing v' and by adding all vertices in any smallest path between r and q', excluding r, that only have vertices not adjacent to w and passes only by original and auxiliary vertices. Since P is an induced path and we removed one vertex and added only one induced path that has either 1 or 3 vertices to create P', we have that P' is an induced path with length $\geq 3k+1$ where all of its vertices have degree 3, except for q', which has degree 2. Therefore, by Lemma 1, we have that $t(G') \geq 3k + 2$.

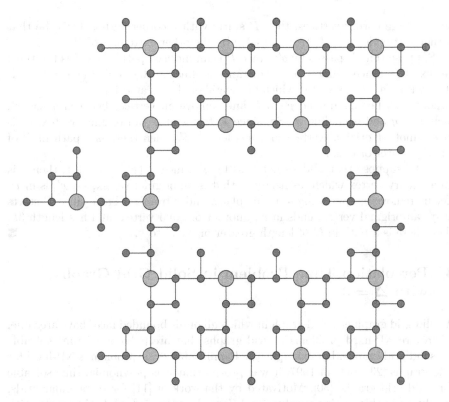

Fig. 3. Grid graph resulting from the reduction applied to the grid graph of the Fig. 1.

Now, suppose that G is a grid graph with maximum degree 3 such that, when we apply the reduction to G to create G', we have that $t(G') \geq 3k + 2$. Let us prove that G has a path of length $\geq k$. Since $t(G') \geq 3k + 2$, applying the Lemma 1, we have that G' has an induced path P where either all vertices in $V(P)$ have degree 3 and $|E(P)| \geq 6k + 2$ or all vertices in $V(P)$ have degree 3, except for one of his extremities, which has degree 2, and $|E(P)| \geq 3k + 1$.

Firstly, suppose that G' has an induced path P where all vertices in $V(P)$ have degree 3 and $|E(P)| \geq 6k + 2$. Since, the only vertices that have degree 3 are the original and auxiliary vertices and for each three consecutive vertices in P there is one original vertex and two auxiliary vertices, it is easy to see that P has at least $k + 1$ original vertices and, thus, there is a path in G of length at least k.

Finally, suppose that G' has an induced path P where all vertices in $V(P)$ have degree 3, except for one of his extremities, which has degree 2, and $|E(P)| \geq 3k+1$. It is enough to analyze the case $|E(P)| = 3k+1$ because, if $|E(P)| > 3k+1$, any subpath of P of length $3k + 1$ that starts at the extremity of P that have degree 2 is an induced path where all of his vertices have degree 3, except for one of his extremities, which has degree 2, and has length $3k + 1$. So, let us say that P starts in the vertex that has degree 2. Since the only vertices that have degree 2 are corner vertices, then P starts with a corner vertex. Let q be that corner vertex, let $q' = f^{-1}(q)$ and let v be the other extremity of P.

Suppose that P passes by q'. Since P is an induced path, then q' is the third vertex of P. Since q and q' are at distance 2 of each other and $|E(P)| = 3k + 1$, then v is an auxiliary vertex which his neighbor that is an original vertex, say v', is not in P. Let us append v' to P and remove all vertices between q and q', including q and excluding q'. So, since P starts at q', an original vertex, ends in v', another original vertex, and has length $3k$, then there is a path in G of length greater or equal to k.

Now, suppose that P does not pass by q'. Since $|E(P)| = 3k + 1$, then v is an auxiliary vertex which his neighbor that is an original vertex, say v', is in P. Let us remove q, appending q' in his place, and v from P. Thus, since P starts at q', an original vertex, ends in v', another original vertex, and has length $3k$, then there is a path in G of length greater or equal to k.   ∎

3   Percolation Time Problem in Solid Grid Graphs with $\Delta = 3$

A solid grid graph is a grid graph in which all of his bounded faces have area one. There are NP-hard problems in grid graphs that are polynomial time solvable for solid grid graphs. For example, the hamiltonian cycle problem is NP-hard for grid graphs [23], but, in 1997, it was proved that it is polynomial time solvable for solid grid graphs [30]. Motivated by the work of [11] for rectangular grids, we obtain in this section a polynomial time algorithm for solid grid graphs with maximum degree 3. However, the Percolation Time Problem for solid grid graphs with maximum degree 4 is still open.

Theorem 3. *For any solid grid graph G with $\Delta = 3$, $t(G)$ can be found in $O(n^2)$ time.*

Proof (sketch of the proof). If a solid grid graph has $\Delta = 3$, then, since it is $K_{1,4}$-free, it becomes a graph formed only by ladders L_k, which are grid graphs with dimensions $2 \times k$, and by paths, possibly linking these ladders by the vertices in their extremities. Let the extremities of a ladder be the four vertices that have only two neighbors in the ladder and let all the other vertices be the vertices internal to the ladder. In Fig. 4, there is an example of solid grid graph with $\Delta = 3$.

To find the percolation time of G, according to Lemma 1, it is enough to find both the longest induced path that starts with a degree 2 vertex and, then, passes only by vertices with degree 3, and the longest induced path that passes only by vertices with degree 3. Thus, since all of G's bounded faces have area one and, besides the ladders, G is composed only by paths, the only difficulty to calculate $t(G)$ is to find the longest induced paths in the ladders between any two extremities that passes only by vertices with degree 3. However, one can easily calculate the longest induced paths between any two extremities of a

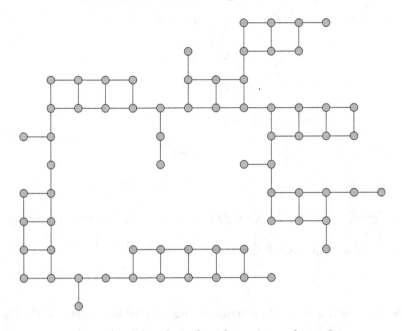

Fig. 4. A solid grid graph with maximum degree 3.

ladder L_k: if the two extremities are neighbors, the length of the longest induced paths between them is 1; if the two extremities are at distance $k-1$, the length of the longest induced paths between them is $(k-t) + 2 \cdot \lfloor (k-t+1)/4 \rfloor - 1 + t$; if the two extremities are at distance k, the length of the longest induced paths between them is $(k-t) + 2 \cdot \lfloor (k-t-1)/4 \rfloor + t$, where t is how many of the two others extremities have degree 2.

So, first, we will transform G in a weighted graph G' where G' is the same graph as G only with all the ladders replaced by weighted K_4's, where the weight of an edge between two vertices in a K_4 represents the length of a longest induced path between the corresponding extremities of the ladders in G that passes only by vertices with degree 3. The weight of all the other edges is 1. The Fig. 5 represents the transformation applied in the graph of the Fig. 4. Note that there is exactly one induced path between any two vertices in G', which length is equal to the longest induced path between the same two vertices in G. It is not hard to see that this transformation from G to G' can be done in linear time.

In Algorithm 1, let $w(u, v)$ be the weight of the edge (u, v). The algorithm, for each vertex $u \in V(G')$ such that $d_G(u) \geq 2$, calls the function LongestInducedPathFrom, which do a Depth-First Search to find the longest induced path in G' from u such that the last vertex is the only vertex in the path that either has degree ≤ 2, besides perhaps the vertex u, or is in the neighborhood of a vertex already in the path, and, then, it subtracts the length of the found path by one. This is necessary because a longest induced path from some vertex u in G can end in a vertex v internal to a ladder, but internal vertices of

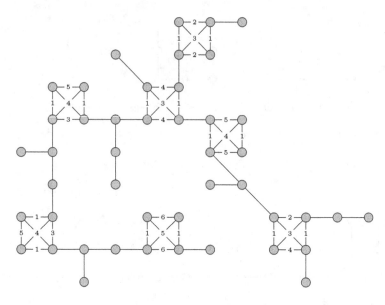

Fig. 5. The resulting graph of the transformation applied to the graph in the Fig. 4.

a ladder are not represented in G'. However, if that happens, since all vertices internal to a ladder have degree 3, then v must be adjacent to some vertex at the extremity of the ladder that has degree 2.

In any case, the resulting length corresponds to the length of the longest induced path in G beginning in u, which last vertex has degree 3 and is not in the neighborhood of any vertex already in the path. Then, it compares all these values, according to the Lemma 1, to find $t(G)$.

Since there is only one induced path between any two vertices in G', we have that the recursive function LongestInducedPathFrom takes the same time as any Depth-First Search algorithm. Thus, since $m = O(n)$, the Function LongestInducedPathFrom takes $O(n)$ time. Therefore, the Algorithm 1 takes $O(n^2)$ time.   ■

4   Percolation Time Problem in Graphs with Bounded Max Degree $\Delta \geq 4$

In Sect. 2 (Theorem 1), we proved that the Percolation Time Problem is polynomial time solvable in grid graphs with $\Delta(G) \leq 3$ for $k = O(\log n)$. In this section, we prove that this not happen for general graphs with fixed maximum degree $\Delta \geq 4$, unless P = NP.

Theorem 4. *Let $\Delta \geq 4$ be fixed. Deciding whether $t(G) \geq k$ is NP-complete for graphs with bounded maximum degree Δ and any $k \geq \log_{\Delta-2} n$.*

Algorithm 1. Algorithm that finds $t(G)$ for any solid grid graph G with $\Delta = 3$

Algorithm MaximumTimeSolidGridΔ3(G)

$\quad$ $G' = \text{Transform}(G)$

$\quad$ maxPercTime $= 0$

$\quad$ **forall the** $u \in V(G')$ *such that* $d_G(u) \geq 2$ **do**

$\quad\quad$ **if** $d_G(u) = 2$ **then**

$\quad\quad\quad$ percTimeU $= \text{LongestInducedPathFrom}(G', u) + 1$

$\quad\quad$ **else**

$\quad\quad\quad$ percTimeU $= \lfloor (\text{LongestInducedPathFrom}(G', u) + 2)/2 \rfloor$

$\quad\quad$ **if** *maxPercTime $<$ percTimeU* **then**

$\quad\quad\quad$ maxPercTime $=$ percTimeU

$\quad$ **return** *maxPercTime*

Proof (sketch of the proof). We obtain a reduction from the variation of the **SAT** problem where each clause has exactly three literals, each variable appears in at most four clauses [29].

Given M clauses $\mathcal{C} = \{C_1, \ldots, C_M\}$ on N variables $X = \{x_1, \ldots, x_N\}$ as an instance of **SAT**, we denote the three literals of C_i by $\ell_{i,1}$, $\ell_{i,2}$ and $\ell_{i,3}$. Note, since any variable can only appear in at most 4 clauses, that $N/3 \leq M \leq 4N/3$. So, first, let us show how to construct a graph G with maximum degree Δ. For each clause C_i of $\mathcal{C}$, add to G a gadget as the one in Fig. 6. Then, for each pair of literals $\ell_{i,a}, \ell_{j,b}$ such that one is the negation of the other, add a vertex $y_{(i,a),(j,b)}$ and link it to either $w_{i,a}^A$ or $w_{i,a}^B$ and either $w_{j,b}^A$ or $w_{j,b}^B$, but always respecting the restriction where each one of the vertices $w_{i,a}^A$, $w_{i,a}^B$, $w_{j,b}^A$ and $w_{j,b}^B$ can only have degree at most 4. Since each variable can appear in at most 4 clauses, it is always possible to do that. Let Y be the set of all vertices $y_{(i,a),(j,b)}$ created this way. Notice that $y = |Y| \leq 4N$.

Then, add the maximum full $(\Delta - 2)$-ary tree with root z such that the number of leaves is less than y and, then, add a new vertex adjacent vertex of degree one to each vertex in the tree. Let T be the tree that we have just added and t be the set of vertices that we just added. After that, link each leaf to at least one and at most $\Delta - 2$ vertices in Y. Thus, each vertex in the tree has degree Δ, except for the leaves, which have degree at most Δ, and z, which has degree $\Delta - 1$. Note that $t = 2 \cdot |V(T)| \leq 2 \cdot \frac{(\Delta-2)y-1}{\Delta-3} \leq 16N$ and $height(T) = \lceil \log_{\Delta-2} y \rceil$.

Let $c = (\Delta-2)^{-8}$, $\alpha = xc$, where $x = \max(41 + \lceil c \rceil, \lceil 1/c \rceil)$, $r = \lceil \log_{\Delta-2}(4N\alpha) \rceil - \lceil \log_{\Delta-2} y \rceil$ and $\beta = 4N\alpha/c - (39M + y + t + 2 + 2r)$. With some work, one can prove that both r and β are non-negative integers.

If $r > 0$, add a path with r vertices, link one end to z and let q be the other end. Also, add a new neighbor of degree one to each vertex that belongs to the path. Let P' be the set of vertices in this path and his neighbors of degree one. If $r = 0$, let $q = z$.

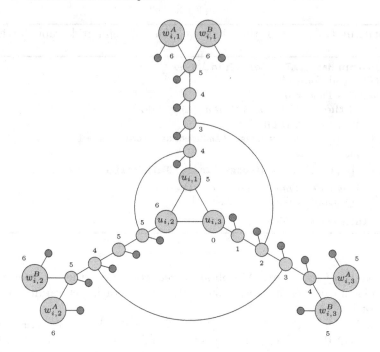

Fig. 6. Gadget with infection times for each clause C_i.

Finally, add a path with $\beta + 2$ vertices and link one end to q. Let P be the set of vertices in this path and let x be the vertex in P that is adjacent to q. By our construction, since $\Delta \geq 4$, we have that G is a graph in which every vertex has degree at most Δ.

Notice that any percolating set must contain a vertex of $\{u_{i,1}, u_{i,2}, u_{i,3}\}$ for each clause C_i of C and all vertices that have degree 1. Thus, following similar arguments presented in [24], it is possible to prove that the maximum percolation time of the vertex z is $height(T) + 7$ if and only if C is satisfiable, which implies that the maximum percolation time of the vertex x is $\lceil \log_{\Delta-2}(4N\alpha) \rceil + 8$ if and only if C is satisfiable.

Then, we have that C is satisfiable if and only if $t(G) \geq \lceil \log_{\Delta-2}(4N\alpha) \rceil + 8$, but, since $n = |V(G)| = 39M + y + t + 2 + 2r + \beta$, then $4N\alpha = c \cdot n$. Therefore, since $c = (\Delta - 2)^{-8}$, C is satisfiable if and only if $t(G) \geq \lceil \log_{\Delta-2} n \rceil$. ∎

5   Fixed Parameter Tractability of the Percolation Time Problem

We say that a decision problem is *fixed parameter tractable* (or just *fpt*) on some parameter Ψ if there exists an algorithm (called *fpt-algorithm*) that solves the problem in time $f(\Psi) \cdot n^{O(1)}$, where n is the size of the input and f is an arbitrary function depending only on the parameter Ψ.

In this section, we show that the Percolation Time Problem is fixed parameter tractable for the parameter $tw(G) + k$, for the parameter $\Delta(G) + k$ and for the parameter $q(G)$, where $tw(G)$ is the treewidth of the graph and $q(G)$ is the minimum $q \geq 4$ such that G is a $(q, q - 4)$-graph, which is a graph such that every subset of at most q vertices induces at most $q - 4$ P_4's (cographs have $q(G) = 4$ and P_4-sparse graphs have $q(G) = 5$). These theorems will imply that, if k is fixed, then deciding if $t(G) \geq k$ is linear time solvable for graphs with bounded treewidth or bounded maximum degree. Moreover, they will imply that determining the maximum percolation time is polynomial time solvable for $(q, q - 4)$-graphs with fixed q.

Theorem 5. *Percolation Time Problem is fixed parameter tractable with parameter $tw(G) + k$.*

Proof (sketch of the proof). A consequence of the Courcelle's theorem [18,21] states that, if a decision problem on graphs can be expressed in a Monadic Second Order (MSO) sentence φ, then this problem is fixed parameter tractable in the parameter $tw(G) + |\varphi|$. Moreover, the running time is linear on the size of the input. The Percolation Time Problem can be expressed by the following MSO-sentence:

$$maxtime_k := \exists w, X_0, X_1, \ldots, X_k \; \forall x \left(x \in X_k \right) \wedge \left(\bigwedge_{0 \leq i < k} (x \in X_i \rightarrow x \in X_{i+1}) \right) \wedge$$

$$\wedge \left(\bigwedge_{0 \leq i < k} (x \in X_{i+1} \backslash X_i) \rightarrow \exists y, z(Exy \wedge Exz \wedge (y \in X_i) \wedge (z \in X_i)) \right) \wedge \left(w \in X_k \backslash X_{k-1} \right),$$

where X_i represents the set of vertices infected at time i, Exy is true if xy is an edge (and false, otherwise) and $\wedge$ is the *and* operator. This MSO sentence asserts that all vertices are infected in time k, that a vertex infected in time i remains infected in time $i + 1$, that a vertex infected in time $i + 1$, but not infected in time i, has two neighbors infected in time i, and that there exists a vertex w infected in time k but not infected in time $k - 1$.   ∎

Theorem 6. *Percolation Time Problem is fixed parameter tractable with parameter $\Delta(G) + k$. Moreover, for fixed Δ, the Percolation Time Problem is polynomial time solvable in graphs with bounded maximum degree Δ for $k = \log_\Delta O(\log n)$, if $\Delta \geq 4$, and for $k = O(\log n)$, if $\Delta = 3$.*

Proof (sketch of the proof). Let $\Delta = \Delta(G)$ and let $u \in V(G)$. Then $|N_{\leq k}(u)| \leq \Delta^k$ and, consequently, the power set $2^{N_{\leq k}(u)}$ has $2^{|N_{\leq k}(u)|} \leq 2^{\Delta^k}$ sets. We claim that $t(G) \geq k$ if and only if there is a vertex u and a percolating set $S \supseteq N_{\geq k}(u)$ such that $t(G, S, u) = k$.

If $t(G) \geq k$, then there is a percolating set S' that infects some vertex u at time k. In [24], it was proved that, given a graph G, a set $Q \subseteq V(G)$ and a vertex $z \in V(G) \backslash S$, if $t(G, Q, w) \geq k$, then $t(G, Q, w) \geq t(G, Q \cup \{z\}, w) \geq k$, for any k and any $w \in N_{\geq k}(z)$. Then, applying this result once for each vertex in $N_{\geq k}(u)$, the percolating set $S = S' \cup N_{\geq k}(u)$ infects u also at time k.

On the other hand, if there is a percolating set $S \supseteq N_{\geq k}(u)$ such that $t(G, S, u) = k$, for some vertex u, then, trivially, $t(G) \geq k$. Then the claim is true.

Therefore, since for each vertex u and set $S' \subseteq N_{\leq k-1}(u)$, it takes $O(km)$ time to know whether the set $S' \cup N_{\geq k}(u)$ infects u at time k, this equivalence gives us an algorithm that decides whether $t(G) \geq k$ in time $n \cdot O(m + km \cdot 2^{\Delta^k}) = O(2^{\Delta^k} k\Delta \cdot n^2)$, since $m = O(\Delta n)$. Notice that, if $k = \log_\Delta O(\log n)$, then the time is polynomial in n. Moreover, if $\Delta = 3$, by Theorem 1, we are done.   ∎

Finally, we prove the fixed parameter tractability for the parameter $q(G)$. In 2014, Campos et al. [14] proved that determining the minimum percolating set is fixed parameter tractable on the parameter $q(G)$. Here, we prove the following.

Theorem 7. *Percolation Time Problem is fixed parameter tractable on parameter $q(G)$. Moreover, $t(G) \leq q(G) + 3$ for every graph G.*

To prove this theorem, we use a graph decomposition, called *primeval decomposition*, which is based on some graph operations: *union, join, spider* and *p-component*. Below we define these operations and present the lemmas used to obtain the maximum percolation time. Because of space restrictions, we omit the proofs.

The *union* $G = G_1 \cup G_2$ of two graph G_1 and G_2 is the graph such that $V(G) = V(G_1) \cup V(G_2)$ and $E(G) = E(G_1) \cup E(G_2)$. The *join* $G = G_1 \vee G_2$ is the graph obtained from $G_1 \cup G_2$ by joining every vertex of G_1 to every vertex of G_2. A *spider* (R, K, S) is a graph $G = (R \cup K \cup S, E)$ such that $K = \{k_1, \ldots, k_p\}$ and $S = \{s_1, \ldots, s_p\}$, for $p \geq 2$, induce a clique and a stable set, respectively; either s_i is adjacent to k_j if and only if $i = j$ (a thin spider), or s_i is adjacent to k_j if and only if $i \neq j$ (a thick spider); and every vertex of R is adjacent to each vertex of K and non-adjacent to each vertex of S.

Lemma 2 (union, join and spider). *Let G, G_1 and G_2 be graphs. If $G = G_1 \cup G_2$, then $t(G) = \max\{t(G_1), t(G_2)\}$. If $G = G_1 \vee G_2$ and G_1 and G_2 have at least two vertices each, then $t(G) \leq 3$. If $G = G_1 \vee G_2$ and G_2 has exactly one vertex, then $t(G) = diameter(G_1) + 1$, if either G_1 is disconnected or contains three vertices u, v, w such that $dist(u, w) = dist(v, w) = diameter(G_1)$ and there is no neighbor of u and v in a minimum path from u to w; otherwise, $t(G) = diameter(G_1)$. If G is a spider, then $t(G) \leq 3$. In all the three last cases, $t(G)$ can be found in the worst case in $O(mn^2)$ time.*

A graph is *p-connected* if, for every partition of the vertex set into two parts A and B, there is a crossing P_4 (with vertices of A and B). A p-connected graph is *separable* if it has a particular bipartition (H_1, H_2) such that every crossing P_4 $wxyz$ satisfies $x, y \in H_1$ and $w, z \in H_2$ (such a bipartition is unique [3]). A *p-component* of a graph is a maximal p-connected subgraph.

Given an arbitrary graph G' and a separable p-connected graph H with separation (H_1, H_2), let $G' \uplus H$ be the graph obtained from $G' \cup H$ by joining every vertex of G' to every vertex of H_1 and and to no vertex of H_2. Note that every spider $(\emptyset, K, S)$ is a separable p-connected graph with separation (K, S).

In [3], it was proved an important structural result for $(q, q-4)$-graphs. If G is a $(q, q-4)$-graph, then either $G = G' \cup G''$, or $G = G' \vee G''$, or $G = G' \uplus H$, or G has less than q vertices, where G' and G'' are $(q, q-4)$-graphs and H is either a spider (R, K, S) with $R = \emptyset$, or a separable p-connected $(q, q-4)$-graph with less than q vertices. This characterization leads to a graph decomposition, called *primeval decomposition*, which can be obtained in linear time $O(m+n)$ [4,9].

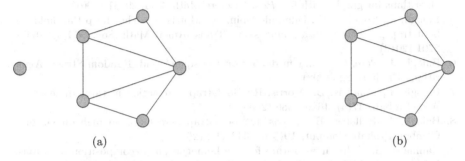

(a)   (b)

Fig. 7. The two possibilities for the graph $H^+[C]$.

Lemma 3 (p-component). *Let G' be a $(q, q-4)$-graph and let H be a separable p-connected $(q, q-4)$-graph with separation (H_1, H_2). If $G = G' \uplus H$, then $t(G) = t(H^+)$, where H^+ is the graph obtained from H by adding a set C of $x = \min\{|V(G')|, 6\}$ new vertices linked to all vertices of H_1, not linked to any vertices in H_2, and, if $x \leq 6$, let $H^+[C]$ and G' be isomorphic graphs. If $x \geq 7$:*

- *if G' is a clique then let $H^+[C]$ be a clique K_x;*
- *if G' is a stable set then let $H^+[C]$ also be a stable set;*
- *Otherwise, let $H^+[C]$ be isomorphic to the graph in Fig. 7(a), if G' is not connected, and let $H^+[C]$ be isomorphic to the graph in Fig. 7(b), if G' is connected.*

As a consequence, if H has less than q vertices (fixed $q \geq 4$), since $|V(H^+)| \leq q + 5$, then $t(G) \leq q + 3$ and $t(G)$ can be obtained in constant time $\leq 2^q q$ (by checking all subsets of vertices of H^+). The two lemmas above, together with the primeval decomposition of $(q, q-4)$-graphs, imply a polynomial time algorithm to determine the maximum percolation time of a $(q, q-4)$-graph, for fixed q, in $O(mn^2)$ time. This also implies that the Percolation Time Problem is fixed parameter tractable for the parameter $q(G)$.

References

1. Amini, H.: Bootstrap percolation in living neural networks. J. Stat. Phys. **141**(3), 459–475 (2010)
2. Araújo, R., Sampaio, R., Santos, V., Szwarcfiter, J.: The convexity of induced paths of order three and applications: complexity aspects. Discrete Appl. Math. (2015, to appear)
3. Babel, L., Olariu, S.: On the structure of graphs with few P_4's. Discrete Appl. Math. **84**, 1–13 (1998)
4. Babel, L., Kloks, T., Kratochvíl, J., Kratsch, D., Müller, H., Olariu, S.: Efficient algorithms for graphs with few P_4's. Discrete Math. **235**, 29–51 (2001)
5. Balogh, J., Bollobás, B., Duminil-Copin, H., Morris, R.: The sharp threshold for bootstrap percolation in all dimensions. Trans. Amer. Math. Soc. **364**(5), 2667–2701 (2012)
6. Balogh, J., Pete, G.: Random disease on the square grid. Random Struct. Algorithms **13**, 409–422 (1998)
7. Balogh, J., Bollobás, B., Morris, R.: Bootstrap percolation in three dimensions. Ann. Probab. **37**(4), 1329–1380 (2009)
8. Balogh, J., Bollobás, B., Morris, R.: Bootstrap percolation in high dimensions. Combin. Probab. Comput. **19**(5–6), 643–692 (2010)
9. Baumann, S.: A linear algorithm for the homogeneous decomposition of graphs. Report No. M-9615, Zentrum Mathematik. Technische Universität München (1996)
10. Benevides, F., Campos, V., Dourado, M.C., Sampaio, R.M., Silva, A.: The maximum time of 2-neighbour bootstrap percolation: algorithmic aspects. In: The Seventh European Conference on Combinatorics, Graph Theory and Applications, Series CRM, vol. 16, pp. 135–139. Scuola Normale Superiore (2013)
11. Benevides, F., Przykucki, M.: Maximum percolation time in two-dimensional bootstrap percolation. Submitted (2014). http://arxiv.org/abs/1310.4457v1
12. Benevides, F., Przykucki, M.: On slowly percolating sets of minimal size in bootstrap percolation. Electron. J. Comb. **20**(2), 46 (2013)
13. Bollobás, B., Holmgren, C., Smith, P.J., Uzzell, A.J.: The time of bootstrap percolation with dense initial sets. Ann. Probab. **42**(4), 1337–1373 (2014)
14. Campos, V., Sampaio, R., Silva, A., Szwarcfiter, J.: Graphs with few P4s under the convexity of paths of order three. Discrete Appl. Math. (2015, to appear). http://dx.doi.org/10.1016/j.dam.2014.05.005
15. Chalupa, J., Leath, P.L., Reich, G.R.: Bootstrap percolation on a Bethe lattice. J. Phys. C **12**(1), 31–35 (1979)
16. Chen, N.: On the approximability of influence in social networks. SIAM J. Discrete Math. **23**(3), 1400–1415 (2009)
17. Coelho, E.M.M., Dourado, M.C., Sampaio, R.M.: Inapproximability results for graph convexity parameters. In: Kaklamanis, C., Pruhs, K. (eds.) WAOA 2013. LNCS, vol. 8447, pp. 97–107. Springer, Heidelberg (2014)
18. Courcelle, B., Makowsky, J., Rotics, U.: On the fixed parameter complexity of graph enumeration problems definable in monadic second order logic. Discrete Appl. Math. **108**, 23–52 (2001)
19. Dreyer, P.A., Roberts, F.S.: Irreversible k-threshold processes: graph-theoretical threshold models of the spread of disease and of opinion. Discrete Appl. Math. **157**(7), 1615–1627 (2009)
20. Fey, A., Levine, L., Peres, Y.: Growth rates and explosions in sandpiles. J. Stat. Phys. **138**, 143–159 (2010)

21. Flum, J., Grohe, M.: Parameterized Complexity Theory. Springer, Heidelberg (2010)
22. Holroyd, A.E.: Sharp metastability threshold for two-dimensional bootstrap percolation. Probab. Theor. Relat. Fields **125**(2), 195–224 (2003)
23. Itai, A., Papadimitriou, C.H., Szwarcfiter, J.L.: Paths in grid graphs. SIAM J. Comput. **11**, 676–686 (1982)
24. Marcilon, T., Nascimento, S., Sampaio, R.: The maximum time of 2-neighbour bootstrap percolation: complexity results. In: Kratsch, D., Todinca, I. (eds.) WG 2014. LNCS, vol. 8747, pp. 372–383. Springer, Heidelberg (2014)
25. Morris, R.: Minimal percolating sets in bootstrap percolation. Electron. J. Comb. **16**(1), 20 (2009)
26. Papadimitriou, C.H., Vazirani, U.V.: On two geometric problems related to the travelling salesman problem. J. Algorithms **5**(2), 231–246 (1984)
27. Przykucki, M.: Maximal percolation time in hypercubes under 2-bootstrap percolation. Electron. J. Comb. **19**(2), 41 (2012)
28. Riedl, E.: Largest minimal percolating sets in hypercubes under 2-bootstrap percolation. Electron. J. Comb. **17**(1), 13 (2010)
29. Tovey, C.A.: A simplified NP-complete satisfiability problem. Discrete Appl. Math. **8**(1), 85–89 (1984)
30. Umans, C., Lenhart, W.: Hamiltonian cycles in solid grid graphs. In: Proceedings of the 38th Annual Symposium on Foundations of Computer Science (FOCS), 1997, Miami, USA, pp. 496–505. IEEE Computer Society, Washington, DC (1997)

Design and Analysis

Minimum Eccentricity Shortest Paths in Some Structured Graph Classes

Feodor F. Dragan and Arne Leitert[(✉)]

Department of Computer Science, Kent State University, Kent, OH, USA
{dragan,aleitert}@cs.kent.edu

Abstract. We investigate the *Minimum Eccentricity Shortest Path* problem in some structured graph classes. It asks for a given graph to find a shortest path with minimum eccentricity. Although it is NP-hard in general graphs, we demonstrate that a minimum eccentricity shortest path can be found in linear time for distance-hereditary graphs (generalizing the previous result for trees) and in $\mathcal{O}(n^3 m)$ time for chordal graphs.

1 Introduction

The *Minimum Eccentricity Shortest Path* problem asks for a given graph $G = (V, E)$ to find a shortest path P such that for each other shortest path Q, $\mathrm{ecc}_G(P) \leq \mathrm{ecc}_G(Q)$ holds. Here, the eccentricity of a set $S \subseteq V$ in G is $\mathrm{ecc}_G(S) = \max_{u \in V} d_G(u, S)$. This problem was introduced in [7]. It may arise in determining a "most accessible" speedy linear route in a network and can find applications in communication networks, transportation planning, water resource management and fluid transportation. It was also shown in [6,7] that a minimum eccentricity shortest path plays a crucial role in obtaining the best to date approximation algorithm for a minimum distortion embedding of a graph into the line. Specifically, every graph G with a shortest path of eccentricity r admits an embedding f of G into the line with distortion at most $(8r + 2) \, \mathrm{ld}(G)$, where $\mathrm{ld}(G)$ is the minimum line-distortion of G (see [7] for details). Furthermore, if a shortest path of G of eccentricity r is given in advance, then such an embedding f can be found in linear time.

Those applications motivate investigation of the Minimum Eccentricity Shortest Path problem in general graphs and in particular graph classes. Fast algorithms for it will imply fast approximation algorithms for the minimum line distortion problem. Existence of low eccentricity shortest paths in structured graph classes will imply low approximation bounds for those classes. For example, all AT-free graphs (hence, all interval, permutation, cocomparability graphs) enjoy a shortest path of eccentricity at most 1 [4], all convex bipartite graphs enjoy a shortest path of eccentricity at most 2 [6].

In [7], the Minimum Eccentricity Shortest Path problem was investigated in general graphs. It was shown that its decision version is NP-complete (even for graphs with vertex degree at most 3). However, there are efficient approximation

© Springer-Verlag Berlin Heidelberg 2016
E.W. Mayr (Ed.): WG 2015, LNCS 9224, pp. 189–202, 2016.
DOI: 10.1007/978-3-662-53174-7_14

algorithms: a 2-approximation, a 3-approximation, and an 8-approximation for the problem can be computed in $\mathcal{O}(n^3)$ time, in $\mathcal{O}(nm)$ time, and in linear time, respectively. Furthermore, a shortest path of minimum eccentricity r in general graphs can be computed in $\mathcal{O}(n^{2r+2}m)$ time. Paper [7] initiated also the study of the Minimum Eccentricity Shortest Path problem in special graph classes by showing that a minimum eccentricity shortest path in trees can be found in linear time. In fact, every diametral path of a tree is a minimum eccentricity shortest path.

In this paper, we design efficient algorithms for the Minimum Eccentricity Shortest Path problem in distance-hereditary graphs and in chordal graphs. We show that the problem can be solved in linear time for distance-hereditary graphs (generalizing the previous result for trees) and in $\mathcal{O}(n^3m)$ time for chordal graphs.

Note that our Minimum Eccentricity Shortest Path problem is close but different from the *Central Path* problem in graphs introduced in [13]. It asks for a given graph G to find a path P (not necessarily shortest) such that any other path of G has eccentricity at least $\mathrm{ecc}_G(P)$. The Central Path problem generalizes the Hamiltonian Path problem and therefore is NP-hard even for chordal graphs [12]. Our problem is polynomial time solvable for chordal graphs.

2   Notions and Notations

All graphs occurring in this paper are connected, finite, unweighted, undirected, loopless and without multiple edges. For a graph $G = (V, E)$, we use $n = |V|$ and $m = |E|$ to denote the cardinality of the vertex set and the edge set of G. $G[S]$ denotes the *induced subgraph* of G with the vertex set S.

The *length* of a path from a vertex v to a vertex u is the number of edges in the path. The *distance* $d_G(u, v)$ of two vertices u and v is the length of a shortest path connecting u and v. The distance between a vertex v and a set $S \subseteq V$ is defined as $d_G(v, S) = \min_{u \in S} d_G(u, v)$. The *eccentricity* $\mathrm{ecc}_G(v)$ of a vertex v is $\max_{u \in V} d_G(u, v)$. For a set $S \subseteq V$, its eccentricity is $\mathrm{ecc}_G(S) = \max_{u \in V} d_G(u, S)$. If no ambiguity arises, we will omit the subscript G. For a vertex pair s, t, a shortest (s, t)-path P has *minimal eccentricity*, if there is no shortest (s, t)-path Q with $\mathrm{ecc}(Q) < \mathrm{ecc}(P)$. Two vertices x and y are called *mutually furthest* if $d_G(x, y) = \mathrm{ecc}(x) = \mathrm{ecc}(y)$. A vertex u is k-*dominated* by a vertex v (by a set $S \subset V$), if $d_G(u, v) \leq k$ ($d_G(u, S) \leq k$, respectively).

The *diameter* of a graph G is $\mathrm{diam}(G) = \max_{u, v \in V} d_G(u, v)$. The diameter $\mathrm{diam}_G(S)$ of a set $S \subseteq V$ is defined as $\max_{u, v \in S} d_G(u, v)$. A pair of vertices x, y of G is called a *diametral pair* if $d_G(x, y) = \mathrm{diam}(G)$. In this case, every shortest path connecting x and y is called a *diametral path*.

For a vertex $v \in V$, $N(v) = \{u \in V \mid uv \in E\}$ is called the *open neighborhood*, and $N[v] = N(v) \cup \{v\}$ the *closed neighborhood* of v. $N^r[v] = \{u \in V \mid d_G(u, v) \leq r\}$ denotes the *disk* of radius r around vertex v. Additionally, $L_r^{(v)} = \{u \in V \mid d_G(u, v) = i\}$ denotes the vertices with distance r from v. For two vertices u and v, $I(u, v) = \{w \mid d_G(u, v) = d_G(u, w) + d_G(w, v)\}$

is the *interval* between u and v. The set $S_i(s,t) = L_i^{(s)} \cap I(u,v)$ is called a *slice* of the interval from u to v. For any set $S \subseteq V$ and a vertex v, $\Pr(v,S) = \{u \in S \mid d_G(u,v) = d_G(v,S)\}$ denotes the *projection* of v on S.

A *chord* in a path is an edge connecting two non-consecutive vertices of the path. A set of vertices S is a *clique* if all vertices in S are pairwise adjacent. A graph is *chordal* if every cycle with at least four vertices has a chord. A graph is *distance-hereditary* if the distances in any connected induced subgraph are the same as they are in the original graph. For more definitions of these classes and relations between them see [2].

3   A Linear-Time Algorithm for Distance-Hereditary Graphs

Distance-hereditary graphs can be defined as graphs where each chordless path is a shortest path [10]. Several interesting characterizations of distance-hereditary graphs in terms of metric and neighborhood properties, and forbidden configurations were provided by BANDELT and MULDER [1], and by D'ATRI and MOSCARINI [5]. The following proposition lists the basic information on distance-hereditary graphs that is needed in what follows.

Proposition 1 ([1,5]). *For a graph G the following conditions are equivalent:*

(1) G is distance-hereditary;

(2) The house, domino, gem (see Fig. 1) and the cycles C_k of length $k \geq 5$ are not induced subgraphs of G;

(3) For an arbitrary vertex x of G and every pair of vertices $u,v \in L_k^{(x)}$, that are in the same connected component of the graph $G[V \setminus L_{k-1}^{(x)}]$, we have
$$N(v) \cap L_{k-1}^{(x)} = N(u) \cap L_{k-1}^{(x)}.$$

(4) (4-point condition) For any four vertices u,v,w,x of G at least two of the following distance sums are equal: $d_G(u,v) + d_G(w,x)$; $d_G(u,w) + d_G(v,x)$; $d_G(u,x) + d_G(v,w)$. If the smaller sums are equal, then the largest one exceeds the smaller ones at most by 2.

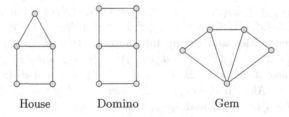

House   Domino   Gem

Fig. 1. Forbidden induced subgraphs in a distance-hereditary graph.

As a consequence of statement (3) of Proposition 1 we get.

Corollary 1. *Let* $P := P(s,t)$ *be a shortest path in a distance-hereditary graph* G *connecting vertices* s *and* t, *and* w *be an arbitrary vertex of* G. *Let* a *be a vertex of* $\Pr(w,P)$ *that is closest to* s, *and let* b *be a vertex of* $\Pr(w,P)$ *that is closest to* t. *Then* $d_G(a,b) \leq 2$ *and there must be a vertex* w' *in* G *adjacent to both* a *and* b *and at distance* $d_G(w,P) - 1$ *from* w.

As a consequence of statement (4) of Proposition 1 we get.

Corollary 2. *Let* x,y,v,u *be arbitrary vertices of a distance-hereditary graph* G *with* $v \in I(x,u)$, $u \in I(y,v)$, *and* $d_G(u,v) > 1$, *then* $d_G(x,y) = d_G(x,v) + d_G(v,u) + d_G(u,y)$. *That is, if two shortest paths share ends of length at least 2, then their union is a shortest path.*

Proof. Consider distance sums $S_1 := d_G(x,v) + d_G(u,y)$, $S_2 := d_G(x,y) + d_G(u,v)$ and $S_3 := d_G(x,u) + d_G(v,y)$. Since $d_G(x,u) + d_G(v,y) = d_G(x,v) + d_G(u,y) + 2\,d_G(u,v)$, we have $S_3 > S_1$. Then, either $S_2 = S_3$ or $S_1 = S_2$ and $S_3 - S_1 \leq 2$. If the latter is true, then $2 \geq S_3 - S_1 = d_G(x,v) + d_G(u,y) + 2\,d_G(u,v) - d_G(x,v) - d_G(u,y) = 2\,d_G(v,u) > 2$ and a contradiction arises. Thus, $S_2 = S_3$ and we get $d_G(x,y) = d_G(x,v) + d_G(v,u) + d_G(u,y)$. $\qquad\square$

Lemma 1. *Let* x,y *be a diametral pair of vertices of a distance-hereditary graph* G, *and* k *be the minimum eccentricity of a shortest path in* G. *If for some shortest path* $P = P(x,y)$, *connecting* x *and* y, $\mathrm{ecc}(P) > k$ *holds, then* $\mathrm{diam}(G) = d_G(x,y) \geq 2k$. *Furthermore, if* $d_G(x,y) = 2k$ *then there is a shortest path* P^* *between* x *and* y *with* $\mathrm{ecc}(P^*) = k$.

Proof. Consider a vertex v with $d_G(v,P) > k$. Let x' be a vertex of $\Pr(v,P)$ closest to x, and y' be a vertex of $\Pr(v,P)$ closest to y. By Corollary 1, $d_G(x',y') \leq 2$ and there must be a vertex v' in G adjacent to both x' and y' and at distance $d_G(v,P) - 1$ from v. Let $P(x,x')$ and $P(y',y)$ be subpaths of P connecting vertices x,x' and vertices y,y', respectively. Consider also an arbitrary shortest path $Q(v,v')$ connecting v and v' in G. By choices of x' and y', no chords in G exist in paths $P(x,x') \cup Q(v',v)$ and $P(y,y') \cup Q(v',v)$. Hence, those paths are shortest in G. Since x,y is a diametral pair, we have $d_G(x,x') + d_G(x',y') + d_G(y',y) = d_G(x,y) \geq d_G(x,v) = d_G(x,x') + 1 + d_G(v',v)$. That is, $d_G(y',y) \geq d_G(v',v) + 1 - d_G(x',y')$. Similarly, $d_G(x',x) \geq d_G(v',v) + 1 - d_G(x',y')$. Combining both inequalities and taking into account that $d_G(v,v') \geq k$, we get $d_G(x,y) = d_G(x,x') + d_G(x',y') + d_G(y',y) \geq 2k + 2 - d_G(x',y') \geq 2k$. Furthermore, we have $d_G(x,y) \geq 2k + 1$ if $d_G(x',y') = 1$ and $d_G(x,y) \geq 2k + 2$ if $d_G(x',y') = 0$. Also, if $d_G(x,y) = 2k$ then $d_G(x',y') = 2$, $d_G(v,v') = k$, $d_G(x,x') = d_G(y,y') = k - 1$ and $d_G(v,x) = d_G(v,y) = 2k$.

Now assume that $d_G(x,y) = 2k$. Consider sets $S = \{w \in V \mid d_G(x,w) = d_G(y,w) = k\}$ and $F_{x,y} = \{u \in V \mid d_G(u,x) = d_G(u,y) = 2k\}$. Let $c \in S$ be a vertex of S that k-dominates the maximum number of vertices in $F_{x,y}$. Consider a shortest path P^* connecting vertices x and y and passing through vertex c.

We will show that $ecc(P^*) = k$. Let x' (y') be the neighbor of c in subpath of P^* connecting c with x (with y, respectively).

Assume there is a vertex v in G such that $d_G(v, P^*) > k$. As in the first part of the proof, one can show that $d_G(v, x') = d_G(v, y') = k + 1$, i.e., $x', y' \in Pr(v, P^*)$ and $d_G(v, P^*) = k + 1$. Furthermore, $d_G(v, x) = d_G(v, y) = 2k$, i.e., $v \in F_{x,y}$. Also, vertex v', that is adjacent to x', y' and at distance k from v, must belong to S. Since $d_G(v, c) > k$ but $d_G(v, v') = k$, by choice of c, there must exist a vertex $u \in F_{x,y}$ such that $d_G(u, c) \le k$ and $d_G(u, v') > k$. Since $d_G(u, y) = d_G(u, x) = 2k$, $d_G(u, c)$ must equal k and both $d_G(u, x')$ and $d_G(u, y')$ must equal $k + 1$.

Since $d_G(v, u) \le \text{diam}(G) = 2k$ and $d_G(v, y') = d_G(v, x') = k + 1 = d_G(u, x') = d_G(u, y')$, we must have a chord between vertices of a shortest path $P(v, v')$ connecting v with v' and vertices of a shortest path $P(u, c)$ connecting u with c. If no chords exist or only chord cv' is present, then $d_G(v, u) \ge 2k + 1$, contradicting with $\text{diam}(G) = 2k$. So, consider a chord ab with $a \in P(v, v')$, $b \in P(u, c)$, $ab \ne cv'$, and $d_G(a, v') + d_G(b, c)$ is minimum. We know that $d_G(a, v') = d_G(b, c)$ must hold since $d_G(u, v') > k = d_G(u, c)$ and $d_G(v, c) > k = d_G(v, v')$. To avoid induced cycles of length $k \ge 5$, $d_G(a, v') = d_G(b, c) = 1$ must hold. But then, vertices a, b, c, x', v' form either an induced cycle C_5, when c and v' are not adjacent, or a house, otherwise. Note that, by distance requirements, edges bv', ca, bx', and ax' are not possible.

Contradictions obtained show that such a vertex v with $d_G(v, P^*) > k$ is not possible, i.e., $ecc(P^*) = k$. □

Lemma 2. *In every distance-hereditary graph there is a minimum eccentricity shortest path $P(s, t)$ where s and t are two mutually furthest vertices.*

Proof. Let k be the minimum eccentricity of a shortest path in G. Let $Q := Q(s, t) = (s = v_0, v_1, \ldots, v_i, \ldots, v_q = t)$ be a shortest path of G of eccentricity k with maximum q, that is, among all shortest paths with eccentricity k, Q is a longest one. Assume, without loss of generality, that t is not a vertex most distant from s. Let $i \le q$ be the smallest index such that subpath $Q(s, v_i) = (v_0, v_1, \ldots, v_i)$ of Q has also the eccentricity k. By choice of i, there must exist a vertex v in G which is k-dominated only by vertex v_i of $Q(s, v_i)$, i.e., $Pr(v, Q(s, v_i)) = \{v_i\}$ and $d_G(v, Q(s, v_i)) = k$. Let $P(v, v_i)$ be an arbitrary shortest path of G connecting v with v_i. By choice of i, no vertex of $P(v, v_i) \setminus \{v_i\}$ is adjacent to a vertex of $Q(s, v_i) \setminus \{v_i\}$. Hence, path obtained by concatenating $Q(s, v_i)$ with $P(v_i, v)$ is chordless and, therefore, shortest in G, and has eccentricity k, too. Note that v is now a most distant vertex from s, i.e., $d_G(s, v) = ecc(s)$. Since $d_G(s, v) > d_G(s, t)$, a contradiction with maximality of q arises. □

The main result of this section is the following.

Theorem 1. *Let x, y be a diametral pair of vertices of a distance-hereditary graph G, and k be the minimum eccentricity of a shortest path in G. Then, there is a shortest path P between x and y with $ecc(P) = k$.*

Proof. We may assume that for some shortest path P' connecting x and y, $\mathrm{ecc}(P') > k$ holds (otherwise, there is nothing to prove). Then, by Lemma 1, we have $d(x, y) \geq 2k$.

Let $Q := Q(s, t) = (s = v_0, v_1, \ldots, v_i, \ldots, v_q = t)$ be a shortest path of G of eccentricity k such that s and t are two mutually furthest vertices (see Lemma 2). Consider projections of x and y to Q. We distinguish between three cases: $\mathrm{Pr}(x, Q)$ is completely on the left of $\mathrm{Pr}(y, Q)$ in Q; $\mathrm{Pr}(x, Q)$ and $\mathrm{Pr}(y, Q)$ have a common vertex w; and the remaining case (see Corollary 1) when $\mathrm{Pr}(x, Q) = \{v_{i-1}, v_{i+1}\}$ and $\mathrm{Pr}(y, Q) = \{v_i\}$ for some index i.

Case 1: $\mathrm{Pr}(x, Q)$ is completely on the left of $\mathrm{Pr}(y, Q)$ in Q.

Let x' be a vertex of $\mathrm{Pr}(x, Q)$ closest to t and y' a vertex of $\mathrm{Pr}(y, Q)$ closest to s. Consider an arbitrary shortest path $P(x, x')$ of G connecting vertices x and x', an arbitrary shortest path $P(y', y)$ of G connecting vertices y' and y, and a subpath $Q(x', y')$ of $Q(s, t)$ between vertices x' and y'. We claim that the path P of G obtained by concatenating $P(x, x')$ with $Q(x', y')$ and then with $P(y', y)$ is a shortest path of eccentricity k.

Indeed, by choice of x', no edge connecting a vertex in $P(x, x') \setminus \{x'\}$ with a vertex in $Q(x', y') \setminus \{x'\}$ can exist in G. Similarly, no edge connecting a vertex in $P(y', y) \setminus \{y'\}$ with a vertex in $Q(x', y') \setminus \{y'\}$ can exist in G. Since we also have $d_G(x, y) \geq 2k$, $d_G(x, Q) \leq k$ and $d_G(y, Q) \leq k$, no edge connecting a vertex in $P(y', y) \setminus \{y'\}$ with a vertex in $P(x, x') \setminus \{x'\}$ can exist in G. Hence, chordless path $P = P(x, x') \cup Q(x', y') \cup P(y', y)$ is a shortest path of G.

Consider now an arbitrary vertex v of G. We want to show that $d_G(v, P) \leq k$. Since $\mathrm{ecc}(Q) = k$, $d_G(v, Q) \leq k$. Consider the projection of v to Q. We may assume that $\mathrm{Pr}(v, Q) \cap Q(x', y') = \emptyset$ and, without loss of generality, that vertices of $\mathrm{Pr}(v, Q)$ are closer to s than vertex x'. Let v' be a vertex of $\mathrm{Pr}(v, Q)$ closest to x'. As before, by choices of v' and y', paths $P(y, y') \cup Q(y', v')$ and $P(v, v') \cup Q(y', v')$ are chordless and, therefore, are shortest paths of G (here $P(v, v')$ is an arbitrary shortest path of G connecting v with v'). Since $d_G(v', y') \geq 2$, by Corollary 2, $d_G(v, y) = d_G(v, v') + d_G(v', y') + d_G(y', y)$. Hence, from $d_G(x, y) \geq d_G(y, v)$, $d_G(x, y) = d_G(x, x') + d_G(x', y)$ and $d_G(v, y) = d_G(v, x') + d_G(x', y)$, we obtain $d_G(v, x') \leq d_G(x, x') \leq k$.

Case 2: $\mathrm{Pr}(x, Q)$ and $\mathrm{Pr}(y, Q)$ have a common vertex w.

In this case, we have $d_G(x, y) \leq d_G(x, w) + d_G(y, w) \leq k + k = 2k$. Earlier we assumed also that $d_G(x, y) \geq 2k$. Hence, $\mathrm{diam}(G) = d_G(x, y) = 2k$ and the statement of the theorem follows from Lemma 1.

Case 3: Remaining case when $\mathrm{Pr}(x, Q) = \{v_{i-1}, v_{i+1}\}$ and $\mathrm{Pr}(y, Q) = \{v_i\}$ for some index i.

In this case, we have $d_G(x, y) \leq d_G(x, v_{i-1}) + 1 + d_G(v_i, y) \leq 2k + 1$. By Lemma 1, we can assume that $\mathrm{diam}(G) = d_G(x, y) = 2k + 1$, i.e., $d_G(x, v_{i-1}) = d_G(x, v_{i+1}) = d_G(v_i, y) = k$.

Let $Q(s, v_{i-1})$ and $Q(t, v_{i+1})$ be subpaths of Q connecting vertices s and v_{i-1} and vertices t and v_{i+1}, respectively. Pick an arbitrary shortest path $P(y, v_i)$ connecting y with v_i. Since no chords are possible between $Q(s, v_i) \setminus \{v_i\}$ and

$P(y, v_i) \setminus \{v_i\}$ and between $Q(t, v_i) \setminus \{v_i\}$ and $P(y, v_i) \setminus \{v_i\}$, we have $d_G(y, t) = d_G(y, v_i) + d_G(v_i, t) = k + d_G(v_i, t)$ and $d_G(y, s) = d_G(y, v_i) + d_G(v_i, s) = k + d_G(v_i, s)$. Inequalities $d_G(x, y) \geq d_G(y, t)$ and $d_G(x, y) \geq d_G(y, s)$ imply $d_G(v_{i+1}, t) \leq d_G(v_{i+1}, x) = k$ and $d_G(v_{i-1}, s) \leq d_G(v_{i-1}, x) = k$. If both $d_G(v_{i+1}, t)$ and $d_G(v_{i-1}, s)$ equal k, then $d_G(s, t) = 2k + 2$ contradicting with $\operatorname{diam}(G) = 2k + 1$. Hence, we may assume, without loss of generality, that $d_G(v_{i-1}, s) \leq k - 1$. We will show that shortest path $P := P(x, v_{i+1}) \cup P(v_i, y)$ has eccentricity k (here, $P(x, v_{i+1})$ is an arbitrary shortest path of G connecting x with v_{i+1}).

Consider a vertex v in G and assume that $\Pr(v, Q)$ is strictly contained in $Q(t, v_{i+1})$. Denote by v' the vertex of $\Pr(v, Q)$ that is closest to s. Let $P(v, v')$ be an arbitrary shortest path connecting v and v'. As before, $P(v, v') \cup Q(v', s)$ is a chordless path and therefore $d_G(v, s) = d_G(v, v_{i+1}) + d_G(v_{i+1}, s)$. Since t is a most distant vertex from s, $d_G(s, v) \leq d_G(s, t)$. Hence, $d_G(v, v_{i+1}) + d_G(v_{i+1}, s) = d_G(s, v) \leq d_G(s, t) = d_G(s, v_{i+1}) + d_G(v_{i+1}, t)$, i.e., $d_G(v, v_{i+1}) \leq d_G(v_{i+1}, t) \leq k$.

Consider a vertex v in G and assume now that $\Pr(v, Q)$ is strictly contained in $Q(s, v_{i-1})$. Denote by v' the vertex of $\Pr(v, Q)$ that is closest to t. Let $P(v, v')$ be an arbitrary shortest path connecting v and v'. Again, $P(v, v') \cup Q(v', t)$ is a chordless path and therefore $d_G(v, t) = d_G(v, v_i) + d_G(v_i, t)$. Since s is a most distant vertex from t, $d_G(t, v) \leq d_G(s, t)$. Hence, $d_G(v, v_i) + d_G(v_i, t) = d_G(t, v) \leq d_G(s, t) = d_G(s, v_i) + d_G(v_i, t)$, i.e., $d_G(v, v_i) \leq d_G(v_i, s) \leq k$.

Thus, all vertices of G are k-dominated by $P(x, v_{i+1}) \cup P(v_i, y)$.   □

It is known [8] that a diametral pair of a distance-hereditary graph can be found in linear time. Hence, according to Theorem 1, to find a shortest path of minimum eccentricity in a distance-hereditary graph in linear time, one needs to efficiently extract a best eccentricity shortest path for a given pair of end-vertices. In what follows, we demonstrate that, for a distance-hereditary graph, such an extraction can be done in linear time as well.

We will need few auxiliary lemmas.

Lemma 3. *In a distance-hereditary graph G, for each pair of vertices s and t, if x is on a shortest path from v to $\Pi_v = \Pr(v, I(s, t))$ and $d_G(x, \Pi_v) = 1$, then $\Pi_v \subseteq N(x)$.*

Proof. Let p and q be two vertices in Π_v and $d_G(v, \Pi_v) = r$. By statement (3) of Proposition 1, $N(p) \cap L_{r-1}^{(v)} = N(q) \cap L_{r-1}^{(v)}$. Thus, each vertex x on a shortest path from v to Π_v with $d_G(x, \Pi_v) = 1$ (which is in $N(p) \cap L_{r-1}^{(v)}$ by definition) is adjacent to all vertices in Π_v, i.e., $\Pi_v \subseteq N(x)$.   □

Lemma 4. *In a distance-hereditary graph G, let $S_i(s, t)$ and $S_{i+1}(s, t)$ be two consecutive slices of an interval $I(s, t)$. Each vertex in $S_i(s, t)$ is adjacent to each vertex in $S_{i+1}(s, t)$.*

Proof. Consider statement (3) of Proposition 1 from perspective of t. Thus, $S_i(s, t) \subseteq N(v)$ for each vertex $v \in S_{i+1}(s, t)$. Additionally, from perspective of s, $S_{i+1}(s, t) \subseteq N(u)$ for each vertex $u \in S_i(s, t)$.   □

Lemma 5. *In a distance-hereditary graph G, if a projection $\Pi_v = \Pr(v, I(s,t))$ intersects two slices of an interval $I(s,t)$, each shortest (s,t)-path intersects Π_v.*

Proof. Because of Lemma 3, there is a vertex x with $N(x) \supseteq \Pi_v$ and $d_G(v,x) = d_G(v, \Pi_v) - 1$. Thus, Π_v intersects at most two slices of interval $I(s,t)$ and those slices have to be consecutive, otherwise x would be a part of the interval. Let $S_i(s,t)$ and $S_{i+1}(s,t)$ be these slices. Note that $d_G(s,x) = i + 1$. Thus, by statement (3) of Proposition 1, $N(x) \cap S_i(s,t) = N(u) \cap S_i(s,t)$ for each $u \in S_{i+1}(s,t)$. Therefore, $S_i(s,t) \subseteq \Pi_v$, i.e., each shortest path from s to t intersects Π_v.   □

From the lemmas above, we can conclude that, for determining a shortest (s,t)-path with minimal eccentricity, a vertex v is only relevant if $d_G(v, I(s,t)) = \mathrm{ecc}(I(s,t))$ and the projection of v on the interval $I(s,t)$ only intersects one slice. Algorithm 1 uses this.

Algorithm 1. Computes a shortest (s,t)-path P with minimal eccentricity for a given distance-hereditary graph G and a vertex pair s,t.

Input: A distance-hereditary graph $G = (V,E)$ and two distinct vertices s and t.

Output: A shortest path P from s to t with minimal eccentricity.

1 Compute the sets $V_i = \{v \mid d_G(v, I(s,t)) = i\}$ for $1 \le i \le \mathrm{ecc}(I(s,t))$.

2 Each vertex $v \notin I(s,t)$ gets a pointer $g(v)$ initialised with $g(v) := v$ if $v \in V_1$, and $g(v) := \varnothing$ otherwise.

3 **for** $i := 2$ **to** $\mathrm{ecc}(I(s,t))$ **do**

4   For each $v \in V_i$, select a vertex $u \in V_{i-1} \cap N(v)$ and set $g(v) := g(u)$.

5 **foreach** $v \in V_{\mathrm{ecc}(I(s,t))}$ **do**

6   If $N(g(v))$ intersects only one slice of $I(s,t)$, flag $g(v)$ as *relevant*.

7 Set $P := \{s,t\}$.

8 **for** $i := 1$ **to** $d_G(s,t) - 1$ **do**

9   Find a vertex $v \in S_i(s,t)$ for which the number of *relevant* vertices in $N(v)$ is maximal.

10   Add v to P.

Lemma 6. *For a distance-hereditary graph G and an arbitrary vertex pair s,t, Algorithm 1 computes a shortest (s,t)-path with minimal eccentricity in linear time.*

Proof. The loop in line 3 determines for each vertex v outside of the interval $I(s,t)$ a *gate vertex* $g(v)$ such that $N(g(v)) \supseteq \Pr(v, I(s,t))$ and $d_G(v, I(s,t)) = d_G(v, g(v)) + 1$ (see Lemma 3). From Lemmas 5 and 4, it follows that for a vertex v which is not in $V_{\mathrm{ecc}(I(s,t))}$ or its projection to $I(s,t)$ is intersecting two slices of $I(s,t)$, $d_G(v, P(s,t)) \le \mathrm{ecc}(I(s,t))$ for every shortest path $P(s,t)$ between s and t. Therefore, line 6 only marks $g(v)$ if $v \in V_{\mathrm{ecc}(I(s,t))}$ and its projection $\Pr(v, I(s,t))$ intersects only one slice. Because only one slice is intersected and each vertex in a slice is adjacent to all vertices in the consecutive slice (see Lemma 4), in each slice the vertex of an optimal (of minimum eccentricity) path P can be selected independently from the preceding vertex. If a vertex x of a slice $S_i(s,t)$ has the maximum number of *relevant* vertices in $N(x)$, then x is good to put in P. Indeed, if x dominates all relevant vertices adjacent to vertices of $S_i(s,t)$, then x is a perfect choice to put in P. Else, any vertex y of a slice $S_i(s,t)$ is a good vertex to put in P. Hence, P is optimal if the number of *relevant* vertices adjacent to P is maximal. Thus, the path selected in line 8 to line 10 is optimal. □

Running Algorithm 1 for a diametral pair of vertices of a distance-hereditary graph G, by Theorem 1, we get a shortest path of G with minimum eccentricity. Thus, we have proven the following result.

Theorem 2. *A shortest path with minimum eccentricity of a distance-hereditary graph $G = (V, E)$ can be computed in $\mathcal{O}(|V| + |E|)$ total time.*

4   A Polynomial-Time Algorithm for Chordal Graphs

In what follows, we will show that the minimum eccentricity shortest path problem for chordal graphs can be solved in polynomial time.

For distance-hereditary graphs, we were able to show that there is a shortest path with minimum eccentricity between a diametral pair of vertices. This is not always the case for chordal graphs. Consider the graph G given in Fig. 2. The only diametral path in G is from s to w. Because of u, it has eccentricity 3. However, a shortest path from s to v containing t has eccentricity 2 which is optimal for G.

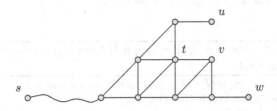

Fig. 2. A chordal graph for which no diametral path has the optimal eccentricity.

To find an optimal path, we create a simpler graph H for given start and end vertices s and t of a chordal graph G. Then, each shortest path P from s to t in H has eccentricity at most 2. Additionally, if P has minimal eccentricity in H, the corresponding path in G also has minimal eccentricity. Repeating this for each vertex pair s, t in G, we can find the minimum eccentricity shortest path of G.

The following lemmas allow us to create H.

Lemma 7 ([3]). *For every chordal graph G and any two of its vertices s and t, each slice $S_i(s,t)$ is a clique.*

Corollary 3. *For each shortest path P from s to t in a chordal graph G,*

$$\text{ecc}(I(s,t)) \leq \text{ecc}(P) \leq \text{ecc}(I(s,t)) + 1.$$

Lemma 8 ([9]). *Let G be a chordal graph. If for two distinct vertices x, y in a disk $N_G^r[v]$ there is a path P connecting them with $P \cap N_G^r[v] = \{x, y\}$, then x and y are adjacent.*

Lemma 9. *Let G be a chordal graph. For each vertex $v \notin I(s,t)$, if a projection $\Pi_v = \text{Pr}(v, I(s,t))$ is not a clique, then each shortest path from s to t intersects Π_v.*

Proof. Because Π_v is not a clique, there are two distinct vertices $u_i \in S_i(s,t) \cap \Pi_v$ and $u_j \in S_j(s,t) \cap \Pi_v$ which are not adjacent to each other. Consider an arbitrary shortest path Q from s to t and two vertices $q_i \in S_i(s,t) \cap Q$ and $q_j \in S_j(s,t) \cap Q$. Because each slice is a clique (see Lemma 7), there is a path $Q' = \{u_i\} \cup Q(q_i, q_j) \cup \{u_j\}$ from u_i to u_j. Note that Π_v is the intersection of $I(s,t)$ with the disk $N_G^r[v]$ (for $r = d_G(v, I(s,t))$). Thus, if Q and Π_v do not intersect, then $Q' \cap N_G^r[v] = \{u_i, u_j\}$. However, because u_i and u_j are not adjacent, this contradicts with Lemma 8. Therefore, Q and Π_v intersect.   □

The conclusion from Corollary 3 and Lemma 9 is that a vertex v is only relevant for determining a minimal eccentricity shortest path from s to t, if $d_G(v, I(s,t)) = \text{ecc}(I(s,t))$ and the projection of v on $I(s,t)$ intersects at most two slices. Therefore, we can create a graph H for a given chordal graph G using Algorithm 2. We call H a *hedgehog graph* for G.

Algorithm 2. Creates a hedgehog graph H from a given chordal graph G for its vertex pair s, t.

Input: A chordal graph $G = (V_G, E_G)$, and a vertex pair s, t.
Output: A hedgehog graph $H = (V_H, E_H)$.

1 Initialise $V_H := \emptyset$ and $E_H := \emptyset$.
2 Add $I_G(s,t)$ to H, i.e. $V_H := V_H \cup I_G(s,t)$ and
  $E_H := E_H \cup \{uv \in E_G \mid u, v \in I_G(s,t)\}$.
3 **foreach** $v \in V_G$ with $d_G(v, I_G(s,t)) = \text{ecc}_G(I_G(s,t))$ **do**
4 | If $\text{Pr}_G(v, I_G(s,t))$ intersects at most two slices of $I_G(s,t)$, then create a new
  | vertex $g(v)$, add to H, and connect it with every vertex in $\text{Pr}_G(v, I_G(s,t))$,
  | i.e. $V_H := V_H \cup \{g(v)\}$ and $E_H := E_H \cup \{ug(v) \mid u \in \text{Pr}_G(v, I_G(s,t))\}$.

Theorem 3. *For a chordal graph G and a vertex pair s, t, let H be the hedgehog graph of G created by Algorithm 2. A shortest (s, t)-path P in H has eccentricity 1 if and only if P has eccentricity $ecc_G(I_G(s, t))$ in G.*

Proof. Assume $ecc_H(P) = 1$. Therefore, for all vertices $g(v) \in V_H \setminus I_H(s, t)$, P intersects the projection $\Pr_H(g(v), I_H(s, t))$. Based on the construction of H, $\Pr_H(g(v), I_H(s, t)) = \Pr_G(v, I_G(s, t))$. Thus, $d_G(v, P) = ecc_G(I_G(s, t))$ for all $v \in V_G$ with $d_G(v, I_G(s, t)) = ecc_G(I_G(s, t))$ and projection $\Pr_G(v, I_G(s, t))$ intersecting at most two slices of $I_G(s, t)$. For all other $v \in V_G$, $d_G(v, P) \leq ecc_G(I_G(s, t))$ follows from Corollary 3 and Lemma 9. Thus, $ecc_G(P) = ecc_G(I_G(s, t))$.

Assume $ecc_H(P) > 1$. Thus, there is a vertex $v \in V_H \setminus I_H(s, t)$ such that P does not intersect the projection $\Pr_H(g(v), I_H(s, t))$. Therefore, $d_G(v, P) > d_G(v, I_G(s, t)) = ecc_G(I_G(s, t))$. □

For the analysis of the complexity of Algorithm 2, we assume that the distance between any two vertices can be determined in constant time (i.e., the distance matrix of the graph is given). Computing the interval $I_G(s, t)$ and $ecc_G(I_G(s, t))$ can be done in $\mathcal{O}(m)$ total time. For a given vertex v, $d_G(v, I_G(s, t))$ (line 3) and $\Pr_G(v, I_G(s, t))$ (line 4) can be calculated in $\mathcal{O}(n)$ time by determining the distance to all vertices in $I_G(s, t)$. Repeating this for all vertices in V_G leads to a total runtime of $\mathcal{O}(n^2)$.

After generating H, we need to determine if there is a shortest path from s to t in H with eccentricity 1.

Algorithm 3. Finds a shortest (s, t)-path with minimal eccentricity in a hedgehog graph H of a chordal graph.

Input: A hedgehog graph $H = (V, E)$ of a chordal graph and a vertex pair s, t.
Output: A shortest path P from s to t with minimal eccentricity.
1 For each vertex v and each edge e in H, set $\omega(v) := 0$ and $\omega(e) := 0$.
2 **foreach** $u \notin I_H(s, t)$ **do**
3   **if** $N_H(u) \subseteq S_i(s, t)$ *(for some i)* **then**
4   Set $\omega(v) := \omega(v) + 1$ for all $v \in N_H(u)$.
5   **else**
6   **foreach** $vw \in E$ with $d_H(u, vw) = 1$ *and* $d_H(s, u) = d_H(s, v) + 1 = d_H(s, w)$ **do**
7             $\omega(vw) := \omega(vw) + 1$

8 Find a shortest path P from s to t such that the sum of all vertex and edge weights of P is maximal.

Lemma 10. *For a given vertex pair s, t and the corresponding hedgehog graph H, Algorithm 3 determines a shortest (s, t)-path of H with minimal eccentricity (one or two) in $\mathcal{O}(nm)$ time.*

Proof (Correctness). Let P be an arbitrary shortest path from s to t in H. We say $\omega(P) = \sum_{u \in P} \omega(u) + \sum_{vw \in E, vw \in P} \omega(vw)$ is the total weight of P. Because H is based on a chordal graph, each slice of $I_H(s,t)$ is a clique. Thus, each path from s to t has eccentricity at most 2. To proof the correctness of the algorithm, we will show that the total weight of P is equal to the number of vertices adjacent to P which are not part of the interval, i.e. $\omega(P) = |N_H[P] \setminus I_H(s,t)|$.

The vertices of H can be partitioned into the following three sets: $I_H(s,t)$, V_1 containing all vertices x whose $N_H(x)$ intersects only one slice of $I_H(s,t)$, and V_2 containing all vertices x whose $N_H(x)$ intersects two slices of $I_H(s,t)$.

For each vertex $u \in V_1$ the weight of every of its neighbors v is increased by 1 (line 4). Thus, $\omega(v) = |N_H(v) \cap V_1|$. Note that for $v, v' \in I_H(s,t)$, $d_G(s,v) \neq d_G(s,v')$ implies $N_H(v) \cap N_H(v') \cap V_1 = \emptyset$. Therefore, $\sum_{v \in P} \omega(v)$ is the number of vertices in V_1 which are adjacent to P.

Let $u \in V_2$ be a vertex such that $N_H(u)$ intersects the slices $S_i(s,t)$ and $S_{i+1}(s,t)$. Then the weight of all edges vw from $S_i(s,t)$ to $S_{i+1}(s,t)$ which intersect $N_H(u)$ is increased by 1 (line 7). Because the weight of an edge vw is only increased if $d_H(s,u) = d_H(s,v) + 1 = d_H(s,w)$, $\omega(vw) = |N_H(vw) \cap V_2 \cap L_{i+1}^{(s)}|$. Therefore, $\sum_{vw \in E, vw \in P} \omega(vw)$ is the number of vertices in V_2 which are adjacent to P.

It follows that each vertex in $N_H[P] \setminus I_H(s,t)$ is counted exactly once for the total weight of P. Therefore, P has eccentricity 1 if and only if $\omega(P) = |V_1 \cup V_2|$. $\square$

Proof (Complexity). Initialising the vertex and edge weights can be done in linear time. For a vertex $u \notin I_H(s,t)$, line 4 only updates the neighborhood of u. Thus, the total runtime for line 4 is $\mathcal{O}(m)$.

Line 7 can be implemented in $\mathcal{O}(nm)$ time as follows. Let $N_H(u)$ intersect the slices $S_i(s,t)$ and $S_{i+1}(s,t)$. First, update $\omega(vw')$, for all vertices $v \in S_i(s,t) \cap N_H(u)$ and all $w' \in S_{i+1}(s,t) \cap N_H(v)$. Also mark v as visited. Then, update $\omega(v'w)$ for all vertices $w \in S_{i+1}(s,t) \cap N_H(u)$ and all $v' \in S_i(s,t) \cap N_H(w)$ where v' is not marked as visited. Last, remove all marks from all vertices $v \in S_i(s,t) \cap N_H(u)$. For a given u, this runs in $\mathcal{O}(m)$ time. Thus, the total runtime for line 7 is in $\mathcal{O}(nm)$.

Finding a shortest path P such that $\omega(P)$ is maximal can be easily done in linear time. Therefore, the overall runtime of Algorithm 3 is in $\mathcal{O}(nm)$. $\square$

Using the methods described above, we can now construct Algorithm 4 to compute a minimum eccentricity shortest path in chordal graphs.

Theorem 4. *Algorithm 4 computes a minimum eccentricity shortest path for a given chordal graph in $\mathcal{O}(n^3 m)$ time.*

Proof (Correctness). The algorithm creates a hedgehog graph $H(s,t)$ for each vertex pair s, t (line 3). Then it determines a shortest path $P(s,t)$ from s to t in $H(s,t)$ with minimal eccentricity (line 4). By Theorem 3, $P(s,t)$ is also a shortest path with minimal eccentricity from s to t in G. Therefore, for at least one pair s, t, the selected path $P(s,t)$ is a minimum eccentricity shortest path of G. Such a path is selected in line 5. $\square$

Algorithm 4. Finds a shortest path P with minimum eccentricity for a given chordal graph G.

Input: A chordal graph $G = (V, E)$.
Output: A shortest path P with minimum eccentricity.
1 Calculate the pairwise distance for all vertices.
2 **foreach** $s, t \in V$ **do**
3 | Create a hedgehog graph $H(s, t)$ of G for s and t using Algorithm 2.
4 | Find a shortest path $P(s, t)$ with minimal eccentricity in $H(s, t)$ using
   | Algorithm 3.
5 Among all shortest paths $P(s, t)$, select one for which $\text{ecc}_G(P(s, t))$ is minimal.

Proof (Complexity). Calculating the pairwise distances between vertices (line 1) can be done in $\mathcal{O}(nm)$ time. This allows to extract the distance between any two vertices in constant time. Thus, $H(s, t)$ can be created in $\mathcal{O}(n^2)$ time. By Lemma 10, finding a path with minimal eccentricity in $H(s, t)$ runs in $\mathcal{O}(nm)$ time. Therefore, the overall runtime for line 3 and line 4 is in $\mathcal{O}(n^3 m)$. The total runtime for determining the eccentricities of all calculated paths to select the minimum is in $\mathcal{O}(n^2 m)$. Thus, the algorithm runs in $\mathcal{O}(n^3 m)$ time.   □

5   Conclusion

We have investigated the Minimum Eccentricity Shortest Path problem for distance-hereditary graphs and for chordal graphs. For distance-hereditary graphs, we were able to present a linear time algorithm. For chordal graphs, we gave an $\mathcal{O}(n^3 m)$ time algorithm.

The main reason for the large difference in the run-times of the two algorithms is that the second one iterates over all vertex pairs of a chordal graph. We know that, for general graphs, the problem remains NP-complete even if a start-end vertex pair is given (see the reduction in [7]). Also, we have shown that there is a shortest path with minimum eccentricity between every diametral pair of vertices of a distance-hereditary graph (Theorem 1). This leads to the following question: How hard is it to determine the start and end vertices of an optimal path? This question applies to general graphs as well as to special graph classes like chordal graphs.

Another interesting question is, for which other graph classes the problem remains NP-complete or can be solved in polynomial time. The NP-completeness proof in [7] uses a reduction from SAT. There is a planar version of 3-SAT (see [11]). Does this imply that the problem remains NP-complete for planar graphs?

Acknowledgement. This work was partially supported by the NIH grant R01 GM103309.

References

1. Bandelt, H.-J., Mulder, H.M.: Distance-hereditary graphs. J. Comb. Theor. Ser. B **41**, 182–208 (1986)
2. Brandstädt, A., Le, V.B., Spinrad, J.: Graph Classes: A Survey. SIAM, Philadelphia (1999)
3. Chang, G.J., Nemhauser, G.L.: The k-domination and k-stability problems on sunfree chordal graphs. SIAM J. Algebraic Discrete Meth. **5**, 332–345 (1984)
4. Corneil, D.G., Olariu, S., Stewart, L.: Linear time algorithms for dominating pairs in asteroidal triple-free graphs. SIAM J. Comput. **28**, 292–302 (1997)
5. D'Atri, A., Moscarini, M.: Distance-hereditaxy graphs, Steiner trees and connected domination. SIAM J. Comput. **17**, 521–538 (1988)
6. Dragan, F.F., Köhler, E., Leitert, A.: Line-distortion, bandwidth and path-length of a graph. In: Ravi, R., Gørtz, I.L. (eds.) SWAT 2014. LNCS, vol. 8503, pp. 158–169. Springer, Heidelberg (2014)
7. Dragan, F.F., Leitert, A.: On the minimum eccentricity shortest path problem. In: Dehne, F., Sack, J.-R., Stege, U. (eds.) WADS 2015. LNCS, vol. 9214, pp. 276–288. Springer, Heidelberg (2015)
8. Dragan, F.F., Nicolai, F.: LexBFS-orderings of distance-hereditary graphs with application to the diametral pair problem. Discrete Appl. Math. **98**, 191–207 (2000)
9. Faber, M., Jamison, R.E.: Convexity in graphs and hypergraphs. SIAM J. Algebraic Discrete Methods **7**, 433–444 (1986)
10. Howorka, E.: A characterization of distance-hereditary graphs. Quart. J. Math. Oxford Ser. **2**(28), 417–420 (1977)
11. Lichtenstein, D.: Planar formulae and their uses. SIAM J. Comput. **11**, 329–343 (1982)
12. Müller, H.: Hamiltonian circuits in chordal bipartite graphs. Discrete Math. **156**, 291–298 (1996)
13. Slater, P.J.: Locating central paths in a graph. Transp. Sci. **16**, 1–18 (1982)

Approximating Source Location and Star Survivable Network Problems

Guy Kortsarz[1] and Zeev Nutov[2(✉)]

[1] Rutgers University, Camden, USA
guyk@camden.rutgers.edu
[2] The Open University of Israel, Raanana, Israel
nutov@openu.ac.il

Abstract. In Source Location (SL) problems the goal is to select a minimum cost source set $S \subseteq V$ such that the connectivity (or flow) $\psi(S, v)$ from S to any node v is at least the demand d_v of v. In many SL problems $\psi(S, v) = d_v$ if $v \in S$, so the demand of nodes selected to S is completely satisfied. In a variant suggested recently by Fukunaga [7], every node v selected to S gets a "bonus" $p_v \leq d_v$, and $\psi(S, v) = p_v + \kappa(S \setminus \{v\}, v)$ if $v \in S$ and $\psi(S, v) = \kappa(S, v)$ otherwise, where $\kappa(S, v)$ is the maximum number of internally disjoint (S, v)-paths. While the approximability of many SL problems was seemingly settled to $\Theta(\ln d(V))$ in [20], for his variant on undirected graphs Fukunaga achieved ratio $O(k \ln k)$, where $k = \max_{v \in V} d_v$ is the maximum demand. We improve this by achieving ratio $\min\{p^* \ln k, k\} \cdot O(\ln k)$ for a more general version with node capacities, where $p^* = \max_{v \in V} p_v$ is the maximum bonus. In particular, for the most natural case $p^* = 1$ we improve the ratio from $O(k \ln k)$ to $O(\ln^2 k)$. To derive these results, we consider a particular case of the Survivable Network (SN) problem when all edges of positive cost form a star. We obtain ratio $O(\min\{\ln n, \ln^2 k\})$ for this variant, improving over the best ratio known for the general case $O(k^3 \ln n)$ of Chuzhoy and Khanna [3].

In addition, we show that directed SL with unit costs is $\Omega(\log n)$-hard to approximate even for $0, 1$ demands, while SL with uniform demands can be solved in polynomial time. Finally, we obtain a logarithmic ratio for a generalization of SL where we also have edge-costs and flow-cost bounds $\{b_v : v \in V\}$, and require that the minimum cost of a flow of value d_v from S to every node v is at most b_v.

1 Introduction

In Source Location (SL) problems, the goal is to select a minimum cost source set $S \subseteq V$ such that the connectivity from S to any node v is at least the demand d_v of v. Formally, the generic version of this problem is as follows.

© Springer-Verlag Berlin Heidelberg 2016
E.W. Mayr (Ed.): WG 2015, LNCS 9224, pp. 203–218, 2016.
DOI: 10.1007/978-3-662-53174-7_15

Source Location (SL)

Instance: A graph $G = (V, E)$ with node-costs $c = \{c_v : v \in V\}$, *connectivity demands* $d = \{d_v : v \in V\}$, and a *source connectivity function* $\psi : 2^V \times V \rightarrow \mathbb{Z}_+$, where $\mathbb{Z}_+$ denotes the set of non-negative integers.

Objective: Find a minimum cost source node set $S \subseteq V$ such that $\psi(S, v) \geq d_v$ for every $v \in V$.

Several source connectivity functions ψ appear in the literature. To avoid considering many cases, we suggest two generic types, that include previous particular cases.

Definition 1. *An integer set-function f on a groundset U is submodular if $f(A) + f(B) \geq f(A \cap B) + f(A \cup B)$ for all $A, B \subseteq U$, and f is non-decreasing if $f(A) \leq f(B)$ for all $A \subseteq B \subseteq U$.*

Definition 2. *Let $G = (V, E)$ be a graph with node-capacities $\{q_u : u \in V\}$. For $S \subseteq V$ and $v \in V$ the (S, v)-q-connectivity $\lambda_G^q(S, v)$ is the maximum number of edge-disjoint paths from $S \setminus \{v\}$ to v in G such that every node $u \in V$ is an internal node in at most q_u paths. Given connectivity bonuses $\{p_u \geq q_u : u \in V\}$, the (S, v)-(p, q)-connectivity $\lambda_G^{p,q}(S, v)$ is defined by: $\lambda_G^{p,q}(S, v) = p_v + \lambda_G^q(S, v)$ if $v \in S$, and $\lambda_G^{p,q}(S, v) = \lambda_G^q(S, v)$ otherwise.*

We will say that a source connectivity function $\psi(S, v)$ is submodular if for every $v \in V$, the function $f_v(S) = \psi(S, v)$ is submodular and non-decreasing; $\psi(S, v)$ is survivable if it is of the type $\psi(S, v) = \lambda_G^{p,q}(S, v)$. The concept of q-connectivity is essentially "mixed connectivity" (the case $q_u \in \{0, k\}$) introduced by Frank, Ibaraki, and Nagamochi [5], while (p, q)-connectivity combines it with the connectivity function introduced recently by Fukunaga [7] (the case $q \equiv 1$). The case of arbitrary node capacities includes additional connectivity versions compared to [7], e.g., the edge-connectivity case.

It is not hard to see that every survivable source connectivity function $\psi(S, v)$ is submodular (see Sect. 4), but the inverse is not true in general. This gives only two types of SL problems.

Submodular SL: The connectivity function $\psi(S, v)$ is submodular.
Survivable SL: The connectivity function $\psi(S, v)$ is survivable.

We list four source connectivity functions that appear in the literature. All of them are submodular, and three of them are also survivable. Given an SL instance let $k = \max_{v \in V} d_v$ denote the maximum demand, and in the case of Survivable SL let $p^* = \max_{u \in V} p_u$ denote the maximum connectivity bonus and $q^* = \min_{u \in V} q_u$ denote the minimum node capacity. In what follows assume that $1 \leq q_u \leq p_u \leq k$ for all $u \in V$, and thus $1 \leq p^* \leq k$ and $1 \leq q^* \leq k$ holds.

1. λ-SL: $\lambda_G(S, v)$ is the maximum number of pairwise edge-disjoint (S, v)-paths if $v \notin S$ and $\lambda_G(S, v) = \infty$ otherwise.
   This is Survivable SL with $p_u = q_u = k$ for every $u \in V$.

2. κ-SL: $\kappa(S,v)$ is the maximum number of (S,v)-paths no two of which have a common node in $V \setminus (S \cup v)$ if $v \notin S$, and $\kappa(S,v) = \infty$ otherwise.
3. $\hat{\kappa}$-SL: $\hat{\kappa}(S,v)$ is the maximum number of (S,v)-paths no two of which have a common node in $V \setminus \{v\}$ if $v \notin S$, and $\hat{\kappa}(S,v) = \infty$ otherwise.
   This is Survivable SL with $p_u = k$ and $q_u = 1$ for every $u \in V$.
4. κ'-SL: $\kappa'(S,v) = \hat{\kappa}(S,v)$ if $v \notin S$ and $\kappa'(S,v) = p_v + \hat{\kappa}(S \setminus \{v\}, v)$ if $v \in S$.
   This is Survivable SL with $q_u = 1$ for every $u \in V$.

Table 1. Previous approximation ratios and lower bounds for SL problems. GC and UC stand for general and uniform costs, GD and UD stand for general and uniform demands, respectively.

c,d	λ $(p,q \equiv k)$		κ	
	Undirected	Directed	Undirected	Directed
GC,GD	$\Theta(\ln d(V))$ [2,20]	$\Theta(\ln d(V))$ [2,20]	$\Theta(\ln d(V))$ [2,20]	$\Theta(\ln d(V))$ [2,20]
GC,UD	in P [1]	$O(\ln d(V))$ [2]	$O(\ln d(V))$ [2]	$O(\ln d(V))$ [2]
UC,GD	in P [1]	$O(\ln d(V))$ [2]	$O(\ln d(V))$ [2]	$O(\ln d(V))$ [2]
UC,UD	in P [22]	in P [10]	$O(\ln d(V))$ [2]	$O(\ln d(V))$ [2]
	$\hat{\kappa}$ $(p \equiv k, q \equiv 1)$		κ' $(q \equiv 1)$	
GC,GD	$\Theta(\ln d(V))$ [20] $O(k \ln k)$ [7]	$\Theta(\ln d(V))$ [20]	$O(\ln d(V))$ [7] $O(k \ln k)$ [7]	$O(\ln d(V))$ [7]
GC,UD	in P [17]	in P [17]		
UC,GD	$O(\ln d(V))$ [20] $O(k)$ [9]	$O(\ln d(V))$ [20]		
UC,UD	in P [17]	in P [17]		

The known approximability status of SL problems with source connectivity functions $\lambda, \kappa, \hat{\kappa}, \kappa'$, is summarized in Table 1; see also a survey in [16]. The approximability of $\lambda, \kappa, \hat{\kappa}$-SL problems was settled to $O(\ln d(V))$ in [20] (where $d(V) = \sum_{v \in V} d_v$), while Fukunaga [7] showed that undirected κ'-SL admits ratio $O(k \ln k)$. We prove the following.

Theorem 1. Submodular SL *admits ratio* $O(\ln d(V))$. *Undirected* Survivable SL *admits ratio* $\min\{p^* \ln k, k\} \cdot O(\ln(k/q^*))$; *furthermore, if* $q^* = k$ *(this is the edge-connectivity case) then the ratio is exactly* k.

Theorem 1 has several consequences. While ratio $O(\ln(d(V))$ was known for source connectivity functions $\lambda, \kappa, \hat{\kappa}$ [20], our proof of a more general result is simpler and shorter than the proof of each particular case. For undirected graphs, the second part of Theorem 1 implies that Survivable SL problems admit ratio $O(k \ln(k/q^*))$ if $p^* \geq k/\ln k$ (e.g., $p^* = k$ in λ-SL and $\hat{\kappa}$-SL), and ratio $O(p^* \ln k \ln(k/q^*))$ if $p^* < k/\ln k$ (e.g., κ'-SL with $p^* = 1$). In the case of λ-SL we have $q^* = k$ which implies ratio exactly k. We note that ratio k for λ-SL can be achieved by decomposing the problem into k problems with demands in $\{0, \ell\}$, $\ell = 1 \ldots, k$; each of these problems can be solved in polynomial time. However, this algorithm is just a particular case of our algorithm.

Summarizing, we get the following result for connectivity functions λ, κ'.

Corollary 1. λ-SL *admits ratio* k *and* κ'-SL *admits ratio* $O(p^* \ln^2 k)$.

To prove Theorem 1, we consider the following known problem.

Survivable Network (SN)

Instance: A graph $G = (V, E)$ with edge-costs $\{c_e : e \in E\}$ and node capacities $\{q_u : u \in V\}$, and connectivity requirements $r = \{r_{sv} : sv \in D\}$ on a set D of demand edges on V.

Objective: Find a minimum-cost subgraph G' of G such that $\lambda_{G'}^q(s, v) \geq r_{sv}$ for every $sv \in D$.

Let $k = \max_{sv \in D} r_{sv}$ denote the maximum requirement. For $q \equiv k$ we get the edge-connectivity version which admits ratio 2 due to Jain [11], while for $q \equiv 1$ we get the node-connectivity version. SN admits a folklore ratio $O(|D|)$, and for directed graphs no better ratio is known. Undirected SN admits ratios $O\left(k^3 \log n\right)$ [3] for edge-costs, and $O\left(k^4 \log^2 n\right)$ for node-costs [18,23], and has an $\Omega(\max\{\{k^{1/4}, |D|^{1/6}\})$ approximation lower bound [14]. We consider the following particular case of SN, studied previously in [7,13].

Star-SN: the set F of edges in E of positive cost is a star with center a.

The Star-SN problem was defined in [13], where it was shown to admit ratio $O(\ln n)$ for unit edge-costs. The study of this problem in [13] is motivated by the observation that directed SN instances when (V, F) is a complete graph with unit edge costs (so called **Connectivity Augmentation** problem) can be reduced to Star-SN with a loss of a factor of 2 in the approximation ratio. Fukunaga [7] observed that κ'-SL is a special case of Star-SN. Hence the Star-SN problem is important, as it generalizes several well known problems, and it is also a particular interesting case of the SN problem. Our results for Star-SN substantially improve over the best known ratios. These results are of independent interest, as they show that Star-SN admits much better ratios than general SN.

Theorem 2. Star-SN *admits approximation ratios* $O(\ln n)$ *for directed graphs, and* $O(\min\{\ln n, \ln k \ln(k/q^*)\})$ *for undirected graphs.*

We further study SL problems and prove the following.

Theorem 3. *Directed* Survivable SL *for* $k = 1$ *and unit costs is* $\Omega(\log n)$-*hard to approximate. Directed/undirected* κ'-SL *with uniform demands and with* $p \equiv 1$ *can be solved in polynomial time.*

Finally, we consider the following generalization of Survivable SL. Given an instance of Survivable SL and edge-costs $c = \{c_e : e \in E\}$, let $\mu_G^{p,q}(S, v)$ denote the minimum cost of an edge set $F \subseteq E$ such that $\lambda_{(V,F)}^{p,q}(S, v) \geq d_v$, where $\mu_G^{p,q}(S, v) = \infty$ if no such edge set F exists (namely, if $\lambda_G^{p,q}(S, v) < d_v$).

Survivable SL with Flow-Cost Bounds

Instance: As in Survivable SL, but in addition we are also given edge-costs $\{c_e : e \in E\}$ and flow-cost bounds $\{b_v \leq c(E) : v \in V\}$.

Objective: As in Survivable SL, with an additional constraint $\mu_G^{p,q}(S, v) \leq b_v$ for every $v \in V$.

Theorem 4. Survivable SL with Flow-Cost Bounds *admits approximation ratio* $H(d(V)) + H(nc(E) - b(V))$.

Theorem 4 will be proved in the full version, due to space limitation.

2 Relations Between SL and SN Problems

To explain the relation between SL and SN problems it would be convenient to consider the augmentation version of the SN problem, with arbitrary connectivity functions and allowing node-costs. Given a function $w = \{w_u : u \in U\}$ on a groundset U and $U' \subseteq U$, let $w(U') = \sum_{u \in U'} w_u$. If w is a cost function on U and I is an edge-set on U, then the cost (or the node-costs) $w(I)$ of I is the cost of the set of the endnodes of I. Formally, we define the problem we need as follows.

Network Augmentation (NA)
Input: A graph $G = (V, E)$, an edge-set F on V, a cost function c on F or on V, connectivity requirements $r = \{r_{sv} : sv \in D\}$ on a set D of demand edges on V, and a family $\{f_{sv} : 2^F \to \mathbb{Z}_+ : sv \in D\}$ of connectivity functions.
Output: A min-cost edge-set $I \subseteq F$ such that $f_{sv}(I) \geq r_{sv}$ for every $sv \in D$.

Note that here the connectivity functions $f_{sv}(I)$ differ from the source connectivity functions in SL problems. As in the case of SL problems, we consider two types of NA problems:

Submodular NA: connectivity functions $f_{sv}(I)$ are submodular and non-decreasing.
Survivable NA: connectivity functions are $f_{sv}(I) = \lambda_{G+I}^q(s, v)$.

SN is a particular case of Survivable NA when $E = \emptyset$, but for edge-costs the problems are equivalent. Here a survivable connectivity function may not be submodular; indeed, we will obtain a logarithmic ratio for Submodular NA, while Survivable NA has a polynomial approximation threshold. To see this, consider the following simple example: $V = \{s, u, v\}$, $E = \emptyset$, $F = \{su, uv\}$, and $f(I) = f_{sv}(I) = \lambda_{(V,I)}(s, v)$ is just the edge-connectivity function. Let $A = \{su\}$ and $B = \{sv\}$. Then $f(A) = f(B) = 0$ and $f(A \cup B) = 1$, and the submodular inequality in Definition 1 does not hold. However, we will show that if F is a star, then in the case of directed graphs every survivable connectivity function is submodular.

Let Rooted NA be a particular case of NA when D is a star with center s. As we shall see, SL is equivalent to the node-costs version of the following particular case of both Star-NA and Rooted NA.

Centered-NA: D, F are both stars with a common center s.

Fukunaga [7] made an important observation that κ'-SL is equivalent (via an approximation ratio preserving reduction) to Survivable Centered-NA with *edge-costs* and $q \equiv 1$. Here we further observe the following. For an edge-set/graph J let $\delta_J(X)$ denote the set of edges in J from X to $V \setminus X$.

Lemma 1. *For both directed and undirected graphs,* Survivable SL *is equivalent to* Survivable Centered-NA *with node-costs such that* $\delta_G(s) = \emptyset$ *and* $c(s) = 0$.

Proof. Given a Survivable SL instance construct a Survivable Centered-NA instance as follows: add to G a new node s of cost 0, and for every $v \in V$ set $r_{sv} = d_v$ and put p_v edges from s to v into F. Conversely, given a Survivable Centered-NA instance construct a Survivable SL instance as follows. Remove s from G, and for every $v \in V$ set p_v to be the number of edges in F from s to v and $d_v = r_{sv}$. In both directions, it is easy to see that S is a solution to the Survivable SL instance, if, and only if, the edge set I of all edges in F from s to S is a solution to the Survivable Centered-NA instance, and clearly I and S have the same node-cost.   □

It is not hard to see that for Survivable Star-NA, approximation ratio ρ for directed graphs implies ratio ρ for undirected graphs. This is achieved by a standard reduction of bidirecting the edges of the undirected instance, removing the directed edges entering the center a, and solving the problem on the obtained directed instance. The same reduction works for Submodular Star-NA problems. We omit the somewhat standard proof details.

The best known ratios for Survivable NA are $O(k^3 \log n)$ for edge-costs [3], and $O\left(k^4 \log^2 n\right)$ for node-costs [18,23]. The best known ratio for Survivable Rooted NA are $O(k \log k)$ for edge-costs [18] and $O\left(k^2 \log n\right)$ for node-costs [18,23], and no better ratios were known even for Survivable Centered-NA, see [7] where ratio $O(k \log k)$ for undirected Survivable Centered-NA was deduced in two ways: from the ratio $O(k \log k)$ for Survivable Rooted NA [18], and via iterative rounding. Our results for Star-NA, that imply Theorems 1 and 2, are summarized in the following three statements.

Let $H(j)$ denote the jth Harmonic number. The following lemma says that Submodular Star-NA problems admit approximation ratio that is logarithmic in terms of certain parameters α and β. These parameters are the maximum total increase (namely, the sum of the increases) in connectivity of all pairs in D as a result of taking a single edge (the parameter α) or a single node (the parameter β) to the solution.

Lemma 2. *For directed graphs,* Submodular NA *with edge costs admits ratio* $H(\alpha)$, *and* Submodular Star-NA *with node costs admits ratio* $H(\beta)$, *where*

$$\alpha = \max_{e \in F} \sum_{sv \in D} [\min\{f_{sv}(\{e\}), r_{sv}\} - f_{sv}(\emptyset)]$$

$$\beta = \max_{z \in V} \sum_{sv \in D} [\min\{f_{sv}(\delta_F(z)), r_{sv}\} - f_{sv}(\emptyset)].$$

The next lemma says that Survivable Star-NA is a particular case of Submodular Star-NA, and thus the previous lemma can be applied. Moreover, the lemma bounds the parameters α and β as above in terms of the Survivable Star-NA instance ingredients r, D, and F.

Lemma 3. *For directed graphs, any* Survivable Star-NA *problem is a* Submodular NA *problem, for which* $\alpha \le |D|$ *and* $\beta \le \min\{r(D), p^*|D|\}$ *holds, where here* p^* *denotes the maximum number of parallel edges in* F.

The above two lemmas imply ratio no better than $O(\ln|D|) = O(\ln n)$ for Survivable Star-NA. The next theorem, which is our main technical contribution, says that for undirected graphs we can achieve ratio roughly $O(p^* \ln^2 k)$, which may be much better that $O(\ln n)$ if the maximum requirement $k = \max_{sv \in D} r_{sv}$ and the maximum number p^* of parallel edges in F are small.

Theorem 5. *Undirected* Survivable Star-NA *admits ratio* $O(\ln k \ln(k/q^*))$ *for edge-costs and* $\min\{p^* \ln k, k\} \cdot O(\ln(k/q^*))$ *for node-costs; furthermore, in the case of node costs and* $q^* = k$ *the ratio is exactly* k.

The above three statements imply Theorem 2; combined with Lemma 1 they also imply Theorem 1. Our ratios for Star-NA and SL are summarized Table 2.

Table 2. Approximation ratios for Star-NA and SL problems proved in this paper.

	Submodular		Survivable	
	directed	*undirected*	*directed*	*undirected*
Star-NA (edge-costs)	$H(\alpha)$	$H(\alpha)$	$H(|D|)$	$H(|D|)$ $O(\ln k \ln(k/q^*))$
Star-NA (node-costs)	$H(\beta)$	$H(\beta)$	$H(\min\{r(D), p^*|D|\})$	$H(\min\{r(D), p^*|D|\})$ $\min\{p^* \ln k, k\} \cdot$ $O(\ln(k/q^*))$
SL	$H(d(V))$	$H(d(V))$	$H(\min\{d(V), p^*|V|\})$	$H(\min\{d(V), p^*|V|\})$ $\min\{p^* \ln k, k\} \cdot$ $O(\ln(k/q^*))$

We briefly mention the techniques we use to prove these statements. Lemma 2 is essentially an easy application of the greedy algorithm of Wolsey [24] for the Submodular Cover problem. Parts of Lemma 3 were implicitly proved in [13], but our proof is both more general and substantially simpler. Our main technical contribution is Theorem 5. To prove this theorem, we consider the augmentation version of Survivable Star-NA with edge-costs where the goal is to increase the connectivity by one between the pairs in D. Using LP-scaling we show that ratio ρ for the augmentation version implies ratio $O(\rho \ln k)$ for the edge-costs version of the general problems, and ratio $\min\{p^* \ln k, k\} \cdot O(\rho)$ for the node-costs version. Then we design an $O(\ln(k/q^*))$-approximation algorithm for the augmentation version. This is achieved by formulating the augmentation problem as a Biset-Family Edge-Cover problem, reducing the later problem to the problem of finding a minimum cost vertex cover in a hypergraph, and using a theorem from [19] to show that the maximum degree in the obtained hypergraph is $O\left((k/q^*)^2\right)$.

3   Directed Submodular NA Problems (Lemma 2)

All graphs in this and the next sections are assumed to be directed. To prove
Lemma 2 we use a result due to Wolsey [24] about a performance of a greedy
algorithm for submodular covering problems. In a generic covering problem we
are given by a value oracle two set functions on a groundset U: a cost-function
$c : 2^U \rightarrow \mathbb{R}$ and a progress function $g : 2^U \rightarrow \mathbb{Z}$. The goal is to find $S \subseteq U$
of minimum cost such that $g(S) = g(U)$. The Submodular Cover problem is a
special case when the function g is submodular and non-decreasing, and $c(S) =$
$\sum_{v \in S} c(v)$ for some $c : U \rightarrow \mathbb{R}^+$. Wolsey [24] proved that then, the greedy
algorithm, that starts with $S = \emptyset$ and as long as $g(S) < g(U)$ repeatedly adds
to A an element $u \in U \setminus S$ with maximum $\frac{g(S \cup \{u\}) - g(S)}{c_u}$, has approximation
ratio $H (\max_{u \in U} g(\{u\}) - g(\emptyset))$.

We start with the case of edge-costs. Then the function g is defined in the
same way as in [13, 20]: $U = F$ and for $I \subseteq F$

$$g(I) = \sum_{sv \in D} \min\{f_{sv}(I), r_{sv}\}.$$

It is not hard to verify that g is non-decreasing, and that I is a feasible solution
to an NA instance if and only if $g(I) = g(F) = r(D)$. Also, for any $e \in F$

$$g(\{e\}) - g(\emptyset) = \sum_{sv \in D} [\min\{f_{sv}(\{e\}), r_{sv}\} - f_{sv}(\emptyset)].$$

We show that g is submodular. It is known (c.f. [21]) that if h is submodular, then $\min\{h, r\}$ is submodular for any constant r. Thus the function
$h_{sv}(I) = \min\{f_{sv}(I), r_{sv}\}$ is submodular. As a sum of submodular functions is
also submodular, we obtain that g is submodular.

Now let us consider node-costs. For $S \subseteq V$ let F_S denote the set of edges in
F from a to S, and let $f'_{sv}(S) = f_{sv}(F_S)$. We have $U = V$ and for $S \subseteq V$ let

$$g'(S) = \sum_{sv \in D} \min\{f'_{sv}(S), r_{sv}\}.$$

As in the edge-costs case, it is not hard to verify that g' is non-decreasing and
that S is a feasible solution to an NA instance if and only if $g'(S) = g'(V) = r(D)$.
Also, for any $z \in V$

$$g'(\{z\}) - g'(\emptyset) = \sum_{sv \in D} [\min\{f_{sv}(\delta_F(z)), r_{sv}\} - f_{sv}(\emptyset)].$$

We show that g' is submodular. We claim that the submodularity of $f(I)$
implies that $f'(S)$ is submodular. This is not true in general, but holds if F is
a star, and hence for Star-NA instances. More precisely, it is not hard to verify
the following statement, that finishes the proof of Lemma 2.

Lemma 4. *Let (V, F) be a graph and let f be a submodular set function on F. If F is a star with center a, then the set function $f'(S) = f(F_S)$ defined on $V \setminus \{a\}$ is also submodular.*

Proof. Let $A, B \subseteq V \setminus \{a\}$. It is easy to see that since F is a star then

$$F_A \cap F_B = F_{A \cap B} \qquad F_A \cup F_B = F_{A \cup B}.$$

Thus by the definition of f' and the submodularity of f we have

$$\begin{aligned} f'(A) + f'(B) &= f(F_A) + f(F_B) \geq f(F_A \cap F_B) + f(F_A \cup F_B) \\ &= f(F_{A \cap B}) + f(F_{A \cup B}) = f'(A \cap B) + f'(A \cup B). \end{aligned} \qquad \square$$

4 Survivable Star-NA is a Submodular NA Problem (Lemma 3)

We start by showing that in the case of edge-costs, directed Survivable Star-NA is a particular case of Submodular NA. Let $s, v \in V$ and let $f : 2^F \to \mathbb{Z}$ be defined by $f(I) = \lambda^q_{G+I}(s, v)$, $I \subseteq F$. It is easy to see that f is non-decreasing and we prove that if F is a star then f is submodular. For that, we use the following known characterization of submodularity, c.f. [21]:

    A set-function f on F is submodular if, and only if

$$f(I_0 \cup \{e\}) + f(I_0 \cup \{e'\}) \geq f(I_0) + f(I_0 \cup \{e, e'\}) \quad \forall I_0 \subset F, e, e' \in F \setminus I_0$$

Let us fix $I_0 \subseteq F$. Revising our notation to $G \leftarrow G + I_0$, $F \leftarrow F \setminus I_0$, and denoting $h(I) = f(I_0 \cup I) - f(I_0)$, we get that f is submodular if, and only if

$$h(\{e\}) + h(\{e'\}) \geq h(\{e, e'\}) \quad \forall e, e' \in F.$$

In our setting, F is a star and $h(I) = \lambda^q_{G+I}(s, v) - \lambda^q_G(s, v)$ is the increase in the (s, v)-q-connectivity as a result of adding I to G. Thus $0 \leq h(I) \leq |I|$ for any $I \subseteq F$, so $0 \leq h(\{e, e'\}) \leq 2$. If $h(\{e, e'\}) = 0$, then we are done; if $h(\{e, e'\}) = 1$, then we need to show that $h(\{e\}) = 1$ or $h(\{e'\}) = 1$; and if $h(\{e, e'\}) = 2$, then we need to show that $h(\{e\}) = 1$ and $h(\{e'\}) = 1$. We prove a general statement, that implies the above; it says that if an augmenting edge set I is a star that increases the st-connectivity by h, then there are h edges in I that cover all minimum st-cuts, and thus each of these edges increases the st-connectivity by 1.

Lemma 5. *Let $G = (V, E)$ be a directed graph with node capacities $\{q_v : v \in V\}$, let I be a set of edges on V disjoint to E such that I is a star with center a, let $s, t \in V$, and let $h = \lambda^q_{G+I}(s, t) - \lambda^q_G(s, t)$. Then there is $J \subseteq I$ of size $|J| \geq h$ such that $\lambda^q_{G+\{e\}}(s, t) = \lambda^q_G(s, t) + 1$ for every $e \in J$.*

Proof. Since we consider directed graphs, it is sufficient to prove the lemma for the case of edge-connectivity. For that, apply the following standard reduction that eliminates node capacities: replace every $v \in V \setminus \{s, t\}$ by two nodes v^{in}, v^{out} connected by q_v parallel edges from v^{in} to v^{out} and replace every $uv \in E \cup I$ by an edge from u^{out} to v^{in}. Hence we will prove the lemma for the edge connectivity function λ.

Let us say that $S \subseteq V$ is *tight* if $s \in S$, $t \notin S$, and $|\delta_G(S)| = \lambda_G(s, t)$, namely, if $\delta_G(S)$ is a minimum st-cut. Let $\mathcal{F}$ be the family of tight sets. By Menger's Theorem $\mathcal{F}$ is non-empty. It is also known that $\mathcal{F}$ is a ring family, namely, the intersection of all the sets in $\mathcal{F}$ is nonempty, and if $X, Y \in \mathcal{F}$ then $X \cap Y, X \cup Y \in \mathcal{F}$. Thus $\mathcal{F}$ has a unique inclusion-minimal set $S_{\min}$ and a unique inclusion-maximal set $S_{\max}$, and $S_{\min} \subseteq S_{\max}$ holds.

Let $J = \{av \in I : a \in S_{\min}, v \in V \setminus S_{\max}\}$ be the set of edges in I that go from $S_{\min}$ to $V \setminus S_{\max}$. Each edge in J covers all members in $\mathcal{F}$, hence by Menger's Theorem $\lambda_{G+\{e\}}(s, t) = \lambda_G(s, t) + 1$ for every $e \in J$.

It remains to prove that $|J| \geq h$. We claim that since I is a star, then $\lambda_{G+I}(s, t) \leq \lambda_G(s, t) + |J|$, hence $|J| \geq \lambda_{G+I}(s, t) - \lambda_G(s, t) = h$. Note that from Menger's Theorem we have

$$\lambda_{G+I}(s, t) \leq \lambda_G(s, t) + |\delta_I(S_{\min})| \qquad \lambda_{G+I}(s, t) \leq \lambda_G(s, t) + |\delta_I(S_{\max})|$$

The first inequality implies that if $\delta_I(S_{\min}) = \emptyset$, then $\lambda_{G+I}(s, t) = \lambda_G(s, t)$, and thus we are done. Else, $a \in S_{\min}$. In this case $J = \delta_I(S_{\max})$, since I is a star. Then the second inequality implies $\lambda_{G+I}(s, t) \leq \lambda_G(s, t) + |J|$, as claimed. $\square$

Note that Lemma 5 does not hold if I is an arbitrary edge set. To see this, consider the following example (this is the example given at the beginning of Sect. 2): $V = \{s, u, t\}$, $E = \emptyset$, and $I = \{su, ut\}$. Then $h = \lambda_{G+I}(s, t) - \lambda_G(s, t) = 1 - 0 = 1$, but $\lambda_{G+\{e\}}(s, t) = 0$ for every $e \in I$.

We now bound the parameters α and β. The bound $\beta \leq r(D)$ is obvious, while the other bounds on α and β follow from the simple observation that for any $s, v \in V$, the set-function on F defined by $f(I) = \lambda_{G+I}^q(s, v)$ has the following properties: $f(\{e\}) \leq 1$ for any $e \in F$ and $f(\delta_F(z)) \leq |\delta_F(z)| \leq p^*$ for any $z \in V$.

The proof of Lemma 3 is now complete.

5   Undirected Survivable Star-NA (Theorem 5)

All graphs in this and the next section are assumed to be undirected. We start by considering the edge-costs case, and then will show that it implies the node-costs case by reductions. We need several definitions.

Definition 3. *An ordered pair $\mathbb{A} = (A, A^+)$ of subsets of a groundset V is called a* biset *if $A \subseteq A^+$; A is the* inner part *and A^+ is the* outer part *of $\mathbb{A}$, and $\partial \mathbb{A} = A^+ \setminus A$ is the* boundary *of $\mathbb{A}$. An edge e covers a biset $\mathbb{A}$ if it has one endnode in A and the other in $V \setminus A^+$. For a biset $\mathbb{A}$ and an edge-set/graph J let $\delta_J(\mathbb{A})$ denote the set of edges in J covering $\mathbb{A}$.*

Given an instance of Survivable NA and a biset $\mathbb{A}$ on V, let the requirement of $\mathbb{A}$ be $r(\mathbb{A}) = \max\{r_{uv} : uv \in \delta_D(\mathbb{A})\}$ if $\delta_D(\mathbb{A}) \neq \emptyset$ and $r(\mathbb{A}) = 0$ otherwise. By the q-connectivity version of Menger's Theorem (c.f. [12]), $I \subseteq F$ is a feasible solution to an Survivable NA instance if, and only if, $|\delta_I(\mathbb{A})| \geq h(\mathbb{A})$ for every bisets $\mathbb{A}$ on V, where h is a biset-function defined by

$$h(\mathbb{A}) = \max\{r(\mathbb{A}) - (q(\partial\mathbb{A}) + |\delta_G(\mathbb{A})|), 0\} \tag{1}$$

Let $\mathcal{P}_h$ denote the polytope of "fractional edge-covers" of h, namely,

$$\mathcal{P}_h = \{x \in \mathbb{R}^F : x(\delta_F(\mathbb{A})) \geq h(\mathbb{A}) \ \forall \text{ biset } \mathbb{A} \text{ on } V, \ 0 \leq x_e \leq 1 \ \forall e \in F\}.$$

Let $\tau(h)$ denote the optimal value of a standard LP-relaxation for edge covering h by a minimum cost edge set, namely, $\tau(h) = \min\left\{\sum_{e \in F} c_e x_e : x \in \mathcal{P}_h\right\}$.

As an intermediate problem, we consider Survivable NA instances when we seek to increase the connectivity by 1 for every $uv \in D$, namely, when $r_{uv} = \lambda_G^q(u, v) + 1$ for all $uv \in D$.

D-Survivable NA (the edge-costs version)
Input: A graph $G = (V, E)$ with node-capacities $\{q_v : v \in V\}$, an edge set F on V, a cost function c on F, and a set D of demand edges on V.
Output: Find a min-cost edge-set $I \subseteq E$ such that $\lambda_{G+I}^q(u, v) \geq \lambda_G^q(u, v) + 1$ for all $uv \in D$.

Given a D-Survivable NA instance, let us say that a biset $\mathbb{A}$ is *tight* if $h(\mathbb{A}) = 1$, where h is defined by (1). The D-Survivable NA problem is equivalent to the problem of finding a minimum cost edge-cover of the family $\mathcal{F} = \{\mathbb{A} : h(\mathbb{A}) = 1\}$ of tight bisets. Thus the following generic problem includes D-Survivable NA.

Biset-Family Edge-Cover
Input: A graph (V, F) with edge-costs and a biset family $\mathcal{F}$ on V.
Output: Find a min-cost $\mathcal{F}$-cover $I \subseteq F$.

For a biset-family $\mathcal{F}$ let $\tau(\mathcal{F})$ denote the optimal value of a standard LP-relaxation for edge covering $\mathcal{F}$ by a minimum cost edge set, namely, $\tau(\mathcal{F}) = \tau(h)$ where h is defined by $h(\mathbb{A}) = 1$ if $\mathbb{A} \in \mathcal{F}$ and $h(\mathbb{A}) = 0$ otherwise.

The following statement considers the factor invoked by applying the so called "backward augmentation" method due to [8]. Some parts of this statement are known, but we will provide a proof for completeness of exposition.

Proposition 1. *Suppose that D-Survivable Star-NA with edge-costs admits a polynomial time algorithm that computes a solution of cost at most $\rho(k)\tau(\mathcal{F})$, where $\mathcal{F}$ is the family of tight bisets. Then Survivable Star-NA admits a polynomial time algorithm that computes a solution I such that:*

- *For edge-costs, $c(I) \leq \tau(h) \cdot \sum_{\ell=1}^{k} \frac{\rho(\ell)}{k-\ell+1}$, where h is defined by (1).*
- *For node-costs, $c(I) \leq \mathsf{opt} \cdot \sum_{\ell=1}^{k} \rho(\ell) \cdot \min\left\{\frac{p^*}{k-\ell+1}, 1\right\}$.*

Proof. We start with the edge-costs case. Consider the following sequential algorithm. Start with $I = \emptyset$. At iteration $\ell = 1, \ldots, k$, add to I and remove from F an edge-set $I_\ell \subseteq F$ that increases by 1 the q-connectivity of $G + I$ on the set of demands

$$D_\ell = \{sv : \lambda^q_{G+I}(s,v) = r(s,v) - k + \ell - 1, sv \in D\},$$

by covering the corresponding biset-family $\mathcal{F}_\ell$ using the ρ-approximation algorithm. After iteration ℓ, we have $\lambda^q_{G+I}(s,v) \geq r(s,v) - k + \ell$ for all $sv \in D$. Consequently, after k iterations $\lambda^q_{G+I}(s,v) \geq r(s,v)$ holds for all $sv \in D$, thus the computed solution is feasible. The approximation ratio follows from the following two observations.

(i) $c(I_\ell) \leq \rho(\ell) \cdot \tau(\mathcal{F}_\ell)$. This is so since $\lambda(s,v) \leq \ell - 1$ for every $sv \in D_\ell$, hence the maximum requirement at iteration ℓ is at most ℓ.

(ii) $\tau(\mathcal{F}_\ell) \leq \frac{\tau(h)}{k - \ell + 1}$. To see this, note that if $\mathbb{A} \in \mathcal{F}_\ell$ and $x \in \mathcal{P}_h$ then $x(\delta(\mathbb{A})) \geq k - \ell + 1$, by Menger's Theorem. Thus $x/(k - \ell + 1)$ is a feasible solution for the LP-relaxation for edge-covering $\mathcal{F}_\ell$, of value $c \cdot x/(k - \ell + 1)$.

Consequently, $c(I) = \sum_{\ell=1}^{k} c(I_\ell) \leq \sum_{\ell=1}^{k} \rho(\ell) \cdot \frac{\tau(h)}{k - \ell + 1} = \tau(h) \cdot \sum_{\ell=1}^{k} \frac{\rho(\ell)}{k - \ell + 1}$.

Now let us consider the case of node-costs. Then we convert node-costs into edge-costs by assigning to every edge $e = av$ the cost $c'(e) = c(v)$. Let opt' denote the optimal solution value of the edge-costs instance obtained. Clearly, $\mathrm{opt} \leq \mathrm{opt}' \leq p^* \cdot \mathrm{opt}$. Note that any inclusion minimal solution to a D-Survivable NA instance has no parallel edges. This implies that $c(I_\ell) \leq \rho(\ell) \cdot \mathrm{opt}$ and that $c(I_\ell) = c'(I_\ell)$. The latter implies $c(I_\ell) = c'(I_\ell) \leq \rho(\ell) \cdot \frac{\mathrm{opt}'}{k - \ell + 1} \leq \rho(\ell) \cdot \mathrm{opt} \cdot \frac{p^*}{k - \ell + 1}$, and the statement for the node-costs case follows.   □

In the next section we prove the following theorem, that together with Proposition 1 finishes the proof of Theorem 5.

Theorem 6. *Undirected D-Survivable Star-NA with edge-costs admits a polynomial time algorithm that computes a solution I of cost $\tau(\mathcal{F}) \cdot O(\ln(k/q^*))$. Furthermore, if D is a star then $c(I) \leq \tau(\mathcal{F}) \cdot H\left(2\left\lfloor \frac{k-1}{q^*} \right\rfloor + 1\right)$.*

6   Proof of Theorem 6

Recall that D-Survivable NA reduces to Biset-Family Edge-Cover with $\mathcal{F}$ being the family of tight bisets; in the case of rooted requirements, when D is a star with center s, it is sufficient to cover the biset-family

$$\mathcal{F}^s = \{\mathbb{A} \in \mathcal{F} : s \in V \setminus A^+\}.$$

Biset-families arising from Survivable NA instances have some special properties, that are summarized in the following definitions.

Definition 4. *The intersection and the union of two bisets $\mathbb{A}, \mathbb{B}$ is defined by $\mathbb{A} \cap \mathbb{B} = (A \cap B, A^+ \cap B^+)$ and $\mathbb{A} \cup \mathbb{B} = (A \cup B, A^+ \cup B^+)$. The biset $\mathbb{A} \setminus \mathbb{B}$ is defined by $\mathbb{A} \setminus \mathbb{B} = (A \setminus B^+, A^+ \setminus B)$. We write $\mathbb{A} \subseteq \mathbb{B}$ and say that $\mathbb{B}$ contains $\mathbb{A}$ if $A \subseteq B$ and $A^+ \subseteq B^+$. Let $\mathcal{C}_\mathcal{F}$ denote the inclusion-minimal bisets in $\mathcal{F}$.*

Definition 5. *Two bisets* $\mathbb{A}, \mathbb{B}$ *covered by an edge-set* D *are* D-*independent if for any* $xx', yy' \in D$ *such that* xx' *covers* $\mathbb{A}$ *and* yy' *covers* $\mathbb{B}$, $\{x, x'\} \cap \partial \mathbb{B} \neq \emptyset$ *or* $\{y, y'\} \cap \partial \mathbb{A} \neq \emptyset$; *otherwise,* $\mathbb{A}, \mathbb{B}$ *are* D-*dependent. We say that a biset family* $\mathcal{F}$ *is* D-*uncrossable if* D *covers* $\mathcal{F}$ *and if for any* D-*dependent* $\mathbb{A}, \mathbb{B} \in \mathcal{F}$ *the following holds:*

$$\mathbb{A} \cap \mathbb{B}, \mathbb{A} \cup \mathbb{B} \in \mathcal{F} \text{ or } \mathbb{A} \setminus \mathbb{B}, \mathbb{B} \setminus \mathbb{A} \in \mathcal{F}. \tag{2}$$

Similarly, given a set $T \subseteq V$ *of terminals, we say that* $\mathbb{A}, \mathbb{B}$ *are* T-*independent if* $A \cap T \subseteq \partial \mathbb{B}$ *or if* $B \cap T \subseteq \partial \mathbb{A}$, *and* $\mathbb{A}, \mathbb{B}$ *are* T-*dependent otherwise. We say that* $\mathcal{F}$ *is* T-*uncrossable if* T *covers the set-family of the inner parts of* $\mathcal{F}$, *and if (2) holds for any* T-*dependent* $\mathbb{A}, \mathbb{B} \in \mathcal{F}$.

A biset-family $\mathcal{F}$ is symmetric if $\mathbb{A} \in \mathcal{F}$ implies $(V \setminus A^+, V \setminus A) \in \mathcal{F}$. We will use the the following statement, that was implicitly proved in [19].

Lemma 6 ([19]). *The family* $\mathcal{F}$ *of tight bisets is symmetric and* D-*uncrossable; if* D *is a star with leaf-set* T *then* $\{\mathbb{A} \in \mathcal{F} : s \notin A^+\}$ *is* T-*uncrossable.*

For a biset-family $\mathcal{F}$ let $\gamma_{\mathcal{F}} = \max\{|\partial \mathbb{A}| : \mathbb{A} \in \mathcal{F}\}$ denote the maximum size of the boundary of a biset in $\mathcal{F}$. Note that if $\mathcal{F}$ is the family of tight bisets then $\gamma_{\mathcal{F}} \leq (k-1)/q^*$. Given an instance of **Biset-Family Edge-Cover**, we will assume that the family $\mathcal{C}$ of the inclusion members of $\mathcal{F}$ can be computed in polynomial time. We note that for $\mathcal{F}$ being the family of tight bisets, this step can be implemented in polynomial time, c.f. [19]. Under this assumption, we prove the following generalization of Theorem 6.

Theorem 7. *For edge/node-costs,* **Biset-Family Edge-Cover** *with* F *being a star admits a polynomial time algorithm that computes a cover* I *of* $\mathcal{F}$ *such that:*

(i) $c(I) \leq H\left((4\gamma_C + 1)^2\right) \cdot \tau(\mathcal{F})$ *if* $\mathcal{F}$ *is symmetric and* D-*uncrossable.*

(ii) $c(I) \leq H(2\gamma_C + 1) \cdot \tau(\mathcal{F})$ *if* $\mathcal{F}$ *is* T-*uncrossable and* $a \in V \setminus X^+$ *for all* $\mathbb{A} \in \mathcal{F}$.

In the rest of this section we prove Theorem 7.

Definition 6. *A set* $U \subseteq V$ *of nodes is a* $\mathcal{C}$-*transversal of a hypergraph (set-family)* $\mathcal{C}$ *on* V *if* U *intersects every set in* $\mathcal{C}$; *if* $\mathcal{C}$ *is a biset-family then* U *should intersect the inner part of every member of* $\mathcal{C}$. *Given node costs* $\{c_v : v \in V\}$, *let* $t^*(\mathcal{C})$ *denote the minimum value of a fractional* $\mathcal{C}$-*transversal, namely:*

$$t^*(\mathcal{C}) = \min\left\{\sum_{v \in V} c_v x_v : x(C) \geq 1 \ \forall C \in \mathcal{C}, \ x(v) \geq 0 \ \forall v \in V\right\}.$$

In [19], the following is proved.

Theorem 8 ([19]). *Let* $\mathcal{C}$ *be the family of the inclusion members of a biset family* $\mathcal{F}$. *Then the maximum degree in the hypergraph* $\{C : \mathbb{C} \in \mathcal{C}\}$ *is at most:*

(i) $(4\gamma_C + 1)^2$ *if* $\mathcal{F}$ *is* D-*uncrossable.*

(ii) $2\gamma_C + 1$ *if* $\mathcal{F}$ *is* T-*uncrossable.*

Given a hypergraph $(V, \mathcal{C})$ with node-costs, the greedy algorithm computes in polynomial time a $\mathcal{C}$-transversal $U \subseteq V$ of cost $c(U) \le H(\Delta(\mathcal{C}))t^*(\mathcal{C})$, where $\Delta(\mathcal{C})$ is the maximum degree of the hypergraph (c.f. [15]).

The following statement is obvious.

Lemma 7. *If an edge-set I covers a biset-family $\mathcal{F}$ then the set of endnodes of I is a transversal of $\mathcal{F}$.*

Lemma 8. *Let $\mathcal{F}$ be a biset family on V and I a star with center a on a transversal $U \subseteq V$ of $\mathcal{F}$. Then I covers $\mathcal{F}$ in each one of the following cases.*

(i) $\mathcal{F}$ is symmetric and $a \notin \Gamma(\mathbb{A})$ for all $\mathbb{A} \in \mathcal{F}$.
(ii) $a \in V \setminus A^+$ for all $\mathbb{A} \in \mathcal{F}$.

Proof. Let $\mathbb{A} \in \mathcal{F}$. Then $a \in A$ or $a \in V \setminus A^+$. If $a \in V \setminus A^+$, then since U is a transversal of $\mathcal{C}$, there is $u \in U \cap A$. If $a \in A$, then if $\mathcal{F}$ is symmetric, then there $u \in U \cap (V \setminus X^+)$. In both cases, there is an edge $au \in I$, and it covers $\mathbb{A}$. $\square$

The algorithm as in Theorem 7, for both edge-costs and node-costs is as follows, where in the case of node-costs we may assume that the cost of a is zero.

---

1. For every $v \in V \setminus \{a\}$, let e_v be the minimum-cost edge incident to v, and in the case of edge-costs define node-costs $c_v = \min_{e \in \delta_F(v)} c_e$ if $\delta_F(v) \ne \emptyset$, and $c_v = \infty$ otherwise.
2. Let $\mathcal{C}$ be the family of the inclusion members of $\mathcal{F}$. With node-costs $\{c_v : v \in V\}$, compute a transversal U of $\mathcal{C}$ of cost $c(U) \le H(\Delta(\mathcal{C}))t^*(\mathcal{C})$.
3. Return $I = \{e_v : v \in U\}$.

The solution computed is feasible by Lemma 8. The approximation ratio follows from Theorem 8 and Lemma 7.

This concludes the proof of Theorem 6, and thus also the proof of Theorem 5.

7   Proof of Theorem 3

Note that in the reduction in Lemma 1 we have the following.

– Uniform demands $d_v = k$ for all $v \in V$ in Survivable SL correspond to requirements $r_{sv} = k$ for all $v \in V \setminus \{s\}$ in Survivable Centered-NA.
– κ'-SL with $p \equiv 1$ corresponds to Survivable Centered-NA with edge costs.
– Unit node-costs in Survivable SL correspond to unit node-costs in Survivable Centered-NA.

Directed Rooted Survivable NA with edge-costs and requirements $r_{sv} = k$ for all $v \in V \setminus \{s\}$ can be solved in polynomial time [6]; this implies that also *undirected* Survivable Centered-NA with edge-costs and requirements $r_{sv} = k$ for all $v \in V \setminus \{s\}$ can be solved in polynomial time. Thus the same holds for κ'-SL with $p \equiv 1$ and uniform demands.

Frank [4] showed that *directed* Survivable Centered-NA with $\delta_G(s) = \emptyset$ and $k = 1$ is NP-hard. Using a slight modification of his reduction we can show that the problem is in fact Set-Cover hard to approximate, and thus is $\Omega(\log n)$-hard to approximate. Given an instance of Set-Cover, where a family A of sets needs to cover a set B of elements, construct the corresponding directed bipartite graph $G' = (A \cup B, E')$, by putting an edge from every set to each element it contains. The graph $G = (V, E)$ is obtained from G' by adding M copies of B, connecting A to each copy in the same way as to B, and adding a new node s. Let $F = \{sv : v \in V\}$, $c(e) = 1$ for every $e \in F$, and $r_{sv} = 0$ if $v \in A$ and $r_{sv} = 1$ otherwise. It is easy to see that if $I \subseteq F$ is a feasible solution to the obtained Survivable Centered-NA instance, then either I corresponds to a feasible solution to the Set-Cover instance, or $|I| \geq M$. The $\Omega(\log n)$-hardness follows for M large enough, say $|M| = (|A| + |B|)^2$, and $|A| = |B|$. Since for $k = 1$ all connectivity functions of Survivable NA are equivalent, we get $\Omega(\log n)$ hardness for directed Survivable NA with $k = 1$ and unit costs.

References

1. Arata, K., Iwata, S., Makino, K., Fujishige, S.: Locating sources to meet flow demands in undirected networks. J. Algorithms **42**, 54–68 (2002)
2. Bar-Ilan, J., Kortsarz, G., Peleg, D.: Generalized submodular cover problems and applications. Theor. Comput. Sci. **250**(1–2), 179–200 (2001)
3. Chuzhoy, J., Khanna, S.: An $O(k^3 \log n)$-approximation algorithms for vertex-connectivity survivable network design. In: FOCS, pp. 437–441 (2009)
4. Frank, A.: Augmenting graphs to meet edge-connectivity requirements. SIAM J. Discrete Math. **5**(1), 25–53 (1992)
5. Frank, A., Ibaraki, T., Nagamochi, H.: On sparse subgraphs preserving connectivity properties. J. Graph Theor. **17**(3), 275–281 (1993)
6. Frank, A., Tardos, E.: An application of submodular flows. Linear Algebra Appl. **114**(115), 329–348 (1989)
7. Fukunaga, T.: Approximating minimum cost source location problems with local vertex-connectivity demands. In: Ogihara, M., Tarui, J. (eds.) TAMC 2011. LNCS, vol. 6648, pp. 428–439. Springer, Heidelberg (2011)
8. Goemans, M., Goldberg, A., Plotkin, S., Shmoys, D., Tardos, E., Williamson, D.: Improved approximation algorithms for network design problems. In: SODA, pp. 223–232 (1994)
9. Ishii, T.: Greedy approximation for source location problem with vertex-connectivity requirements in undirected graphs. J. Discrete Algorithms **7**, 570–578 (2009)
10. Ito, H., Makino, K., Arata, K., Honami, S., Itatsu, Y., Fujishige, S.: Source location problem with flow requirements in directed networks. Optim. Methods Softw. **18**(4), 427–435 (2003)
11. Jain, K.: A factor 2 approximation algorithm for the generalized steiner network problem. Combinatorica **21**(1), 39–60 (2001)
12. Kortsarz, G., Nutov, Z.: Approximating minimum cost connectivity problems. In: Gonzales, T.F. (ed.) Approximation Algorithms and Metaheuristics, Ch. 58 (2007)
13. Kortsarz, G., Nutov, Z.: Tight approximation algorithm for connectivity augmentation problems. J. Comput. Syst. Sci. **74**(5), 662–670 (2008)

14. Laekhanukit, B.: Parameters of two-prover-one-round game and the hardness of connectivity problems. In: SODA, pp. 1626–1643 (2014)

15. Lovász, L.: On the ratio of optimal integral and fractional covers. Discrete Math. **13**, 383–390 (1975)

16. Nagamochi, H., Ibaraki, T.: Algorithmic Aspects of Graph Connectivity, Ch. 9. Cambridge University Press, Cambridge (2008)

17. Nagamochi, H., Ishii, T., Ito, H.: Minimum cost source location problem with vertex-connectivity requirements in digraphs. IPL **80**, 287–294 (2001)

18. Nutov, Z.: Approximating minimum cost connectivity problems via uncrossable bifamilies. Trans. Algorithms **9**(1), 1 (2012)

19. Nutov, Z.: Approximating node-connectivity augmentation problems. Algorithmica **63**(1–2), 398–410 (2012)

20. Sakashita, M., Makino, K., Fujishige, S.: Minimum cost source location problems with flow requirements. Algorithmica **50**, 555–583 (2008)

21. Schrijver, A.: Combinatorial Optimization Polyhedra and Efficiency. Springer, Heidelberg (2004)

22. Tamura, H., Sengoku, M., Shinoda, S., Abe, T.: Location problems on undirected flow networks. IEICE Trans. **E73**, 1989–1993 (1990)

23. Vakilian, A.: Node-weighted prize-collecting network design problems. M.Sc. thesis (2013)

24. Wolsey, L.A.: An analysis of the greedy algorithm for the submodular set covering problem. Combinatorica **2**, 385–393 (1982)

On the Complexity of Computing
the k-restricted Edge-connectivity of a Graph

Luis Pedro Montejano[1] and Ignasi Sau[2($\boxtimes$)]

[1] Département de Mathématiques, Université de Montpellier 2, Montpellier, France
lpmontejano@gmail.com
[2] AlGCo project team, CNRS, LIRMM, Montpellier, France
ignasi.sau@lirmm.fr

Abstract. The k-*restricted edge-connectivity* of a graph G, denoted by $\lambda_k(G)$, is defined as the minimum size of an edge set whose removal leaves exactly two connected components each containing at least k vertices. This graph invariant, which can be seen as a generalization of a minimum edge-cut, has been extensively studied from a combinatorial point of view. However, very little is known about the complexity of computing $\lambda_k(G)$. Very recently, in the parameterized complexity community the notion of *good edge separation* of a graph has been defined, which happens to be essentially the same as the k-restricted edge-connectivity. Motivated by the relevance of this invariant from both combinatorial and algorithmic points of view, in this article we initiate a systematic study of its computational complexity, with special emphasis on its parameterized complexity for several choices of the parameters. We provide a number of NP-hardness and W[1]-hardness results, as well as FPT-algorithms.

Keywords: Graph cut · k-restricted edge-connectivity · Good edge separation · Parameterized complexity · FPT-algorithm · polynomial kernel

1   Introduction

Motivation. The k-restricted edge-connectivity is a graph invariant that has been widely studied in the literature from a combinatorial point of view [1,5, 16,21,30,31]. Since the classical edge-connectivity may not suffice to measure accurately how connected a graph is after deleting some edges, Esfahanian and Hakimi [15] proposed in 1988 the notion of *restricted edge-connectivity*. An edge-cut S is called a *restricted edge-cut* if there are no isolated vertices in $G - S$. The *restricted edge-connectivity* $\lambda'(G)$ is the minimum cardinality over all *restricted edge-cuts* S.

Inspired by the above definition, Fàbrega and Fiol [16] proposed in 1994 the notion of k-*restricted edge-connectivity*, where k is a positive integer, generalizing this notion. An edge-cut S is called a k-*restricted edge-cut* if every component of $G - S$ has at least k vertices. Assuming that G has k-restricted edge-cuts, the

© Springer-Verlag Berlin Heidelberg 2016
E.W. Mayr (Ed.): WG 2015, LNCS 9224, pp. 219–233, 2016.
DOI: 10.1007/978-3-662-53174-7_16

k-restricted edge-connectivity of G, denoted by $\lambda_k(G)$, is defined as the minimum cardinality over all k-restricted edge-cuts of G, i.e.,

$$\lambda_k(G) = \min\{|S| : S \subseteq E(G) \text{ is a } k\text{-restricted edge-cut}\}.$$

Note that for any graph G, $\lambda_1(G)$ is the size of a minimum edge-cut, and $\lambda_2(G) = \lambda'(G)$. A connected graph G is called λ_k-*connected* if $\lambda_k(G)$ exists. Let $[X, Y]$ denote the set of edges between two disjoint vertex sets $X, Y \subseteq V(G)$, and let $\overline{X}$ denote the complement $\overline{X} = V(G) \backslash X$ of vertex set X. It is clear that for any k-restricted cut $[X, \overline{X}]$ of size $\lambda_k(G)$, the graph $G - [X, \overline{X}]$ has exactly two connected components.

Very recently, Chitnis *et al.* [6] defined the notion of *good edge separation* for algorithmic purposes. For two positive integers k and ℓ, a partition $(X, \overline{X})$ of the vertex set of a connected graph G is called a (k, ℓ) -*good edge separation* if $|X|, |\overline{X}| > k$, $|[X, \overline{X}]| \leq \ell$, and both $G[X]$ and $G[\overline{X}]$ are connected. That is, it holds that $\lambda_k(G) \leq \ell$ if and only if G admits a $(k - 1, \ell)$-good edge separation. Thus both notions, which have been defined independently and for which there existed no connection so far, are essentially the same.

Good edge separations turned out to be very useful for designing parameterized algorithms for cut problems [6], by using a technique known as *recursive understanding*, which basically consists in breaking up the input graph into highly connected pieces in which the considered problem can be *efficiently* solved. It should be mentioned that Kawarabayashi and Thorup [22] had defined before a very similar notion for *vertex-cuts* and introduced the idea of recursive understanding. This technique has also been subsequently used in [9, 23].

Very little is known about the complexity of computing the k-restricted edge-connectivity of a graph, in spite of its extensive study in combinatorics. In this article we initiate a systematic analysis on this topic, with special emphasis on the parameterized complexity of the problem. In a nutshell, the main idea is to identify relevant parameters of the input of some problem, and study how the running time of an algorithm solving the problem depends on the chosen parameters. See [12, 17, 27] for introductory textbooks to this area.

Our results. We consider two problems concerning the k-restricted edge-connectivity of a graph. Namely, given a connected graph G and an integer k, determining whether G is λ_k-connected or not, and determining the value of $\lambda_k(G)$, if it exists. This latter problem, which we call RESTRICTED EDGE-CONNECTIVITY (REC for short), can be seen as a generalization of computing a MINIMUM CUT in a graph, which is polynomial-time solvable [28]. In Sect. 2 we prove that it is NP-hard, even restricted to λ_k-connected graphs. In Sect. 3 we study the parameterized complexity of the REC problem. More precisely, given a connected graph G and two integers k and ℓ, we consider the problem of determining whether $\lambda_k(G) \leq \ell$. Existing results concerning good edge separations imply that the problem is FPT when parameterized by k and ℓ. We prove that it is W[1]-hard when parameterized by k, and that it is unlikely to admit polynomial kernels when parameterized by ℓ. Moreover, we prove that deciding whether a graph is λ_k-connected is FPT when parameterized by k. Finally, in Sect. 4 we also consider the maximum degree Δ of the input graph as a parameter, and we prove that deciding whether a graph is λ_k-connected remains NP-complete in graphs with $\Delta \leq 5$, and that the REC problem is FPT when parameterized by k and Δ. Note that this implies, in particular, that the REC problem parameterized by k is FPT in graphs of bounded degree. Table 1 summarizes the results of this article.

Table 1. Summary of our results, where Δ denotes the maximum degree of the input graph G, and NPc (resp. NPh) stands for NP-complete (resp. NP-hard). The symbol '$\star$' denotes that the problem is not defined for that parameter.

Problem	Classical complexity	Parameterized complexity with parameter				
		$k + \ell$	k	ℓ	$k + \Delta$	$\ell + \Delta$
Is G λ_k-connected ?	NPc, even if $\Delta \leq 5$ (Theorem 6)	FPT (Theorem 2 by [6])	FPT (Theorem 4)	$\star$	FPT (Theorem 7)	$\star$
$\lambda_k(G) \leq \ell$?	NPh, even if G is λ_k-connected (Theorem 1)	FPT (Theorem 2 by [6])	W[1]-hard (Theorem 3)	No poly kernels (Theorem 5)	FPT (Theorem 7)	?

Remarks and further research. In view of Table 1, the main open question is whether the REC problem is FPT when parameterized by ℓ. In this direction, it is worth noting that for the MINIMUM BISECTION problem, which is strongly related to the REC problem with parameter ℓ, the non-existence of polynomial kernels [29] was known before the problem was recently proved to be FPT [9]. In fact, it was proved in [9] that the following more general problem, which is even closer to the REC problem, is FPT with parameter ℓ: Given a graph G and two integers k and ℓ, decide whether there exists a partition of $V(G)$ into A and B such that $|A| = k$ and there are at most ℓ edges between A and B.

Even when considering the combined parameter $\ell + \Delta$, the parameterized complexity of the REC problem is open. Our intuition is that adding Δ as a parameter may not make the problem much easier, as for MINIMUM BISECTION, in terms of approximability the problem is as hard in 3-regular graphs as in general graphs [2].

Other interesting questions are determining the existence of polynomial kernels for the REC problem with parameter $k + \ell$, improving the bound on the maximum degree in Theorem 6, and studying the (parameterized) complexity of the REC problem in planar graphs and other sparse graph classes.

Finally, let us mention that Bonsma, Ueffing and Volkmann [5] defined an extension of the minimum edge-degree of a graph for an integer $k \geq 2$, called the *minimum k-edge-degree*. This notion turns out to be related to the k-restricted edge-connectivity of a graph. Due to this relation, in Appendix C we study the parameterized complexity of computing the minimum k-edge-degree of a graph.

Notation. We use standard graph-theoretic notation; see for instance [10]. For a graph G, let $\Delta(G)$ denote its maximum degree, and for a vertex v, its degree in G is denoted by $d_G(v)$. If $S \subseteq V(G)$, we define $G - S = G[V(G) \setminus S]$, and if $S \subseteq E(G)$, we define $G - S = (V(G), E(G) \setminus S)$. Unless stated otherwise, throughout the article n denotes the number of vertices of the input graph of the problem under consideration. We will always assume that the input graphs are connected.

2   Preliminary Results

Clearly, any connected graph G is λ_1-connected, and $\lambda_1(G)$ can be computed in polynomial time by a MINIMUM CUT algorithm (cf. [28]). However, for $k \geq 2$, there exist infinitely many connected graphs which are not λ_k-connected, such as the graphs containing a cut vertex u such that every component of $G - u$ has at most $k - 1$ vertices (these graphs are called *flowers* in the literature [5], and correspond exactly to stars when $k = 2$). Moreover, the problem of determining whether a graph is λ_k-connected is hard. Indeed, given a graph G, if n is even and $k = n/2$, by [13, Theorem 2.2] it is NP-complete to determine whether G contains two vertex-disjoint connected subgraphs of order $n/2$ each. We can summarize this discussion as follows.

Remark 1. *Determining whether a connected graph is λ_k-connected is NP-complete when k is part of the input.*

In Sect. 4 we will strengthen the above hardness result to the case where the maximum degree of the input graph is at most 5.

In this article we will be interested in the following optimization problem.

RESTRICTED EDGE-CONNECTIVITY (REC)
  Instance: A connected graph $G = (V, E)$ and a positive integer k.
  Output: $\lambda_k(G)$, or a correct report that G is not λ_k-connected.

Note that Remark 1 implies that the above problem is NP-hard. Furthermore, even if the input graph G is guaranteed to be λ_k-connected, computing $\lambda_k(G)$ remains hard, as shown by the following theorem.

Theorem 1. *The REC problem is NP-hard restricted to λ_k-connected graphs.*

PROOF: We prove it for n even and $k = n/2$. The reduction is from the MINIMUM BISECTION problem[1] restricted to connected 3-regular graphs, which is known to be NP-hard [3]. Given a 3-regular connected graph G with even number of vertices as instance of MINIMUM BISECTION, we construct from it an instance G' of REC by adding two non-adjacent universal vertices v_1 and v_2. Note that G' is $\lambda_{n/2}$-connected, since any bipartition of $V(G')$ containing v_1 and v_2 in different parts induces two connected subgraphs.

We claim that v_1 and v_2 should necessarily belong to different connected subgraphs in any optimal solution in G'. Indeed, let (V_1, V_2) be a bipartition of $V(G)$

---

[1] Given a graph G with even number of vertices, the MINIMUM BISECTION problem consists in partitioning $V(G)$ into two equally-sized parts minimizing the number of edges with one endpoint in each part.

such that $|[V_1, V_2]| = \lambda_{n/2}(G')$, and assume for contradiction that $v_1, v_2 \in V_1$. Since G is connected, there is a vertex $u \in V_2$ with at least one neighbor in $V_1 \backslash \{v_1, v_2\}$. Let $V_1' := V_1 \cup \{u\} \backslash \{v_2\}$ and $V_2' := V_2 \cup \{v_2\} \backslash \{u\}$, and note that both $G[V_1']$ and $G[V_2']$ are connected. Since u has at least one neighbor in $V_1 \backslash \{v_1, v_2\}$, G is 3-regular, and v_1 and v_2 are non-adjacent and adjacent to all other vertices of G', it can be checked that $|[V_1', V_2']| \leq |[V_1, V_2]| - 1 = \lambda_{n/2}(G') - 1$, contradicting the definition of $\lambda_{n/2}(G')$.

Therefore, solving the REC problem in G' corresponds exactly to solving the MINIMUM BISECTION problem in G, concluding the proof.   □

3   A Parameterized Analysis

The NP-hardness results of the previous section naturally lead to considering parameterized versions of the problem. In this section we consider the following three distinct parameterizations.

PARAMETERIZED RESTRICTED EDGE-CONNECTIVITY (p-REC)
> **Instance:** A connected graph $G = (V, E)$ and two integers k and ℓ.
> **Parameter 1:** The integers k and ℓ.
> **Parameter 2:** The integer k.
> **Parameter 3:** The integer ℓ.
> **Question:** $\lambda_k(G) \leq \ell$?

As mentioned in the introduction, determining whether $\lambda_k(G) \leq \ell$ corresponds exactly to determining whether G admits a $(k - 1, \ell)$-good edge separation. This latter problem has been recently shown to be solvable in time $2^{O(\min\{k,\ell\} \log(k+\ell))} \cdot n^3 \log n$ by Chitnis et al. [6, Lemma II.2].

Theorem 2 (Chitnis et al. [6]). *The* p-REC *problem is* FPT *when parameterized by both k and ℓ.*

We would like to note that any improvement on the running time of the algorithm behind Theorem 2 would answer an open question raised in [7,8], and would have direct consequences and improve the algorithms described in [6,9,23].

As pointed out in [15,20], the p-REC problem can be solved in time $O^*(n^{2k})$. Roughly, the idea is to guess two sets of k vertices inducing a connected subgraph, contract them into two vertices s and t, and then call a polynomial-time MINIMUM CUT algorithm between s and t (cf. [28]). In other words, it is in XP when parameterized by k. The following theorem shows that this is essentially the best algorithm we can hope for when the parameter is only k.

Theorem 3. *The* p-REC *problem is* W[1]-*hard when parameterized by k.*

PROOF: We reduce from k-CLIQUE, which is known to be W[1]-hard [12]. The parameterized reduction is the same as the one given by Downey et al. in [11, Theorem 2] to show the W[1]-hardness of the CUTTING k VERTICES FROM A GRAPH problem, only the analysis changes.

Let $G = (V, E)$ be an n-vertex graph for which we wish to determine whether it has a k-clique. We construct a graph G' as follows:

(1) We start with a clique C of size n^3 and n *representative* vertices corresponding bijectively with the vertices of G.
(2) Every representative vertex v is connected to $n^2 - d_G(v)$ arbitrary vertices of C.
(3) If $uv \in E(G)$ then $uv \in E(G')$.

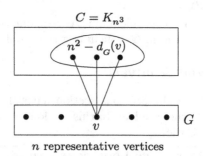

$$C = K_{n^3}$$

$$n^2 - d_G(v)$$

G

v

n representative vertices

Fig. 1. Illustration of the graph G' in the proof of Theorem 3.

See Fig. 1 for an illustration of G'. Consider $\ell = kn^2 - 2\binom{k}{2}$ and take $k \leq n/2$. We claim that G has a k-clique if and only if G' is a YES-instance of **p-REC**.

Suppose first that $K \subseteq V(G)$ is a k-clique in G. Obviously, K is connected in G' and has k vertices. On the other hand, $G' - K$ is also connected with at least $n^3 - |K| > k$ vertices. Finally, it is straightforward to check that $||[K, V(G')\backslash K]|| = kn^2 - 2\binom{k}{2} = \ell$.

In the other direction, suppose G' has a k-restricted edge-cut with at most ℓ edges, i.e., there exists $K \subseteq V(G')$ such that $G[K]$ and $G' - K$ are connected, $|K| \geq k$, $|V(G')\backslash K| \geq k$, and $||[K, V(G')\backslash K]|| \leq \ell$. Two cases need to be distinguished.

Case 1. $C \cap K = \emptyset$. Then every vertex of K must be a representative vertex. Hence $||[K, V(G')\backslash K]|| = |K|n^2 - 2|E(G'[K])|$ since every representative vertex has degree n^2. As by hypothesis $||[K, V(G')\backslash K]|| \leq \ell = kn^2 - 2\binom{k}{2}$, it follows that $|E(G'[K])| = \binom{k}{2}$, hence K must be a k-clique.

Case 2. $C \cap K \neq \emptyset$. Note that for every bipartition (C_1, C_2) of C we have that $||[C_1, C_2]|| \geq n^3 - 1 > \ell$. Now suppose $C \cap (V(G')\backslash K) \neq \emptyset$ and consider the bipartition $C_1 = C \cap K$ and $C_2 = C \cap (V(G')\backslash K)$ of C. Then $||[K, V(G')\backslash K]|| \geq ||[C_1, C_2]|| \geq n^3 - 1 > \ell$, a contradiction. Therefore, we have that $C \cap (V(G')\backslash K) = \emptyset$. The proof concludes by applying Case 1 to $V(G')\backslash K$ instead of K. $\qquad\square$

In contrast to Theorem 3 above, we now prove that the problem of determining whether a connected graph G is λ_k-connected (which is NP-complete by Remark 1) is FPT when parameterized by k. The proof uses the technique of *splitters* introduced by Naor *et al.* [26], which has also been recently used for designing parameterized algorithms in [6,9,23]. Our main tool is the following lemma.

Lemma 1 (Chitnis *et al.* [6]). *There exists an algorithm that given a set U of size n and two integers $a, b \in [0, n]$, outputs in time $2^{O(\min\{a,b\} \cdot \log(a+b))} \cdot n \log n$ a set $\mathcal{F} \subseteq 2^U$ with $|\mathcal{F}| = 2^{O(\min\{a,b\} \cdot \log(a+b))} \cdot \log n$ such that for every two sets $A, B \subseteq U$, where $A \cap B = \emptyset$, $|A| \leq a$, and $|B| \leq b$, there exists a set $S \in \mathcal{F}$ with $A \subseteq S$ and $B \cap S = \emptyset$.*

Theorem 4. *Given a connected graph G and a positive integer k, the problem of determining whether G is λ_k-connected is* FPT *when parameterized by k.*

PROOF: We use the easy property that G is λ_k-connected if and only if G contains two vertex-disjoint trees T_1 and T_2 such that $|V(T_1)| \geq k$ and $|V(T_2)| \geq k$. In order to apply Lemma 1, we take $U = V(G)$ and $a = b = k$, obtaining in time $k^{O(k)} \cdot n \log n$ the desired family $\mathcal{F}$ of subsets of vertices of G. Now, if such trees T_1 and T_2 exist, then necessarily there exists a set $S \in \mathcal{F}$ such that $V(T_1) \subseteq S$ and $V(T_2) \cap S = \emptyset$. Therefore, in order to determine whether G is λ_k-connected or not, it suffices to check, for each set $S \in \mathcal{F}$, whether both $G[S]$ and $G - S$ contain a connected component with at least k vertices. (Note that for each such set $S \in \mathcal{F}$, this can be done in linear time.) Indeed, if such a set S exists, then clearly G is λ_k-connected. Otherwise, by the property of the family $\mathcal{F}$, G does not contain two disjoint trees T_1 and T_2 of size k each, and therefore G is not λ_k-connected. □

Concerning the parameterized complexity of the **p-REC** problem, in view of Theorems 2 and 3, it just remains to settle the case when the parameter is ℓ only. While we cannot prove whether the **p-REC** problem is FPT or not when parameterized by ℓ, we are able to prove that it does not admit polynomial kernels, assuming that coNP $\subseteq$ NP/poly.

Theorem 5. *Unless* coNP $\subseteq$ NP/poly, *the* **p-REC** *problem does not admit polynomial kernels when parameterized by ℓ.*

PROOF: The proof is strongly inspired by the one given by van Bevern *et al.* [29, Theorem 3] to prove that the MINIMUM BISECTION problem does not admit polynomial kernels, which in turn resembles the proof given by Garey *et al.* [19] to prove the NP-hardness of MINIMUM BISECTION. The main difference with respect to the proof given in [29] is that we need to make the appropriate modifications to guarantee that both parts left out by the edge-cut are *connected*, which is not an issue in the MINIMUM BISECTION problem.

We will first rule out the existence of polynomial kernels for the generalization of **p-REC** where the edges have non-negative integer weights, and the objective is to decide whether the input graph can be partitioned into two connected subgraphs with at least k vertices each by removing a set of edges whose total weight does not exceed ℓ. We call this problem EDGE-WEIGHTED **p-REC**. Then it will just remain to get rid of the edge weights. This is done in Appendix A.

As shown by Bodlaender *et al.* [4], in order to prove that EDGE-WEIGHTED **p-REC** does not admit polynomial kernels when parameterized by ℓ (assuming that coNP $\subseteq$ NP/poly), it is sufficient to define a *cross composition* from an NP-hard problem to EDGE-WEIGHTED **p-REC**. In our case, the NP-hard problem is

MAXIMUM CUT (see [18]), which is defined as follows. Given a graph $G = (V, E)$ and an integer p, one has to decide whether V can be partitioned into two sets A and B such that there are at least p edges with an endpoint in A and an endpoint in B.

A cross composition from MAXIMUM CUT to EDGE-WEIGHTED p-REC parameterized by ℓ consists in a polynomial-time algorithm that, given t instances $(G_1, p_1), \ldots, (G_t, p_t)$ of MAXIMUM CUT, constructs an instance (G^*, k, ℓ) of EDGE-WEIGHTED p-REC such that (G^*, k, ℓ) is a YES-instance if and only if one of the t instances of MAXIMUM CUT is a YES-instance, and such that ℓ is polynomially bounded as a function of $\max_{1 \leq i \leq t} |V(G_i)|$. Similarly to [29], we may safely assume that t is odd, that for each $1 \leq i \leq t$ we have $|V(G_i)| =: n$ and $p_i =: p$, and that $1 \leq p \leq n^2$.

Given $(G_1, p), \ldots, (G_t, p)$, we create G^* as follows. Let $w_1 := 5n^2$ and $w_2 := 5$. For each graph $G_i = (V_i, E_i)$ add to G^* the vertices in V_i and a clique V_i' with $|V_i| = n$ vertices whose edges have weight w_1. Add an edge of weight w_1 between each vertex in V_i and each vertex in V_i'. For each pair of vertices $u, v \in V_i$, add the edge $\{u, v\}$ to G^* with weight $w_1 - w_2$ if $\{u, v\} \in E_i$, and with weight w_1 otherwise. Let s_i^1, s_i^2 be two arbitrary distinct vertices in V_i', which we call *link* vertices. For $1 \leq i \leq t - 1$, add two edges with weight 1 between s_i^1 and s_{i+1}^1 and between s_i^2 and s_{i+1}^2, and two edges with weight 1 between s_t^1 and s_1^1 and between s_t^2 and s_1^2. This completes the construction of G^*; see Fig. 2 for an illustration. These $2t$ edges among distinct V_i''s are called *chain* edges. Finally, we set $k := |V(G^*)|/2$ and $\ell := w_1 n^2 - w_2 p + 4$. Note that k is *not* polynomially bounded in terms of n, but this is not a problem since the parameter we consider is ℓ, which is bounded by $5n^4$. This construction can clearly be performed in polynomial time in $t \cdot n$. We claim that (G^*, k, ℓ) is a YES-instance of EDGE-WEIGHTED p-REC if and only if there exists $i \in \{1, \ldots, t\}$ such that (G_i, p) is a YES-instance of MAXIMUM CUT.

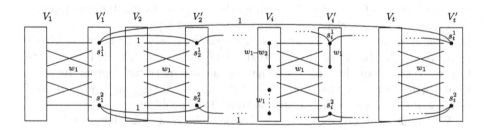

Fig. 2. Illustration of the graph G^* in the proof of Theorem 5.

Assume first that there exists $i \in \{1, \ldots, t\}$ such that (G_i, p) is a YES-instance of MAXIMUM CUT. Assume without loss of generality that $i = 1$, and let $V_1 = A \uplus B$ such that there are at least p edges in E_1 between A and B. We proceed to partition $V(G^*)$ into two equally-sized sets A' and B' such that both $G^*[A']$ and $G^*[B']$ are connected, and such that the total weight of the edges in G^* with one endpoint in A' and one endpoint in B' is at most ℓ. The set A' contains $V_1 \cap A$, any set of $|B|$ vertices in V_1 containing exactly one of s_1^1 and s_1^2 (this is possible since $1 \leq |B| \leq n-1$), and $\bigcup_{i=2}^{\lceil t/2 \rceil} V_i \cup V_i'$. Then $B' = V(G^*) \setminus A'$. Since t is odd, $|A'| = |B'|$. Let us now see that $G^*[A']$ is connected. As $1 \leq |A| \leq n-1$, the set $V_1 \cap A'$ is connected to $V_1' \cap A'$, which is connected to V_2' since A' contains exactly one of the link vertices s_1^1 and s_1^2. The graph $G^*[\bigcup_{i=2}^{\lceil t/2 \rceil} V_i \cup V_i']$ is clearly connected because of the chain edges, which implies that $G^*[A']$ is indeed connected. The proof for the connectivity of $G^*[B']$ is similar, using that $V_1' \cap B'$ is connected to $V_1' \cap B'$ since $1 \leq |B| \leq n-1$, which is in turn connected to V_t' since B' contains exactly one of the link vertices s_1^1 and s_1^2. Finally, let us show that the total weight of the edges between A' and B' is at most ℓ. Note first that two chain edges incident to V_1' and two chain edges incident to $V_{\lceil t/2 \rceil}'$ belong to the cut defined by A' and B', and no other chain edge belongs to the cut. Beside the chain edges, only edges in the graph $G^*[V_1 \cup V_1']$ are cut. Note that $G^*[V_1 \cup V_1']$ is a clique on $2n$ vertices and each of $(V_1 \cup V_1') \cap A'$ and $(V_1 \cup V_1') \cap B'$ contains n vertices. Since (G_1, p) is a YES-instance of MAXIMUM CUT, at least p of the edges of weight $w_1 - w_2$ belong to the cut. Therefore, the total weight of the cut is at most $w_1 n^2 - w_2 p + 4 = \ell$.

Conversely, assume that for all $i \in \{1, \ldots, t\}$, (G_i, p) is a NO-instance of MAXIMUM CUT, and we want to prove that (G^*, k, ℓ) is a NO-instance of EDGE-WEIGHTED $\mathbf{p}$-REC. Let $A \uplus B$ be a partition of $V(G^*)$ such that $|A| = |B|$ and both $G^*[A]$ and $G^*[B]$ are connected, and such that the weight of the edges between A and B is minimized among all such partitions. For $1 \leq i \leq t$, we let $a_i := |(V_i \cup V_i') \cap A|$. Since for $1 \leq i \leq t$, (G_i, p) is a NO-instance of MAXIMUM CUT, any bipartition of V_i cuts at most $p-1$ edges. Therefore, the total weight of the edges between $(V_i \cup V_i') \cap A$ and $(V_i \cup V_i') \cap B$ is at least $w_1 a_i (2n - a_i) - (p-1) w_2$.

Since t is odd, necessarily at least one of the graphs G_i is cut by $A \uplus B$. Assume first that exactly one graph G_i is cut by $A \uplus B$. Since $|A| = |B|$, we have that $a_i = n$, so the value of the cut is at least $w_1 n^2 - (p-1) w_2 = w_1 n^2 - p w_2 + w_2 > w_1 n^2 - p w_2 + 4 = \ell$, and thus (G^*, k, ℓ) is a NO-instance of EDGE-WEIGHTED $\mathbf{p}$-REC.

We claim that there is always exactly one graph G_i cut by $A \uplus B$. Assume for contradiction that it is not the case, that is, that there are two strictly positive values a_i, a_j for some $i \neq j$. By symmetry between A and B, we may assume that $a_i + a_j \leq 2n$. The total weight of the edges cut in $G^*[V_i \cup V_i']$ and $G^*[V_j \cup V_j']$ is at least

$$w_1 a_i (2n - a_i) - (p-1) w_2 + w_1 a_j (2n - a_j) - (p-1) w_2 = 2n w_1 (a_i + a_j) - w_1 (a_i^2 + a_j^2) - 2 w_2 (p-1).$$

Now we construct another solution of EDGE-WEIGHTED $\mathbf{p}$-REC in G^* where the $a_i + a_j$ vertices are cut in only one of $V_i \cup V_i'$ and $V_j \cup V_j'$, say $V_i \cup V_i'$ (note that this is possible since $a_i + a_j \leq 2n$). The connectivity of each of the two parts of the newly obtained bipartition $A' \uplus B'$ of G^* can be guaranteed as follows. If $a_i + a_j = 2n$, then $V_i \cup V_i'$ is entirely contained in A' or B', and both parts are clearly connected. If $a_i + a_j < 2n$, then we choose $(V_i \cup V_i') \cap A'$ such that it contains exactly one of s_i^1 and s_i^2, which ensures the connectivity of both $G^*[A']$ and $G^*[B']$. Taking into account that each V_i' has four incident chain edges, the total weight of the edges cut in $G^*[V_i \cup V_i']$ and $G^*[V_j \cup V_j']$ by the new solution is at most

$$w_1(a_i + a_j)(2n - a_i - a_j) + 8 =$$
$$2nw_1(a_i + a_j) - w_1(a_i^2 + a_j^2) - 2w_1a_ia_j + 8.$$

That is, the weight of the cut defined by $A \uplus B$ minus the weight of the cut defined by $A' \uplus B'$ is at least

$$-2w_2(p - 1) + 2w_1a_ia_j - 8 =$$
$$2(w_1a_ia_j - w_2(p - 1) - 4) \geq$$
$$2(w_1 - w_2(n^2 - 1) - 4) > 0,$$

where we have used that $a_i, a_j \geq 1$, $p \leq n^2$, $w_1 = 5n^2$, and $w_2 = 5$. In other words, $A' \uplus B'$ defines a cut of strictly smaller weight, contradicting the definition of $A \uplus B$. $\qquad\square$

4   Considering the Maximum Degree as a Parameter

Towards understanding the parameterized complexity of the REC problem, one may wonder whether considering the maximum degree of the input graph as an extra parameter turns the problem easier (this is a classical approach in parameterized complexity, see for instance [24,25]). We first prove that, from a classical complexity point of view, bounding the degree of the input graph does not turn the problem easier. Before stating the hardness result, we need the define the 3-DIMENSIONAL MATCHING problem, 3DM for short.

An instance of 3DM consists of a set $W = R \cup B \cup Y$, where R, B, Y are disjoint sets with $|R| = |B| = |Y| = m$, and a set of triples $T \subseteq R \times B \times Y$. The question is whether there exists a matching $M \subseteq T$ covering W, i.e., $|M| = m$ and each element of $W = R \cup B \cup Y$ occurs in exactly one triple of M.

An instance of 3DM can be represented by a bipartite graph $G_I = (W \cup T, E_I)$, where $E_I = \bigcup_{t=(r,b,y)\in T} \{\{r, t\}, \{b, t\}, \{y, t\}\}$.

It is known that 3DM is NP-complete even if each element of W appears in 2 or 3 triples only [13,14]. In [13, Theorem 2.2] it is proved that partitioning a graph G into two connected subgraphs of equal size is NP-hard, using a reduction from 3DM. It is worth noting that the graph constructed in the NP-hardness reduction contains only two vertices of degree greater than five. In Theorem 6 we appropriately modify the reduction of [13, Theorem 2.2] so that the constructed graph has maximum degree at most 5. The proof can be found in Appendix B.

Theorem 6. *Determining whether a connected graph G is λ_k-connected is NP-complete when k is part of the input, even if the maximum degree of G is 5.*

In order to understand to which extent the vertices of high degree make the complexity of computing the restricted edge-connectivity of a graph hard, we also consider the maximum degree of the input graph as a parameter for the p-REC problem.

Theorem 7. *The* **p-REC** *problem is* FPT *when parameterized by k and the maximum degree Δ of the input graph.*

PROOF: The algorithm is based on a simple exhaustive search. We use the property that, for any graph G and any two integers k, ℓ, $\lambda_k(G) \leq \ell$ if and only if G contains two vertex-disjoint trees T_1 and T_2 with $|V(T_1)| \geq k$ and $|V(T_2)| \geq k$, such that there exists an edge set S in G with $|S| \leq \ell$ such that in $G - S$ the trees T_1 and T_2 belong to different connected components. Hence, we just have to determine whether these trees exist in G or not. For doing so, for every pair of distinct vertices v_1 and v_2 of G, we exhaustively consider all trees T_1 and T_2 with k vertices containing v_1 and v_2, respectively. Note that the number of such trees is at most Δ^{2k}. For every pair of vertex-disjoint trees T_1 and T_2, we proceed as follows. We contract tree T_1 (resp. T_2) to a single vertex t_1 (resp. t_2), keeping edge multiplicities, and then we run in the resulting graph a polynomial-time MINIMUM CUT algorithm between t_1 and t_2 (cf. [28]). If the size of the returned edge-cut is at most ℓ, then T_1 and T_2 are the desired trees. Otherwise, we continue searching. It is clear that the overall running time of this algorithm is $O(\Delta^{2k} \cdot n^{O(1)})$. $\qquad\square$

A   Dealing with the Edge Weights in the Proof of Theorem 5

As in [29], we show how to convert the instance (G^*, k, ℓ) of EDGE-WEIGHTED p-REC that we just constructed into an equivalent instance of **p-REC** such that the resulting parameter remains polynomial in n. Given (G^*, k, ℓ), we define $(\hat{G}, k, \ell)$ as the instance of **p-REC**, where $\hat{G}$ is an unweighted graph obtained from G^* as follows. We replace each vertex v of G^* with a clique C_v of size $w_1 + \ell + 1$, and for each edge $\{u, v\}$ of G^* with weight w, we add w pairwise disjoint edges between the cliques C_u and C_v. Since no cut of size at most ℓ in $\hat{G}$ can separate a clique C_v introduced for a vertex v, it follows that (G^*, k, ℓ) is a YES-instance of EDGE-WEIGHTED p-REC if and only if $(\hat{G}, k, \ell)$ is a YES-instance of **p-REC**. Finally, it is clear that the desired cut size ℓ is still polynomial in n.

B   Proof of Theorem 6

Given an instance (W, T) of 3DM with $W = R \cup B \cup Y$, $|R| = |B| = |Y| = m$, and $T \subseteq R \times B \times Y$ such that each element of W appears in 2 or 3 triples only, we define an n-vertex graph $G = (V, E)$ with maximum degree 5 as follows (see Fig. 3 for an illustration).

The set of vertices of G is

$$V = W \cup T \cup T_a \cup T_b \cup \mathcal{P} \cup \{a\},$$

where $T_a = \{t_1^a \ldots, t_{|T|}^a\}$, $T_b = \{b = t_1^b, t_2^b \ldots, t_{|T|}^b\}$, $T = \{t_1 \ldots, t_{|T|}\}$ is the set of triples, and $\mathcal{P} = \bigcup_{\sigma \in W \cup T_b \cup \{a\}} P_\sigma$, where $P_\sigma = \{(\sigma, t) : t = 1, \ldots, n_\sigma\}$ with $n_a = (3m + |T|)n_b + 5m - |T| - 1$, $n_b = 2m^3$, and $n_\sigma = n_b$ for every $\sigma \in W \cup T_b$.

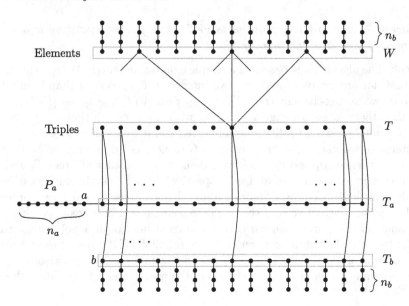

Fig. 3. Construction of the graph G in the proof of Theorem 6, with $\Delta(G) = 5$.

The set of edges of G is

$$E = E_I \cup E_{T_a} \cup E_{T_b} \cup E_{T^+} \bigcup_{\sigma \in W \cup T_b \cup \{a\}} E_\sigma,$$

where $E_{T_a} = \{\{t_i^a, t_{i+1}^a\} : 1 \leq i \leq |T| - 1\}$, $E_{T_b} = \{\{t_i^b, t_{i+1}^b\} : 1 \leq i \leq |T| - 1\}$, $E_{T^+} = \{\{t_i, t_i^a\}, \{t_i, t_i^b\} : 1 \leq i \leq |T|\}$, and $E_\sigma = \{\{\sigma, (\sigma, 1)\}\} \cup \{\{(\sigma, t), (\sigma, t + 1)\} : 1 \leq t \leq n_\sigma - 1\} \cup \{a, t_1^a\}$ for every $\sigma \in W \cup T_b \cup \{a\}$.

Note that the maximum degree of G is indeed 5. Since $n = 1 + 3m + 3|T| + n_a + (3m + |T|)n_b$, we can observe that

$$n = 2(n_a + 1 + 2|T| - m).$$

Next, we show that for $k = n/2$, G is YES-instance of the REC problem if and only if T contains a matching covering W.

One direction is easy. Suppose first that T contains a matching M covering W. Let $S = \{a\} \cup P_a \cup T_a \cup (T \backslash M)$. It is straightforward to check that $|S| = n/2$ and that $G[S]$, $G[V \backslash S]$ are both connected.

Conversely, suppose that G can be partitioned into 2 connected subgraphs $G[S]$, $G[V \backslash S]$ with $|S| = n/2$. We can assume that $a \in S$, and then it follows that $P_a \subseteq S$. Now $|S \backslash (P_a \cup \{a\})| = 2|T| - m < 2m^3 = n_b$ since $|T| \leq m^3$. As $P_\sigma \subseteq S$ if and only if $\sigma \in S \cap (W \cup T_b)$, then $S \cap (W \cup T_b) = \emptyset$ since $|S \backslash (P_a \cup \{a\})| < n_b$ and $|P_\sigma| = n_b$ for every $\sigma \in W \cup T_b$. Hence $S \backslash (P_a \cup \{a\}) \subseteq T \cup T_a$. Let $M = (V \backslash S) \cap T$. Then $|M| \leq m$ since $|S \backslash (P_a \cup \{a\})| = 2|T| - m$. Finally, as $G[V \backslash S]$ is connected and $W \cup T_b \subseteq V \backslash S$, it follows that $|M| \geq m$. Hence $|M| = m$ and M must be a matching covering W.

C   Computing the Minimum k-edge-degree

As it has been already mentioned, a graph may not have k-restricted edge-cuts. In fact, Esfahanian and Hakimi [15] showed that each connected graph G of order $n \geq 4$ except a star, is λ_2-connected and satisfies $\lambda_2(G) \leq \xi(G)$, where $\xi(G)$ is the *minimum edge-degree* of G defined as

$$\xi(G) = \min\{d_G(u) + d_G(v) - 2 : uv \in E(G)\}.$$

Bonsma, Ueffing and Volkmann [5] defined an extension of the minimum edge-degree of a graph G for an integer $k \geq 2$, called the *minimum k-edge-degree*, as follows:

$$\xi_k(G) = \min\{|[X, \overline{X}]| : |X| = k \text{ and } G[X] \text{ is connected}\}.$$

They proved that $\lambda_k(G) \leq \xi_k(G)$ for $1 \leq k \leq 3$ and all graphs G aside from a class of exceptions for $k = 3$ determined in [5]. Also in the same paper, the authors give a number of examples, which show that $\lambda_k(G) \leq \xi_k(G)$ is not true in general for $k \geq 4$. In 2005, Zhang and Yuan [31] proved that, except for the class of flowers, graphs with minimum degree greater than or equal to $k - 1$ are λ_k-connected. Moreover, for the same class of graphs they showed that $\lambda_k(G) \leq \xi_k(G)$ (recall that a graph G with $|V(G)| \geq 2k$ is called a *flower* if it contains a cut vertex u such that every component of $G - u$ has order at most $k - 1$).

Therefore, given the relation between the invariants λ_k and ξ_k, we are interested in the following parameterized problem.

PARAMETERIZED MINIMUM EDGE-DEGREE (p-MED)
   Instance: A graph $G = (V, E)$ and two positive integers k and ℓ.
Parameter 1: The integers k and ℓ.
Parameter 2: The integer k.
Parameter 3: The integer ℓ.
   Question: $\xi_k(G) \leq \ell$?

Similarly to what happens with the **p-REC** problem, the **p-MED** problem is FPT with parameters k and ℓ, and W[1]-hard with parameter k.

Theorem 8. *The* **p-MED** *problem is* FPT *when parameterized by k and ℓ.*

PROOF: The proof is based on a simple application of the splitters technique. Note that given G, k, ℓ, if $\xi_k(G) \leq \ell$ then G contains a vertex set X of size k whose neighborhood in $G - X$, say N_X, has size at most ℓ. We apply Lemma 1 with $U = V(G)$, $a = k$, and $b = \ell$, obtaining in time $(k+\ell)^{O(k+\ell)} \cdot n^{O(1)}$ a family $\mathcal{F}$ of subsets of $V(G)$ with $|\mathcal{F}| = (k + \ell)^{O(k+\ell)} \cdot \log n$. If $\xi_k(G) \leq \ell$, then there exists $S \in \mathcal{F}$ containing X and disjoint from N_X. Therefore, it suffices to check, for every $S \in \mathcal{F}$, whether $G[S]$ contains a connected component X with $|X| = k$ and $|[X, \overline{X}]| \leq \ell$.   □

Theorem 9. *The* **p**-MED *problem is* W[1]-*hard when parameterized by* k.

PROOF: The reduction is inspired by the one given in [24] to prove the W[1]-hardness of the CUTTING ℓ VERTICES problem. Given an instance (G, k) of k-CLIQUE such that G is r-regular for some $r \geq k$ (the k-CLIQUE problem is easily seen to remain W[1]-hard with this assumption [24]), we define an instance of **p**-MED as (G, k, ℓ), with $\ell := k(r - k + 1)$. It is then clear that G has a clique of size k if and only if it has a vertex subset X such that $|X| = k$, $G[X]$ is connected, and $|[X, \overline{X}]| \leq k(r - k + 1)$.   □

We leave as an open problem the parameterized complexity of the **p**-MED problem with parameter ℓ.

References

1. Balbuena, C., Carmona, A., Fàbrega, J., Fiol, M.A.: Extraconnectivity of graphs with large minimum degree and girth. Discrete Math. **167**, 85–100 (1997)
2. Berman, P., Karpinski, M.: Approximation hardness of bounded degree MIN-CSP and MIN-BISECTION. Electron. Colloquium Comput. Complex. 8(26) (2001)
3. Berman, P., Karpinski, M.: Approximation hardness of bounded degree MIN-CSP and MIN-BISECTION. In: Widmayer, P., Triguero, F., Morales, R., Hennessy, M., Eidenbenz, S., Conejo, R. (eds.) ICALP 2002. LNCS, vol. 2380, pp. 623–632. Springer, Heidelberg (2002)
4. Bodlaender, H.L., Jansen, B.M.P., Kratsch, S.: Kernelization lower bounds by cross-composition. SIAM J. Discrete Math. **28**(1), 277–305 (2014)
5. Bonsma, P., Ueffing, N., Volkmann, L.: Edge-cuts leaving components of order at least three. Discrete Math. **256**(1), 431–439 (2002)
6. Chitnis, R.H., Cygan, M., Hajiaghayi, M., Pilipczuk, M., Pilipczuk, M.: Designing FPT algorithms for cut problems using randomized contractions. In: Proceedings of the 53rd Annual IEEE Symposium on Foundations of Computer Science (FOCS), pp. 460–469 (2012)
7. Cygan, M., Fomin, F., Jansen, B.M., Kowalik, L., Lokshtanov, D., Marx, D., Pilipczuk, M., Pilipczuk, M.: Open problems from School on Parameterized Algorithms and Complexity (2014). http://fptschool.mimuw.edu.pl/opl.pdf
8. Cygan, M., Kowalik, L., Pilipczuk, M.: Open problems from Update Meeting on Graph Separation Problems (2013). http://worker2013.mimuw.edu.pl/slides/update-opl.pdf
9. Cygan, M., Lokshtanov, D., Pilipczuk, M., Pilipczuk, M., Saurabh, S.: Minimum bisection is fixed parameter tractable. In: Proceedings of the 46th ACM Symposium on Theory of Computing (STOC), pp. 323–332 (2014)
10. Diestel, R.: Graph Theory, 3rd edn. Springer, Berlin (2005)
11. Downey, R.G., Estivill-Castro, V., Fellows, M.R., Prieto, E., Rosamond, F.A.: Cutting up is hard to do: the parameterized complexity of k-cut and related problems. Electron. Notes Theor. Comput. Sci. **78**, 209–222 (2003)
12. Downey, R.G., Fellows, M.R.: Parameterized Complexity. Springer, New York (1999)
13. Dyer, M.E., Frieze, A.M.: On the complexity of partitioning graphs into connected subgraphs. Discrete Appl. Math. **10**, 139–153 (1985)

14. Dyer, M.E., Frieze, A.M.: Planar 3DM is NP-complete. J. Algorithms **7**(2), 174–184 (1986)
15. Esfahanian, A.-H., Hakimi, S.L.: On computing a conditional edge-connectivity of a graph. Inf. Process. Lett. **27**(4), 195–199 (1988)
16. Fàbrega, J., Fiol, M.A.: Extraconnectivity of graphs with large girth. Discrete Math. **127**, 163–170 (1994)
17. Flum, J., Grohe, M.: Parameterized Complexity Theory. Springer, Heidelberg (2006)
18. Garey, M.R., Johnson, D.S.: Computers and Intractability: A Guide to the Theory of NP-Completeness. W. H. Freeman and Co., New York (1979)
19. Garey, M.R., Johnson, D.S., Stockmeyer, L.J.: Some simplified NP-complete graph problems. Theoret. Comput. Sci. **1**(3), 237–267 (1976)
20. Holtkamp, A.: Connectivity in Graphs and Digraphs. Maximizing vertex-, edge- and arc-connectivity with an emphasis on local connectivity properties. Ph.D. thesis, RWTH Aachen University (2013)
21. Holtkamp, A., Meierling, D., Montejano, L.P.: k-restricted edge-connectivity in triangle-free graphs. Discrete Appl. Math. **160**(9), 1345–1355 (2012)
22. Kawarabayashi, K., Thorup, M.: The minimum k-way cut of bounded size is fixed-parameter tractable. In: Proceedings of the 52nd Annual Symposium on Foundations of Computer Science (FOCS), pp. 160–169 (2011)
23. Kim, E.J., Oum, S., Paul, C., Sau, I., Thilikos, D.M.: The List Allocation Problem and Some of its Applications in Parameterized Algorithms. Manuscript submitted for publication (2015). http://www.lirmm.fr/~sau/Pubs/LA.pdf
24. Marx, D.: Parameterized graph separation problems. Theor. Comput. Sci. **351**(3), 394–406 (2006)
25. Marx, D., Pilipczuk, M.: Everything you always wanted to know about the parameterized complexity of subgraph isomorphism (but were afraid to ask). In: Proceedings of the 31st International Symposium on Theoretical Aspects of Computer Science (STACS), pp. 542–553 (2014)
26. Naor, M., Schulman, L.J., Srinivasan, A.: Splitters and near-optimal derandomization. In: Proceedings of the 36th Annual Symposium on Foundations of Computer Science (FOCS), pp. 182–191 (1995)
27. Niedermeier, R.: Invitation to Fixed-Parameter Algorithms. Oxford University Press, Oxford (2006)
28. Stoer, M., Wagner, F.: A simple min-cut algorithm. J. ACM **44**(4), 585–591 (1997)
29. van Bevern, R., Feldmann, A.E., Sorge, M., Suchý, O.: On the parameterized complexity of computing graph bisections. In: Brandstädt, A., Jansen, K., Reischuk, R. (eds.) WG 2013. LNCS, vol. 8165, pp. 76–87. Springer, Heidelberg (2013)
30. Yuan, J., Liu, A.: Sufficient conditions for λ_k-optimality in triangle-free graphs. Discrete Math. **310**, 981–987 (2010)
31. Zhang, Z., Yuan, J.: A proof of an inequality concerning k-restricted edge-connectivity. Discrete Math. **304**, 128–134 (2005)

Computational Geometry

Weak Unit Disk and Interval Representation of Graphs

M.J. Alam[1], S.G. Kobourov[1], S. Pupyrev[1,2(✉)], and J. Toeniskoetter[1]

[1] Department of Computer Science, University of Arizona, Tucson, AZ, USA
spupyrev@gmail.com
[2] Institute of Mathematics and Computer Science, Ural Federal University,
Yekaterinburg, Russia

Abstract. We study a variant of intersection representations with unit balls: unit disks in the plane and unit intervals on the line. Given a planar graph and a bipartition of the edges of the graph into *near* and *far* edges, the goal is to represent the vertices of the graph by unit-size balls so that the balls for two adjacent vertices intersect if and only if the corresponding edge is near. We consider the problem in the plane and prove that it is NP-hard to decide whether such a representation exists for a given edge-partition. On the other hand, we show that series-parallel graphs (which include outerplanar graphs) admit such a representation with unit disks for any near/far bipartition of the edges. The unit-interval on the line variant is equivalent to threshold graph coloring, in which context it is known that there exist girth-3 planar graphs (even outerplanar graphs) that do not admit such coloring. We extend this result to girth-4 planar graphs. On the other hand, we show that all triangle-free outerplanar graphs and all planar graphs with maximum average degree less than 26/11 have such a coloring, via unit-interval intersection representation on the line. This gives a simple proof that all planar graphs with girth at least 13 have a unit-interval intersection representation on the line.

1 Introduction

Intersection graphs of various geometric objects have been extensively studied for their many applications [17]. A graph is a d-dimensional *unit ball graph* if its vertices are represented by unit-size balls in $\mathbb{R}^d$, and an edge exists between two vertices if and only if the corresponding balls intersect. Unit ball graphs are called *unit disk graphs* when $d = 2$ and *unit interval graphs* when $d = 1$. In this paper we study *weak unit ball graphs*: given a graph G whose edges have been partitioned into "near" and "far" sets, we wish to assign unit balls to the vertices of G so that, for an edge (u, v) of G, the balls representing u and v intersect if the edge (u, v) is near and do not intersect if the edge (u, v) is far. Note that if (u, v) is not an edge of G, then the balls of u and v may or may not intersect. We refer to such graphs as *weak unit disk* $(d = 2)$ and *weak unit interval graphs* $(d = 1)$. A geometric representation of such graphs (particularly, a mapping of the vertices to unit balls in $\mathbb{R}^2$ or $\mathbb{R}$), is called a *weak unit disk representation*

© Springer-Verlag Berlin Heidelberg 2016
E.W. Mayr (Ed.): WG 2015, LNCS 9224, pp. 237–251, 2016.
DOI: 10.1007/978-3-662-53174-7_17

or a *weak unit interval representation*; see Fig. 1. Near edges are shown as thick line segments and far edges are dashed line segments and we use this convention to distinguish near/far edges in the rest of the paper. Unit disk representations allow us to represent the edges of a graph by spatial proximity, which is intuitive from the point of view of human perception. Weak unit disk graphs also allow to arbitrarily forbid edges between certain pairs of vertices, which is useful in representation of "almost" unit disk graphs. It has been shown that weak unit interval graphs can be used to compute *unit-cube contact representations* of planar graphs [5,18].

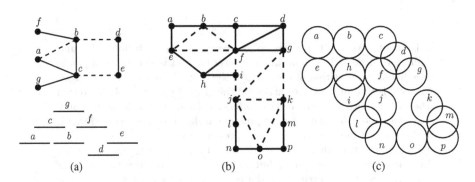

Fig. 1. (a) A graph with an edge-labeling and its weak unit interval representation. (b, c) A graph with an edge-labeling and its weak unit disk representation. In the figures we indicate near edges with solid lines and far edges with dashed lines.

Unit disk graphs have been extensively studied for their application to wireless sensor and radio networks. In such a network each sensor or radio can be modeled as a device with a unit size broadcast range, which naturally induces a unit disk graph by adding an edge whenever two ranges intersect. This setting makes it easy to study various practical problems. For example, in the *frequency assignment problem* the goal is to assign frequencies to radio towers so that nearby towers do not interfere with each other [15]. A weakness of the unit disk model is that it does not allow for interference between nodes (e.g., due to geography) and it does not account for the possibility that a pair of nodes may not be able to communicate (e.g., due to technological barriers). One attempt to address this issue are *quasi unit disk graphs* [19], where each vertex is represented by a pair of concentric disks, one of radius r, $0 < r < 1$, and the other of radius 1. In this model, two vertices are connected by an edge if their radius-r disks overlap, and do not have an edge if their radius-1 disks do not overlap. The remaining edges are in or out of the graph on a case by case basis. In the weak unit disk model such problems can be dealt with by simply deleting edges between nodes which are nearby but whose ranges do not overlap (e.g., because they are separated by a mountain range). This gives us more flexibility than quasi unit disk graphs.

Formally, an *edge-labeling* of a graph $G = (V, E)$ is a map $\ell : E \to \{N, F\}$. If $(u, v) \in E$, then (u, v) is called *near* if $\ell(u, v) = N$, and otherwise (u, v) is called *far*. In a unit disk (interval) representation I, each vertex $v \in V$ is represented as a disk (interval) centered at the point $I(v) \in \mathbb{R}^2$ ($\mathbb{R}$). We denote by $\|I(u) - I(v)\|$ the distance between the points $I(u)$ and $I(v)$, and by a slight abuse of notation, we also refer to $I(v)$ as the disk (interval) representing $v \in V$. A weak unit disk (interval) representation of G with respect to ℓ is a representation I such that for each edge $(u, v) \in E$, $\|I(u) - I(v)\| \leq t$ if and only if $\ell(u, v) = N$, for some fixed unit $t > 0$ (in other words, the disks and intervals have diameter t). Unless otherwise stated, we assume $t = 1$. We say that a graph is a *total weak unit disk (interval)* graphs if it has an appropriate representation for all possible edge-labelings.

Related Work: Weak unit ball graphs can be seen as a form of graph drawing/labeling where a notion of "closeness" between vertices is used to define edges, from a given set of permissible edges. There are many classes of graphs defined on some notion of vertex closeness. For example, *proximity graphs* are those that can be drawn in the plane such that every pair of adjacent vertices satisfies some fixed notion of closeness, whereas every pair of non-adjacent vertices satisfy some notion of farness [20]. Examples of proximity graphs are Gabriel graphs, Delaunay triangulations, and relative neighborhood graphs. *Gabriel graphs*, defined in the context of categorizing biological populations [13], can be embedded in the plane so that for every pair of vertices (u, v), the disk with u and v as antipodal points contains no other vertex if and only if (u, v) is an edge. Recently, Evans et al. [10] studied *region of influence graphs*, where each pair of vertices u, v in the plane is assigned a region $R(u, v)$, and there is an edge if and only if $R(u, v)$ contains no vertices, except possibly u and v. They generalize this class of graphs to *approximate proximity graphs*, where there are parameters $\epsilon_1 > 0$ and $\epsilon_2 > 0$, such that a vertex other than u or v is contained in $R(u, v)$, scaled by $1/(1 + \epsilon_1)$, if and only if (u, v) is an edge; the region $R(u, v)$, scaled by $1 + \epsilon_2$, is empty if and only if (u, v) is not an edge. However there is a significant difference between the notion of proximity graphs and the notion of weak unit ball graphs. In proximity graphs the notion of closeness is defined by two groups, namely adjacent and non-adjacent pairs of vertices, whereas for weak unit ball graphs, there are three groups. Specifically, the near and far edges in the input graph G represent vertex pairs with closeness and farness requirements, while all nonadjacent vertex pairs in G have no requirement on proximity. Thus proximity graphs is more restricted than the weak unit ball graphs, in that they can be modeled by weak unit ball graphs where the input graph is the complete graph K_n.

Weak unit ball representability in 1D is related to the recently introduced *threshold-coloring* problem [1] and we show that these two problems are in fact equivalent. In this variant of graph coloring, integer colors are assigned to the vertices so that endpoints of near edges differ by less than a given threshold, while endpoints of far edges differ by more than the threshold. Deciding whether a graph is threshold-colorable with respect to a given partition of edges into near and far is equivalent to the graph sandwich problem for unit-interval-representability, which is known to be NP-hard [14]. Hence, deciding whether a graph admits a weak unit interval representation with respect to a given edge-labeling is also NP-hard. In fact, along the lines of argument used in [1], one can prove that recognizing weak unit ball graphs with a given edge-labeling in any dimension $d = 1, 2, \ldots$ is equivalent to the graph sandwich problem for unit-ball-representability in dimension d. Note that the problem of recognizing weak unit interval graphs is different than recognizing unit interval graphs, which can be done in linear time [11]. It is known that planar graphs with girth (the length of a shortest cycle in the graph) at least 10 are always threshold-colorable. Several Archimedean lattices (which correspond to tilings of the plane by regular polygons), and some of their duals, the Laves lattices, are also threshold-colorable [2] for any edge-labeling. Hence, these graph classes are weak unit interval graphs.

Unit interval graphs are also related to threshold and difference graphs. In *threshold graphs* there exists a real number S and for every vertex v there is a real weight a_v so that (v, w) is an edge if and only if $a_v + a_w \geq S$ [21]. A graph is a *difference graph* if there is a real number S and for every vertex v there is a real weight a_v so that $|a_v| < S$ and (v, w) is an edge if and only if $|a_v - a_w| \geq S$ [16]. Note that for both these classes the existence of an edge is completely determined by the threshold S, while in our setting the edges defined by the threshold (size of the ball) must also belong to the original (not necessarily complete) graph. Threshold-colorability is also related to the *integer distance graph* representation [9,12]. An integer distance graph is a graph with the set of integers as vertex set and with an edge joining two vertices u and v if and only if $|u - v| \in D$, where D is a subset of the positive integers.

Our Results: We introduce the notion of weak unit disk and interval representations. While finding representations with unit intervals is equivalent to threshold-coloring where some results are already known, the problem of weak unit disk representability is new. We first show that recognizing weak unit disk graphs is NP-hard. Note that the NP-hardness of the unit interval variant follows from the results in [1].

We then consider subclasses of planar graphs that admit weak unit disk (interval) representation. We show that every 2-reducible graph (as defined later) has a weak unit disk representation for any edge-labeling. In particular, any series-parallel graph (which includes all outerplanar graphs) has a weak unit disk representation for any edge-labeling. For representation with unit intervals, it follows from [1] that all planar graphs with girth at least 10 are total weak unit interval graphs. We generalize the result by proving that graphs of bounded maximum average degree have weak unit interval representations for any given edge-labeling. In the other direction, we construct an example of a planar girth-4 graph which is not a total weak unit interval graph, improving on the earlier girth-3 example. Further, we show that dense planar graphs do not always admit weak unit interval graph representation.

Finally we study outerplanar graphs. It is known that some outerplanar graphs with girth 3 are not total weak unit interval graphs, and our example of girth-4 graph is not outerplanar. Thus, a natural question in this context is whether every girth-4 outerplanar graph admits weak unit interval representation for any edge-labeling. We show that this is indeed the case.

2 Weak Unit Disk Graph Representations

First we consider the complexity of recognizing weak unit disk graphs.

Lemma 1. *It is NP-hard to decide if a graph G with an edge-labeling ℓ admits a weak unit disk representation, even if the edges labeled N induce a planar subgraph.*

Proof. It is known that deciding whether a planar graph is a unit disk graph is NP-hard [6]. Let n be the number of vertices of G, and define an edge-labeling ℓ of K_n by setting $\ell(e) = N$ if and only if e is an edge of G. Clearly, a unit disk representation of G is also a weak unit disk representation of K_n with respect to ℓ and vice versa.   □

Note that Lemma 1 only proves NP-hardness, and the problem of deciding whether a graph with an edge-labeling has a weak unit disk representation is not known to be in NP. The obvious approach is to use a weak unit disk representation as a polynomial size certificate. Unfortunately, it has recently been showed that unit disks graphs on n vertices may require $2^{2^{\Theta(n)}}$ bits for a unit disk representation with integer coordinates [22].

Unit Disk Representation of Outerplanar and Related Graphs

Note that the class of weak unit disk graphs strictly contains the class of weak unit interval graphs. For example, in Fig. 2, we provide a weak unit disk representation of the sungraph for a particular edge-labeling, which does not admit a weak unit interval representation. Our main goal here is to prove that every series-parallel graph is a total weak unit disk graph. To this end, we study a larger class of graphs, called *2-reducible* graphs [25]. A simple graph G is a *2-reducible* graph if one of the following holds:

1. G is an independent set;
2. G has an edge (u, v) such that v has degree at most 2, and the graph obtained by contracting (u, v) and removing parallel edges is a 2-reducible graph.

Note that 2-reducible graphs are a subclass of *2-degenerate* graphs, which are graphs where every subgraph has a vertex with degree at most 2 [23]. For example, the graph obtained by subdividing one edge of K_4 is a 2-degenerate graph, but not a 2-reducible graph.

Theorem 1. *Every 2-reducible graph is a total weak unit disk graph.*

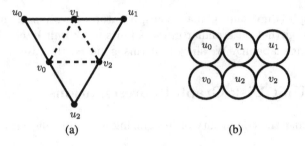

Fig. 2. (a) The sungraph has no weak unit interval representation, but (b) it has a weak unit disk representation. Near/far edges are indicated with solid/dashed line segments.

Proof. We prove the theorem by induction on the number of edges in a graph. Assume the inductive hypothesis that every 2-reducible graph G with m edges has a weak unit disk representation I with respect to any edge-labeling ℓ so that (i) the disks have diameter $t = 2$ and (ii) for every edge (x, y) of G, $1 < \|I(x) - I(y)\| < 4$. The base case $m = 0$ is trivial, so assume that the claim holds for $m > 0$ and for any 2-reducible graph G' with $m' < m$ edges. Now consider an arbitrary 2-reducible graph G with m edges and an arbitrary edge-labeling ℓ of G.

If G has a vertex of degree 1, then the desired representation can be constructed by removing the vertex and considering a representation for the resulting graph. Hence, we assume that G has no degree-1 vertices. Then G has a vertex v with exactly two neighbors u and w, such that contracting the edge (u, v) results in a 2-reducible graph G'. We adopt the equivalent convention that, instead of contracting (u, v), we delete v and add the edge (u, w) if it is not already present. Clearly in G' the number of edges $m' < m$. Thus by the inductive hypothesis, G' has a weak unit disk representation I' with respect to the edge-labeling ℓ restricted to the edges of G' (if edge (u, w) does not belong to G, give it an arbitrary label). Furthermore, $1 < \|I'(u) - I'(w)\| < 4$. Without loss of generality, assume that $I'(u) = (0,0) = p$ (say) and $I'(w) = (d,0) = q$ (say), where $d > 0$. Then $1 < d < 4$. We construct a representation I of G by setting $I(x) = I'(x)$ for every vertex $x \neq v$. To compute the value of $I(v)$, consider the following four cases, based on the values of $\ell(u, v)$ and $\ell(w, v)$.

Case 1: $\ell(u, v) = N$, $\ell(w, v) = N$. If $\ell(u, w) = N$, i.e., the disks for u and w intersect each other, then set $I(v)$ to be the apex of an equilateral triangle with pq as a side. The disk for v then intersect both the disks for u and w. Otherwise, if $\ell(u, w) = F$, set $I(v) = (0, d/2)$. Then $\|I(u) - I(v)\| = \|I(w) - I(v)\| = d/2$ and since $d < 4$, $d/2 < 2$. However since $\ell(u, w) = F$, we have $d > 2$; hence $d/2 > 1$.

Case 2: $\ell(u, v) = N$, $\ell(w, v) = F$. Set $I(v)$ to be $(0, -t)$, where $1 < t < 2$.

Case 3: $\ell(u, v) = F$, $\ell(w, v) = N$. Set $I(v)$ to be $(0, t)$, where $d + 1 < t < d + 2$.

Case 4: $\ell(u,v) = F$, $\ell(w,v) = F$. Set $I(v)$ to be the apex of an isosceles triangle with height h and with pq as the base, where $2 < h < 3$. Then $\|I(u) - I(v)\| = \|I(w) - I(v)\| = d' = \sqrt{h^2 + (d/2)^2}$. Thus $d' > h > 2$, so that the disks for u and w do not intersect with the disk for v. Furthermore $d < 4$ and $h < 3$ imply $d' < 4$. $\qquad\qquad\square$

Series-parallel graphs are defined as the graphs that do not have K_4 as a minor [8]. Hence by definition, these graphs are closed under edge contraction. It is also well-known that a series-parallel graph is subgraph of a 2-tree, which is 2-reducible, and that every outerplanar graph is a subgraph of a series parallel graph. Thus, by Theorem 1, we have the following corollary.

Corollary 1. *Every outerplanar and series-parallel graph is a total weak unit disk graph.*

3   Weak Unit Interval Graph Representations

In this section we study weak unit interval representability. Given a graph $G = (V, E)$, an edge-labeling $\ell : E \to \{N, F\}$, and integers $r > 0$, $t \geq 0$, G is said to be (r, t)-*threshold-colorable* with respect to ℓ if there exists a coloring $c : V \to \{1, \ldots, r\}$ such that for each edge $(u, v) \in E$, $|c(u) - c(v)| \leq t$ if and only if $\ell(u, v) = N$. The coloring c is known as a *threshold-coloring* [1]. It is easy to see that threshold-coloring is a special case of weak unit disk representation when restricted to unit interval representation. As defined, threshold coloring requires integer coordinates for the vertices. The next lemma shows that this requirement does not significantly affect the correspondence between the two problems.

Lemma 2. *A graph G has a weak unit interval representation for an edge-labeling ℓ if and only if G is (r, t)-threshold-colorable with respect to ℓ for integers $r > 0$, $t \geq 0$.*

Proof. We first show that a threshold-coloring c with respect to an edge-labeling ℓ yields a weak interval representation for ℓ. Indeed if c is an (r, t)-threshold-coloring of G with respect to ℓ, then for each vertex u of G, define an interval $I(u)$, which is centered at the point $c(u)$ and has length t. Then for any pair of vertices u, v of G, the intervals $I(u)$ and $I(v)$ intersect if and only if $|c(u) - c(v)| \leq t$. Since c is a threshold coloring, for any edge (u, v), $|c(u) - c(v)| \leq t$ if and only if $\ell(u, v) = N$. Thus the set of intervals defines a weak interval representation of G for ℓ.

For the other direction, let I be a weak unit interval representation of G with respect to ℓ. We can then find a threshold coloring of G from I as follows. Let $c(u)$ be the center of the interval $I(u)$ for each vertex u and let t be the length of the intervals in I. Then by definition, for each edge (u, v) of G, $|c(u) - c(v)| \leq t$ if and only if $\ell(u, v) = N$. However, the centers $c(u)$ and the length t are not necessarily integers. We now modify the representation so that the centers $c(u)$

and the length t are all integers, while the weak interval representation property is maintained. First increase the length of each interval by some $\epsilon > 0$ so that no two intervals intersect each other only at their endpoints. Choose ϵ so that the intervals have rational lengths. Next perturb the center of each interval by some $\epsilon' < \epsilon/2$ so that each interval is centered at a rational point. Note that for any two intervals $I(u)$, $I(v)$, we have that $I(u)$ and $I(v)$ intersect each other after these modification if and only if they intersected each other before the modification. Finally scale the representation so that the center of each interval is an integer, and the length of the intervals is also an integer. Then the centers of the intervals in the modified representation give a threshold-coloring (although r and t may be large). □

Since deciding threshold-colorability is NP-complete [1], so is the recognition problem for weak unit interval graphs.

Lemma 3. *It is NP-complete to decide if a graph with an edge-labeling admits a weak unit interval representation*

Next, we study weak unit interval representation for some graph classes. We first present a method for representing graphs, which admit a decomposition into a forest and a 2-independent set. By $G[U]$ we mean the subgraph of G induced by the vertex set $U \subseteq V$. Recall that a subset $\mathcal{I}$ of vertices in a graph G is called *independent* if $G[\mathcal{I}]$ has no edges. Similarly, $\mathcal{I}$ is called *2-independent* if the shortest path in G between any two vertices of $\mathcal{I}$ has length greater than 2. Such decompositions have been applied to other graph coloring problems [2,3,24].

Lemma 4. *Suppose $G = (\mathcal{I} \cup \mathcal{F}, E)$ is a graph such that $\mathcal{I}$ is 2-independent, $G[\mathcal{F}]$ is a forest, and $\mathcal{I} \cap \mathcal{F} = \emptyset$. Then G is a total weak unit interval graph.*

Proof. We assume that all intervals in the proof are centered at integer coordinates and have length $t = 1$. Suppose $\ell : E \rightarrow \{N, F\}$ is an arbitrary edge-labeling. For each $v \in \mathcal{I}$, set $I(v) = 0$. Each vertex in $G[\mathcal{F}]$ is assigned a point from $\{-2, -1, 1, 2\}$ as follows. Choose a component T of $G[\mathcal{F}]$, and select a root vertex w of T. If w is far from a neighbor in $\mathcal{I}$, set $I(w) = 2$; otherwise, $I(w) = 1$. Now perform breadth first search on T, assigning an interval for each vertex as it is traversed. When we reach a vertex $u \neq w$, it has one neighbor x in T which has been processed, and at most one neighbor $v \in \mathcal{I}$. If v exists, we choose the interval $I(u) = 1$ if $\ell(u, v) = N$, and $I(u) = 2$ otherwise. Then, if the label of edge (u, x) is not satisfied by $\|I(u) - I(x)\|$, we multiply $I(u)$ by -1. If v does not exist, choose $I(u) = 1$ or -1 to satisfy the edge (u, x). By repeating the procedure on each component of $G[\mathcal{F}]$, we construct a representation of G. □

Recall that the *maximum average degree* of a graph G is the maximum of the average degree of each of its subgraphs $H = (V_H, E_H)$, and it is given by $\mathrm{mad}(G) = \max(2|E_H|/|V_H|)$, where the maximum is taken over all subgraphs of G. It is known that every planar graph G of maximum average degree $\mathrm{mad}(G)$ strictly less than $\frac{26}{11}$ can be decomposed into a 2-independent set and a forest [7]. Hence,

Fig. 3. Decomposition of a graph into a nearly 2-independent set (red vertices) and a forest (black vertices and edges). Thin blue are $\mathcal{I}$-edges. (Color figure online)

Theorem 2. *Every planar graph G with $\mathrm{mad}(G) < \frac{26}{11}$ is a total weak unit interval graph.*

We also note that a planar graph with girth g satisfies $\mathrm{mad}(G) < \frac{2g}{g-2}$ [4]. Therefore, a planar graph with girth at least 13 always has a weak unit interval representation.

Next we present a generalization of Lemma 4, suitable for graphs which have an independent set that is in some sense nearly 2-independent. The strategy is to delete certain edges so the independent set becomes 2-independent, obtain a unit interval representation using Lemma 4, and then modify it so that it is a representation of the original graph. Formally, let $\mathcal{I}$ be an independent set in a graph G. Suppose that for every vertex $v \in \mathcal{I}$, there is at most one vertex $u \in \mathcal{I}$ such that the distance between v and u in G is 2. Also suppose that there is only one path with two edges connecting v to u. Then we call $\mathcal{I}$ *nearly 2-independent*. The pair $\{u, v\}$ is called an $\mathcal{I}$-*pair*, and the edges of the path (u, x, v) connecting u and v are called $\mathcal{I}$-*edges*, which are associated with the $\mathcal{I}$-pair $\{u, v\}$; see Fig. 3.

Lemma 5. *Let $G = (\mathcal{I} \cup \mathcal{F}, E)$ be a graph, where $\mathcal{I}$ is a nearly 2-independent set, $G[\mathcal{F}]$ is a forest and $\mathcal{I} \cap \mathcal{F} = \emptyset$. Then G has a total weak unit interval graph.*

Proof. Assume that all intervals in the proof are centered at integer coordinates and have size $t = 3$. Suppose that $\ell : E \to \{N, F\}$ is an arbitrary edge-labeling of G. Let $E' \subseteq E$ be a set such that for each $\mathcal{I}$-pair $\{u, v\}$, exactly one of the $\mathcal{I}$-edges associated with $\{u, v\}$ belongs to E'. Let $G' = (V, E - E')$. Then clearly $\mathcal{I}$ is 2-independent in G' and $G'[\mathcal{F}]$ is a forest; by Lemma 4, there exists a weak unit interval representation I' of G' for ℓ.

We now modify I' to construct a weak unit interval representation I of G with respect to ℓ. First, for each vertex $v \in V$, set $I(v) = 0$ if $I'(v) = 0$, $I(v) = 2$ if $I'(v) = 1$, and $I(v) = 5$ if $I'(v) = 2$ (if $I'(v)$ is negative, do the same but set $I(v)$ negative). It is clear that I is a weak unit interval representation of G'. Now, let $(x, y) \in E'$. One of these vertices, say x, is in $\mathcal{I}$ so $I(x) = 0$, and $I(y) \in \{-5, -2, 2, 5\}$. Without loss of generality assume that $I(y) > 0$; the case where $I(y) < 0$ is symmetric. Now it is possible that $\ell(x, y) = N$ but $\|I(x) - I(y)\| > 3$ or that $\ell(x, y) = F$ but $\|I(x) - I(y)\| \leq 3$. In the first case, we must have $I(y) = 5$. We modify I so that $I(x) = 1$ and $I(y) = 4$. Note that y is still near to vertices with intervals centered at 2 or 5, and far from vertices with intervals centered at less than 1. Similarly, x is still close to the intervals at -2, 0, or 2, but far from -5 and 5. Thus all the edges of $E - E'$ are satisfied by the modification of I, and additionally the edge (x, y) is satisfied. In the second case, we have $I(y) = 2$. We modify I so that $I(x) = -1$ and $I(y) = 3$. As before, no edges which disagreed with the edge-labeling still disagree with the edge-labeling.

Since $\mathcal{I}$ is nearly 2-independent, our modifications to the representation I will not affect non-local vertices, as every vertex in $\mathcal{I}$ is adjacent to at most one edge of E'. $\qquad \square$

Weak Unit Interval Representation of Outerplanar Graphs

It is known [1] that some outerplanar graphs containing triangles are not total weak unit interval graphs, e.g., the sungraph in Fig. 2. Hence, we study weak unit interval representability of triangle-free outerplanar graphs. We start with a claim for girth 5.

Lemma 6. *An outerplanar graph with girth 5 is a total weak unit interval graph.*

Proof. We prove that girth-5 outerplanar graphs may be decomposed into a forest and a 2-independent set using induction on the number of internal faces. The result will follow from Lemma 4. The claim is trivial for a single internal face, so assume that it is true for all girth-5 outerplanar graphs with $k \geq 1$ internal faces. Let G be a girth-5 outerplanar graph with $k + 1$ internal faces. Since G is outerplanar, it must have at least one face $f = (v_1, \ldots, v_l)$, $l \geq 5$, such that every vertex of f except v_1, v_l is of degree 2. Consider the graph G' obtained by deleting $v_2, \ldots, v_{l-1}$. The vertices of G' have a decomposition into a 2-independent set $\mathcal{I}$ and a set $\mathcal{F}$ such that $G'[\mathcal{F}]$ is a forest. Now we will add the vertices $v_2, \ldots, v_{l-1}$ to either $\mathcal{I}$ or $\mathcal{F}$ so that $\mathcal{I}$ is a 2-independent set in G, and $G[\mathcal{F}]$ is a forest. If either of v_1, v_l belongs to $\mathcal{I}$, then add all the remaining vertices to $\mathcal{F}$. Otherwise, add v_3 to $\mathcal{I}$ and the rest to $\mathcal{F}$. Since v_1, v_l are not in $\mathcal{I}$, v_3 has distance at least 3 from any other element of $\mathcal{I}$. $\qquad \square$

Next our goal is to show that a triangle-free outerplanar graph G always has a weak unit interval representation for any edge-labeling. We assume that all intervals are centered at integer coordinates and we use intervals of size $t = 2$. Our strategy is to find a representation of G by a traversal in a depth-first search manner of its weak dual graph G^* (the planar dual minus the outerface). We find intervals for all the vertices in each interior face of G as it is traversed in G^*. Since we are considering triangle-free graphs, this implies that we take an induced path $P_n = (u_1, u_2, \ldots, u_n)$ of G, $n \geq 4$, where the two end vertices u_1 and u_n are already processed and we need to assign unit intervals to the internal vertices $u_2, \ldots, u_{n-1}$ of P_n. Note that this path P_n along with the edge (u, v) forms an internal face of G. We additionally maintain the invariant in our representation that for each edge (u, v) of G, $\|I(u) - I(v)\| \leq 6$. For a particular edge-labeling ℓ of $P_n = (u_1, \ldots, u_n)$, call a pair of coordinates x, y feasible if there is a weak unit interval representation I of P_n for ℓ with $t = 2$, where

$I(u_1) = x$, $I(u_n) = y$, and for any $i \in \{1, \ldots, n-1\}$, $\|I(u_i) - I(u_{i+1})\| \leq 6$. We first need the following three claims.

Claim 1. *For any value of $x \in \{2, 3, -2, -3\}$, the pair $0, x$ is feasible for any edge-labeling ℓ of $P_3 = (u_1, u_2, u_3)$.*

Proof. Without loss of generality, we may assume that $x > 0$. We compute a desired weak unit interval representation I with $t = 2$ for P_3 with respect to ℓ as follows. Assign $I(u_1) = 0$ and $I(u_3) = x$. Assign $I(u_2)$ in such a way that $|I(u_2)| = 2$ if $\ell(u_1, u_2) = N$, and $|I(u_2)| = 3$ if $\ell(u_1, u_2) = F$. Then choose the sign of $I(u_2)$ to be the same as $I(u_3)$ if $\ell(u_2, u_3) = N$, and the opposite of $I(u_3)$ if $\ell(u_2, u_3) = F$. □

Claim 2. *For any edge-labeling of $P_3 = (u_1, u_2, u_3)$, either $0, 4$ or $0, 6$ are feasible.*

Proof. We compute a desired weak unit interval representation I with $t = 2$ for ℓ as follows. If $\ell(u_1, u_2) = l(u_2, u_3) = N$, then $I(u_1) = 0$, $I(u_2) = 2$, and $I(u_3) = 4$. Otherwise, assign $I(u_1) = 0$, $I(u_3) = 6$, and $I(u_2) = 2, 3$ or 4 when $(l(u_1, u_2), l(u_2, u_3))$ have values (N, F), (F, F), and (F, N), respectively. □

Claim 3. *For any integer value of $x \in [-6, 6]$, the pair $0, x$ is feasible for any edge-labeling of $P_n = (u_1, u_2, \ldots, u_n)$, $n \geq 4$.*

Proof. Without loss of generality, let $x \geq 0$. Consider first the case for $n = 4$. Take a particular edge-labeling ℓ of P_4. For any integer value of $0 \leq x \leq 5$, there is at least one number $y \in \{2, 3, -2, -3\}$ and at least one number $z \in \{2, 3, -2, -3\}$ such that $|x - y| \leq 2$ and $2 < |x - z| \leq 6$. In particular, it suffices to choose for $x = 0$, $y = 2$, $z = 3$; for $x = 1, 2, 3, 4$, $y = 2$, $z = -2$ and for $x = 5$, $y = 3$, $z = 2$. Thus if $0 \leq I(u_4) \leq 5$, and regardless of whether $\ell(u_3, u_4)$ is N or F, one can choose a value for $I(u_3)$ from $\{2, 3, -2, -3\}$ respecting both the edge-labeling of (u_3, u_4) and the property that $\|I(u_3) - I(u_4)\| \leq 6$. Then by Claim 1, 0 and x is feasible for the edge-labeling ℓ of P_4. A similar argument shows that if $\ell(u_3, u_4) = F$, then 0 and $x = 6$ is feasible. On the other hand, if $x = 6$ and $\ell(u_3, u_4) = N$, then both 4 and 6 are valid choices for $I(u_3)$. By Claim 2, 0 and 6 is feasible for any edge-labeling ℓ of P_4.

Consider now the case with $n > 4$. Then assign coordinates $I(u_1) = 0$, $I(u_n) = x$ and for $i \in \{n-1, \ldots, 4\}$, assign $I(u_i) \in [-6, 6]$ such that it respects both $\ell(u_i, u_{i+1})$ and the property that $\|I(u_i) - I(u_{i+1})\| \leq 6$. Then a similar argument as that for $n = 4$ can be used to extend this representation to u_2 and u_3. □

The next corollary immediately follows from Claim 3.

Corollary 2. *Any pair x, y with $|x - y| \leq 6$, is feasible for any edge-labeling of $P_n = (u_1, u_2, \ldots, u_n)$, $n \geq 4$.*

Theorem 3. *Every triangle-free outerplanar graph is a total weak unit interval graph.*

Proof. If G is not 2-connected, we augment it in the following way. Let v be a cut vertex of G and let $H_1, \ldots, H_k$ be the 2-connected components of G containing v. For $i \in \{1, \ldots, k-1\}$, let u be a neighbor of v in H_i, and w be a neighbor of v in H_{i+1}. Add the path (u, x, w), where x is a new vertex. Clearly, any weak unit interval representation of the new 2-connected graph is also a weak unit interval representation of G, and the new graph is outerplanar with girth 4.

Now let G be a 2-connected triangle-free outerplanar graph with $n > 4$ vertices embedded in the plane with every vertex on the outerface, and let ℓ be an arbitrary edge-labeling of G. We next compute a weak unit interval representation of G for ℓ. The proof is by induction on the number of vertices in G, with the n-vertex cycle as a base case. Assume the inductive hypothesis that every triangle-free outerplanar graph with fewer than n vertices is a total weak unit interval graph. Further, assume that for such a graph G' with any edge-labeling ℓ', there is a weak unit interval representation of G' for ℓ' where any two neighbor vertices u and v satisfy $\|I(u) - I(v)\| \leq 6$. Clearly if G has at least two cycles, then G has a path $P_k = (u_1, \ldots, u_k)$, $k \geq 4$ with $\deg(u_i) = 2$ for some $1 < i < k$. The theorem follows from the inductive hypothesis and Corollary 2. $\square$

Planar Graphs without Weak Unit Interval Representations

Planar graphs with high edge density may not have weak unit interval representations. First we prove the result for a *wheel graph*, defined as W_n, $n \geq 4$, formed by adding an edge from a vertex v_1 to every vertex of an $(n-1)$-cycle $(v_2, \ldots, v_n, v_2)$.

Lemma 7. *A wheel graph is not a total weak unit interval graph.*

Proof. Define an edge-labeling ℓ of W_n by $\ell(v_2, v_n) = F$, $\ell(v_1, v_i) = F$ for $3 \leq i \leq n - 1$, and every other edge labeled N; see Fig. 4(a). Suppose I is a weak unit interval representation of W_n with respect to ℓ. Since only one edge of the triangle (v_1, v_2, v_n) is far, $I(v_1) \neq I(v_2)$, hence assume that $I(v_1) < I(v_2)$. For $3 \leq i \leq n$, if $I(v_{i-1}) > I(v_1)$, we have $I(v_i) > I(v_1)$, since $\ell(v_{i-1}, v_i) = N$ and either $\ell(v_1, v_{i-1})$ or $\ell(v_1, v_i)$ is F. Then $I(v_1) < I(v_2) \leq I(v_1) + 1$, and $I(v_1) < I(v_n) \leq I(v_1) + 1$, contradicting that $\ell(v_2, v_n) = F$ and I is weak interval representation. $\square$

Using Lemma 7, it is easy to see that any maximal planar graph with $|V| \geq 4$ is not a weak unit interval graph. Indeed, consider such a graph $G = (V, E)$ and a vertex $v \in V$; the neighborhood $N(v) = \{u \mid (v, u) \in E\}$ together with v induces a wheel subgraph. The observation leads to the following theorem.

Theorem 4. *Any planar graph G with $\mathrm{mad}(G) \geq \frac{11}{2}$ is not a total weak unit interval graph.*

Proof. To prove the claim, we show that a total weak unit interval planar graph has at most $\lfloor 11|V|/4 \rfloor - 6$ edges.

Consider a vertex v of a weak unit interval planar graph $G = (V, E)$ and assume it is embedded in the plane. The neighborhood of v is acyclic; otherwise

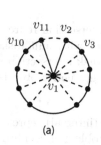 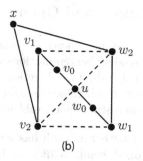

Fig. 4. (a) A wheel graph W_{11} with an edge-labeling, that has no weak unit interval representation. (b) A girth-4 graph with an edge-labeling, that has no weak unit interval representation.

v and its neighborhood induce a wheel, which by Lemma 7 is not a weak unit interval graph. Thus the number of edges between any two neighbors of v is at most $\deg(v) - 1$, where $\deg(v)$ is the degree of v. Denote the number for a vertex v by $s(v)$. Consider the sum, $S = \sum_v s(v)$, taken over all vertices of G. It is easy to see that $S \leq 2|E| - |V|$.

Let T and $\overline{T}$ be the sets of triangular and non-triangular faces in an embedding of G. For each triangle $x \in T$ each of the edges in x is counted once in S. Thus, $2|E| - |V| \geq 3|T| \Rightarrow |T| \leq (2|E| - |V|)/3$. Counting both sides of the edges we get $2|E| \geq 3|T| + 4|\overline{T}| \Rightarrow |T| + |\overline{T}| \leq (2|E| + |T|)/4 \leq (8|E| - |V|)/12$, since $|T| \leq (2|E| - |V|)/3$. Thus, from Euler's formula $|V| - |E| + |T| + |\overline{T}| = 2$, we have $|V| - |E| + (8|E| - |V|)/12 \geq 2 \Rightarrow |E| \leq 11|V|/4 - 6$. □

In [1] all examples of graphs without threshold-coloring (and thus not total weak unit interval graphs) have girth 3. We strengthen the bound by proving the following.

Lemma 8. *There exist planar girth-4 graphs that are not total weak unit interval graphs.*

Proof. Consider the graph in Fig. 4(b) with the given edge-labeling. Suppose there exists a weak unit interval representation I. Without loss of generality suppose that $I(w_2) > I(u)$. Let us consider two cases. First, suppose $I(v_2) < I(u)$. Since the edges (u, v_2) and (u, w_2) are labeled F, it must be that $I(v_2) < I(u) - 1$ and $I(u) + 1 < I(w_2)$. Then vertex x must be represented by an interval near to both of these, which is impossible since $||I(v_2) - I(w_2)|| > 2$.

Otherwise $I(v_2) > I(u)$. Then $I(v_1) \geq I(v_2) - 1 > I(u)$, and $I(u) < I(w_2)$ implies that $I(v_1) < I(w_2)$. Similarly, $I(w_1) < I(v_2)$. Now, either $I(w_2) \leq I(v_2)$, or $I(v_2) < I(w_2)$. In the first case, w_2 is near to v_1 since $I(v_1) < I(w_2) \leq I(v_2)$ and $||I(v_1) - I(v_2)|| \leq 1$. The second case leads to a similar contradiction. □

4   Conclusion and Open Problems

In this paper we introduced the concept of weak intersection representation of graphs and studied representations of planar graphs with unit disks and unit intervals. A natural future direction is to consider weak intersection representations for other graph classes and/or with different geometric objects. Next we list several interesting open problems.

1. Deciding whether a graph has a weak unit disk (interval) representation for a given edge-labeling is NP-hard. However, the problem of deciding whether a graph is a total weak unit disk (interval) graph is open, and it remains open when restricted to planar graphs. Note that the class of total weak unit disk (interval) planar graphs is not closed under taking minors, as subdividing each edge of a planar graph three times results in a planar graph with girth at least 10, which is a total weak unit interval graph.
2. Tightening the lower and upper bounds for maximum average degree of total weak unit interval planar graphs, given in Theorems 2 and 4, is a challenging open problem. Based on extensive computer experiments, we conjecture that there are no total weak unit interval graphs with more than $2|V| - 3$ edges.
3. We considered planar graphs, but little is known for general graphs. In particular, it would be interesting to find out whether the edge density of total weak unit disk (interval) graphs is always bounded by a constant.
4. We proved that a graph has a weak unit interval representation for an edge-labeling ℓ if and only if it is (r, t)-threshold-colorable with respect to ℓ for integers $r > 0$, $t \geq 0$, but in the proof, the values of r and t can be arbitrarily large. It would be interesting to bound the values of r and t for any n-vertex graph.

Acknowledgments. We thank Michalis Bekos, Gasper Fijavz, and Michael Kaufmann for productive discussions about several variants of these problems.

References

1. Alam, M.J., Chaplick, S., Fijavž, G., Kaufmann, M., Kobourov, S.G., Pupyrev, S.: Threshold-coloring and unit-cube contact representation of graphs. In: Brandstädt, A., Jansen, K., Reischuk, R. (eds.) WG 2013. LNCS, vol. 8165, pp. 26–37. Springer, Heidelberg (2013)
2. Alam, M.J., Kobourov, S.G., Pupyrev, S., Toeniskoetter, J.: Happy edges: threshold-coloring of regular lattices. In: Ferro, A., Luccio, F., Widmayer, P. (eds.) FUN 2014. LNCS, vol. 8496, pp. 28–39. Springer, Heidelberg (2014)
3. Albertson, M.O., Chappell, G.G., Kierstead, H.A., Kündgen, A., Ramamurthi, R.: Coloring with no 2-colored P4. Electron. J. Combin. **11**(1), R26 (2004)
4. Borodin, O., Kostochka, A., Nešetřil, J., Raspaud, A., Sopena, E.: On the maximum average degree and the oriented chromatic number of a graph. Dis. Math. **206**(1), 77–89 (1999)

5. Bremner, D., Evans, W., Frati, F., Heyer, L., Kobourov, S.G., Lenhart, W.J., Liotta, G., Rappaport, D., Whitesides, S.H.: On representing graphs by touching cuboids. In: Didimo, W., Patrignani, M. (eds.) GD 2012. LNCS, vol. 7704, pp. 187–198. Springer, Heidelberg (2013)
6. Breu, H., Kirkpatrick, D.G.: Unit disk graph recognition is NP-hard. Comput. Geom. **9**(1), 3–24 (1998)
7. Bu, Y., Cranston, D.W., Montassier, M., Raspaud, A., Wang, W.: Star coloring of sparse graphs. J. Graph. Theory **62**(3), 201–219 (2009)
8. Duffin, R.: Topology of series-parallel networks. J. Math. Anal. Appl. **10**, 303–318 (1965)
9. Eggleton, R., Erdös, P., Skilton, D.: Colouring the real line. J. Comb. Theory, Ser. B **39**(1), 86–100 (1985)
10. Evans, W., Gansner, E.R., Kaufmann, M., Liotta, G., Meijer, H., Spillner, A.: Approximate proximity drawings. In: Speckmann, B. (ed.) GD 2011. LNCS, vol. 7034, pp. 166–178. Springer, Heidelberg (2011)
11. Herrera de Fegueiredo, C.M., Meidanis, J., Picinin de Mello, C.: A linear-time algorithm for proper interval graph recognition. Inf. Process. Lett. **56**(3), 179–184 (1995)
12. Ferrara, M., Kohayakawa, Y., Rödl, V.: Distance graphs on the integers. Comb. Probab. Comput. **14**(1), 107–131 (2005)
13. Gabriel, K.R., Sokal, R.R.: A new statistical approach to geographic variation analysis. Syst. Biol. **18**(3), 259–278 (1969)
14. Golumbic, M.C., Kaplan, H., Shamir, R.: Graph sandwich problems. J. Algorithms **19**(3), 449–473 (1995)
15. Hale, W.K.: Frequency assignment: theory and applications. Proc. IEEE **68**(12), 1497–1514 (1980)
16. Hammer, P.L., Peled, U.N., Sun, X.: Difference graphs. Dis. App. Math. **28**(1), 35–44 (1990)
17. Hliněný, P., Kratochvíl, J.: Representing graphs by disks and balls (a survey of recognition-complexity results). Discrete Math. **229**(1), 101–124 (2001)
18. Kleist, L., Rahman, B.: Unit contact representations of grid subgraphs with regular polytopes in 2D and 3D. In: Duncan, C., Symvonis, A. (eds.) GD 2014. LNCS, vol. 8871, pp. 137–148. Springer, Heidelberg (2014)
19. Kuhn, F., Wattenhofer, R., Zollinger, A.: Ad hoc networks beyond unit disk graphs. Wireless Netw. **14**(5), 715–729 (2008)
20. Liotta, G.: Proximity drawings. In: Tamassia, R. (ed.) Handbook of Graph Drawing and Visualization. Chapman & Hall/CRC, Boca Raton (2007)
21. Mahadev, N.V., Peled, U.N.: Threshold Graphs and Related Topics. North Holland, Amsterdam (1995)
22. McDiarmid, C., Müller, T.: Integer realizations of disk and segment graphs. J. Comb. Theory, Ser. B **103**(1), 114–143 (2013)
23. Thomassen, C.: Decomposing a planar graph into degenerate graphs. J. Comb. Theory, Ser. B **65**(2), 305–314 (1995)
24. Timmons, C.: Star coloring high girth planar graphs. Electron. J. Comb. **15**(1), R124 (2008)
25. Wiegers, M.: Recognizing outerplanar graphs in linear time. In: Tinhofer, G., Schmidt, G. (eds.) WG 1986. LNCS, vol. 246, pp. 165–176. Springer, Heidelberg (1987)

Simultaneous Visibility Representations of Plane *st*-graphs Using L-shapes

William S. Evans[1], Giuseppe Liotta[2], and Fabrizio Montecchiani[2]([⊠])

[1] University of British Columbia, Vancouver, Canada
will@cs.ubc.ca
[2] Università degli Studi di Perugia, Perugia, Italy
{giuseppe.liotta,fabrizio.montecchiani}@unipg.it

Abstract. Let $\langle G_r, G_b \rangle$ be a pair of plane *st*-graphs with the same vertex set V. A simultaneous visibility representation with L-shapes of $\langle G_r, G_b \rangle$ is a pair of bar visibility representations $\langle \Gamma_r, \Gamma_b \rangle$ such that, for every vertex $v \in V$, $\Gamma_r(v)$ and $\Gamma_b(v)$ are a horizontal and a vertical segment, which share an end-point. In other words, every vertex is drawn as an *L*-shape, every edge of G_r is a vertical visibility segment, and every edge of G_b is a horizontal visibility segment. Also, no two L-shapes intersect each other. An L-shape has four possible rotations, and we assume that each vertex is given a rotation for its L-shape as part of the input. Our main results are: (i) a characterization of those pairs of plane *st*-graphs admitting such a representation, (ii) a cubic time algorithm to recognize them, and (iii) a linear time drawing algorithm if the test is positive.

1 Introduction

Let G_r and G_b be two plane graphs with the same vertex set. A *simultaneous embedding (SE)* of $\langle G_r, G_b \rangle$ consists of two planar drawings, Γ_r of G_r and Γ_b of G_b, such that every edge is a simple Jordan arc, and every vertex is the same point both in Γ_r and in Γ_b. The problem of computing SEs has received a lot of attention in the Graph Drawing literature, partly for its theoretical interest and partly for its application to the visual analysis of dynamically changing networks on a common (sub)set of vertices. For example, it is known that any two plane graphs with the same vertex set admit a SE where the edges are polylines with at most two bends, which are sometimes necessary [8]. If the edges are straight-line segments, the representation is called a *simultaneous geometric embedding (SGE)*, and many graph pairs do not have an SGE: a tree and a path [1], a planar graph and a matching [6], and three paths [5]. On the positive side, the discovery of graph pairs that have an SGE is still a fertile research topic. The reader can refer to the survey by Bläsius, Kobourov and Rutter [21] for references and open problems.

Research supported in part by the MIUR project AMANDA "Algorithmics for MAssive and Networked DAta", prot. 2012C4E3KT_001 and NSERC Canada.

E.W. Mayr (Ed.): WG 2015, LNCS 9224, pp. 252–265, 2016.
DOI: 10.1007/978-3-662-53174-7_18

Only a few papers study simultaneous representations that adopt a drawing paradigm different from SE and SGE. A seminal paper by Jampani and Lubiw initiates the study of *simultaneous intersection representations (SIR)* [16]. In an intersection representation of a graph, each vertex is a geometric object and there is an edge between two vertices if and only if the corresponding objects intersect. Let $\langle G_r, G_b \rangle$ be two graphs that have a subgraph in common. A SIR of $\langle G_r, G_b \rangle$ is a pair of intersection representations where each vertex in $G_r \cap G_b$ is mapped to the same object in both realizations. Polynomial-time algorithms for testing the existence of SIRs for chordal, comparability, interval, and permutation graphs have been presented [4,15,16].

We introduce and study a different type of simultaneous representation, where each graph is realized as a *bar visibility representation* and two segments representing the same vertex share an end-point. A bar visibility representation of a plane graph G is an embedding preserving drawing Γ where the vertices of G are non-overlapping horizontal segments, and two segments are joined by a vertical visibility segment if and only if there exists an edge in G between the two corresponding vertices (see, e.g., [18,22]). A visibility segment has thickness $\epsilon > 0$ and does not intersect any other segment.

A *simultaneous visibility representation with L-shapes* of $\langle G_r, G_b \rangle$ is a pair of bar visibility representations $\langle \Gamma_r, \Gamma_b \rangle$ such that for every vertex $v \in V$, $\Gamma_r(v)$ and $\Gamma_b(v)$ are a horizontal and a vertical segment that share an end-point. In other words, every vertex is an L-shape, and every edge of G_r (resp., G_b) is a vertical (resp., horizontal) visibility segment. Also, no two L-shapes intersect. A simultaneous visibility representation with L-shapes of $\langle G_r, G_b \rangle$ where the rotation of the L-shape of each vertex in V is defined by a function $\Phi : V \to \mathcal{H} = \{ \mathsf{L}, \mathsf{\lrcorner}, \mathsf{\urcorner}, \mathsf{\Gamma} \}$, is called a Φ-*LSVR* in the following. While this definition does not assume any particular direction on the edges of G_r (resp., G_b), the resulting representation does induce a bottom-to-top (resp., left-to-right) *st*-orientation. In this paper, we assume that G_r and G_b are directed and this direction must be preserved in the visibility representation. Also, the two graphs have been augmented with distinct (dummy) sources and sinks. More formally, $G_r = (V \cup \{s_r, t_r\}, E_r)$ and $G_b = (V \cup \{s_b, t_b\}, E_b)$ are two plane st-graphs with sources s_r, s_b, and sinks t_r, t_b.

In terms of readability, this kind of simultaneous representation has the following advantages: (i) The edges are depicted as straight-line segments (as in SGE) and the edge-crossings are rectilinear; (ii) The edges of the two graphs are easy to distinguish, since they consistently flow from bottom to top for one graph and from left to right for the other graph. Having rectilinear crossing edges is an important benefit in terms of readability, as shown in [14], which motivated a relevant amount of research on right-angle crossing (RAC) drawings, see [9] for a survey.

Our main contribution is summarized by the following theorem.

Theorem 1. *Let G_r and G_b be two plane st-graphs defined on the same set of n vertices V and with distinct sources and sinks. Let $\Phi : V \to \mathcal{H} = \{ \mathsf{L}, \mathsf{\lrcorner}, \mathsf{\urcorner}, \mathsf{\Gamma} \}$.*

There exists an $O(n^3)$-time algorithm to test whether $\langle G_r, G_b \rangle$ admits a Φ-LSVR. Also, in the positive case, a Φ-LSVR can be computed in $O(n)$ time.

This result relates to previous studies on topological rectangle visibility graphs [20] and transversal structures (see, e.g., [12,13,17,19]). Also, starting from a Φ-LSVR of $\langle G_r, G_b \rangle$, we can compute a *simultaneous RAC embedding* of the two graphs with at most two bends per edge, improving the general upper bound by Bekos *et al.* [3] for those pairs of graphs that can be directed and augmented to admit a Φ-LSVR.

The proof of Theorem 1 is based on a characterization described in Sect. 3, which allows for an efficient testing algorithm presented in Sect. 4. Due to space restrictions, some proofs are omitted or only sketched in the text; full proofs can be found in [11].

2   Preliminaries

A graph $G = (V, E)$ is *simple*, if it contains neither loops nor multiple edges. We consider simple graphs, if not otherwise specified. A *drawing* Γ of G maps each vertex of V to a point of the plane and each edge of E to a Jordan arc between its two end-points. We only consider *simple drawings*, i.e., drawings such that the arcs representing two edges have at most one point in common, which is either a common end-vertex or a common interior point where the two arcs properly cross. A drawing is *planar* if no two arcs representing two edges cross. A planar drawing subdivides the plane into topologically connected regions, called *faces*. The unbounded region is called the *outer face*. A *planar embedding* of a graph is an equivalence class of planar drawings that define the same set of faces. A graph with a given planar embedding is a *plane* graph. For a non-planar drawing, we can still derive an embedding considering that the boundary of a face may consist also of edge segments between vertices and/or crossing points of edges. The unbounded region is still called the outer face.

A graph is *biconnected* if it remains connected after removing any one vertex. A directed graph (a digraph for short) is biconnected if its underlying undirected graph is biconnected. The *dual graph* D of a plane graph G is a plane multigraph whose vertices are the faces of G with an edge between two faces if and only if they share an edge. If G is a digraph, D is also a digraph whose dual edge e^* for a primal edge e is conventionally directed from the face, $left_G(e)$, on the left of e to the face, $right_G(e)$, on the right of e. Since we also use the opposite convention, we let $D^\rightarrow$ (resp., $D^\leftarrow$) be the dual whose edges cross the primal edges from left to right (resp., right to left).

A *topological numbering* of a digraph is an assignment, X, of numbers to its vertices such that $X(u) < X(v)$ for every edge (u, v). A graph admits a topological numbering if and only if it is acyclic. An acyclic digraph with a single source s and a single sink t is called an *st-graph*. In such a graph, for every vertex v, there exists a directed path from s to t that contains v [22]. A *plane st-graph* is an *st*-graph that is planar and embedded such that s and t are on the boundary of the outer face. In any st-graph, the presence of the edge (s,t) guarantees that the graph is biconnected. In the following we consider st-graphs that contain the edge (s,t), as otherwise it can be added without violating planarity. Let G be a plane st-graph, then for each vertex v of G the incoming edges appear consecutively around v, and so do the outgoing edges. Vertex s only has outgoing edges, while vertex t only has incoming edges. This is a particular transversal structure (see Sect. 3) known as a *bipolar orientation* [18,22]. Each face f of G is bounded by two directed paths with a common origin and destination, called the *left path* and *right path* of f. For all vertices v and edges e on the left (resp., right) path of f, we let $right_G(v) = right_G(e) = f$ (resp., $left_G(v) = left_G(e) = f$).

Tamassia and Tollis [22] proved the following lemma.

Lemma 1 [22]. *Let G be a plane st-graph and let $D^{\rightarrow}$ be its dual graph. Let u and v be two vertices of G. Then exactly one of the following four conditions holds:* (i) *G has a path from u to v, or* (ii) *from v to u;* (iii) *$D^{\rightarrow}$ has a path from $right_G(u)$ to $left_G(v)$, or* (iv) *from $right_G(v)$ to $left_G(u)$.*

Let v be a vertex of G, then denote by $B(v)$ (resp., $T(v)$) the set of vertices that can reach (resp., can be reached from) v. Also, denote by $L(v)$ (resp., $R(v)$) the set of vertices that are to the left (resp., to the right) of every path from s to t through v. By Lemma 1, these four sets partition the vertices of $G \setminus \{v\}$. In every planar drawing of G, they are contained in four distinct regions of the plane that share point v. The vertices of $B(v)$ are in the region delimited by the leftmost and the rightmost paths from s to v, while the vertices of $T(v)$ are in the region delimited by the leftmost and the rightmost paths from v to t. Edge (s,t) separates the two regions containing the vertices of $L(v)$ and $R(v)$, as in Fig. 1. Refer to [7] for further details.

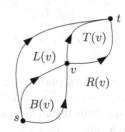

Fig. 1. Vertex sets $B(v)$, $T(v)$, $L(v)$, and $R(v)$ and their corresponding regions of the plane.

3   Characterization

A *transversal structure* of a plane graph G, is a coloring and an orientation of the inner edges (i.e., those edges that do not belong to the outer face) of the graph that obey some local and global conditions. Transversal structures have been widely studied and important applications have been found. Bipolar orientations (also known as *st-orientations*) of plane graphs have been used to compute bar visibility representations [18,22]. Further applications can be found in [12,13,17,19], see also [10] for a survey.

To characterize those pairs of graphs that admit a Φ-LSVR, we introduce a new transversal structure for the union of the two graphs (which may be non-planar) and show that it is in bijection with the desired representation. In what follows $G_r = (V_r = V \cup \{s_r, t_r\}, E_r)$ and $G_b = (V_b = V \cup \{s_b, t_b\}, E_b)$ are two plane st-graphs with duals $D_r^{\rightarrow}$ and $D_b^{\leftarrow}$, respectively.

Definition 1. *Given* $\Phi : V \to \mathcal{H} = \{ \llcorner , \lrcorner , \urcorner , \ulcorner \}$, *a* (4-polar) Φ-transversal *is a drawing of a directed (multi)graph on the vertex set* $V \cup \{ s_r, t_r, s_b, t_b \}$ *whose edges are partitioned into red edges, blue edges, and the four special edges* (s_r, s_b), (s_b, t_r), (t_r, t_b), *and* (t_b, s_r) *forming the outer face, in clockwise order. In addition, the* Φ-transversal *obeys the following conditions:*

c1. *The red (resp., blue) edges induce an st-graph with source* s_r *(resp.,* s_b*) and sink* t_r *(resp.,* t_b*).*

c2. *For every vertex* $u \in V$, *the clockwise order of the edges incident to* u *forms four non-empty blocks of monochromatic edges, such that all edges in the same block are either all incoming or all outgoing with respect to* u. *The four blocks are encountered around* u *depending on* $\Phi(u)$ *as in the following table.*

$\llcorner \Rightarrow$	$\lrcorner \Rightarrow$	$\urcorner \Rightarrow$	$\ulcorner \Rightarrow$
$f_b(u)=right_b(u)$ $f_r(u)=left_r(u)$	$f_b(u)=right_b(u)$ $f_r(u)=right_r(u)$	$f_b(u)=left_b(u)$ $f_r(u)=right_r(u)$	$f_b(u)=left_b(u)$ $f_r(u)=left_r(u)$

c3. *Only blue and red edges may cross and only if blue crosses red from left to right.*

A pair of plane st-graphs $\langle G_r, G_b \rangle$ admits a Φ-transversal if there exists a Φ-transversal G_{rb} such that restricting $G_{rb} \setminus \{s_b, t_b\}$ to the red edges realizes the planar embedding G_r and restricting $G_{rb} \setminus \{s_r, t_r\}$ to the blue edges realizes the planar embedding G_b.

Let u be a vertex of V, then the edges of a single color enter and leave u by the same face in the embedding of the other colored graph. In other words, as condition **c2** indicates, $\Phi(u)$ defines the face of G_b (resp., G_r), denoted by $f_b(u)$ (resp., $f_r(u)$), by which the edges of G_r (resp., G_b) incident to u enter and leave u, in the Φ-transversal. Also, condition **c3** implies that edges $\{(s_r, s_b), (s_b, t_r), (t_r, t_b), (t_b, s_r)\}$ are not crossed, because they are not colored.

In the remainder of this section we will prove the next theorem.

Theorem 2. *Let* G_r *and* G_b *be two plane st-graphs defined on the same set of vertices* V *and with distinct sources and sinks. Let* $\Phi : V \to \mathcal{H} = \{ \llcorner , \lrcorner , \urcorner , \ulcorner \}$. *Then* $\langle G_r, G_b \rangle$ *admits a* Φ-LSVR *if and only if it admits a* Φ-transversal.

The necessity of the Φ-transversal is easily shown. Let $\langle \Gamma_r, \Gamma_b \rangle$ be a Φ-LSVR of $\langle G_r, G_b \rangle$ with two additional horizontal bars at the bottommost and topmost sides of the drawing that represent s_r and t_r, and two additional vertical bars at the leftmost and rightmost sides of the drawing that represent s_b and t_b. From such a representation we can compute a Φ-transversal G_{rb} as follows. Since the four vertices s_r, t_r, s_b, and t_b are represented by the extreme bars in the drawing, these four vertices belong to the outer face, and the four edges on the outer face can be added without crossings. Also, we color red all inner edges represented by vertical visibilities (directed from bottom to top), and blue all inner edges represented by horizontal visibilities (directed from left to right). To see that

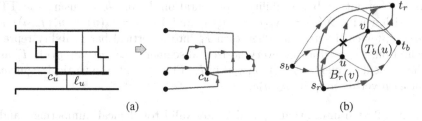

Fig. 2. (a) The replacement of the L-shape, ℓ_u, for vertex u with its corner point c_u and the drawing of u's adjacent edges with 2 bends per edge when constructing a Φ-transversal from a Φ-LSVR. Only ℓ_u's visibilities are shown. (b) Illustration for the proof of Lemma 3: the case when u is in $B(v)$ and v is in $T(u)$.

conditions **c1**, **c2** and **c3** are satisfied, let G_{rb} be a polyline drawing computed as follows. Let c_u be the corner of the L-shape, ℓ_u, representing vertex u. For every edge (u, v), replace its visibility segment by a polyline from c_u to c_v that has two bends, both contained in the visibility segment and each at distance δ from a different one of its endpoints, for an arbitrarily small, fixed $\delta > 0$. See Fig. 2(a). Finally, replace every L-shape ℓ_u with its corner c_u. Since each bar visibility representation preserves the embedding of the input graph, **c1** is respected. Also, **c2** and **c3** are clearly satisfied by the embedding derived from G_{rb}. We remark that, by construction, each edge is represented by a polyline with two bends and two edges cross only at right angles; this observation will be used in Sect. 5.

To prove sufficiency, assume $\langle G_r, G_b \rangle$ admits a Φ-transversal G_{rb}. We present an algorithm, Φ**LSVRDrawer**, that takes as input G_{rb} and returns a Φ-LSVR $\langle \Gamma_r, \Gamma_b \rangle$ of $\langle G_r, G_b \rangle$. We first recall the algorithm by Tamassia and Tollis (TT in the following) to compute an embedding-preserving bar visibility representation of a plane st-graph G [7,22]:

1. Compute the dual $D^{\rightarrow}$ of G.
2. Compute a pair of topological numberings Y of G and X of $D^{\rightarrow}$.
3. Draw each vertex v as a horizontal bar with y-coordinate $Y(v)$ and between x-coordinates $X(left_G(v))$ and $X(right_G(v)) - \epsilon$.
4. Draw each edge $e = (u, v)$ as a vertical segment at x-coordinate $X(left_G(e))$, between y-coordinates $Y(u)$ and $Y(v)$, and with thickness ϵ.

We are now ready to describe algorithm Φ**LSVRDrawer**.

Step 1: Compute the dual graphs $D_r^{\rightarrow}$ of G_r and $D_b^{\leftarrow}$ of G_b.
Step 2: Compute a pair of topological numberings n_r of G_r and n_b of G_b.
Step 3: Compute a pair of topological numberings n_r^* of $D_r^{\rightarrow}$ and n_b^* of $D_b^{\leftarrow}$.
Step 4: Compute a bar visibility representation Γ_r of G_r by using the TT algorithm with $X(u) = X_r(u) = n_r^*(u)$ and $Y(u) = Y_r(u) = n_b^*(f_b(u)) + n_r(u)\delta$, for each vertex u. Also, shift the horizontal segment for each vertex u to the left by $n_b(u)\delta$.

Step 5: Compute a bar visibility representation Γ_b' of G_b by using the TT algorithm with $X(u) = X_b(u) = n_b^*(u)$ and $Y(u) = Y_b(u) = n_r^*(f_r(u)) + n_b(u)\delta$, for each vertex u. Then turn Γ_b' into a vertical bar visibility representation, Γ_b, by drawing every horizontal segment $((x_0, y), (x_1, y))$ in Γ_b' as the vertical segment $((y, x_0), (y, x_1))$ in Γ_b. Finally, shift the vertical segment for each vertex u up by $n_r(u)\delta$.

Lemma 2 guarantees that Y_r and Y_b are valid topological numberings, and thus, that Γ_r and Γ_b are two bar visibility representations. Also, Lemma 3 ensures the union of Γ_r and Γ_b is a Φ-LSVR. The shifts performed at the end of **Steps 4–5** are to prevent the bars of two L-shapes from coinciding. The value $\delta > 0$ is chosen to be less than ϵ and less than the smallest difference between distinct numbers divided by the largest number from any topological numbering n_r, n_b, n_r^*, or n_b^*. This choice of δ guarantees that all visibilities are preserved after the shift, and that no new visibilities are introduced.

Lemma 2. *Y_r is a valid topological numbering of G_r and Y_b is a valid topological numbering of G_b.*

Proof. Let (u, v) be a red edge from u to v. We know that $n_r(u) < n_r(v)$. Let $e_0, e_1, \ldots, e_k$ be the blue edges crossed by (u, v) in G_{rb}. Due to conditions **c2** and **c3**, there exists a path $\{f_b(u) = right_b(e_0),\ left_b(e_0) = right_b(e_1),\ \ldots, left_b(e_{k-1}) = right_b(e_k),\ left_b(e_k) = f_b(v)\}$ in $D_b^\leftarrow$. Thus, we also know that $n_b^*(f_b(u)) \leq n_b^*(f_b(v))$. Since $Y_r(u) = n_b^*(f_b(u)) + n_r(u)\delta$ and $\delta > 0$, it follows that $Y_r(u) < Y_r(v)$. A symmetric argument shows $Y_b(u) < Y_b(v)$ if (u, v) is a blue edge. $\qquad\square$

Lemma 3. *Each vertex u of V is represented by an L-shape ℓ_u in $\langle \Gamma_r, \Gamma_b \rangle$ as defined by the function Φ. Also no two L-shapes intersect each other.*

Proof. Suppose $\Phi(u) = \llcorner$, as the other cases are similar. Then, $f_b(u) = right_b(u)$ and $f_r(u) = left_r(u)$. The horizontal bar representing u in Γ_r is the segment $[p_0(u), p_1(u)]$, where the two points $p_0(u)$ and $p_1(u)$ are $p_0(u) = (n_r^*(left_r(u)) + n_b(u)\delta,\ Y_r(u))$, and $p_1(u) = (n_r^*(right_r(u)) + n_b(u)\delta,\ Y_r(u))$. Note that $n_r^*(left_r(u)) < n_r^*(right_r(u))$. The vertical bar representing u in Γ_b is the segment $[q_0(u), q_1(u)]$, where the two points $q_0(u)$ and $q_1(u)$ are $q_0(u) = (Y_b(u),\ n_b^*(right_b(u)) + n_r(u)\delta)$, and $q_1(u) = (Y_b(u),\ n_b^*(left_b(u)) + n_r(u)\delta)$. Note that $n_b^*(right_b(u)) < n_b^*(left_b(u))$. Since $Y_r(u) = n_b^*(f_b(u)) + n_r(u)\delta = n_b^*(right_b(u)) + n_r(u)\delta$, the bottom coordinate of the vertical bar representing u matches the y-coordinate of horizontal bar representing u. Since $Y_b(u) = n_r^*(f_r(u)) + n_b(u)\delta = n_r^*(left_r(u)) + n_b(u)\delta$, the left coordinate of the horizontal bar representing u matches the x-coordinate of the vertical bar representing u. Thus the two bars form the L-shape $\llcorner$.

We now show that no two L-shapes properly intersect each other. Suppose by contradiction that the vertical bar of a vertex u, properly intersects the horizontal bar of a vertex v. Based on Φ, the vertical bar of u involved in the intersection is either a left vertical bar or a right vertical bar, and it is drawn at x-coordinate

$n_r^*(left_r(u)) + n_b(u)\delta$ or $n_r^*(right_r(u)) + n_b(u)\delta$, respectively. Suppose it is a left vertical bar, as the other case is symmetric. Since u's vertical bar properly intersects v's horizontal bar, we know by construction that $n_r^*(left_r(v)) + n_b(v)\delta < n_r^*(left_r(u)) + n_b(u)\delta < n_r^*(right_r(v)) + n_b(v)\delta$. Proper intersection implies that these inequalities are strict, that there is a path in the red dual $D_r^{\rightarrow}$ from $left_r(v)$ to $left_r(u)$ to $right_r(v)$, and that the three faces are distinct. This implies that u belongs either to $B_r(v)$ or to $T_r(v)$, and it lies in the corresponding regions of the plane, with $f_r(u)$ (and hence the start/end of curves representing blue edges incident to u) inside the region. Similarly, by considering the blue dual $D_b^{\leftarrow}$, $n_b^*(right_b(u)) + n_r(u)\delta < n_b^*(f_b(v)) + n_r(v)\delta < n_b^*(left_b(u)) + n_r(u)\delta$, we know that v belongs either to $B_b(u)$, or to $T_b(u)$, and it lies in the corresponding regions of the plane, with $f_b(v)$ (and hence the start/end of curves representing red edges incident to v) inside the region. No matter which region, $B_r(v)$ or $T_r(v)$, vertex u lies in, or which region, $B_b(u)$ or $T_b(u)$, vertex v lies in, the directed boundary of the blue region ($B_b(u)$ or $T_b(u)$) containing v crosses the directed boundary of the red region ($B_r(v)$ or $T_r(v)$) containing u from right to left. This either violates **c3** (if edges of the boundaries cross) or it violates **c2** (if the boundaries share a vertex). See also Fig. 2(b).   □

Theorem 3. *Let G_r and G_b be two plane st-graphs defined on the same set of n vertices V and with distinct sources and sinks. Let $\Phi : V \rightarrow \mathcal{H} = \{ \llcorner, \lrcorner, \urcorner, \ulcorner \}$. If $\langle G_r, G_b \rangle$ admits a Φ-transversal, then algorithm ΦLSVRDrawer computes a Φ-LSVR of $\langle G_r, G_b \rangle$ in $O(n)$ time.*

Proof. Lemmas 2 and 3 imply that ΦLSVRDrawer computes a Φ-LSVR of $\langle G_r, G_b \rangle$. Computing the dual graphs and the four topological numberings (**Steps 1–3**), as well as computing the two bar visibility representations and shifting each segment (**Steps 4–5**), can be done in $O(n)$ time, as shown in [7, 22].   □

4   Testing Algorithm

In this section we show how to test whether two plane st-graphs with the same set of vertices admit a Φ-LSVR for a given function ϕ. In [11] it is shown a pair of graphs $\langle G_r, G_b \rangle$ that does not admit a Φ-LSVR for any function Φ. This emphasizes the need for an efficient testing algorithm. Our algorithm exploits the interplay between the primal of the blue (red) graph and the dual of the red (blue) graph. Given the circular order of the edges around each vertex imposed by the function ϕ, we aim to compute a suitable path in the red dual for each blue edge. Such paths will then be used to route the blue edges. Finally, we check that no two blue edges cross.

We first introduce a few definitions. Let G and $D^{\rightarrow}$ be a plane st-graph and its dual. Let f and g be two faces of G that share an edge $e = (x, z)$ of G, such that e belongs to the right (resp., left) path of f (resp., g). Let e^* be the dual edge in $D^{\rightarrow}$ corresponding to e. Let w be a vertex on the right path of f (or, equivalently, on the left path of g). Then w is *cut from above* (resp., *below*) by e^*, if w precedes z (resp., succeeds x) along the right path of f, i.e., all vertices that precede z (including x) are cut from above, while all vertices that succeed x (including z) are cut from below by e^*.

Let G_r and G_b be a pair of plane st-graphs with the same vertex set V and with distinct sources and sinks. Let $\Phi : V \to \mathcal{H} = \{ \llcorner, \lrcorner, \urcorner, \ulcorner \}$. Recall that, for a given vertex u of G_b, with the notation $L_b(u)$, $R_b(u)$, $T_b(u)$ and $B_b(u)$ we represent the set of vertices to the left, to the right, that are reachable from, and that can reach u in G_b, respectively (see Sect. 2). Then consider an edge $e = (u, v)$ of G_b and a path[1] $\pi_e = \{ f_r(u) = f_0, e_0^*, f_1, \ldots, f_{k-1}, e_{k-1}^*, f_r(v) = f_k \}$ in $D_r^\to$, where f_i $(0 \le i \le k)$ are the faces traversed by the path, and e_i^* $(0 \le i < k)$ are the dual edges used by the path to go from f_i to f_{i+1}. Path π_e is a *traversing path* for e, if $\pi_e = \{ f_r(u) = f_r(v) \}$, or for all $0 \le i < k$ and all vertices w in the right path of f_i:

p1. If $w \in L_b(u)$ then w is cut from below by e_i^*. See Fig. 3(a).

p2. If $w \in R_b(u)$ then w is cut from above by e_i^*.

p3. If $w \in B_b(u)$, then either $\Phi(w) = \llcorner$ or $\Phi(w) = \ulcorner$. Also, if $\Phi(w) = \llcorner$ (resp., $\Phi(w) = \ulcorner$) then w is cut from above (resp., below) by e_i^*. See Fig. 3(b).

p4. If $w \in T_b(v)$ (and thus $w \in T_b(u)$), then either $\Phi(w) = \lrcorner$ or $\Phi(w) = \urcorner$. Also, if $\Phi(w) = \lrcorner$ (resp., $\Phi(w) = \urcorner$) then w is cut from above (resp., below) by e_i^*. See Fig. 3(c).

p5. If $w \in T_b(u)$ and $w \notin T_b(v)$, then let (u, z) be an edge having u as end-vertex and belonging to a path from u to w. If (u, z) is to the left (resp., right) of any path from s_b to t_b through (u, v), then w is cut from below (resp., above) by e_i^*. See Fig. 3(d).

We now show that if $\langle G_r, G_b \rangle$ admits a Φ-transversal, then for each blue edge (the same argument would apply for red edges) there exists a unique traversing path.

Lemma 4. *Let G_r and G_b be two plane st-graphs with the same vertex set V and with distinct sources and sinks. Let $\Phi : V \to \mathcal{H} = \{ \llcorner, \lrcorner, \urcorner, \ulcorner \}$. If $\langle G_r, G_b \rangle$ admits a Φ-transversal, then for every edge e of G_b there is a unique traversing path π_e in $D_r^\to$.*

Proof. If $\langle G_r, G_b \rangle$ admits a Φ-transversal G_{rb}, then for every edge $e = (u, v)$ of G_b there exists a path $\pi_e = \{ f_r(u) = f_0, e_0^*, f_1, \ldots, f_{k-1}, e_{k-1}^*, f_r(v) = f_k \}$ in $D_r^\to$, which is the path used by e to go from $f_r(u)$ to $f_r(v)$ in G_{rb}.

If f_0 and f_k coincide, then π_e is a traversing path. Otherwise, if f_0 and f_k coincide and π_e is not a traversing path, we would have a cycle $\pi_e = \{ f_0 = f_k, \ldots, f_0 = f_k \}$, which is not possible since $D_r^\to$ is acyclic, being the dual of a plane st-graph.

If f_0 and f_k do not coincide, let w be a vertex in the right path of f_i. First, if w belongs to $L_b(u)$, then it is cut from below. Otherwise, if w was cut from above, since edge $e = (u, v)$ cannot cross the right path of f_i twice (by condition **c3**), it would belong to $R_b(u)$, a contradiction with the fact that the embedding

---

[1] Since $D_r^\to$ is a multigraph, to uniquely identify π_e we specify the edges that are traversed.

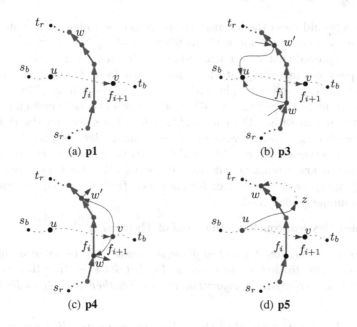

Fig. 3. Illustration for the properties of a traversing path π_e. (Color figure online)

of G_b is preserved. Thus **p1** is respected by π_e. With a symmetric argument we can also prove **p2**.

Suppose now that w belongs to $B_b(u)$. Then $f_r(w) = f_i = left_r(w)$, otherwise if $f_r(w) = f_{i+1} = right_r(w)$, an edge of the blue path from w to u would violate **c3**. In other words, either $\Phi(w) = \mathsf{L}$ or $\Phi(w) = \mathsf{\Gamma}$. Furthermore, if $\Phi(w) = \mathsf{L}$, then w must be cut from above, while if $\Phi(w) = \mathsf{\Gamma}$, then w must be cut from below, as otherwise an edge of the blue path from s_b to w must enter the region delimited by the blue path from w to u, part of the blue edge (u, v), and part of the (red) right path of f_i, which violates the planarity of the embedding of G_b or condition **c3**. Thus, π_e obeys to **p3**.

If w belongs to both $T_b(u)$ and $T_b(v)$, then $f_r(w) = f_{i+1} = right_r(w)$, otherwise if $f_r(w) = f_i = left_r(w)$, an edge of the blue path from v to w would violate **c3**. In other words, either $\Phi(w) = \mathsf{J}$ or $\Phi(w) = \mathsf{\daleth}$. Furthermore, if $\Phi(w) = \mathsf{J}$, then w must be cut from above, while if $\Phi(w) = \mathsf{\daleth}$, then w must be cut from below, as otherwise an edge of the blue path from w to t_b must exit the region delimited by the blue path from v to w, part of the blue edge (u, v), and part of the (red) right path of f_i, which again violates the planarity of the embedding of G_b or condition **c3**. Thus, π_e obeys to **p4**.

Let w belong to $T_b(u)$ but not to $T_b(v)$, and let (u, z) be an edge having u as end-vertex and belonging to a path from u to w (there exists at least one since $w \in T_b(u)$). Notice that z cannot coincide with v as $w \notin T_b(v)$. If (u, z) is to the left (resp., right) of any path from s_b to t_b through (u, v), then w is cut from below (resp., above) by e_i^*. Otherwise, recall that $w \notin T_b(v)$, the blue path from u to w would cross the leftmost (resp., rightmost) blue path from s_b to t_b through (u, v), a contradiction with the planarity of G_b. Hence, π_e obeys to **p5**.

So far we proved that π_e is a traversing set. To prove that π_e is unique, note that any possible traversing path for e must start from f_0 and leave this face. Hence, any vertex w on the right path of f_0 must be cut from either above or below, according to properties **p1**–**p5** (which cover all possible cases for w). The only edge that can satisfy the cut condition for all vertices on the right path of f_0, is an edge e_0^* whose corresponding red primal edge, denoted by (x, z), is such that all vertices on the right path of f_0 above x must be cut from below and all those below z must be cut from above. Clearly, this edge is unique. By repeatedly applying this argument for each face f_i $(0 \le i < k)$, the traversing path π_e is uniquely identified. $\square$

The next theorem concludes the proof of Theorem 1.

Theorem 4. *Let G_r and G_b be two plane st-graphs with the same set of n vertices V and with distinct sources and sinks. Let $\Phi : V \to \mathcal{H} = \{\llcorner, \lrcorner, \urcorner, \ulcorner\}$. There exists an $O(n^3)$-time algorithm to test whether $\langle G_r, G_b \rangle$ admits a Φ-transversal.*

Proof sketch. Our testing algorithm aims to compute (if it exists) a Φ-transversal G_{rb} for $\langle G_r, G_b \rangle$. We first fix the circular order of the edges restricted to the blue edges (resp., red edges) around each vertex u of G_{rb} to satisfy **c1** and to maintain the planar embedding of G_b (resp., G_r). We then fix the circular order of the blue edges with respect to the red edges around each vertex u of G_{rb} to satisfy **c2** (i.e., to obey $\Phi(u)$). Then, we first check if for every blue edge e there exists a traversing path π_e; if so, we verify that by routing every blue edge e through π_e no two blue edges cross each other. If this procedure succeeds then $\langle G_r, G_b \rangle$ admits Φ-transversal G_{rb}. Indeed, by construction, the resulting embedding of G_{rb} satisfies conditions **c1**, **c2** and **c3** and it is such that restricting $G_{rb} \setminus \{s_b, t_b\}$ to the red edges realizes the planar embedding G_r and restricting $G_{rb} \setminus \{s_r, t_r\}$ to the blue edges realizes the planar embedding G_b. Otherwise, either there exists a blue edge with no traversing path, or two traversing paths are such that the two corresponding edges of G_b cross if routed through them. In the first case $\langle G_r, G_b \rangle$ does not admit a Φ-transversal by Lemma 4. In the second case, since the traversing paths are unique, condition **c1** cannot be satisfied, and again $\langle G_r, G_b \rangle$ does not admit a Φ-transversal.

The testing algorithm works in two phases as follows.

Phase 1. For every edge $e = (u, v) \in E_b$. If $f_r(u) = f_r(v)$, we have found a traversing path. Otherwise, we label each vertex on the right path of $f_r(u)$, by A if it must be cut from above or by B if it must be cut from below, according to properties **p1**–**p5**. Then we check if the sequence of labels along the path is a nonzero number of A's followed by a nonzero number of B's. If so, then the dual edge of the traversing path is the one whose corresponding primal edge has the two end-vertices with different labels (which is unique). If this is not the case, then a traversing path for e does not exist. In the positive case, we add the dual edge we found and the next face we reach through this edge to π_e and we iterate the algorithm until we reach either $f_r(v)$ or the outer face of $D_r^{\rightarrow}$. In the former case π_e is a traversing path for e, while in the latter case, since the edges of the outer face of G_{rb} cannot be crossed by definition of Φ-transversal, we have that again no traversing path can be found.

Phase 2. We now check that by routing every edge $e \in E_b$ through its corresponding traversing path π_e, no two of these edges cross each other. Consider the dual graph $D_r^{\rightarrow}$, which is a plane st-graph. Construct a planar drawing Γ of $D_r^{\rightarrow}$. Consider any two traversing paths π_e and π'_e, which corresponds to two paths in Γ, and let $e = (u, v)$ and $e' = (w, z)$ be the two corresponding edges of G_b. Denote by $\hat{\pi}_e = \{u\} \cup \pi_e \cup \{v\}$ and $\hat{\pi}'_e = \{w\} \cup \pi'_e \cup \{z\}$ the two enriched paths. Enrich Γ by adding the four edges $(x, f_r(x))$, where $x \in \{u, v, w, z\}$, in a planar way respecting the original embedding of G_b. Consider now the subdrawing Γ' of Γ induced by $\hat{\pi}_e \cup \hat{\pi}'_e$. If e and e' cross each other, then $\pi_e \cap \pi'_e$ cannot be empty. Moreover, the intersection $\pi_e \cap \pi'_e$ must be a single subpath, as otherwise the two traversing paths would not be unique. Let f be the first face and let g be the last face in this subpath. Let e_u be the incoming edge of f that belongs to the subpath of $\hat{\pi}_e$ from u to f; and let e_w be the incoming edge of f that belongs to the subpath of $\hat{\pi}'_e$ from w to f. Also, let e_v be the outgoing edge of g that belongs to the subpath of $\hat{\pi}_e$ from g to v; and let e_z be the outgoing edge of g that belongs to the subpath of $\hat{\pi}'_e$ from g to z. Then e and e' cross if and only if walking clockwise along $\pi_e \cup \pi'_e$ from f to g and back to f these four edges are encountered in the circular order e_u, e_z, e_v, e_w. Note that, e_u and e_w may coincide if $u = w$, and similarly for e_v and e_z. $\qquad\square$

5   Final Remarks and Open Problems

In this paper we have introduced and studied the concept of simultaneous visibility representation with L-shapes of two plane st-graphs. We remark that it is possible to include in our theory the case when the vertices can also be drawn as rectangles. Nevertheless, this would not enlarge the class of representable pairs of graphs. In fact, for every vertex v drawn as a rectangle $\mathcal{R}_v$, we can replace $\mathcal{R}_v$ with any L-shape by keeping only two adjacent sides of $\mathcal{R}_v$ in the drawing and prolonging the visibilites incident to the removed sides of $\mathcal{R}_v$. The converse is not true. Indeed, roughly speaking, L-shapes can be nested, whereas rectangles cannot. To give an example, if a vertex v must see a vertex u both vertically and horizontally, this immediately implies that the two corresponding rectangles need to overlap, while two L-shapes could instead be nested. Several extensions of the model introduced in this paper can also be studied, e.g., the case where every edge is represented by a T-shape, or more generally by a +-shape.

Our results can also be used to shed more light on the problem of computing a simultaneous RAC embedding (SRE) [2,3]. Given two planar graphs with the same vertex set, an SRE is a simultaneous embedding where crossings between edges of the two graphs occur at right angles. Argyriou et al. proved that it is always possible to construct an SRE with straight-line edges of a cycle and a matching, while there exist a wheel graph and a cycle that do not admit such a representation [2]. This motivated recent results about SRE with bends along the edges. Namely, Bekos *et al.* show that two planar graphs with the same vertex set admit an SRE with at most six bends per edge in both graphs [3]. We observe that any pair of graphs that admit a simultaneous visibility representation with L-shapes also admits an SRE with at most two bends per edge. This is obtained with the technique used in Sect. 3 to compute a Φ-transversal from a Φ-LSVR, see Fig. 2(a). Thus, a new approach to characterize graph pairs that have SREs with at most two bends per edge is as follows: Given two planar graphs with the same vertex set, add to each of them a unique source and a unique sink, and look for two *st*-orientations (one for each of the two graphs) and a function Φ such that the two graphs admit a Φ-LSVR. In [11], we show an alternative proof of another result by Bekos *et al.* that a wheel graph and a matching admit an SRE with at most two bends for each edge of the wheel, and no bends for the matching edges [3].

Three questions that stem from this paper are whether the time complexity of the testing algorithm in Sect. 4 can be improved; what is the complexity of deciding if two given plane *st*-graphs admit a Φ-LSVR for some function Φ, which is not part of the input; and what is the complexity of deciding if two undirected graphs admit a Φ-LSVR for some function Φ.

References

1. Angelini, P., Geyer, M., Kaufmann, M., Neuwirth, D.: On a tree and a path with no geometric simultaneous embedding. J. Graph Algorithms Appl. **16**(1), 37–83 (2012)
2. Argyriou, E.N., Bekos, M.A., Kaufmann, M., Symvonis, A.: Geometric RAC simultaneous drawings of graphs. J. Graph Algorithms Appl. **17**(1), 11–34 (2013)
3. Bekos, M.A., van Dijk, T.C., Kindermann, P., Wolff, A.: Simultaneous drawing of planar graphs with right-angle crossings and few bends. In: Rahman, M.S., Tomita, E. (eds.) WALCOM 2015. LNCS, vol. 8973, pp. 222–233. Springer, Heidelberg (2015)
4. Bläsius, T., Rutter, I.: Simultaneous pq-ordering with applications to constrained embedding problems. In: Khanna, S. (ed.) SODA 2013, pp. 1030–1043. SIAM (2013)
5. Braß, P., Cenek, E., Duncan, C.A., Efrat, A., Erten, C., Ismailescu, D., Kobourov, S.G., Lubiw, A., Mitchell, J.S.B.: On simultaneous planar graph embeddings. Comput. Geom. **36**(2), 117–130 (2007)
6. Cabello, S., van Kreveld, M.J., Liotta, G., Meijer, H., Speckmann, B., Verbeek, K.: Geometric simultaneous embeddings of a graph and a matching. J. Graph Algorithms Appl. **15**(1), 79–96 (2011)
7. Di Battista, G., Eades, P., Tamassia, R., Tollis, I.G.: Graph Drawing. Prentice Hall, Englewood Cliffs (1999)
8. Di Giacomo, E., Liotta, G.: Simultaneous embedding of outerplanar graphs, paths, and cycles. Int. J. Comput. Geom. Appl. **17**(2), 139–160 (2007)
9. Didimo, W., Liotta, G.: The crossing angle resolution in graph drawing. In: Pach, J. (ed.) Thirty Essays on Geometric Graph Theory, pp. 167–184. Springer, New York (2012)

10. Eppstein, D.: Regular labelings and geometric structures. In: CCCG 2010, pp. 125–130 (2010)
11. Evans, W.S., Liotta, G., Montecchiani, F.: Simultaneous visibility representations of plane *st*-graphs usingL-shapes. arXiv (2015). http://arxiv.org/abs/1505.04388
12. Felsner, S.: Rectangle and square representations of planar graphs. In: Pach, J. (ed.) Thirty Essays on Geometric Graph Theory, pp. 213–248. Springer (2013)
13. Fusy, É.: Transversal structures on triangulations: a combinatorial study and straight-line drawings. Discr. Math. **309**(7), 1870–1894 (2009)
14. Huang, W., Hong, S.-H., Eades, P.: Effects of crossing angles. In: PacificVis 2008, pp. 41–46. IEEE (2008)
15. Jampani, K.R., Lubiw, A.: Simultaneous interval graphs. In: Cheong, O., Chwa, K.-Y., Park, K. (eds.) ISAAC 2010, Part I. LNCS, vol. 6506, pp. 206–217. Springer, Heidelberg (2010)
16. Jampani, K.R., Lubiw, A.: The simultaneous representation problem for chordal, comparability and permutation graphs. J. Graph Algorithms Appl. **16**(2), 283–315 (2012)
17. Kant, G., He, X.: Regular edge labeling of 4-connected plane graphs and its applications in graph drawing problems. Theor. Comput. Sci. **172**(1–2), 175–193 (1997)
18. Rosenstiehl, P., Tarjan, R.E.: Rectilinear planar layouts and bipolar orientations of planar graphs. Discr. Comput. Geom. **1**, 343–353 (1986)
19. Schnyder, W.: Embedding planar graphs on the grid. In: Johnson, D.S. (ed.) SODA 1990, pp. 138–148. SIAM (1990)
20. Streinu, I., Whitesides, S.: Rectangle visibility graphs: Characterization, construction, and compaction. In: Alt, H., Habib, M. (eds.) STACS 2003. LNCS, vol. 2607, pp. 26–37. Springer, Heidelberg (2003)
21. Tamassia, R.: Simultaneous embedding of planar graphs. In: Handbook of Graph Drawing and Visualization. CRC Press, Boca Raton (2013)
22. Tamassia, R., Tollis, I.G.: A unified approach to visibility representations of planar graphs. Discr. Comput. Geom. **1**(1), 321–341 (1986)

An Abstract Approach to Polychromatic Coloring: Shallow Hitting Sets in ABA-free Hypergraphs and Pseudohalfplanes

Balázs Keszegh[1] and Dömötör Pálvölgyi[2(✉)]

[1] Alfréd Rényi Institute of Mathematics, Budapest, Hungary
keszegh.balazs@renyi.mta.hu
[2] Institute of Mathematics, Eötvös University, Budapest, Hungary
dom@cs.elte.hu

Abstract. The goal of this paper is to give a new, abstract approach to cover-decomposition and polychromatic colorings using hypergraphs on ordered vertex sets. We introduce an abstract version of a framework by Smorodinsky and Yuditsky, used for polychromatic coloring halfplanes, and apply it to so-called *ABA-free hypergraphs*, which are a generalization of *interval graphs*. Using our methods, we prove that $(2k-1)$-uniform ABA-free hypergraphs have a polychromatic k-coloring, a problem posed by the second author. We also prove the same for hypergraphs defined on a point set by pseudohalfplanes. These results are best possible.

We also introduce several new notions that seem to be important for investigating polychromatic colorings and ϵ-nets, such as *shallow hitting sets*. We pose several open problems related to them. For example, is it true that given a finite point set S on a sphere and a set of halfspheres $\mathcal{F}$, such that $\{S \cap F \mid F \in \mathcal{F}\}$ is a Sperner family, we can select an $R \subset S$ such that $1 \leq |F \cap R| \leq 2$ holds for every $F \in \mathcal{F}$?

1 Introduction

The study of proper and polychromatic colorings of geometric hypergraphs has attracted much attention, not only because this is a very basic and natural theoretical problem but also because such problems often have important applications. One such application area is resource allocation, e.g., battery consumption in sensor networks. Moreover, the coloring of geometric shapes in the plane is related to the problems of cover-decomposability, conflict-free colorings and ϵ-nets; these problems have applications in sensor networks and frequency assignment as well as other areas. For surveys on these and related problems see [13,18].

In a (primal) *geometric hypergraph polychromatic coloring* problem, we are given a natural number k, a set of points and a collection of regions in $\mathbb{R}^d$, and our goal is to k-color the points such that every region that contains at least

Research supported by Hungarian Scientific Research Fund (OTKA), under grant NN 102029, NK 78439, PD 104386, PD 108406, by the János Bolyai Research Scholarship of the Hungarian Academy of Sciences.

E.W. Mayr (Ed.): WG 2015, LNCS 9224, pp. 266–280, 2016.
DOI: 10.1007/978-3-662-53174-7_19

$m(k)$ points contains a point of every color, where m is some function that we try to minimize. We call such a coloring a *polychromatic k-coloring*. In a *dual* geometric hypergraph polychromatic coloring problem, our goal is to k-color the regions such that every point which is contained in at least $m(k)$ regions is contained in a region of every color. In other words, in the dual version our goal is to *decompose* an $m(k)$-fold covering of some point set into k coverings. The primal and the dual versions are equivalent if the underlying regions are the translates of some fixed set. For the proof of this statement and an extensive survey of results related to *cover-decomposition*, see e.g., [13]. Below we mention some of these results, stated in the equivalent primal form.

The most general result about translates of polygons is that given a fixed convex polygon, there exists a c (that depends only on the polygon) such that any *finite* point set has a polychromatic k-coloring such that any translate of the fixed convex polygon that contains at least $m(k) = c \cdot k$ points contains a point of every color [6]. Non-convex polygons for which such a finite $m(k)$ (for any $k \geq 2$) exists have been classified [14,17].

As it was shown recently [16], there is no such finite $m(2)$ for convex sets with a smooth boundary, e.g., for the translates of a disc. However, it was also shown in the same paper that for the translates of any *unbounded* convex set $m(2) = 3$ is sufficient. In this paper we extend this result to every k, showing that $m(k) = 2k - 1$ is an optimal function for unbounded convex sets.

For homothets of a given shape the primal and dual problems are not equivalent. For homothets of a triangle (a case closely related to the case of translates of octants [9,10]), there are several results, the current best are $m(k) = O(k^{4.53})$ in the primal version [2,11] and $m(k) = O(k^{5.53})$ in the dual version [3]. For the homothets of other convex polygons, in the dual case there is no finite $m(2)$ [12], and in the primal case only conditional results are known [11], namely, that the existence of a finite $m(2)$ implies the existence of an $m(k)$ that grows at most polynomially in k. In fact, it is even possible that for *any* polychromatic coloring problem $m(k) = O(m(2))$.

For other shapes, cover-decomposability has been studied less, in these cases the investigation of polychromatic-colorings is motivated rather by conflict-free colorings or ϵ-nets. Most closely related to our paper, coloring halfplanes for small values were investigated in [5,7,8], and polychromatic k-colorings in [19]. We generalize all the (primal and dual) results of the latter paper to pseudohalf-planes. Note that translates of an unbounded convex set form a set of pseudo-halfplanes, thus the above mentioned result about unbounded convex sets is a special case of this generalization to pseudohalfplanes.

Besides generalizing earlier results, our contribution is a more abstract app-roach to the above problems. Namely, we introduce the notion of *ABA-free families* (see Definition 1) and *shallow hitting sets* (see Definition 5).

1.1 Definitions and Statements of Main Results

Definition 1. *A hypergraph* $\mathcal{H}$ *with an ordered vertex set is called* ABA-free *if H does not contain two hyperedges A and B for which there are three vertices* $x < y < z$ *such that* $x, z \in A \setminus B$ *and* $y \in B \setminus A$.

A hypergraph with an unordered vertex set is ABA-free if its vertices have an ordering with which the hypergraph is ABA-free.[1]

Remark 1. ABA-free hypergraphs were first defined in [16] under the name *special shift-chains*, as they are a special case of *shift-chains* introduced in [15].

Example 2. *An* interval hypergraph *is a hypergraph whose vertices are some points of* $\mathbb{R}$, *and hyperedges are some intervals from* $\mathbb{R}$, *with the incidences preserved. By definition, every interval hypergraph is ABA-free and in fact every ABA-free hypergraph is similar, in a way, to an interval hypergraph.*

Example 3 [16]. *Let S be a set of points in the plane with different x-coordinates and let C be a convex set that contains a vertical halfline. Define a hypergraph* $\mathcal{H}$ *whose vertex set is the x-coordinates of the points of S. A set of numbers X is a hyperedge of* $\mathcal{H}$ *if there is a translate of C such that the x-coordinates of the points of S contained in the translate is exactly X. The hypergraph* $\mathcal{H}$ *defined this way is ABA-free.*

Example 4. *Let S be a set of points in the plane in general position. Define a hypergraph* $\mathcal{H}$ *whose vertex set is the x-coordinates of the points of S. A set of numbers X is a hyperedge of* $\mathcal{H}$ *if there is a positive halfplane H (i.e., that contains a vertical positive halfline) such that the x-coordinates of the points of S contained in H is exactly X. The hypergraph* $\mathcal{H}$ *defined this way is ABA-free.*

The above examples show how to reduce geometric problems to abstract problems about ABA-free hypergraphs. Observe that given an S, by choosing an appropriately big parabola, all hyperedges defined by positive halfplanes is also defined by some translate of this parabola, thus the first example is more general than the second. Even more, as we will see later in Sect. 3, finite ABA-free hypergraphs have an equivalent geometric representation with graphic pseudoline arrangements (sets are defined by the regions above the pseudolines, for the definitions and details see Sect. 3) and both translates of the boundary of an unbounded convex set and lines in the plane form graphic pseudoline arrangements, showing again that the above examples are special cases of ABA-free hypergraphs.

To study polychromatic coloring problems, we also introduce the following definition, which is implicitly used in [19], but deserves to be defined explicitly as it seems to be important in the study of polychromatic colorings.

---

[1] While it might seem that using the same notion for ordered and unordered hypergraphs leads to confusion as by forgetting the ordering of an ordered hypergraph it might become ABA-free, from the context it will always be perfectly clear what we mean.

Definition 5. *A set R is a c-shallow hitting set of the hypergraph $\mathcal{H}$ if for every $H \in \mathcal{H}$ we have $1 \le |R \cap H| \le c$.*

Actually, almost all our results are based on shallow hitting sets.

Observation 6. *An induced subhypergraph of an ABA-free hypergraph is also ABA-free.*

Our main results and the organization of the rest of this paper is as follows.

In Sect. 2 we prove (following closely the ideas of Smorodinsky and Yuditsky [19]) that every $(2k - 1)$-uniform ABA-free hypergraph has a polychromatic coloring with k colors. We then observe that the dual of this problem is equivalent to the primal, which implies that the edges of every $(2k - 1)$-uniform ABA-free hypergraph can be colored with k colors, such that if a vertex v is in a subfamily $\mathcal{H}_v$ of at least $m(k) = 2k - 1$ of the edges of $\mathcal{H}$, then $\mathcal{H}_v$ contains a hyperedge from each of the k color classes.

In Sect. 3 we give an abstract equivalent definition (using ABA-free hypergraphs) of hypergraphs defined by pseudohalfplanes, and we prove that given a finite set of points S and a pseudohalfplane arrangement $\mathcal{H}$, we can k-color S such that any pseudohalfplane in $\mathcal{H}$ that contains at least $m(k) = 2k - 1$ points of S contains all k colors. Both results are sharp. Note that these results imply the same for hypergraphs defined by unbounded convex sets.

In Sect. 5 we discuss dual and other versions of the problem. For example we prove that given a pseudohalfplane arrangement $\mathcal{H}$, we can k-color $\mathcal{H}$ such that if a point p belongs to a subfamily $\mathcal{H}_p$ of at least $m(k) = 3k - 2$ of the pseudohalfplanes of $\mathcal{H}$, then $\mathcal{H}_p$ contains a pseudohalfplane from each of the k color classes. This result might not be sharp, the best known lower bound for $m(k)$ is $2k - 1$ [19]. We also discuss consequences about ϵ-nets on pseudohalfplanes.

We denote the symmetric difference of two sets, A and B, by $A \triangle B$, the complement of a hyperedge F by $\bar{F}$ and for a family $\mathcal{F}$ we use $\bar{\mathcal{F}} = \{\bar{F} | F \in \mathcal{F}\}$. We will suppose (unless stated otherwise) that all hypergraph and point sets are finite, and denote the smallest (resp. largest) element of an ordered set H by $\min(H)$ (resp. $\max(H)$).

2   ABA-free Hypergraphs

Suppose we are given an ABA-free hypergraph $\mathcal{H}$ on n vertices. As the hypergraph is ABA-free, for any pair of sets $A, B \in \mathcal{H}$ either there are $a < b$ such that $a \in A \setminus B$ and $b \in B \setminus A$, or there are $b < a$ such that $a \in A \setminus B$ and $b \in B \setminus A$, or none of them, but not both as that would contradict ABA-freeness.

Define $A < B$ if and only if there are $a < b$ such that $a \in A \setminus B$ and $b \in B \setminus A$. Define $A \le B$ if and only if either $A = B$ (as sets) or $A < B$. By the above observation, this is well-defined, and it gives a partial ordering of the sets.

Definition 7. *A vertex a is skippable if there exists an $A \in \mathcal{H}$ such that $\min(A) < a < \max(A)$ and $a \notin A$. In this case we say that A skips a. A vertex a is unskippable if there is no such A.*

Observation 8. *If a vertex a is unskippable in some ABA-free hypergraph $\mathcal{H}$, then after adding the one-element edge $\{a\}$ to $\mathcal{H}$, it remains ABA-free.*

Note that the following two lemmas show that the unskippable vertices of an ABA-free hypergraph behave similarly to vertices on the convex hull of a hypergraph on a point set defined by halfplanes. These two lemmas make it possible to use the framework of [19] on ABA-free hypergraphs.

Lemma 9. *If $\mathcal{H}$ is finite ABA-free, then every $A \in \mathcal{H}$ contains an unskippable vertex.*

Remark 2. Note that finiteness is needed, as the hypergraph whose vertex set is $\mathbb{Z}$ and edge set is $\{\mathbb{Z} \setminus \{n\} \mid n \in \mathbb{Z}\}$ is ABA-free without unskippable vertices.

Proof. Take an arbitrary set $A \in \mathcal{H}$, suppose that it does not contain an unskippable vertex, we will reach a contradiction. Call $a \in A$ *rightskippable* if there is a $B \in \mathcal{H}$ rightskipping a, that is for which $a \in A \setminus B$ and there are $b_1, b_2 \in B$ such that $b_1 < a < b_2$ where $b_2 \in B \setminus A$.

If A contains no unskippable vertex, $\max(A)$ must be rightskippable (any set skipping $\max(A)$ must also rightskip $\max(A)$). Also, $\min(A)$ cannot be rightskippable, as otherwise A and the set B rightskipping $\min(A)$ would violate ABA-freeness (we would get $b_1 < \min(A) < b_2$ where $b_1, b_2 \in B \setminus A, \min(A) \in A \setminus B$). Therefore we can take the largest $a \in A$ that is not rightskippable. By the assumption, it is skipped by a set, call it B, i.e., $b_1 < a < b_2$ where $b_1, b_2 \in B \not\ni a$. Moreover, wlog., suppose that b_2 is the smallest element of B which is bigger than a. Since a is not rightskippable, $b_2 \in A$ must also hold. As $b_2 \in A$ is rightskippable, there is a C such that $c_1 < b_2 < c_2$ where $c_1, c_2 \in C$ and $b_2 \notin C, c_2 \notin A$. Wlog., suppose c_1 is the largest element of C which is smaller than b_2. If $c_1 < a$, then C would rightskip a, a contradiction. Thus, $b_1 < a \leq c_1$, and from the choice of b_2 we conclude that $c_1 \notin B$. As $c_2 \notin A$, also $c_2 \notin B$, otherwise B would rightskip a. Putting all together, we get $c_1 < b_2 < c_2$, thus B and C contradict ABA-freeness.

Recall that a hypergraph is called *Sperner* if no two of its sets (i.e., hyperedges) contain each other, we further assume in it that the hypergraph is non-empty, i.e., it contains at least one edge.

Lemma 10. *If $\mathcal{H}$ is finite, ABA-free and Sperner, then any minimal hitting set of $\mathcal{H}$ that contains only unskippable vertices is 2-shallow.*

Proof. Let R be a minimal hitting set of unskippable vertices. Assume to the contrary that there exists a set A such that $|A \cap R| \geq 3$. Let $l = \min(A \cap R)$ and $r = \max(A \cap R)$. There exists a third vertex $l < a < r$ in $A \cap R$. We claim that $R' = R \setminus \{a\}$ hits all sets of $\mathcal{H}$, contradicting its minimality. Assume on the contrary that R' is disjoint from some $B \in \mathcal{H}$. As R must hit B, we have $R \cap B = \{a\}$. If there is a $b \in B \setminus A$ such that $l < b < r$, that would contradict the ABA-free property. If there is a $b \in B$ such that $b < l < a$ or $a < r < b$, that would contradict that l and r are unskippable. Thus $B \subset A$, contradicting that $\mathcal{H}$ is Sperner.

Now we present the framework of [19] modified for ABA-free hypergraphs. Our algorithm to give a polychromatic k-coloring of the vertices of an ABA-free hypergraph with edges of size at least $2k - 1$ is as follows.

Algorithm 11. *At the beginning we are given an ABA-free hypergraph $\mathcal{H}$ with edges of size at least $2k - 1$.*

Repeat $k - 1$ times ($i = 1, \ldots, k - 1$) the **general step i of the algorithm:**
At the beginning of step i we have an ABA-free hypergraph $\mathcal{H}$ with edges of size at least $2k - 2i + 1$. If any hyperedge contains another, then delete the bigger hyperedge. Repeat this until no hyperedge contains another, thus making our hypergraph Sperner. Next, take the set of all unskippable vertices, which is a hitting set by Lemma 9 and delete vertices from this set until it becomes a minimal hitting set R. By Lemma 10 R is a 2-shallow hitting set, color its vertices with the i-th color. Delete these vertices from $\mathcal{H}$ (the edges of the new hypergraph are the ones induced by the remaining vertices). Using Observation 6, at the end of this step i, we have an ABA-free hypergraph with edges of size at least $2k - 2i - 1$.

After $k - 1$ iterations of the above, we are left with a 1-uniform hypergraph whose vertices we can color with the k-th color.

Algorithm 11 implies the following theorem.

Theorem 12. *Given a finite ABA-free $\mathcal{H}$ we can color its vertices with k colors such that every $A \in \mathcal{H}$ whose size is at least $2k - 1$ contains all k colors.*

Notice that the above theorem is sharp, as taking $\mathcal{H}$ to be all subsets of size $2k - 2$ from $2k - 1$ vertices, in any coloring of the vertices, one color must occur at most once and is thus missed by some hyperedge.

We state another corollary of Lemma 9 that we need later. Before that, we need another simple claim.

Proposition 13. *Suppose we insert a new vertex, v, somewhere into the (ordered) vertex set of an ABA-free hypergraph, $\mathcal{H}$, and add v to every edge that contains a vertex before and another vertex after v, then we get an ABA-free hypergraph.*

Proof. We show that if in the new hypergraph, $\mathcal{H}'$, two hyperedges A' and B' violate ABA-freeness, then we can find two hyperedges A and B in the original hypergraph, $\mathcal{H}$, that also violate ABA-freeness, which would be a contradiction. We define $A = A' \setminus \{v\}$ and $B = B' \setminus \{v\}$. If both A' and B' contain or do not contain v, then by definition A and B also violate the condition. If, say, $v \notin A'$ and $v \in B'$, then without loss of generality we can suppose that all the vertices of $A = A'$ are before v. This means that if there are $x < y < z$ such that $x, z \in A' \setminus B'$ and $y \in B' \setminus A'$, then necessarily $v = z$. But as B' has an element z' that is bigger than v, we have $x, z' \in A \setminus B$ and $y \in B \setminus A$, a contradiction.

Lemma 14. *If $\mathcal{H}$ is ABA-free, $A \in \mathcal{H}$, then there is a vertex $a \in A$ such that $\mathcal{H} \cup \{A \setminus \{a\}\}$ is also ABA-free.*

Proof. If $|A| = 1$, then trivially $\mathcal{H}$ can be extended with $\emptyset$. If $|A| > 1$, then we proceed by induction on the size of A. Using Lemma 9, there is an unskippable vertex $v \in A$. Delete this vertex from $\mathcal{H}$ to obtain some ABA-free $\mathcal{H}_v$ and let $A_v = A \setminus \{v\}$. Using induction on A_v, there is an $A'_v = A_v \setminus \{a\}$ such that $\mathcal{H}_v \cup \{A'_v\}$ is also ABA-free. We claim that with $A' = A'_v \cup \{v\} = A \setminus \{a\}$, the family $\mathcal{H} \cup \{A'\}$ is also ABA-free.

Notice that adding back v to $\mathcal{H}_v$ is very similar to the operation of Proposition 13, as v is unskippable in $\mathcal{H}$. The only difference is that we might also have to add it to some further hyperedges, ending in or starting at v. But a hyperedge that contains v cannot violate the ABA-free condition with A', since it also contains v, so the corresponding hyperedges in $\mathcal{H}_v$ would also violate the ABA-free condition.

Notice that with the repeated application of Lemma 14 we can extend any ABA-free hypergraph, such that in any set A there is a vertex a for which $\{a\}$ is a singleton edge, implying that a is unskippable in A. Thus in fact Lemma 14 is equivalent to Lemma 9. Moreover, in Sect. 3, in the more general context of pseudohalfplanes, it will be the abstract equivalent of a known and important property of pseudohalfplanes.

We prove another interesting property of ABA-free hypergraphs before which we need the following definition.

Definition 15. *The dual of a hypergraph $\mathcal{H}$, denoted by $\mathcal{H}^*$, is such that its vertices are the hyperedges of $\mathcal{H}$ and its hyperedges are the vertices of $\mathcal{H}$ with the same incidences as in $\mathcal{H}$.*

Proposition 16. *If $\mathcal{H}$ is ABA-free, then its dual $\mathcal{H}^*$ is also ABA-free with respect to some ordering of its vertices.*

Proof. Take the partial order "$<$" of the hyperedges of $\mathcal{H}$ and extend this arbitrarily into a total order $<^*$. We claim that $\mathcal{H}^*$ is ABA-free if its vertices are ordered with respect to $<^*$. To check the condition, suppose for a contradiction that $H_x <^* H_y <^* H_z$ and $a \in (H_x \cap H_z) \setminus H_y$ and $b \in H_y \setminus (H_x \cup H_z)$. Without loss of generality, suppose that $a < b$. But in this case $H_z < H_y$ holds, contradicting $H_y <^* H_z$.

Corollary 17. *The edges of every $(2k - 1)$-uniform ABA-free hypergraph can be colored with k colors, such that if a vertex v is in a subfamily $\mathcal{H}_v$ of at least $m(k) = 2k - 1$ of the edges of $\mathcal{H}$, then $\mathcal{H}_v$ contains a hyperedge from each of the k color classes.*

Corollary 18. *Any $(2k-1)$-fold covering of a finite point set with the translates of an unbounded convex planar set is decomposable into k coverings.*

3   Pseudohalfplanes

Here we extend a result of Smorodinsky and Yuditsky [19]. A *pseudoline arrangement* is a finite collection of simple curves in the plane in general position (i.e., no three curves have a common point) such that any two are either disjoint or intersect once and in the intersection point they cross. Some further definitions and well-known results about pseudoline arrangements are collected below, which can be found in [1]. We also recommend [4] where generalizations of classical theorems are proved for *topological affine planes*.

An *infinite* pseudoline arrangement is such that cutting a pseudoline in two, both parts are unbounded. A curve is *graphic* if it is the graph of a function, i.e., an x-monotone infinite curve that intersects every vertical line of the plane. A *graphic* pseudoline arrangement is such that every curve is graphic. We say that two pseudoline arrangements are *equivalent* if there is a bijection between their pseudolines such that the order in which a pseudoline intersects the other pseudolines remains the same. A *pseudohalfplane arrangement* is an infinite pseudoline arrangement, with a side of each pseudoline selected (note that the two sides of a pseudoline are well-defined regions in this case).

Facts About Pseudoline Arrangements

   I. (Levi Enlargement Lemma) Given a pseudoline arrangement, any two points of the plane can be connected by a new pseudoline (if they are not connected already).
  II. Given a pseudoline arrangement, we can find a(n infinite) pseudoline arrangement in which every pair of pseudolines intersects exactly once, and the order in which a pseudoline intersects the other pseudolines remains the same (ignoring the new intersections).
 III. Given an infinite pseudoline arrangement, we can find an equivalent graphic pseudoline arrangement.

From these facts it follows that in the definition of a pseudohalfplane we can (and will) suppose that the underlying pseudoline arrangement is a graphic pseudoline arrangement.

Notice that ABA-free hypergraphs are in a natural bijection with (graphic) pseudoline arrangements and sets of points, such that each hyperedge corresponds to the subset of points *above* a pseudoline.

Proposition 19. *Given in the plane a set of points S (with all different x-coordinates) and a graphic pseudoline arrangement L, define the hypergraph $\mathcal{H}_{S,L}$ with vertex set S such that for each pseudoline $l \in L$ the set of points above l is a hyperedge of $\mathcal{H}_{S,L}$. Then $\mathcal{H}_{S,L}$ is ABA-free with the order on the vertices defined by the x-coordinates.*

Conversely, given an ABA-free hypergraph $\mathcal{H}$, there exists a set of points S and a graphic pseudoline arrangement L such that $\mathcal{H} = \mathcal{H}_{S,L}$.

Proof. The first part is almost trivial, suppose that there are two hyperedges A, B in $\mathcal{H}_{S,L}$ having an ABA-sequence on the vertices corresponding to the points $a, b, c \in S$. The pseudolines corresponding to the hyperedges A and B are denoted by ℓ_A and ℓ_B. The pseudoline ℓ_A intersects the vertical line through a below a, the vertical line through b above b and the vertical line through c above c, while ℓ_A intersects these in the opposite way (above/below/above). Thus these

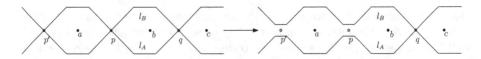

Fig. 1. Redrawing a lens to decrease the number of intersections

lines must intersect in the vertical strip between a and b and also in the strip between b and c, thus having two intersections, a contradiction.

The second part of the proof is also quite natural. Given an ABA-free hypergraph $\mathcal{H}(V, E)$ with an ordering on V, we want to realize it with a planar point set S and a graphic pseudoline arrangement L. Let S be $|V|$ points on the x axis corresponding to the vertices in V such that the order on V is the same as the order given by the x-coordinates on S. From now on we identify the vertices of V with the corresponding points of V.

For a given $A \in \mathcal{H}$ it is easy to draw an ℓ_A graphic curve for which the points of S above ℓ_A are exactly in A. Draw a pseudoline ℓ_A for every $A \in \mathcal{H}$, such that there are finitely many intersections among these pseudolines, all of them crossings. What we get is an arrangement of graphic curves, but it can happen that they intersect more than twice. Now among such drawings take one which has the minimal number of intersections, we claim that this is a pseudoline arrangement.

Assume on the contrary, that there are two curves ℓ_A and ℓ_B intersecting (at least) twice. Let two consecutive (in the x-order) intersection points be p and q, where p has smaller x-coordinate than q. Without loss of generality, ℓ_A is above ℓ_B close to the left of p and close to the right of q, while ℓ_A is below ℓ_B in the open vertical strip between p and q. This structure is usually called a lens, and we want to eliminate it in a standard way, decreasing the number of intersections. We can change the part of ℓ_A and ℓ_B to the left of p (and to the right from the intersection p' next to and left of p if there is any) and change their drawing locally around p (and p' if it exists) such that we get rid of the intersection at p, see Fig. 1. If there are no points of S between ℓ_A and ℓ_B and to the left of p (and to the right of p'), then this redrawing does not change the hyperedges defined by ℓ_A and ℓ_B, so we get a representation of $\mathcal{H}$ with less intersections, a contradiction. Thus there is a point $(p' <)a < p$ below ℓ_A and above ℓ_B. Similarly, there must be a point $p < b < q$ above ℓ_A and below ℓ_B and finally a point $q < c$ below ℓ_A and above ℓ_B, otherwise we could redraw the pseudolines with less intersections. These three points $a < b < c$ contradict the ABA-freeness of $\mathcal{H}$ as by the definition of the pseudolines, $b \in A \setminus B$ and $a, c \in B \setminus A$.

From these, it follows that the hypergraphs defined by points contained in pseudohalfplanes have the following structure.

Definition 20. *A hypergraph* $\mathcal{H}$ *on an ordered set of points* S *is called a* pseudo-halfplane-hypergraph *if there exists an ABA-free hypergraph* $\mathcal{F}$ *on* S *such that* $\mathcal{H} \subset \mathcal{F} \cup \bar{\mathcal{F}}.$

Note that $\bar{\mathcal{F}}$ is also ABA-free with the same ordering of the points. We refer to the edges of a pseudohalfplane-hypergraph also as pseudohalfplanes.

Using Lemma 14 on a hyperedge of a pseudohalfplane-hypergraph, we get the following.

Proposition 21. *Given a pseudohalfplane-hypergraph* $\mathcal{H}$, *and an edge* A *of* $\mathcal{H}$, *we can add a new hyperedge* A' *contained completely in* A *that contains all but one of the points of* A, *such that* $\mathcal{H}$ *remains a pseudohalfplane-hypergraph.*

In the geometric setting this corresponds to the known and useful fact that given a pseudohalfplane arrangement and a finite set of points A contained in the pseudohalfplane H, we can add a new pseudohalfplane H' contained completely in H that contains all but one of the points of A.

Now we show how to extend Theorem 12 to pseudohalfplane arrangements, i.e., to the case when the points of S *below* a line also define a hyperedge.

Theorem 22. *Given a finite set of points* S *and a pseudohalfplane arrangement* $\mathcal{H}$, *we can color* S *with* k *colors such that any pseudohalfplane in* $\mathcal{H}$ *that contains at least* $2k - 1$ *points of* S *contains all* k *colors. Equivalently, the vertices* S *of a finite pseudohalfplane-hypergraph can be colored with* k *colors such that any hyperedge containing at least* $2k - 1$ *points contains all* k *colors.*

We remark that a similar statement is not true for the union of two arbitrary ABA-free hypergraphs (instead of an ABA-free hypergraph and its complement), in fact the union of two arbitrary ABA-free hypergraphs might not be 2-colorable, see [16] for such a construction.

4 Proof of Theorem 22

Proof. Our proof is completely about the abstract setting, yet it translates naturally to the geometric setting, also the figures illustrate the geometric interpretations.

By definition there exists an ABA-free $\mathcal{F}$ such that $\mathcal{H} \subset \mathcal{F} \cup \bar{\mathcal{F}}$. Call $\mathcal{U} = \mathcal{H} \cap \mathcal{F}$ the upsets and $\mathcal{D} = \mathcal{H} \cap \bar{\mathcal{F}}$ the downsets, observe that both $\mathcal{U}$ and $\mathcal{D}$ are ABA-free.

Further, the unskippable vertices of $\mathcal{U}$ (resp. $\mathcal{D}$) are called top (resp. bottom) vertices. The top and bottom vertices are called the unskippable vertices of $\mathcal{H}$. Recall that by adding these unskippable vertices as one-element edges to $\mathcal{H}$, $\mathcal{H}$ remains to be a pseudohalfplane-hypergraph, as we can extend $\mathcal{F}$ and $\bar{\mathcal{F}}$ with the appropriate hyperedge (this is a convenient way of thinking about top/bottom vertices in the geometric setting, as seen later in the figures).

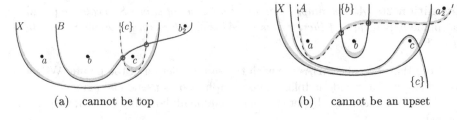

(a)   cannot be top   (b)   cannot be an upset

Fig. 2. Proof of Lemma 24

Observation 23. *If x is top and X is a downset and $x \in X$, then X contains all vertices that are bigger or all vertices that are smaller than x. The same holds if x is bottom, X is an upset and $x \in X$.*

Lemma 24. *If $\mathcal{H}$ is a finite Sperner pseudohalfplane-hypergraph, then any minimal hitting set of $\mathcal{H}$ that contains only unskippable vertices is 2-shallow.*

Proof. Let R be a minimal hitting set of unskippable vertices. Suppose for a contradiction that $\{a, b, c\} \subset R \cap X$ and $a < b < c$ for some $X \in \mathcal{H}$. Without loss of generality, suppose that b is top. As R is minimal, let B be a set for which $B \cap R = \{b\}$. From Observation 23 it follows that B is an upset.

First suppose that X is an upset. As $B \not\subset X$, take a $b_2 \in B \setminus X$. As B and X are both upsets and thus have the ABA-free property, we have $b_2 < a$ or $c < b_2$. Without loss of generality, we can suppose $c < b_2$. If c is top, $\{c\}$ and B violate ABA-freeness. See Fig. 2a. If c is bottom, then using Observation 23, X contains all the vertices that are smaller than c. Take a set $A \not\subset X$ for which $A \cap R = \{a\}$. This set must contain an $a_2 \in A \setminus X$ and so we must have $c < a_2$. If A is an upset, as it does not contain b and recall $a < b < a_2$, A and $\{b\}$ violate ABA-freeness. See Fig. 2b. If A is a downset, as it does not contain c and recall $a < c < a_2$, A and $\{c\}$ violate ABA-freeness, both cases lead to a contradiction.

The case when X is a downset is similar. Using Observation 23 for X and $\{b\}$ we can suppose without loss of generality that X contains all vertices that are smaller than b. Take a set $A \not\subset X$ for which $A \cap R = \{a\}$ and an $a_2 \in A \setminus X$. As X contains all vertices smaller than b, we have $b < a_2$. A cannot be an upset, as then it would contain b, so it is a downset. If $b < a_2 < c$, then A and X would violate ABA-freeness, thus we must have $c < a_2$. This means c cannot be bottom, so it is top. Using Observation 23, X contains all the vertices that are smaller than c. But then $B \setminus X$ must have an element that is bigger than c, contradicting the ABA-freeness of B and $\{c\}$.

From here the rest of the proof is the same. Our algorithm to give a k-coloring of the vertices of $\mathcal{H}$ such that every pseudohalfplane of size at least $2k - 1$ contains all k colors is as follows. Using Proposition 21, it is enough to consider pseudohalfplanes of size exactly $2k - 1$. Apply Lemma 24 to select a 2-shallow hitting set R and color its vertices with the first color. Delete these vertices and apply induction on k.

5   The Dual Problem and Signed ABA-free Hypergraphs

We are also interested in coloring pseudohalfplanes with k colors such that all points that are covered many times will be contained for each k colors in a pseudohalfplane of that color. For example, we can also generalize the dual result about coloring halfplanes of [19] to pseudohalfplanes.

Theorem 25. *Given a pseudohalfplane arrangement $\mathcal{H}$, we can color $\mathcal{H}$ with k colors such that if a point p belongs to a subset $\mathcal{H}_p$ of at least $3k - 2$ of the pseudohalfplanes of $\mathcal{H}$ then $\mathcal{H}_p$ contains a pseudohalfplane of every color.*

Theorem 25 follows from Theorem 29, that we will state and prove later.

However, instead of coloring pseudohalfplanes, we stick to coloring points with respect to pseudohalfplanes and work with *dual hypergraphs*, where the vertex-hyperedge incidences are preserved, but vertices become hyperedges and hyperedges become vertices. Since we have already seen in Sect. 3 the equivalence of our abstract definition and the standard definition of a pseudohalfplane arrangement, we can use the well-known properties of the dual arrangement (see, e.g., [1]) to obtain the following.

Proposition 26. *A* dual pseudohalfplane-hypergraph *is a hypergraph $\mathcal{H}$ on an ordered set of vertices S such that there exists a set $X \subset S$ and an ABA-free hypergraph $\mathcal{F}$ on S such that the edges of $\mathcal{H}$ are the edges $F \Delta X$ for every $F \in \mathcal{F}$.*

Proof. Using Definition 20, let $\mathcal{F}$ be an ABA-free hypergraph that represents the original pseudohalfplane-hypergraph, that is, every pseudohalfplane is equal to a set $F \in \mathcal{F}$ or to a set $F \in \bar{\mathcal{F}}$. Using Proposition 16, the dual of $\mathcal{F}$ is an ABA-free hypergraph, $\mathcal{F}^*$, whose vertices $\{v_F : F \in \mathcal{F}\}$ correspond to the edges of $\mathcal{F}$ (ordered in some way) and whose edges $\{f_p : p \in S\}$ correspond to the vertices of $\mathcal{F}$, with incidence relations preserved, i.e., $f_p = \{v_F : p \in F\}$. Now we add also the set of vertices $\{v_{\bar{F}} : \bar{F} \in \mathcal{F}\}$ corresponding to the edges of $\bar{\mathcal{F}}$. Each vertex $v_{\bar{F}}$ is put right after vertex v_F in the order. The edges change in the following way. As in $\mathcal{F}$ we have $p \notin F \in \mathcal{F}$ if and only if $p \in \bar{F}$, in the dual the corresponding edge is $h_p = \{v_F : p \in F\} \cup \{v_{\bar{F}} : p \in \bar{F}\}$ contains exactly one of v_F and $v_{\bar{F}}$. First, without loss of generality, we can suppose that for every $A \in \mathcal{F}$, $\mathcal{H}$ contains at least one of $A \in \mathcal{F}$ and $\bar{A} \in \bar{\mathcal{F}}$, otherwise we can delete A from $\mathcal{F}$ too. Further, we can suppose that $\mathcal{H}$ contains exactly one of $A \in \mathcal{F}$ and $\bar{A} \in \bar{\mathcal{F}}$, as if it contains both, we can add another copy A' of A to $\mathcal{F}$ ($\mathcal{F}$ is then a(n ABA-free) multihypergraph) and regard $\bar{A}$ as $\bar{A}'$. This way it can never happen that both $A \in \mathcal{H}$ and $\bar{A} \in \mathcal{H}$, thus in the dual only one of the corresponding vertices are present. Thus, we can relabel to w_A the one vertex that is present in $\mathcal{H}$ among v_A and $v_{\bar{A}}$. After the relabeling we have $V = \{w_A : A \in \mathcal{F}\}$. Denote by X the set of vertices of V for which $w_A = v_{\bar{A}}$. Now an arbitrary edge $h_p = \{v_F : p \in F\} \cup \{v_{\bar{F}} : p \in \bar{F}\} = \{w_F : p \in F \cap \bar{X}\} \cup \{w_F : p \in \bar{F} \cap X\} = f'_p \Delta X$, where Δ denotes the symmetric difference of two sets and $f'_p = \{w_F : v_F \in f_P\}$, i.e., f_p injected in the natural way into the relabeled set V. These f'_p define the same (up to this projection) ABA-free hyperedge $\mathcal{G}$ as $\mathcal{F}^*$.

Now we define a common generalization of the primal and dual definitions.

Definition 27. *A signed pseudohalfplane-hypergraph is a hypergraph $\mathcal{H}$ on an ordered set of vertices S such that there exists a set $X \subset S$ and an ABA-free hypergraph $\mathcal{F}$ on S such that the edges of $\mathcal{H}$ are some subset of $\{F \Delta X, \bar{F} \Delta X \mid F \in \mathcal{F}\}$.*

It is easy to see that the dual of such a signed pseudohalfplane-hypergraph is also a signed pseudohalfplane-hypergraph, just like in Proposition 16. Furthermore, there is a nice geometric representation of such hypergraphs; $\mathcal{H}$ is a signed pseudohalfplane-hypergraph if and only if there is a set of points, S, on the surface of a sphere and a *pseudohalfsphere arrangement* $\mathcal{F}$ on the sphere such that the incidences among S and $\mathcal{F}$ give $\mathcal{H}$. (Here we omit the exact definition of pseudohalfsphere arrangements, which are a generalization of a collection of some halfspheres of a sphere. The interested reader can find it in [1].)

Another popular geometric representation on the plane, adding *signs* to lines and points, is the following. The vertices correspond to a set of points in the plane together with a direction (up or down), and the edges correspond to a set of (x-monotone) pseudolines with a sign ($+$ or $-$). The hyperedge corresponding to a positive pseudoline is the set of points that point *towards* the pseudoline, while the hyperedge corresponding to a negative pseudoline is the set of points that point *away* from the pseudoline. Positive pseudolines correspond to $\mathcal{F}$, negative pseudolines to $\bar{\mathcal{F}}$, up points correspond to X and down points correspond to $\bar{X}$. With this interpretation, ABA-free hypergraphs have only $+$ and up signs, pseudohalfplane-hypergraphs have $\pm$ and up signs, dual pseudohalfplane-hypergraphs have $+$ and up/down signs.

In the next table we summarize the best known results about these hypergraphs, with respect to how many points each edge has to contain to have a polychromatic k-coloring and the values of the smallest c for which there exists a c-shallow hitting set for Sperner families.

	Polychromatic k-coloring	Shallow hitting set
ABA-free hypergraphs	$2k - 1$ (Theorem 12)	2 (Lemma 10)
Pseudohalfplane-hypergraphs	$2k - 1$ (Theorem 22)	2 (Lemma 24)
Dual pseudohalfplane-hypergraphs	$\leq 3k - 2$ (Theorem 25)	≤ 3 (Theorem 29)
Signed pseudohalfplane-hypergraphs	$\leq 4k - 3$ (Corollary 28)	?

We conjecture that even Sperner pseudohalfsphere arrangements have a 2-shallow hitting set, which would also imply (using the framework described above Theorem 12) that any family whose sets have size at least $2k - 1$ admits a polychromatic k-coloring, but we could not even prove for any constant c that a c-shallow hitting set exists.

As we can find a polychromatic k-coloring of the points of X and $\bar{X}$ independently with respect to the sets of $\mathcal{F}$ and $\bar{\mathcal{F}}$, respectively, of size at least $2k - 1$ using Theorem 22, the following is true.

Corollary 28. *Given a finite set of points S on the sphere and a pseudohalf-sphere arrangement $\mathcal{H}$, we can color S with k colors such that any pseudohalfsphere in $\mathcal{H}$ that contains at least $4k - 3$ points of S contains all k colors. Equivalently, the vertices S of a finite signed pseudohalfplane-hypergraph can be colored with k colors such that any hyperedge containing at least $4k - 3$ points contains all k colors.*

To finish, we first need to prove the following theorem, which, using the usual framework, will imply Theorem 25.

Theorem 29. *Every Sperner dual pseudohalfplane hypergraph has a 3-shallow hitting set.*

The proof of this result follows again closely the argument of [19] and the interested reader can find it in the full version of the paper.

Acknowledgement. We would like to thank the anonymous referees for their several useful suggestions and comments.

References

1. Björner, A., Las Vergnas, M., Sturmfels, B., White, N., Ziegler, G.: Oriented Matroids Encyclopedia of Mathematics and Its Applications 46. Cambridge University Press, Cambridge (1999)
2. Cardinal, J., Knauer, K., Micek, P., Ueckerdt, T.: Making triangles colorful. J. Comput. Geom. **4**, 240–246 (2013)
3. Cardinal, J., Knauer, K., Micek, P., Ueckerdt, T.: Making octants colorful and related covering decomposition problems. In: Proceedings of SODA 2014, pp. 1424–1432 (2014)
4. Dhandapani, R., Goodman, J.E., Holmsen, A., Pollack, R., Smorodinsky, S.: Convexity in topological affine planes. Discrete Comput. Geom. (DCG) **38**, 243–257 (2007)
5. Fulek, R.: Coloring geometric hypergraph defined by an arrangement of half-planes. In: Proceedings of CCCG 2010, pp. 71–74 (2010)
6. Gibson, M., Varadarajan, K.: Decomposing coverings and the planar sensor cover problem. In: Proceedings of FOCS 2009, pp. 159–168 (2009)
7. Keszegh, B.: Weak conflict free colorings of point sets and simple regions. In: Proceedings of CCCG 2007, pp. 97–100 (2007)
8. Keszegh, B.: Coloring half-planes and bottomless rectangles. Comput. Geom. Theory Appl. **45**(9), 495–507 (2012). Elsevier
9. Keszegh, B., Pálvölgyi, D.: Octants are cover decomposable. Discrete Comput. Geom. **47**(3), 598–609 (2012)
10. Keszegh, B., Pálvölgyi, D.: Octants are cover decomposable into many coverings. Comput. Geom. Theory Appl. **47**(5), 585–588 (2014)
11. Keszegh, B., Pálvölgyi, D.: Convex polygons are self-coverable. Discrete Comput. Geom. **51**(4), 885–895 (2014)
12. Kovács, I.: Indecomposable coverings with homothetic polygons. arXiv:1312.4597

13. Pach, J., Pálvölgyi, D., Tóth, G.: Survey on decomposition of multiple coverings. In: Bárány, I., Böröczky, K.J., Fejes Tóth, G., Pach, J. (eds.) Geometry-Intuitive, Discrete, and Convex. Bolyai Society Mathematical Studies, vol. 24, pp. 219–257. Springer, Heidelberg (2014)
14. Pálvölgyi, D.: Indecomposable coverings with concave polygons. Discrete Comput. Geom. **44**(3), 577–588 (2010)
15. Pálvölgyi, D.: Decomposition of Geometric Set Systems and Graphs. Ph.D. thesis (2010). arXiv:1009.4641
16. Pálvölgyi, D.: Indecomposable coverings with unit discs (2013). arXiv:1310.6900
17. Pálvölgyi, D., Tóth, G.: Convex polygons are cover-decomposable. Discrete Comput. Geom. **43**(3), 483–496 (2010)
18. Smorodinsky, S.: Conflict-free coloring and its applications. In: Bárány, I., Böröczky, K.J., Fejes Tóth, G., Pach, J. (eds.) Geometry-Intuitive, Discrete, and Convex. Bolyai Society Mathematical Studies, vol. 24. Springer, Heidelberg (2014)
19. Smorodinsky, S., Yuditsky, Y.: Polychromatic coloring for half-planes. J. Comb. Theory Ser. A **119**(1), 146–154 (2012)

Unsplittable Coverings in the Plane

János Pach[1] and Dömötör Pálvölgyi[2]([✉])

[1] EPFL, Lausanne and Rényi Institute, Budapest, Hungary
pach@cims.nyu.edu
[2] Institute of Mathematics, Eötvös University, Budapest, Hungary
dom@cs.elte.hu

Et tu mi fili, Brute?
(Julius Caesar)

Abstract. A system of sets forms an *m-fold covering* of a set X if every point of X belongs to at least m of its members. A 1-fold covering is called a *covering*. The problem of splitting multiple coverings into several coverings was motivated by classical density estimates for *sphere packings* as well as by the *planar sensor cover problem*. It has been the prevailing conjecture for 35 years (settled in many special cases) that for every plane convex body C, there exists a constant $m = m(C)$ such that every *m*-fold covering of the plane with translates of C splits into 2 coverings. In the present paper, it is proved that this conjecture is false for the unit disk. The proof can be generalized to construct, for every m, an unsplittable m-fold covering of the plane with translates of any open convex body C which has a smooth boundary with everywhere *positive curvature*. Somewhat surprisingly, *unbounded* open convex sets C do not misbehave, they satisfy the conjecture: every 3-fold covering of any region of the plane by translates of such a set C splits into two coverings. To establish this result, we prove a general coloring theorem for hypergraphs of a special type: *shift-chains*. We also show that there is a constant $c > 0$ such that, for any positive integer m, every m-fold covering of a region with unit disks splits into two coverings, provided that every point is covered by *at most $c2^{m/2}$* sets.

1 Introduction

Let $\mathcal{C}$ be a family of sets in $\mathbb{R}^d$, and let $P \subseteq \mathbb{R}^d$. We say that $\mathcal{C}$ is an *m-fold covering of P* if every point of P belongs to at least m members of $\mathcal{C}$. A 1-fold covering is called a *covering*. Clearly, the union of m coverings is an m-fold covering. We will be mostly interested in the case when P is a large region or the whole space $\mathbb{R}^d$.

The authors were completely convinced that the unit disk does not misbehave. Research was supported by Hungarian Scientific Research Fund EuroGIGA Grant OTKA NN 102029 and PD 104386, by Swiss National Science Foundation Grants 200020-144531 and 200021-137574. This work started in 1986, when the second author was still in kindergarden [27].

© Springer-Verlag Berlin Heidelberg 2016
E.W. Mayr (Ed.): WG 2015, LNCS 9224, pp. 281–296, 2016.
DOI: 10.1007/978-3-662-53174-7_20

Sphere packings and coverings have been studied for centuries, partially because of their applications in crystallography, Diophantine approximation, number theory, and elsewhere. The research in this field has been dominated by density questions of the following type: What is the most "economical" (i.e., least dense) m-fold covering of space by unit balls or by translates of a fixed convex body? It is suggested by many classical results and physical observations that, at least in low-dimensional spaces, the optimal arrangements are typically periodic, and they can be split into several lattice-like coverings [14,15]. Does a similar phenomenon hold for all sufficiently "thick" multiple coverings, without any assumption on their densities?

About 15 years ago, a similar problem was raised for *large scale ad hoc sensor networks*; see Feige *et al.* [13], Buchsbaum *et al.* [6]. In the – by now rather extensive – literature, it is usually referred to as the *sensor cover problem*. In its simplest version it can be phrased as follows. Suppose that a large region P is monitored by a set of sensors, each having a circular range of unit radius and each powered by a battery of unit lifetime. Suppose that every point of P is within the range of at least m sensors, that is, the family of ranges of the sensors, C, forms an m-fold covering of P. If C can be split into k coverings $C_1, \ldots, C_k$, then the region can be monitored by the sensors for at least k units of time (Fig. 1). Indeed, at time i, we can switch on all sensors whose ranges belong to C_i $(1 \leq i \leq k)$. We want to maximize k, in order to guarantee the longest possible service. Of course, the first question is the following.

Problem 1 (Pach, 1980 [31]). Is it true that every m-fold covering of the plane with unit disks splits into two coverings, provided that m is sufficiently large?

In a long unpublished manuscript, Mani and Pach [27] claimed that the answer to this question was in the affirmative with $m \leq 33$. Pach [35] warned that this "has never been independently verified." Winkler [42] even conjectured that the statement is true with $m = 4$. For more than 30 years, the prevailing conjecture has been that for any open plane *convex body* (i.e., bounded convex set) C, there exists a positive integer $m = m(C)$ such that every m-fold covering of the plane with translates of C splits into two coverings. This conjecture was proved in [32] for centrally symmetric convex polygons C. It took almost 25 years to generalize this statement to all convex polygons [38,40]. Moreover, it was proved by Aloupis *et al.* [3] and Gibson and Varadarajan [19] that in these cases, for every integer k, every at least bk-fold covering splits into k coverings, where $b = b(C)$ is a suitable positive constant. See [33,34,36], for surveys.

Here we disprove the above conjecture by giving a negative answer to Problem 1.

Theorem 1. *For every positive integer m, there exists an m-fold covering of the plane with open unit disks that cannot be split into 2 coverings.*

Our construction can be generalized as follows.

Theorem 2. *Let C be any open plane convex set, which has two parallel supporting lines with positive curvature at their points of tangencies. Then, for every*

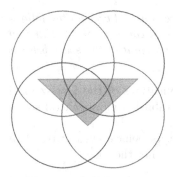
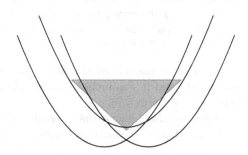

(a) The disks form a 2-fold covering of the green triangle. The colors give a split into 2 coverings. It follows, however, from the proof of Theorem 1, that for any m there is an m-fold covering of a triangle by disks that does not split.

(b) The parabolas form a 2-fold covering of the green triangle, but no matter how we 2-color them, there will be a point not covered by one of the colors. It follows, however, from Theorem 4 and a standard compactness argument, that any 3-fold covering of a closed triangle by the translates of an open parabola splits into 2 coverings.

Fig. 1. Two simple examples.

positive integer m, there exists an m-fold covering of the plane with translates of C that cannot be split into 2 coverings.

As was mentioned above, for every open convex polygon Q, there exists a smallest positive integer $m(Q)$ such that every $m(Q)$-fold covering of the plane with translates of Q splits into 2 coverings. We have that $\sup m(Q) = \infty$, where the sup is taken over all convex polygons Q. Otherwise, we could approximate the unit disk with convex n-gons with n tending to infinity. By compactness, we would conclude that the unit disk C satisfies $m(C) < +\infty$, which contradicts Theorem 1.

Problem 2. Does there exist, for any $n > 3$, an integer $m(n)$ such that every convex n-gon Q satisfies $m(Q) \leq m(n)$?

For any triangle T, there is an affine transformation of the plane that takes it into an equilateral triangle T_0. Therefore, we have $m(T) = m(T_0)$ and $m(3)$ is finite. For $n = 4$, Problem 2 is open.

In spite of our sobering negative answer to Problem 1 and its analogues in higher dimensions (cp. [27]), there are important classes of multiple coverings such that all of their members are splittable. According to our next, somewhat counter-intuitive result, for example, any m-fold covering of $\mathbb{R}^d$ with unit balls can be split into 2 coverings, provided that no point of the space is covered by too many balls. (We could innocently believe that heavily covered points make it only easier to split an arrangement.)

Theorem 3. *For every $d \geq 2$, there exists a positive constant c_d with the following property. For every positive integer m, any m-fold covering of $\mathbb{R}^d$ with unit balls can be split into two coverings, provided that no point of the space belongs to more than $c_d 2^{m/d}$ balls.*

Theorem 3 was one of the first geometric applications of the Lovász local lemma [10], and it was included in [2]. Here, we establish a more general statement (see Theorem 5.2).

One may also believe that *unbounded* convex sets behave even worse than the bounded ones. It turns out, however, that this is not the case.

Theorem 4. *Let C be an unbounded open convex set and let P be a finite set of points in the plane. Then every 3-fold covering of $P \subset \mathbb{R}^2$ with translates of C can be split into two coverings of P.*

In fact, using a standard compactness argument, Theorem 4 also holds if P is any *compact* set in the plane. However, Theorem 4 does not generalize to higher dimensions. Indeed, it follows from the proof of Theorem 1 that, for every positive integer m, there exists a finite family $\mathcal{C}$ of open unit disks in the plane and a finite set $P \subset \mathbb{R}^2$ such that $\mathcal{C}$ is an m-fold covering of P that cannot be split into two coverings. Consider now an unbounded convex cone C' in $\mathbb{R}^3$, whose intersection with the plane $\mathbb{R}^2$ is an open disk. Take a system of translates of C' such that their intersections with the plane coincide with the members of $\mathcal{C}$. These cones form an m-fold covering of P that cannot be split into two coverings.

For interesting technical reasons, the proof of Theorem 4 becomes much easier if we restrict our attention to multiple coverings of the *whole plane*. In fact, in this case, we do not even have to consider *multiple* coverings! Moreover, the statement remains true in higher dimensions.

Proposition 5. *Let C be an unbounded line-free open convex set in $\mathbb{R}^d$. Then every covering of $\mathbb{R}^d$ with translates of C can be split into two, and hence into infinitely many, coverings.*

The reason why we assume here that C is *line-free* (i.e., does not contain a full line) is the following. If C contains a straight line, then it can be obtained as the direct product of a line l and a $(d-1)$-dimensional open convex set C'. Any arrangement $\mathcal{C}$ of translates of C in $\mathbb{R}^d$ is combinatorially equivalent to the $(d-1)$-dimensional arrangement of translates of C', obtained by cutting $\mathcal{C}$ with a hyperplane orthogonal to l. In particular, the problem whether an m-fold covering of $\mathbb{R}^d$ with translates of C can be split into two coverings reduces to the respective question about m-fold coverings of $\mathbb{R}^{d-1}$ with translates of C'.

Proposition 5 is false already in the plane without the assumption that C is open. However, every 2-fold covering of the plane with translates of an unbounded C can be split into two coverings. We omit the proof as it reduces to a simple claim about intervals.

However, in higher dimensions, the similar claim is false.

Theorem 6. *There is a bounded convex set $C' \subset \mathbb{R}^3$ with the following property. One can construct a family of translates of $C = C' \times [0, \infty) \subset \mathbb{R}^4$ which covers every point of $\mathbb{R}^4$ infinitely many times, but which cannot be split into two coverings.*

Our construction is based on an example of Naszódi and Taschuk [30], and explores the fact that the boundary of C' can be rather "erratic." We do not know whether sufficiently thick coverings of $\mathbb{R}^3$ by translates of an unbounded line-free convex set can be split into two coverings or not.

In the sequel, we will study the equivalent *"dual" form* of the above questions. Consider a family $C = \{C_i : i \in I\}$ of translates of a set $C \subset \mathbb{R}^d$ that form an m-fold covering of $P \subseteq \mathbb{R}^d$. Suppose without loss of generality that C contains the origin 0. For every $i \in I$, let c_i denote the point of C_i that corresponds to $0 \in C$. In other words, we have $C = \{C + c_i : i \in I\}$. Assign to each $p \in P$ a translate of $-C$, the reflection of C about the origin, by setting $C_p^* = -C + p$. Observe that

$$p \in C_i \iff c_i \in C_p^*.$$

In particular, the fact that C forms an m-fold covering of P is equivalent to the following property: Every member of the family $C^* = \{C_p^* : p \in P\}$ contains at least m elements of $\{c_i : i \in I\}$. Thus, Theorem 1 can be rephrased in the following *dual* form.

Theorem 1'. *For every $m \geq 2$, there is a set of points $P^* = P^*(m)$ in the plane with the property that every open unit disk contains at least m elements of P^*, and no matter how we color the elements of P^* with two colors, there exists a unit disk such that all points in it are of the same color.*

A set system *not* satisfying this condition is said to have *property B* (in honor of Bernstein) or is *2-colorable* (see [9,29,39]). Generalizations of this notion are related to *conflict-free colorings* [12] and have strong connections, e.g., to the theory of ε-*nets, geometric set covers* and to *combinatorial game theory* [1,18,21,34,41].

The rest of this paper is organized as follows. In the next three sections, we prove Theorem 1' in 3 steps. In Sect. 2, we exhibit a family of non-2-colorable m-uniform hypergraphs $\mathcal{H}(k,l)$. In Sect. 3, we construct planar "realizations" of these hypergraphs, where the vertices correspond to points and the (hyper)edges to unit disks, preserving the incidence relations. In Sect. 4, we extend this construction, without violating the colorability condition, so that every disk contains at least m points. The proof of a more general version of Theorem 3, using the Lovász local lemma, can be found in Sect. 5. Finally, in Sect. 6 we make some concluding remarks and mention a couple of open problems.

The proof of Theorem 2, a generalization of Theorem 1 to bounded plane convex bodies with a smooth boundary, and the proofs of our results related to multiple coverings with *unbounded* convex sets, Theorem 4, Proposition 5, and Theorem 6, can be found in the full version of the paper available online.

2   A Family of Non-2-colorable Hypergraphs $\mathcal{H}(k,l)$

In this section we define, for any positive integers k and l, an abstract hypergraph $\mathcal{H}(k,l)$ with vertex set $V(k,l)$ and edge set $E(k,l)$. The hypergraphs $\mathcal{H}(k,l)$ are defined recursively. The edge set $E(k,l)$ will be the disjoint union of two sets, $E(k,l) = E_R(k,l) \cup E_B(k,l)$, where the subscripts R and B stand for red and blue. All edges belonging to $E_R(k,l)$ will be of size k, all edges belonging to $E_B(k,l)$ will be of size l. In other words, $\mathcal{H}(k,l)$ is the union of a k-uniform and an l-uniform hypergraph. If $k = l = m$, we get an m-uniform hypergraph (Fig. 2).

Definition 2.1. *Let k and l be positive integers.*

1. *For $k = 1$, let $V(1,l)$ be an l-element set.*
   Set $E_R(1,l) := V(1,l)$ and $E_B(1,l) := \{V(1,l)\}$.
2. *For $l = 1$, let $V(k,1)$ be a k-element set.*
   Set $E_R(k,1) := \{V(k,1)\}$ and $E_B(k,1) := V(k,1)$.
3. *For any $k,l > 1$, we pick a new vertex p, called the* root, *and let*

$$V(k,l) := V(k-1,l) \cup V(k,l-1) \cup \{p\},$$

$$E_R(k,l) := \{e \cup \{p\} : e \in E_R(k-1,l)\} \cup E_R(k,l-1),$$

$$E_B(k,l) := E_B(k-1,l) \cup \{e \cup \{p\} : e \in E_B(k,l-1)\}.$$

By recursion, we obtain that

$$|V(k,l)| = \binom{k+l}{k} - 1,$$

$$|E_R(k,l)| = \binom{k+l-1}{k}, \quad |E_B(k,l)| = \binom{k+l-1}{l},$$

$$|E(k,l)| = |E_R(k,l)| + |E_B(k,l)| = \binom{k+l}{k}.$$

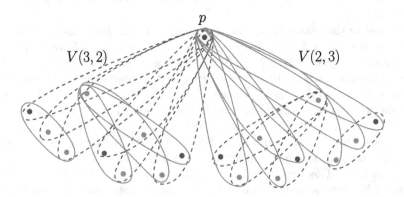

Fig. 2. The hypergraph $\mathcal{H}(3,3)$ with (arbitrarily) 2-colored vertices. There is a blue (dashed) set with 3 blue vertices or a red (solid) set with 3 red vertices. (Color figure online)

Lemma 2.2 ([37]). *For any positive integers k, l, the hypergraph $\mathcal{H}(k, l)$ is not 2-colorable. Moreover, for every coloring of $V(k, l)$ with red and blue, there is an edge in $E_R(k, l)$ such that all of its k vertices are red or an edge in $E_B(k, l)$ such that all of its l vertices are blue.*

For completeness, here we include the proof of Lemma 2.2 from [37]. The induction on two parameters, k and l, is similar to the proof of Ramsey's theorem by Erdős and Szekeres [11].

Proof. We will prove that for every coloring of $V(k, l)$ with red and blue, there is an edge in $E_R(k, l)$ such that all of its k vertices are red or an edge in $E_B(k, l)$ such that all of its l vertices are blue.

Suppose first that $k = 1$. If any vertex in $V(1, l)$ is red, then it is a red singleton edge in $\mathcal{H}(1, l)$. If all vertices in $V(1, l)$ are blue, then the (only) edge $V(1, l) \in E_B(1, l)$ contains only blue points. Analogously, the assertion is true if $l = 1$.

Suppose next that $k, l > 1$. Assume without loss of generality that the root p is red. Consider the subhypergraph $\mathcal{H}(k - 1, l) \subset \mathcal{H}(k, l)$ induced by the vertices in $V(k - 1, l)$. If it has a monochromatic red edge $e \in E_R(k - 1, l)$, then $e \cup \{p\} \in E_R(k, l)$ is red. If there is a monochromatic blue edge in $E_B(k - 1, l)$, then we are again done, because it is also an edge in $E_B(k, l)$.

For other interesting properties of the hypergraphs $\mathcal{H}(k, l)$ related to hereditary discrepancy, see Matoušek [28].

3   Geometric Realization of the Hypergraphs $\mathcal{H}(k, l)$

The aim of this section is to establish the following weaker version of Theorem 1'.

Theorem 1". *For every $m \geq 2$, there exists a finite point set $P = P(m) \subset \mathbb{R}^2$ and a finite family of unit disks $\mathcal{C} = \mathcal{C}(m)$ with the property that every member of $\mathcal{C}$ contains at least m elements of P, and no matter how we color the elements of P with two colors, there exists a disk in $\mathcal{C}$ such that all points in it are of the same color.*

We *realize* the hypergraph $\mathcal{H}(k, l)$ defined in Sect. 2 with points and disks. The vertex set $V(k, l)$ is mapped to a point set $P(k, l) \subset \mathbb{R}^2$, and the edge sets, $E_R(k, l)$ and $E_B(k, l)$, to families of open unit disks, $\mathcal{C}_R(k, l)$ and $\mathcal{C}_B(k, l)$, so that a vertex belongs to an edge if and only if the corresponding point is contained in the corresponding disk. The geometric properties of this realization are summarized in the following lemma.

Given two unit disks C, C', let $d(C, C')$ denote the distance between their centers. We fix an orthogonal coordinate system in the plane so that we can talk about the *topmost* and the *bottommost* points of a disk.

Lemma 3.1. *For any positive integers k, l and for any $\varepsilon > 0$, there is a finite point set $P = P(k, l)$ and a finite family of open unit disks $\mathcal{C}(k, l) = \mathcal{C}_R(k, l) \cup \mathcal{C}_B(k, l)$ with the following properties.*

1. *Any disk $C \in C_R(k,l)$ (resp. $C_B(k,l)$) contains precisely k (resp. l) points of P.*
2. *For any coloring of P with red and blue, there is a disk in $C_R(k,l)$ such that all of its points are red or a disk in $C_B(k,l)$ such that all of its point are blue. In fact, P and $C(k,l)$ realize the abstract hypergraph $\mathcal{H}(k,l)$ in the above sense.*
3. *For the coordinates (x,y) of any point from P, we have $-\varepsilon < x < \varepsilon$ and $-\varepsilon^2 < y < \varepsilon^2$.*
4. *For the coordinates (x,y) of the center of any disk from $C_R(k,l)$, we have $-\varepsilon < x < \varepsilon$ and $-\varepsilon^2 < y - 1 < \varepsilon^2$.*
5. *For the coordinates (x,y) of the center of any disk from $C_B(k,l)$, we have $-\varepsilon < x < \varepsilon$ and $-\varepsilon^2 < y + 1 < \varepsilon^2$.*
6. *The topmost and the bottommost points of a disk $C \in C(k,l)$ are not covered by the closure of any other member of $C(k,l)$.*

Looking at our construction from "far away" the two families C_R and C_B look like two touching disks, with all points of P very close to the touching point. The segments connecting the centers of disks from different families are almost vertical with all members of C_R lying "above" all members of C_B. We prove the lemma by induction. Most conditions are needed for the induction to go through. Condition 6 is an exception: it will be used in Sect. 4.

Proof. We give a recursive construction. We can assume that $\varepsilon < 1/10$. It is easy to see that, for $k = 1$ or $l = 1$, there exists such a family of unit disks for any $\varepsilon > 0$, see Fig. 3(a). The family $C(2,2)$ is depicted in Fig. 3(b), where the main idea of the induction may already be visible.

Suppose that $k,l \geq 2$ and we have already constructed $P(k-1,l)$ and $C(k-1,l)$, and $P(k,l-1)$ and $C(k,l-1)$, for some $\varepsilon(k-1,l) < \varepsilon/100$ and $\varepsilon(k,l-1) < \varepsilon/100$, respectively. To obtain $P(k,l)$, we place the root p of $\mathcal{H}(k,l)$ into the origin $(0,0)$, and we shift (translate) $P(k-1,l)$ and $P(k,l-1)$ into new positions such that their roots are at $(-\varepsilon/3, -\varepsilon^2/10)$ and $(\varepsilon/3, \varepsilon^2/10)$, respectively. With a slight abuse of notation, the shifted copies will also be denoted $P(k-1,l)$ and $P(k,l-1)$. See Fig. 3. In this way, it is guaranteed that for the coordinates (x,y) of any point of P, we have

$$-\varepsilon < -(\varepsilon/3 + \varepsilon(k-1,l) + \varepsilon(k,l-1)) < x < \varepsilon/3 + \varepsilon(k-1,l) + \varepsilon(k,l-1) < \varepsilon$$

and

$$-\varepsilon^2 < -(\varepsilon^2/10 + \varepsilon^2(k-1,l) + \varepsilon^2(k,l-1)) < y < \varepsilon^2/3 + \varepsilon^2(k-1,l) + \varepsilon^2(k,l-1) < \varepsilon^2.$$

Thus, **property 3** of the lemma holds.

The family $C(k,l)$ is defined as the union of two previously defined families, $C(k-1,l)$ and $C(k,l-1)$, translated by the same vectors as $P(k-1,l)$ and, resp. $P(k,l-1)$ were. Again, we use the same symbols to denote the translated copies. To verify **properties 4 and 5**, we only have to repeat the above calculations, with the y-coordinates being shifted 1 higher (resp. 1 lower).

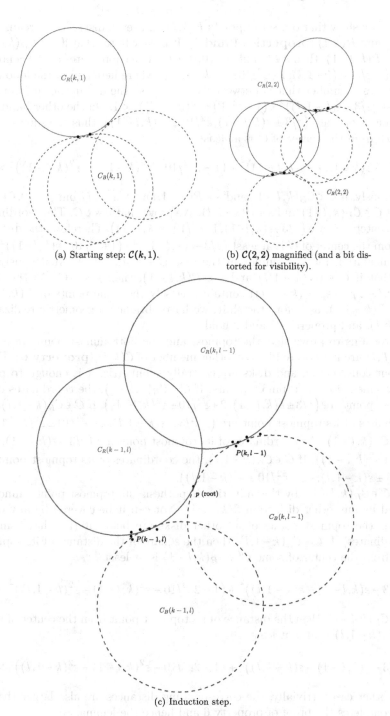

(a) Starting step: $C(k, 1)$.

(b) $C(2, 2)$ magnified (and a bit distorted for visibility).

(c) Induction step.

Fig. 3. The construction.

Now we show that our set of points $P(k,l)$ and set of disks $\mathcal{C}(k,l)$ realize the hypergraph $\mathcal{H}(k,l)$ (**properties 1 and 2**). It is easy to see that if $C \in \mathcal{C}_R(k-1,l)$ and $s \in P(k,l-1)$, then $s \notin C$ but $p = (0,0) \in C$. The coordinates of the center of C are $\left(-\varepsilon/3 \pm \varepsilon(k-1,l), 1 - \varepsilon^2/10 \pm \varepsilon^2(k-1,l)\right)$ (where here and in the following, $\pm z$ denotes a number that is between $-z$ and z), so the distance of p from C is at most $(\varepsilon/3 + \varepsilon(k-1,l))^2 + (1 - \varepsilon^2/10 + \varepsilon^2(k-1,l))^2 < 1$. On the other hand, the coordinates of s are $\left(\varepsilon/3 \pm \varepsilon(k,l-1), \varepsilon^2/10 \pm \varepsilon^2(k,l-1)\right)$, thus the square of its distance from the center of C is at least

$$\left(2\varepsilon/3 - \varepsilon(k-1,l) - \varepsilon(k,l-1)\right)^2 + \left(1 - 2\varepsilon^2/10 - \varepsilon^2(k-1,l) - \varepsilon^2(k,l-1)\right)^2 > 1.$$

Analogously, if $C \in \mathcal{C}_B(k,l-1)$ and $s \in P(k-1,l)$, then $s \notin C$ but $p = (0,0) \in C$.

Let $C \in \mathcal{C}_R(k,l-1)$ and $s \in P(k-1,l)$. We prove that $p, s \notin C$. The coordinates of the center of C are $\left(\varepsilon/3 \pm \varepsilon(k,l-1), 1 + \varepsilon^2/10 \pm \varepsilon(k,l-1)\right)$. Therefore, the distance of p from the center of C is at least $(\varepsilon/3 - \varepsilon(k,l-1))^2 + (1 + \varepsilon^2/10 - \varepsilon(k,l-1))^2 > 1$. The calculation for s is similar in the case $C \in \mathcal{C}_R(k-1,l)$. Analogously, we have that if $C \in \mathcal{C}_B(k-1,l)$ and $s \in P(k,l-1)$, then $p, s \notin C$. As the disks in $\mathcal{C}(k,l-1)$ (resp. $\mathcal{C}(k-1,l)$) contain precisely the same points of $P(k,l-1)$ (resp. $P(k-1,l)$, as before the shift, we have obtained a geometric realization of $\mathcal{H}(k,l)$, and properties 1 and 2 hold.

It remains to prove that the topmost and the bottommost points of a disk $C \in \mathcal{C}(k,l)$ are not covered by any other member of $\mathcal{C}(k,l)$ (**property 6**). Using that our construction and disks are centrally symmetric, it is enough to prove the statement for the topmost points. If $C \in \mathcal{C}_R(k,l-1)$, the coordinates of its topmost point are $\left(\varepsilon/3 \pm \varepsilon(k,l-1), 2 + \varepsilon^2/10 \pm \varepsilon^2(k,l-1)\right)$. If $C \in \mathcal{C}_R(k-1,l)$, the coordinates of its topmost point are $\left(-\varepsilon/3 \pm \varepsilon(k-1,l), 2 - \varepsilon^2/10 \pm \varepsilon^2(k-1,l)\right)$. If $C \in \mathcal{C}_B(k,l-1)$, the coordinates of its topmost point are $\left(\varepsilon/3 \pm \varepsilon(k,l-1), -2 + \varepsilon^2/10 \pm \varepsilon^2(k,l-1)\right)$. If $C \in \mathcal{C}_B(k-1,l)$, the coordinates of its topmost point are $\left(-\varepsilon/3 \pm \varepsilon(k-1,l), -2 - \varepsilon^2/10 \pm \varepsilon^2(k-1,l)\right)$.

If $C \in \mathcal{C}_R(k,l-1)$, by the induction hypothesis, its topmost point cannot be covered by any other disk from $\mathcal{C}(k,l-1)$. Nor can it be covered by any other disk, as the topmost points of all other disks are below it (i.e., have smaller y-coordinates). If $C \in \mathcal{C}_R(k-1,l)$, then the square of the distance of its topmost point from the center of some $C' \in \mathcal{C}_R(k,l-1)$ is at least

$$\left(2\varepsilon/3 - \varepsilon(k,l-1) - \varepsilon(k-1,l)\right)^2 + \left(1 - 2\varepsilon^2/10 - \varepsilon^2(k,l-1) - \varepsilon^2(k-1,l)\right)^2 > 1.$$

If $C \in \mathcal{C}_B(k,l-1)$, then the distance of its topmost point from the center of some $C' \in \mathcal{C}_R(k-1,l)$ is also at least

$$\left(2\varepsilon/3 - \varepsilon(k,l-1) - \varepsilon(k-1,l)\right)^2 + \left(1 - 2\varepsilon^2/10 - \varepsilon^2(k,l-1) - \varepsilon^2(k-1,l)\right)^2 > 1.$$

In all other cases, trivially, the corresponding distances are also larger than 1. This completes the proof of property 6 and hence the lemma.

4   Adding points to P – Proof of Theorem 1'

In this section, we extend the proof of Theorem 1" to establish Theorem 1' (which is equivalent to Theorem 1). Note that the only difference between Theorems 1" and 1' is that in the latter it is also required that every unit disk of the plane contains at least m elements of the point set $P^* = P^*(m)$. The set $P = P(m, m)$ constructed in Lemma 3.1, does not satisfy this condition. In order to fix this, we will add all points *not* in $\cup\mathcal{C}(m, m)$ to the set P (or rather a sufficiently dense discrete subset of $\mathbb{R}^2 \smallsetminus \cup\mathcal{C}(m, m)$). In order to show that the resulting set P^* meets the requirements of Theorem 1', all we have to show is the following.

Lemma 4.1. *No (open) unit disk $C \notin \mathcal{C}(k, l)$ is entirely contained in $\cup\mathcal{C}(k, l)$.*

For future purposes, we prove this statement in a slightly more general form. In what follows, we only assume that C is an open convex body with a unique topmost point t and a unique bottommost point b, which divide the boundary of C into two *closed arcs*. They will be referred to as the *left boundary arc* and a *right boundary arc*.

Definition 4.2. *A collection $\mathcal{C}$ of translates of C is said to be* exposed *if the topmost and bottommost points of its members do not belong to the closure of any other member of $\mathcal{C}$.*

By the last condition in Lemma 3.1, the collections of disks $\mathcal{C}(k, l)$ constructed in the previous section are exposed. We prove the following generalization of Lemma 4.1.

Lemma 4.3. *Let $\mathcal{C}$ be a finite exposed collection of translates of an open convex body C with unique topmost and bottommost points. If $C \notin \mathcal{C}$, then $C \not\subseteq \cup\mathcal{C}$.*

For the proof, we need a simple observation.

Proposition 4.4. *If the right boundary arcs of two translates of C intersect, then the closure of one of the translates must contain the topmost or bottommost point of the other.*

Proof. Let C_1 and C_2 be the two translates, and let γ_i denote the closed convex curve formed by the right boundary arc of C_i and the straight-line segment connecting its two endpoints (the topmost and the bottommost points of C_i). The curves γ_1 and γ_2 are translates of each other, and since they intersect, they must *cross* twice. (At a *crossing*, one curve comes from the exterior of the other, then it shares an arc with it, which may be a single point, and enters the interior.) It cannot happen that both crossings occur between the right boundary arcs, because they are convex and translates of each other. Therefore, one of the two crossings involves the straight-line segment of one the curves, say, γ_1. But since the condition is that the right boundary arcs intersect, one of the two endpoints of this straight-line segment, either the topmost or the bottommost point of C_1, lies in the closure of C_2.

Proof (of Lemma 4.3). Suppose, for contradiction, that $C \subsetneq \cup \mathcal{C}$. By removing some members of $\mathcal{C}$ if necessary, we can assume that $\mathcal{C}$ is a *minimal* collection of translates that covers C. Then C must have a point which belongs to (at least) three translates, $C_1, C_2, C_3 \in \mathcal{C}$. None of the topmost and bottommost points of these translates can be covered by C, otherwise, it would also be covered by another member of $\mathcal{C}$, contradicting the assumption that $\mathcal{C}$ is exposed.

Thus, C intersects either the left or the right boundary arc of every C_i. Without loss of generality, suppose that C intersects the right boundary arcs of C_1 and C_2. These right boundary arcs must intersect inside C, otherwise $C_1 \cap C \subseteq C_2 \cap C$ or $C_2 \cap C \subseteq C_1 \cap C$, and $\mathcal{C}$ would not be minimal. Therefore, we can apply Proposition 4.4 to conclude that one of them must contain the topmost or bottommost point of the other.

Remark 1. In the construction described in Lemma 3.1, every disk in $\mathcal{C}(m, m)$ contains at most $|P(m, m)| < 2^{2m}$ points. At the last stage, we added many new points to P. We can keep the maximum number of points of P lying in a unit disk bounded from above by a function $f(m)$. What is the best upper bound? The bound given by our construction depends on $\varepsilon(m, m) \le 100^{-2m} \varepsilon(1, 1)$.

5   Bounded Coverings

We prove Theorem 3 in a somewhat more general form. For the proof we need the following consequence of the Lovász local lemma.

Lemma 5.1 (Erdős-Lovász [10]). *Let $k, m \ge 2$ be integers. If every edge of a hypergraph has at least m vertices and every edge intersects at most $k^{m-1}/4(k-1)^m$ other edges, then its vertices can be colored with k colors so that every edge contains at least one vertex of each color.*

Let $\mathcal{C}$ be a class of subsets of $\mathbb{R}^d$. Given n members $C_1, \ldots, C_n$ of $\mathcal{C}$, assign to each point $x \in \mathbb{R}^d$ a *characteristic vector* $c(x) = (c_1(x), \ldots, c_n(x))$, where $c_i(x) = 1$ if $x \in C_i$ and $c_i(x) = 0$ otherwise. The number of distinct characteristic vectors shows how many "pieces" $C_1, \ldots, C_n$ cut the space into. The *dual shatter function* of $\mathcal{C}$, denoted by $\pi_{\mathcal{C}}^*(n)$, is the maximum of this quantity over all n-tuples $C_1, \ldots, C_n \in \mathcal{C}$. For example, when $\mathcal{C}$ is the family of open *balls* in $\mathbb{R}^d$, it is well known that

$$\pi_{\mathcal{C}}^*(n) \le \binom{n-1}{d} + \sum_{i=0}^{d} \binom{n}{i} \le n^d, \tag{1}$$

provided that $2 \le d \le n$.

Theorem 5.2. *Let $\mathcal{C}$ be a class of open sets in $\mathbb{R}^d$ with diameter at most D and volume at least v. Let $\pi(n) = \pi_{\mathcal{C}}^*(n)$ denote the dual shatter function of $\mathcal{C}$, and let B^d denote the unit ball in $\mathbb{R}^d$. Then, for every positive integer m, any m-fold covering of $\mathbb{R}^d$ with members of $\mathcal{C}$ splits into two coverings, provided that no point of the space is covered more than $\frac{v}{(2D)^d VolB^d} \pi^{-1}(2^{m-3})$ times, where $VolB^d$ is the volume of B^d.*

Proof. Given an m-fold covering of $\mathbb{R}^d$ in which no point is covered more than M times, define a hypergraph $\mathcal{H} = (V, E)$, as follows. Let V consist of all members of $\mathcal{C}$ that participate in the covering. To each point $x \in \mathbb{R}^d$, assign a (hyper)edge $e(x)$: the set of all members of the covering that contain x. (Every edge is counted only once.) Since every point x is covered by at least m members of $\mathcal{C}$, every edge $e(x) \in E$ consists of at least m points.

Consider two edges $e(x), e(y) \in E$ with $e(x) \cap e(y) \neq \varnothing$. Then there is a member of $\mathcal{C}$ that contains both x and y, so that y must lie in the ball $B(x, D)$ of radius D around x. Hence, all members of the covering that contain y lie in the ball $B(x, 2D)$ of radius $2D$ around x. Since the volume of each of these members is at least v, and no point of $B(x, 2D)$ is covered more than M times, we obtain that $B(x, D)$ can be intersected by at most $M VolB(x, 2D)/v = M(2D)^d VolB^d/v$ members of the covering. By the definition of the dual shatter functions, those members of the covering that intersect $B(x, D)$ cut $B(x, D)$ into at most $\pi(M(2D)^d VolB^d/v)$ pieces, each of which corresponds to an edge of $\mathcal{H}$. Therefore, for the maximum number N of edges of $\mathcal{H}$ that can intersect the same edge $e(x) \in E$, we have

$$N \leq \pi(M(2D)^d VolB^d/v).$$

According to Lemma 5.1 (for $k = 2$), in order to show that the covering can be split into two, i.e., the hypergraph $\mathcal{H}$ is 2-colorable, it is sufficient to assume that $N \leq 2^{m-3}$. Comparing this with the previous inequality, the result follows.

In the special case where $\mathcal{C}$ is the class of unit balls in $\mathbb{R}^d$, we have $v = VolB^d$, $D = 2$, and, in view of (1), $\pi^{-1}(z) \geq z^{1/d}$. Thus, we obtain Theorem 3 with $c_d = 2^{-2d-3/d}$.

If we want to decompose an m-fold covering into $k > 2$ coverings, then the above argument shows that it is sufficient to assume that

$$\pi(M(2D)^d VolB^d/v) \leq k^{m-1}/4(k-1)^m.$$

In case of unit balls, this holds for $M \leq c_{k,d}(1 + \frac{1}{k-1})^{m/d}$ with $c_{k,d} = k^{-1/d}4^{-d-1/d}$.

Two sets are *homothets* of each other if one can be obtained from the other by a dilation with positive coefficient followed by a translation. It is easy to see [20] that for $d = 2$, the dual shatter function of the class $\mathcal{C}$ consisting of all homothets of a fixed convex set C is at most $n^2 - n + 2 \leq n^2$, for every $n \geq 2$. In this case, Theorem 5.2 immediately implies

Corollary 5.3. *Every m-fold covering $\mathcal{C}$ of the plane with homothets of a fixed convex set can be decomposed into two coverings, provided that no point of the plane belongs to more than $2^{(m-11)/2}$ members of $\mathcal{C}$.*

Naszódi and Taschuk [30] constructed a convex set C in $\mathbb{R}^3$ such that the dual shatter function of the class of all translates of C cannot be bounded from above by any polynomial of n. Therefore, for translates of C, the above approach breaks down. We do not know how to generalize Theorem 3 from balls to arbitrary convex bodies in $\mathbb{R}^d$, for $d \geq 3$.

For some related combinatorial results, see Bollobás et al. [5].

6   Open Problems and Concluding Remarks

Theorem 2 states that, if C is a plane convex body with two antipodal points at which the curvature is positive, then for every m, there exists an m-fold covering of $\mathbb{R}^2$ with translates of C that does not split into two coverings. We also know that this statement is false for any convex polygon. But what happens if C "almost satisfies" the condition concerning the antipodal point pair?

Problem 3. Does there exist an integer m such that every m-fold covering of $\mathbb{R}^2$ with translates of an open semidisk splits into two coverings?

Another question, which surprisingly is widely open even in a completely abstract setting, is the following.

Problem 4. Suppose that for a body C, there is an integer m such that every m-fold covering of $\mathbb{R}^d$ with translates of C splits into two coverings. Does it follow that for every $k > 2$, there is an integer m_k such that every m_k-fold covering of $\mathbb{R}^d$ with translates of C splits into k coverings? Is it true that (for the smallest such m_k) even $m_k = O_C(k)$?

According to Theorem 4, every 3-fold cover of a finite point set by the translates of an unbounded open convex set splits into two coverings. Keszegh and Pálvölgyi [25] recently extended this theorem to splitting any $(3k - 2)$-fold so-called *pseudohalfplane* arrangement into k coverings. The $3k - 2$ can probably be always improved to $2k - 1$, which was done in special cases, e.g., for translates of an unbounded open convex set.

As was stated in the introduction, for every triangle (in fact, for every convex polygon) C, there is an integer $m(C)$ such that every m-fold covering of the plane with *translates* of C splits into two coverings. Keszegh and the Pálvölgyi [22] extended this theorem to m-fold coverings with *homothets* of a triangle. (Two sets are homothets of each other if one can be obtained from the other by a dilation with positive coefficient followed by a translation.) Using the idea of the proof of our Theorem 1, Kovács [26] has recently showed that the analogous statement is false for homothets of any convex polygon with more than 3 sides. For further results about decomposition of multiple coverings, see [4,5,7,8,19,23,24].

Acknowledgment. The authors are deeply indebted to Professor Peter Mani, who passed away in 2013, for many interesting conversations about the topics, and his ideas reflected in the long unpublished manuscript [27]. It was the starting point and an important source for the present work.

The authors would also like to thank Radoslav Fulek, Balázs Keszegh, and Géza Tóth for their many valuable remarks.

References

1. Alon, N.: A non-linear lower bound for planar epsilon-nets. Discrete Comput. Geom. **47**(2), 235–244 (2012)
2. Alon, N., Spencer, J.H.: The Probabilistic Method. Wiley-Interscience Series in Discrete Mathematics and Optimization, 3rd edn. Wiley, Hoboken, NJ (2008)
3. Aloupis, G., Cardinal, J., Collette, S., Langerman, S., Orden, D., Ramos, P.: Decomposition of multiple coverings into more parts. Discrete Comput. Geom. **44**(3), 706–723 (2010)
4. Asinowski, A., et al.: Coloring hypergraphs induced by dynamic point sets and bottomless rectangles. In: Dehne, F., Solis-Oba, R., Sack, J.-R. (eds.) WADS 2013. LNCS, vol. 8037, pp. 73–84. Springer, Heidelberg (2013)
5. Bollobás, B., Pritchard, D., Rothvoß, T., Scott, A.: Cover-decomposition and polychromatic numbers. SIAM J. Discrete Math. **27**(1), 240–256 (2013)
6. Buchsbaum, A.L., Efrat, A., Jain, S., Venkatasubramanian, S., Yi, K.: Restricted strip covering, the sensor cover problem. In: Proceedings of the Eighteenth Annual ACM-SIAM Symposium on Discrete Algorithms (SODA 2007), pp. 1056–1063 (2007)
7. Cardinal, J., Knauer, K., Micek, P., Ueckerdt, T.: Making triangles colorful. J. Comput. Geom. **4**(1), 240–246 (2013)
8. Cardinal, J., Knauer, K., Micek, P., Ueckerdt, T.: Making octants colorful and related covering decomposition problems. SIAM J. Discrete Math. **28**(4), 1948–1959 (2014)
9. Erdős, P.: On a combinatorial problem. Nordisk Mat. Tidskr. **11**, 5–10 (1963). 40
10. Erdős, P., Lovász, L.: Problems and results on 3-chromatic hypergraphs and some related questions. In: Infinite and finite sets (Colloq., Keszthely 1973; dedicated to P. Erdős on his 60th birthday), Vol. II, pp. 609–627. Colloq. Math. Soc. János Bolyai, 10
11. Erdős, P., Szekeres, G.: A combinatorial problem in geometry. Compos. Math. **2**, 463–470 (1935)
12. Even, G., Lotker, Z., Ron, D., Smorodinsky, S.: Conflict-free colorings of simple geometric regions with applications to frequency assignment in cellular networks. SIAM J. Comput. **33**(1), 94–136 (2003)
13. Feige, U., Halldórsson, M.M., Kortsarz, G.: Approximating the domatic number. SIAM J. Comput. **32**(1), 172–195 (2002)
14. Fejes Tóth, G.: New results in the theory of packing and covering. In: Gruber, P., Wills, J. (eds.) Convexity and Its Applications, pp. 318–359. Birkhäuser, Basel (1983)
15. Tóth, G., Kuperberg, W.: A survey of recent results in the theory of packing and covering. In: Pach, J. (ed.) New Trends in Discrete and Computational Geometry. Algorithms and Combinatorics, vol. 10, pp. 251–279. Springer, Heidelberg (1993)
16. Fulek, R.: Personal communication (2010). See also in [36]
17. Fulek, R., Hubai, T., Keszegh, B., Nagy, Z., Rothvoß, T., Vizer, M.: Personal communication (2010)
18. Gebauer, H., Gebauer, H.: Disproof of the neighborhood conjecture with implications to SAT. Combinatorica **32**(5), 573–587 (2012)
19. Gibson, M., Varadarajan, K.: Optimally decomposing coverings with translates of a convex polygon. Discrete Comput. Geom. **46**(2), 313–333 (2011)
20. Grünbaum, B.: Venn diagrams and independent families of sets. Math. Mag. **48**, 12–23 (1975)

21. Haussler, D., Welzl, E.: ε-nets and simplex range queries. Discrete Comput. Geom. **2**(2), 127–151 (1987)
22. Keszegh, B., Pálvölgyi, D.: Octants are cover-decomposable. Discrete Comput. Geom. **47**(3), 598–609 (2012)
23. Keszegh, B., Pálvölgyi, D.: Octants are cover-decomposable into many coverings. Comput. Geom. **47**(5), 585–588 (2014)
24. Keszegh, B., Pálvölgyi, D.: Convex polygons are self-coverable. Discrete Comput. Geom. **51**(4), 885–895 (2014)
25. Keszegh, B., Pálvölgyi, D.: An abstract approach to polychromatic coloring: shallow hitting sets in ABA-free hypergraphs and pseudohalfplanes. arXiv:1410.0258
26. Kovács, I. Indecomposable coverings with homothetic polygons. arXiv: 1312.4597
27. Mani-Levitska, P., Pach, J.: Decomposition problems for multiple coverings with unit balls, manuscript (1986). Parts of the manuscript are available at http://www.math.nyu.edu/~pach/publications/unsplittable.pdf
28. Matoušek, J.: The determinant bound for discrepancy is almost tight. Proc. Amer. Math. Soc. **141**(2), 451–460 (2013)
29. Miller, E.W.: On a property of families of sets. C. R. Soc. Sci. Varsovie **30**, 31–38 (1937)
30. Naszódi, M., Taschuk, S.: On the transversal number and VC-dimension of families of positive homothets of a convex body. Discrete Math. **310**(1), 77–82 (2010)
31. Pach, J.: Decomposition of multiple packing and covering. In: Diskrete Geometrie, vol. 2. Kolloq. Math. Inst. Univ. Salzburg, pp. 169–178 (1980)
32. Pach, J.: Covering the plane with convex polygons. Discrete Comput. Geom. **1**(1), 73–81 (1986)
33. Pach, J., Pálvölgyi, D., Tóth, G.: Survey on decomposition of multiple coverings. In: Geometry–intuitive, discrete, and convex, 219–257, Bolyai Soc. Math. Stud., 24, János Bolyai Math. Soc., Budapest (2013)
34. Pach, J., Tardos, G., Tóth, G.: Indecomposable coverings. Canad. Math. Bull. **52**(3), 451–463 (2009)
35. Pach, J., Tóth, G.: Decomposition of multiple coverings into many parts. Comput. Geom. **42**(2), 127–133 (2009)
36. Pálvölgyi, D.: Decomposition of geometric set systems and graphs, Ph.D. thesis, EPFL, Lausanne (2010). arXiv:1009.4641
37. Pálvölgyi, D.: Indecomposable coverings with concave polygons. Discrete Comput. Geom. **44**(3), 577–588 (2010)
38. Pálvölgyi, D., Tóth, G.: Convex polygons are cover-decomposable. Discrete Comput. Geom. **43**(3), 483–496 (2010)
39. Radhakrishnan, J., Srinivasan, A.: Improved bounds and algorithms for hypergraph 2-coloring. Random Struct. Algorithms **16**(1), 4–32 (2000)
40. Tardos, G., Tóth, G.: Multiple coverings of the plane with triangles. Discrete Comput. Geom. **38**(2), 443–450 (2007)
41. Varadarajan, K.: Weighted geometric set cover via quasi-uniform sampling, in STOC 2010–Proceedings of the 2010 ACM International Symposium on Theory of Computing pp. 641–647. ACM, New York (2010)
42. Winkler, P.: Mathematical mind-benders, p. 137. A K Peters, Wellesley, MA: Winkler, P.: Puzzled: covering the plane. Commun. ACM **52**(11), 112(2009)

Structural Graph Theory

Induced Minor Free Graphs: Isomorphism and Clique-width

Rémy Belmonte[1], Yota Otachi[2(✉)], and Pascal Schweitzer[3]

[1] Department of Architectural Engineering, Kyoto University, Kyoto, Japan
remy.belmonte@gmail.com
[2] School of Information Science,
Japan Advanced Institute of Science and Technology, Nomi, Japan
otachi@jaist.ac.jp
[3] RWTH Aachen University, Aachen, Germany
schweitzer@informatik.rwth-aachen.de

Abstract. Given two graphs G and H, we say that G contains H as an induced minor if a graph isomorphic to H can be obtained from G by a sequence of vertex deletions and edge contractions. We study the complexity of GRAPH ISOMORPHISM on graphs that exclude a fixed graph as an induced minor. More precisely, we determine for every graph H that GRAPH ISOMORPHISM is polynomial-time solvable on H-induced-minor-free graphs or that it is isomorphism complete. Additionally, we classify those graphs H for which H-induced-minor-free graphs have bounded clique-width. Those two results complement similar dichotomies for graphs that exclude a fixed graph as an induced subgraph, minor or subgraph.

1 Introduction

Remaining unresolved, the algorithmic problem GRAPH ISOMORPHISM persists as a fundamental graph theoretic challenge which, despite generating ongoing interest, has neither been shown to be NP-hard nor polynomial-time solvable. The problem asks whether two given graphs are structurally the same, that is, whether there exists an adjacency and non-adjacency preserving map from the vertices of one graph to the vertices of the other graph.

Related work. In the absence of a result determining the complexity of the general problem, considerable effort has been put into classifying the isomorphism problem restricted to graph classes as being polynomial time tractable or polynomial time equivalent to the general problem, i.e., GI-complete. Most graph classes considered in these efforts are graph classes that are closed under some basic operations. Operations that are typically considered are edge contraction, vertex deletion and edge deletion. A class of graphs closed under all of these operations is said to be minor closed and can also be described as a class of graphs avoiding a set of forbidden minors. As shown by Ponomarenko, the GRAPH ISOMORPHISM problem can be solved in polynomial time on H-minor

© Springer-Verlag Berlin Heidelberg 2016
E.W. Mayr (Ed.): WG 2015, LNCS 9224, pp. 299–311, 2016.
DOI: 10.1007/978-3-662-53174-7_21

free graphs for any fixed graph H [24]. This implies prior results on solvability of graphs of bounded treewidth, planar graphs and bounded genus. The result on minor closed graph classes was recently extended by Grohe and Marx to H-topological minor free graphs [11], and Lokshtanov, Pilipczuk, Pilipczuk and Saurabh [18] showed that the problem is actually fixed-parameter tractable on graphs of bounded treewidth, an important class of minor-free graphs. When a graph class is only required to be closed under some of the above named operations, isomorphism on such a graph class can sometimes be polynomial-time solvable and sometimes be isomorphism complete. We say that a graph G is H-free if it does not contain the graph H as an induced subgraph. When forbidding one induced subgraph, it is known that GRAPH ISOMORPHISM can be solved in polynomial time on H-free graphs if H is an induced subgraph of P_4 (the path on 4 vertices) and GI-complete otherwise (see [2]). For two forbidden induced subgraphs such a classification into graph isomorphism complete and polynomial-time solvable cases turns out to be more complicated [17, 25]. In the case where we consider forbidden subgraphs (i.e., also allowing edge and vertex deletions) there is a complete dichotomy for the computational complexity of GRAPH ISOMORPHISM on classes characterized by a finite set of forbidden subgraphs, while there are intermediate classes defined by infinitely many forbidden subgraphs [20] (assuming that graph isomorphism is not polynomial time solvable).

Our results. In this paper we consider graph classes closed under edge contraction and vertex deletion (but not necessarily under edge deletion). The corresponding graph containment relation is called induced minor. More precisely, a graph H is an *induced minor* of a graph G if H can be obtained from G by repeated vertex deletion and edge contraction. If no induced minor of G is isomorphic to H, we say that G is H-induced-minor-free. We consider graph classes characterized by one forbidden induced minor, and on these classes we study the computational complexity of the GRAPH ISOMORPHISM problem and whether the value of the parameter clique-width is bounded by some universal constant c_H. The isomorphism problem for such classes was first considered by Ponomarenko [24] for the case where H is connected. In that paper two choices for the graph H play a crucial role, namely choosing H to be the gem and choosing H to be co-$(P_3 \cup 2K_1)$ (see Fig. 1). Forbidding either of these graphs as induced minor yields a graph class with an isomorphism problem solvable in polynomial time. However, to show polynomial time solvability for the gem, the proof of [24], due to a common misunderstanding concerning the required preconditions, incorrectly relies on a technique of [14] to reduce the problem to the 3-connected case (see Subsect. refsubsection:gem). To clarify the situation, we provide a proof that avoids this reduction and instead use a reduction of the problem to the 2-connected case for which we provide a polynomial time isomorphism test. To extend Ponomarenko's theorem to the disconnected case, we provide a reduction structurally different from the ones used previously, allowing us to treat the case where H consists of a cycle with an added isolated vertex. Overall we extend Ponomarenko's results to obtain the following theorem (see Fig. 1 for the graphs that are mentioned).

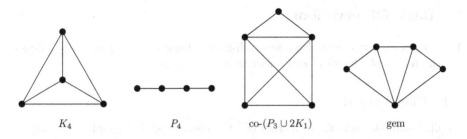

Fig. 1. The graphs K_4, P_4, co-$(P_3 \cup 2K_1)$ and the gem.

Theorem 1.1. *Let H be a graph. The* GRAPH ISOMORPHISM *problem on H-induced-minor-free graphs is polynomial-time solvable if H is complete or an induced subgraph of* co-$(P_3 \cup 2K_1)$ *or the gem, and GI-complete otherwise.*

Our proofs rely on structural descriptions that also allow us to determine exactly which classes characterized by one forbidden induced minor have bounded clique-width.

Theorem 1.2. *Let H be a graph. The clique-width of the H-induced-minor-free graphs is bounded if and only if H is an induced subgraph of* co-$(P_3 \cup 2K_1)$ *or the gem.*

While it is still open whether GRAPH ISOMORPHISM is polynomial time solvable for graphs of bounded clique-width[1], our theorems are in accordance with the seemingly reoccurring pattern that the isomorphism problem for graphs of bounded clique-width is polynomial time solvable, while there are graph classes with unbounded clique width on which GRAPH ISOMORPHISM is polynomial-time solvable. For example, planar graphs, interval graphs, and permutation graphs are such graph classes. Additionally, note that H-free graphs have bounded clique-width if and only H is an induced subgraph of P_4 [8] and that H-minor-free graphs have bounded clique-width if and only if H is planar [15]. Recently, Dabrowski and Paulusma gave a dichotomy for the clique-width of bipartite H-free graphs [7], and initiated the study of clique-width on graphs that forbid two graphs as induced subgraphs [8].

Structure of the paper. We first summarize well known observations about induced-minor-free graphs, isomorphism and clique-width (Sect. 2). We then consider classes that are characterized by one forbidden induced minor of size at most 5 (Sect. 3). Finally we show that the observations of Sects. 2 and 3 resolve all cases with forbidden induced minors of size at least 6 (Sect. 4). In this paper all graphs that are considered are finite. Throughout the paper, we use standard notation and terminology from Diestel [10].

---

[1] Since the acceptance of this paper for the publication in the conference proceedings, a preprint has become available addressing the graph isomorphism problem for graphs of bounded clique width [12].

2   Basic Observations

In this section, we summarize a few well-known basic observations about clique-width and graph classes closed under induced minors.

2.1   Clique-Width

In [6], Courcelle and Olariu introduced the clique-width of graphs as a way of measuring the complexity of minimal separators in a graph. Similarly to graphs of bounded treewidth, it has been shown that a large class of problems can be solved efficiently on graphs of bounded clique-width [5]. However, while GRAPH ISOMORPHISM has long been known to be solvable in polynomial time on graphs of bounded treewidth [24], it is not currently known whether the problem is tractable on graphs of bounded clique-width.

For any given graph G, the clique-width of G, denoted by $cw(G)$, is defined as the minimum number of labels needed to construct G by means of the following 4 operations: (i) creation of a new vertex v with label i; (ii) forming the disjoint union of two labeled graphs G_1 and G_2; (iii) joining by an edge every vertex labeled i to every vertex labeled j, where $i \neq j$; (iv) renaming label i to label j. In the remainder of the paper, we will be using the following well-known observations to derive upper bounds or lower bounds on the value of clique-width of H-induced-minor-free graphs. See e.g., [13] for an overview of clique-width.

Theorem 2.1 ([6]). *Let G be a graph and $\overline{G}$ its edge complement, then* $cw(G) \leq 2 \cdot cw(\overline{G})$.

Theorem 2.2 ([19]). *Let G be a graph and S a subset of the vertices of G. We have* $cw(G - S) \leq cw(G) \leq 2^{|S|}(cw(G - S) + 1) - 1$.

Let G be a graph and u a vertex of G. The *local complementation* of G at u is the graph obtained from G by replacing the subgraph induced by the neighbors of u with its edge complement. The following observation follows from the well-known facts that for any graph G, we have $rw(G) \leq cw(G) \leq 2^{rw(G)+1} - 1$ (see [23]), where rw denotes the rank-width, and that rank-width remains constant under local complementations [22].

Observation 2.3. *Let G and G' be two graphs such that G' can be obtained from G by a sequence of local complementations, then* $cw(G) \leq 2^{cw(G')+1} - 1$.

Theorem 2.4 ([4]). *Let G and G' be two graphs such that G' can be obtained from G by a sequence of edge subdivisions, i.e., replacing edges with paths of length 2. Then* $cw(G) \leq 2^{cw(G')+1} - 1$.

Theorem 2.5 ([1,19]). *Let G be a graph and $\mathcal{B}$ the set of its biconnected components. It holds that* $cw(G) \leq t + 2$, *where* $t = \max_{B \in \mathcal{B}}\{cw(B)\}$.

Finally, note that for any graph G, the clique-width of G is at most $3 \cdot 2^{tw(G)-1}$, where $tw(G)$ denotes the treewidth of G [3].

2.2 Some Tractable Cases

Lemma 2.6. *If H is a complete graph, then* GRAPH ISOMORPHISM *for H-induced-minor-free graphs can be solved in polynomial time.*

Lemma 2.7. *Let H be a complete graph K_k. The H-induced-minor-free graphs have bounded clique-width if and only if $k \leq 4$.*

Note that the lemma above is used to prove Theorem 1.2, but K_4 is not explicitly mentioned in the statement, due to the fact that K_4 is an induced subgraph of co-$(P_3 \cup 2K_2)$.

Lemma 2.8. *If H is an induced subgraph of P_4 then* GRAPH ISOMORPHISM *for the H-induced-minor-free graphs can be solved in linear time.*

It is well known that P_4-free graphs are exactly the graphs of clique-width at most 2 (see [15]).

2.3 Some Intractable Cases

A *split partition* (C, I) of a graph G is a partition of $V(G)$ into a clique C and an independent set I. A *split graph* is a graph admitting a split partition. We say a split graph is of *restricted split type* if it has a split partition (C, I) such that each vertex in I has at most two neighbors in C. Note that a non-complete split graph of restricted split type has minimum degree at most 2. The classes of co-bipartite graphs and restricted split graphs are closed under vertex deletions and edge contractions, and thus under induced minors. As also argued in [17,24], the standard graph-isomorphism reductions to split graphs and co-bipartite graphs explained in [2] imply the following lemmas.

Lemma 2.9. *If H is not of restricted split type or H is not co-bipartite, then* GRAPH ISOMORPHISM *for the H-induced-minor-free graphs is GI-complete.*

The reductions used in the lemma can be achieved by performing edge subdivisions and subgraph complementation. *Subgraph complementation* is the operation of complementing the edges of an induced subgraph. The clique-width of graphs in the class obtained by applying subgraph complementation a constant number of times is bounded if and only if it is bounded for graphs in the original class [15]. Together with Theorem 2.4, this implies that restricted split graphs and co-bipartite graphs obtained by the reductions from general graphs have unbounded clique-width.

Corollary 2.10. *If H is not of restricted split type or H is not co-bipartite, then the H-induced-minor-free graphs have unbounded clique-width.*

3 Graphs on at Most 5 Vertices

In this section we study graph classes characterized by a forbidden induced minor H that has at most 5 vertices.

3.1 The Graph $K_3 \cup K_1$

We show that GRAPH ISOMORPHISM is GI-complete on graphs that do not contain $K_3 \cup K_1$ as an induced minor. Additionally, we show that these graphs have unbounded clique-width.

Theorem 3.1. *The* GRAPH ISOMORPHISM *problem is isomorphism complete on graphs that do not contain $K_3 \cup K_1$ as an induced minor.*

Theorem 3.2. *The class of graphs that do not contain $K_3 \cup K_1$ as an induced minor does not have bounded clique-width.*

3.2 The Gem

We now consider the class of graphs that do not contain the gem as an induced minor (see Fig. 1). In [24] this class is also considered, however, there is an issue with the proof for the fact that the isomorphism problem of graphs in this class is polynomial-time solvable. More precisely, a common misunderstanding of how the reduction to three connected components by Hopcroft and Tarjan [14] is to be applied has happened. Indeed, the techniques of Hopcroft and Tarjan do not show that graph isomorphism in a graph class $\mathcal{C}$ polynomial-time reduces to graph isomorphism of 3-connected components in $\mathcal{C}$, even if $\mathcal{C}$ is a minor closed graph class. If this were the case then the class of split graphs of restricted type would be polynomial-time solvable since the only 3-connected graphs of this type are complete graphs. Additionally to $\mathcal{C}$ being minor closed, for the techniques to be applicable it is necessary to solve the edge-colored isomorphism problem for 3-connected graphs in $\mathcal{C}$. However, edge-colored isomorphism is already GI-complete on complete graphs.

We now provide a proof that isomorphism of graphs not containing the gem as an induced minor is polynomial-time solvable without alluding to 3-connectivity. For this we first need to extend the structural considerations for such graphs performed in [24] for 3-connected graphs to biconnected graphs.

Let C be a subgraph of G. We say a vertex v in a vertex set $M \subseteq V(G) \backslash C$ has *exclusive attachment* with respect to C among the vertices of M if $N(v) \cap C \neq \emptyset$ but there is no vertex $v' \in M \setminus \{v\}$ with $(N(v) \cap C) \cap (N(v') \cap C) \neq \emptyset$. That is, no other vertex of M shares a neighbor in C with v.

Lemma 3.3. *Let G be a biconnected gem-induced-minor-free graph. Suppose C is a biconnected subgraph of G with at least 3 vertices and M is a component of $G - C$ such that $N(M) \cap C \neq C$. If $v \in M$ is a vertex with $|N(v) \cap C| = 1$ then v has exclusive attachment.*

Lemma 3.4. *Let G be a biconnected gem-induced-minor-free graph. Suppose C is a biconnected subgraph of G and M is a component of $G - C$ with $N(M) \cap C \neq C$ and $|N(M) \cap C| \leq 3$. If there is no vertex x in M with $|N(x) \cap C| = 1$ then every vertex of M has a neighbor in C, and M is a P_4-free graph.*

We call a vertex of a biconnected graph G a *branching vertex* if it has degree at least 3.

Lemma 3.5. *Let G be a biconnected gem-induced-minor-free graph that contains the path P_4 as an induced subgraph. Then at least one of the following two options holds:*

- *G has an induced subgraph H which is a path containing at most 2 inner vertices (of the path) that are branching vertices of G such that $G - H$ is disconnected, or*
- *G has an induced subgraph H which is a cycle containing at most 3 branching vertices of G such that for every connected component M of $G - H$ we have $N(M) \cap H \neq H$.*

Let G be a graph with induced subgraphs H and K. We say that G is *sutured* from H and K along $V \subseteq V(H)$ and $V' \subseteq V(K)$ if G is obtained in the following way. First we require that $|V| = |V'|$. We also require that $V(H) \cap V(K) = V \cap V'$. The graph G must then be formed from the (not necessarily disjoint) union $K \cup H$ in the following way. We add edges that form a perfect matching between vertices in $V \setminus V'$ and $V' \setminus V$. Finally we may subdivide the edges in the matching an arbitrary number of times, see Fig. 2. (Note that if H and K are not induced subgraphs of G then it is not necessarily the case that $E(K[V \cap V']) = E(H[V \cap V'])$ but this will not be relevant in the following.)

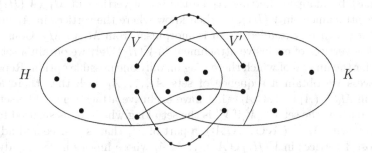

Fig. 2. A suture of two graphs H and K.

Lemma 3.6. *Let G be a biconnected gem-induced-minor-free graph. There exists an induced subgraph H of G which is isomorphic to either a path or a cycle, contains at most 4 branching vertices, and such that for every component M of $G - H$ the following holds: the graph $G[M \cup H]$ is sutured from H and some graph K along V and V' such that $K \setminus V'$ is P_4-free. Moreover $|V'| \leq 4$ and every vertex of $K - V'$ has a neighbor in V'.*

Theorem 3.7. *The* GRAPH ISOMORPHISM *problem can be solved in polynomial time on gem-induced-minor-free graphs.*

Proof. It is folklore that graph isomorphism of a hereditary graph class C reduces to isomorphism of vertex-colored biconnected graphs in C (see for example [9] or [21]). We thus assume that the input graphs are colored and biconnected. If G is such a biconnected graph, we search for an induced subgraph H that satisfies the assumptions of Lemma 3.6, that is, H is a path or a cycle with at most 4 branching vertices such that for every component M of $G - H$ we know that $G[M \cup H]$ is a suture of H with a graph K along sets V and V' such that $K \setminus V'$ is P_4-free. Moreover $|V'| \leq 4$ and every vertex in $K - V'$ has a neighbor in V'. Each H is determined by the branch vertices, the leaves (if H is a path) and choices of the paths of non-branching vertices connecting such vertices.

Now suppose G_1 and G_2 are biconnected input graphs to the isomorphism problem. Since there are only polynomially many possible choices for H, we can find an induced graph H_1 in G_1 with said properties and test for every H_2 in G_2 whether there is an isomorphism that maps H_1 to H_2. To do so we iterate over all isomorphisms φ from H_1 to H_2, there are only polynomially many, and check whether such an isomorphism extends to an isomorphism from G_1 to G_2. To check whether such an isomorphism extends, it suffices to know which component M_1 of $G_1 - H_1$ can be mapped isomorphically to which component M_2 of $G_2 - H_2$ such that the isomorphism can be extended to an isomorphism from $G_1[H_1 \cup M_1]$ to $G_2[H_2 \cup M_2]$ such that H_1 is mapped to H_2 in agreement with φ.

Note that the mapping φ determines how vertices with exclusive attachment in H_1 must be mapped. Letting A_1 be the set of vertices in $M_1 \setminus V(H_1)$ with exclusive attachment in $V(H_1)$, we then know where the vertices in A_1 must be mapped if φ can be extended to an isomorphism from M_1 to M_2. Considering in turn the vertices of exclusive attachment in $V(H_1) \cup A_1$ we obtain a set A_2 of vertices for the images of which there is again only one possible option. Repeating this process we obtain a sequence of sets $A_1, \ldots, A_t$ such that there are no vertices in $M_0 - (A_1 \cup \ldots \cup A_t)$ that have exclusive attachment. (The set $A_1 \cup \ldots \cup A_t$ contains the set $V' \setminus V$ if V' is the set along which M_0 is sutured to H_1.) We are left with $M_1 - (A_1 \cup \ldots \cup A_t)$, a part of M_1 that is P_4-free and adjacent to at most 4 vertices in $V(H_1) \cup A_1 \cup \ldots \cup A_t$ whose images have already been determined.

The isomorphism problem for vertex-colored P_4-free graphs is solvable in polynomial time (see [25]) and thus the problem for graphs obtained from P_4-free graphs by adding a bounded number of vertices can be solved in polynomial time ([16, Theorem 1]). Using this algorithm the theorem follows.   □

Theorem 3.8. *If H is an induced subgraph of the gem, then the H-induced-minor-free graphs have bounded clique-width.*

3.3   The Graph co-$(P_3 \cup 2K_1)$

In the following we will analyze the graphs that do not contain an induced minor isomorphic to co-$(P_3 \cup 2K_1)$, the graph obtained from K_5 by removing two incident edges. While it has already been shown in [24] that isomorphism for such graphs reduces to isomorphism of graphs not containing the gem (and is thus polynomially solvable), we provide a refinement of the proof in [24] for this. We do this to obtain a finer structural description of these graphs, allowing us to also bound the clique-width in the graph class.

Suppose G is a co-$(P_3 \cup 2K_1)$-induced-minor-free graph. If G does not have a K_t minor for some fixed t then G is in particular in the minor closed graph class of K_t-minor free graphs, and, as described in the introduction, the isomorphism problem can be solved in polynomial time for such graphs. Our strategy is thus to find a K_t minor and use this to analyze the structure of G. In general, of course, there is no constant bound on the number of vertices required to form a K_t minor. However in a co-$(P_3 \cup 2K_1)$-induced-minor-free graph there is such a bound. We call a K_t minor *compact* if every bag has at most 2 vertices.

Lemma 3.9. *If a co-$(P_3 \cup 2K_1)$-induced-minor-free graph G has a K_t minor for $t \geq 5$ then G has a compact K_t minor.*

Proof. Let $M_1, \ldots, M_t$ be the bags of a K_t minor in G such that the M_i are inclusion minimal with respect to forming a K_t minor. That is, removing a vertex from one of the M_i yields a minor different from K_t. We analyze the structure of the minor. We say a vertex v is adjacent to a bag M_j if there exists a vertex $v' \in M_j$ that is adjacent to v.

For a vertex $v \in M_i$ define $\mathrm{Mdeg}(v) = |\{M_j \mid j \neq i, N(v) \cap M_j \neq \emptyset\}|$ to be the number of bags different from M_i adjacent to v. Using several steps we will show that $\mathrm{Mdeg}(v) \geq t - 2$ for all $v \in M_1 \cup M_2 \cup \cdots \cup M_t$. We first argue that in the case $\mathrm{Mdeg}(v) > 1$ then $\mathrm{Mdeg}(v) \geq t - 2$. Indeed, if $\mathrm{Mdeg}(v) > 1$ then consider the minor obtained by removing all vertices from M_i different from v. If $\mathrm{Mdeg}(v) < t - 2$ we can choose 2 bags which have vertices adjacent to v and two bags which do not have such vertices. Using these bags and the vertex v we obtain the forbidden induced minor co-$(P_3 \cup 2K_1)$. We call vertices with $\mathrm{Mdeg}(v) = 0$ *inner vertices*, those with $\mathrm{Mdeg}(v) = 1$ *low degree vertices* and we call vertices with $\mathrm{Mdeg}(v) \geq t - 2$ *high degree vertices*. Next we argue that there are at most 2 high degree vertices in each bag. First, observe that if M_i contains a vertex v such that $\mathrm{Mdeg}(v) = t - 1$, then v is the only vertex in M_i and we are done. Therefore we may assume that every vertex v in M_i satisfies $\mathrm{Mdeg}(v) \leq t - 2$ and M_i contains at least 2 high degree vertices. Then, there are at least 2 such vertices, we can pick two vertices v, v' in M_i which are not adjacent to exactly the same bags, such that v is a high degree vertex and there is a path from v to v' in M_i that does not contain any other high degree vertex. Since every bag different from M_i is adjacent to v or v', removing all vertices different from v and v' and not lying on the path yields a K_t minor. Since the bags $M_1, \ldots, M_t$ were chosen to be minimal, we conclude that there are at most 2 high degree vertices in each bag.

We further argue that there is no low degree vertex in M_i. Indeed, in case there is at least one low degree vertex in M_i, we can choose a low degree vertex $v \in M_i$ and a vertex $v' \in M_i$ adjacent to a bag M_j with $j \neq i$ such that v

is not adjacent to M_j and such that there exists a path in M_i of inner vertices connecting v and v'. We remove all vertices in M_i different from v and v' and not on said path connecting them. We then move the vertex v' from M_i to M_j. We obtain the induced minor co-$(K_{1,t-3} \cup 2K_1)$, which contains co-$(P_3 \cup 2K_1)$ since $t \geq 5$.

Finally we argue that there are no inner vertices. Indeed, by minimality we can assume that every inner vertex v lies on a path between two high degree vertices v_1 and v_2, say. We again remove all vertices different from v_1 and v_2 not on the path. We then move v_1 to an adjacent bag M_j and v_2 to an adjacent bag $M_{j'}$ such that $j \neq j'$. This is possible since the vertices have high degree. Again we obtain a forbidden induced minor co-$(K_{1,t-3} \cup 2K_1)$ as above.

Since there are only high degree vertices and since each bag can only contain two such vertices, the minimal minor is compact. □

Lemma 3.10. *If G is a biconnected co-$(P_3 \cup 2K_1)$ induced-minor-free graph and M is a compact K_t minor with $t \geq 5$ then $G - M$ is $(K_2 \cup K_1)$-free.*

Corollary 3.11. *If a biconnected co-$(P_3 \cup 2K_1)$-induced-minor-free graph G has a K_8 minor then G is $(K_2 \cup K_1)$-free.*

Since the gem is biconnected, and thus every occurrence of a gem as induced minor must occur within a biconnected component of a graph, the corollary is a refinement of Ponomarenko's result [24] that says that if a co-$(P_3 \cup 2K_1)$-induced-minor-free graph G has a $K_{2^{18}+4}$-minor then it does not contain a gem as induced minor.

Theorem 3.12. *The Graph isomorphism problem for co-$(P_3 \cup 2K_1)$-induced-minor-free graphs can be solved in polynomial time.*

To show that the co-$(P_3 \cup 2K_1)$-induced-minor-free graphs have bounded clique-width, we need the following fact, which was indirectly proven by van't Hof et al. in the proof of Theorem 9 in [26].

Theorem 3.13. *For any graph F and for any planar graph H, there exists a constant $c_{F,H}$ such that an F-minor-free graph of treewidth at least $c_{F,H}$ has H as an induced minor.*

Theorem 3.14. *If H is an induced subgraph of co-$(P_3 \cup 2K_1)$, then the H-induced-minor-free graphs have bounded clique-width.*

3.4   The Remaining Graphs on at Most 5 Vertices

Now we study the remaining small graphs of at most five vertices. We show that every case here can be reduced to some case we have already solved.

Lemma 3.15. *Let H be a non-complete graph of 5 vertices. If H is neither co-$(P_3 \cup 2K_1)$ nor the gem, then GRAPH ISOMORPHISM for H-induced-minor-free graphs is GI-complete.*

Lemma 3.16. *Let H be a graph of at most 4 vertices. The* GRAPH ISOMOR-PHISM *problem for H-induced-minor-free graphs is polynomial-time solvable if H is an induced subgraph of either co-$(P_3 \cup 2K_1)$ or P_4. Otherwise, it is GI-complete.*

The two lemmas above together imply the following theorem.

Theorem 3.17. *Let H be a non-complete graph of at most 5 vertices. Then* GRAPH ISOMORPHISM *for H-induced-minor-free graphs is polynomial-time solvable if H is an induced subgraph of co-$(P_3 \cup 2K_1)$ or the gem; otherwise, it is GI-complete.*

The reductions we used above in order to show GI-completeness preserve the property that the clique-width is unbounded (see Subsect. 2.3). Thus we have the following corollary.

Corollary 3.18. *Let H be a non-complete graph of at most 5 vertices. Then the H-induced-minor-free graphs have bounded clique-width if and only if H is an induced subgraph of co-$(P_3 \cup 2K_1)$ or the gem.*

4 Non-complete Graphs on at Least 6 Vertices

In this section, we show that if H is not a complete graph and has at least six vertices, then GRAPH ISOMORPHISM for the H-induced-minor-free graphs is GI-complete.

Lemma 4.1. *If H is non-complete and contains a clique of size 5, then* GRAPH ISOMORPHISM *for H-induced-minor-free graphs is GI-complete.*

Theorem 4.2. *If H is a non-complete graph of size at least 6, then* GRAPH ISOMORPHISM *for H-induced-minor-free graphs is GI-complete.*

Since the reductions that we used above in order to show GI-completeness preserve the property that the clique-width is unbounded (see Subsect. 2.3), we have the following corollary.

Corollary 4.3. *If H is a non-complete graph of size at least 6, then H-induced-minor-free graphs have unbounded clique-width.*

References

1. Boliac, R., Lozin, V.V.: On the Clique-width of graphs in hereditary classes. In: Bose, P., Morin, P. (eds.) ISAAC 2002. LNCS, vol. 2518, pp. 44–54. Springer, Heidelberg (2002)
2. Booth, K.S., Colbourn, C.J.: Problems polynomially equivalent to graph isomorphism. Technical report CS-77-04, Computer Science Department, University of Waterloo (1979)

3. Corneil, D.G., Rotics, U.: On the relationship between clique-width and treewidth. SIAM J. Comput. **34**(4), 825–847 (2005)
4. Courcelle, B.: Clique-width and edge contraction. Inf. Process. Lett. **114**, 42–44 (2014)
5. Courcelle, B., Makowsky, J.A., Rotics, U.: Linear time solvable optimization problems on graphs of bounded clique-width. Theor. Comput. Syst. **33**(2), 125–150 (2000)
6. Courcelle, B., Olariu, S.: Upper bounds to the clique width of graphs. Discrete Appl. Math. **101**, 77–114 (2000)
7. Dabrowski, K.K., Paulusma, D.: Classifying the clique-width of H-Free bipartite graphs. In: Cai, Z., Zelikovsky, A., Bourgeois, A. (eds.) COCOON 2014. LNCS, vol. 8591, pp. 489–500. Springer, Heidelberg (2014)
8. Dabrowski, K.K., Paulusma, D.: Clique-width of graph classes defined by two forbidden induced subgraphs. In: Paschos, V.T., Widmayer, P. (eds.) CIAC 2015. LNCS, vol. 9079, pp. 167–181. Springer, Heidelberg (2015)
9. Datta, S., Limaye, N., Nimbhorkar, P., Thierauf, T., Wagner, F.: Planar graph isomorphism is in log-space. In: IEEE Conference on Computational Complexity, pp. 203–214 (2009)
10. Diestel, R.: Graph Theory, Electronic edn. Springer, Heidelberg (2005)
11. Grohe, M., Marx, D.: Structure theorem and isomorphism test for graphs with excluded topological subgraphs. In: STOC, pp. 173–192 (2012)
12. Grohe, M., Schweitzer, P.: Isomorphism testing for graphs of bounded rank width. CoRR, abs/1505.03737 (2015). http://arxiv.org/abs/1208.0142
13. Hlinený, P., Oum, S., Seese, D., Gottlob, G.: Width parameters beyond tree-width and their applications. Comput. J. **51**(3), 326–362 (2008)
14. Hopcroft, J.E., Tarjan, R.E.: Isomorphism of planar graphs. In: Complexity of Computer Computations, pp. 131–152 (1972)
15. Kamiński, M., Lozin, V.V., Milanič, M.: Recent developments on graphs of bounded clique-width. Discrete Appl. Math. **157**, 2747–2761 (2009)
16. Kratsch, S., Schweitzer, P.: Isomorphism for graphs of bounded feedback vertex set number. In: Kaplan, H. (ed.) SWAT 2010. LNCS, vol. 6139, pp. 81–92. Springer, Heidelberg (2010)
17. Kratsch, S., Schweitzer, P.: Graph isomorphism for graph classes characterized by two forbidden induced subgraphs. In: Golumbic, M.C., Stern, M., Levy, A., Morgenstern, G. (eds.) WG 2012. LNCS, vol. 7551, pp. 34–45. Springer, Heidelberg (2012)
18. Lokshtanov, D., Pilipczuk, M., Pilipczuk, M., Saurabh, S.: Fixed-parameter tractable canonization and isomorphism test for graphs of bounded treewidth. In: FOCS 2014, pp. 186–195 (2014)
19. Lozin, V.V., Rautenbach, D.: On the band-, tree- and clique-width of graphs with bounded vertex degree. SIAM J. Discrete Math. **18**, 195–206 (2004)
20. Otachi, Y., Schweitzer, P.: Isomorphism on subgraph-closed graph classes: a complexity dichotomy and intermediate graph classes. In: Cai, L., Cheng, S.-W., Lam, T.-W. (eds.) ISAAC 2013. LNCS, vol. 8283, pp. 111–118. Springer, Heidelberg (2013)
21. Otachi, Y., Schweitzer, P.: Reduction techniques for graph isomorphism in the context of width parameters. In: Ravi, R., Gørtz, I.L. (eds.) SWAT 2014. LNCS, vol. 8503, pp. 368–379. Springer, Heidelberg (2014)
22. Oum, S.: Rank-width and vertex-minors. J. Comb. Theor. Ser. B **95**(1), 79–100 (2005)

23. Oum, S., Seymour, P.D.: Approximating clique-width and branch-width. J. Comb. Theor. Ser. B **96**(4), 514–528 (2006)
24. Ponomarenko, I.N.: The isomorphism problem for classes of graphs closed under contraction. Zapiski Nauchnykh Seminarov Leningradskogo Otdeleniya Matematicheskogo Instituta **174**, 147–177 (1988). Russian. English Translation in Journal of Soviet Mathematics **55**, 1621–1643 (1991)
25. Schweitzer, P.: Towards an isomorphism dichotomy for hereditary graph classes. In: STACS, vol. 30, pp. 689–702 (2015)
26. van't Hof, P., Kamiński, M., Paulusma, D., Szeider, S., Thilikos, D.M.: On graph contractions and induced minors. Discrete Appl. Math. **160**, 799–809 (2012)

On the Complexity of Probe and Sandwich Problems for Generalized Threshold Graphs

Fernanda Couto[1]([⊠]), Luerbio Faria[2], Sylvain Gravier[3], Sulamita Klein[1,4], and Vinicius F. dos Santos[5]

[1] PESC - COPPE, Universidade Federal do Rio de Janeiro, Rio de Janeiro, Brazil
nandavdc@gmail.com
[2] IM, Universidade Estadual do Rio de Janeiro, Rio de Janeiro, Brazil
luerbio@cos.ufrj.br
[3] IF, Université Joseph Fourier, Saint Martin d'hères, France
sylvain.gravier@ujf-grenoble.fr
[4] IM, Universidade Federal do Rio de Janeiro, Rio de Janeiro, Brazil
sula@cos.ufrj.br
[5] DECOM, CEFET-MG, Belo Horizonte, Brazil
vinicius.santos@gmail.com

Abstract. A cograph is a graph without induced P_4. A graph G is (k, ℓ) if its vertex set can be partitioned into at most k independent sets and ℓ cliques. Threshold graphs are cographs-$(1, 1)$. We proved recently that cographs-$(2, 1)$ are their generalization and, as threshold graphs, they can be recognized in linear time. GRAPH SANDWICH PROBLEMS FOR PROPERTY Π (Π-SP) were defined by Golumbic et al. as a natural generalization of RECOGNITION PROBLEMS. PARTITIONED PROBE PROBLEMS are particular cases of GRAPH SANDWICH PROBLEMS. In this paper we show that, similarly to PROBE THRESHOLD GRAPHS and PROBE COGRAPHS, PROBE COGRAPHS-$(2, 1)$ and PROBE JOIN OF TWO THRESHOLDS are recognizable in polynomial time. In contrast, although COGRAPH-SP and THRESHOLD-SP are polynomially solvable problems, we prove that COGRAPH-$(2, 1)$-SP and JOIN OF TWO THRESHOLDS -SP are NP-complete problems.

Keywords: Graph sandwich problems · Cograph-$(2, 1)$ · Join of two threshold graphs

1 Introduction

Interval Probe Graphs were introduced by Zhang [1] in 1994 as a new graph theoretic model and used in [2] and in [3] to model certain problems in physical mapping of DNA. We will use a well-known generalization of this concept (as surveyed in [4]):

This work was partially supported by CAPES, CNPq, FAPERJ and FAPEMIG.

E.W. Mayr (Ed.): WG 2015, LNCS 9224, pp. 312–324, 2016.
DOI: 10.1007/978-3-662-53174-7_22

Definition 1. *Let $\mathcal{G}$ be a class of graphs. A graph $G = (V, E)$ is a probe graph if its vertex set can be partitioned into a set of probes P and an independent set of nonprobes N, such that G can be embedded in a graph of $\mathcal{G}$ by adding edges between certain nonprobes.*

If the partition of the vertex set into *probes* P and *nonprobes* N is an input data, then we call G a *partitioned probe graph of $\mathcal{G}$* if G can be embedded into a graph of $\mathcal{G}$ by adding some edges between nonprobe vertices. We call a graph H obtained from G by adding some edges between vertices of N an *embedding* of G. We denote a partitioned graph as $G = (P + N, E)$, and when this notation is used it is understood that we work with the partitioned problem. Moreover, we refer to a probe problem for a class $\mathcal{G}$ as PROBE $\mathcal{G}$. When we want to refer the partitioned version, we use the notation PP-$\mathcal{G}$.

In 1995, as a natural generalization of RECOGNITION PROBLEMS, Golumbic, Kaplan and Shamir [5] introduced the concept of a new decision problem: GRAPH SANDWICH PROBLEMS, which we formulate below:

GRAPH SANDWICH PROBLEM FOR PROPERTY Π (Π-SP)

Input: Two graphs $G^1 = (V, E^1)$ and $G^2 = (V, E^2)$ such that $E^1 \subseteq E^2$.
Question: Is there a graph $G = (V, E)$ satisfying property Π and such that $E^1 \subseteq E \subseteq E^2$?

If such a graph exists, it is called *sandwich graph*. Each edge in E^1 is a *forced edge*, while each edge of E^2 is called *optional edge*. Then, every edge that is not in E^2 is considered a *forbidden edge*. We will denote by $G^3 = (V, E^3)$ the complement graph of G^2, and each edge in E^3 will be a forbidden one. We can then define GRAPH SANDWICH PROBLEMS accordingly to this perspective.

GRAPH SANDWICH PROBLEM FOR PROPERTY Π (Π-SP)

Input: A triple (V, E^1, E^3), where $E^1 \cap E^3 = \emptyset$.
Question: Is there a graph $G = (V, E)$ satisfying Π such that $E^1 \subseteq E$ and $E \cap E^3 = \emptyset$?

This will be the definition we will adopt from now on when dealing with Π-SP.

We have a clear relation between PARTITIONED PROBE PROBLEMS and GRAPH SANDWICH PROBLEMS: the last generalizes the former. Thus, if the partitioned version of a probe problem for a class $\mathcal{G}$ is known to be NP-complete, so will be the GRAPH SANDWICH PROBLEM for this class. Conversely, if the GRAPH SANDWICH PROBLEM is polynomially solvable, then the PARTITIONED PROBE PROBLEM is also in P.

Perfect Graphs attract a lot of attention in Graph Theory. In the seminal paper of GRAPH SANDWICH PROBLEMS [5], Golumbic et al. worked only with subclasses of *perfect graphs*, for instance, *chordal graphs, cographs, threshold graphs* and *split graphs*, for which, except for chordal graphs, GRAPH SAND-WICH PROBLEMS are polynomially solvable. They left some open problems, for example when Π is "to be strongly chordal" or "to be chordal bipartite". Both were proved to be NP-complete in [6] and [7], respectively. But one problem they left open, still remains open: PERFECT GRAPH SANDWICH PROBLEM.

In this work, we are particularly interested in one well-known subclass of perfect graphs: cographs.

Definition 2 (Corneil et al. [8]). *A cograph can be defined recursively as follows:*

1. *The trivial graph K_1 is a cograph;*
2. *If $G_1, G_2, \ldots, G_p$ are cographs, then $G_1 \cup G_2 \cup \ldots \cup G_p$ is a cograph;*
3. *If G is a cograph, then $\bar{G}$ is a cograph.*

There are at least 6 equivalent forms of characterizing a cograph [8], but one of the well known is the characterization by forbidden subgraphs.

Theorem 1 (Corneil et al. [8]). *A cograph is a graph without induced P_4, i.e. induced paths with 4 vertices.*

Corneil in 1985 [9], presented the first, but not the only one, linear time algorithm to recognize cographs [10,11].

Threshold graphs are a special case of cographs and split graphs. More formally, a graph is a threshold graph if and only if it is both a cograph and a split graph. Introduced by Chvátal and Hammer in 1977 [12], Theorem 2 characterizes them.

Theorem 2 (Chvátal and Hammer [12]). *For every graph G, the following three conditions are equivalent:*

1. *G is threshold;*
2. *G has no induced subgraph isomorphic to $2K_2, P_4$ or C_4;*
3. *There is an ordering $v_1, v_2, \ldots, v_n$ of vertices of G and a partition of $\{v_1, v_2, \ldots v_n\}$ into disjoint subsets P and Q such that:*
   - *Every $v_j \in P$ is adjacent to all vertices v_i with $i < j$,*
   - *Every $v_j \in Q$ is adjacent to none of the vertices v_i with $i < j$.*

Thus, threshold graphs can be constructed from a trivial graph K_1 by repeated applications of the following two operations:

1. Addition of a single isolated vertex to the graph.
2. Addition of a single dominating vertex to the graph, i.e. a single vertex that is adjacent to each other vertex.

In [13–15], Brandstädt et al. defined a special class of graphs named (k, ℓ)-*graphs*, i.e., graphs whose vertex set can be partitioned into at most k independent sets and ℓ cliques: a generalization of split graphs, which can be described as $(1,1)$-graphs. Moreover, they proved that the recognition problem for this class of graphs is NP-complete for k or ℓ at least 3 and polynomial, otherwise. We already know how to fully classify (k, ℓ)GRAPH SANDWICH PROBLEMS with respect to the solution complexity, for integers k, ℓ: the problem is NP-complete for $k + \ell \geq 3$ and polynomial, otherwise [16].

For cograph-(k, ℓ), there is a characterization by forbidden subgraphs [17]. Recently, we provided a structural characterization and decomposition for cographs-$(2, 1)$ which leads us to a linear time algorithm to recognize this class of graphs, a generalization of threshold graphs, accordingly to the characterization [18]. Before presenting this previous result, we make some helpful definitions.

Definition 3. *A biclique is a complete bipartite graph.*

Definition 4. *The union of two graphs $G = (V_G, E_G)$ and $H = (V_H, E_H)$ is the union of their vertex and edge sets: $G \cup H = (V_G \cup V_H, E_G \cup E_H)$.*

Definition 5. *The disjoint union of two graphs $G = (V_G, E_G)$ and $H = (V_H, E_H)$ is the union of their vertex and edge sets when V_G and V_H are disjoint: $G + H = (V_G + V_H, E_G + E_H)$.*

Definition 6. *The join $G \oplus H$ of two graphs $G = (V_G, E_G)$ and $H = (V_H, E_H)$ is their graph union with all the edges that connect the vertices of G with the vertices of H, i.e., $G \oplus H = (V_G \cup V_H, E_G \cup E_H \cup \{uv : u \in V_G, v \cup V_H\})$.*

Theorem 3 (Couto et al. [18]). *Let G be a graph. Then the following are equivalent.*

1. *G is a cograph-$(2,1)$.*
2. *G can be partitioned into a collection of maximal bicliques $B = \{B_1, \ldots, B_l\}$ and a clique K such that $B_i = (X_i, Y_i)$ and $V(K)$ is the union of non-intersecting sets K^1 and K^2 such that the following properties hold.*
   (a) *There are no edges between vertices of B_i and B_j for $i \neq j$;*
   (b) *Let $L(v)$ be the list of bicliques in the neighborhood of v, $\forall v \in V$.*
      $K^1 = \{v \in K | N(v) \cap B \subseteq B_1\} = K^{1,1} \cup K^{1,2}$ *and*
      $K^2 = \{v \in K | L(v) \geq 2, B_i \in L(v) \Leftrightarrow B_i \subseteq N(v)\}$, *where*
      $K^{1,1} = \{v \in K^1 | vx \in E(G), \forall x \in X_1\}$ *and*
      $K^{1,2} = K^1 \setminus K^{1,1}$ *and it holds that $uy \in E(G), \forall u \in K^{1,2}$ and $y \in Y_1$;*
   (c) *$G[X_1 \cup Y_1 \cup K^{1,1} \cup K^{1,2}]$ is the join of threshold graphs $(K^{1,1}, Y_1)$ and $(K^{1,2}, X_1)$;*
   (d) *There is an ordering $v_1, v_2, \ldots, v_{|K^2|}$ of K^2's vertices such that $N(v_i) \subseteq N(v_j)$, $\forall i \leq j$ and $N(v) \subseteq N(v_1)$, $\forall v \in K^1$.*
3. *G is either a join of two threshold graphs or it can be obtained from the join of two threshold graphs by the applications of any sequence of the following operations:*
   – *Disjoint union with a biclique;*
   – *Join with a single vertex.*

After providing this characterization, the question about PROBE COGRAPH-$(2,1)$ and PP-JOIN OF TWO THRESHOLDS complexities arose. In this paper we prove that, similarly to PROBE COGRAPH and PP-THRESHOLD, these two problems are polynomially solvable. Our next step was to consider the GRAPH SANDWICH PROBLEM for both classes. Surprisingly, we got two new examples of the non-monotony of GRAPH SANDWICH PROBLEMS. Moreover, when dealing with Theorem 3 to solve COGRAPH-$(2,1)$-SP, another interesting and motivating question arose: Given a property Π for which Π-SP is known to be polynomially solvable, is $(\Pi \oplus \Pi)$-SP also in P?

In this paper, besides of recognizing in polynomial time PROBE COGRAPHS-$(2,1)$ and PP-JOIN OF TWO THRESHOLDS, we answer the question above negatively and we present the first NP-complete $(\Pi \oplus \Pi)$-SP, with Π-SP in P: JOIN OF TWO THRESHOLDS GRAPH SANDWICH PROBLEM. With this result in hands, we show, in Sect. 6, that COGRAPH-$(2,1)$-SP is also NP-complete. Both results corroborate the fact that GRAPH SANDWICH PROBLEMS are not monotone.

2   Partitioned Probe Join of Two Threshold Graphs

The goal of this section is to prove that PARTITIONED PROBE JOIN OF TWO THRESHOLD GRAPHS can be recognized in polynomial time. This result is application of cograph-$(2,1)$'s decomposition [18]. It will be used later in Sect. 3.

We define PP-JOIN OF TWO THRESHOLD GRAPHS as follows:

PARTITIONED PROBE JOIN OF TWO THRESHOLD GRAPHS (PP-JTT)
Input: A graph $G = (V, E)$ such that $V = P + N$ where N is an independent set called *non-probe set* and P is the *probe set.*
Question: Is there an edge set E' whose edges have both extremes in N and so that $G' = (V, E \cup E')$ is a join of two threshold graphs?

Next, we present some results and definitions from the literature.

Theorem 4 (Chvátal and Hammer [12]). *Let $G = (V, E)$ be a graph. Then G is a threshold graph if and only if every induced subgraph has an isolated vertex or a universal vertex.*

Definition 7 (H.N. de Ridder [19]). *A vertex x is probe threshold if either*

1. *$x \in P$ and x is isolated or universal, or*
2. *$x \in N$ and x is either isolated or x is adjacent to all probes of G.*

Theorem 5 (H.N. de Ridder [19]). *A partitioned graph $G = (P + N, E)$ is a probe threshold graph if and only if every induced subgraph has a probe threshold vertex.*

Similarly, we define:

Definition 8. *Given a graph $G = (V, E)$, a vertex $x \in V$ is probe universal if either*

1. *$x \in P$ and x is universal, or*
2. *$x \in N$ and x is adjacent to all probes of G.*

Definition 9. *Given a graph $G = (V, E)$, a pair of vertices (x, y) is a probe JTT pair if either*

1. $x \in P$, $y \in P$ such that $V = N(x) + N(y)$ with $N_P(x)$ and $N_P(y)$ completely adjacent, $N_N(x)$ and $N_P(y)$ completely adjacent, $N_N(y)$ and $N_P(x)$ completely adjacent; or

2. $x \in N$, $y \in N$ such that $P = N(x) + N(y)$ with $N(x)$ and $N(y)$ completely adjacent; or

3. $x \in P$, $y \in N$ such that $P = N_P(x) + N(y)$ with $N(x)$ and $N(y)$ completely adjacent and $V \setminus (N(x) \cup N(y))$ and $N_P(x)$ completely adjacent.

Lemma 1. *Let H_1 and H_2 be instances for* PP-THRESHOLD *such that there are embeddings H_1' and H_2' for them. If H_1 or H_2 has a probe threshold vertex, then $H_1 \oplus H_2$ has a probe universal vertex.*

Algorithm 1. Algorithm for solving PP-JTT

```
 1  begin
 2      while G has a probe universal vertex v do
 3          V = V \ {v}

 4      for each probe JTT pair (x, y) ∈ V do
 5          if x, y ∈ P then
 6              P ∩ V_{H_1} = N_P(y)
 7              P ∩ V_{H_2} = N_P(x)
 8              N ∩ V_{H_1} = N_N(y)
 9              N ∩ V_{H_2} = N_N(x)

10          if x ∈ N and y ∈ P then
11              P ∩ V_{H_1} = N_P(y)
12              P ∩ V_{H_2} = N_P(x)
13              N ∩ V_{H_1} = N_N(y)
14              N ∩ V_{H_2} = N \ V_{H_1}

15          if x, y ∈ N then
16              N ∩ V_{H_1} = {x}
17              N ∩ V_{H_2} = {y}
18              P ∩ V_{H_1} = N(y)
19              P ∩ V_{H_2} = N(x)
20              for each vertex v ∈ N \ {x, y} do
21                  if N(v) = P ∩ V_{H_1} then
22                      N ∩ V_{H_2} = N ∩ V_{H_2} ∪ {v}
23                  else
24                      if N(v) = P ∩ V_{H_2} then
25                          N ∩ V_{H_1} = N ∩ V_{H_1} ∪ {v}
26                      else
27                          Go back to line 4

28          if G[V_{H_1}] = H_1 and G[V_{H_2}] = H_2 are PP-THRESHOLD then
29              return G is a PP-JTT

30      return G is not a PP-JTT
```

Proof. It follows from the definitions.

Lemma 2. *Let $G = P + N$ be a graph without a probe universal vertex. If G is* PP-JTT, *then G has a probe JTT pair.*

Lemma 3. *If $G = P + N$ is* PP-JTT *without a probe universal vertex and has a probe JTT pair (x, y), then, after placing x and y, there is only one way of partitioning G into two feasible instances for* PP-THRESHOLD H_1 *and H_2, i.e., instances that can have embeddings.*

We can solve PP-JTT following Algorithm 1.

Now, we can state Theorem 6.

Theorem 6. PARTITIONED PROBE JOIN OF TWO THRESHOLDS *can be solved in polynomial time.*

3   Partitioned Probe Cographs-$(2, 1)$

In this section we work with PARTITIONED PROBE COGRAPHS-$(2, 1)$, which can be formulated as follows:

PARTITIONED PROBE COGRAPHS-$(2, 1)$ (PP - COGRAPH-$(2, 1)$)

Input: A graph $G = (V, E)$ such that $V = P + N$ where N is an independent set called the *non-probe set* and P is the *probe set.*
Question: Is there an edge set E' whose edges have both extremes in N and such that $G' = (V, E \cup E')$ is a cograph-$(2, 1)$?

In this section we prove that PP-COGRAPH-$(2, 1)$ is solvable in polynomial time. In order to make it happen, we need the definition below.

Definition 10. *An induced subgraph $B_i = (P_i + N_i, E_i)$ of a partitioned graph $G = (P + N, E)$ is called probe-biclique if $B_i[P_i]$ is a biclique (X_i, Y_i) and $B_i[N_i]$ is an independent set (X_i', Y_i') such that $N_i(x) = Y_i, \forall x \in X_i'$ and $N_i(y) = X_i, \forall y \in Y_i'$.*

Algorithm 2. Algorithm for solving PP-COGRAPH-$(2, 1)$

---

1 **begin**
2   **while** *G has a probe universal vertex v or an isolated probe-biclique B_i* **do**
3   **if** *G has a probe universal vertex v* **then**
4          $V = V \setminus \{v\}$
5   **if** *G has an isolated probe-biclique B_i* **then**
6          $V = V \setminus V(B_i)$

7   **if** *G is a* PP-JTT *graph* **then**
8   **return** *G is a* PP-COGRAPH-$(2, 1)$
9   **return** *G is not a* PP-COGRAPH-$(2, 1)$

Next, we present Algorithm 2 which is based on the decomposition presented in item 3 of Theorem 3. It also uses some features of Sect. 2.

Theorem 7. PARTITIONED PROBE COGRAPH-$(2,1)$ *can be solved in polynomial time.*

4 Probe Cographs-$(2,1)$

In this section we prove that recognizing a PROBE COGRAPHS-$(2,1)$ is a polynomial time solvable problem.

Theorem 8 deals with the unpartitioned version of PROBE COGRAPHS and proves that, given a graph G, then G has a polynomial number of feasible partitions.

Theorem 8 (Chandler et al. [20]). *Let G and $\overline{G}$ be connected and assume that G is not a cograph. Then G is a* PROBE COGRAPH *if and only if there are two non-adjacent vertices x and y in G such that G is a probe cograph with probe set $P = N(x) + N(y)$ and non-probe set $N = V \setminus P$.*

In addition, Theorem 9 proves that it is possible to reduce the PROBE COGRAPH to the PP-COGRAPH in linear time.

Theorem 9 (Chandler et al. [20]). *The problem of recognizing* PROBE COGRAPHS *can be reduced to the problem of recognizing* PARTITIONED PROBE COGRAPHS *in $O(n+m)$ time.*

These previous results allow us to state the following Theorem.

Theorem 10. *The recognition of* PROBE COGRAPHS-$(2,1)$ *is a polynomial time solvable problem.*

5 Join of Two Thresholds Graph Sandwich Problem

In this section we prove that, even knowing that THRESHOLD GRAPH SANDWICH PROBLEM is polynomially solvable [5], JOIN OF TWO THRESHOLDS GRAPH SANDWICH PROBLEM is NP-complete.

Proposition 1 is the key to prove that THRESHOLD-SP is a polynomial time solvable problem and it will be very helpful in this work.

Proposition 1 (Golumbic, Kaplan and Shamir [5]). *Let (V, E^1, E^3) be a threshold sandwich instance and let $v \in V$ be an isolated vertex in G^1 or in G^3. There is a threshold sandwich for (V, E^1, E^3) if and only if there is a threshold sandwich for $(V, E^1, E^3)_{V \setminus \{v\}}$.*

Next we define JOIN OF TWO THRESHOLD GRAPH SANDWICH PROBLEM. We remark we are considering that one of the thresholds of the join might be empty.

JOIN OF TWO THRESHOLD GRAPH SANDWICH PROBLEM (JTT-SP)

Input: A triple (V, E^1, E^3), where $E^1 \cap E^3 = \emptyset$.
Question: Is there a graph $G = (V, E)$ which is a join of two threshold graphs such that $E^1 \subseteq E$ and $E \cap E^3 = \emptyset$?

To prove the main result of this section, stated below, we make a polynomial time reduction from the NP-complete problem MONOTONE NAE3SAT [21], which can be formulated as follows:

MONOTONE NOT ALL EQUAL 3-SATISFIABILITY (MONOTONE NAE3SAT)

Input: A pair (X, C), where $X = \{x_1, \ldots, x_n\}$ is the set of variables, and $C = \{c_1, \ldots, c_m\}$ is the collection of clauses over X such that each clause $c \in C$ has exactly 3 positive literals.
Question: Is there a truth assignment for X such that each clause has at least one true and one false literal ?

Theorem 11. JOIN OF TWO THRESHOLDS GRAPH SANDWICH PROBLEM *is an NP-complete problem.*

Proof. In order to reduce MONOTONE NAE3SAT to JOIN OF TWO THRESHOLDS-SP we first construct a particular instance (V, E^1, E^3) of JTT-SP, from a generic instance (X, C) of MONOTONE NAE3SAT. Second, in Lemma 4 we prove that if there is a sandwich graph which is the join of two threshold graphs for (V, E^1, E^3), then there is a truth assignment satisfying each clause of (X, C) such that in each clause we have at least one true and one false literal. Finally, in Lemma 5 we prove that if there is a truth assignment satisfying each clause of (X, C), an instance of MONOTONE NAE3SAT, then there is a sandwich graph for (V, E^1, E^3) which is the join of two threshold graphs.

Remark 1. Let $G = (V, E)$ be a sandwich graph for (V, E^1, E^3) which is the join of two threshold graphs H_1, H_2. If $e = xy \in E^3$, then x, y are both either in H_1 or in H_2.

Remark 2. If G is the join of two threshold graphs, then G is a cograph. Therefore, G has no induced P_4.

Remark 3. If G is the join of two threshold graphs H_1, H_2 and G has an induced C_4 $\{a, b, c, d, a\}$, then $\{a, b, c, d, a\}$ cannot be entirely contained in H_1 or in H_2.

Construction of the particular instance (V, E^1, E^3) for JTT-SP:

Variable gadget

- Vertices: For each variable $x_i \in X$, $i \in \{1, \ldots, n\}$ add vertices $x_1^i, y_1^i, x_2^i, y_2^i$. For each time that a variable $x_i \in X$, $i \in \{1, \ldots, n\}$ figures in a clause c_j, $j \in \{1, \ldots, m\}$, add vertices c_i^j, d_i^j, h_i^j.

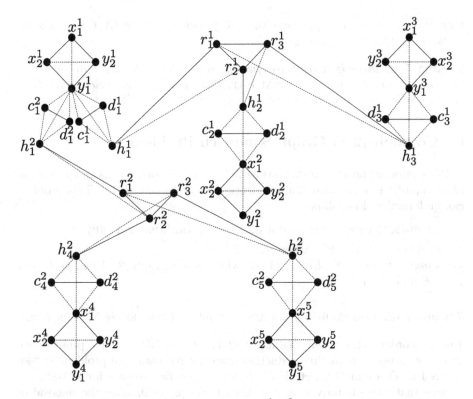

Fig. 1. Example of a particular instance (V, E^1, E^3) of JTT-SP obtained from the instance of MONOTONE NAE 3SAT: $I = (X, C) = (\{x_1, x_2, x_3, x_4, x_5\}, (x_1 \vee x_2 \vee x_3) \wedge (x_1 \vee x_4 \vee x_5))$. Solid edges are forced E^1-edges, dashed edges are forbidden E^3-edges and omitted edges are optional edges.

- For $i \in \{1, \ldots, n\}$ and $j \in \{1, \ldots, m\}$, add the following forced edges: $\{x_1^i x_2^i, x_1^i y_2^i, x_2^i y_1^i, y_1^i y_2^i, c_i^j d_i^j\}$.
- For $i \in \{1, \ldots, n\}$ and $j \in \{1, \ldots, m\}$, add the following forbidden edges: $\{x_1^i y_1^i, x_2^i y_2^i, y_1^i c_i^j, y_1^i d_i^j, y_1^i h_i^j, c_i^j h_i^j, d_i^j h_i^j\}$

Clause gadget

- Vertices: For each clause $c_j \in C$, $j \in \{1, \ldots, m\}$, add vertices r_1^j, r_2^j, r_3^j.
- For $j \in \{1, \ldots, m\}$ add the following forced edges: $\{r_1^j r_2^j, r_1^j r_3^j, r_2^j, r_3^j\}$.
  For each clause $c_j = (l_1^j \vee l_2^j \vee l_3^j) \in C$, add to E^1 edges $h_1^j r_1^j, h_2^j r_2^j, h_3^j r_3^j$.
- Forbidden edges:
  For each clause $c_j = (l_1^j \vee l_2^j \vee l_3^j) \in C$, add to E^3 edges $h_1^j r_2^j, h_2^j r_3^j, h_3^j r_1^j$.

See Fig. 1 as an example.

Lemma 4. *If there is a sandwich graph $G = (V, E)$ which is the join of two threshold graphs for the particular instance (V, E^1, E^3) constructed above,* *then there is a truth assignment satisfying each clause of (X, C), a generic instance of* MONOTONE NAE3SAT.

Lemma 5. *If there is a truth assignment satisfying each clause of (X, C), a generic instance of* MONOTONE NAE3SAT, *then there is a sandwich graph $G = (V, E)$ which is a join of two threshold graphs for (V, E^1, E^3).*

6   Cograph-(2,1) Graph Sandwich Problem

In this section we prove that COGRAPH-$(2, 1)$-SP is NP-complete as an application of cograph-$(2, 1)$ structural characterization we presented in [18]. This problem can be formulated as follows:

COGRAPH-$(2, 1)$ GRAPH SANDWICH PROBLEM (COGRAPH-$(2, 1)$-SP)
Input: A triple (V, E^1, E^3), where $E^1 \cap E^3 = \emptyset$.
Question: Is there a graph $G = (V, E)$ which is a cograph-$(2, 1)$ and such that $E^1 \subseteq E$ and $E \cap E^3 = \emptyset$?

Theorem 12. COGRAPH-$(2, 1)$ GRAPH SANDWICH PROBLEM *is NP-complete.*

Proof. In order to prove that COGRAPH-$(2, 1)$-SP is an NP-complete problem, we will make a polynomial time reduction from the NP-complete problem JTT-SP (proved in Theorem 11). Let (V, E^1, E^3) be a generic instance for JTT-SP. We assume that there is no vertex u such that $N_{G^3}(u) = \emptyset$, since the removal of this kind of vertex does not affect the property of being a PP-JTT. First, we will construct a particular instance (V', E'^1, E'^3) for COGRAPH-$(2, 1)$-SP. Second, we prove that if there is a join of two thresholds sandwich graph for (V, E^1, E^3), then there is a cograph-$(2, 1)$ sandwich graph for (V', E'^1, E'^3). Third, we prove that if there is a cograph-$(2, 1)$ sandwich graph for (V', E'^1, E'^3), then there is a join of two thresholds sandwich graph for (V, E^1, E^3).

Construction of the particular instance (V', E'^1, E'^3) for COGRAPH-$(2, 1)$-SP:

- $V' = V \cup \{a, b, c, d\}$,
- $E'^1 = E^1 \cup \{ab, bc, cd, da\} \cup \{xy | x \in V, y \in \{a, b\}\}$
- $E'^3 = E^3 \cup \{ac, bd\}$

The next Lemmas complete Theorem 12's proof.

Lemma 6. *Let $G = (V, E)$ be a graph with a universal vertex u. G is a join of two thresholds graph if and only if $G \setminus \{u\}$ is a join of two thresholds graph.*

Lemma 7. *If there is a join of two thresholds sandwich graph for (V, E^1, E^3), then there is a cograph-$(2, 1)$ sandwich graph for (V', E'^1, E'^3).*

Lemma 8. *If there is a cograph-$(2, 1)$ sandwich for (V', E'^1, E'^3), then there is a join of two thresholds sandwich graph for (V, E^1, E^3).*

7   Conclusions

In this paper, we proved that, similarly to PROBE COGRAPH PROBLEM and PARTITIONED PROBE THRESHOLD PROBLEM, PROBE COGRAPH-$(2, 1)$ PROBLEM and PARTITIONED PROBE JOIN OF TWO THRESHOLDS PROBLEM are polynomially solvable. In contrast, we showed that, although THRESHOLD-SP and COGRAPH GRAPH SANDWICH PROBLEM are polynomially solvable problems [5], JOIN OF TWO THRESHOLDS GRAPH SANDWICH PROBLEM and consequently COGRAPH-$(2, 1)$ GRAPH SANDWICH PROBLEM are NP-complete ones, contradicting all natural feelings around two well-known classes of graphs.

References

1. Zhang, P.: Probe interval graph and its application to physical mapping of DNA. Manuscript (1994)
2. Zhang, P., Schon, E.A., Fisher, S.G., Cayanis, E., Weiss, J., Kistler, S., Bourne, P.E.: An algorithm based on graph theory for the assembly of contigs in physical mapping of DNA. CABIOS **10**, 309–317 (1994)
3. Zhang, P., Ye, X., Liao, L., Russo, J., Fisher, S.G.: Integrated mapping package - a physical mapping software tool kit. Genomics **55**, 78–87 (1999)
4. Chandler, D.B., Chang, M.-S., Kloks, T., Liu, J., Peng, S.-L.: Probe Graph Classes. Online Manuscript (2012)
5. Golumbic, M.C., Kaplan, H., Shamir, R.: Graph sandwich problems. J. Algorithm **19**(3), 449–473 (1995)
6. Figueiredo, C.M.F., Faria, L., Klein, S., Sritharan, R.: On the complexity of the sandwich problems for strongly chordal graphs and chordal bipartite graphs. Theor. Comput. Sci. **381**, 57–67 (2007)
7. Sritharan, R.: Chordal bipartite completion of colored graphs. Discrete Math. **308**, 2581–2588 (2008)
8. Corneil, D.G., Lerchs, H., Burlingham, L.S.: Complement reducible graphs. Discrete App. Math. **3**(3), 163–174 (1981)
9. Corneil, D.G., Perl, Y., Stewart, L.K.: A linear recognition algorithm for cographs. SIAM J. Comput. **14**(4), 926–934 (1985)
10. Bretscher, A., Corneil, D.G., Habib, M., Paul, C.: A simple linear time lexbfs cograph recognition algorithm. In: 29th WG Workshop, pp. 119–130 (2003)
11. Habib, M., Paul, C.: A simple linear time algorithm for cograph recognition. Discrete Appl. Math. **145**(2), 183–197 (2005)
12. Chvátal, V., Hammer, P.L.: Aggregation of inequalities in integer programming. In: Korte, B.H., Hammer, P.L., Johnson, E.L., Nemhauser, G.L. (eds.) Studies in Integer Programming, vol. 1, Annals of Discrete Mathematics, pp. 145–162. Elsevier (1977)
13. Brandstädt, A.: Partitions of graphs into one or two independent sets and cliques. Discrete Math. **152**(1–3), 47–54 (1996)
14. Brandstädt, A.: Corrigendum. Discrete Math. **186**, 295 (2005)
15. Brandstädt, A., Le, V.B., Szymczak, T.: The complexity of some problems related to graph 3-colorability. Discrete Appl. Math. **89**(1–3), 59–73 (1998)
16. Dantas, S., de Figueiredo, C.M., Faria, L.: On decision and optimization (k, l)-graph sandwich problems. Discrete Appl. Math. **143**, 155–165 (2004)

17. Bravo, R., Klein, S., Nogueira, L.: Characterizing (k, ℓ)-partitionable cographs. ENDM **22**, 277–280 (2005)
18. Couto, F., Faria, L., Gravier, S., Klein, S., dos Santos, V.F.: Structural characterization, decomposition for cographs-$(2, 1)$, $(1, 2)$: a natural generalization of threshold graphs. Electron. Notes Discrete Math. **50**, 133–138 (2015)
19. de Ridder, H.N.: On probe classes of graphs. Ph.D. thesis, Rostock (2007)
20. Chandler, D.B., Chang, M.-S., Kloks, T., Peng, S.-L.: Probe Graphs. Online Manuscript (2009)
21. Garey, M.R., Johnson, D.S.: Computers and Intractability: A Guide to the Theory of NP-Completeness. W. H. Freeman & Co., New York (1979)

Colouring and Covering Nowhere Dense Graphs

Martin Grohe[1], Stephan Kreutzer[2], Roman Rabinovich[2], Sebastian Siebertz[2], and Konstantinos Stavropoulos[1(✉)]

[1] RWTH Aachen University, Aachen, Germany
{grohe,stavropoulos}@informatik.rwth-aachen.de
[2] Technical University Berlin, Berlin, Germany
{stephan.kreutzer,roman.rabinovich,sebastian.siebertz}@tu-berlin.de

Abstract. In [9] it was shown that nowhere dense classes of graphs admit sparse *neighbourhood covers* of small degree. We show that a monotone graph class admits sparse neighbourhood covers if and only if it is nowhere dense. The existence of such covers for nowhere dense classes is established through bounds on so-called *weak colouring numbers.*

The core results of this paper are various lower and upper bounds on the weak colouring numbers and other, closely related generalised colouring numbers. We prove tight bounds for these numbers on graphs of bounded tree width. We clarify and tighten the relation between the expansion (in the sense of "bounded expansion" [15]) and the various generalised colouring numbers. These upper bounds are complemented by new, stronger exponential lower bounds on the generalised colouring numbers. Finally, we show that computing weak r-colouring numbers is NP-complete for all $r \geq 3$.

1 Introduction

Nowhere dense classes of graphs have been introduced by Nešetřil and Ossona de Mendez [15,16] as a general model of "sparse" graph classes. They include and generalise many other natural sparse graph classes, among them all classes of bounded degree, classes of bounded genus, classes defined by excluded (topological) minors, and classes of bounded expansion. It has been demonstrated in several papers, e.g., [2,9,15] that nowhere dense graph classes have nice algorithmic properties; many problems that are hard in general can be solved (more) efficiently on nowhere dense graph classes. As a matter of fact, nowhere dense classes are a natural limit for the efficient solvability of a wide class of problems [6,9,12].

In [9], it was shown that nowhere dense classes of graphs admit sparse *neighbourhood covers*. Neighbourhood covers play an important role in the study of distributed network algorithms and other application areas (see, for example, [17]). The neighbourhood covers developed in [9] combine low radius and low degree making them interesting for the applications outlined above. In this paper, we prove a (partial) converse to the result of [9]: we show that monotone graph classes (that is, classes closed under taking subgraphs) are nowhere dense if and only if they admit sparse neighbourhood covers.

© Springer-Verlag Berlin Heidelberg 2016
E.W. Mayr (Ed.): WG 2015, LNCS 9224, pp. 325–338, 2016.
DOI: 10.1007/978-3-662-53174-7_23

Nowhere denseness has turned out to be a very robust property of graph classes with various seemingly unrelated characterisations (see [8,15]), among them characterisations through so-called *generalised colouring numbers*. These are particularly relevant in the algorithmic context, because the existence of sparse neighbourhood covers for nowhere dense classes is established through such colouring numbers—the *weak r-colouring numbers*, to be precise—and the value of these numbers is directly related to the degree of the neighbourhood covers. Besides the weak r-colouring numbers $\mathrm{wcol}_r(G)$ of graphs G we study the *r-colouring numbers* $\mathrm{col}_r(G)$ and the *r-admissibility numbers* $\mathrm{adm}_r(G)$. The two families of colouring numbers where introduced by Kierstead and Yang in [11], and the admissibility numbers go back to Kierstead and Trotter in [10] and were generalised by Dvořák in [5]. All these numbers generalise the *degeneracy*, a.k.a. *colouring number*, which is defined to be the minimum d such that there is a linear order of the vertices of G in which every vertex has at most d smaller neighbours. The name "colouring number" comes from the fact that graphs of degeneracy d have a proper $d + 1$ colouring which can be computed efficiently by a simple greedy algorithm. For the generalised r-colouring numbers, instead of smaller neighbours of a vertex we count smaller vertices reachable by certain paths of length r; the numbers differ by the kind of paths of length r considered. We observe that with growing r the colouring numbers converge to the treewidth of the graph.

The core results of this paper are various upper and lower bounds for these families of colouring numbers. In particular, we prove tight bounds for $\mathrm{wcol}_r(G)$ for graphs G of bounded tree width. We clarify and tighten the relation between the expansion (in the sense of "bounded expansion" [15]) and the various generalised colouring numbers and use it to prove that for every $\varepsilon > 0$, $\mathrm{wcol}_r(G) = \mathcal{O}(r^{(1+\varepsilon)r})$ for graph classes of constant expansion, e.g. for classes that exclude a (topological) minor. These upper bounds are complemented by new, stronger exponential lower bounds on the generalised colouring numbers. The lower bounds can already be achieved on graph classes of bounded degree. As mentioned above, the bounds on the weak colouring numbers of graph classes are directly related to the sparseness of the neighbourhood covers. Finally, we show that computing weak r-colouring numbers is NP-complete for all $r \geq 3$.

After giving some graph theoretic background in Sect. 2, we prove our various bounds on the generalised colouring numbers in Sects. 3, 4 and 5. Section 6 is devoted to sparse neighbourhood covers, and the NP-completeness result for the weak colouring numbers is proved in Sect. 7.

2   Generalised Colouring Numbers

Our notation from graph theory is standard, we refer the reader to [3] for background. All graphs in this paper are finite and simple, i.e. they do not have loops or multiple edges between the same pair of vertices. A class of graphs is *monotone* if it is closed under subgraphs. The *radius* $\mathrm{rad}(G)$ of G is $\min_{u \in V(G)} \max_{v \in V(G)} \mathrm{dist}^G(u, v)$. By $N_r^G(v)$ we denote the *r-neighbourhood* of v in G, i.e. the set of vertices of distance at most r from v in G.

We represent a linear order on $V(G)$ as an injective function $L : V(G) \to \mathbb{N}$ and write $\Pi(G)$ for the set of all linear orders on $V(G)$.

Vertex u is *weakly r-reachable* from v with respect to the order L, if there is a path P of length $0 \leq \ell \leq r$ from v to u such that $L(u) \leq L(w)$ for all $w \in V(P)$. Let $\mathrm{WReach}_r[G, L, v]$ be the set of vertices that are weakly r-reachable from v with respect to L. If furthermore, all inner vertices w of P satisfy $L(v) < L(w)$, then u is called *strongly r-reachable* from v. Let $\mathrm{SReach}_r[G, L, v]$ be the set of vertices that are strongly r-reachable from v with respect to L.

The *r-admissibility* $\mathrm{adm}_r[G, L, v]$ of v with respect to L is the maximum size k of a family $\{P_1, \ldots, P_k\}$ of paths of length at most r in G that start in v, end at a vertex w with $L(w) \leq L(v)$ and satisfy $V(P_i) \cap V(P_j) = \{v\}$ for $1 \leq i \neq j \leq k$. As we can always let the paths end in the first vertex smaller than v, we can assume that the internal vertices of the paths are larger than v. Note that $\mathrm{adm}_r[G, L, v]$ is an integer, whereas $\mathrm{WReach}_r[G, L, v]$ and $\mathrm{SReach}_r[G, L, v]$ are sets of vertices.

The *weak r-colouring number* $\mathrm{wcol}_r(G)$, the *r-colouring number* $\mathrm{col}_r(G)$, and the *r-admissibility* $\mathrm{adm}_r(G)$ are defined as

$$\mathrm{wcol}_r(G) = \min_{L \in \Pi(G)} \max_{v \in V(G)} |\mathrm{WReach}_r[G, L, v]|,$$

$$\mathrm{col}_r(G) = \min_{L \in \Pi(G)} \max_{v \in V(G)} |\mathrm{SReach}_r[G, L, v]|,$$

$$\mathrm{adm}_r(G) = \min_{L \in \Pi(G)} \max_{v \in V(G)} \mathrm{adm}_r[G, L, v].$$

It follows from the definitions that, for all $r \in \mathbb{N}$, $\mathrm{adm}_r(G) \leq \mathrm{col}_r(G) \leq \mathrm{wcol}_r(G)$. Furthermore, $\mathrm{adm}_1(G) \leq \mathrm{adm}_2(G) \leq \ldots \leq \mathrm{adm}_n(G)$, $\mathrm{wcol}_1(G) \leq \mathrm{wcol}_2(G) \leq \ldots \leq \mathrm{wcol}_n(G) = \mathrm{td}(G)$ (where $\mathrm{td}(G)$ is the treedepth of G, see e.g. [15]) and $\mathrm{col}_1(G) \leq \mathrm{col}_2(G) \leq \ldots \leq \mathrm{col}_n(G) = \mathrm{tw}(G)$ (where $\mathrm{tw}(G)$ is the treewidth of G).

To see that $\mathrm{col}_n(G) = \mathrm{tw}(G)$, note that treewidth can be characterised by elimination orders. An *elimination order* of a graph G is a linear order L on $V(G)$ with which we associate a sequence of graphs G_i. Let $V(G) = \{1, \ldots, n\}$ and $L(i) < L(j)$ for $i < j$, then $G_0 = G$ and for $0 < i \leq n$, $V(G_i) = V(G_{i-1}) \backslash \{i\}$ and $E(G_i) = \left(E(G_{i-1}) \backslash \{\{i, j\} : j \leq n\} \right) \cup \{\{\ell, j\} : \{\ell, i\}, \{i, j\} \in E(G_{i-1})\}$, i.e. we eliminate vertex i and make a clique out of the neighbours of i in G_{i-1}. The width of the elimination order is one plus the maximum size of a clique over all G_i. The elimination width of G is the minimum width over all possible widths of elimination orders of G. It is well known that the treewidth of G is equal to its elimination width. Let L' be the reverse to L. An easy induction shows that the neighbours of a vertex i in G_{i-1} are exactly those of $\mathrm{SReach}_n[G, L', i]$. It follows that $\mathrm{col}_n(G) = \mathrm{tw}(G)$.

Furthermore, it was shown that the generalised colouring numbers are strongly related, i.e. $\mathrm{col}_r(G) \leq (\mathrm{adm}_r(G) - 1) \cdot (\mathrm{adm}_r(G) - 2)^{r-1} + 1$ and $\mathrm{wcol}_r(G) \leq \mathrm{adm}_r(G)^r$ (see for example [5], but note that in that work, paths of length 0 are not considered for the r-admissibility).

3   Admissibility and Expansion

For $r \in \mathbb{N}$, an *r-subdivision* of a graph H is obtained from H by replacing edges by pairwise internally disjoint paths of length at most $r+1$. If a graph G contains a $2r$-subdivision of H as a subgraph, then H is a *topological depth-r minor* of G, written $H \preceq_r^t G$. Recall that H is a *topological minor* of G (we write $H \preceq^t G$) if some subdivision of H is a subgraph of G, that is, if $H \preceq_r^t G$ for some $r \in \mathbb{N}$.

The *edge density* of a graph G is $\varepsilon(G) = |E(G)|/|V(G)|$. Note that the average degree of G is $2\varepsilon(G)$. A graph is *k-degenerate* if every subgraph has a vertex of degree at most k. The maximum of the edge densities of all $H \preceq_r^t G$ is known as the *topological greatest reduced average density* $\widetilde{\nabla}_r(G)$ of G with rank r.

A class $\mathcal{C}$ of graphs is *nowhere dense* if for all $\varepsilon > 0$ and all $r \in \mathbb{N}$ there is an $n_0 \in \mathbb{N}$ such that all n-vertex graphs $G \in \mathcal{C}$ with at least n_0 vertices satisfy $\widetilde{\nabla}_r(G) \leq n^\varepsilon$. $\mathcal{C}$ is said to have *bounded expansion* if for every r there is a $c(r)$ such that $\widetilde{\nabla}_r(G) \leq c(r)$ for all $G \in \mathcal{C}$. It is easy to see that all classes of bounded expansion are nowhere dense; the converse does not hold. We say that $\mathcal{C}$ has *constant expansion* if there is a constant c such that $\widetilde{\nabla}_r(G) \leq c$ for all $r \in \mathbb{N}$ and $G \in \mathcal{C}$.

The following theorem implies improvements of previous results from Kierstead and Yang [11] and Zhu [20] to the exponent of their upper bounds for colouring numbers and the weak colouring numbers.

Theorem 3.1. *Let G be a graph and let $r \in \mathbb{N}$. Then* $\mathrm{adm}_r(G) \leq 2r\big(\widetilde{\nabla}_{r-1}(G)\big)^3$.

Every class that excludes a topological minor has constant expansion. This includes familiar classes such as classes of bounded degree, bounded genus, and bounded tree width. We obtain the following corollary.

Corollary 3.2. *Let $\mathcal{C}$ be a graph class that excludes some fixed graph as a topological minor. Then for all $G \in \mathcal{C}$ we have $\mathrm{adm}_r(G) = \mathcal{O}(r)$ and for every $\varepsilon > 0$, $\mathrm{wcol}_r(G) = \mathcal{O}(r^{(1+\varepsilon)r})$.*

For the proof of Theorem 3.1 we need a lemma which is a variation of a result of Dvořák [5]. For a set $S \subseteq V(G)$ and $v \in S$, let $b_r(S, v)$ be the maximum number k of paths $P_1, \ldots, P_k$ of length at most r from v to S with internal vertices in $V(G) \setminus S$ and with $V(P_i) \cap V(P_j) = \{v\}$ for $1 \leq i \neq j \leq k$.

Lemma 3.3 [5]**.** *For all graphs G and $r \in \mathbb{N}$, there exists a set $S \subseteq V(G)$ such that $b_r(S, v) = \mathrm{adm}_r(G)$ for all $v \in S$.*

Proof. Assume that all $S \subseteq V(G)$ contain a vertex v such that $b_r(S, v) < \mathrm{adm}_r(G)$. We construct an order $L(v_1) < L(v_2) < \ldots < L(v_n)$ of $V(G)$ as follows. If $v_{i+1}, \ldots, v_n$ have already been ordered, choose v_i such that if $S_i = \{v_1, \ldots, v_i\}$, then $b_r(S_i, v_i)$ is minimal. Clearly, the r-admissibility of the resulting order is one of the values $b_r(S_i, v_i)$ occuring in its construction. This implies $\mathrm{adm}_r(G) < \mathrm{adm}_r(G)$, a contradiction.   □

Proof (of Theorem 3.1). Let G be a graph with $\widetilde{\nabla}_{r-1}(G) \leq c$, and let $\ell := 2rc^3 + 1$. Suppose for contradiction that $\mathrm{adm}_r(G) > \ell$. By Lemma 3.3, there exists a set S such that $b_r(S, v) > \ell$ for all $v \in S$. For $v \in S$, let $\mathcal{P}_v$ be a set of paths from v to S witnessing this, and let $s := |S|$.

Choose a maximal set $\mathcal{P}$ of pairwise internally vertex-disjoint paths of length at most $2r - 1$ connecting pairs of vertices from S whose internal vertices belong to $V(G) \setminus S$ such that each pair of vertices is connected by at most one path. Let H be the graph with vertex set S and edges between all vertices $v, w \in S$ connected by a path in $\mathcal{P}$. Then $H \preceq^t_{r-1} G$ and hence $|\mathcal{P}| = |E(H)| \leq s \cdot c$. Let M be the set of all internal vertices of the paths in $\mathcal{P}$, and let $m := |M|$. Then $m \leq s \cdot c \cdot (2r - 2)$.

Note that we not only have $H \preceq^t_{r-1} G$, but also $H' \preceq^t_{r-1} G$ for all $H' \subseteq H$. Thus $\varepsilon(H') \leq c$, and therefore H' has a vertex of degree at most $c/2$. We claim that H contains an independent set R of size $\lfloor s/(c/2 + 1) \rfloor$. We can iteratively build the set as follows. Choose a vertex v of minimum degree and add it to R. Delete v and all its neighbours from $V(H)$ and continue inductively. Clearly the resulting set R is independent in H. As all subgraphs of H have a vertex of degree at most $c/2$, we delete at most $c/2 + 1$ vertices from S in each step. Hence R has size at least $\lfloor s/(c/2 + 1) \rfloor$.

For every $v \in S$, we let $\mathcal{Q}_v$ be the set of initial segments of paths in $\mathcal{P}_v$ from v to a vertex in $(M \cup S) \setminus \{v\}$ with all internal vertices in $V(G) \setminus (M \cup S)$. Observe that for $u, v \in R$ the paths in $\mathcal{Q}_v$ and $\mathcal{Q}_u$ are internally disjoint, because if $Q \in \mathcal{Q}_u$ and $Q' \in \mathcal{Q}_v$ had an internal vertex in common, then $Q \cup Q'$ would contain a path of length at most $2r - 2$ that is internally disjoint from all paths in $\mathcal{P}$, contradicting the maximality of $\mathcal{P}$.

Let G' be the union of all paths in $\mathcal{P}$ and all paths in $\mathcal{Q}_v$ for $v \in R$, and let H' be obtained from G' by contracting all paths in $\bigcup_{v \in R} \mathcal{Q}_v$ to single edges. Then $H' \preceq^t_{r-1} G$.

We have $|V(H')| \leq s + m \leq s + s \cdot c \cdot (2r - 2) \leq s \cdot c \cdot (2r - 1)$ and at least $|E(H')| \geq \lfloor s/(c/2 + 1) \rfloor \cdot \ell \geq s \cdot \ell/c$ edges. Thus $\varepsilon(H') \geq \ell/2rc^2 > c$. A contradiction. $\qquad\square$

4 The Colouring Numbers of Graphs of Bounded Treewidth

A *tree decomposition* of a graph G is a pair (T, X), where T is a tree, $X = (X_t : t \in V(T))$, is a family of subsets of V(G) (called bags) such that (i) $\bigcup_{t \in V(T)} X_t = V(G)$, (ii) for every edge $\{u, v\}$ of G there exists $t \in V(T)$ with $u, v \in X_t$ and (iii) if $r, s, t \in V(T)$ and s is on the path of T between r and t, then $X_r \cap X_t \subseteq X_s$.

A graph has *treewidth* at most k if it admits a tree decomposition (T, X) such that $|X_t| \leq k + 1$ for each $t \in V(T)$ and we write $\mathrm{tw}(G)$ for the treewidth of G. We assume familiarity with the basic theory of tree decompositions as in [3].

It is well known that a graph of treewidth k has a tree decomposition (T, X) of width k such that for every $\{s, t\} \in E(T)$ we have $|X_s \setminus X_t| \leq 1$. We call such decompositions *smooth*. The following separation property of tree decompositions is well known.

Lemma 4.1. *If $r, s, t \in V(T)$, $u \in X_r$ and $v \in X_t$ and s is on the path of T between r and t, then every path from u to v in G uses a vertex contained in X_s.*

For a tree decomposition (T, X) of G and a node $s \in V(T)$ we define a partial order $L^{T,s}$ on $V(T)$ demanding that $L^{T,s}(t) \le L^{T,s}(u)$ if t lies on the path from s to u (i.e. $L^{T,s}$ is the standard tree order where s is minimum).

Theorem 4.2. *Let $\mathrm{tw}(G) \le k$. Then $\mathrm{wcol}_r(G) \le \binom{r+k}{k}$.*

Proof. Let (T, X) be a smooth tree decomposition of G of width at most k. Since if G' is a subgraph of G, then $\mathrm{wcol}_r(G') \le \mathrm{wcol}_r(G)$, w.l.o.g. we may assume that G is edge maximal of treewidth k, i.e. each bag induces a clique in G. We choose an arbitrary root s of T and let L' be some linear extension of $L^{T,s}$. For every $v \in V(G)$, let t_v be the unique node of T such that $L'(t_v) = \min\{L'(t)|v \in X_t\}$ and define a linear ordering L of $V(G)$ such that: (i) $L'(t_v) < L'(t_u) \Rightarrow L(v) < L(u)$, and (ii) if $L'(t_v) = L'(t_u)$ (which is possible in the root bag X_s), break ties arbitrarily.

Fix some $v \in V(G)$ and let $w \in \mathrm{WReach}_r[G, L, v]$. By Lemma 4.1 and the definition of L, it is immediate that t_w lies on the path from t_v to s in T. Let $u \in X_{t_v}$ be such that $L(u) \le L(u')$ for all $u' \in X_{t_v}$. If $t_v = s$, then $|\mathrm{WReach}_r[G, L, v]| \le k+1$ and we are done. Otherwise, as the decomposition is smooth, $L'(t_u) < L'(t_v)$. We define two subgraphs G_1 and G_2 of G as follows. The graph G_1 is induced by the vertices from the bags between s and t_u, i.e. by the set $\bigcup \{X_t \in V(T) : L^{T,s}(t) \le L^{T,s}(t_u)\}$. The graph G_2 is induced by $\bigcup \{X_t \in V(T) : L^{T,s}(t_v) \le L^{T,s}(t) < L^{T,s}(t_u)\} \setminus V(G_1)$.

Let L_i be the restriction of L to $V(G_i)$, for $i = 1, 2$, respectively. We claim that if $w \in \mathrm{WReach}_r[G, L, v]$, then $w \in \mathrm{WReach}_{r-1}[G_1, L_1, u] \cup \mathrm{WReach}_r[G_2, L_2, v]$. To see this, let $P = (v = v_1, \ldots, v_\ell = w)$ be a shortest path between v and w of length $\ell \le r$ such that $L(w)$ is minimum among all vertices of $V(P)$.

We claim that $L(v_1) > \ldots > L(v_\ell)$ (and call P a decreasing path). This implies in particular that all t_{v_i} lie on the path from t_v to s and that $L^{T,s}(t_{v_1}) \ge \ldots \ge L^{T,s}(t_{v_\ell})$ (non-equality may only hold in the last step, if we take a step in the root bag).

Assume that the claim does not hold and let i be the first position with $L(v_i) < L(v_{i+1})$. It suffices to show that we can find a subsequence (which is also a path in G) $Q = v_i, v_j, \ldots, v$ of P with $j > i+1$. By definition of $t_{v_{i+1}} =: t$, X_t contains v_i. (Indeed, there is an edge between v_i and v_{i+1}, which must be contained in some bag, but v_{i+1} appears first in X_t counting from the root and each bag induces a clique in G). Let t' be the parent node of t. $X_{t'}$ also contains v_i, as the decomposition is smooth and v_{i+1} is the unique vertex that joins X_t. But by Lemma 4.1, $X_{t'}$ is a separator that separates v_{i+1} from all vertices smaller than v_{i+1}. We hence must visit another vertex v_j from $X_{t'}$ in order to finally reach v. We can hence shorten the path as claimed.

If $L(w) \leq L(u)$, then P goes through X_{t_u} by Lemma 4.1. Let u' be the first vertex of P that lies in X_{t_u}. We show that there is a shortest path from v to u' that uses u as the second vertex. By assumption, $v \neq u$. If $\{v, u'\} \in E(G)$, then $\{v, u'\}$ must be contained in some bag $X_{t'}$. By definition of t_v, $t' = t_v$, as t_v is the first node of T on the path from s to t_v containing v. By definition of u and because (T, X) is smooth, u is the only vertex from t_v that appears in t_u. Thus $u' = u$, so the shortest path from v to u' uses u. If the distance between v and u' is at least 2, a shortest path can be chosen as v, u, u'. Indeed $u \in X_{t_u} \cap X_{t_v}$ and every bag induces a clique by assumption.

It follows that if $L(w) \leq L(u)$ and $w \in \mathrm{WReach}_r[G, L, v]$, then there is a shortest path from v to w that uses u as the second vertex. Thus $w \in \mathrm{WReach}_{r-1}[G_1, L_1, u]$, as P is decreasing.

If $L(w) > L(u)$, then P never visits vertices of G_1. If P lies completely in G_2, we have $w \in \mathrm{WReach}_r[G_2, L_2, v]$. If P leaves G_2, it visits vertices of G that are contained only in bags strictly below t_v. However, this is impossible, as P is decreasing.

Hence $|\mathrm{WReach}_r[G, L, v]| \leq |\mathrm{WReach}_{r-1}[G_1, L_1, u]| + |\mathrm{WReach}_r[G_2, L_2, v]|$. The treewidth of G_2 is at most $k - 1$, as we removed u from every bag. More precisely, the tree decomposition (T^2, X^2) of G_2 of width at most $k - 1$ is the restriction of (T, X) to G_2, i.e. we take tree nodes t contained between t_u and t_v (including t_v and not including t_u) and define $X_t^2 = X_t \cap V(G_2)$.

For the induction base, recall that $\mathrm{wcol}_1(G)$ equals the degeneracy of G plus one and that every graph of treewidth $\leq k$ is k-degenerate. Furthermore, wcol_r of a tree is at most $r + 1$ (using any linearisation of the standard tree order).

We recursively define the following numbers $w(k, r)$ such that $w(k, 1) = k + 1$ for $k \geq 1$, $w(1, r) = r + 1$ for $r \geq 1$ and $w(k, r) = w(k, r - 1) + w(k - 1, r)$ for $k, r > 1$. By our above argumentation, $|\mathrm{WReach}_r[G, L, v]| \leq w(k, r)$. We observe that this is the recursive definition of the binomial coefficients and conclude that $|\mathrm{WReach}_r[G, L, v]| \leq \binom{r+k}{k}$. $\qquad\square$

The proof of Theorem 4.2 gives rise to a construction of a class of graphs that matches the upper bound proven there. We construct a graph of treewidth k and weak r-colouring number $\binom{k+r}{k}$ whose tree decomposition has a highly branching host tree. This enforces a path in the tree from the root to a leaf that realises the recursion from the proof of Theorem 4.2.

Theorem 4.3. *There is a family of graphs G_r^k with $\mathrm{tw}(G_r^k) = k$, such that $\mathrm{wcol}_r(G_r^k) = \binom{r+k}{k}$. In fact, for all $r' \leq r$, $\mathrm{wcol}_{r'}(G_r^k) = \binom{r'+k}{k}$.*

Proof. Fix r, k and let $c = \binom{r+k}{k}$. We define graphs $G(k', r')$ for all $r' \leq r, k' \leq k$ and corresponding tree decompositions $T(k', r') = (T(k', r'), X(k', r'))$ of $G(k', r')$ of width k' with a distinguished root $s(T(k', r'))$ by induction on k' and r'. We will show that $\mathrm{wcol}_{r'}(G(k', r')) \leq \binom{r'+k'}{k'}$. We guarantee several invariants for all values of k' and r' which will give us control over a sufficiently large part of any order that witnesses $\mathrm{wcol}_{r'}(G(k', r')) \leq \binom{r'+k'}{k'}$.

1. There is a bijection $f : V(T(k', r')) \rightarrow V(G(k', r'))$ such that $f(s(T(k', r')))$ is the unique vertex contained in $X_{s(T(k',r'))}$ and if t is a child of t' in $T(k', r')$, then $f(t)$ is the unique vertex of $X_t \setminus X_{t'}$. Hence any order defined on $V(T)$ directly translates to an order of $V(G)$ and vice versa.
2. In any order L of $V(G(k', r'))$ which satisfies $\text{wcol}_r(G(k', r')) \leq c$, there is some root-leaf path $P = t_1, \ldots, t_m$ such that $L(f(t_1)) < \ldots < L(f(t_m))$.
3. Every bag of $T(k', r')$ contains at most $k' + 1$ vertices.

It will be convenient to define the tree decompositions first and to define the corresponding graphs as the unique graphs induced by the decomposition in the following sense. For a tree T and a family of finite and non-empty sets $(X_t)_{t \in V(T)}$ such that if $z, s, t \in V(T)$ and s is on the path of T between z and t, then $X_z \cap X_t \subseteq X_s$, we define the graph *induced* by $(T, (X_t)_{t \in V(T)})$ as the graph G with $V(G) = \bigcup_{t \in V(T)} X_t$ and $\{u, v\} \in E(G)$ if and only if $u, v \in X_t$ for some $t \in V(T)$. Then $(T, (X_t)_{t \in V(T)})$ is a tree decomposition of G.

For $k' \geq 1, r' = 1$, let $T(k', r') =: T$ be a tree of depth $k' + 1$ and branching degree c with root s. Let $L^{T,s}$ be the natural partial tree order. Let $f: V(T) \rightarrow V$ be a bijection to some new set V. We define $X_t := \{f(t) : L^{T,s}(t') \leq L^{T,s}(t)\}$. Let $G(k', r')$ be the graph induced by the decomposition. The first and the third invariants clearly hold. For the second invariant, consider a simple pigeonhole argument. For every non-leaf node t, the vertex $f(t)$ has c neighbours $f(t')$ in the child bags $X_{t'}$ of t. Hence some $f(t')$ must be larger in the order. This guarantees the existence of a path as required.

For $k' = 1, r' \geq 1$, let $T(k', r') =: T$ be a tree of depth $r' + 1$ and branching degree c with root s and let f be as before. Let $X_s := \{f(s)\}$ and for each $t' \in V(T)$ with parent $t \in V(T)$ let $X_{t'} := \{f(t), f(t')\}$. Let $G(k', r')$ be the graph induced by the decomposition. All invariants hold by the same arguments as above. Note that G_1^1 is the same graph in both constructions and is hence well defined.

Now assume that $G(k', r' - 1)$ and $G(k' - 1, r')$ and their respective tree decompositions have been defined. Let $T(k', r')$ be the tree which is obtained by attaching c copies of $T(k' - 1, r')$ as children to each leaf of $T(k', r' - 1)$. We define the bags that belong to the copy of $T(k', r' - 1)$, exactly as those of $T(k', r' - 1)$. To every bag of a copy of $T(k' - 1, r')$ which is attached to a leaf z, we add $f'(z)$ (where f' is the bijection from $T(k', r' - 1)$). Let $G(k', r')$ be the graph induced by the decomposition.

It is easy to see how to obtain the new bijection f on the whole graph such that it satisfies the invariant. It is also not hard to see that each bag contains at most $k' + 1$ vertices. For the second invariant, let $P_1 = t_1, \ldots, t_m$ be some root-leaf path in $T(k', r' - 1)$ which is ordered such that $L(f(t_1)) < \ldots < L(f(t_m))$. Let $v = f(t_m)$ be the unique vertex in the leaf bag in which P_1 ends. By the same argument as above, this vertex has many neighbours s' such that $f^{-1}(s')$ is a root of a copy of $T(k' - 1, r')$. One of them must be larger than v. In appropriate copy we find a path P_2 with the above property by assumption. We attach the paths to find the path $P = t_1 \ldots t_\ell$ in $T(k', r')$.

We finally show that $\text{WReach}_{r'}[G(k',r'),L,f(t_\ell)] = c$. This is again shown by an easy induction. Using the notation of the proof of Theorem 4.2, we observe that the graph G_1 is isomorphic to $G(k',r'-1)$ in $G(k',r')$ and G_2 is isomorphic to $G(k'-1,r')$. Furthermore we observe that the number of vertices reached in these graphs are exactly $w(k',r'-1)$ and $w(k'-1,r')$, so that the upper bound is matched. Similarly one shows that $\text{wcol}_{r'}(G(k,r)) = \binom{r'+k}{k}$. The theorem follows by letting $G_r^k := G(k,r)$. □

It is proven in [11,15] that for every graph G, $\text{wcol}_r(G) \leq (\text{col}_r(G))^r$. To our knowledge, there is no example in the literature that verifies the exponential gap between wcol_r and col_r. As $\text{col}_r(G) \leq \text{tw}(G)$ and G_r^k contains a $k+1$-clique, Theorem 4.3 provides an example that is close to an affirmative answer for arbitrarily large generalised colouring numbers, in a rather uniform manner.

Corollary 4.4. *For every $k \geq 1$, $r \geq 1$, there is a graph G_r^k such that for all $1 \leq r' \leq r$ we have $\text{col}_{r'}(G_r^k) = k+1$ and $\text{wcol}_{r'}(G_r^k) \geq \left(\frac{\text{col}_{r'}(G_r^k)}{r'}\right)^{r'}$.*

5   High-Girth Regular Graphs

We want to explore if assuming constant expansion for a graph class (such as classes excluding a topological minor) results to polynomial colouring numbers. To this end, we would expect such classes to have exponential weak colouring numbers, but it is much more unclear what the case for their colouring numbers is (which can have an exponential gap with the weak colouring number). Surprisingly, we prove that, in fact, even classes of bounded degree (which are of the simplest classes that can exclude a topological minor) can't have polynomial colouring numbers. For this section, we let $n := |V(G)|$.

Theorem 5.1. *Let G be a d-regular graph of girth at least $4g+1$, where $d \geq 7$. Then for every $r \leq g$, $\text{col}_r(G) \geq \frac{d}{2}\left(\frac{d-2}{4}\right)^{2^{\lfloor \log r \rfloor}-1}$.*

Proof. For an ordering L of G, let $R_r(v) = \text{SReach}_r[G,L,v] \setminus \text{SReach}_{r-1}[G,L,v]$ and $U_r = \sum_{v \in V(G)} |R_r(v)|$.

Suppose that $r \leq g$ and notice that for $u, w \in R_r(v)$, we have that either $u \in R_{2r}(w)$ or $w \in R_{2r}(u)$. Therefore, every vertex $v \in V(G)$ contributes at least $\binom{|R_r(v)|}{2}$ times to U_{2r}. Moreover, since $r \leq g$, for every u, w with $u \in R_{2r}(w)$ there is at most one vertex $v \in V(G)$ such that $u, w \in R_r(v)$ (namely the middle vertex of the unique (u,v)-path of length $2r$ in G). It follows that for every $r \leq g$, $U_{2r} \geq \sum_{v \in V(G)} \binom{|R_r(v)|}{2} = \frac{1}{2}\sum_{v \in V(G)} |R_r(v)|^2 - \frac{1}{2}\sum_{v \in V(G)} |R_r(v)| \geq \frac{1}{2n}\left(\sum_{v \in V(G)} |R_r(v)|\right)^2 - \frac{1}{2}U_r = \frac{1}{2n}U_r^2 - \frac{1}{2}U_r$ where for the second inequality we have used the Cauchy-Schwarz inequality.

Let $c_r = \frac{U_r}{n}$. Then for every $r \leq g$, we obtain $c_{2r} \geq \frac{1}{2}c_r(c_r - 1)$. But, $U_1 = \sum_{v \in V(G)} |\text{SReach}_1[G,L,v] \setminus \{v\}| = \frac{1}{2}dn$, so that $c_1 = \frac{d}{2} > 3$, since $d \geq 7$. By induction and because $c_{2r} \geq \frac{1}{2}c_r(c_r-1)$, for every $r = 2^{r'} \leq g$ we have

$c_{2r} \geq c_r \geq 3$, therefore $c_r \geq c_1 = \frac{d}{2}$. Again because $c_{2r} \geq \frac{1}{2}c_r(c_r - 1)$, for every $r = 2^{r'} \leq g$ we have $c_{2r} \geq \frac{1}{2}c_r^2 - \frac{1}{2}c_r \geq \frac{1}{2}c_r^2 - \frac{1}{4}c_r^2 = \frac{d-2}{2d}c_r^2$. Then for every $r = 2^{r'} \leq g$, it easily follows that $c_r \geq \frac{d}{2}\left(\frac{d-2}{4}\right)^{r-1}$.

Finally, let $C_r = \frac{1}{n}\sum_{v \in V(G)}|\text{SReach}_r[G, L, v]|$. Then, $C_r = \sum_{i=1}^{r} c_i$, hence $C_r \geq c_{2\lfloor \log r \rfloor} \geq \frac{d}{2}\left(\frac{d-2}{4}\right)^{2^{\lfloor \log r \rfloor}-1}$, and hence for every $r \leq g$ there exists a vertex $v_r \in V(G)$ such that $|\text{SReach}_r[G, L, v_r]| \geq \frac{d}{2}\left(\frac{d-2}{4}\right)^{2^{\lfloor \log r \rfloor}-1}$. Since L was arbitrary, the theorem follows.   □

Unfortunately, our proof above makes sense only if $d \geq 7$, which is also best possible with this approach, since for $d \leq 6$, we have $c_1 \leq 3$. Then for the recurrence relation $c_{2r} = \frac{1}{2}c_r^2 - \frac{1}{2}c_r$, we get $c_{2^i} \leq c_{2^{i-1}}$ for every i and we clearly cannot afford to have c_{2^i} non-increasing. Somewhat better constants can be achieved if in the estimation of c_r one uses that $c_{2^i} \geq c_{2^{i_0}}$, for $i \geq i_0 > 0$, instead of the relation $c_{2^i} \geq c_1$, as in our proof. Since $d \geq 7$ would be still the best that we would be able to do, we adopted the simpler approach for easier readability.

Actually, by combining a known result for the ∇_r (which stands for the expansion function defined through the minor resolution, instead of the topolological minor resolution that we use here for our purposes) of high-girth regular graphs ([4,15] Exercise 4.2) and ([20], Lemma 3.3), we get exponential lower bounds for the weak colouring number of high-girth d-regular graphs, already for $d \geq 3$. In particular, for a 3-regular graph G of high enough girth, $\text{wcol}_r(G) \geq 3 \cdot 2^{\lfloor r/4 \rfloor - 1}$. The methods above can be extended to get appropriate bounds in terms of their degree for regular graphs of higher degree, but by adopting a more straightforward approach, we get better bounds for high-girth d-regular graphs for $d \geq 4$.

Theorem 5.2. *Let G be a d-regular graph of girth at least $2g + 1$, where $d \geq 4$. Then for every $r \leq g$,*

$$\text{wcol}_r(G) \geq \frac{d}{d-3}\left(\left(\frac{d-1}{2}\right)^r - 1\right).$$

Proof. Let L be an ordering of G. For $u, v \in V(G)$ with $d(u, v) \leq r$, let P_{uv} be the unique (u, v)-path of length at most r, due to the girth of G. Let $S_r = \sum_{v \in V(G)}|Q_r(v)|$, where $Q_r(v) = \text{WReach}_r[G, L, v] \setminus \text{WReach}_{r-1}[G, L, v]$. For $r \leq g - 1$, a vertex $u \in Q_r(v)$ and $w \in N(v) \setminus V(P_{uv})$, it holds that either $w \in Q_{r+1}(u)$ or $u \in Q_{r+1}(w)$. Notice that $|N(v) \setminus V(P_{uv})| = d - 1$ and that P_{vu} and P_{uw} are unique. Therefore, every pair of vertices v, u with $u \in Q_r(v)$ corresponds to at least $d - 1$ pairs of vertices u, w with $u \in Q_{r+1}(w)$ or $w \in Q_{r+1}(u)$ and hence contributes at least $d - 1$ times to S_{r+1}. Since every path of length $r + 1$ contains exactly two subpaths of length r, we have for every $r \leq g - 1$ that $2S_{r+1} \geq (d-1)S_r$. Let $w_r = \frac{S_r}{n}$. Then, for every $r \leq g - 1$ we have $w_{r+1} \geq \frac{d-1}{2}w_r$.

But, $\sum_{v \in V(G)}|\text{WReach}_1[G, L, v] - v| = \frac{1}{2}dn$, so that $w_1 = \frac{d}{2}$. It easily follows that for every $r \leq g$, $w_r \geq \frac{d}{2}\left(\frac{d-1}{2}\right)^{r-1}$.

Finally, let $W_r = \frac{1}{n}\sum_{v \in V(G)} |\text{WReach}_r[G, L, v]|$. Then, $W_r = \sum_{i=1}^r w_i \geq \sum_{i=1}^r \frac{d}{2}\left(\frac{d-1}{2}\right)^{i-1} = \frac{d}{d-3}\left(\left(\frac{d-1}{2}\right)^r - 1\right)$, and hence for every $r \leq g$ there exists a vertex $v_r \in V(G)$ such that $\text{WReach}_r[G, L, v_r] \geq \frac{d}{d-3}\left(\left(\frac{d-1}{2}\right)^r - 1\right)$. Since L was arbitrary, the theorem follows.   □

Remark 5.3. *Notice that for every d-regular graph G and every radius r, we have $\text{adm}_r(G) \leq \Delta(G) + 1 = d + 1$, so by Theorem 5.1 for every $d \geq 7$ and every $r \leq g$, the d-regular graphs of girth at least $2g + 1$ verify the exponential gap between $\text{adm}_r, \Delta(G)$ and $\text{col}_r, \text{wcol}_r$ of the known relations from Sect. 2.*

6 Neighbourhood Covers

Neighbourhood covers of small radius and small size play a key role in the design of many data structures for distributed systems. For references about neighbourhood covers, we refer the reader to [1].

For $r \in \mathbb{N}$, an *r-neighbourhood cover* $\mathcal{X}$ of a graph G is a set of connected subgraphs of G called *clusters*, such that for every vertex $v \in V(G)$ there is some $X \in \mathcal{X}$ with $N_r(v) \subseteq X$.

The *radius* $\text{rad}(\mathcal{X})$ of a cover $\mathcal{X}$ is the maximum radius of any of its clusters. The *degree* $d^{\mathcal{X}}(v)$ of v in $\mathcal{X}$ is the number of clusters that contain v. A class $\mathcal{C}$ *admits sparse neighbourhood covers* if for every $r \in \mathbb{N}$, there exists $c \in \mathbb{N}$ such that for all $\varepsilon > 0$, there is $n_0 \in \mathbb{N}$ such that for all $G \in \mathcal{C}$ of order at least n_0, there exists an r-neighbourhood cover of radius at most $c \cdot r$ and degree at most $|V(G)|^\varepsilon$. For any graph G, one can construct an r-neighbourhood cover of radius $2r - 1$ and degree $2k \cdot |V(G)|^{1/r}$ and asymptotically these bounds cannot be improved [19].

Theorem 6.1 (Theorem 16.2.4 of [17,19]). *For every r and $k \geq 3$, there exist infinitely many graphs G for which every r-neighbourhood cover of radius at most k has degree $\Omega(|V(G)|^{1/k})$.*

For restricted classes of graphs, better covers exist. The most general results are that a class excluding a complete graph on t vertices as a minor admits an r-neighbourhood cover of radius $\mathcal{O}(t^2 \cdot r)$ and degree $2^{\mathcal{O}(t)} t!$ [1] and the following result from [9].

Theorem 6.2 [9]. *Let $\mathcal{C}$ be a nowhere dense class of graphs. There is a function f such that for all $r \in \mathbb{N}$ and $\varepsilon > 0$ and all graphs $G \in \mathcal{C}$ with $|V(G)| \geq f(r, \varepsilon)$, there exists an r-neighbourhood cover of radius at most 2r and maximum degree at most $|V(G)|^\varepsilon$. More precisely, if $\text{wcol}_{2r}(G) = d$, then there exists an r-neighbourhood cover of radius at most 2r and maximum degree at most d.*

We show that for monotone classes the converse is also true. We first observe that the lower bounds in Theorem 6.1 come from a well known somewhere dense class.

Lemma 6.3. *Let $d \geq 1, k \geq 2$ and let G be a graph of girth at least $k + 1$ and edge density at least d. Then every 1-neighbourhood cover of radius at most k has degree at least d.*

Lemma 6.4 [13]. *Let $r \geq 5$. There are infinitely many graphs G of girth at least $4r$ with edge density at least $c_0 \cdot |V(G)|^{1/(3(r-1))}$ for some constant $c_0 > 0$.*

Theorem 6.5. *If $\mathcal{C}$ is somewhere dense and monotone, then $\mathcal{C}$ does not admit sparse neighbourhood covers.*

Proof. Let $\mathcal{C}$ be somewhere dense. Then for some s, all graphs H are topological depth-s minors of a graph $G \in \mathcal{C}$. Assume towards a contradiction that $\mathcal{C}$ admits a sparse neighbourhood cover. Then for every $G \in \mathcal{C}$ there is an $r \cdot s$-neighbourhood cover of radius $c \cdot s \cdot r$ (for some constant c) which for every $\varepsilon > 0$ has degree at most $|V(G)|^\varepsilon$ if G is sufficiently large. Fix some $r \geq 5$.

Claim 1. If an s-subdivision of H admits an $r \cdot s$-neighbourhood cover of radius $c \cdot r \cdot s$ and degree d, then H admits an r-neighbourhood cover of radius $c \cdot r \cdot s$ and degree d.

Proof. Let G be an s-subdivision of H and let $\mathcal{X}$ be an $r \cdot s$-neighbourhood cover of G. Let $\mathcal{Y}$ be the *projected cover* which for every $X \in \mathcal{X}$ has a cluster
$$Y(X) := X \cap V(H)$$
Then $\mathcal{Y}$ is an r-neighbourhood cover of radius crs and degree d: Clearly, every $Y(X)$ is connected and has radius at most crs. Let $v \in V(G)$. There is a cluster $X \in \mathcal{X}$ such that $N_{rs}^G(v) \subseteq X$. Then $N_r^H(v) = N_{rs}^G(v) \cap V(H) \subseteq X \cap V(H) = Y(X)$. Finally, the degree of $\mathcal{Y}$ is at most d, as every vertex v of H is exactly in those clusters $Y(X)$ with $v \in X$. This proves the claim.   ⊣

Let H be a large graph of girth greater than $c \cdot r \cdot s$ with edge density $d = c_0 \cdot |V(H)|^{1/(crs)}$ for some constant c_0. Such H exists by Lemma 6.4 and H does not admit an $r \cdot s$-neighbourhood cover of radius $c \cdot r \cdot s$ and degree d by Lemma 6.3. As $\mathcal{C}$ is monotone, an s-subdivision of H is a graph $G \in \mathcal{C}$ with $|V(G)| \leq |V(H)| + s \cdot |E(H)| \leq 2c_0 s |V(H)|^{1+1/(crs)}$.

By assumption, G admits an $r \cdot s$-neighbourhood cover of radius at most $c \cdot r \cdot s$ and degree at most $|V(G)|^\varepsilon$ for $\varepsilon = 1/(2crs)$ if G is large enough. It follows from Claim 1 that H has a cover of radius $c \cdot r \cdot s$ and degree at most

$$|V(G)|^\varepsilon \leq \left(2c_0 s |V(H)|^{1+1/(crs)} \right)^\varepsilon = (2c_0 s)^\varepsilon \cdot |V(H)|^{\varepsilon + \varepsilon/(crs)}$$
$$< c_0 |V(H)|^{2\varepsilon} = c_0 |V(H)|^{1/(crs)}$$

for sufficiently large H. A contradiction.   □

We were informed by Nešetřil and Ossona de Mendez [14] that they found a similar characterisation also for classes of bounded expansion.

7 The Complexity of Computing $\mathrm{wcol}_r(G)$

Unlike computing the degeneracy of a graph G, i.e. $\mathrm{wcol}_1(G) + 1$, deciding whether $\mathrm{wcol}_r(G) = k$ turns out to be NP-complete for all $r \geq 3$. The case $r = 2$ remains an open question. Clearly, the problem is in NP, hence it remains to show NP-hardness. The proof is a straightforward modification of a proof of Pothen [18], showing that computing a minimum elimination tree height problem is NP-complete. It is based on a reduction from the NP-complete problem BALANCED COMPLETE BIPARTITE SUBGRAPH (BCBS, problem GT24 of [7]): given a bipartite graph G and a positive integer k, decide whether there are two disjoint subsets $W_1, W_2 \subseteq V(G)$ such that $|W_1| = |W_2| = k$ and such that $u \in W_1, v \in W_2$ implies $\{u, v\} \in E(G)$. For a graph G, let $\bar{G}$ be its complement graph.

Lemma 7.1. *Let $G = (V_1 \cup V_2, E)$ be a bipartite n-vertex graph and let $k \in \mathbb{N}$. Then G has a balanced complete bipartite subgraph with partitions W_1, W_2 of size k if and only if $\mathrm{wcol}_r(\bar{G}) = \mathrm{wcol}_3(\bar{G}) \leq n - k$ for all $r \geq 3$.*

Proof. $\bar{G}$ is the complement of a bipartite graph, i.e. V_1 and V_2 induce complete subgraphs in $\bar{G}$ and there are possibly further edges between vertices of V_1 and V_2. Thus, for any two vertices u, v which are connected in $\bar{G}$ by a path P, there is a subpath of P between u and v of length at most 3. Hence $\mathrm{wcol}_r(\bar{G}) = \mathrm{wcol}_3(\bar{G})$ for any $r \geq 3$ and it suffices to show that G has a balanced complete bipartite subgraph with partitions W_1, W_2 of size k if and only if $\mathrm{wcol}_3(\bar{G}) = n - k$.

First assume that there are sets $W_1 \subseteq V_1, W_2 \subseteq V_2$ with $|W_1| = |W_2| = k$ and such that for all $u \in W_1, v \in W_2$ there is an edge $\{u, v\} \in E(G)$. Let L be some order which satisfies $L(u) < L(v)$ if $u \in V(\bar{G}) \setminus (W_1 \cup W_2)$ and $v \in W_1 \cup W_2$ and $L(v) < L(w)$ if $v \in W_1$ and $w \in W_2$. Then any vertex from $V(\bar{G}) \setminus (W_1 \cup W_2)$ weakly reaches at most $n - 2k$ vertices and any vertex from W_i for $1 \leq i \leq 2$ weakly reaches at most $n - k$ vertices.

Now let L be an order with $\mathrm{WReach}_3[\bar{G}, L, v] \leq n - k$ for all $v \in V(G)$. Assume without loss of generality that $V(G) = \{v_1, v_2, \ldots, v_n\}$ with $L(v_i) < L(v_{i+1})$ for all $i < n$. Denote by $\bar{G}_i$ the subgraph $\bar{G}[\{v_i, \ldots, v_n\}]$ and let $V_1^i := V(\bar{G}_i) \cap V_1$ and $V_2^i := V(\bar{G}_i) \cap V_2$. Let $\ell \geq 1$ be minimal such that there is no edge between V_1^ℓ and V_2^ℓ in $\bar{G}$. It exists because one of V_1^n or V_2^n is empty. Clearly, V_1^ℓ and V_2^ℓ induce a complete bipartite graph in G. Let $j_1 := |V_1^\ell|$ and $j_2 := |V_2^\ell|$. We show that $j_1, j_2 \geq k$. It is easy to see that $\mathrm{WReach}_3[\bar{G}, L, w_1] \leq \ell + j_1$ for the maximal element $w_1 \in V_1^\ell$ and $\mathrm{WReach}_3[\bar{G}, L, w_2] \leq \ell + j_2$ for the maximal element $w_2 \in V_2^\ell$. We have $j_1 + j_2 = n - \ell$ and, without loss of generality, $\ell + j_1 \leq \ell + j_2 \leq n - k$. Hence $j_1 \leq j_2 \leq n - \ell - k = j_1 + j_2 - k$, which implies both $j_1 \geq k$ and $j_2 \geq k$. $\square$

The above reduction is polynomial time computable, so we obtain the following theorem.

Theorem 7.2. *Given a graph G and $k, r \in \mathbb{N}, r \geq 3$, it is NP-complete to decide whether $\mathrm{wcol}_r(G) = k$.*

References

1. Abraham, I., Gavoille, C., Malkhi, D., Wieder, U.: Strong-diameter decompositions of minor free graphs. In: Proceedings of the Nineteenth Annual ACM Symposium on Parallel Algorithms and Architectures, pp. 16–24. ACM (2007)
2. Dawar, A., Kreutzer, S.: Domination problems in nowhere-dense graph classes. In: Kannhan, R., Kumar, K.N. (eds.) Proceedings of the 29th Conference on Foundations of Software Technology and Theoretical Computer Science. LIPIcs, vol. 4, pp. 157–168. Schloss Dagstuhl - Leibniz-Zentrum fuer Informatik (2009)
3. Diestel, R.: Graph Theory, 3rd edn. Springer, Heidelberg (2005)
4. Diestel, R., Rempel, C.: Dense minors in graphs of large girth. Combinatorica 25(1), 111–116 (2004)
5. Dvořák, Z.: Constant-factor approximation of the domination number in sparse graphs. Eur. J. Comb. 34(5), 833–840 (2013)
6. Dvořák, Z., Král', D., Thomas, R.: Deciding first-order properties for sparse graphs. J. ACM (2013). to appear
7. Garey, M.R., Johnson, D.S.: Computers and Intractability, vol. 29. W.H. freeman, New York (2002)
8. Grohe, M., Kreutzer, S., Siebertz, S.: Characterisations of nowhere dense graphs. In: Seth, A., Vishnoi, N.K. (eds.) Proceedings of the 32nd IARCS Annual Conference on Foundations of Software Technology and Theoretical Computer Science. LIPIcs, vol. 24, pp. 21–40. Schloss Dagstuhl - Leibniz-Zentrum fuer Informatik (2013)
9. Grohe, M., Kreutzer, S., Siebertz, S.: Deciding first-order properties of nowhere dense graphs. In: Proceedings of the 46th ACM Symposium on Theory of Computing, pp. 89–98 (2014)
10. Kierstead, H.A., Trotter, W.T.: Planar graph coloring with an uncooperative partner. DIMACS Ser. Discrete Math. Theor. Comput. Sci. 9, 85–93 (1993)
11. Kierstead, H.A., Yang, D.: Orderings on graphs and game coloring number. Order 20(3), 255–264 (2003)
12. Kreutzer, S.: Algorithmic meta-theorems. In: Esparza, J., Michaux, C., Steinhorn, C. (eds.) Finite and Algorithmic Model Theory, London Mathematical Society Lecture Note Series, chap. 5, pp. 177–270. Cambridge University Press (2011)
13. Lazebnik, F., Ustimenko, V.A., Woldar, A.J.: A new series of dense graphs of high girth. Bull. Am. Math. Soc. 32(1), 73–79 (1995)
14. Nešetřil, J., Ossona de Mendez, P.: Characterization of nowhere dense classes and classes with bounded expansion by coverings. (to appear)
15. Nešetřil, J., Ossona de Mendez, P.: Sparsity. Springer, Heidelberg (2012)
16. Nešetřil, J., Ossona de Mendez, P.: On nowhere dense graphs. Eur. J. Comb. 32(4), 600–617 (2011)
17. Peleg, D.: Distributed computing. SIAM Monogr. Discrete Math. Appl. 5 (2000)
18. Pothen, A.: The complexity of optimal elimination trees. Pennsylvania State University, Department of Computer Science (1988)
19. Thorup, M., Zwick, U.: Approximate distance oracles. J. ACM (JACM) 52(1), 1–24 (2005)
20. Zhu, X.: Colouring graphs with bounded generalized colouring number. Discrete Math. 309(18), 5562–5568 (2009)

Parity Linkage and the Erdős-Pósa Property of Odd Cycles Through Prescribed Vertices in Highly Connected Graphs

Felix Joos[✉]

Institut für Optimierung und Operations Research,
Universität Ulm, Helmholtzstraße 18, 89081 Ulm, Germany
felix.joos@uni-ulm.de

Abstract. We show the following for every sufficiently connected graph G, any vertex subset S of G, and given integer k: there are k disjoint odd cycles in G containing each a vertex of S or there is set X of at most $3k - 3$ vertices such that $G - X$ does not contain any odd cycle that contains a vertex of S. We prove this via an extension of Kawarabayashi and Reed's result about parity-k-linked graphs (Combinatorica 29, 215–225). From this result it is easy to deduce several other well known results about the Erdős-Pósa property of odd cycles in highly connected graphs. This strengthens results due to Thomassen (Combinatorica 21, 321–333), and Rautenbach and Reed (Combinatorica 21, 267–278), respectively. Furthermore, we consider algorithmic consequences of our results.

Keywords: Cycles · Packing · Covering

AMS Subject Classification: 05C70

1 Introduction

We consider only finite and simple graphs. A family $\mathcal{F}$ of graphs has the *Erdős-Pósa property* if there is a function $f : \mathbb{N} \mapsto \mathbb{N}$ such that for every positive integer k and every graph G, the graph G contains k disjoint subgraphs from $\mathcal{F}$ or there is a set X of vertices of G with $|X| < f(k)$ such that $G - X$ contains no subgraph from $\mathcal{F}$. This notion has been introduced because Erdős and Pósa proved that the family of cycles has the Erdős-Pósa property [7]. It is one facet of the duality between packing and covering in graphs, which is one of the most fundamental concepts in graph theory. There is a huge number of results about families of graphs which have the Erdős-Pósa property. For example, Birmelé, Bondy, and Reed [1] verified it for the family of cycles of length at least ℓ for some integer ℓ and Robertson and Seymour [19] showed it for the family of graphs that contain a fixed planar graph as a minor.

In contrast, the family of odd cycles does not have the Erdős-Pósa property. In particular, there is a sequence of graphs $(G_n)_{n \in \mathbb{N}}$ such that G_n does not

© Springer-Verlag Berlin Heidelberg 2016
E.W. Mayr (Ed.): WG 2015, LNCS 9224, pp. 339–350, 2016.
DOI: 10.1007/978-3-662-53174-7_24

contain two disjoint odd cycles, all odd cycles are of length $\Omega(\sqrt{n})$, and every set that intersects all odd cycles has cardinality at least $\Omega(\sqrt{n})$ [17].

However, Thomassen [21] proved that the family of odd cycles has the Erdős-Pósa property if we restrict ourselves to graphs with high connectivity. Rautenbach and Reed [16] improved Thomassen's connectivity bound from a double-exponential to linear one. Later, Kawarabayashi and Reed [11] lowered this bound to $24k$, and Kawarabayashi and Wollan [13] improved this further to $\frac{31}{2}k$.

More than 50 years ago, Dirac [5] showed that in every k-connected graph G, there is a cycle containing any prescribed set of k vertices. Later, Bondy and Lovász [2] extended Dirac's result and proved among other results along this line that for every k-connected non-bipartite graph G, there is an odd cycle containing any prescribed set of $k - 1$ vertices.

If one asks for many disjoint cycles through a prescribed set S of vertices it is natural to start with disjoint cycles each containing at least one element of S. We call such cycles S-cycles. Pontecorvi and Wollan [15] showed that the class of S-cycles has the Erdős-Pósa property with $f(k) = O(k \log k)$, which improved the quadratic bound from [10]. Bruhn et al. [3] proved that the class of all S-cycles of length at least ℓ has the Erdős-Pósa property with $f(k, \ell) = O(\ell k \log k)$. For $S = V(G)$, these results equal the Erdős-Pósa property for cycles and cycles of length at least ℓ, respectively.

Although, the Erdős-Pósa property does not hold for odd cycles, it is proved in [9] that a half-integral version for the Erdős-Pósa property of odd S-cycles holds. This generalizes a result of Reed [17], who proved the case $S = V(G)$.

In this paper we continue the study of S-cycles by showing the following theorem. We say a set of vertices X is an *odd cycle cover* of G if $G - X$ is bipartite. As mentioned above, the results in [11,13,16] show that linear connectivity ensures that a graph has k vertex disjoint odd cycles or an odd cycle cover of size $2k - 2$. We show that a sufficiently connected graph has k vertex disjoint odd S-cycles for any prescribed vertex set S of at least k vertices or has an odd cycle cover of size $3k - 3$.

Moreover, the bound of $3k - 3$ is tight for any connectivity and this can be seen as follows. Let G arise from a large complete bipartite graph with bipartition (A, B) by adding the edges of a clique on $2k - 1$ vertices to A and the edges of a clique on k vertices to B. Let S be the set of k vertices in B containing the k-clique. There do not exist k disjoint odd S-cycles, nor an odd cycle cover of size $3k - 4$ in G.

Theorem 1. *For any integer k, any $50k$-connected graph G, and any subset S of at least k vertices of G, at least one of the following statements hold:*

1. *G contains k disjoint odd S-cycles.*
2. *There is a set X with $|X| \leq 3k - 3$ such that $G - X$ is bipartite.*

In fact, we prove more detailed results than Theorem 1. These results in turn imply the already known result that every $50k$-connected graph G that fails to have k disjoint odd cycles contains an odd cycle cover of size $2k - 2$, which corresponds to the case $S = V(G)$ in [11,13,16].

It is not difficult to see that there are arbitrarily highly connected graphs that contain k disjoint odd cycles and an odd cycle cover of size less than $2k - 2$. In this paper, we present an equivalent condition for $50k$-connected graphs for having k disjoint odd cycles and deduce the known Erdős-Pósa-type result from this result.

The following decision problem is closely related to Theorem 1:

> *Given an integer k, a $50k$-connected graph G and a set of vertices S.*
> *Does G contain k disjoint odd S-cycles?*

We present an algorithm running in $O(f(k)nm)$-time answering this question, where n and m denote the number of vertices and edges of G, respectively.

Now, we turn our attention to linkage problems which appear to have a very close relation to the results mentioned above. A graph is k-*linked* if for every set of distinct vertices $\{s_1, \ldots, s_k, t_1, \ldots, t_k\}$ there are disjoint paths $P_1, \ldots, P_k$ such that P_i connects s_i and t_i. Moreover, a graph G is *parity-k-linked* if it is k-linked and we can additionally specify whether the length of each P_i should be odd or even for every $1 \le i \le k$.

There are several results of the form if G is $g_1(k)$-connected, then G is k-linked. The best result is due to Thomas and Wollan [20] who proved that $g_1(k) = 10k$ suffices. They even proved the following stronger result.

Theorem 2 ([20]). *Every $2k$-connected graph G with at least $5k|V(G)|$ edges is k-linked.*

Analogously, there are results of the form if G is $g_2(k)$-connected and without an odd cycle cover of size $4k - 4$, then G is parity-k-linked. In particular, Kawarabayashi and Reed [11] proved the following.

Theorem 3 ([11]). *Every $50k$-connected graph without an odd cycle cover of size $4k - 4$ is parity-k-linked.*

The condition of having no small odd cycle cover is necessary and best possible – there are graphs of arbitrarily high connectivity and with an odd cycle cover of size $4k - 4$ that are not parity-k-linked. For example, consider a large complete bipartite graph G with bipartition (A, B) where we add to A the edges of a clique on $2k - 1$ vertices and we add to B the edges of a clique on $2k$ vertices minus a perfect matching.

One can apply Theorem 3 almost directly to obtain that every $50k$-connected graph G without an odd cycle cover of size $4k - 4$ has k disjoint odd S-cycles for any set S of at least k vertices. However, the bound on the size of the odd cycle cover is not optimal. In this paper we prove a stronger version of Theorem 3, reprove the Erdős-Pósa property for odd cycles for $50k$-connected graphs, and as the main result of this paper, we prove Theorem 1.

In addition, we prove several results on the way that may be of independent interest.

The paper is organized as follows. In Sect. 2 we deal with the results concerning the parity-k-linkage, in Sect. 3 we prove the results about the Erdős-Pósa property for odd S-cycles, and in Sect. 4 we discuss algorithmic consequences of our results.

2   Highly Parity Linked Graphs

In the next theorem we explicitly characterize the obstruction for a $50k$-connected graph and a set $S = \{s_1, \ldots, s_k, t_1, \ldots, t_k\}$ of $2k$ distinct vertices for not having k disjoint $P_1, \ldots, P_k$ paths of prescribed length parity where P_i connects s_i and t_i.

Before we state the theorem, we introduce some definitions. A *partition* (A, B) of G is partition of the vertex set into two sets A and B. For a partition (A, B) of G, we denote by $G_{A,B}$ the graph $G[A] \cup G[B]$. A partition (A, B) of G is *nice* if there is a minimum odd cycle cover X of G for which $(A \setminus X, B \setminus X)$ is a bipartition of $G - X$ such that a vertex of X is in A if and only if it has more neighbors in $B \setminus X$ than in $A \setminus X$. We say that a minimum odd cycle cover X *induces* a nice partition (A, B) of G if $(A \setminus X, B \setminus X)$ is a bipartition such that a vertex of X is in A if and only if it has more neighbors in $B \setminus X$ than in $A \setminus X$. Note that every minimum odd cycle cover induces a nice partition.

Let (A, B) be nice partition of G and let $S = \{s_1, \ldots, s_k, t_1, \ldots, t_k\}$ be a set of $2k$ distinct vertices. A *parity breaking matching* for S (with respect to the partition (A, B)) is a matching M such that $M \subseteq E(G_{A,B})$ and there is no edge $rr' \in M$ with $r \in \{s_i, t_i\}$ and $r' \in \{s_j, t_j\}$ for $i \neq j$.

Theorem 4. *Let $k \in \mathbb{N}$. Let G be a $50k$-connected graph and let $S = \{s_1, \ldots, s_k, t_1, \ldots, t_k\}$ be a set of $2k$ distinct vertices. Exactly one of the following two statements holds.*

1. *G contains k disjoint paths $P_1, \ldots, P_k$ of any prescribed parity such that P_i connects s_i and t_i.*
2. *For all nice partitions of G, there is no parity breaking matching for S of size k.*

There are plenty of consequences of Theorem 4. Firstly, it is easy to see that it implies Theorem 5.

Theorem 5. *Let $k \in \mathbb{N}$ and let G be a $50k$-connected graph. Exactly one of the following two statements holds.*

1. *G is k-parity linked.*
2. *There is a set $\{s_1, \ldots, s_k, t_1, \ldots, t_k\} \subset V(G)$ such that for all nice partitions of G, there is no parity breaking matching of size k.*

Secondly, later we deduce Theorem 3. The third consequence (Corollary 1) shows that the bound "$4k - 4$" in Theorem 3 can be strengthened to "$2k - 2$" if $\{s_1, \ldots, s_k, t_1, \ldots, t_k\}$ is an independent set. Note that both bounds "$4k - 4$" and "$2k - 2$" are best possible, respectively. As a fourth consequence we prove the Erdős-Pósa property for odd S-cycles (Theorem 1) in Sect. 3.

We say that G is *parity-k-linked restricted to independent sets* if for every independent set of $2k$ vertices $\{s_1, \ldots, s_k, t_1, \ldots, t_k\}$, there are disjoint paths $P_1, \ldots, P_k$ such that P_i connects s_i and t_i and we can choose whether the length of P_i is odd or even.

Corollary 1. *Let $k \in \mathbb{N}$ and let G be a $50k$-connected graph. At least one of the following statements holds.*

1. *G is parity-k-linked restricted to independent sets.*
2. *There is a set X of $2k - 2$ vertices such that $G - X$ is bipartite.*

Next, we mention two results needed in the proof of Theorem 4. The first result is basically due to Mader and there is a slightly improved version in the textbook of Diestel on page 13 [4].

Lemma 1 (Mader [14]). *If G is a graph such that $\frac{|E(G)|}{|V(G)|} \geq 2k$, then G contains a $(k + 1)$-connected graph H such that $\frac{|E(H)|}{|V(H)|} \geq \frac{|E(G)|}{|V(G)|} - k$.*

Hence, if $\delta(G) \geq 12k$, then G contains a $2k$-connected subgraph H such that $\frac{|E(H)|}{|V(H)|} \geq 5k$. Using Theorem 2, this implies in turn that the subgraph H is k-linked.

Another result which is used in the proof of Theorem 4 is due to Geelen et al. For a graph G and a set of vertices Z, a Z-*path* is a path P such that $V(P) \cap Z$ are exactly the end vertices of P.

Theorem 6 (Geelen et al. [8]). *For any set Z of vertices of a graph G and any positive integer ℓ at least one of the following statements holds;*

- *there are ℓ disjoint odd Z-paths or*
- *there is a vertex set X of order at most $2\ell - 2$ such that $G - X$ contains no odd Z-path.*

We proceed with the proof of Theorem 4 that leads to the several consequences mentioned above.

Proof (Proof of Theorem 4). Suppose the first statement holds. Let (A, B) be some nice partition of G. Let $P_1, \ldots, P_k$ be disjoint paths where P_i is a s_i, t_i-path and we choose the parity of P_i to be even if exactly one vertex of $\{s_i, t_i\}$ belongs to A and odd otherwise. Thus P_i contains at least one edge m_i in $E(G_{A,B})$. Therefore, $\{m_1, \ldots, m_k\}$ is a parity breaking matching for S of size k.

Next, suppose the second statement does not hold; that is, there is a nice partition (A, B) of G with a parity breaking matching $M = \{m_1, \ldots, m_k\}$. If an edge of M covers a vertex of $\{s_i, t_i\}$, let $m_i = x_i y_i$ be this edge and choose x_i, y_i such that $s_i = x_i$ or $t_i = y_i$. Let X be a minimum odd cycle cover of G that induces the nice partition (A, B).

Suppose first that $|X| < 8k$. By the definition of a nice partition and the fact that G is $50k$-connected, we know that every vertex in $a \in A$ has at least $20k$ neighbors in B and so at least $12k$ in $B \setminus X$. The same holds vice versa for the vertices in B. Therefore, we can find a set of $4k$ distinct vertices $\bigcup_{i=1}^{k} \{s_i', t_i', x_i', y_i'\} \subset V(G) \setminus (X \cup \bigcup_{i=1}^{k} \{s_i, t_i, x_i, y_i\})$ such that z' is a neighbor of z for $z \in \bigcup_{i=1}^{k} \{s_i, t_i, x_i, y_i\}$ (symbolically written) and exactly one vertex of the set $\{z_i, z_i'\}$ belongs to A.

Let $G' = G - (X \cup \bigcup_{i=1}^{k} \{s_i, t_i, x_i, y_i\})$. Thus G' is $38k$-connected and bipartite. In addition, by Theorem 2, we obtain that G' is $2k$-linked.

For every i we proceed as follows. If our choice of the parity of P_i shall respect the parity naturally given by the sides of the partition (A, B), then let P_i' be a path connecting s_i' and t_i' in G' and let P_i be the conjunction of $s_i s_i'$, the path P_i', and $t_i' t_i$. Otherwise, let P_i' be a path connecting s_i' and x_i' and P_i'' be a path connecting t_i' and y_i'. If $\{s_i, t_i\} \cap \{x_i, y_i\} = \emptyset$, then let P_i be the conjunction of $s_i s_i'$, the path P_i', the path $x_i' x_i y_i y_i'$, the path P_i'', and $t_i' t_i$. If $s_i = x_i$ and $t_i \neq y_i$, then let P_i be the conjunction of $s_i y_i y_i'$, the path P_i'', and $t_i' t_i$. If $s_i \neq x_i$ and $t_i = y_i$, then let P_i be the conjunction of $s_i s_i'$, the path P_i', and $x_i' x_i t_i$ Finally, if $s_i = x_i$ and $t_i = y_i$, then let $P_i = s_i t_i$.

As mentioned above, because G' is $2k$-linked, we can choose the corresponding paths $P_1', P_1'', \dots, P_k', P_k''$ in G' to be pairwise disjoint and hence also $P_1, \dots, P_k$ are pairwise disjoint.

It remains to show that if X has size at least $8k$, then the first statement holds. This part of the proof can basically be found in [11]. However, we change some arguments which leads to a shorter proof. Let $G' = G - \{s_1, \dots, s_k, t_1, \dots, t_k\}$ and let (A', B') be a partition of G' such that $|E(G_{A', B'})|$ is minimized. Note that $\delta(G') \geq 24k$. By Lemma 1, there is a $4k$-connected subgraph H of G' with $|E(H)| \geq 10k|V(H)|$. Moreover, by Theorem 2, the graph H is $2k$-linked. Let (A_H, B_H) be the (unique) bipartition of H such that $A_H \subset A'$.

Theorem 6 guarantees a set Y with $|Y| \leq 6k - 6$ that intersects all odd A_H-paths in G' or $3k$ disjoint odd A_H-paths in G'. Suppose that there is a set Y of at most $6k - 6$ vertices such that $G' - Y$ contains no odd A_H-path. For a contradiction, we assume that $G' - Y$ is not bipartite. Thus there is an odd cycle C in $G' - Y$. Since G' is $48k$-connected, $G' - Y$ is 2-connected. Hence there are two disjoint A_H-C-paths in G'. Note that the length of these paths could be zero. Nevertheless, combining these two paths with one part of the cycle C leads to an odd A_H-path, which is a contradiction. This in turn implies that $S \cup Y$ is an odd cycle cover of G of size at most $8k - 6$, which is a contradiction to the assumption $|X| \geq 8k$. Thus Theorem 6 implies the existence of $3k$ odd A_H-paths.

Let P be one of these $3k$ odd A_H-path. There is a natural partition of $E(P)$ into H-paths. Because P is an odd A_H-path, there is a subpath P' of P such that P' is an odd H-path and both endvertices of P' lie in the same side of the bipartition of H or P' is an even H-path and exactly one endvertex of P' lies in A_H.

Therefore, there is a set $\mathcal{Q}$ of $3k$ disjoint H-paths $Q_1, \dots, Q_{3k}$ where the length of Q_i is odd if both endvertices lie in the same side of the bipartition of H and even otherwise.

Since G is $50k$-connected, there is a set of $2k$ disjoint paths $\mathcal{P} = \{P_1, \dots, P_{2k}\}$ connecting $\{s_1, \dots, s_k, t_1, \dots, t_k\}$ and H. Choose these paths such that they intersect as few as possible paths from $\mathcal{Q}$. Under this condition choose these paths such that their edge intersection with $\mathcal{Q}$ is as large as possible. The latter condition implies that if $Q \in \mathcal{Q}$ has nonempty intersection with a path in $\mathcal{P}$ – let z', z'' be the endvertices of Q, and let P be first path that intersects Q seen from the direction of z' – then P follows the path Q up to z' beside the case that P is the only path intersecting Q, and P follows Q to z''. Hence for every $Q \in \mathcal{Q}$ that intersects a path in $\mathcal{P}$, there is at least one path $P \in \mathcal{P}$ such that there is vertex z that is an endvertex of P and Q. Therefore, the paths in $\mathcal{P}$ intersect at most $2k$ paths in $\mathcal{Q}$ and hence there is a collection $\mathcal{Q}' = \{Q'_1, \ldots, Q'_k\} \subset \mathcal{Q}$ of k paths such that $Q \cap P = \emptyset$ for $P \in \mathcal{P}$ and $Q \in \mathcal{Q}'$.

Since H is $2k$-linked, we can find the desired disjoint paths of specified parity connecting s_i and t_i by using the paths $\mathcal{P}$ and then either directly linking the ends in H of the paths belonging to s_i and t_i or by using the path Q'_i in between.   □

For a graph G, let a set of vertices X of G be a *vertex cover* of G if every edge is incident to at least one vertex of X. Let the *vertex cover number* $\tau(G)$ of G be the least number k such that G has a vertex cover X with $|X| = k$. Since a vertex cover has to contain at least one vertex of every edge in a matching M, we have on the one hand $|M| \leq \tau(G)$ for every matching M in G. On the other hand, we observe the following.

> If M is a maximal matching of G, then the vertices covered by M form a vertex cover of G and hence $\tau(G) \leq 2|M|$.   (1)

A graph G is τ-*critical* if $\tau(G - e) < \tau(G)$ and $\tau(G - v) < \tau(G)$ for every edge $e \in E(G)$ and every vertex $v \in V(G)$. A result of Erdős and Gallai [6] says if G is τ-critical, then $\tau(G) \geq |V(G)|/2$.

Having these definitions in mind we reprove Kawabarayashi's and Reed's result and directly afterwards Corollary 1.

Proof (Proof of Theorem 3). Suppose that X is a minimum odd cycle cover and $|X| \geq 4k - 3$. We show by induction on k that G contains a parity breaking matching for S. Let (A, B) be a nice partition induced by X. Note that X is a minimum vertex cover of $G_{A,B}$.

Suppose $k = 1$. Since $|X| \geq 1$, the graph $G_{A,B}$ must contain an edge e and e is a parity breaking matching.

Hence we may assume that $k \geq 2$. Because $2k < 4k - 3$, there is a vertex r such that $r \in X \setminus \{s_1, \ldots, s_k, t_1, \ldots, t_k\}$. Since X is minimum, r has a neighbor r' in $G_{A,B}$. If $r' \notin \{s_1, \ldots, s_k, t_1, \ldots, t_k\}$, we have $\tau(H - \{s_1, t_1, r, r'\}) \geq 4(k-1) - 3$ and combining rr' and the induction hypothesis, we conclude that G contains a parity breaking matching M. Now we may assume that, by symmetry, $r' = s_1$. However, using the same argument again, there is a parity breaking matching for S. Applying Theorem 4 completes the proof.   □

Proof (Proof of Corollary 1). Suppose that X is a minimum odd cycle cover and $|X| \geq 2k - 1$. Fix some independent set $S = \{s_1, \ldots, s_k, t_1, \ldots, t_k\}$ of $2k$ distinct vertices. We show by induction on k that G contains a parity breaking matching for S. Let (A, B) be a nice partition induced by X. Note that X is a minimum vertex cover of $G_{A,B}$.

In the following we show that there is a parity breaking matching of size k. We proceed by induction on k. If $k = 1$, then $G_{A,B}$ contains an edge because $|X| \geq 1$ and this edge is a parity breaking matching of size 1.

Suppose $G_{A,B} - S$ contains no edges. This implies that $G_{A,B}$ is bipartite with bipartition $(V(G_{A,B}) \setminus S, S)$. By König's Theorem, the matching number of $G_{A,B}$ equals the vertex cover number and hence $G_{A,B}$ contains a matching M of size $2k - 1$. Let N be the matching obtained from M by deleting one of the matching edges $s_i p$ and $t_i q$ if both exist in M. Therefore, $|N| \geq k$ and every set of k edges of N is a parity breaking matching of size k.

In the following we may assume that $G_{A,B} - S$ contains edges. Suppose s_i is an isolated vertex in $G_{A,B}$. Let $e = uv$ be an edge incident to t_i if such an edge exists otherwise let e be an edge in $G_{A,B} - S$. By induction, $\tau(G_{A,B} - \{u, v, s_i, t_i\}) \geq 2k - 3$ and thus there exits a parity breaking matching M of size $k - 1$. Combining M and e leads to the desired matching.

Therefore, we may assume that every vertex in S has a neighbor in $G_{A,B}$. Induction can also be applied if $|X| \geq 2k$ by deleting s_1, t_1, and a neighbor of s_1 from $G_{A,B}$. Thus we may assume $|X| = 2k - 1$.

Let $G'_{A,B}$ be the induced subgraph of $G_{A,B}$ which is obtained from $G_{A,B}$ by deleting all isolated vertices from $G_{A,B}$. We may assume that the $\tau(G'_{A,B} - e) < \tau(G'_{A,B})$ for every $e \in E(G'_{A,B})$ and $\tau(G'_{A,B} - r) < \tau(G'_{A,B})$ for every $r \in V(G'_{A,B}) \setminus S$. Moreover, if $\tau(G'_{A,B} - s_i) = \tau(G'_{A,B})$, then let r be a neighbor of t_i and the statement follows by induction because $\tau(G'_{A,B} - \{s_i, t_i, r\}) \geq \tau(G'_{A,B}) - 2$. This implies that $G'_{A,B}$ is a τ-critical graph.

Since S is an independent set, the complement of an independent set is a vertex cover, and $|X| = 2k - 1$, we conclude that $\tau(G'_{A,B}) < n(G'_{A,B})/2$. However, this contradicts a theorem of Erdős and Gallai mentioned before. $\square$

3 Odd Cycles Through Prescribed Vertices

In this section we present several results concerning the Erdős-Pósa property of odd S-cycles in highly connected graphs. This extends the results concerning the Erdős-Pósa property of odd cycles in highly connected graphs. Furthermore, assuming a slightly higher connectivity, we show how known results follow easily from Theorem 4.

Lemma 2. *Let $k \in \mathbb{N}$ and let G be a $50k$-connected graph. Let S be a set of k vertices. Exactly one of the following statements holds.*

1. *G contains k disjoint odd S-cycles.*
2. *For all nice partitions (A, B) of G, there is no matching M of size k in $G_{A,B}$ where an edge in M covers at most one vertex of S.*

In addition, if one nice partition of G has a matching as in 2, then all nice partitions have such a matching.

Proof. Suppose first that G contains k disjoint S-cycles and let (A, B) be a partition of G. Since every odd cycle contains at least one edge of $E(G_{A,B})$ and by picking exactly one of these edges from every odd S-cycle, we get a matching of size k in $G_{A,B}$ where one edge does not cover two vertices of S, which implies that the second statement does not hold.

Next, we suppose that the second statement does not hold and let (A, B) be a nice partition such that there is a matching M of size k in $G_{A,B}$ and one edge in M does not cover two vertices of S.

Let $T = \{t_1, \ldots, t_k\}$ be a set of vertices distinct from S and distinct from the vertices covered by M. By Theorem 4, there are disjoint paths $P_1, \ldots, P_k$ with prescribed parity such that P_i connects s_i and t_i because M is a parity breaking matching for $\{s_1, \ldots, s_k, t_1, \ldots, t_k\}$.

Let $(\tilde{A}, \tilde{B})$ be any partition of G. By suitably choosing the parity of P_i, every path uses at least one edge in $G_{\tilde{A},\tilde{B}}$ and hence there is a matching in $G_{\tilde{A},\tilde{B}}$ of size at least k such that each edge covers at most one vertex of S.

For every s_i, add to G a vertex s_i' such that $N(s_i) = N(s_i')$ and denote this new graph by G'. Note that G' is $50k$-connected. Let (A', B') be a nice partition of G'. As mentioned above, in the subgraph G of G' for every partition $(\tilde{A}, \tilde{B})$, the graph $G_{\tilde{A},\tilde{B}}$ contains a matching of size k such that each edge covers at most one vertex of S, in particular, this holds for the partition induced by (A', B'). Let M be such a matching of size k. Note that M does not cover a vertex of the set $\{s_1', \ldots, s_k'\}$. Thus M is a parity breaking matching for $S \cup \{s_1', \ldots, s_k'\}$ in G'. By Theorem 4, there are disjoint paths $P_1', \ldots, P_k'$ of odd length where P_i' joins s_i and s_i'.

By identifying s_i and s_i', this in turn implies the existence of k disjoint odd S-cycles $C_1, \ldots, C_k$ in G where C_i contains s_i. $\square$

After having proved Lemma 2, it is not difficult to prove the Erdős-Pósa property of odd S-cycles in highly connected graphs.

Theorem 7. *Let $k \in \mathbb{N}$ and let G be a $50k$-connected graph. Let S be a set of k vertices. At least one of the following statements holds.*

1. *G contains k disjoint odd S-cycles.*
2. *There is a set X with $|X| = 2k - 2 + \tau(G[S])$ such that $G - X$ is bipartite.*

Proof. Let X be a minimum odd cycle cover. We may assume that $|X| \geq 2k - 1 + \tau(G[S])$. Let (A, B) be nice partition of G and let Y be a minimum vertex cover of S. Since X is a minimum vertex cover of $G_{A,B}$, we conclude $\tau(G_{A,B} - Y) \geq 2k - 1$. Using (1), this in turn implies the existence of a matching M of size k in $G_{A,B} - Y$ such that no vertex of M covers more than one vertex of S. Using Lemma 2, this implies the existence of k disjoint odd S-cycles in G. $\square$

Note that $\tau(G[S]) \leq k - 1$ and thus $2k - 2 + \tau(G[S]) \leq 3k - 3$. Moreover, the bound "$2k - 2 + \tau(G[S])$" is sharp for every possible value of $\tau(G[S])$ no matter how large the connectivity of G is. To see this, let G arise from a large complete bipartite graph with bipartition (A, B) by adding the edges of a clique on $2k - 1$ vertices to A and the edges of a clique on τ vertices to B for some $1 \leq \tau \leq k$. Let S be a set of k vertices in B containing the τ-clique. Hence $\tau(G[S]) = \tau - 1$, there do not exist k disjoint odd S-cycles, and there is no set X of $2k - 3 + \tau(G[S])$ vertices such that $G - X$ is bipartite.

In Theorem 7 we require that $|S| = k$. However, it is not difficult to extend Theorem 7 to arbitrary subsets of $V(G)$ as follows.

On the one hand, if $|S| < k$, then G does not contain k disjoint odd S-cycles and one can remove (the small set) S from G to obtain a graph without an odd S-cycle. Of course, one cannot hope for a small set X such that $G - X$ is bipartite because G could simply be a large clique.

On the other hand if $|S| \geq k$, then let $S' \subset S$ such that $|S'| = k$. In the case that G does not contain k disjoint odd S'-cycles, Theorem 7 implies the existence of a set X of $3k - 3$ vertices such that $G - X$ is bipartite, in particular, $G - X$ does not contain odd S-cycles. In addition, this proves Theorem 1.

We complete the paper with two results concerning the case $S = V(G)$.

Corollary 2. *Let $k \in \mathbb{N}$ and let G be a 50k-connected graph. Exactly one of the following statements holds.*

1. *G contains k disjoint odd cycles.*
2. *For every nice partition (A, B), the graph $G_{A,B}$ does not contain a matching of size k.*

Proof. If the first statement holds, then the second does clearly not hold.

Suppose that the second statement does not hold. We may assume that G contains an independent set of cardinality k, otherwise there are k disjoint triangles in G. This can be seen as follows: Let x be any vertex in G. The neighborhood of x contains an edge. Combining this edge with x leads to a triangle. Repeating this argument k times, leads to the first statement. Thus G contains an independent set S of cardinality k and we apply Lemma 2 to obtain k disjoint odd $(S$-$)$cycles in G. $\qquad\square$

The following corollary is already proven by Thomassen [21] and Rautenbach and Reed [16] with a higher connectivity bound. Later Kawarabayashi and Reed [11] and Kawarabayashi and Wollan [13] improved this bound to $24k$ and $\frac{31}{2}k$, respectively.

Corollary 3. *Let $k \in \mathbb{N}$ and let G be a 50k-connected graph. At least one of the following statements holds.*

1. *G contains k disjoint odd cycles.*
2. *G contains a set X of $2k - 2$ vertices such that $G - X$ is bipartite.*

Proof. Suppose the second statement does not hold. Let (A, B) a nice partition of G induced by a minimum odd cycle cover X. Note that X is a minimum vertex cover of $G_{A,B}$ and since $|X| \geq 2k - 1$ by our assumption and by applying (1), the graph $G_{A,B}$ contains a matching of size k. Because this holds for every minimum odd cycle cover and so for every nice partition of G, the statement follows by the previous corollary. $\qquad\square$

Clearly, assuming $50k$-connectivity in our results in not the best function in terms of k one can hope for. However, it is essentially best possible in that sense as one can easily construct graphs that show that linear connectivity in k is necessary. It would be interesting to know which connectivity is needed to ensure that our results hold. It even seems possible that the approach via a parity-k-linkage theorem cannot lead to the best connectivity bound.

4 Algorithmic Consequences

The problem of finding k disjoint odd cycles in a graph is known to be NP-complete if k is part of the input. Moreover, it is fixed parameter tractable (an algorithm is announced in [12]).

We also aim for an FPT-algorithm parameterized in k. Let us consider the following problem:

> Given an integer k, a $50k$-connected graph G and a set of vertices S. Does G contain k disjoint odd S-cycles?
> $\hspace{3em}$ (2)

We briefly describe how our results lead to an algorithm for problem (2). Trivially, we assume that $|S| \geq k$. In [18] Reed, Smith and Vetta present an algorithm for deciding whether a graph has an odd cycle cover of size at most p that runs in $O(g(p)mn)$-time for some exponential function g. In addition, if this is the case, the algorithm computes such an odd cycle cover. In order to solve our problem, we use this algorithm to decide whether G has an odd cycle cover of size at most $3k - 3$. If this is not the case, then by Theorem 1 the graph G has k disjoint odd S-cycles. Suppose now that G has an odd cycles cover of size at most $3k - 3$. It is straightforward to construct a nice partition (A, B) of G. Next we check whether $G_{A,B}$ has a matching M of size k and a set $S' \subseteq S$ such that $|S'| = k$ and an edge of M covers at most one vertex of S'. This in turn answers (2) by applying Lemma 2. Clearly, this algorithm runs in $O(f(k)nm)$-time.

References

1. Birmelé, E., Bondy, J.A., Reed, B.: The Erdős-Pósa property for long circuits. Combinatorica **27**, 135–145 (2007)
2. Bondy, J.A., Lovász, L.: Cycles through specified vertices of a graph. Combinatorica **1**, 117–140 (1981)
3. Bruhn, H., Joos, F., Schaudt, O.: Long cycles through prescribed vertices have the Erdős-Pósa property (2014). arXiv:1412.2894
4. Diestel, R.: Graph Theory. GTM, vol. 173, 4th edn. Springer, Heidelberg (2010)
5. Dirac, G.A.: In abstrakten Graphen vorhandene vollständige 4-Graphen und ihre Unterteilungen. Math. Nachr. **22**, 61–85 (1960)
6. Erdős, P., Gallai, T.: On the minimal number of vertices representing the edges of a graph. Magyar Tud. Akad. Mat. Kutató Int. Közl. **6**, 181–203 (1961)
7. Erdős, P., Pósa, L.: On the maximal number of disjoint circuits of a graph. Publ. Math. Debrecen **9**, 3–12 (1962)

8. Geelen, J., Gerards, B., Reed, B., Seymour, P., Vetta, A.: On the odd-minor variant of Hadwiger's conjecture. J. Comb. Theor. (Series B) **99**, 20–29 (2009)

9. Kakimura, N., Kawarabayashi, K.: Half-integral packing of odd cycles through prescribed vertices. Combinatorica **33**, 549–572 (2013)

10. Kakimura, N., Kawarabayashi, K., Marx, D.: Packing cycles through prescribed vertices. J. Comb. Theor. (Series B) **101**, 378–381 (2011)

11. Kawarabayashi, K., Reed, B.: Highly parity linked graphs. Combinatorica **29**, 215–225 (2009)

12. Kawarabayashi, K., Reed, B.: Odd cycle packing [extended abstract]. In: Proceedings of the 2010 ACM International Symposium on Theory of Computing, STOC 2010, pp. 695–704. ACM, New York (2010)

13. Kawarabayashi, K., Wollan, P.: Non-zero disjoint cycles in highly connected group labelled graphs. J. Comb. Theor. (Series B) **96**, 296–301 (2006)

14. Mader, W.: Existenz n-fach zusammenhängender Teilgraphen in Graphen genügend grosser Kantendichte. Abh. Math. Sem. Univ. Hamburg **37**, 86–97 (1972)

15. Pontecorvi, M., Wollan, P.: Disjoint cycles intersecting a set of vertices. J. Comb. Theor. (Series B) **102**, 1134–1141 (2012)

16. Rautenbach, D., Reed, B.: The Erdős-Pósa property for odd cycles in highly connected graphs. Combinatorica **21**, 267–278 (2001)

17. Reed, B.: Mangoes and blueberries. Combinatorica **19**, 267–296 (1999)

18. Reed, B., Smith, K., Vetta, A.: Finding odd cycle transversals. Oper. Res. Lett. **32**, 299–301 (2004)

19. Robertson, N., Seymour, P.: Graph minors. V. Excluding a planar graph. J. Comb. Theor. (Series B) **41**, 92–114 (1986)

20. Thomas, R., Wollan, P.: An improved linear edge bound for graph linkages. Eur. J. Comb. **26**, 309–324 (2005)

21. Thomassen, C.: The Erdős-Pósa property for odd cycles in graphs of large connectivity. Combinatorica **21**, 321–333 (2001)

Well-quasi-ordering Does Not Imply Bounded Clique-width

Vadim V. Lozin[1], Igor Razgon[2(✉)], and Viktor Zamaraev[3]

[1] Mathematics Institute, University of Warwick, Coventry CV4 7AL, UK
V.Lozin@warwick.ac.uk
[2] Department of Computer Science and Information Systems,
Birkbeck, University of London, London, UK
igor@dcs.bbk.ac.uk
[3] Mathematics Institute, University of Warwick, Coventry CV4 7AL, UK
V.Zamaraev@warwick.ac.uk

Abstract. We present a hereditary class of graphs of unbounded clique-width which is well-quasi-ordered by the induced subgraph relation. This result provides the negative answer to a question asked by Daligault, Rao and Thomassé in [3].

1 Introduction

Well-quasi-ordering (WQO) is a highly desirable property and frequently discovered concept in mathematics and theoretical computer science [6,10]. One of the most remarkable recent results in this area is the proof of Wagner's conjecture stating that the set of all finite graphs is well-quasi-ordered by the minor relation [13]. This is, however, not the case for the induced subgraph relation, since the set of cycles $\{C_n | n \geq 3\}$ forms an infinite antichain with respect to this relation. On the other hand, the induced subgraph relation may become a well-quasi-order when restricted to graphs in particular classes, such as cographs [4] or k-letter graphs [12]. It is interesting to observe that in both examples we deal with graphs of bounded clique-width, which is another property of great importance in mathematics and computer science. Moreover, the same is true for all available examples of hereditary graph classes which are well-quasi-ordered by the induced subgraph relation (see e.g. [9]). This raises an interesting question whether the clique-width is always bounded for graphs in well-quasi-ordered hereditary classes. This question was formally stated as an open problem (Problem 6) by Daligault, Rao and Thomassé in [3]. In the present paper, we answer former question negatively by exhibiting a hereditary class of graphs of unbounded clique-width which is well-quasi-ordered by the induced subgraph relation.

Remark. The proposed result *does not* resolve Conjecture 5 of [3] asserting that 2-well-quasi-orderability of a hereditary class implies its bounded cliquewidth. Indeed, being 2-well-quasi-ordered is a stronger (more constrained)

This research has been supported by EPSRC grant EP/LO20408/1.

© Springer-Verlag Berlin Heidelberg 2016
E.W. Mayr (Ed.): WG 2015, LNCS 9224, pp. 351–359, 2016.
DOI: 10.1007/978-3-662-53174-7_25

property than just being well-quasi-ordered. Therefore, the existence of a well-quasi-ordered class of unbounded cliquewidth says nothing about existence of such a 2-well-quasi-ordered class. In fact, we believe that the class considered in this paper is not 2-well-quasi-ordered.

Our result shows that it is generally non-trivial to determine whether a given problem definable in Monadic Second Order (MSO) logic is polynomially solvable on a WQO class, since unboundedness of clique-width does not allow a straightforward application of Courcelle et al.'s theorem [2]. This makes the WQO classes an interesting object to study from the algorithmic perspective.

Graphs in the class introduced in this paper are *dense* (in particular, they are P_7-free). The density is a necessary condition, because an earlier result [1] shows that for sparse graph classes (those where a large biclique is forbidden as a subgraph) well quasi-orderability by induced subgraphs imply bounded treewidth (and hence bounded clique-width). We believe that the result of [1] can be strengthened by showing that well quasi-orderability by induced subgraphs in sparse classes implies bounded pathwidth (and hence *linear* clique-width [7]). Our result proved in the present paper shows a stark contrast between dense and sparse graphs in this context.

The rest of the paper is structured as follows. In Sect. 2 we define the class of graphs studied in this paper and state the main result. The unboundedness of clique-width and well-quasi-orderability by induced subgraphs is proved in Sects. 3 and 4, respectively. We use standard graph-theoretic notation as e.g. in [5]. The notions of clique-width and well-quasi-ordering are introduced in respective sections where they are actually used.

2   The Main Result

In this section, we define the class $\mathcal{D}$, which is the main object of the paper, and state the main result.

Let P be a path with vertex set $\{1, \ldots, n\}$ with two vertices i and j being adjacent if and only if $|i - j| = 1$. For vertex i, the largest 2^k that divides i is called the *power* of i and is denoted by $q(i)$. For example, $q(5) = 1, q(6) = 2, q(8) = 8, q(12) = 4$. Add edges to P that connect i and j whenever $q(i) = q(j)$. We denote the graph obtained in this way by D_n. Figure 1 illustrates graph D_{16}.

Clearly, the edges $E(D_n) \setminus E(P)$ form a set of disjoint cliques and we call them *power cliques*. If a power clique Q contains a vertex i with $q(i) = 2^k$ we say that Q *corresponds to* 2^k. We call P the *body* of D_n, the edges of $E(P)$ the *path edges*, and the edges of $E(D_n) \setminus E(P)$ the *clique edges*. The class $\mathcal{D}$ is the set of all graphs D_n and all their induced subgraphs. In what follows we prove that

- clique-width of graphs in $\mathcal{D}$ is unbounded (Sect. 3),
- graphs in $\mathcal{D}$ are well-quasi-ordered by the induced subgraph relation (Sect. 4).

These two facts imply the following conclusion, which is the main result of the paper.

Theorem 1. *Within the family of hereditary graph classes, there exist classes of unbounded clique-width which are well-quasi-ordered by the induced subgraph relation.*

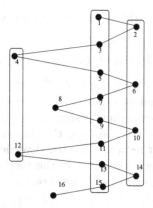

Fig. 1. Graph D_{16}. To avoid shading the picture with many edges, cliques are represented as rectangular boxes.

3   Clique-Width is Unbounded in $\mathcal{D}$

The clique-width of a graph G, denoted $cwd(G)$, is the minimum number of labels needed to construct the graph by means of the four graph operations: creation of a new vertex, disjoint union of two labeled graphs, connecting vertices with specified labels i and j, and renaming label i to label j. Every graph G can be constructed by means of these four operations, and the process of the construction can be described either by an algebraic expression or by a rooted binary tree, whose leaves correspond to the vertices of G, the root corresponds to G and the internal nodes correspond to the union operations.

Given a graph G and a subset $U \subset V(G)$, we denote by $\overline{U}$ the set $V(G) - U$. We say that two vertices $x, y \in U$ are U-*similar* if $N(x) \cap \overline{U} = N(y) \cap \overline{U}$, i.e. if x and y have the same neighbourhood outside of U. Clearly, the U-similarity is an equivalence relation and we denote the number of similarity classes of U by $\mu_G(U)$. Also, we denote

$$\mu(G) = \min_{\frac{1}{3}n \leq |U| \leq \frac{2}{3}n} \mu_G(U),$$

where $n = |V(G)|$. Our proof of the main result of this section is based on the following lemma.

Lemma 1. *For any graph G, $\mu(G) \leq cwd(G)$.*

Proof. Let T be an *optimal* decomposition tree, t a node of T and U_t the set of vertices of G that are leaves of the subtree of T rooted at t. It is known (see e.g. [11]) that $cwd(G) \geq \mu_G(U_t)$ for any node t of T. According to a well known folklore result, the binary tree T has a node t such that $\frac{1}{3}|V(G)| \leq |U_t| \leq \frac{2}{3}|V(G)|$, in which case $\mu_G(U_t) \geq \mu(G)$. Hence the lemma.   □

Let $U \subseteq V(D_n)$, and let P be the body of D_n. We denote by P^U the subgraph of P induced by U. In other words, P^U is obtained from $D_n[U]$ by removing the clique edges. Since P is a path, P^U is a graph every connected component of which is a path.

Lemma 2. *If P^U has $c + 1$ connected components, then $\mu_{D_n}(U) \geq c/2$.*

Proof. In the i-th connected component of P^U, $i \leq c$, we choose the last vertex (listed along the path P) and denote it by u_i. The next vertex of P, denoted $\overline{u}_i$, belongs to $\overline{U}$. This creates a matching of size c with edges $(u_i, \overline{u}_i)$. Note that none of $(u_i, \overline{u}_j)$ is a path edge for $i < j$. Among the chosen vertices of U at least half have the same parity. Their respective matched vertices of $\overline{U}$ have the opposite parity. Since the clique edges connect only the vertices of the *same* parity, we conclude that at least $c/2$ vertices of U have pairwise different neighbourhoods in $\overline{U}$, i.e. $\mu_{D_n}(U) \geq c/2$. □

Note that if P^U has c connected components, then $P^{\overline{U}}$ has at least $c - 1$ connected components. Therefore, in light of Lemma 2, it remains to consider the case where both P^U and $P^{\overline{U}}$ have a limited number of connected components. Taking into account the definition of $\mu(G)$ and Lemma 1 we can assume that both U and $\overline{U}$ are 'large' (between one third and two third of $|V(G)|$), and hence each of P^U and $P^{\overline{U}}$ has a 'large' connected component. In order to address this case we use the following lemma which states that a large number of power cliques intersecting both U and $\overline{U}$ implies a large value of $\mu_{D_n}(U)$.

Lemma 3. *If there exist c different power cliques $Q_1, \ldots, Q_c$ each of which*

(1) corresponds to a power of 2 greater than 1 and
(2) intersects both U and $\overline{U}$

then $\mu_{D_n}(U) \geq c$.

Proof. Let u_i and $\overline{u}_i$ be some vertices in Q_i, which belong to U and $\overline{U}$, respectively. Since all the vertices in $M = \{u_1, \overline{u}_1, \ldots, u_c, \overline{u}_c\}$ are even and two even vertices are adjacent in D_n if and only if they belong to the same power clique, M induces a matching in D_n with edges $(u_i, \overline{u}_i)$, $i = 1, \ldots, c$. This implies that $u_1, \ldots, u_c$ have pairwise different neighbourhoods in $\overline{U}$, that is $\mu_{D_n}(U) \geq c$. □

The only remaining ingredient to prove the main result of this section is the following lemma.

Lemma 4. *Let c be a constant and P' a subpath of P of length at least 2^{c+1}. Then P' intersects each of the power cliques corresponding to $2^1, \ldots, 2^c$.*

Proof. The statement easily follows from the fact that for a fixed k vertices v with $q(v) = 2^k$ are of the form $v = 2^k(2p + 1)$. That is, they occur in P with period 2^{k+1}. □

Now we are ready to prove the main result of this section.

Theorem 2. *Let n and c be natural numbers such that $n \geq 3((2c+1)(2^{c+1}-1)+1)$. Then $cwd(D_n) \geq c$ and hence the clique-width of graphs in $\mathcal{D}$ is unbounded.*

Proof. Let U be an arbitrary subset of vertices of D_n, such that $\frac{n}{3} \leq |U| \leq \frac{2n}{3}$. Note that the choice of U implies that the cardinalities of both U and $\overline{U}$ are at least $\frac{n}{3} \geq (2c+1)(2^{c+1}-1)+1$.

If P^U has at least $2c+1$ connected components, then by Lemma 2 $\mu_{D_n}(U) \geq c$. Otherwise P^U has less than $2c+1$ connected components and $P^{\overline{U}}$ has less than $2c+2$ connected components. By the pigeonhole principle, both graphs have connected components of size at least 2^{c+1}. Clearly, these connected components are disjoint subpaths of P. By Lemma 4, the power cliques corresponding to $2^1, \ldots, 2^c$ intersect both U and $\overline{U}$, and hence, by Lemma 3, $\mu_{D_n}(U) \geq c$.

Since U has been chosen arbitrarily, we conclude that $\mu(D_n) \geq c$, and therefore, by Lemma 1, $cwd(D_n) \geq c$, as required.   □

4 $\mathcal{D}$ is WQO by Induced Subgraphs

A binary relation $\leq$ on a set W is a *quasi-order* (also known as *preorder*) if it is reflexive and transitive. Two elements $x, y \in W$ are said to be comparable with respect to $\leq$ if either $x \leq y$ or $y \leq x$. Otherwise, x and y are incomparable. A set of pairwise comparable elements is called a *chain* and a set of pairwise incomparable elements an *antichain*. A quasi-order $(W, \leq)$ is a *well-quasi-order* (WQO) if it contains neither infinite strictly decreasing chains nor infinite antichains.

In this section, we show that graphs in $\mathcal{D}$ are well-quasi-ordered by the induced subgraph relation. In the proof we apply the celebrated Higman's lemma [8] which can be stated as follows.

For an arbitrary set M, let M^* be the set of all finite sequences of elements of M. Any quasi-order $\leq$ on M defines a quasi-order $\preceq$ on M^* as follows: $(a_1, \ldots, a_m) \preceq (b_1, \ldots, b_n)$ if and only if there is an order-preserving injection $f : \{a_1, \ldots, a_m\} \rightarrow \{b_1, \ldots, b_n\}$ with $a_i \leq f(a_i)$ for each $i = 1, \ldots, m$.

Lemma 5 [8]. *If $(M, \leq)$ is a WQO, then $(M^*, \preceq)$ is a WQO.*

Obviously, the induced subgraph relation contains no infinite strictly decreasing chains. Therefore, to prove that this relation is a WQO on $\mathcal{D}$ we need to show that for each infinite sequence $\mathcal{G} = G_1, G_2 \ldots$ of graphs in $\mathcal{D}$ there are i, j such that G_i is an induced subgraph of G_j.

We recall that $V(D_n)$ is the set of integers $1, 2, \ldots, n$ listed along the body of D_n and any graph in $\mathcal{D}$ is an induced subgraph of D_n with some n. Among all possible sets of integers inducing a graph (isomorphic to) $G \in \mathcal{D}$ we pick *one* (arbitrarily) and identify $V(G)$ with this set.

Any set of consecutive integers will be called an *interval* and any subgraph of D_n induced by an interval will be called a *factor*. The number of elements in an interval inducing a factor is called the *length* of the factor. If a graph $G \in \mathcal{D}$ is not a factor, its vertex set can be split into maximal intervals and we call the subgraphs of G induced by these intervals *factor-components* of G. The set of all factor-components of G will be denoted $\mathcal{F}(G)$.

Lemma 6. *If $\mathcal{G}$ contains graphs with arbitrarily long factor-components, then $\mathcal{G}$ is not an antichain.*

Proof. Pick an arbitrary G_i and let n be the smallest number such that G_i is an induced subgraph of D_n. By our assumption, there is G_j with factor-component F of length at least $5n$. Let us show that D_n is an induced subgraph of G_j. By the transitivity of the induced subgraph relation, this will imply that G_i is an induced subgraph of G_j.

Let 2^k be the smallest power of 2 larger than n. Clearly, $2^{k+1} \leq 4n$. Hence, by Lemma 4, there is a vertex y among the first $4n$ vertices of F with $q(y) = 2^k$. Let F' be the factor induced by the vertices of F starting at $y+1$. Since F is of length at least $5n$ and y is among the first $4n$ vertices of F, the length of F' is at least n. Thus we can define an injective function $f : V(D_n) \to V(F')$ as follows: $f(z) = y+z$ for $1 \leq z \leq n$. We claim that f is an induced subgraph isomorphism from D_n to a subgraph of G_j. Clearly, $f(z+1) = f(z)+1$ for $1 \leq z < n$, hence it remains to verify that adjacencies and non-adjacencies are preserved for vertices z_1, z_2 of D_n such that $z_2 > z_1 + 1$. Clearly, in this case z_1 and z_2 are adjacent if and only if $q(z_1) = q(z_2)$. Moreover, since $f(z_2) > f(z_1)+1$, $f(z_2)$ and $f(z_1)$ are adjacent if and only if $q(f(z_1)) = q(f(z_2))$. Below we prove that $q(f(z)) = q(z)$ for $1 \leq z \leq n$ and hence $q(z_1) = q(z_2)$ if and only if $q(f(z_1)) = q(f(z_2))$, implying the lemma.

Indeed, $f(z) = y + z = 2^k p + 2^{k_1} p_1$, where $2^{k_1} = q(z)$ and p, p_1 are odd numbers. Since $2^{k_1} \leq n < 2^k$, $k_1 < k$ and hence $y + z$ can be written as $2^{k_1}(2^{k-k_1}p + p_1)$. Since $k > k_1$, 2^{k-k_1} is even and hence $2^{k-k_1}p + p_1$ is odd. Consequently, $q(y + z) = 2^{k_1}$, as required. □

From now on, we assume the length of factor-components of graphs in $\mathcal{G}$ is bounded by some constant $c = c(\mathcal{G})$. In what follows we prove that in this case $\mathcal{G}$ is not an antichain as well.

Let F be a factor. We say that a vertex u of F is *maximal* if $q(u) \geq q(v)$ for each vertex v of F different from u.

Lemma 7. *Every factor F of D_n contains precisely one maximal vertex.*

Proof. Suppose that F contains two maximal vertices $2^k p$ and $2^k(p+r)$ for some odd number p and even number $r \geq 2$. Then F also contains the vertex $2^k(p+1)$. Clearly $p+1$ is an even number and hence $q(2^k(p+1)) \geq 2^{k+1}$, which contradicts the maximality of 2^k. □

In light of Lemma 7, we denote the unique maximal vertex of F by $m(F)$. Also, let $s(F)$ be the smallest vertex of F.

Now we define two equivalence relations on the set of factor graphs as follows. We say that two factors F_1 and F_2 are

- *t-equivalent* if they are of the same length and $m(F_1) - s(F_1) = m(F_2) - s(F_2)$,
- *ℓ-equivalent* if $q(m(F_1)) = q(m(F_2))$.

We denote by L_i the ℓ-equivalence class such that $q(m(F)) = 2^i$ for every factor F in this class. We also order the t-equivalence classes (arbitrarily) and denote by T_j the j-th class in this order.

Lemma 8. *Let F be a factor of length at most c. Let v be a vertex of F different from its maximal vertex $m = m(F)$. Then $q(v) = q(|m - v|)$ and, in particular, $q(v) < c$.*

Proof. We can assume without loss of generality that $v > m$. Let k_1, p_1, k_2, p_2 be such that $m = 2^{k_1} p_1$ and $v - m = 2^{k_2} p_2$, with p_1, p_2 being odd numbers. Observe that $k_2 < k_1$. Indeed, otherwise $v = 2^{k_1} p_1 + 2^{k_2} p_2 = 2^{k_1}(p_1 + 2^{k_2 - k_1} p_2)$, where $p_1 + 2^{k_2 - k_1} p_2$ is a natural number. Therefore, $q(v) \geq 2^{k_1} = q(m)$ in contradiction either to the maximality of m or to Lemma 7.

Consequently, $v = 2^{k_1} p_1 + 2^{k_2} p_2 = 2^{k_2}(2^{k_1 - k_2} p_1 + p_2)$, where $2^{k_1 - k_2} p_1 + p_2$ is an odd number because of $2^{k_1 - k_2} p_1$ being even. Hence, $q(v) = 2^{k_2} = q(v - m)$.

Finally, since the length of F is at most c, we conclude that $v - m < c$, and therefore $q(v) = q(v - m) < c$. $\qquad \square$

Corollary 1. *Let F be a factor of length at most c. Let m be a vertex of F with $q(m) \geq c$. Then m is the maximal vertex of F.*

Corollary 2. *Let F_1, F_2 be two t-equivalent factors. Then there exists an isomorphism f from F_1 to F_2 such that:*

(a) $f(m(F_1)) = m(F_2)$;
(b) $q(f(v)) = q(v)$ for all $v \in V(F_1)$ except possibly for $m(F_1)$.

Proof. We claim that the function f that maps the i-th vertex of factor F_1 (starting from the smallest) to the i-th vertex of factor F_2 is the desired isomorphism. Indeed, property (a) follows from the condition that the factors are t-equivalent. Now property (a) together with Lemma 8 implies property (b). Finally, since adjacency between vertices in a factor is completely determined by their adjacency in the body and by their powers, we conclude that f is, in fact, isomorphism. $\qquad \square$

For a graph $G \in \mathcal{D}$, we denote by $G_{i,j}$ the set of factor-components of G in $L_i \cap T_j$, and define a binary relation $\leq$ on graphs of $\mathcal{D}$ as follows: $G \leq H$ if and only if $|G_{i,j}| \leq |H_{i,j}|$ for all i and j (clearly in this definition one can be restricted to non-empty sets $G_{i,j}$).

Finally, for a constant $c = c(\mathcal{G})$ we slightly modify the definition of $\leq$ to $\leq_c$ as follows. We say that a mapping $h : \mathbb{N} \to \mathbb{N}$ is c-preserving if it is injective and $h(i) = i$ for all $i \leq \lfloor \log c \rfloor$. Then $G \leq_c H$ if and only if there is a c-preserving mapping h such that $|G_{i,j}| \leq |H_{h(i),j}|$ for all i and j.

The importance of the binary relation $\leq_c$ is due to the following lemma.

Lemma 9. *Suppose the length of factor-components of G and H is bounded by c and $G \leq_c H$, then G is an induced subgraph of H.*

Proof. We say that a factor F is *low-powered* if $F \in L_i$, for some $i \leq \lfloor \log c \rfloor$, i.e. $q(m(F)) \leq c$.

It can be easily checked that the definition of $\leq_c$ implies the existence of an injective function $\phi : \mathcal{F}(G) \to \mathcal{F}(H)$ that possesses the following properties:

(1) ϕ maps each of the factors in $\mathcal{F}(G)$ to a t-equivalent factor in $\mathcal{F}(H)$;
(2) $F \in \mathcal{F}(G)$ is a low-powered factor if and only if $\phi(F)$ is;
(3) ϕ preserves power of the maximal vertex for each of the low-powered factors, i.e. $q(m(F)) = q(m(\phi(F)))$ for every low-powered factor $F \in \mathcal{F}(G)$;
(4) for any two factors $F_1, F_2 \in \mathcal{F}(G)$, $q(m(F_1)) = q(m(F_2))$ if and only if $q(m(\phi(F_1))) = q(m(\phi(F_2)))$.

To show that G is an induced subgraph of H we define a witnessing function that maps vertices of a factor $F \in \mathcal{F}(G)$ to vertices of $\phi(F) \in \mathcal{F}(H)$ according to an isomorphism described in Corollary 2. This mapping guarantees that a factor F of G is isomorphic to the factor $\phi(F)$ of H. Therefore it remains to check that adjacency relation between vertices in different factors is preserved under the defined mapping.

Note that adjacency between two vertices in different factors is determined entirely by powers of these vertices. Moreover, Corollary 2 and property (3) of ϕ imply that our mapping preserves powers of all vertices except possibly maximal vertices of power more that c. Therefore in order to complete the proof we need only to make sure that in graph G a maximal vertex m of a factor F with $q(m) > c$ is adjacent to a vertex v in a factor different from F if and only if the corresponding images of m and v are adjacent in H.

Taking into account Corollary 1 we derive that a maximal vertex with $q(m) > c$ is adjacent to a vertex v in a different factor if and only if v is maximal and $q(m) = q(v)$. Now the desired conclusion follows from Corollary 2 and properties (2) and (4) of function ϕ.   □

Lemma 10. *The set of graphs in $\mathcal{D}$ in which factor-components have size at most c is well-quasi-ordered by the $\leq_c$ relation.*

Proof. We associate with each graph $G \in \mathcal{D}$ containing no factor-component of size larger than c a matrix $M_G = m(i, j)$ with $m(i, j) = |G_{i,j}|$.

Each row of this matrix corresponds to an ℓ-equivalence class and we delete any row corresponding to L_i with $i > \lfloor \log c \rfloor$ which is empty (contains only 0s). This leaves a finite amount of rows (since G is finite).

Each column of M_G corresponds to a t-equivalence class and we delete all columns corresponding to t-equivalence classes containing factors of size larger than c (none of these classes has a factor-component of G). This leaves precisely $\binom{c+1}{2}$ columns in M_G.

We define the relation $\preceq_c$ on the set $\mathcal{M}$ of matrices constructed in this way as follows. For $M_1, M_2 \in \mathcal{M}$ we say that $M_1 \preceq_c M_2$ if and only if there is a c-preserving mapping β such that $m_1(i, j) \leq m_2(\beta(i), j)$ for all i and j.

It is not difficult to see that if $M_{G_1} \preceq_c M_{G_2}$, then $G_1 \leq_c G_2$. Therefore, if $\preceq_c$ is a well-quasi-order, then $\leq_c$ is a well-quasi-order too. The well-quasi-orderability of matrices follows by repeated applications of Higman's lemma. First, we split each matrix $M \in \mathcal{M}$ into two sub-matrices M' and M'' so that M' contains the first $\lfloor \log c \rfloor$ rows and M'' contains the remaining rows. Let $\mathcal{M}' = \{M' | M \in \mathcal{M}\}$ and $\mathcal{M} = \{M'' | M \in \mathcal{M}\}$

To see that the set of matrices $\mathcal{M}'$ is WQO we apply Higman's lemma twice. First, the set of rows is WQO since each of them is a finite word over the alphabet of non-negative integers (which is WQO by the ordinary arithmetic $\leq$ relation). Second, the set of matrices is WQO since each of them is a finite word over the alphabet of rows. Similarly, the set of matrices $\mathcal{M}''$ is WQO.

Note that in both applications of Lemma 5 to $\mathcal{M}'$ and in the first application to $\mathcal{M}''$, we considered sets of sequences of the *same length*. Hence, in this, case, Higman's lemma in fact implies the existence of two sequences one of them is *coordinate-wise* smaller than the other, exactly what we need in these cases.

Finally, the set of matrices $\mathcal{M}$ is WQO since each of them is a word of two letters over the alphabet $\mathcal{M}' \cup \mathcal{M}''$ which is WQO.   □

Combining Lemmas 6 and 10, we obtain the main result of this section.

Theorem 3. $\mathcal{D}$ *is* WQO *by the induced subgraph relation.*

References

1. Atminas, A., Lozin, V.V., Razgon, I.: Well-quasi-ordering, tree-width and sub-quadratic properties of graphs. CoRR, abs/1410.3260 (2014)
2. Courcelle, B., Makowsky, J.A., Rotics, U.: Linear time solvable optimization problems on graphs of bounded clique-width. Theory Comput. Syst. **33**(2), 125–150 (2000)
3. Daligault, J., Rao, M., Thomassé, S.: Well-quasi-order of relabel functions. Order **27**(3), 301–315 (2010)
4. Damaschke, P.: Induced subgraphs and well-quasi-ordering. J. Graph Theor. **14**(4), 427–435 (1990)
5. Diestel, R.: Graph Theory, 3rd edn. Springer-Verlag, Heidelberg (2005)
6. Finkel, A., Schnoebelen, P.: Well-structured transition systems everywhere!. Theor. Comput. Sci. **256**(1–2), 63–92 (2001)
7. Gurski, F., Wanke, E.: On the relationship between NLC-width and linear NLC-width. Theor. Comput. Sci. **347**(1–2), 76–89 (2005)
8. Higman, G.: Ordering by divisibility in abstract algebras. Proc. London Math. Soc. **2**, 326–336 (1952)
9. Korpelainen, N., Lozin, V.V.: Two forbidden induced subgraphs and well-quasi-ordering. Discrete Math. **311**(16), 1813–1822 (2011)
10. Kruskal, J.B.: The theory of well-quasi-ordering: a frequently discovered concept. J. Comb. Theory Ser. A **13**(3), 297–305 (1972)
11. Lozin, V.V., Rautenbach, D.: The relative clique-width of a graph. J. Comb. Theory Ser. B **97**(5), 846–858 (2007)
12. Petkovsek, M.: Letter graphs and well-quasi-order by induced subgraphs. Discrete Math. **244**(1–3), 375–388 (2002)
13. Robertson, N., Seymour, P.D.: Graph minors XX. Wagner's conjecture. J. Comb. Theory Ser. B **92**(2), 325–357 (2004)

A Slice Theoretic Approach for Embedding Problems on Digraphs

Mateus de Oliveira Oliveira[(✉)]

Institute of Mathematics - Academy of Sciences of the Czech Republic,
Prague, Czech Republic
mateus.oliveira@math.cas.cz

Abstract. We say that a digraph H can be covered by k paths if there exist k directed paths $\mathfrak{p}_1, \mathfrak{p}_2, \ldots, \mathfrak{p}_k$ such that $H = \cup_{i=1}^{k} \mathfrak{p}_i$. In this work we devise parameterized algorithms for embedding problems on digraphs in the setting in which the host digraph G has *directed* pathwidth w and the pattern digraph H can be covered by k paths. More precisely, we show that the subgraph isomorphism, subgraph homeomorphism, and two other related embedding problems can each be solved in time $2^{O(k \cdot w \log k \cdot w)} \cdot |H|^{O(k \cdot w)} \cdot |G|^{O(k \cdot w)}$. We note in particular that for constant values of w and k, our algorithm runs in polynomial time with respect to the size of the pattern digraph H. Therefore for the classes of digraphs considered in this work our results yield an exponential speedup with respect to the best general algorithm for the subgraph isomorphism problem which runs in time $O^*(2^{|H|} \cdot |G|^{tw(H)})$ (where $tw(H)$ is the *undirected* treewidth of H), and an exponential speedup with respect to the best general algorithm for the subgraph homeomorphism problem which runs in time $|G|^{O(|H|)}$.

Keywords: Directed pathwidth · Subgraph isomorphism · Subgraph homeomorphism · Slice languages

1 Introduction

The problem of determining whether a structure can be embedded into another is ubiquitous in mathematics and computer science. In the context of graph theory, two notoriously hard variants of embedding problems have been extensively studied. These are the subgraph isomorphism problem [1,9] and the subgraph homeomorphism problem [10,12,14,15]. In this work we provide new parameterized algorithms for embedding problems on digraphs. Our parameters are the *directed* pathwidth w of the host digraph G and the minimum number k of paths necessary to cover all edges and vertices of the pattern digraph H. A distinctive feature of our algorithms is that they work in polynomial time with respect to the number of vertices of the pattern digraph H whenever the two parameters w and k are held constant.

The notion of *directed* pathwidth is one of the earliest examples of directed width measure. According to Barát [3] this concept was introduced by Reed,

© Springer-Verlag Berlin Heidelberg 2016
E.W. Mayr (Ed.): WG 2015, LNCS 9224, pp. 360–372, 2016.
DOI: 10.1007/978-3-662-53174-7_26

Robertson and Seymour in the mid-nineties with the aim of developing an algorithmic metatheory suited for digraphs. There are several features that make the directed pathwidth of a digraph occupy a special place among directed width measures. First, this notion can be defined very naturally, via a slightly modification of the traditional notion of *undirected* pathwidth. Second, *directed* pathwidth is a generalization of its undirected counterpart. Indeed, if G is an undirected graph, and G' is obtained from G by replacing each of its edges by a pair of directed edges oriented in opposite directions, then the *directed* pathwidth of G' coincides with the *undirected* pathwidth of G [3]. Third, this generalization is strict. Digraphs of constant *directed* pathwidth may already have simultaneously unbounded *undirected* treewidth and clique-width. As a standard example for this fact, consider the $n \times n$ grid in which all horizontal edges are directed from left to right and all vertical edges are oriented upwards. This grid has directed pathwidth 0, since any DAG has directed pathwidth 0. Nevertheless this grid has *undirected* treewidth $\Omega(n)$ and clique-width $\Omega(n)$. This fact implies in particular that powerful algorithmic metatheorems such as those introduced in [2,4,5] cannot be applied to classes of digraphs of constant directed pathwidth. Fourth, many interesting combinatorial problems have been shown to be solvable in polynomial time on digraphs of constant directed pathwidth [8]. Fifth, directed pathwidth asymptotically lower bounds other important directed width measures such as, *cycle rank* [13], *DAG-depth ddp(G)* [11] and *K-width Kw(G)* [11]. Finally, given any constant w, one can determine in polynomial time whether a digraph G has directed pathwidth at most w, and if this is the case, one can construct a directed path decomposition in polynomial time [16].

The goal of this work is to extend the algorithmic theory of digraphs of constant pathwidth by addressing embedding problems in which the pattern digraph H is given in the input. We say that a digraph $H = (V, E)$ can be covered by k paths if there exist k directed paths $\mathfrak{p}_1, \mathfrak{p}_2, \ldots, \mathfrak{p}_k$ in H such that $\mathfrak{p}_i = (V_{\mathfrak{p}_i}, E_{\mathfrak{p}_i})$, $V = \cup_{i=1}^k V_i$ and $E = \cup_{i=1}^k E_i$. Our main result states that for each constants w and k, the subgraph isomorphism problem, the subgraph homeomorphism problem, and two other related embedding problems can be solved in polynomial time whenever the host digraph G has directed pathwidth w and the pattern digraph H can be covered by k paths.

Theorem 1 (Main Theorem). *Let G be a digraph of directed pathwidth w and H be a digraph that can be covered by k paths. Each of the following four problems can be solved in time $2^{O(k \cdot w \log k \cdot w)} \cdot |H|^{O(k \cdot w)} \cdot |G|^{O(k \cdot w)}$.*

1. *Determine whether H is isomorphic to a subgraph of G.*
2. *Determine whether a subdivision of H is isomorphic to a subgraph of G.*
3. *Determine whether H is isomorphic to a subdivision of a subgraph of G.*
4. *Determine whether a subdivision of H is isomorphic to a subdivision of a subgraph of G.*

Problem 1 is the classic subgraph isomorphism problem. The best parameterized algorithm for general graphs runs in time $O^*(2^{|H|} \cdot |G|^{tw(H)})$ where $tw(H)$ denotes the *undirected* treewidth of H [1,9]. We note that this running time depends exponentially on the number of vertices in H, whereas in our result, for constant values of k and w, the dependence is polynomial on the number of vertices of H.

Problem 2 is the classic subgraph homeomorphism problem. If the pattern graph H is not fixed, this problem is NP-complete even if both graphs have constant *undirected* treewidth [14]. For undirected graphs, the problem can be solved in time $f(|H|) \cdot |G|^3$ for some function $f(|H|)$ [12]. In other words, in the undirected case, the problem is fixed parameter tractable with respect to the size of the pattern graph H. For directed graphs however, it can be shown that the subgraph homeomorphism problem is NP-complete even if the pattern graph H consists only of two disjoint edges [10]. Thus not even an algorithm running in time $|G|^{O(|H|)}$ is expected to exist for general digraphs assuming $P \neq NP$. If the host graph has *directed* treewidth t then the problem can be solved in time $|G|^{O(t \cdot |H|)}$ using techniques from [15]. Note that in our algorithm the exponent of $|G|$ does not depend on the size of H, but only on k and w.

In Problems 3 and 4 we address two interesting variants of embedding problems for digraphs that are related to notions of embedding encountered in topology. Intuitively, Problem 3 asks whether the host digraph G can be slightly deformed so that it can accomodate the pattern digraph H. In Problem 4 we address a graph theoretic analog of the topological homeomorphism problem. Intuitively, we ask whether a deformed version of the pattern digraph H can be embedded into a deformed version of the host digraph G.

1.1   Proof Techniques

We will prove Theorem 1 using the framework of slice languages. This framework was introduced in [6,7] and used to solve several problems in the partial order theory of concurrency. Subsequently, slice languages were generalized to the context of digraphs and used to provide the first algorithmic metatheorem for digraphs of constant *directed* pathwidth [8]. In this work we will prove Theorem 1 by developing new machinery for the manipulation of slice languages.

Intuitively, a slice is a digraph $\mathbf{S}$ with special in-frontier and out-frontier vertices which can be used for composition. A slice $\mathbf{S}_1$ can be glued to a slice $\mathbf{S}_2$ if the out-frontier of $\mathbf{S}_1$ can be coherently matched with the in-frontier of $\mathbf{S}_2$. In this case, the glueing gives rise to a bigger slice $\mathbf{S}_1 \circ \mathbf{S}_2$ which is obtained by matching the out-frontier of $\mathbf{S}_1$ with the in-frontier of $\mathbf{S}_2$. A sequence $\mathbf{U} = \mathbf{S}_1 \mathbf{S}_2 \dots \mathbf{S}_n$ where each two consecutive slices can be glued is called a slice decomposition. After gluing each two consecutive slices in $\mathbf{U}$ we obtain a digraph $\overset{\circ}{\mathbf{U}} = \mathbf{S}_1 \circ \mathbf{S}_2 \circ \dots \circ \mathbf{S}_n$. Therefore, slices may be regarded as the basic constituents of digraphs in the same way that letters are the basic constituents of words. We may define infinite families of digraphs via finite automata that concatenate slices. We call these automata, *slice automata*. In order to define infinite families of digraphs, we associate two languages with each slice automaton $\mathcal{A}$. The first, the slice language $\mathcal{L}(\mathcal{A})$, is simply the set of all sequences of slices accepted by $\mathcal{A}$. The second, the graph language $\mathcal{L}_\mathcal{G}(\mathcal{A})$, is the set of all digraphs obtained by glueing the slices in each sequence accepted by $\mathcal{A}$.

In [8] we introduced the notion of z-saturated slice automaton, a notion that will be used recurrently in this work. We showed that given a digraph G of *directed* pathwidth z, and a z-saturated slice automaton $\mathcal{A}$ generating only digraphs that are the union of k paths, one can determine in time $|G|^{O(k \cdot z)} \cdot |\mathcal{A}|^{O(1)}$ whether some subgraphs of G is isomorphic to some digraph in the graph language $\mathcal{L}_\mathcal{G}(\mathcal{A})$ represented by $\mathcal{A}$. The caveat is that defining interesting families of digraphs via z-saturated slice automata is a difficult task. To circumvent this difficulty, we recurred to the monadic second order logic of graphs with edge set quantifications (MSO$_2$ Logic). We showed that for each given MSO$_2$ sentence φ, one can automatically construct a z-saturated slice automaton $\mathcal{A}(\varphi, k, z)$ whose graph language $\mathcal{L}_\mathcal{G}(\mathcal{A}(\varphi, k, z))$ consists precisely of the set of digraphs that at the same time are the union of k paths and satisfy φ. Using this result we were able to show how to solve in polynomial time several natural and interesting counting problems on digraphs of constant *directed* pathwidth.

Unfortunately, the construction of the automaton $\mathcal{A}(\varphi, k, z)$ given in [8] takes time at least exponential on $|\varphi|$. Additionally, assuming that P $\neq$ PSPACE, there is no algorithm that takes an MSO$_2$ formula φ and constructs the automaton $\mathcal{A}(\varphi, k, z)$ in polynomial time. This is due to the fact that the model checking problem for MSO logic is PSPACE-complete even for strings [17], which can be easily modeled in terms of graphs. Now suppose we want to solve the subgraph homeomorphism problem using the construction in [8]. Even if it is easy to define an MSO$_2$ formula φ_H^{sub} whose models are precisely the subdivisions of a digraph H, the construction of the automaton $\mathcal{A}(\varphi_H^{sub}, k, z)$ using the translation from formulas to slice automata given in [8] would take exponential time on $|\varphi_H^{sub}|$ and hence exponential time on $|H|$. Note that a simple counting argument implies that $|\varphi_H^{sub}| = \Omega(|H|)$ for almost all digraphs H.

One of the main technical contributions of this work is to show that for each digraph H that can be covered by k paths one can construct in time $2^{O(k \cdot z \log k \cdot z)} \cdot |H|^{O(k \cdot z)}$ a z-saturated slice automaton $\mathcal{A}(H, k, z)$ whose graph language consists only of H itself, i.e., $\mathcal{L}_\mathcal{G}(\mathcal{A}(H, k, z)) = \{H\}$. Another contribution is to show that given any z-saturated slice automaton $\mathcal{A}$ one can construct in time $2^{O(k \cdot z \log k \cdot z)} |\mathcal{A}|^{O(1)}$ a slice automaton $\mathbf{sdiv}(\mathcal{A})$ whose graph language $\mathcal{L}_\mathcal{G}(\mathcal{A})$ consists precisely of the subdivisions of digraphs in $\mathcal{L}_\mathcal{G}(\mathcal{A})$. In particular we have that $\mathcal{L}_\mathcal{G}(\mathbf{sdiv}(\mathcal{A}(H, k, z)))$ has also $2^{O(k \cdot z \log k \cdot z)} \cdot |H|^{O(k \cdot z)}$ states. To prove these results we will need to introduce new slice theoretic machinery. In particular we will define several non-boolean operations that will allow us to perform surgical transformations on all digraphs represented by a slice language. In this respect our work also represents a contribution to formal language theory. In particular we believe that our techniques can be used in other contexts where the goal is to develop graph theoretic algorithms via the manipulation of infinite families of graphs.

2 Directed Pathwidth and Zig-Zag Number

A *directed path decomposition* [3] of a digraph $G = (V, E)$ is a sequence $\mathcal{X} = (X_1, X_2, \ldots, X_r)$ of subsets of vertices G satisfying the following conditions.

(i) $\bigcup_{i=1}^{r} X_i = V$,

(ii) If $i < j < l$ then $X_i \cap X_l \subseteq X_j$,

(iii) For each edge $e \in E$ there exist $i, j \in \{1, \ldots, r\}$ with $i \leq j$ such that the source of e is in X_i and the target of e is in X_j.

The width of $\mathcal{X}$ is defined as $\max\{|X_i| : 1 \leq i \leq n\} - 1$. The *directed pathwidth* of a digraph G, denoted $ddp(G)$, is defined as the minimum width over all possible directed path decompositions of G. As a matter of comparison we observe that the classic notion of *undirected* pathwidth is recovered when Condition (iii) is replaced by

(iii') For each edge $e \in E$ there exists $i \in \{1, \ldots, r\}$ such that both endpoints of e lie in X_i.

Next, we define the notion of *z-topological ordering* which can be used to establish a connection between digraphs of constant *directed* pathwidth and slice languages. Let $G = (V, E)$ be a directed graph. For subsets of vertices $V_1, V_2 \subseteq V$ we let $E(V_1, V_2)$ denote the set of edges with one endpoint in V_1 and another endpoint in V_2. We say that a linear ordering $\omega = (v_1, v_2, \ldots, v_n)$ of the vertices of V is a *z-topological ordering* of G if for every *directed* simple path $\mathfrak{p} = (V_\mathfrak{p}, E_\mathfrak{p})$ in G and every $i \in \{1, \ldots, n-1\}$, we have that $|E_\mathfrak{p} \cap E(\{v_1 \ldots, v_i\}, \{v_{i+1}, \ldots, v_n\})| \leq z$. In other words, ω is a *z-topological ordering* if for each $i \in \{1, \ldots, n-1\}$, each *directed* simple path of G enters or leaves the set $\{v_1, \ldots, v_i\}$ at most z times. We say that a digraph G has *zig-zag number z* if some ordering ω of the vertices of G is a *z-topological ordering*.

Theorem 2 ([8]). *Let G be a digraph of directed pathwidth w. Then one can construct in time $|G|^{O(w)}$ a z-topological ordering $\omega = (v_1, \ldots, v_n)$ for G such that $z \leq 2w + 1$.*

3   Slices and Slice Languages

A slice $\mathbf{S} = (V, E, \rho, \xi, s, t, [C, I, O])$ is a digraph comprising a set of vertices V, a set of edges E, a vertex labeling function $\rho : V \to \Gamma_1$ for some finite set of symbols Γ_1, an edge labeling function $\xi : E \to \Gamma_2$ for some finite set of symbols Γ_2 and total functions $s, t : E \to V$ associating with each edge $e \in E$ a source vertex e^s and a target vertex e^t. Alternatively, we say that e^s and e^t are the endpoints of e. The vertex set V is partitioned into three disjoint subsets: an in-frontier I, a center C, and an out-frontier O. A slice is subject to the following restrictions:

1. The frontier vertices of $\mathbf{S}$ are labeled by ρ with natural numbers in such a way that no two vertices in the same frontier are labeled with the same number.
2. Each frontier vertex in $I \cup O$ is the endpoint of exactly one edge.
3. No edge has both endpoints in the same frontier.

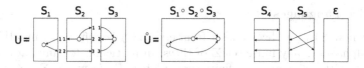

Fig. 1. A unit decomposition $\mathbf{U} = \mathbf{S}_1\mathbf{S}_2\mathbf{S}_3$ and the digraph $\overset{\circ}{\mathbf{U}} = \mathbf{S}_1 \circ \mathbf{S}_2 \circ \mathbf{S}_3$ which is obtained by gluing each two adjacent slices in $\mathbf{U}$. All slices in $\mathbf{U}$ are normalized. The in-frontier of $\mathbf{S}_1$ is empty and the out-frontier of $\mathbf{S}_3$ is empty. The slices $\mathbf{S}_4$ and $\mathbf{S}_5$ and ε are permutation slices. $\mathbf{S}_4$ is additionally an identity slice. ε is the empty slice.

From now on we let the source and target functions be implicit when defining a slice. When referring to a slice $\mathbf{S} = (V, E, \rho, \xi)$ with frontiers (I, O) we mean that $\mathbf{S}$ has in-frontier I and out-frontier O. We say that a slice $\mathbf{S}$ with frontiers (I, O) is normalized if $\rho(I) = \{1, \ldots, |I|\}$ and $\rho(O) = \{1, \ldots, |O|\}$. Non-normalized slices will play an important role later in this section when we introduce the notion of sub-decomposition. Let $i \in \rho(I)$. We say that a slice $\mathbf{S}$ is a *unit slice* if $\mathbf{S}$ has at most one center vertex. Unit slices will play a very important role in this work. We denote by $e(I, i)$ the unique edge that has as one endpoint the vertex of I labeled with i. Analogously, $e(O, i)$ denotes the unique edge that has as one endpoint the vertex in O labeled with i. A slice $\mathbf{S}_1 = (V_1, E_2, \rho_1, \xi_1)$ with frontiers (I_1, O_1) can be glued to a slice $\mathbf{S}_2 = (V_2, E_2, \rho_2, \xi_2)$ with frontiers (I_2, O_2) provided the following conditions are satisfied.

1. $\rho_1(O_1) = \rho_2(I_2)$,
2. for each $i \in \rho(O_1)$, $\xi_1(e(O_1, i)) = \xi_2(e(I_2, i))$,
3. for each $i \in \rho(O_1)$, either the target of $e(O_1, i)$ lies in O_1 and the source of $e(I_2, i)$ in I_2, or the source of $e(O_1, i)$ lies in O_1 and the target of $e(I_2, i)$ in I_2.

Intuitively, $\mathbf{S}_1$ can be glued to $\mathbf{S}_2$ if for each $i \in \rho(O_1)$, the edge $e(O_1, i)$ can be matched with the edge $e(I_2, i)$ in such a way that the two edges agree both in labeling (Condition 2) and direction (Condition 3). If $\mathbf{S}_1$ can be glued to $\mathbf{S}_2$, then we let $e(i, \mathbf{S}_1, \mathbf{S}_2)$ denote the edge that is obtained by fusing $e(O_1, i)$ with $e(I_2, i)$. More precisely, if the target of $e(O_1, i)$ lies in O_1 then we set $e(\mathbf{S}_1, \mathbf{S}_2, i)^s = e(O_1, i)^s$ and $e(\mathbf{S}_1, \mathbf{S}_2, i)^t = e(I_2, i)^t$. Otherwise, if the source of $e(O_1, i)$ lies in O_1, then we set $e(\mathbf{S}_1, \mathbf{S}_2, i)^s = e(I_2, i)^s$ and $e(\mathbf{S}_1, \mathbf{S}_2, i)^t = e(O_1, i)^t$. If $\mathbf{S}_1$ can be glued to $\mathbf{S}_2$ then the gluing gives rise to the slice $\mathbf{S}_1 \circ \mathbf{S}_2 = (V_3, E_3, \rho_3, \xi_3)$ with frontiers (I_1, O_2) where the vertex set is $V_3 = (V_1 \cup V_2)\backslash(O_1 \cup I_2)$, and the edge set is $E_3 = [(E_1 \cup E_2)\backslash\{e(O_1, i), e(I_2, i) \mid i \in \rho(O_1)\}] \cup \{e(\mathbf{S}_1, \mathbf{S}_2, i) \mid i \in \rho(O_1)\}$.

The labels of vertices and edges are inherited from the slice they come from. More precisely for $j \in \{1, 2\}$, $\rho_3|_{V_3 \cap V_j} = \rho_j|_{V_3 \cap V_j}$, $\xi_3|_{E_3 \cap E_j} = \xi_j|_{E_3 \cap E_j}$ and $\xi(e(\mathbf{S}_1, \mathbf{S}_2, i)) = \xi_1(e(O_1, i))$ for each $i \in \rho_1(O_1)$. We note that in the glueing process the vertices belonging to the glued frontiers O_1 and I_2 are deleted, while the center vertices of both slices and the vertices in I_1 and O_2 remain intact.

3.1   Slice Languages

The width $\mathbf{w(S)}$ of a slice $\mathbf{S}$ with frontiers (I, O) is defined as $\max\{|I|, |O|\}$. A slice alphabet is any finite set Σ of slices. In particular for any natural numbers c, q, ν with $c \leq q$, and any finite sets of labels Γ_1, Γ_2 we let $\Sigma(c, q, \nu, \Gamma_1, \Gamma_2)$ be the slice alphabet formed by all slices of width at most c, with at most ν center vertices, whose center vertices are labeled with elements from Γ_1, edges are labeled with elements from Γ_2 and whose frontier vertices are labeled with numbers in $\{1, \ldots, q\}$. We write $\Sigma(c, q, \Gamma_1, \Gamma_2)$ as an abbreviation for $\Sigma(c, q, 1, \Gamma_1, \Gamma_2)$ and $\Sigma(c, \Gamma_1, \Gamma_2)$ for the set of all unit normalized slices in $\Sigma(c, c, 1, \Gamma_1, \Gamma_2)$.

If Σ is a slice alphabet, we denote by Σ^* the free monoid generated by Σ. In other words Σ^* is simply the set of all sequences of slices taken from Σ. The operation of the monoid is simply concatenation and should not be confused with glueing. The identity of the monoid is simply the empty string λ for which $\mathbf{S}\lambda = \mathbf{S} = \lambda\mathbf{S}$ and should not be confused with the empty slice ε. We say that a slice is *initial* if its in-frontier is empty, and *final* if its out-frontier is empty. A *slice decomposition* is a sequence $\mathbf{D} = \mathbf{S}_1\mathbf{S}_2 \ldots \mathbf{S}_n$ of slices such that $\mathbf{S}_1$ is initial, $\mathbf{S}_n$ is final and such that $\mathbf{S}_i$ can be glued to $\mathbf{S}_{i+1}$ for each $i \in \{1, \ldots, n-1\}$. The width $\mathbf{w(D)}$ of $\mathbf{D}$ is defined as the maximum width of a slice in $\mathbf{D}$. We let $\mathcal{L}(\Sigma)$ denote the set of all slice decompositions in Σ^*. A *slice language* is any subset $\mathcal{L} \subseteq \mathcal{L}(\Sigma)$. Any slice decomposition $\mathbf{D} = \mathbf{S}_1\mathbf{S}_2 \ldots \mathbf{S}_n$ in a slice language $\mathcal{L}$ gives rise to a digraph $\mathring{\mathbf{D}} = \mathbf{S}_1 \circ \mathbf{S}_2 \circ \ldots \circ \mathbf{S}_n$ which is obtained by gluing each two consecutive slices in $\mathbf{D}$. Thus slice languages may be regarded as a syntactic way of representing possibly infinite families of digraphs. Namely, the graph language derived from $\mathcal{L}$ is defined as

$$\mathcal{L}_{\mathcal{G}} = \{\mathring{\mathbf{D}} \mid \mathbf{D} \in \mathcal{L}\}. \tag{1}$$

In this work we will only be concerned with slice languages that can be effectively represented. In particular we will be concerned with the class of regular slice languages, which are those languages that can be represented via finite automata over slice alphabets. We call these automata *slice automata*.

Definition 3 (Slice Automaton). *A slice automaton over a slice alphabet Σ is a finite automaton $\mathcal{A} = (Q, \mathfrak{R}, Q_0, F)$ where Q is a set of states, $Q_0 \in Q$ is a set of initial states, $F \subseteq Q$ is a set of final states, and $\mathfrak{R} \subseteq Q \times \Sigma \times Q$ is a transition relation such that the following conditions are satisfied for every $r, r', r'' \in Q$ and every $\mathbf{S} \in \Sigma$.*

1. *if $(r, \mathbf{S}, r') \in \mathfrak{R}$ and $r \in Q_0$ then $\mathbf{S}$ is an initial slice,*
2. *if $(r, \mathbf{S}, r') \in \mathfrak{R}$ and $r' \in F$, then $\mathbf{S}$ is a final slice,*
3. *if $(r, \mathbf{S}, r') \in \mathfrak{R}$ and $(r', \mathbf{S}', r'') \in \mathfrak{R}$, then $\mathbf{S}$ can be glued to $\mathbf{S}'$.*

We denote by $\mathcal{L}(\mathcal{A})$ the slice language accepted by $\mathcal{A}$ and by $\mathcal{L}_{\mathcal{G}}(\mathcal{A})$ the graph language derived from $\mathcal{L}(\mathcal{A})$ according to Eq. 1. For instance, in Fig. 2 we depict a slice automaton representing an infinite family of digraphs.

Given a slice alphabet Σ, we denote by $\mathcal{A}(\Sigma)$ the minimum deterministic slice automaton generating the slice language $\mathcal{L}(\Sigma)$ consisting of all slice decompositions over Σ. It can be easily shown that $\mathcal{A}(\Sigma)$ has size $O(|\Sigma|)$.

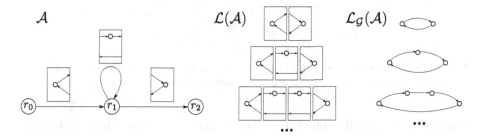

Fig. 2. A slice automaton $\mathcal{A}$, its slice language $\mathcal{L}(\mathcal{A})$ and the graph language $\mathcal{L}_{\mathcal{G}}(\mathcal{A})$ which consists of all cycles of size at least two.

3.2 Sub-slices and Sub-decompositions

A sub-slice of $\mathbf{S}$ is a subgraph $\mathbf{S}'$ of $\mathbf{S}$ which is itself a slice. Note that the numbering of the frontier vertices of $\mathbf{S}'$ is inherited from the numbering of the frontier vertices of $\mathbf{S}$. Thus even if $\mathbf{S}$ is a normalized slice, $\mathbf{S}'$ may not be normalized. Let $\mathbf{D} = \mathbf{S}_1\mathbf{S}_2\ldots\mathbf{S}_n$ be a slice decomposition. A sub-decomposition of $\mathbf{D}$ is a decomposition $\mathbf{D}' = \mathbf{S}_1'\mathbf{S}_2'\ldots\mathbf{S}_n'$ such that $\mathbf{S}_i'$ is a sub-slice of $\mathbf{S}_i$ for each $i \in \{1,\ldots,n\}$. Note that even if $\mathbf{D}$ is a normalized slice decomposition, a sub-decomposition $\mathbf{D}'$ of $\mathbf{D}$ may not be normalized (Fig. 3).

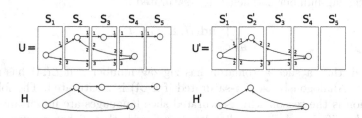

Fig. 3. A digraph H and one of its unit decompositions $\mathbf{U}$. The digraph H' is a subgraph of H and the unit decomposition $\mathbf{U}'$ is a sub-decomposition of $\mathbf{U}$. Note that some slices in $\mathbf{U}'$ are not normalized. For instance, $\mathbf{S}_3'$ is not normalized because even though it has width 2, its frontier vertices are labeled with numbers from $\{2,3\}$ and not from $\{1,2\}$. Finally, note that since a sub-decomposition has the same number of slices of the original decomposition, some of the slices occurring in it may be permutation slices or even the empty slice, e.g., $\mathbf{S}_3'$ and $\mathbf{S}_5'$.

3.3 z-Saturated Slice Languages

Recall that a unit slice is a slice with at most one center vertex. A *unit decomposition* is a slice decomposition $\mathbf{U} = \mathbf{S}_1\mathbf{S}_2\ldots\mathbf{S}_n$ in which all slices are unit slices. Most of the time we will use the letter $\mathbf{U}$, possibly with subscripts, to denote unit decompositions. We will denote by $\overset{\circ}{\mathbf{U}}$ the graph $\mathbf{S}_1 \circ \mathbf{S}_2 \circ \ldots \circ \mathbf{S}_n$ derived from a unit decomposition $\mathbf{U} = \mathbf{S}_1\mathbf{S}_2\ldots\mathbf{S}_n$. We say that $\mathbf{U}$ is a unit

Fig. 4. Two unit decompositions of the cycle on 4 vertices. The unit decomposition (i) has zig-zag number 2 while the unit decomposition (ii) has zig-zag number 3.

decomposition of a digraph G if $G = \overset{\circ}{\mathbf{U}}$. In general a digraph may have many unit decompositions (Fig. 4).

Let H be a digraph with n vertices, $\omega = (v_1, v_2, \ldots, v_n)$ be a linear ordering of the vertices of H and $\mathbf{U} = \mathbf{S}_1 \mathbf{S}_2 \ldots \mathbf{S}_m \in \mathcal{L}(\mathbf{\Sigma})$ be a unit decomposition of H for some $m \geq n$. We say that $\mathbf{U}$ is compatible with ω if there exists a sequence $j_1 < j_2 < \ldots < j_n$ of indices such that the vertex v_i of H is the center vertex of $\mathbf{S}_{j_i}$. Observe that we need to use double sub-indexes j_k because m may be greater than n, since the unit decomposition $\mathbf{U}$ may have some slices with no center vertex. We say that $\mathbf{U}$ has zig-zag number z if $\mathbf{U}$ is compatible with a z-topological ordering of H. Intuitively, $\mathbf{U}$ has zig-zag number z if every simple path in the digraph $\overset{\circ}{\mathbf{U}}$ crosses each frontier of each slice of $\mathbf{U}$ at most z times (Fig. 4). A slice language $\mathcal{L}$ has zig-zag number z if each unit decomposition in $\mathcal{L}$ has zig-zag number at most z. A slice language $\mathcal{L} \subseteq \mathcal{L}(\mathbf{\Sigma})$ is z-saturated if for each digraph $H \in \mathcal{L}$ all unit decompositions of H of zig-zag number z are in $\mathcal{L}$. More precisely, let $\mathbf{ud}(H, \mathbf{\Sigma}, z) \subseteq \mathcal{L}(\mathbf{\Sigma})$ denote the set of all unit decompositions of H of zig-zag number z. Then $\mathcal{L}$ is z-saturated if

$$\bigcup_{H \in \mathcal{L}_\mathcal{G}} \mathbf{ud}(H, \mathbf{\Sigma}, z) \subseteq \mathcal{L}. \tag{2}$$

We say that a slice automaton has zig-zag number z if $\mathcal{L}(\mathcal{A})$ has zig-zag number z. Analogously, $\mathcal{A}$ is z-saturated if $\mathcal{L}(\mathcal{A})$ is z-saturated. The following proposition is the reason why z-saturated slice languages are important for us. It says that if $\mathcal{L}$ and $\mathcal{L}'$ are slice languages such that $\mathcal{L}$ has zig-zag number z and $\mathcal{L}'$ is z-saturated, then the graph language $[\mathcal{L} \cap \mathcal{L}']_\mathcal{G}$ represented by the intersection $\mathcal{L} \cap \mathcal{L}'$ is precisely the intersection of the graph languages $\mathcal{L}_\mathcal{G}$ and $\mathcal{L}_{\mathcal{G}'}$ represented by $\mathcal{L}$ and $\mathcal{L}'$ respectively.

Proposition 4 ([8]). *Let $\mathcal{L}$ and $\mathcal{L}'$ be slice languages over $\mathbf{\Sigma}$. Let $\mathcal{L}$ have zig-zag number z and $\mathcal{L}'$ be z-saturated. Then $(\mathcal{L} \cap \mathcal{L}')_\mathcal{G} = \mathcal{L}_\mathcal{G} \cap \mathcal{L}'_\mathcal{G}$.*

We observe that Proposition 4 is not true in general if none of the slice languages is z-saturated. As a simple example, let $\mathcal{L} = \{\mathbf{U}\}$ and $\mathcal{L}' = \{\mathbf{U}'\}$ where $\mathbf{U}$ and $\mathbf{U}'$ are two distinct unit decompositions of the same graph H. Then $\mathcal{L}_\mathcal{G} \cap \mathcal{L}'_\mathcal{G} = \{H\}$ but $\mathcal{L} \cap \mathcal{L}' = \emptyset$.

4   High Level Proof of Theorem 1

In this section we will state four results relating regular slice languages with subgraphs and subdivisions of digraphs. Subsequently, we will use these results to prove Theorem 1. The proofs of these four results can be found in the full version of this paper. We start by stating a lemma that says that given any digraph G of zig-zag number z, one can construct in polynomial time a slice automaton of zig-zag number z whose graph language contains all subgraphs of G that are the union of k paths.

Lemma 5. *Let $G = (V, E)$ be a digraph and $\mathbf{U} \in \mathcal{L}(\Sigma(|E|, \Gamma_1, \Gamma_2))$ be a normalized unit decomposition of G of zig-zag number z. Then one can construct in time $|G|^{O(k \cdot z)}$ a slice automaton $\mathcal{A}(\mathbf{U}, k \cdot z)$ over $\Sigma(k \cdot z, \Gamma_1, \Gamma_2)$ whose graph language $\mathcal{L}_{\mathcal{G}}(\mathcal{A}(\mathbf{U}, k \cdot z))$ contains all subgraphs of G that are the union of k paths.*

We observe that the slice automaton $\mathcal{A}(G, k \cdot z)$ constructed in Lemma 5 is not necessarily z-saturated. We also note that $\mathcal{L}_{\mathcal{G}}(\mathcal{A}(G, k \cdot z))$ may contain digraphs that are isomorphic to subgraphs of G that are not the union of k paths. What is important in Lemma 5 is that whenever a subgraph H of G can be covered by k paths, there will be a unit decomposition $\mathbf{U}$ in $\mathcal{L}(\mathcal{A}(G, k \cdot z))$ such that $\mathring{\mathbf{U}} = H$.

The next lemma says that given any slice automaton $\mathcal{A}$ of zig-zag number z, one can efficiently construct another slice automaton $\mathbf{div}(\mathcal{A})$ of zig-zag number z whose graph language $\mathcal{L}_{\mathcal{G}}(\mathbf{div}(\mathcal{A}))$ is the set of all subdivisions of digraphs in $\mathcal{L}_{\mathcal{G}}(\mathcal{A})$. In other words, Lemma 6 below gives us a way to transform a concrete representation of an infinite set of digraphs into a concrete representation of the set of subdivisions of these digraphs.

Lemma 6. *Let $\mathcal{A}$ be a slice automaton over $\Sigma(c, \Gamma_1, \Gamma_2)$ of zig-zag number z. Then one can construct in time $O(c \cdot |\mathcal{A}|)$ a slice automaton $\mathbf{div}(\mathcal{A})$ of zig-zag number z such that*

$$\mathcal{L}_{\mathcal{G}}(\mathbf{div}(\mathcal{A})) = \{H \mid \exists H' \in \mathcal{L}_{\mathcal{G}}(\mathcal{A}) \text{ such that } H \text{ is a subdivision of } H'\}.$$

We say that $\mathbf{div}(\mathcal{A})$ is the subdivision of $\mathcal{A}$. We observe that even if the slice automaton $\mathcal{A}$ is z-saturated, the slice automaton $\mathbf{div}(\mathcal{A})$ is not necessarily z-saturated. The next theorem says that if H can be covered by k paths, then one can construct in polynomial time a normalized slice automaton generating precisely the unit decompositions of H of zig-zag number at most z.

Theorem 7. *Let H be a digraph that can be covered by k paths. One can construct in time $2^{O(k \cdot z \cdot \log k \cdot z)} \cdot |E|^{O(k \cdot z)}$ a slice automaton $\mathcal{A}(H, k, z)$ generating the following z-saturated slice language over $\Sigma(k \cdot z, \Gamma_1, \Gamma_2)$:*

$$\mathcal{L}(\mathcal{A}(H, k, z)) = \{\mathbf{U} \mid \mathbf{U} \text{ has zig-zag number } z, \mathring{\mathbf{U}} = H\} \tag{3}$$

We observe that the graph H in Theorem 7 is fixed. Thus, while the slice language $\mathcal{L}(\mathcal{A}(H, k, z))$ consists of all unit decompositions of H of zig-zag number at most z, the graph language $\mathcal{L}_{\mathcal{G}}(\mathcal{A}(H, k, z))$ is simply the singleton $\{H\}$.

Finally, the next theorem says that given any z-saturated slice automaton $\mathcal{A}$ one can construct in polynomial time a z-saturated slice automaton $\mathbf{sdiv}(\mathcal{A})$ such that the graph language of $\mathbf{sdiv}(\mathcal{A})$ consists of all subdivisions of digraphs in $\mathcal{L}_{\mathcal{G}}(\mathcal{A})$.

Theorem 8. *Let $\mathcal{A}$ be a z-saturated slice automaton over $\Sigma(c, \Gamma_1, \Gamma_2)$. Then one can construct in time $2^{O(c \log c)} \cdot |\mathcal{A}|$ a z-saturated slice automaton $\mathbf{sdiv}(\mathcal{A})$ such that*

$$\mathcal{L}_\mathcal{G}(\mathbf{sdiv}(\mathcal{A})) = \{H \mid H \text{ is a subdivision of a digraph in } \mathcal{L}_\mathcal{G}(\mathcal{A})\} \quad (4)$$

There are two important differences between the slice automaton $\mathbf{sdiv}(\mathcal{A})$ of Theorem 8 and the slice automaton $\mathbf{div}(\mathcal{A})$ of Lemma 6. The first is that $\mathbf{sdiv}(\mathcal{A})$ preserves z-saturation, while $\mathbf{div}(\mathcal{A})$ does not. The second is that Theorem 8 can only be applied if $\mathcal{A}$ is z-saturated, while Lemma 6 only requires that $\mathcal{A}$ has zig-zag number z.

Now we are in a position to prove Theorem 1 using Lemmas 5 and 6, and Theorems 7 and 8. We will use these two lemmas and two theorems to reduce the problems stated in Theorem 1 to a problem of non-emptiness of intersection between a slice language of zig-zag number z and a z-saturated slice language.

Proof of Theorem 1. Let $G = (V, E)$ be a host digraph with n vertices and directed pathwidth w, and let H be a pattern digraph that can be covered by k paths. Applying Theorem 2 we can construct in time $|G|^{O(w)}$ a z-topological ordering $\omega = (v_1, v_2, \ldots, v_n)$ of G where $z \leq 2w + 1$. Using ω we can construct in linear time a normalized unit decomposition $\mathbf{U} = \mathbf{S}_1 \mathbf{S}_2 \ldots \mathbf{S}_n$ of G of zig-zag number z in which v_i is the center vertex of $\mathbf{S}_i$.

1. By Lemma 5, the graph language $\mathcal{L}_\mathcal{G}(\mathcal{A}(\mathbf{U}, k \cdot z))$ of the slice automaton $\mathcal{A}(\mathbf{U}, k \cdot z)$ contains all subgraphs of G that are the union of k paths. The pattern digraph H can be covered by k paths. Additionally, by Theorem 7, the graph language $\mathcal{L}_\mathcal{G}(\mathcal{A}(H, k, z))$ is simply the singleton $\{H\}$. Since $\mathcal{A}(\mathbf{U}, k \cdot z)$ has zig-zag number z and $\mathcal{A}(H, k, z)$ is z-saturated, Proposition 4 implies that H is a subgraph of G if and only if $\mathcal{L}(\mathcal{A}(\mathbf{U}, k \cdot z) \cap \mathcal{A}(H, k, z))$ is non-empty. Since $\mathcal{A}(\mathbf{U}, k \cdot z)$ has $|G|^{O(k \cdot z)}$ states and $\mathcal{A}(H, k, z)$ has $2^{O(k \cdot z \log k \cdot z)} \cdot |H|^{O(k \cdot z)}$ states, we have that the non-emptiness of intersection of these two automata can be determined in time $2^{O(k \cdot z \log k \cdot z)} \cdot |H|^{O(k \cdot z)} \cdot |G|^{O(k \cdot z)}$.

2. By Theorem 8, a digraph H' belongs to the graph language $\mathcal{L}_\mathcal{G}(\mathbf{sdiv}(\mathcal{A}(H, k, z)))$ if and only if H' is a subdivision of H. Observe that, since H can be covered by k paths, each of its subdivisions is also the union of k paths. Since $\mathbf{sdiv}(\mathcal{A}(H, k, z))$ is z-saturated, by Proposition 4, we have that a subdivision of H is a subgraph of G if and only if the slice language $\mathcal{L}(\mathbf{sdiv}(\mathcal{A}(H, k, z)) \cap \mathcal{A}(\mathbf{U}, k \cdot z))$ is non-empty. Since $\mathbf{sdiv}(\mathcal{A}(H, k, z))$ has $2^{O(k \cdot z \log k \cdot z)} \cdot |H|^{O(k \cdot z)}$ states, we have that the non-emptiness of the intersection above can be decided in time $2^{O(k \cdot z \log k \cdot z)} \cdot |H|^{O(k \cdot z)} \cdot |G|^{O(k \cdot z)}$.

3. By Lemma 6, the graph language $\mathcal{L}_\mathcal{G}(\mathbf{div}(\mathcal{A}(\mathbf{U}, k \cdot z)))$ consists of all subdivisions of subgraphs of G that are the union of k paths. Since $\mathbf{div}(\mathcal{A}(\mathbf{U}, k \cdot z))$ has zig-zag number z and $\mathcal{A}(H, k, z)$ is z-saturated, Proposition 4 implies that a subdivision of H is isomorphic to a subgraph of G if and only if the slice language $\mathcal{L}(\mathcal{A}(H, k, z) \cap \mathbf{div}(\mathcal{A}(\mathbf{U}, k \cdot z)))$ is non-empty. Since $\mathbf{div}(\mathcal{A}(\mathbf{U}, k \cdot z))$ has $|G|^{O(k \cdot z)}$ states, we have that the non-emptiness of this intersection can be tested in time $2^{O(k \cdot z \log k \cdot z)} \cdot |H|^{O(k \cdot z)} \cdot |G|^{O(k \cdot z)}$.

4. Since $\mathbf{sdiv}(\mathcal{A}(H, k, z))$ is z-saturated and $\mathbf{div}(\mathcal{A}(\mathbf{U}, k, z))$ has zig-zag number z, by Proposition 4 we have that a subdivision of H is isomorphic to a subdivision of a subgraph of G if and only if the slice language of the intersection $\mathbf{div}(\mathcal{A}(\mathbf{U}, k \cdot z)) \cap \mathbf{sdiv}(\mathcal{A}(H, k, z))$ is non-empty. By similar arguments as in the last three items, we have that the intersection can be tested in time $2^{O(k \cdot z \log k \cdot z)} \cdot |H|^{O(k \cdot z)} \cdot |G|^{O(k \cdot z)}$.

5 Conclusion

In this work we showed how to solve in polynomial time four variants of embedding problems in which the host digraph has constant directed pathwidth, and the pattern digraph can be covered by a constant number of directed paths. Indeed we reduced all four problems to the problem of intersection of slice languages. To prove our main results we introduced new slice theoretic machinery that may be of independent interest. In particular, the concept of z-saturation, together with the new operations we defined on slice languages may have the potential to shed new light on algorithmic questions on digraphs for which other techniques such as dynamic programming do not provide an evident solution.

Acknowledgements. I gratefully acknowledge financial support from the European Research Council, ERC grant agreement 339691, within the context of the project Feasibility, Logic and Randomness (FEALORA).

References

1. Alon, N., Yuster, R., Zwick, U.: Color-coding. J. ACM (JACM) **42**(4), 844–856 (1995)
2. Arnborg, S., Lagergren, J., Seese, D.: Easy problems for tree-decomposable graphs. J. Algorithms **12**(2), 308–340 (1991)
3. Barát, J.: Directed path-width and monotonicity in digraph searching. Graphs and Combinatorics **22**(2), 161–172 (2006)
4. Courcelle, B.: Graph rewriting: an algebraic and logic approach. In: Handbook of Theoretical Computer Science, Chap. 5, pp. 194–242. Elsevier, Amsterdam (1990)
5. Courcelle, B., Makowsky, J.A., Rotics, U.: Linear time solvable optimization problems on graphs of bounded clique-width. Theor. Comput. Syst. **33**(2), 125–150 (2000)
6. de Oliveira Oliveira, M.: Hasse diagram generators and Petri nets. Fundamenta Informaticae **105**(3), 263–289 (2010)
7. de Oliveira Oliveira, M.: Canonizable partial order generators. In: Dediu, A.-H., Martín-Vide, C. (eds.) LATA 2012. LNCS, vol. 7183, pp. 445–457. Springer, Heidelberg (2012)
8. de Oliveira Oliveira, M.: Subgraphs satisfying MSO properties on z-topologically orderable digraphs. In: Gutin, G., Szeider, S. (eds.) IPEC 2013. LNCS, vol. 8246, pp. 123–136. Springer, Heidelberg (2013)
9. Fomin, F.V., Lokshtanov, D., Raman, V., Saurabh, S., Rao, B.V.R.: Faster algorithms for finding and counting subgraphs. J. Comput. Syst. Sci. **78**(3), 698–706 (2012)

10. Fortune, S., Hopcroft, J.E., Wyllie, J.: The directed subgraph homeomorphism problem. Theor. Comput. Sci. **10**, 111–121 (1980)
11. Ganian, R., Hliněný, P., Kneis, J., Langer, A., Obdržálek, J., Rossmanith, P.: On digraph width measures in parameterized algorithmics. In: Chen, J., Fomin, F.V. (eds.) IWPEC 2009. LNCS, vol. 5917, pp. 185–197. Springer, Heidelberg (2009)
12. Grohe, M., Kawarabayashi, K.-i., Marx, D., Wollan, P.: Finding topological subgraphs is fixed-parameter tractable. In: STOC, pp. 479–488. ACM (2011)
13. Gruber, H.: Digraph complexity measures and applications in formal language theory. Discrete Math. Theor. Comput. Sci. **14**(2), 189–204 (2012)
14. Gupta, A., Nishimura, N., Proskurowski, A., Ragde, P.: Embeddings of k-connected graphs of pathwidth k. Discrete Appl. Math. **145**(2), 242–265 (2005)
15. Johnson, T., Robertson, N., Seymour, P.D., Thomas, R.: Directed tree-width. J. Comb. Theor. Ser. B **82**(1), 138–154 (2001)
16. Tamaki, H.: A polynomial time algorithm for bounded directed pathwidth. In: Kolman, P., Kratochvíl, J. (eds.) WG 2011. LNCS, vol. 6986, pp. 331–342. Springer, Heidelberg (2011)
17. Vardi, M.Y.: The complexity of relational query languages. In: Proceedings of the Fourteenth Annual ACM Symposium on Theory of Computing, pp. 137–146. ACM (1982)

Decomposition Theorems for Square-free 2-matchings in Bipartite Graphs

Kenjiro Takazawa$^{(\boxtimes)}$

Department of Industrial and Systems Engineering, Faculty of Science
and Engineering, Hosei University, Tokyo 184-8584, Japan
takazawa@hosei.ac.jp

Abstract. A square-free 2-matching in an undirected graph is a simple
2-matching without cycles of length four. In bipartite graphs, the maximum square-free 2-matching problem is well-solved. Previous results
include min-max theorems, polynomial combinatorial algorithms, poly-
hedral description with dual integrality, and discrete convex structure.

In this paper, we further investigate the structure of square-free
2-matchings in bipartite graphs to present new decomposition theorems,
which serve as an analogue of the Dulmage-Mendelsohn decomposition
for bipartite matchings and the Edmonds-Gallai decomposition for nonbipartite matchings. We exhibit two canonical minimizers of the set function in the min-max formula, and a characterization of the maximum
square-free 2-matchings with the aid of these canonical minimizers.

Keywords: Matching theory · Square-free 2-matching · Dulmage-
Mendelsohn decomposition · Edmonds-Gallai decomposition

1 Introduction

For a simple undirected graph $G = (V, E)$ and a positive integer k, a subset M
of E is a 2-*matching* if each vertex in V has at most two incident edges in M,
and a 2-matching M is called C_k-*free* if it does not contain a cycle of length k or
less. The maximum C_k-*free* 2-*matching problem* is a problem of finding a C_k-free
2-matching of maximum size for given G and k. If a 2-matching has size $|V|$,
which shall be maximum, then it is called a 2-*factor*.

The larger k becomes, the closer a C_k-free 2-factor becomes to a Hamilton cycle. Indeed, a main motivation of investigating the maximum C_k-free
2-matching problem is that it is a relaxation of the Hamilton cycle problem.
Utilizing matching theory has been one of the most effective approaches to
the Hamilton cycle problem and the traveling salesman problem (TSP), and
C_k-free 2-factors would provide a tighter relaxation of Hamilton cycles to 2-
factors. For instance, the most standard linear programming relaxation of the
TSP due to Dantzig, Fulkerson and Johnson [11] is exactly a fractional 2-
matching polytope with subtour elimination constraints added. Clearly, a C_k-
free 2-matching corresponds to an integer vector in the 2-matching polytope

© Springer-Verlag Berlin Heidelberg 2016
E.W. Mayr (Ed.): WG 2015, LNCS 9224, pp. 373–387, 2016.
DOI: 10.1007/978-3-662-53174-7_27

satisfying the constraints arising from the subtours of length at most four. Thus, analysis of C_k-free 2-matchings shall result in obtaining better lower bounds and designing approximation algorithms for the TSP. In particular, for the graphic traveling salesman problem and the minimum 2-edge connected spanning subgraph problem, C_k-free 2-matchings are directly applied to designing approximation algorithms. For these two problems, if a C_k-free 2-factor is found, then $(1 + 2/(k + 1))$-approximation immediately follows. For a more elaborated use of C_k-free 2-factors, see [6,7,9,25,40].

The complexity of the maximum C_k-free 2-matching problem varies due to k. As stated above, the case $k \geq |V|/2$ contains the Hamilton cycle problem and hence is NP-hard, while the case $k \leq 2$ is exactly the classical maximum simple 2-matching problem and hence is polynomially solvable. Moreover, Papadimitriou proved NP-hardness for the case $k \geq 5$ (see [8]), whereas Hartvigsen [19] proposed a combinatorial algorithm for the case $k = 3$. The case $k = 4$ is left open.

The weighted version of the maximum C_k-free 2-matching problem is also of interest. The NP-hardness of the case $k \geq 5$ follows from that of the unweighted version, while the case $k = 2$ is the classical maximum-weight simple 2-matching problem and hence polynomially solvable. A nontrivial result is due to Vornberger [42], who proved the NP-hardness of the case $k = 4$. The maximum-weight C_3-free 2-matching problem is still open.

In bipartite graphs, C_4-free 2-matchings are often referred to as *square-free 2-matchings*, and in the present paper we mainly use this terminology. About fifteen years after the above basic results, in 1999, Hartvigsen [20] proposed a Tutte-type theorem characterizing bipartite graphs admitting a square-free 2-factor and a combinatorial algorithm. Király [26] gave a precise description and proof of the Tutte-type theorem, and extended it to a min-max formula. Since then, the maximum C_k-free 2-matching problem for the case $k = 3, 4$ has been studied actively. Frank [15] introduced the $K_{t,t}$-free t-matching problem in bipartite graphs, which is a generalization of the square-free 2-matching problem in bipartite graphs, and presented a min-max formula. After that, a full version [21] of [20] followed, and Pap [37] also gave a combinatorial algorithm for the maximum square-free 2-matching problem in bipartite graphs, which slightly differs from Hartvigsen's algorithm and is extended to the maximum $K_{t,t}$-free t-matching problem in bipartite graphs (see also [36]). Babenko [1] improved the time complexity of Pap's algorithm. We remark here that the min-max formula in [15] differs from the formula in [21,26]. We will give a detailed comparison of these two min-max formulas in Sect. 3.

For the weighted version of the maximum C_k-free 2-matching problem in bipartite graphs, NP-hardness is proved for the case $k \geq 6$ by Geelen [18] and for the case $k = 4$ by Király (see [15]). On the other hand, for the case $k = 4$ and the edge weight satisfies a property that the weight is vertex-induced on every square, Makai [33] presented a linear programming formulation with dual integrality. This formulation implies polynomial solvability via the ellipsoid method, and a combinatorial algorithm for this case was given by Takazawa [39].

Discrete convex structure of the C_k-free 2-matchings was first studied by Cunningham [10], who proved that the set of the degree sequences of the C_k-free 2-matchings is a *jump system* [5] for the case $k \leq 3$, and is not necessarily a jump system for the case $k \geq 5$. We remark that this result is consistent with the polynomial solvability of the maximum C_k-free 2-matching problem. For the case $k = 4$, Cunningham conjectured that the degree sequences of the C_4-free 2-matchings form a jump system, and later this was proved by Kobayashi, Szabó and Takazawa [29]. In [29], it is also proved that the weighted square-free 2-matchings in bipartite graphs induce an *M-concave function on a constant-parity jump system* [34] if and only if the edge weight is vertex-induced on every square, which is also consistent with polynomial solvability.

Through these results, one could assert that the maximum square-free 2-matching problem in bipartite graphs is indeed a well-solved case of the maximum C_k-free 2-matching problem. Apart from bipartite graphs, in subcubic graphs C_3- or C_4-free 2-matchings become tractable as well. See [2, 3, 22, 23, 27, 28, 30, 42] for progress in subcubic graphs.

The purpose of the present paper is to deepen the theory of C_k-free 2-matchings by investigating the structure of the square-free 2-matchings in bipartite graphs. First we exhibit that the two min-max formulas in [21, 26] and [15] are essentially different in a sense that a vertex set minimizing the set function in one formula is not necessarily a minimizer in the other. We then establish decomposition theorems for square-free 2-matchings in bipartite graphs, which serve as an analogue for the Dulmage-Mendelsohn decomposition for matchings in bipartite graphs [12, 13] and the Edmonds-Gallai decomposition for matchings in nonbipartite graphs [14, 16, 17]. Here we focus on the min-max formula in [21, 26], and we prove that two minimizers found by the algorithm in [21] are canonical in some sense. With these two minimizers, we can characterize the structure of the maximum square-free 2-matchings. We can know, e.g., which vertices have degree two for every maximum square-free 2-matching, and which edges belong in some maximum square-free 2-matching. These theorems suggest that the maximum square-free 2-matching problem has similarity to the maximum matching problem in both bipartite and nonbipartite graphs.

The rest of the paper is organized as follows. In Sect. 2, we review basic theorems for matchings in bipartite and nonbipartite graphs, such as the min-max theorems and the Dulmage-Mendelsohn and Edmonds-Gallai decompositions. In Sect. 3, we compare two min-max formulas for the maximum square-free 2-matching problem in bipartite graphs, and review Hartvigsen's algorithm [21]. Our decomposition theorems for square-free 2-matchings in bipartite graphs appear in Sect. 4.

2   Min-Max and Decomposition Theorems for Matchings

In this section, we review the basic results of matchings in bipartite graphs and nonbipartite matchings such as the min-max formulas, the Dulmage-Mendelsohn decomposition, and the Edmonds-Gallai decomposition. For more detailed discussion, the readers are referred to [24, 32, 35, 38].

Let $G = (V, E)$ be a simple undirected graph with vertex set V and edge set E. For $X \subseteq V$, the complement of X is denoted by $\bar{X}$, i.e., $\bar{X} = V \setminus X$. For $X \subseteq V$ and $F \subseteq E$, let $F[X]$ denote the set of edges in F spanned by X. A graph $G' = (V', E')$ is a *subgraph* of G if $V' \subseteq V$ and $E' \subseteq E$. For a subgraph H of G, the vertex and edge sets of H are denoted by $V(H)$ and $E(H)$, respectively. For $X \subseteq V$, let $G[X] = (X, E[X])$, the subgraph induced by X. For $F \subseteq E$ and two disjoint vertex subsets $X, Y \subseteq V$, let $F[X, Y]$ denote the set of all edges in F connecting X and Y. Let $G[X, Y] = (X \cup Y, E[X, Y])$. If G is bipartite, we often denote $G = (V^+, V^-; E)$, where $\{V^+, V^-\}$ is a partition of V and every edge in E connects V^+ and V^-. For $X \subseteq V$, let $X^+ = X \cap V^+$ and $X^- = X \cap V^-$.

For $F \subseteq E$ and a vertex $v \in V$, the *degree* of F on v is the number of edges in F incident to v and denoted by $\deg_F(v)$. A subset M of edges is called a *matching* if $\deg_M(v) \leq 1$ for each $v \in V$. Recall that M is a 2-matching if $\deg_M(v) \leq 2$ for each $v \in V$, and a 2-factor if $\deg_M(v) = 2$ for each $v \in V$. More generally, for an integer vector $b \in \mathbf{Z}^V$, an edge subset $M \subseteq E$ is called a *b-matching* if M satisfies $\deg_M(v) \leq b_v$ for each $v \in V$, and a *b-factor* if $\deg_M(v) = b_v$ for each $v \in V$. For $X \subseteq V$, let $b(X) = \sum_{v \in X} b_v$.

We remark that in the literature a b-matching with the above definition is often called a simple b-matching, and in a b-matching multiplicities on edges are allowed. In the present paper, since we only discuss subsets of edges and never put multiplicities on edges, a b-matching always means a simple b-matching, even if the term "simple" is omitted.

We begin with the classical min-max theorem for matchings in bipartite graphs of Kőnig [31]. For a graph $G = (V, E)$, $X \subseteq V$ is called a *vertex cover* if every edge in E is incident to at least one vertex in X.

Theorem 1 ([31]). *For a bipartite graph $G = (V, E)$, the maximum size of a matching is equal to the minimum size of a vertex cover.*

Theorem 1 is extended to the following min-max theorems for b-matchings in bipartite graphs and matchings in nonbipartite graphs. A component of a graph G is called *odd* if it consists of odd number of vertices, and let $o(G)$ denote the number of odd components in G. For $X \subseteq V$, $G - X$ denotes the subgraph obtained from G by deleting X and edges incident to at least one vertex in X.

Theorem 2. *Let $G = (V, E)$ be a bipartite graph and $b \in \mathbf{Z}^V$. The maximum size of a simple b-matching in G is equal to*

$$\min_{X \subseteq V} \{b(\bar{X}) + |E[X]|\}. \tag{1}$$

Theorem 3 (Tutte-Berge formula [4, 41]). *The maximum size of a matching in a graph $G = (V, E)$ is equal to*

$$\frac{1}{2} \min_{X \subseteq V} \{|V| + |X| - o(G - X)\}. \tag{2}$$

We remark that Theorems 2 and 3 are extended to distinct min-max theorems for the maximum square-free 2-matching problem in bipartite graphs, which are described in Sect. 3.

The *Dulmage-Mendelsohn decomposition* [12,13] characterizes the maximum matchings and minimum vertex covers in bipartite graphs. Roughly, a bipartite graph is decomposed into three parts: one part is covered by an arbitrary maximum matching, i.e., an arbitrary maximum matching of the entire graph contains a perfect matching of this part; and the other parts involve vertices missed by at least one maximum matching and vertices which might be matched with those vertices in a maximum matching. By this decomposition, we could know the whole structure of the maximum matchings and minimum covers.

From the rich structure of the Dulmage-Mendelsohn decomposition, in Theorem 4 below, we describe in detail the statements which are related to our decomposition theorems for square-free 2-matchings. We provide this precise description in order to offer a clear comparison of the Dulmage-Mendelsohn decomposition and our decomposition theorems for square-free matchings, and make the present paper self-contained.

Call an edge *admissible* if it belongs to some maximum matching. For $X \subseteq V$, let $\Gamma(X)$ denote the set of vertices in $V \setminus X$ adjacent to at least one vertex in X, i.e., $\Gamma(X) = \{v \in V \setminus X \mid \exists u \in X, uv \in E\}$.

Theorem 4 *For a bipartite graph $G = (V, E)$, let $D \subseteq V$ be the set of vertices which are not covered by at least one maximum matching in G. Then, the following statements hold.*

   (i) $X_1 = \bar{D}^+ \cup \Gamma(D^+)$ and $X_2 = \Gamma(D^-) \cup \bar{D}^-$ are minimum vertex covers.
   (ii) For an arbitrary minimum vertex cover Y, it holds that $X_2^+ \subseteq Y^+ \subseteq X_1^+$ and $X_1^- \subseteq Y^- \subseteq X_2^-$.
   (iii) Each edge in $E[\bar{X}_1^+, X_1^-]$ and $E[X_2^+, \bar{X}_2^-]$ is admissible.
   (iv) $G[X_1^+ \setminus X_2^+, X_2^- \setminus X_1^-]$ has a perfect matching.
   (v) $M \subseteq E$ is a maximum matching in G if and only if it is composed of a maximum matching in $G[\bar{X}_1^+, X_1^-]$, a maximum matching in $G[X_2^+, \bar{X}_2^-]$, and a perfect matching in $G[X_1^+ \setminus X_2^+, X_2^- \setminus X_1^-]$.

Indeed, the Dulmage-Mendelsohn decomposition includes a finer decomposition of $G[X_1^+ \setminus X_2^+, X_2^- \setminus X_1^-]$, distributive lattice structure. For details, see, e.g., [24,32,35].

The *Edmonds-Gallai decomposition* [14,16,17] characterizes a minimizer of (2) which is canonical in some sense, and the structure of maximum matchings in nonbipartite graphs. A component Q in a graph G is called *factor-critical* if $Q - \{v\}$ admits a perfect matching for each vertex v in Q. Roughly, by introducing the concept of factor-critical components, the difficulty of nonbipartite matchings clears up and almost reduces to the bipartite case. Below we provide a detailed description as well, which would also help comparing the Edmonds-Gallai decomposition and our decomposition theorems for square-free 2-matchings.

Theorem 5 (Edmonds-Gallai decomposition [14,16,17]; see also [32]).
*For a graph $G = (V, E)$, let $D \subseteq V$ be the set of vertices which are not covered
by at least one maximum matching, $A \subseteq V \setminus D$ be the set of vertices adjacent
to at least one vertex in D, i.e., $A = \Gamma(D)$, and $C = V \setminus (D \cup A)$. Then the
following statements hold.*

(i) Each component in $G[D]$ is factor-critical.

(ii) $G[C]$ has a perfect matching.

*(iii) In the bipartite graph obtained from G by deleting the vertices in C and
edges in $E[A]$ and by contracting each component of $G[D]$ to one vertex,
for each $X \subseteq A$ it holds that $|\Gamma(X)| > |X|$.*

*(iv) If M is a maximum matching in G, then M contains a matching of size
$(|V(Q)| - 1)/2$ in each component Q of $G[D]$ and a perfect matching of
$G[C]$, and matches all vertices of A with vertices in distinct components of
$G[D]$.*

(v) The maximum size of a matching in G is equal to $(|V| + |A| - o(G[D]))/2$.

3   Min-Max Theorems and Algorithms for Square-Free 2-Matchings in Bipartite Graphs

In the sequel, we work on b-matchings in bipartite graphs, where $b_v \in \{0, 1, 2\}$
for each vertex v. For a bipartite graph $G = (V, E)$ and $b \in \{0, 1, 2\}^V$, a *square*
is a subgraph forming a cycle of length four, and a b-matching in G is called
square-free if it does not contain a cycle of length four. Recall that we never put
multiplicities on edges in dealing with b-matchings, and note that a b-matching
with $b \in \{0, 1, 2\}^V$ is a vertex-disjoint collection of cycles and paths, and the
shortest length of a cycle in a bipartite graph is four.

In this section, we exhibit a comparison of the two min-max theorems
in [21,26] and [15], and review Hartvigsen's algorithm [21] for square-free 2-
matchings in bipartite graphs.

3.1   Min-Max Theorems and Optimality Criteria

In a graph, a component consisting of an edge (resp., a square) is called an
edge-component (resp., *square-component*). For $Z \subseteq V$, denote the number of
square-components in $G[Z]$ by $c(Z)$, and the total number of isolated vertices,
edge-components and square-components in $G[Z]$ by $q(Z)$. For the maximum
square-free 2-matching problem in bipartite graphs, the following two min-max
theorems are established.

Theorem 6 ([21,26]). *Let $G = (V, E)$ be a bipartite graph and $b \in \{0, 1, 2\}^V$.
The maximum size of a square-free b-matching in G is equal to*

$$\min_{Z \subseteq V} \{b(\bar{Z}) + |Z| - q(Z)\}. \tag{3}$$

Theorem 7 ([15]). *Let $G = (V, E)$ be a bipartite graph and $b \in \{0, 1, 2\}^V$. The maximum size of a square-free b-matching in G is equal to*

$$\min_{Z \subseteq V} \{b(\bar{Z}) + |E[Z]| - c(Z)\}. \tag{4}$$

Intuitively, Theorem 6 is close to Theorem 3, as well as Theorem 7 resembles Theorem 2. By putting $b_v = 2$ for each $v \in V$ and $X = \bar{Z}$ in (3), we obtain

$$b(\bar{Z}) + |Z| - q(Z) = 2|X| + |\bar{X}| - q(\bar{X}) = |V| + |X| - q(\bar{X}), \tag{5}$$

which is similar to (2).

Theorems 6 and 7 indeed differ from each other in that a minimizer of (3) does not necessarily minimize (4), and vice versa. See Fig. 1 for an example, where $b_v = 2$ for each vertex v. Observe that the maximum size of a square-free b-matching is six, $Z_1 = \{v_1, v_2, v_3, v_4, v_5, v_6\}$ attains six in (3) and seven in (4), and $Z_2 = \{v_1, v_2, v_3, v_4, v_6\}$ attains six in (4) and seven in (3).

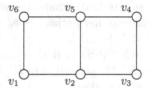

Fig. 1. $Z_1 = \{v_1, v_2, v_3, v_4, v_5, v_6\}$ minimizes (3) and not (4), whereas $Z_2 = \{v_1, v_2, v_3, v_4, v_6\}$ minimizes (4) and not (3).

An advantage of Theorem 7 is that it is extended to a min-max theorem for the maximum $K_{t,t}$-free t-matching problem in bipartite graphs [15], and further to a linear programming formulation with dual integrality for the weighted $K_{t,t}$-free t-matching problem in bipartite graphs, where the edge weight is vertex-induced on each $K_{t,t}$ [33,39]. On the other hand, in this paper we establish a structure theorem (Theorem 10), which is based on Theorem 6 and reveals the existence of some sort of canonical minimizers of (3), as with Theorem 4.

Theorem 6 implies optimality criteria for maximum square-free b-matchings in G and minimizers of (3). For an arbitrary square-free b-matching M in G and an arbitrary $Z \subseteq V$, it holds that $|M[Z]| \leq |Z| - q(Z)$ and $|M[Z, \bar{Z}]| + 2|M[\bar{Z}]| \leq b(\bar{Z})$. Thus, if M is a maximum square-free b-matching and Z minimizes (3), it holds that

$$|M[Z]| = |Z| - q(Z), \tag{6}$$
$$|M[Z, \bar{Z}]| = b(\bar{Z}), \tag{7}$$
$$M[\bar{Z}] = \emptyset. \tag{8}$$

Equation (6) further implies the following property of M and Z.

(*) If a component in $G[Z]$ consists of at least six vertices, then M contains a 2-factor in this component. Otherwise, the component is either a single vertex, a single edge in M, or a single square with three edges in M.

Also, (7) and (8) imply that

$$\deg_M(v) = \deg_{M[Z,\bar{Z}]}(v) = b(v) \quad \text{for each } v \in \bar{Z}. \tag{9}$$

3.2 Hartvigsen's Algorithm

For the maximum square-free 2-matching problem in bipartite graphs, several combinatorial algorithms are designed [1,21,37,39], and they slightly differ from each others. In this paper we discuss Hartvigsen's algorithm [21], since the minimizer Z of (3) found by the algorithm in [21] plays a key role in our decomposition theorem. It is also noteworthy that the minimizer Z of (3) found by the algorithm in [21] is a minimizer of (4) as well, while the minimizers of (4) implied in [1,37,39] do not necessarily minimize (3).

Let us briefly sketch the algorithm. Let $G = (V^+, V^-; E)$ be a bipartite graph and let M be an arbitrary square-free 2-matching in G. In the algorithm, we augment M with the aid of alternating paths. Let $b_v = 2$ for each $v \in V^+ \cup V^-$, $U^+ = \{u \in V^+ \mid \deg_M(u) < b_v\}$ and $U^- = \{v \in V^- \mid \deg_M(v) < b_v\}$. We execute the breadth-first search (BFS) to find a path P from U^+ to U^- such that P starts with an edge in $E \setminus M$, and edges in $E \setminus M$ and in M lie alternately in P. In the BFS, if we reach an edge $e \in E \setminus M$ and a square S such that $\{e\} = E(S) \setminus M = E(S) \cap E(P_e)$, where P_e is the path from U^+ to e obtained in the BFS, we shrink S in the following manner. Let $V(S) = \{v_1^+, v_2^+, v_1^-, v_2^-\}$, where $v_1^+, v_2^+ \in V^+$ and $v_1^-, v_2^- \in V^-$. Then identify v_1^+ and v_2^+ to obtain a new vertex v_S^+, and v_1^- and v_2^- to obtain v_S^-. All edges in $E(S)$ are deleted, and edges incident to v_1^+ or v_2^+ (resp., v_1^- or v_2^-) are connected to v_S^+ (resp., v_S^-). Denote the resulting bipartite graph by $\tilde{G} = (\tilde{V}^+, \tilde{V}^-; \tilde{E})$, and reset $b \in \{1,2\}^{\tilde{V}^+ \cup \tilde{V}^-}$ by

$$b_v := \begin{cases} 1 & \text{if } v = v_S^+ \text{ or } v = v_S^- \text{ for some shrunk square } S, \\ 2 & \text{otherwise.} \end{cases} \tag{10}$$

Now the objective becomes to find a maximum square-free b-matching in $\tilde{G}$. We remark that multiple edges connecting the same pair of vertices may appear in $\tilde{G}$, but M should contain at most one of those edges. Note also that shrunk squares are vertex-disjoint, even if repeated shrinking of squares are executed.

If an alternating path P from U^+ to U^- without such an edge e and a square S is found, then we update $M := M \triangle E(P)$, which is a square-free 2-matching with $|M'| = |M| + 1$.

After augmentation, we execute expanding of each shrunk square, which is the reverse operation of shrinking of a square. Let $\tilde{M}$ be a square-free b-matching in $\tilde{G}$. Then, it is not difficult to see that we can obtain a square-free 2-matching M in G by adding exactly three edges from each shrunk square to $\tilde{M}$.

An entire description of the algorithm is as follows.

ALGORITHM SQUARE-FREE

Input: A bipartite graph $G = (V^+, V^-; E)$.

Output: A maximum square-free 2-matching $M \subseteq E$, and $Z \subseteq V$ minimizing both (3) and (4).

Step 0: Let M be an arbitrary square-free 2-matching in G and $\tilde{G} = G$.

Step 1: In $\tilde{G}$, define b by (10) and let $U^+ = \{u \in \tilde{V}^+ \mid \deg_M(u) < b_u\}$ and $U^- = \{v \in \tilde{V}^- \mid \deg_M(v) < b_v\}$. Construct an auxiliary directed graph $\tilde{G}_M$ from $\tilde{G}$ by orienting the edges in $E \setminus M$ from $\tilde{V}^+$ to $\tilde{V}^-$ and the edges in M from $\tilde{V}^-$ to $\tilde{V}^+$. Execute the BFS from U^+ in $\tilde{G}_M$. For an edge e, denote the path from U^+ to e obtained by the BFS by P_e. If an edge $e \in \tilde{E} \setminus M$ and a square S such that $\{e\} = \tilde{E}(S) \setminus M = \tilde{E}(S) \cap \tilde{E}(P_e)$ are found, then go to Step 2. If a path P from U^+ to U^- without such an edge e and a square S is found, then go to Step 3. Otherwise go to Step 4.

Step 2 (Shrinking): Shrink S and go to Step 1.

Step 3 (Augmentation): Update $M := M \triangle \tilde{E}(P)$, expand all shrunk squares, and then go to Step 1.

Step 4 (Termination):

  Obtaining M. Expand all shrunk squares and return M.

  Obtaining Z. Let $R \subseteq \tilde{V}^+ \cup \tilde{V}^-$ be the set of vertices reachable from U^+ in $\tilde{G}_M$, and let $Z = (\tilde{V}^+ \cap R) \cup (\tilde{V}^- \setminus R)$.

  For each $v \in \tilde{V}^- \setminus R$ which is not contained in any shrunk square, if there exist two edges in M connecting $\tilde{V}^+ \cap R$ and v, then reset $Z := Z \setminus \{v\}$. For each $v_S^- \in \tilde{V}^- \setminus R$ of a shrunk square S, if there exists one edge in M connecting $\tilde{V}^+ \cap R$ and v_S^-, then reset $Z := Z \setminus \{v_S^-\}$. (We remark that $v_S^+ \in Z$ always holds.)

  In expanding each shrunk square S, reset Z by

$$Z := \begin{cases} (Z \setminus \{v_S^+, v_S^-\}) \cup V^+(S) \cup V^-(S) & \text{if } v_S^- \in Z, \\ (Z \setminus \{v_S^+, v_S^-\}) \cup V^+(S) & \text{if } v_S^- \notin Z. \end{cases} \tag{11}$$

  After expanding all shrunk squares, return Z.

As is proved in [21], the output M is a maximum square-free 2-matching and Z minimizes (3). It is also not difficult to check that Z minimizes (4) as well.

Theorem 8. *Let M and Z be outputs of* ALGORITHM SQUARE-FREE. *Then, M is a maximum square-free 2-matching in G, and Z minimizes both (3) and (4).*

4   Decomposition Theorems for Square-Free 2-Matchings in Bipartite Graphs

In this section, we describe our main contribution, structure theorems for square-free 2-matchings in bipartite graphs. Denote the minimizer of (3) obtained by ALGORITHM SQUARE-FREE by Z_1. By replacing the roles of V^+ and V^- in ALGORITHM SQUARE-FREE, we obtain another minimizer of (3), denoted by Z_2. We begin with showing a property of Z_1 and Z_2, which is stronger than (∗).

Proposition 1. *For $i = 1, 2$, it holds that*

- *the components in $G[Z_i]$ is either a single vertex, a single edge, or a single square, and*
- *for an arbitrary maximum square-free 2-matching M, all edges in $E[Z_i]$ except for one edge from each square-component belong to M.*

Proof. We only discuss Z_1, since the same argument applies to Z_2. Since Z_1 and an arbitrary maximum square-free 2-matching M satisfy the property (*), it suffices to prove that $G[Z_1]$ does not have a component with at least six vertices. Suppose to the contrary that $G[Z_1]$ has such a component Q. Denote the maximum square-free 2-matching found by ALGORITHM SQUARE-FREE by M^*. Then we have that $V^+(Q) \subseteq Z_1$ and M^* contains a 2-factor in Q. This implies that the vertices in $V^-(Q)$ cannot belong to Z_1 (see Step 4 of ALGORITHM SQUARE-FREE) regardless of whether Q contains shrunk squares or not, a contradiction. $\square$

In the sequel, we denote the graph and square-free b-matching at the last stage of ALGORITHM SQUARE-FREE, for which neither shrinking nor augmentation is executed, by $\tilde{G} = (\tilde{V}^+, \tilde{V}^-; \tilde{E})$ and $\tilde{M}$. For $X \subseteq V$, let $\tilde{X}$ denote the subset of $\tilde{V}$ corresponding to X.

We are now ready to describe our decomposition theorems. Following the notation of the Edmonds-Gallai decomposition, define $D, A, C \subseteq V$ by

$$D = Z_1^+ \cup Z_2^-, \quad A = \bar{Z}_1^- \cup \bar{Z}_2^+, \quad C = V \setminus (D \cup A). \tag{12}$$

First, the following theorem characterizes the vertex set D.

Theorem 9. *It holds that*

$$D = \{u \in V \mid \exists maximum \ square\text{-}free \ 2\text{-}matching \ M \ such \ that \ \deg_M(u) \le 1\}. \tag{13}$$

Proof. We only discuss the vertices in V^+. The arguments straightforwardly apply to the vertices in V^-.

By (9), $\deg_M(u) = 2$ holds for each $u \in \bar{Z}_1^+$ and an arbitrary maximum square-free 2-matching M. We next show that, for each $u \in Z_1^+$, there exists a maximum square-free 2-matching M such that $\deg_M(u) \le 1$.

Let $u \in Z_1^+$ be a vertex which is not contained in any shrunk square in $\tilde{G}$. In $\tilde{G}_{\tilde{M}}$, there exists a path P from U^+ to u. Let $\tilde{M}' = \tilde{M} \triangle \tilde{E}(P)$ to have $\deg_{\tilde{M}'}(u) \le 1$. By expanding all shrunk squares, we obtain another maximum square-free 2-matching M' from $\tilde{M}'$ with $\deg_{M'}(u) \le 1$.

If $u \in Z_1^+$ is shrunk into v_S^+ for some square S in $\tilde{G}$, let P be a path from U^+ to v_S^+ in $\tilde{G}_{\tilde{M}}$. Again let $\tilde{M}' = \tilde{M} \triangle \tilde{E}(P)$ to have $\deg_{\tilde{M}'}(v_S^+) = 0$, and we can expand all shrunk squares to obtain a maximum square-free 2-matching M' from $\tilde{M}'$ satisfying $\deg_{M'}(u) \le 1$. $\square$

The following theorem corresponds to Theorem 4 (ii), and suggests that the minimizers Z_1 and Z_2 are canonical.

Theorem 10. *For an arbitrary set $Y \subseteq V$ minimizing (3), it holds that $Z_1^+ \subseteq Y^+ \subseteq Z_2^+$ and $Z_2^- \subseteq Y^- \subseteq Z_1^-$.*

Proof. We first prove that $Z_1^+ \subseteq Y^+$. For $u \in Z_1^+$, by Theorem 9 there exists a maximum square-free 2-matching M satisfying $\deg_M(u) \leq 1$. On the other hand, by (9), we have that $\deg_M(v) = 2$ for every $v \in \bar{Y}$. Thus, $u \in Y$ follows.

Next we prove that $Y^- \subseteq Z_1^-$. Suppose to the contrary that there exists $v \in Y^- \setminus Z_1^-$. Let M^* be a maximum square-free 2-matching found by ALGO-RITHM SQUARE-FREE. Since $v \notin Z_1^-$, by (7) there exist two vertices $u_1, u_2 \in Z_1^+$ such that $u_1 v, u_2 v \in M^*$. By the above argument, $u_1, u_2 \in Z_1^+ \subseteq Y^+$ follows, and let Q be a component in $G[Y]$ containing u_1, u_2, v. Since u_1, u_2 are reachable from U^+, it follows that M^* does not contain a 2-factor in Q and thus Q is a square by (*). Let v_0 be the unique vertex in $V^-(Q) \setminus \{v\}$, and without loss of generality let $u_1 v_0 \in M^*$ and $u_2 v_0 \notin M^*$. Since v_0 has no adjacent vertex in $Y^+ \setminus \{u_1, u_2\}$, and so as in $Z_1^+ \setminus \{u_1, u_2\}$, it follows that $v_0 \in Z_1^-$ from (7). This implies that a shrinking involving $u_2 v_0$ should have occurred. If this shrinking also involves v, then it contradicts that $v \notin Z_1^-$ and $v_0 \in Z_1^-$. Otherwise, the shrunk square contains a vertex in $V^+ \setminus \{u_1, u_2\}$ adjacent to v_0, again a contradiction.   □

Theorem 10 implies that, while we defined Z_1 and Z_2 by the output of ALGO-RITHM SQUARE-FREE, Z_1 and Z_2 are uniquely determined. Thus, our decomposition $\{D, A, C\}$ is also uniquely determined.

Finally, the following theorem is a counterpart of the Dulmage-Mendelsohn decomposition for matchings in bipartite graphs (Theorem 4) and the Edmonds-Gallai decomposition for matchings in nonbipartite graphs (Theorem 5).

Theorem 11. *The following statements hold.*

 (i) *The components in $G[D]$ and $G[D, C]$ is either a single vertex, a single edge, or a single square.*
 (ii) *Every edge in $E[D, A]$ is admissible.*
 (iii) *Shrink the squares in $G[D]$ and $G[D, C]$ in the same manner as in ALGO-RITHM SQUARE-FREE to obtain a new graph $G' = (V', E')$, denote the vertex subsets of V' corresponding to D, C by D', C', and define $b' \in \{1, 2\}^{D' \cup C'}$ by*

$$b'_v = \begin{cases} 1 & \text{if } v = v_S^+ \text{ or } v = v_S^- \text{ for some shrunk square } S, \\ & \text{or } v \text{ belongs to an edge-component in } G[D] \text{ or } G[D, C], \quad (14) \\ 2 & \text{otherwise.} \end{cases}$$

Then,
(a) for arbitrary $X \subseteq A$, it holds that $b'(\Gamma(X) \cap D') > 2|X|$, and
(b) $G'[C']$ has a b'-factor.
 (iv) *An arbitrary maximum square-free 2-matching M in G is composed of the following edges:*

(a) in $G[D]$ and $G[D, C]$, M contains the single edge of each edge-component, and exactly three edges from each square-component;

(b) for $u \in A$, M contains two edges connecting u and distinct components in $G[D]$; and

(c) in $G[C]$, $M[C]$ corresponds to a b'-factor in $G'[C']$.

(v) Both $D \cup C^+$ and $D \cup C^-$ minimize both (3) and (4).

Proof. Assertion (v) directly follows from $D \cup C^- = Z_1$ and $D \cup C^+ = Z_2$.

We next prove (i) and (iv)(a). It suffices to deal with $G[D] = G[D^+, D^-]$ and $G[D^+, C^-]$. Let M^* be the maximum square-free 2-matching found by ALGORITHM SQUARE-FREE. By Proposition 1, it suffices to prove that $G[Z_1]$ does not have a square intersecting both D^- and C^-. Suppose to the contrary that $G[Z_1]$ has such a square S. Then, S is not shrunk in $\tilde{G}$, and by (*) one vertex $v \in \tilde{V}^-(S)$ has two incident edges in M^* connecting v and $\tilde{V}(S)^+ \subseteq D^+$. Then v should belong to A^- (see Step 4 of ALGORITHM SQUARE-FREE), a contradiction.

Assertion (iv)(b) is now straightforward from (7) and Assertions (i) and (iv)(a).

We then prove (iii)(b) and (iv)(c). Since $C^+ \subseteq Z_2^+$ and $C^- \subseteq Z_1^-$, it follows from (9) that $\deg_M(v) = 2$ for an arbitrary vertex $v \in C$ and a arbitrary maximum square-free 2-matching M. Since $M[A, C] = \emptyset$ by (iv)(b), Assertions (iii)(b) and (iv)(c) follow from (iv)(a).

Next we prove (iii)(a). It suffices to consider $X \subseteq A^- = \bar{Z}_1^-$. From (iv)(a) and (iv)(b), it is clear that $b'_u \geq |\tilde{M}[\{u\}, X]|$ for each $u \in \Gamma(X) \cap D'$. Suppose that there exists a shortest path P from U^+ to $\tilde{X}$ in $\tilde{G}_{\tilde{M}}$. Denote the last edge of P by uv, where $u \in \tilde{V}^+$ and $v \in \tilde{V}^-$. Then we have that $b_u > |\tilde{M}[\{u\}, \tilde{X}]|$, which implies that $b'(\Gamma(X) \cap D')$ is strictly larger than $2|X|$. Suppose that P does not exist, i.e., all vertices in X are deleted from Z in Step 4. In this case, $b'_u > |\tilde{M}[\{u\}, \tilde{X}]|$ holds for each $u \in \Gamma(X) \cap D'$, since u is reachable from U^+, i.e., $u \in U^+$ or u has an extra edge in M reaching u from U^+, and thus the assertion follows.

Finally we prove (ii). We show that an edge $e \in E[Z_1^+, V^- \setminus Z_1^-]$ is admissible. The same argument applies to edges in $E[V^+ \setminus Z_2^+, Z_2^-]$.

Suppose that $e = uv \in \tilde{E}$, i.e., e does not belong to shrunk squares, where $u \in \tilde{V}^+$ and $v \in \tilde{V}^-$. If $e \in \tilde{M}$, then, from $\tilde{M}$, we obtain a maximum square-free 2-matching containing e by expanding all shrunk squares. If $e \in \tilde{E} \setminus \tilde{M}$, let P be a shortest path from U^+ to u in $\tilde{G}_{\tilde{M}}$, and let $\tilde{M}' = \tilde{M} \triangle \tilde{E}(P)$. Then, from $\tilde{M}'$, we obtain a maximum square-free 2-matching M' in G such that $e \in E \setminus M'$, $\deg_{M'}(u) \leq 1$ and $\deg_{M'}(v) = 2$ by expanding all shrunk squares. By adding e to M' and deleting one of edges incident to v from M', we obtain another maximum square-free 2-matching M'' (we can choose the deleted edge so that M'' does not contain a square).

Suppose that e does not appear in $\tilde{E}$, i.e., $e \in E(S)$ for some shrunk square S. Let P be a path from U^+ to v_S^+. Then $\tilde{M}' = \tilde{M} \triangle \tilde{E}(P)$ is a new square-free b-matching satisfying $|\tilde{M}'| = |\tilde{M}|$ and $\deg_{\tilde{M}'}(v_S^+) = 0$. Now it is not difficult to see that we can add e to $\tilde{M}'$ in expanding S.   □

Here let us describe how Assertions (i) and (iv)(a) in Theorem 11 relates to Theorems 4 and 5. In Assertion (i) in Theorem 11, the components in $G[D]$ are analogue to the components in $G[D]$ in Theorems 4 and 5, which are factor-critical (in Theorem 4, every component in $G[D]$ consists of a single vertex). For a component Q which is either a single vertex, an edge-component or a square-component, the maximum size of a square-free 2-matching in Q is equal to $|V(Q)| - 1$, while the maximum size of a matching in a factor-critical component Q' is $(|V(Q')| - 1)/2$. In particular, if Q is a square-component, for every pair of $u^+ \in V^+(Q)$ and $u^- \in V^-(Q)$ there exists a maximum square-free 2-matching M_Q in Q satisfying $\deg_{M_Q}(u^+) = \deg_{M_Q}(u^-) = 1$ and $\deg_{M_Q}(v) = 2$ for $v \in V(Q) \setminus \{u^+, u^-\}$. This would correspond to the fact that a factor-critical component Q' admits a perfect matching in $Q' - \{v\}$ for each vertex v in Q'. Moreover, by Assertion (iv)(a), an arbitrary maximum square-free 2-matching contains a maximum square-free 2-matching in each component in $G[D]$, as is the case for a maximum matching and the factor-critical components in $G[D]$ in Theorem 5 (iv).

The components in $G[D, C]$ appear in neither the Dulmage-Mendelsohn nor the Edmonds-Gallai decomposition. For edge-components in $G[D, C]$, however, their counterpart indeed exists in bipartite b-matchings, which corresponds to the term $|E[X]|$ in (1). The square-components in $G[D, C]$ are specific to square-free 2-matchings, but again are analogue to the edge-components in $G[X]$ in Theorem 2 in a sense that each square-component S contains three edges from an arbitrary maximum square-free 2-matching by Assertion (iv)(a) and thus it can be shrunk and dealt with just as an edge-component.

With the above analogy in mind, for the other assertions in Theorem 11 it is not difficult to find their counterparts in Theorems 4 and 5.

Acknowledgements. The author is thankful to Satoru Iwata for suggesting this research and Zoltán Király for informing him of the history of the two min-max theorems for the maximum square-free 2-matching problem in bipartite graphs. This work is partially supported by JSPS KAKENHI Grant Numbers 25280004 and 26280001.

References

1. Babenko, M.A.: Improved algorithms for even factors and square-free simple b-matchings. Algorithmica **64**, 362–383 (2012)
2. Bérczi, K., Kobayashi, Y.: An algorithm for $(n - 3)$-connectivity augmentation problem: Jump system approach. J. Comb. Theor. Ser. B **102**, 565–587 (2012)
3. Bérczi, K., Végh, L.A.: Restricted b-matchings in degree-bounded graphs. In: Eisenbrand, F., Shepherd, F.B. (eds.) IPCO 2010. LNCS, vol. 6080, pp. 43–56. Springer, Heidelberg (2010)
4. Berge, C.: Sur le couplage maximum d'un graphe. Comptes Rendus Hebdomadaires Séances l'de Académie de Sciences **247**, 258–259 (1958)
5. Bouchet, A., Cunningham, W.H.: Delta-matroids, jump systems, and bisubmodular polyhedra. SIAM J. Discrete Math. **8**, 17–32 (1995)
6. Boyd, S., Iwata, S., Takazawa, K.: Finding 2-factors closer to TSP tours in cubic graphs. SIAM J. Discrete Math. **27**, 918–939 (2013)

7. Boyd, S., Sitters, R., van der Ster, S., Stougie, L.: The traveling salesman problem on cubic and subcubic graphs. Math. Program. **144**, 227–245 (2014)
8. Cornuéjols, G., Pulleyblank, W.: A matching problem with side conditions. Discrete Math. **29**, 135–159 (1980)
9. Correa, J.R., Larré, O., Soto, J.A.: TSP tours in cubic graphs: beyond 4/3. SIAM J. Discrete Math. **29**, 915–939 (2015)
10. Cunningham, W.H.: Matching, matroids, and extensions. Math. Program. **91**(3), 515–542 (2002)
11. Dantzig, G., Fulkerson, R., Johnson, S.: Solution of a large-scale traveling-salesman problem. Oper. Res. **2**, 393–410 (1954)
12. Dulmage, A.L., Mendelsohn, N.S.: Coverings of bipartite graphs. Can. J. Math. **10**, 517–534 (1958)
13. Dulmage, A.L., Mendelsohn, N.S.: A structure theory of bipartite graphs of finite exterior dimension. Trans. Roy. Soc. Can. Sect. III **53**, 1–13 (1959)
14. Edmonds, J.: Paths, trees, and flowers. Can. J. Math. **17**, 449–467 (1965)
15. Frank, A.: Restricted t-matchings in bipartite graphs. Discrete Appl. Math. **131**, 337–346 (2003)
16. Gallai, T.: Kritische Graphen II. A Magyar Tudományos Akadémia—Matematikai Kutató Intézetének Közleményei **8**, 373–395 (1963)
17. Gallai, T.: Maximale Systeme unabhänginger Kanten. A Magyar Tudományos Akadémia—Matematikai Kutató Intézetének Közleményei **9**, 401–413 (1964)
18. Geelen, J.F.: The C_6-free 2-factor problem in bipartite graphs is NP-complete (1999), unpublished
19. Hartvigsen, D.: Extensions of matching theory. Ph.D. thesis. Carnegie Mellon University (1984)
20. Hartvigsen, D.: The square-free 2-factor problem in bipartite graphs. In: Cornuéjols, G., Burkard, R.E., Woeginger, G.J. (eds.) IPCO 1999. LNCS, vol. 1610, pp. 234–241. Springer, Heidelberg (1999)
21. Hartvigsen, D.: Finding maximum square-free 2-matchings in bipartite graphs. J. Comb. Theor. Ser. B **96**, 693–705 (2006)
22. Hartvigsen, D., Li, Y.: Maximum cardinality simple 2-matchings in subcubic graphs. SIAM J. Discrete Math. **21**, 1027–1045 (2011)
23. Hartvigsen, D., Li, Y.: Polyhedron of triangle-free simple 2-matchings in subcubic graphs. Math. Program. **138**, 43–82 (2013)
24. Iri, M.: Structural theory for the combinatorial systems characterized by submodular functions. In: Pulleyblank, W.R. (ed.) Progress in Combinatorial Optimization, pp. 197–219. Academic Press, New York (1984)
25. Karp, J.A., Ravi, R.: A 9/7-approximation algorithm for graphic TSP in cubic bipartite graphs. In: Jansen, K., Rolim, J., Devanur, N., Moore, C. (eds.) Approximation, Randomization, and Combinatorial Optimization, Algorithms and Techniques, pp. 284–296 (2014)
26. Király, Z.: C_4-free 2-factors in bipartite graphs. Technical report, TR-2001-13, Egerváry Research Group (1999)
27. Kobayashi, Y.: A simple algorithm for finding a maximum triangle-free 2-matching in subcubic graphs. Discrete Optim. **7**, 197–202 (2010)
28. Kobayashi, Y.: Triangle-free 2-matchings and M-concave functions on jump systems. Discrete Appl. Math. **175**, 35–42 (2014)
29. Kobayashi, Y., Szabó, J., Takazawa, K.: A proof of Cunningham's conjecture on restricted subgraphs and jump systems. J. Comb. Theor. Ser. B **102**, 948–966 (2012)

30. Kobayashi, Y., Yin, X.: An algorithm for finding a maximum t-matching excluding complete partite subgraphs. Discrete Optim. **9**, 98–108 (2012)
31. Kőnig, D.: Graphok és matrixok. Matematikai és Fizikai Lapok **38**, 116–119 (1931)
32. Lovász, L., Plummer, M.D.: Matching Theory. AMS Chelsea Publishing, Providence (2009)
33. Makai, M.: On maximum cost $K_{t,t}$-free t-matchings of bipartite graphs. SIAM J. Discrete Math. **21**, 349–360 (2007)
34. Murota, K.: M-convex functions on jump systems: A general framework for min-square graph factor problem. SIAM J. Discrete Math. **20**, 226–231 (2006)
35. Murota, K.: Matrices and Matroids for Systems Analysis. Springer, Heidelberg (2010). softcover edn
36. Pap, G.: Alternating paths revisited II: Restricted b-matchings in bipartite graphs. Technical report, TR-2005-13. Egerváry Research Group (2005)
37. Pap, G.: Combinatorial algorithms for matchings, even factors and square-free 2-factors. Math. Program. **110**, 57–69 (2007)
38. Schrijver, A.: Combinatorial Optimization—Polyhedra and Efficiency. Springer, Heidelberg (2003)
39. Takazawa, K.: A weighted $K_{t,t}$-free t-factor algorithm for bipartite graphs. Math. Oper. Res. **34**, 351–362 (2009)
40. Takazawa, K.: Approximation algorithms for the minimum 2-edge connected spanning subgraph problem and the graph-TSP in regular bipartite graphs via restricted 2-factors, Preprint, RIMS-1826. Kyoto University, Research Institute for Mathematical Sciences (2015)
41. Tutte, W.T.: The factorization of linear graphs. J. Lond. Math. Soc. **22**, 107–111 (1947)
42. Vornberger, O.: Easy and hard cycle covers, Preprint, Universität Paderborn (1980)

Graph Drawing

Saturated Simple and 2-simple Topological Graphs with Few Edges

Péter Hajnal[1], Alexander Igamberdiev[2], Günter Rote[3], and André Schulz[4(✉)]

[1] Bolyai Institute, University of Szeged, Szeged, Hungary
[2] Institut für Mathematische Logik und Grundlagenforschung,
Universität Münster, Münster, Germany
[3] Institut für Informatik, Freie Universität Berlin, Berlin, Germany
[4] FernUniversität in Hagen, Hagen, Germany
andre.schulz@fernuni-hagen.de

Abstract. A *simple topological graph* is a topological graph in which any two edges have at most one common point, which is either their common endpoint or a proper crossing. More generally, in a *k-simple topological graph*, every pair of edges has at most k common points of this kind. We construct *saturated* simple and 2-simple graphs with few edges. These are k-simple graphs in which no further edge can be added. We improve the previous upper bounds of Kynčl, Pach, Radoičić, and Tóth [Comput. Geom., **48**, 2015] and show that there are saturated simple graphs on n vertices with only $7n$ edges and saturated 2-simple graphs on n vertices with $14.5n$ edges. As a consequence, $14.5n$ edges is also a new upper bound for k-simple graphs (considering all values of k). We also construct saturated simple and 2-simple graphs that have some vertices with low degree.

1 Introduction

Let $G = (V, E)$ be a finite simple graph ($E \subseteq V \times V$). A *drawing* of G is a map $\delta \colon V \cup E \to \mathbb{R}^2$ that is one-to-one on $\delta|_V \colon V \to \mathbb{R}^2$, i.e., δ assigns the vertices of the graph to different points of the plane. Furthermore, we require that $\delta|_E \colon E \to \mathcal{C}$, where $\mathcal{C}$ is a set of "nice" non-self-intersecting curves with two boundary points of the plane. For example we might think of $\mathcal{C}$ as the set of all Jordan curves or, more elementary, of the set of all simple polygonal curves. For simplicity, we will not distinguish between an edge and the curve on which it is embedded, and between a vertex and the point on which it is embedded. We assume that for any $e = xy \in E$ the edge $\delta(e)$ is a curve connecting $\delta(x)$ and $\delta(y)$ and it doesn't go through any other vertex, and also that any two different edges meet at finitely many points and any meeting point — that is not a common endvertex — is a proper crossing of the two curves.

The pair (G, δ), i.e., a graph with a drawing, is called a *topological graph*. A topological graph (G, δ) is *simple* if in δ two edges have at most one common point. More generally, the topological graph is called *k-simple* if in δ two edges have at most k common points. For both simple and k-simple graphs we do not

© Springer-Verlag Berlin Heidelberg 2016
E.W. Mayr (Ed.): WG 2015, LNCS 9224, pp. 391–405, 2016.
DOI: 10.1007/978-3-662-53174-7_28

allow self-intersecting edges. A topological graph is a *geometric graph* if all its edges are drawn as straight-line segments. Obviously, every geometric graph is simple, provided that the vertices are placed in general position. Thus, every graph has simple drawings.

For a graph property $\mathcal{T}$, a graph G is $\mathcal{T}$-*saturated* if G has property $\mathcal{T}$, but the addition of any edge joining two non-adjacent vertices of G violates property $\mathcal{T}$. Often structures with property $\mathcal{T}$ are quite hard to grasp, but $\mathcal{T}$-saturated structures might have a more useful character. We direct the interested reader to applications of the saturation technique [1,4,6]. This notion can be naturally extended to hypergraphs. A thorough survey by Faudree, Faudree, and Schmitt [2] discusses the case when property $\mathcal{T}$ is "not having F as a sub(hyper)graph".

In this paper we study *saturated k-simple topological graphs*. These are topological graphs that are k-simple, but no edge can be added without violating the k-simplicity of the drawing. Saturated planar drawings are triangulations and have therefore due to Euler's formula $3n - 6$ edges. Recently, Kynčl, Pach, Radoičić, and Tóth [5] started to investigate saturated simple k-simple graphs. The maximum number of edges a saturated simple topological graph can have is clearly $\binom{n}{2}$, since the geometric graph of K_n with points in convex position is a simple drawing. The more intriguing questions ask about the minimum number of edges for saturated k-simple topological graph. One of the main results of Kynčl et al. [5] is a construction of sparse saturated simple and k-simple topological graphs. We denote by $s_k(n)$ the minimum number of edges a saturated k-simple graph with n vertices can have. Their upper bound on $s_k(n)$ is a linear function of n, for n being the number of vertices; see Table 1 for the bounds obtained by Kynčl et al. [5]. The gap between the best known upper and lower bounds for $s_k(n)$ is quite substantial. We only know that $s_1(n) \geq 1.5n$ and that $s_k(n) \geq n$ [5].

Table 1. Old and new upper bounds for $s_k(n)$, the minimum number of edges in a saturated k-simple graph with n vertices.

k	1	2	3	4	5	6,8,10	7	9, ≥ 11
Old upper bounds [5]	17.5n	16n	14.5n	13.5n	13n	9.5n	10n	7n
New upper bounds	**7n**	**14.5n**						

Our Contribution. We improve the upper bounds for $s_k(n)$ for $k = 1, 2$. We do this by showing that for any positive integer n there exists a saturated simple topological graph with at most $7n$ edges (in Sect. 2), and a saturated 2-simple graph with $14.5n$ edges (in Sect. 3). This result also implies that there are saturated k-simple graphs with $14.5n$ edges for every k. See also Table 1 for a comparison with the old bounds. Our proofs are constructive, i.e., we can give the sparse saturated graphs.

We complete our results by studying *local saturation* of topological graphs. Here, local saturation refers to drawings in which one (or several) vertices have a small vertex degree even though the full drawing might not be the sparsest. Such observations might be helpful in further studies, e.g., if we want to investigate techniques for proving lower bounds that are based on the minimum vertex degree in saturated graphs. We show that there are arbitrarily large saturated simple graphs that have a vertex of degree 4, and saturated simple graphs in which 10 percent of the vertices have degree 5. For saturated 2-simple graphs we can prove that there are drawings with minimum degree 12. The currents lower bounds for $s_k(n)$ are obtained by bounding the minimum vertex degree in saturated k-simple graphs [5]. Our results show the limits of this approach. These results can be found in Sect. 4.

2   Saturated Simple Topological Graph with Few Edges

In this section we give a construction that generates a sparse saturated simple graph. We start with defining a graph G, parametrized by ℓ, with $n = 6\ell$ vertices and $9\ell - 6$ edges. This graph is the backbone of our sparse saturated graph.

The drawing is best visualized on the surface of a long circular cylinder. Figure 1 shows an unrolling of the cylinder into the plane. The cylinder is obtained by cutting the drawing along the two dotted lines and gluing the top and the bottom together. The vertices of the graph are placed in a $3 \times 2\ell$-grid-like fashion. We draw the vertices together in pairs, with each vertex X_i^L on the *left* and the corresponding vertex X_i^R on the *right*, for $X = A, B, C$ and $i = 1, \ldots, \ell$. We refer to the vertices whose label have the subscript i as the i-*th layer*.

G is the union of

- three vertex-disjoint paths of *blue edges* connecting $A_1^L A_2^L \ldots A_\ell^L$, $B_1^L B_2^L \ldots B_\ell^L$, and $C_1^L C_2^L \ldots C_\ell^L$,
- three vertex-disjoint paths of *red edges* connecting $A_1^R A_2^R \ldots A_\ell^R$, $B_1^R B_2^R \ldots B_\ell^R$, and $C_1^R C_2^R \ldots C_\ell^R$, and
- k disjoint cycles of *green edges* connecting $A_i^L B_i^L C_i^L$.

We will first consider the graph G_{RB} that omits the green edges, because this graph is more symmetric: with the exception of the vertices $X_1^{L/R}$ and $X_\ell^{L/R}$ near the boundary, all vertices look identical. Apart from these boundary effects, the drawing has a rotational symmetry, cyclically shifting the labels $A \to B \to C \to A$, a translational symmetry, shifting indices i up or down, and a mirror symmetry, exchanging left with right and blue with red. The green edges destroy this mirror symmetry: there are then two classes of vertices, the blue vertices X_i^L and the red vertices X_i^R.

We will show that the maximum degree in any saturated drawing which extends G_{RB} is 16. The 16 potential neighbors of a typical vertex A_i^L are shown in Fig. 1. This establishes that there are saturated drawings with n vertices and less than $8n$ edges. When the green edges are included, the three dash-dotted edges in Fig. 1 become impossible. Thus, each blue vertex has 13 potential neighbors. The red vertex A_{i+1}^R, which can be taken as a representative of a typical red vertex, loses A_i^L as a potential neighbor. Thus, each red vertex has at most 15

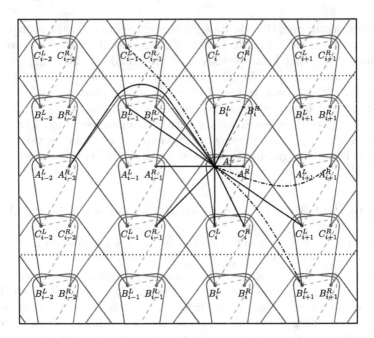

Fig. 1. The graph G on an unrolled cylinder, with the 16 potential neighbors of a vertex. Green edges are dashed. The solid black edges show possible additional edges for the vertex A_i^L. (Color figure online)

potential neighbors. This improves the upper bound for the smallest number of edges in a saturated drawings with n vertices to $7n$.

Theorem 1. *Let $s(n)$ denote the minimum number of edges that a simple saturated drawing with n vertices can have. Then $s(n) \leq 7n$.*

The remainder of this section is devoted to proving the above theorem. We start with the analysis of the graph G_{RB}.

Lemma 1. *The 16 potential neighbors of a typical vertex A_i^L in G_{RB} are all 11 vertices of levels $i-1$ and i ($A_{i-1}^L, B_{i-1}^L, C_{i-1}^L$; $A_{i-1}^R, B_{i-1}^R, C_{i-1}^R$; B_i^L, C_i^L; A_i^R, B_i^R, C_i^R) plus the 5 vertices A_{i-2}^R; $A_{i+1}^L, B_{i+1}^L, C_{i+1}^L$; A_{i+1}^R.*

When any of the neighbors listed above does not exist because $i \leq 2$ or $i = \ell$, the lemma still holds in the sense that the remaining vertices form the set of potential neighbors. In the proofs, when we exclude an edge between, say, levels i and j, our arguments will not use edges outside this range.

In the following we will look at the given drawing of G_{RB} (or G) and argue about the additional edges that can be drawn. The implicit assumption is that these edges cannot cross any given edge more than once. Usually, we will regard a new edge as a directed edge, starting at some vertex and trying to reach another vertex.

A *belt* is a substructure of our drawing. It is formed by the 12 vertices of two successive layers with their 6 edges between them, see Fig. 2. This drawing separates a large face on the left from a large face on the right. More precisely, the belt is defined as the part of the plane (or the cylinder) which lies between these two large faces.

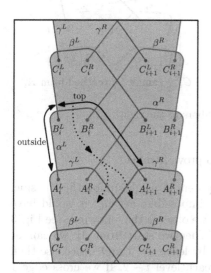
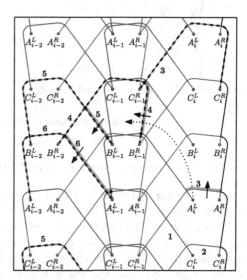

Fig. 2. Escape from a belt is difficult (Lemma 2).

Fig. 3. The situation discussed in the proof of Lemma 1 for left side neighbors.

We denote the six edges of the belt by $\alpha^L, \beta^L, \gamma^L, \alpha^R, \beta^R, \gamma^R$; as shown in Fig. 2. Each edge is cut into six sections by the intersections with the other edges: Two sections are little "stumps" at the end vertices. One section belongs to the boundary between the belt and the *outside*. The remaining three sections form the *top part* of the edge. We say that a new (directed) edge crosses a belt edge *from the outside* or *from the top* if it crosses the boundary part or the top part in the appropriate direction.

Lemma 2. *In a simple drawing that contains G_{RB}, the following holds: (1) If an edge crosses a belt edge from the top or from the outside, it must terminate inside the belt. (2) No edge can cross a belt.*

Proof. We start with the following observation: If an edge crosses α_L from the outside or from the top, and it does not terminate at B_i^L or at B_i^R, then it must later cross γ^L or γ^R from the top. This observation holds symmetrically for α_R instead of α_L, and cyclically for the other four belt edges. Hence, any edge that "enters" the belt from the outside has to continue by crossing another edge from the belt from the top. There is no way to leave the belt without crossing some edge twice. □

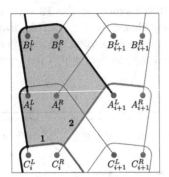
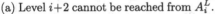

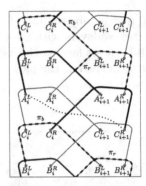

(a) Level $i+2$ cannot be reached from A_i^L.   (b) C_{i+1}^R cannot be reached from A_i^L.

Fig. 4. Restricting the neighbors to the right.

After these preparations, we are ready to prove Lemma 1.

Proof of Lemma 1. Let us first look at the potential neighbors on the left side. A connection from A_i^L to levels $j \leq i - 3$ is impossible, because it would have to cross a belt. For the vertices at level $i - 2$ we observe the following (see Fig. 3 for the edge numbers we are referring to): When we start from A_i^L we cannot cross the right boundary of the belt formed by levels $i - 1$ and i, because then we would have to cross the whole belt to reach level $i - 2$. If we cross edge 1 or 2 from the top, then, by Lemma 2, we are restricted to the belt defined by level $i - 1$ and i. Thus we can regard edge 1 and 2 as closed from the top. (These edges can later be crossed from the bottom.) We successively conclude that the new edges must cross the shadowed parts of the edges 3, 4, 5, and 6. The endpoints $B_{i-2}^R, B_{i-2}^L, A_{i-2}^L$ of the edges 4, 5, and 6 cannot be taken. C_{i-2}^L and C_{i-2}^R are enclosed in a small face delimited by the edges 4, 5, and 6, and cannot be reached. A_{i-2}^R is thus the only reachable vertex of level $i - 2$.

Let us turn to the potential neighbors on the right side. A connection from A_i^L to levels $j \geq i + 3$ is impossible, because it would have to cross a belt. Vertices at level $i + 2$ cannot be reached either, because (i) if we cross the edge forming the left boundary of the belt spanned by the vertices of level i and $i + 1$ we cannot cross this belt anymore and therefore cannot reach level $i + 2$, and (ii) if we cross one of the edges in the face that contains A_i^L from the top (edge labeled 1 and 2 in Fig. 4(a), then, by Lemma 2, we are also restricted to this belt. Thus we are restricted to the shaded region in Fig. 4a.

The vertices B_{i+1}^R and C_{i+1}^R also cannot be neighbors of A_i^L. We discuss the exclusion of C_{i+1}^R as a potential neighbor – the case for B_{i+1}^R is symmetric. The edges incident to A_i^L and C_{i+1}^R, which we call the *closed edges* cannot be crossed. The closed edges are depicted as thicker curves in Fig. 4b. Consider the portion of the red edge π_r that runs between A_i^R and A_{i+1}^R above the closed edges (see Fig. 4(b)). The curve π_r bounds a region below in which the remaining edges bounding this region are parts of the closed edges. Hence, if we enter this region we cannot leave and therefore we cannot cross π_r (see Fig. 4(b)). Let us now consider the partial edge π_b that runs between B_{i+1}^L and B_i^L above the closed edges and π_r. Again, there is a region whose boundary is part of the closed edges and also π_b. To enter and leave this region we have to cross either one of the closed edges or π_r, or we have to cross π_b twice. Since all these options are invalid, we have to avoid this region, and therefore are not allowed to cross π_b. We observe that the closed edges together with π_b and π_r leave A_i^L and C_{i+1}^R in different faces, which shows that these vertices cannot be neighbors unless we cross one edge twice. $\qquad\square$

Now we turn back to G. The additional green edges exclude some of the possible edges from the Lemma 1.

Lemma 3. *1. The* 13 *potential neighbors of a typical vertex A_i^L in G are all 5 vertices of level i (B_i^L, C_i^L; A_i^R, B_i^R, C_i^R), all but one vertex of level $i-1$ (A_{i-1}^L, B_{i-1}^L; $A_{i-1}^R, B_{i-1}^R, C_{i-1}^R$) plus the 3 vertices A_{i-2}^R; A_{i+1}^L, C_{i+1}^R.*
2. The 15 *potential neighbors of a typical vertex A_i^R in G are all 11 vertices of levels i and $i+1$ (B_i^L, C_i^L, A_i^L; B_i^R, C_i^R; $A_{i+1}^L, B_{i+1}^L, C_{i+1}^L$; $A_{i+1}^R, B_{i+1}^R, C_{i+1}^R$) plus the 4 vertices $A_{i-1}^R, B_{i-1}^R, C_{i-1}^R$; A_{i+2}^L.*

The claim immediately follows from the next lemma, whose proof can be found in the full version [3].

Lemma 4. *In a simple extension of G, A_{i+1}^R, C_{i-1}^L and B_{i+1}^L cannot be neighbors of A_i^L.*

As a consequence of Lemma 3 the average degree in a saturated extension of G is at most 14, which proves Theorem 1 when the number n of vertices is a multiple of 6. A more careful analysis reveals that for any $n \geq 12$ that is a multiple of 6, there exists a saturated simple topological graph with n vertices and at most $7n - 30$ edges.

Our construction can be extended to any vertex size by *cloning* some vertices. Take a saturated simple topological graph and any vertex P of it. Next to P we add ρ new copies of P – the clones. Connect the neighbors of P to each clone by edges that are non-intersecting perturbations of the edges incident to P. By this we obtain a simple drawing. A saturation of this drawing can include as additional edges only edges among P and its clones.

For $n \geq 12$, we can write n as $6r + \rho$ where $0 \leq \rho \leq 5$. If $\rho = 0$, we are done. If $\rho \geq 1$, then start with a construction for a saturated simple topological graph with $6r$ vertices. Add ρ clones of its lowest-degree vertex P, and saturate. In our construction, the lowest degree is 7. Cloning such a vertex ρ times adds up to $7\rho + \binom{\rho+1}{2}$ additional edges after saturation. Since $\rho \leq 5$, the number of edges is bounded by

$$7(6r) - 30 + 7\rho + \binom{\rho+1}{2} \leq 7(6r + \rho) - 30 + 15 = 7(6r + \rho) - 15 < 7n$$

The resulting simple topological graph proves Theorem 1 for $n \geq 12$. If $n \leq 11$, then the bound of Theorem 1 holds since even the complete graph has at most $\binom{n}{2} \leq 5n$ edges.

3   Saturated 2-Simple Topological Graphs with Few Edges

3.1   The Grid-Block Configuration

To begin, we study a drawing of 6 edges (three red edges and three black edges) as depicted in Fig. 5. The drawing consists of three disjoint edges representing the red edges r_1, r_2, and r_3. The black edges are drawn, such that one crosses (in order) $r_1, r_2, r_3, r_1, r_2, r_3$, the other $r_2, r_3, r_1, r_2, r_3, r_1$, and the last one $r_3, r_1, r_2, r_3, r_1, r_2$. There are no other crossings in the drawing. Note that the configuration superimposes a grid. We call such an arrangement of edges a *grid-block*. These blocks have been used as so-called $(3, 2)$-grid-blocks by Kynčl et al. as building blocks in their saturated graphs [5].

As done in the previous section we consider the graph as drawn on the cylinder. More precisely, we draw the graph inside a rectangle in which we identify two sides in opposition (*bottom side* and *top side*), while the other sides are named *right side* and *left side*. If an edge uses the transition across the bottom/top edge we say that it *wraps around*. In the following we assume that the grid-blocks are drawn such that only the black edges wrap around. We label every face of the drawing of a grid-block with 2 numbers. These numbers refer to the coordinates of the (dual) superimposed grid, with $(0, 0)$ being the label of the face that contains the two bottom most endpoints of the black edges on the left side. All "vertical" coordinates are considered modulo 3.

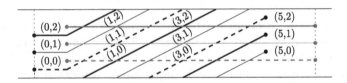

Fig. 5. A grid-block with some labeled faces.

Kynčl et al. observed that every edge connecting the left with the right side of the cylinder has to intersect the edges of grid-block at least 5 times. For our construction we need a stronger statement which is presented in Lemma 5 (proven in the full version [3]).

Lemma 5. *Let γ be a path crossing the grid-block that starts in face $(0, i)$ and ends in face $(5, j)$ and that never visits the faces $(0, \cdot)$, $(5, \cdot)$ again. Then γ can be transformed to a path $\widetilde{\gamma}$ passing through the grid block keeping its endpoints fixed, such that $\widetilde{\gamma}$*

1. *crosses only the edges (with the same or smaller multiplicity) crossed by γ,*
2. *has 5 crossings with the grid-block,*
3. *first walks between the faces $(0, i)$, $0 \le i \le 2$, then crosses some black edges to the right, passing from a face (a, i) to a face $(a + 1, i)$, then crosses some red edges upwards, passing from a face (a, i) to a face $(a + 1, i + 1)$.*

3.2   A Blocking Configuration

We call the building blocks of the following constructions *black block* and *red block*, see Fig. 6(a)-(b). We refer to the edges in the red (black) block as *red edges* (*black edges*). Any two red edges, as well as any two black edges, cross exactly twice. Note that after we have mirrored the red block it is isotopic to the black block.

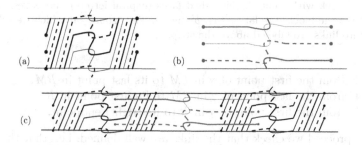

Fig. 6. A black block (a), a red block (b), and a 3-block, formed by consecutive black, red and again black blocks (c). (Color figure online)

We combine two black blocks and a red block as shown in Fig. 6(c) to obtain a drawing that we call a *3-block*. Since the red block differs from the black block only by a reflection, the 3-block built form consecutive black-red-black blocks is a mirror image of the 3-block built from consecutive red-black-red blocks.

The following theorem is the key observation that we need for the construction of the sparse 2-simple drawing.

Theorem 2. *Any path connecting the left with the right sides of the cylinder while passing through the 3-block crosses one of the edges forming the 3-block at least 3 times.*

Before proving the theorem we provide some helpful lemmas. We consider the path as a walk on the graph, dual to the graph of the arrangement of the 3-block. We label some of the faces of the arrangement as shown in Fig. 7. In particular, for $i = 0, 1, 2$, we denote the faces containing the left endpoint of the red edges r_i as L_i, and the faces containing the right endpoint as R_i. The edges of the left black block are named b_i and the edges of the right black block are named b'_i. Finally, let LM_i be the face that contains the right endpoint of b_i, and let RM_i be the face that contains the left endpoint of b'_i. The region spanned by L_0, L_1 and L_2 is denoted by L. We similarly define regions LM, RM and R.

Let γ be a path that passes through the 3-block. To facilitate the analysis we subdivide the path γ into smaller pieces, which we call *links*. The links are defined as follows:

link 1: from the start point (left) of γ to the last point of γ in L,
link 2: from the last point of γ in L to its first point in LM,

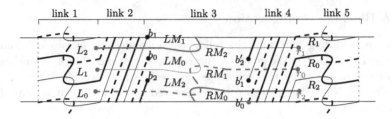

Fig. 7. A 3-block with some distinguished faces (capital letters) and edges. The red edges forming the blocks are labeled b_i, b_i' and r_i. The "zones" at which we subdivide the path into links are labeled above the strip.

link 3: from the first point of γ in LM to its last point in RM,
link 4: from the last point of γ in RM to its first point in R,
link 5: from the first point of γ in R to its (right) endpoint.

Before we proceed we check that the links are well defined, i.e., that the points defining the links appear in order. For the links 1, 3 and 5 this holds trivially, while to check it for links 2 (and, symmetric, 4), we need to prove that the last point in L precedes the first point in LM:

Lemma 6. *No path can visit the regions $L \to LM \to L \to LM$ in this order without crossing some of the edges forming the 3-block at least 3 times.*

Proof. The faces L and LM are separated by a grid-block. Passing through it requires at least 5 crossings of its edges. Any path visiting $L \to LM \to L \to LM$ would cross the grid-block at least 3 times, and hence it would cross the edges of the grid-block at least $3 \times 5 = 15$ times. Since a grid-block is formed by 6 edges, at least one of them will be crossed 3 times or more.   □

We continue by analyzing the path through the 3-block following its links.

Lemma 7. *Any path passing the 3-block from left to right with the last point of link 1 at L_i crosses the edge b_{i+1} at least once or one of the edges b_i and b_{i+2} at least twice at its first link (all indices modulo 3).*

Proof. A path that ends in L_i crosses either b_{i+1} or it crosses b_{i+2} while entering from L_{i+1}. Repeating this argument twice proves the lemma.   □

The following lemma summarizes the behavior of the path on the first two links:

Lemma 8. *Any path γ passing the 3-block that does not intersect any edge 3 times or more crosses the red edges r_j, r_{j+1} before it first visits the region LM at LM_j.*

Proof. We modify the path γ along link 2 following the simplification procedure described in Lemma 5 to get a path $\widetilde{\gamma}$. Lemma 5 also implies that the link 2 of $\widetilde{\gamma}$

consists of exactly 5 "steps": first, $0 \leq h \leq 5$ steps crossing the black edges $\rightarrow$ to the right, followed by $v = 5 - h$ steps crossing red edges $\nearrow$ upward.

Assume that the first point of link 2 of $\tilde{\gamma}$ lies inside the face L_i. Then h horizontal steps of link 2 cross the b_{i+1}, b_i, b_{i-1}, $\ldots$, $b_{i+1-(h-1)}$. Moreover, Lemma 7 guarantees that already link 1 of the path $\tilde{\gamma}$ crossed either b_{i+1} once or one of b_i or b_{i+2} twice. Since $\tilde{\gamma}$ does not cross any of the black edges more than twice, it follows that $h \leq 3$. This, however, shows that $v \geq 2$, which implies that the path $\tilde{\gamma}$ crosses the red edges r_{j+1}, r_j before it reaches the last point of its second link in face LM_j. To finish the proof we recall that the path γ crosses every edge of the 3-block at least as many times as $\tilde{\gamma}$ and that the last points of the link 2 of γ and $\tilde{\gamma}$ coincide.   □

Proof of Theorem 2. We prove by contradiction, namely, we assume that there is a path γ that passes through the 3-block while crossing every edge of the 3-block at most twice. Let LM_j be the face where link 2 ends, and let RM_ℓ be the face where link 4 starts. By Lemma 8 we know that γ crosses r_j and r_{j+1} in link 1 and link 2. Since the structure of the link 4 and 5 coincides with the structure of link 2 and 1 we can apply Lemma 8 also to the last two links. Thus, γ crosses $r_{\ell-1}$, r_ℓ in link 4 and 5. A short case distinction (ℓ might be either j, $j + 1$, or $j + 2$) shows that γ cannot connect endpoints of link 2 and 4 via link 3 without crossing at least one of the red edges 3 times; see Fig. 8. The figure depicts all ways of how to possibly route the path γ in link 3. Each of the possible continuations crosses some of the red edges r_j, r_{j+1}, r_{j-1} twice and is blocked within one of the faces before it reaches the face RM_ℓ. As a consequence the path γ cannot exists.   □

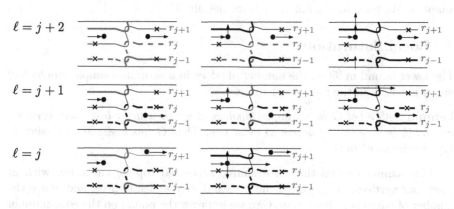

$\ell = j + 2$

$\ell = j + 1$

$\ell = j$

Fig. 8. Each row depicts a case. Black dots inside faces mark the faces LM_j (left) and RM_ℓ (right). Black crosses on red edges mark the edges that are, due to Lemma 8, crossed by the path outside link 3. We color red edges black as soon as they are crossed by the path γ twice and no more crossings are allowed. In the case $\ell = j$ the path can be continued in 3 different directions, in each of them the path is blocked after one step. (Color figure online)

3.3   A Sparse Saturated 2-Simple Drawing

We show now how to combine a sequence of 3-blocks to obtain a 2-simple saturated drawing with few edges.

Theorem 3. *Let $s_2(n)$ denote the minimum number of edges that a 2-simple saturated drawing with n vertices can have. Then $s_2(n) \leq 14.5n$.*

Proof. We consider the drawing that repeats the pattern shown in Fig. 9. The drawing is formed by ℓ consecutive black and red blocks; see Fig. 6. We have 6 vertices per block, thus 6ℓ in total. Clearly, the drawing is 2-simple.

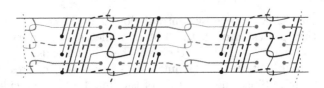

Fig. 9. A 2-simple drawing that does not allow too many edges to be added.

Now we add as many edges as possible without violating the 2-simplicity, so that the drawing becomes saturated (this padding procedure is definitely not unique). Theorem 2 implies that without violating the 2-simplicity any vertex can be connected by an edge only to 29 other vertices. This implies that the maximal number of edges in the resulting saturated 2-simple drawing is less then or equal than $14.5n$. For n not divisible by 6 our construction can be easily adapted. We refer to the full version for details [3].   □

4   Local Saturation

The lower bound in [5] on the number of edges in a saturated simple topological graph is based on the following lemma.

Lemma 9 [5]. *Let G be a simple topological graph with at least four vertices, and let A be a vertex of degree at most two. Then G has a simple extension by an edge incident to A.*

This lemma implies that in a simple saturated topological graph with at least four vertices, every vertex must have degree at least three, and hence the number of edges is at least $1.5n$. Can we improve the bound on the edge number by strengthening the lower bound on the degree? The following considerations establish a limit to this approach: There are saturated graphs with minimum degree four.

We say that a vertex S in a simple topological graph is *saturated* if it cannot be connected to a non-adjacent vertex while maintaining simplicity. The above lemma implies that in a simple topological graph with at least four vertices, a saturated vertex must have degree at least three.

Observation 1. *For any positive integer $n \geq 6$, there is a simple topological graph on n vertices with a saturated vertex of degree four.*

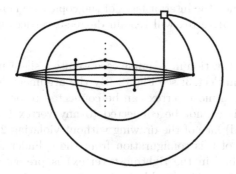

Fig. 10. The boxy vertex of degree four is saturated.

The observation is due to the construction presented in Fig. 10. This example is an extension of the case $n = 6$ from [5, Fig. 2]. The topmost vertex is saturated, since only the straight edges are not incident to that vertex. It is easy to see, that in order to connect a vertex p to the degree four, on has to cross an edge incident to p before reaching p.

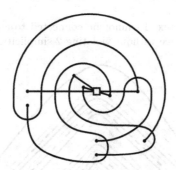

Fig. 11. In the simple topological graph above, the central vertex has degree 5, and it cannot be connected by an edge to any point in the unbounded region while keeping simplicity.

The following lemma presents a construction that realizes small vertex degrees for many vertices.

Lemma 10. *For any positive integer k, there is a saturated simple topological graph on $10k$ vertices with k vertices of degree 5.*

Proof. The main idea of our construction is depicted in Fig. 11. A simple case distinction verifies that no edge can connect the central vertex with a point on the outer face without violating the simplicity of drawing.

Now, take k copies of the drawing in Fig. 11, and place them on the plane next to each other such the interior faces of the copies are non-overlapping. The k copies of the central vertex will remain degree-5 vertices no matter how we saturate the graph.   □

To study local saturation in 2-simple case we use a slight modification of the 3-block introduced in Sect. 3; see Fig. 12. By the arguments given in the proof of Theorem 2 the rightmost vertex can be connected to only 12 other vertices (Fig. 12) and thus it cannot be connected to any vertex that belongs to the leftmost (unbounded) face of the drawing without violating 2-simplicity.

The "unrolling" of this configuration from the cylinder to the plane (with center of the unrolling in the rightmost vertex) is presented in Fig. 13. The central vertex cannot be connected by an edge to any vertex that belongs to the

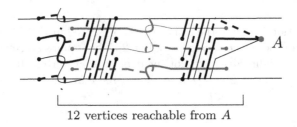

12 vertices reachable from A

Fig. 12. The rightmost vertex A cannot be connected to any vertex that belongs to the leftmost (unbounded) face without violating 2-simplicity.

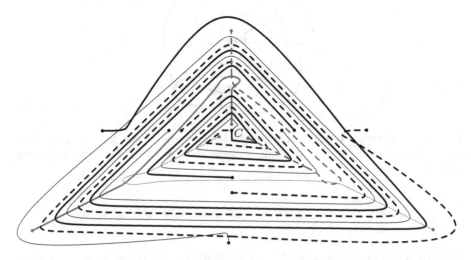

Fig. 13. Unrolling of Fig. 12 to the plane. The central vertex C corresponds to the rightmost vertex A of Fig. 12.

unbounded region without violating 2-simplicity, and so it has degree no larger than 12 in any saturation. After placing k disjoint copies of this construction to the plane next to each other we obtain the following result:

Lemma 11. *For any positive integer k, there is a saturated 2-simple topological graph on $16k$ vertices with k vertices of degree 12.*

Acknowledgments. The first author thanks Géza Tóth for presenting their inspiring results [5] in Szeged and for the encouragement during his investigation. This research was partially initiated at the EuroGIGA *Workshop on Geometric Graphs* in Münster, Germany, 2014, supported by the European Science Foundation (ESF).

References

1. Erdős, P., Hajnal, A., Moon, J.W.: A problem in graph theory. Amer. Math. Monthly **71**, 1107–1110 (1964)
2. Faudree, J.R., Faudree, R.J., Schmitt, J.R.: A survey of minimum saturated graphs. Electron. J. Combin. **18**, #DS19 (2011)
3. Hajnal, P., Igamberdiev, A., Rote, G., Schulz, A.: Saturated simple and 2-simple topological graphs with few edges (2015). http://arxiv.org/abs/1503.01386
4. Kászonyi, L., Tuza, Z.: Saturated graphs with minimal number of edges. J. Graph Theor. **10**(2), 203–210 (1986)
5. Kynčl, J., Pach, J., Radoičić, R., Tóth, G.: Saturated simple and k-simple topological graphs. Comput. Geom. **48**(4), 295–310 (2015)
6. Lovász, L.: Three short proofs in graph theory. J. Comb. Theor. Ser. B **19**(3), 269–271 (1975)

Testing Full Outer-2-planarity in Linear Time

Seok-Hee Hong[1](✉) and Hiroshi Nagamochi[2]

[1] University of Sydney, Sydney, Australia
seokhee.hong@sydney.edu.au
[2] Kyoto University, Kyoto, Japan
nag@amp.i.kyoto-u.ac.jp

Abstract. A graph is 1-planar, if it admits a 1-planar embedding, where each edge has at most one crossing. Unfortunately, testing the *1-planarity* of a graph is known as NP-complete.

This paper initiates the study of the problem of the testing *2-planarity* of a graph, in particular, testing the "full-outer-2-planarity" of a graph. A graph is *outer-2-planar*, if it admits an *outer-2-planar embedding*, that is every vertex is on the outer boundary and no edge has more than two crossings. A graph is *fully-outer-2-planar*, if it admits a *fully-outer-2-planar embedding*, that is an outer-2-planar embedding such that no crossing appears along the outer boundary. We present several structural properties of triconnected outer-2-planar graphs and fully-outer-2-planar graphs, and prove that triconnected fully-outer-2-planar graphs have a constant number of fully-outer-2-planar embeddings. Based on these properties, we present linear-time algorithms for testing the fully-outer-2-planarity of a graph G, whose vertex-connectivity is 1, 2 or at least 3. The algorithm also produces a fully-outer-2-planar embedding of a graph, if it exists. Moreover, we show that every fully-outer-2-planar embedding admits a straight-line drawing.

1 Introduction

A recent research topic in topological graph theory generalises the notion of planarity to *sparse non-planar graphs* with some specific crossings, or with some forbidden crossing patterns. Examples include *k-planar graphs* (i.e., graphs that can be embedded with at most k crossings per edge), *k-quasi-planar graphs* (i.e., graphs that can be embedded without k mutually crossing edges), *RAC graphs* (i.e., graphs that can be embedded with right angle crossings), *fan-crossing-free graphs* (i.e., graphs that can be embedded without fan-crossings), and *fan-planar graphs* (i.e., graphs that can be embedded such that each edge is crossed by a bundle of edges incident to a common vertex) [2,7,9,18,21]. Some mathematical results are known for these graphs. Pach and Toth [21] proved that a 1-planar graph with n vertices has at most $4n - 8$ edges. Agarwal et al. [2] (Ackerman [1]) showed that 3- and 4-quasi-planar graphs have linear number of edges. Fox et al. [11] proved that k-quasi-planar graphs have at most $O(n \log^{1+o(1)} n)$ edges.

For omitted proofs and figures, see the full version of this paper TR [17].

© Springer-Verlag Berlin Heidelberg 2016
E.W. Mayr (Ed.): WG 2015, LNCS 9224, pp. 406–421, 2016.
DOI: 10.1007/978-3-662-53174-7_29

Didimo et al. [9] showed that RAC graphs have at most $4n - 10$ edges. Cheong et al. [7] showed that fan-crossing free graphs have at most $4n - 8$ edges, and Kaufmann and Ueckerdt [7] showed that fan-planar graphs have at most $5n - 10$ edges.

Recently, algorithmics and complexity for such graphs have been investigated. Grigoriev and Bodlaender, and Kohrzik and Mohar proved that testing the 1-planarity of a graph is NP-complete [14,19]. Argyriou et al. proved that testing whether a given graph is a RAC graph is NP-hard [3]. Testing fan-planariy of graphs is NP-hard [6], even if a *rotation system* (i.e., the circular ordering of edges for each vertex) is given [5].

On the positive side, Eades et al. [10] showed that the problem of testing the *maximal 1-planarity* (i.e., addition of an edge destroys 1-planarity) of a graph can be solved in linear time, if a rotation system is given. Hong et al. [15], and Auer et al. [4] independently presented a linear time algorithm for testing the *outer-1-planarity* (i.e., 1-planar graphs with every vertex is on the outer face) of a graph. Bekos et al. presented a polynomial time algorithm for testing the *maximal outer-fan-planarity* (i.e., fan-planar graphs with each vertex is on the outer face and addition of an edge destroys outer-fan-planarity) [5].

This paper initiates the problem of testing the *2-planarity* of a graph, in particular, testing the *fully-outer-2-planarity* of a graph. An embedding γ of a graph G in the plane is *2-planar*, if no edge has more than two crossings. A 2-planar embedding of G is called *outer-2-planar* (O2PE), if each vertex is on the outer boundary. An outer-2-planar embedding of G is called *fully-outer-2-planar* (FO2PE), if no edge crossings appear along the outer boundary. A graph G is *2-planar* (resp., *outer-2-planar, fully-outer-2-planar*) if it admits a 2-planar (resp., *outer-2-planar, fully-outer-2-planar*) embedding (see Fig. 1).

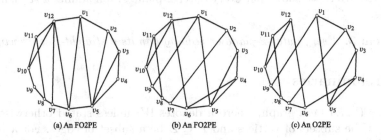

Fig. 1. (a) An FO2PE γ_1 of a biconnected graph G_1; (b) An FO2PE γ_2 of a triconnected graph G_2; (c) An O2PE γ_3 of a triconnected graph G_3.

Note that the problem of testing the outer-2-planarity seems much harder than testing the outer-1-planarity. For example, it was shown that outer-1-planar graphs (say K_4) are indeed planar graphs [4]. However, outer-2-planar graphs (say K_5) are non-planar-graphs. Also there is only one triconnected outer-1-planar graph, K_4, and it has unique outer-1-planar embedding [4,15]. However, we can show that there is a triconnected outer-2-planar graph which has exponentially many outer-2-planar embeddings. Thus, the main thrust of this paper is to devise a linear-time algorithm for testing the full-outer-2-planarity. Moreover, the outer boundary of an FO2PE of a biconnected graph G is a Hamiltonian cycle of G. Note that testing whether a given graph has a Hamiltonian cycle is known to be NP-complete, even for cubic graphs [13].

We first prove several structural properties of outer-2-planar graphs and fully-outer-2-planar graphs. Using these properties, we present a linear-time algorithm for testing the fully-outer-2-planarity of a graph. The following theorem summarizes our main results.

Theorem 1. *There is a linear-time algorithm that tests whether a given graph is fully-outer-2-planar, and computes a fully-outer-2-planar embedding of the graph if it exists.*

We use a connectivity approach to prove Theorem 1. The *oneconnected* case is easy (see Theorem 4), and the *biconnected* case is more involved (see Theorem 5); the main contribution of this paper is to solve the *triconnected* case,. The following theorem is the key to design linear-time algorithm for FO2PE.

Theorem 2. *The number of all FO2PEs of a triconnected graph G is constant, and the set of all FO2PEs of G can be generated in linear time.*

Fary's well-known theorem [12] shows that every plane graph admits a straight-line drawing. Thomassen [22] presented forbidden subgraphs for straight-line drawings of 1-plane graphs. Hong et al. [16] gave a linear-time algorithm to construct a straight-line 1-planar drawing, if it exists. Nagamochi solved the more general problem of straight-line drawability for wider class of embedded graphs [20]. Here we show that every outer-2-plane graph admits a straight-line drawing.

Theorem 3. *Every outer-2-plane embedding admits a straight-line drawing.*

2 Preliminaries

Let $G = (V, E)$ be a graph, where n denotes $|V|$ unless stated otherwise. Let $X, Y \subseteq V$ be subsets of vertices and $F \subseteq E$ be a subset of edges. For a vertex v, let $E(v)$ denote the set of edges vu incident to v, $\deg(v)$ denote the degree $|E(v)|$ of v, $N(v)$ denote the set of neighbors u of v, and $N[v] = N(v) \cup \{v\}$. We may indicate the underlying graph G in these notations in such a way that $E(v)$ is written as $E(v; G)$. Let $G - F$ denote the graph obtained from G by removing the edges in F, and $G - X$ denote the graph obtained from G by removing the vertices in X together with the edges in $\cup_{v \in X} E(v)$. Let G/X denote the graph obtained from a graph G by contracting the vertices in a subset X of vertices into a single vertex, where any resulting loops and multiple edges are removed. A vertex of degree d is called a *degree-d vertex*. A simple cycle of length k is called a *k-cycle*, where a 3-cycle is called a *triangle*.

A *topological graph* or *embedding* γ of a graph G is a representation of a graph (possibly with multiple edges) in the plane, where each vertex is a point and each edge is a Jordan arc between the points representing its endpoints. Two edges *cross* if they have a point in common, other than their endpoints. The point in common is a *crossing*. To avoid pathological cases, standard non-degeneracy conditions apply: (i) two edges intersect at most one point; (ii) an edge does not contain a vertex other than its endpoints; (iii) no edge crosses itself; (iv) edges must not meet tangentially; (v) no three edges share a crossing point; and (vi) no two edges that share an endpoint cross.

For an O2PE γ of a graph $G = (V, E)$, we denote by $\partial\gamma$ the outer boundary of γ, which may pass through a crossing point made by two edges. An edge $e \in E$ is called an *outer* (resp., *inner*) edge of γ if the whole drawing of e is part of $\partial\gamma$ (resp., $\partial\gamma$ passes through only the end-vertices of e). An edge may not be outer or inner when a crossing on it appears along $\partial\gamma$. Let $V_{\partial\gamma}$, $E_{\partial\gamma}$ and $C_{\partial\gamma}$ denote the sets of vertices, outer edges and crossings in $\partial\gamma$.

For two vertices $u, v \in V$, the boundary path traversed from u to v in the clockwise order is denoted by $\partial\gamma[u, v]$. Let $V_{\partial\gamma}[u, v]$, $E_{\partial\gamma}[u, v]$ and $C_{\partial\gamma}[u, v]$ denote the sets of vertices, outer edges and crossings in $\partial\gamma[u, v]$. Also let $V_{\partial\gamma}(u, v] = V_{\partial\gamma}[u, v] - \{u\}$, $V_{\partial\gamma}[u, v) = V_{\partial\gamma}[u, v] - \{v\}$, $V_{\partial\gamma}(u, v) = V_{\partial\gamma}[u, v] - \{u, v\}$. We call the boundary path $\partial\gamma[u, v]$ *crossing-free* if $C_{\partial\gamma}[u, v] = \emptyset$, i.e., it consists of outer edges.

To solve the problem of finding an FO2PE γ of a graph G, we consider the problem with an additional constraint such that a set B of specified edges is required to appear along the boundary; i.e., $B \subseteq E_{\partial\gamma}$, and denote such an instance by (G, B). An FO2PE of γ of G such that $B \subseteq E_{\partial\gamma}$ is called an *FO2PE extension* of (G, B), and an instance (G, B) is called *extendible* if it admits an FO2PE extension.

3 Connected Graphs and Biconnected Graphs

We first observe that we can focus on biconnected graphs to design algorithms for testing the (full) outer-2-planarity.

Theorem 4. *A graph is outer-2-planar (resp., fully-outer-2-planar) if and only if its biconnected components are outer-2-planar (resp., fully-outer-2-planar).*

Thus, in what follows, we treat only *biconnected graphs* G as input. For a permutation $[v_1, v_2, \ldots, v_n]$ of the vertices of a biconnected graph G, let $\gamma = (G, [v_1, v_2, \ldots, v_n])$ denote an embedding of G such that vertices $v_1, v_2, \ldots, v_n$ appear along $\partial\gamma$ in the clockwise manner. We can easily observe that the number of crossings on each edge in an O2PE γ is determined only by the ordering of all vertices along $\partial\gamma$.

Our algorithm for the biconnected case uses the decomposition of a biconnected graph G into triconnected components, also known as the *SPQR tree*, defined by di Battista and Tamassia [8], which can be computed in linear time.

Each triconnected component consists of *real* edges (i.e., edges in the original graph) and *virtual* edges. (i.e., edges introduced during the decomposition process, which represents the other triconnected components, sharing the same virtual edges defined by cut-pairs).

In this paper, we use the SPR tree, a simplified version of the SPQR tree, and treat the SPR tree as a *rooted tree* by choosing an arbitrary node as its root. Each node ν in the SPR tree is associated with a graph called the *skeleton* of ν, denoted by $\sigma(\nu)$, which corresponds to a triconnected component. There are three types of nodes ν in the SPR tree: (i) S-node, where $\sigma(\nu)$ is a simple cycle with at least three vertices; (ii) P-node, where $\sigma(\nu)$ consists of two vertices connected by at least three edges; and (iii) R-node, where $\sigma(\nu)$ is a simple triconnected graph with at least four vertices. The set of virtual edges in the skeleton of a node ν is denoted by $E_{\mathrm{vir}}(\nu)$.

For a given biconnected graph G, we establish a recurrence relationship of FO2PE problem instances (G, B) based on the SPR decomposition of G. In fact we prove that G admits an FO2PE if and only if for each node ν in the SPR decomposition of G, the instance $(\sigma(\nu), E_{\mathrm{vir}}(\nu))$ is extendible. We easily see that for S-nodes ν $(\sigma(\nu), E_{\mathrm{vir}}(\nu))$ are cycles and always extendible.

The following theorem summarizes the main results in this section.

Theorem 5. *A biconnected graph $G = (V, E)$ admits an FO2PE if and only if the following holds: for each P-node ν, $|E_{\mathrm{vir}}(\nu)| \leq 2$; and for each R-node ν, $(\sigma(\nu), E_{\mathrm{vir}}(\nu))$ is extendible. Moreover, there is a linear-time algorithm for constructing an FO2PE of G, if it exists.*

4   Triconnected Graphs

In this section, we prove Theorem 2, i.e., every triconnected graph G has a constant number of FO2PEs, and they can be generated in linear time. By applying Theorem 2 to each triconnected graph $(\sigma(\nu), E_{\mathrm{vir}}(\nu))$ in Theorem 5, we see that FO2PE testing for biconnected graphs can be done in linear time.

4.1   Sketch of Proof for Theorem 2

To prove Theorem 2, we derive a recurrence structure over FO2PE problem instances (G, B) for special local structures B, called "**rims**". We first examine the structure of FO2PEs of triconnected graphs to prove that every FO2PE γ of a triconnected graph G except for some small instances contains a special local structure formed by three or four consecutive vertices along the boundary of γ, called a "rim" (see Fig. 2(a)–(c)). We also prove that a given triconnected graph G can have a constant number of subgraphs that can be a rim of some FO2PE γ of G.

In order to find all FO2PEs of G, we choose each of these subgraphs as a fixed partial embedding B and try to find an extension of instance (G, B) (the union of extensions of (G, B) over all choices of B may contain a duplication of the FO2PE of G). To find all extensions of an instance (G, B) with a rim B, we investigate the structure around the rim B to prove that some vertices or edges adjacent to B must appear next to B along the boundary of *any* extensions of an instance (G, B), and we identify a special subgraph structure called "**frill**" (see Fig. 2(d)), which can be embedded in an FO2PE in two ways (i.e., frill is a locally flippable embedding without destroying the whole FO2PE).

Based on these properties, we replace B with such vertices or edges to be fixed next to B with a new rim B' transforming G to another smaller triconnected graph G' so that (i) each extension γ of (G, B) can be obtained from an extension γ' of the new instance (G', B') by the reverse operation of the transformation, if no frill appears next to B; or (ii) each extension γ of (G, B) can be obtained from an extension γ' of (G', B') by the reverse operation of the transformation, and by the reverse operation followed by an operation for flipping the frill in the embedding, if a frill appears next to B. In (i), there is a one-to-one correspondence between the FO2PEs of (G, B) and those of (G', B'), while in (ii) two FO2PEs will be generated from each FO2PE of (G', B'). This establishes a reduction of an instance (G, B) of a rim B to a smaller instance (G', B') of a rim B', where B and B' may be of different types among $(3, 3)$-rim, $(3, 4)$-rim and 4-rim.

Given a triconnected instance G with n vertices, we directly generate all possible embeddings of it when $n \leq 9$ or reduce it to an instance (G', B) with at most seven vertices by a repeated application of the reduction step otherwise. In the former, there are at most $(9 - 1)!$ FO2PEs of G. In the latter, we will show that the number of times that a "frill" appears during the reduction process is at most 2, which implies that the number of all extensions of the instance (G', B) is at most $2 \cdot 2 \cdot (7 - 1)!$. Then the number of all FO2PEs of a given triconnected instance G is at most $\max\{8!, 4 \cdot 6!\}$, which is constant.

For the time complexity, we use several different transformations to reduce an instance (G, B) to a smaller instance (G', B') and most cases can be easily executed in $O(1)$ time, since in such cases we only need to examine a constant number of vertics or edges adjacent to B to decide which transformation to be applied. However, there are two *special transformations* (see Lemmas 10(iv) and 11(v)), where constructing (G', B') from (G, B) may take $O(n)$ time to collect the necessary information to determine (G', B'). For these cases, we present a sophisticated way of handling the transformation so that the reduction process can still be executed in $O(1)$ time. From the above, we will see that the total running time for generating all FO2PEs of a given triconnected instance G can be done in linear time.

4.2   Structural Results on Triconnected O2PE and FO2PE

We first present structural results on triconnected O2PE.

Lemma 1. *Every O2PE of a triconnected graph G is quasi-planar unless G is* $K_{3,3}$.

Lemma 2. *No triconnected graph G with a vertex of degree ≥ 5 admits an O2PE.*

Lemma 3. *Let $G = (V, E)$ be a triconnected graph which contains K_4 as a subgraph. If G admits an O2PE, then $n \leq 6$.*

By Lemmas 1 and 3, we obtain the following lemma.

Lemma 4. *Let G be a triconnected graph with at least seven vertices. If G admits an O2PE γ, then G contains no subgraph isomorphic to K_4 and γ is quasi-planar.*

For an O2PE γ of a triconnected graph $G = (V, E)$ with $n \geq 7$, the cyclic order $[v_1, v_2, \ldots, v_n]$ of the vertices in $\partial\gamma$ completely determines the embedding γ by Lemma 4. In what follows, an O2PE γ of a graph G is simply denoted by the cyclic order of the vertices in $\partial\gamma$.

For an inner edge uv in an FO2PE γ of a triconnected graph G, there is an edge ab that crosses uv; i.e., ab joins a vertex $a \in V_{\partial\gamma}(u, v)$ and a vertex $b \in V_{\partial\gamma}(v, u)$, since otherwise $\{u, v\}$ would be a cut-pair.

4.3 Identifying a Constant Number of Candidate Partial Embeddings

Let γ be an O2PE of a triconnected graph G. A triangle uvw is called a $(3, 3)$-*rim* (resp., $(3, 4)$-rim) of γ if uv and vw are outer edges in γ and v is a degree-3 (resp., degree-4) vertex. A $(3, 3)$- or $(3, 4)$-rim is called a *3-rim*. A 4-cycle $uvv'w$ is a 4-*rim* of γ if v and v' are degree-3 vertices and uv, vv' and vw are outer edges in γ. A 3- or 4-rim is called a *rim*; see Fig. 2. We show that any FO2PE of a triconnected graph G contains a rim.

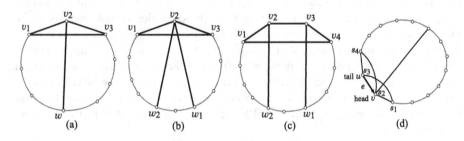

Fig. 2. (a) a $(3,3)$-rim $B = [v_1, v_2, v_3]$ with a degree-3 vertex v_2; (b) a $(3,4)$-rim $B = [v_1, v_2, v_3]$ with a degree-4 vertex v_2; (c) a 4-rim $B = [v_1, v_2, v_3, v_4]$ with degree-3 vertices v_2 and v_3; (d) a frill e with span $[s_1, s_2, s_3, s_4]$ such that s_2 is the head and s_3 is the tail of e.

Lemma 5. *Any FO2PE γ of a triconnected graph G has a rim.*

Our algorithm for constructing an FO2PE of a given triconnected graph G first generates triangles/4-cycles as rims of possible FO2PEs and tries to extend each of the triangles/4-cycles into an FO2PE. By Lemma 2, we can assume that a given triconnected graph G has a maximum degree at most 4. Then there are $O(n)$ triangles and 4-cycles for candidates of rims in an FO2PE of G. The next lemma reduces the number of triangles/4-cycles to be generated as rims of possible FO2PEs to a constant number.

Lemma 6. *Let γ be an FO2PE of a triconnected graph $G = (V, E)$ with $n \geq 10$.*

(i) *Assume that G has a triangle, and let t_1 be a triangle in G. Then $\partial\gamma$ contains a sequence $[u, v, w]$ for the set of vertices u, v and w of some triangle $t' = uvw$ sharing an edge with t_1 (possibly $t' = t_1$) as its subsequence.*

(ii) *Assume that G has no triangle. Then G has a 4-cycle $u_1u_2u_3u_4$ with degree-3 vertices u_2 and u_3 (by Lemma 5). Then $\partial\gamma$ (or its reversal) contains $[u_1, u_2, u_3, u_4]$ (or $[u_3, u_4, u_1, u_2]$ if $\deg(u_4) = \deg(u_1) = 3$) as its subsequence.*

In an FO2PE γ of a triconnected graph G, an outer edge e joining a degree-3 vertex u and a degree-4 vertex v is called a *frill* if γ contains a subsequence $[s_1, s_2, s_3, s_4]$ with $\{s_2, s_3\} = \{u, v\}$ such that $s_1s_2s_3$ and $s_2s_3s_4$ are triangles, where the degree-4 vertex v (resp., degree-3 vertex u) is called the *head* (resp., *tail*) of the frill e (see Fig. 2(d)). We call $[s_1, s_2, s_3, s_4]$ the *span* of frill e. An operation of exchanging the positions of s_2 and s_3 in the cyclic order γ is called *flipping* frill e. It is easy to observe that the cyclic order γ' obtained from γ by flipping a frill is also an FO2PE of G.

Lemma 7. *Let γ be an FO2PE of a triconnected graph $G = (V, E)$ with $n \geq 7$. Then there are at most two frills in γ, and if there are exactly two frills, then their spans share at most one vertex. Moreover flipping a frill in γ never introduces a new frill in the resulting cyclic order γ'.*

We start with a triangle or 4-cycle fixed in Lemma 6 as a rim of a possible FO2PE of G, where the rim is a "partial embedding" of G. For a triangle uvw (resp., a 4-cycle $uvv'w$) in a graph G, the instance where edges uv and vw (resp., uv, vv' and $v'w$) are required to appear as outer edges is given by (G, B) with $B = \{uv, vw\}$ (resp., $B = \{uv, vv', v'w\}$). In what follows, we denote the constraint B simply by a vertex sequence $B = [u, v, w]$ (resp., $B = [u, v, v', w]$).

Our next aim is to design a procedure for constructing a possible FO2PE of G as an extension of the fixed rim. Suppose that Algorithm **EXTEND**(G, B) is a procedure that returns all FO2PE extensions of (G, B). By executing such a procedure to each candidate of rims, we can enumerate all FO2PE of a triconnected graph G in linear time (see [17] for detail).

4.4 Reducing Instances with Fixed Rims

In this section, we prove the following result by designing Algorithm **EXTEND**(G, B).

Lemma 8. *For a triconnected instance* (G, B) *with a fixed rim, the maximum number of FO2PE extensions of* (G, B) *is constant, and all FO2PE extensions of* (G, B) *can be generated in* $O(n)$ *time.*

To prove Theorem 2, it suffices to show Lemma 8. We call an instance (G, B) *triconnected* if G is triconnected. To prove the lemma, we establish a reduction over triconnected instances (G, B) with fixed rims. We try to extend a given partial embedding (G, B) by fixing some other vertices, and simplify the instance with the newly fixed vertices into a triconnected instance (G', B') so that the new instance (G', B') admits an FO2PE extension if and only if so does the original instance.

For an instance (G, B), a sequence $[s_1, s_2, \ldots, s_k]$ is called *inevitable* if any FO2PE extension $\gamma = [v_1, v_2, \ldots, v_n]$ of (G, B) contains the sequence as its subsequence. Given an instance (G, B) with a fixed rim, we identify an inevitable sequence or a frill contained in any FO2PE extension of (G, B) without generating all possible permutations of the vertices in G. Based on the identified local structure of inevitable sequences or frills, we reduce (G, B) to a smaller new instance (G', B') with a new fixed rim B' such that (G, B) is extendible if and only if so is (G', B').

When we construct a new instance $(G' = G/X, B')$ by contracting a vertex subset X in G into a single vertex v^* and setting B' to be the set V' of a new triangle or 4-cycle, we call a vertex $v \in X$ an *attaching point* of (G', B') if each edge $e = uv^* \in E(v^*; G')$ corresponds to an edge $e \in E(v; G)$. We now show how to reduce an instance with a fixed $(3, 3)$-rim.

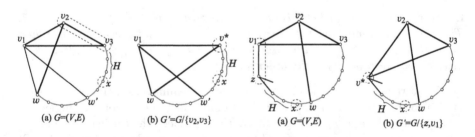

(a) $G=(V,E)$ (b) $G'=G/\{v_2,v_3\}$ (a) $G=(V,E)$ (b) $G'=G/\{z,v_1\}$

Fig. 3. (a) a graph G such that v_1 is a degree-3 vertex adjacent to w; (b) a new instance $(G' = G/\{v_2, v_3\}, B' = [w, v_1, v^*])$ with a new $(3, 3)$-rim of triangle wv_1v^* with a degree-3 vertex v_1; (c) a graph G such that v_1 is a degree-3 vertex not adjacent to w; (d) a new instance $(G' = G/\{z, v_1\}, B' = [v^*, v_2, v_3])$ with a new $(3, 3)$-rim of triangle $v^*v_2v_3$ with a degree-3 vertex v_2.

Lemma 9 ($(3, 3)$-rim reduction). *Let* (G, B) *be a triconnected extendible instance with* $n \geq 7$ *for a fixed* $(3, 3)$-rim $B = [v_1, v_2, v_3]$ *with* $N(v_2) = \{v_1, v_2, w\}$. *Then one of the following conditions* (i) *and* (ii) *holds, and the instance* (G', B') *defined in each condition is triconnected and extendible.*

(i) *Assume that v_1 or v_3, say v_1 is a degree-4 vertex adjacent to w (see Fig. 3(a) and (b)). Then $[w, v_1, v_2, v_3]$ is inevitable to (G, B). Let $G' = G/\{v_2, v_3\}$ and $B' = [w, v_1, v^*]$. Any FO2PE extension of (G, B) is obtained by modifying an FO2PE extension $\gamma' = [u_1 = w, u_2 = v_1, u_3 = v^*, u_4, \ldots, u_{n'}]$ of (G', B') into $\gamma = [w, v_1, v_2, v_3, u_4, \ldots, u_{n'}]$.*

(ii) *Assume that v_1 or v_3, say v_1 is a degree-3 vertex not adjacent to w (see Fig. 3(c) and (d)). Then $[z, v_1, v_2, v_3]$ is inevitable to (G, B). Let $G' = G/\{z, v_1\}$ and $B' = [v^*, v_2, v_3]$. Any FO2PE extension of (G, B) is obtained by modifying any FO2PE extension $\gamma' = [u_1 = v^*, u_2 = v_2, u_3 = v_3, u_4, \ldots, u_{n'}]$ of (G', B') into $\gamma = [z, v_1, v_2, v_3, u_4, \ldots, u_{n'}]$.*

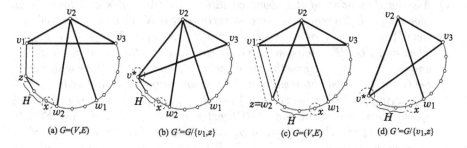

(a) $G=(V,E)$ (b) $G'=G/\{v_1,z\}$ (c) $G=(V,E)$ (d) $G'=G/\{v_1,z\}$

Fig. 4. (a) a graph G such that $z \notin \{w_1, w_2\}$; (b) a new instance $(G' = G/\{z, v_1\}, B' = [v^*, v_2, v_3])$ with a new $(3,4)$-rim of triangle $v^*v_2v_3$ with a degree-4 vertex v_2; (c) a graph G such that $z \in \{w_1, w_2\}$; (d) a new instance $(G' = G/\{z, v_1\}, B' = [v^*, v_2, v_3])$ with a new $(3,3)$-rim of triangle $v^*v_2v_3$ with a degree-3 vertex v_2.

The next lemma shows how to reduce an instance with a fixed $(3,4)$-rim. Note that for an instance $(G, B = [v_1, v_2, v_3])$ with $N(v_2) = \{v_1, v_2, w_1, w_2\}$ for a $(3,4)$-rim, we do not know the order of vertices w_1 and w_2 along the boundary of an FO2PE extension of (G, B).

Lemma 10 ((3,4)-rim reduction). *Let (G, B) be a triconnected extendible instance with $n \geq 7$ for a fixed $(3,4)$-rim $B = [v_1, v_2, v_3]$ with $N(v_2) = \{v_1, v_2, w_1, w_2\}$. Then one of the following conditions (i)–(iv) holds, and the instance (G', B') defined in each condition is triconnected and extendible.*

(i) *Assume that v_1 or v_3, say v_1 is a degree-3 vertex, where $N(v_1) = \{v_2, v_3, z\}$. (See Fig. 4.) Then $[z, v_1, v_2, v_3]$ is inevitable to (G, B). Let $G' = G/\{v_1, z\}$ and $B' = [v^*, v_2, v_3]$. Any FO2PE extension of (G, B) is obtained by modifying an FO2PE extension $\gamma' = [u_1 = v^*, u_2 = v_2, u_3 = v_3, u_4, \ldots, u_{n'}]$ of (G', B') into $\gamma = [z, v_1, v_2, v_3, u_4, \ldots, u_{n'}]$.*

(ii) *Assume that a vertex $v \in \{v_1, v_3\}$ is a degree-4 vertex adjacent to exactly one of w_1 and w_2, say $w \in \{w_1, w_2\}$, and there is a pair of a degree-3 vertex z and a vertex y such that vwz and wzy are triangles. Let $v = v_1$ without loss of generality. (See Fig. 5.) Then any FO2PE extension of (G, B) has zw as* *a frill. Let $G' = G/\{y, z, w, v_1\}$ and $B' = [v^*, v_2, v_3]$. Any FO2PE extension of (G, B) is obtained by modifying an FO2PE extension $\gamma' = [u_1 = v^*, u_2 = v_2, u_3 = v_3, u_4, \ldots, u_{n'}]$ of (G', B') into $\gamma = [y, z, w, v_1, v_2, v_3, u_4, \ldots, u_{n'}]$ and $[y, w, z, v_1, v_2, v_3, u_4, \ldots, u_{n'}]$.*

(iii) *Assume that a vertex $v \in \{v_1, v_3\}$ is a degree-4 vertex adjacent to exactly one of w_1 and w_2, say $w \in \{w_1, w_2\}$, but there is no pair of a degree-3 vertex z and a vertex y such that vwz and wzy are triangles. Let $(v, w) = (v_1, w_2)$ without loss of generality. (See Fig. 6.) Then $[w_2, v_1, v_2, v_3]$ is inevitable to (G, B). Let G' be the graph obtained from G by replacing edges v_1v_3 and v_2w_2 with a new edge w_2v_3, and $B' = [w_2, v_1, v_2, v_3]$. Any FO2PE extension of (G, B) is obtained as an FO2PE extension $\gamma' = [u_1, u_2, u_3, u_4, \ldots, u_{n'}]$ of (G', B').*

(iv) *Assume that none of the above conditions (i)–(iii) holds and there is an edge $z_1z_2 \in E$ between two degree-3 vertices $z_1 \in N(w)$ and $z_2 \in N(w')$ for $\{w, w'\} = \{w_1, w_2\}$ or a degree-4 vertex $z \in N(w_1) \cap N(w_2)$. (See Fig. 7.) Then any FO2PE extension of (G, B) contains exactly one of $[w, z_1, z_2, w']$ and $[w', z_2, z_1, w]$ (or exactly one of $[w, z, w']$ and $[w', z, w]$) as a sequence. Let G' be the graph obtained from G by removing vertex v_2 and adding a new edge w_1w_2, and $B' = [w, z_1, z_2, w']$ (or $B' = [w_1, z, w_2]$). Vertices v_1 and v_3 appear consecutively in any FO2PE extension γ' of (G', B'). Any FO2PE extension of (G, B) is obtained by modifying an FO2PE extension $\gamma' = [u_1 = v_1, u_2 = v_3, u_3, \ldots, u_{n'}]$ of (G', B') into $\gamma = [v_1, v_2, v_3, u_3, \ldots, u_{n'}]$.*

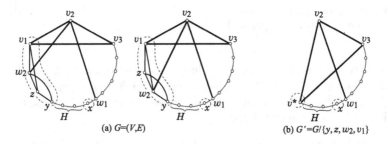

(a) $G=(V,E)$ (b) $G'=G/\{y, z, w_2, v_1\}$

Fig. 5. (a) a graph G such that v_1 is a degree-4 vertex adjacent to exactly one of w_1 and w_2, say w_2, and there is a pair of a degree-3 vertex z and a vertex y such that v_1w_2z and w_2zy are triangles; (b) a new instance $(G' = G/\{y, z, w_2, v_1\}, B' = [v^*, v_2, v_3])$ with a new $(3, 3)$-rim of triangle $v^*v_2v_3$ with a degree-3 vertex v_2.

The next lemma provides how to reduce an instance with a fixed 4-rim. Note that for an instance $(G, B = [v_1, v_2, v_3, v_4])$ with $N(v_2) = \{v_1, v_3, w_2\}$ and $N(v_3) = \{v_2, v_4, w_1\}$ for a 4-rim, we see that w_1 and w_2 appear always in this order after vertices v_1, v_2, v_3, v_4 appear along the boundary of any "quasi-planar" FO2PE extension of (G, B).

Lemma 11 (4-rim reduction). *Let (G, B) be a triconnected extendible instance with $n \geq 7$ for a fixed 4-rim $B = [v_1, v_2, v_3, v_4]$ with $N(v_2) = \{v_1, v_3, w_2\}$*

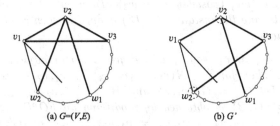

Fig. 6. (a) a graph G such that v_1 is a degree-4 vertex adjacent to exactly one of w_1 and w_2, say w_2, but there is no pair of a degree-3 vertex z and a vertex y such that v_1w_2z and w_2zy are triangles; (b) a new instance $(G', B' = [w_2, v_1, v_2, v_3])$ with a new 4-rim of 4-cycle $w_2v_1v_2v_3$ with degree-3 vertices v_1 and v_2.

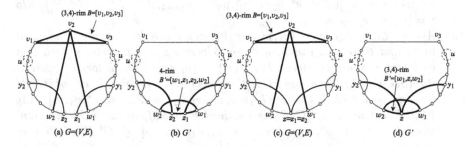

Fig. 7. (a) a graph G such that none of conditions (i)–(iii) in Lemma 10 holds and there is an edge $z_1z_2 \in E$ between two degree-3 vertices $z_1 \in N(w)$ and $z_2 \in N(w')$ for $\{w, w'\} = \{w_1, w_2\}$; (b) a new instance $(G', B' = [w, z_1, z_2, w'])$ with a new 4-rim of 4-cycle wz_1z_2w' with degree-3 vertices z_1 and z_2; (c) a graph G such that none of conditions (i)–(iii) in Lemma 10 holds and there is a degree-4 vertex $z \in N(w_1) \cap N(w_2)$; (d) a new instance $(G', B' = [w_1, z, w_2])$ with a new (3, 4)-rim of 3-cycle wzw' with a degree-3 vertex z.

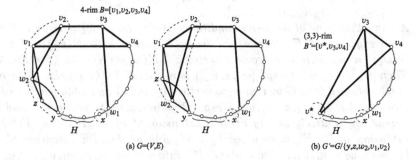

Fig. 8. (a) a graph G such that v_1 is a degree-4 vertex adjacent to w_2, and there is a pair of a degree-3 vertex z and a vertex y such that v_1w_2z and w_2zy are triangles; (b) a new instance $(G' = G/\{y, z, w_2, v_1, v_2\}, B' = [v^*, v_3, v_4])$ with a new (3, 3)-rim of triangle $v^*v_3v_4$ with a degree-3 vertex v_3.

and $N(v_3) = \{v_2, v_4, w_1\}$ (possibly $w_1 = w_2$). Then one of the following conditions (i)–(v) holds, and the instance (G', B') defined in each condition is triconnected and extendible.

(i) Assume that v_1 or v_4, say v_1 is a degree-3 vertex adjacent to neither of w_1 and w_2. Then for $z \in N(v_1) - \{v_2, v_3\}$, $[z, v_1, v_2, v_3, v_4]$ is inevitable to (G, B). Let $G' = G/\{v_1, z\}$ and $B' = [v^*, v_2, v_3, v_4]$. Any FO2PE extension of (G, B) is obtained by modifying an FO2PE extension $\gamma' = [u_1 = v^*, u_2 = v_2, u_3 = v_3, u_4 = v_4, u_5, \ldots, u_{n'}]$ of (G', B') into $\gamma = [z, v_1, v_2, v_3, v_4, u_5, \ldots, u_{n'}]$.

(ii) Assume that for $(v, w) = (v_1, w_2)$ or (v_4, w_1), v is a degree-3 vertex adjacent to w. Let $(v, w) = (v_1, w_2)$ without loss of generality. Then $[w_2, v_1, v_2, v_3, v_4]$ is inevitable to (G, B). Let $G' = G/\{w_2, v_1, v_2\}$ and $B' = [v^*, v_3, v_4]$. Any FO2PE extension of (G, B) is obtained by modifying an FO2PE extension $\gamma' = [u_1 = v^*, u_2 = v_3, u_3 = v_4, u_4, \ldots, u_{n'}]$ of (G', B') into $\gamma = [w_2, v_1, v_2, v_3, v_4, u_4, \ldots, u_{n'}]$.

(iii) Assume that for $(v, w) = (v_1, w_2)$ or (v_4, w_1), v is a degree-4 vertex adjacent to w, and there is a pair of a degree-3 vertex z and a vertex y such that vwz and wzy are triangles. Let $(v, w) = (v_1, w_2)$ without loss of generality. (See Fig. 8.) Then any FO2PE extension $\gamma = [v_1, v_2, \ldots, v_n]$ of (G, B) has zw_2 as a frill. Let $G' = G/\{y, z, w_2, v_1, v_2\}$ and $B' = [v^*, v_3, v_4]$. Any FO2PE extension of (G, B) is obtained by modifying an FO2PE extension $\gamma' = [u_1 = v^*, u_2 = v_3, u_3 = v_4, u_4, \ldots, u_{n'}]$ of (G', B') into $\gamma = [y, z, w_2, v_1, v_2, v_3, v_4, u_4, \ldots, u_{n'}]$ and $[y, w_2, z, v_1, v_2, v_3, v_4, u_4, \ldots, u_{n'}]$.

(iv) Assume that for $(v, w) = (v_1, w_2)$ or (v_4, w_1), v is a degree-4 vertex adjacent to w, but there is no pair of a degree-3 vertex z and a vertex y such that vwz and wzy are triangles. Let $(v, w) = (v_1, w_2)$ without loss of generality. Then $[w_2, v_1, v_2, v_3, v_4]$ is inevitable to (G, B). Let G' be the graph obtained from G by replacing edges v_1v_4 and v_2w_2 with a new edge w_2v_4 and contracting v_1 and v_2 into a single vertex v^*, and $B' = [w_2, v^*, v_3, v_4]$. Any FO2PE extension of (G, B) is obtained by modifying an FO2PE extension $\gamma' = [u_1 = w_2, u_2 = v^*, u_3 = v_3, u_4 = v_4, u_5, \ldots, u_{n'}]$ of (G', B') into $\gamma = [w_2, v_1, v_2, v_3, v_4, u_5, \ldots, u_{n'}]$.

(v) Assume that none of the above conditions (i)–(iv) holds, $w_1 \neq w_2$, and there is an edge $z_1z_2 \in E$ between two degree-3 vertices $z_1 \in N(w_1)$ and $z_2 \in N(w_2)$ (resp., there is a degree-4 vertex $z \in N(w_1) \cap N(w_2)$). (See Fig. 9.) Then $[w_1, z_1, z_2, w_2]$ (resp., $[w_1, z, w_2]$) is inevitable to (G, B). Let G' be the graph obtained from G by removing vertices v_2 and v_3 and adding a new edge w_1w_2, and $B' = [w_1, z_1, z_2, w_2]$ (resp., $B' = [w_1, z, w_2]$). Vertices v_1 and v_4 appear consecutively in any FO2PE extension γ' of (G', B'). Any FO2PE extension of (G, B) is obtained by modifying an FO2PE extension $\gamma' = [u_1 = v_1, u_2 = v_4, u_3, \ldots, u_{n'}]$ of (G', B') into $\gamma = [v_1, v_2, v_3, v_4, u_3, \ldots, u_{n'}]$.

Note that in each of Lemmas 9, 10 and 11, constructing a new instance (G', B') and modifying an FO2PE extension γ' of (G', B') into an FO2PE extension γ of (G, B) can be executed in $O(1)$ since G is a degree-bounded graph and γ can be obtained by inserting a subsequence.

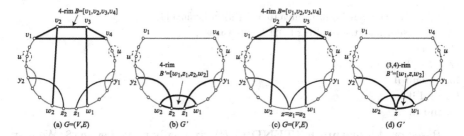

Fig. 9. (a) a graph G such that none of conditions (i)–(iv) in Lemma 11 holds, $w_1 \neq w_2$, and there is an edge $z_1 z_2 \in E$ between two degree-3 vertices $z_1 \in N(w_1)$ and $z_2 \in N(w_2)$; (b) a new instance $(G', B' = [w_1, z_1, z_2, w_2])$ with a new 4-rim of 4-cycle $w_1 z_1 z_2 w_2$ with degree-3 vertices z_1 and z_2; (c) a graph G such that none of conditions (i)–(iv) in Lemma 11 holds, $w_1 \neq w_2$, and there is a degree-4 vertex $z \in N(w_1) \cap N(w_2)$; (d) a new instance $(G', B' = [w_1, z, w_2])$ with a new $(3,4)$-rim of triangle $w_1 z w_2$ with degree-4 vertex z.

The **Algorithm EXTEND**(G, B), which takes a triconnected graph G and a permutation B of vertices in a triangle uvw or a 4-cycle $uvv'w$ with degree-3 vertices v and v', and outputs all FO2PE extensions of (G, B), is described below.

Algorithm EXTEND(G, B)

Input: A triconnected simple graph $G = (V, E)$ with $n \geq 7$ and a permutation B of vertices in a triangle uvw or a 4-cycle $uvv'w$ with degree-3 vertices v and v'.

Output: All FO2PE extensions of (G, B).

1: **if** $n \leq 7$ **then**

2:    Return all FO2PE extensions γ of (G, B) (if any), or Return $\emptyset$ (otherwise);

3: **else**

   /* Partial embedding B is specified as one of the following:

      Case 1: $B = [v_1, v_2, v_3]$ for a triangle $v_1 v_2 v_3$ with a degree-3 vertex v_2,

         where $N(v_2) = \{v_1, v_3, w\}$;

      Case 2: $B = [v_1, v_2, v_3]$ for a triangle $v_1 v_2 v_3$ with a degree-4 vertex v_2,

         where $N(v_2) = \{v_1, v_3, w_1, w_2\}$; and

      Case 3: $B = [v_1, v_2, v_3, v_4]$ for a 4-cycle $v_1 v_2 v_3 v_4$ with degree-3 vertices v_2 and

         v_3, where $N(v_2) = \{v_1, v_3, w_2\}$ and $N(v_3) = \{v_2, v_4, w_1\}$ */

4: **if** Case 1 (resp., Case 2, 3) holds, but none of the conditions (i)–(ii) in Lemma 9

   (resp., (i)–(iv) in Lemma 10 and (i)–(v) in Lemma 11) holds **then**

5:    Return $\emptyset$;

6: **else**

7:    Construct (G', B') according to the the conditions (i)–(ii) in Lemma 9

      (resp., (i)–(v) in Lemmas 10 and 11) currently satisfied by (G, B);

8:    **if** $\Gamma := $**EXTEND**$(G', B') \neq \emptyset$ **then**

9:       Modify each $\gamma' \in \Gamma$ into an FO2PE extension γ of (G, B)

         according to the operation in Lemma 9 (resp., Lemmas 10 and 11), where

         two FO2PE extensions of (G, B) will be constructed from the same γ'

```
        for the cases (ii) in Lemma 10 and (iii) in Lemma 11;
10:        Return all the resulting FO2PE extensions γ
11:    else
12:        Return ∅
13:    end if
14: end if
15: end if.
```

Based on **Algorithm EXTEND**(G, B), we finally prove Lemma 8. We first show that **Algorithm EXTEND**(G, B) correctly delivers all FO2PE extensions of (G, B), if any. In line 8, **Algorithm EXTEND**(G', B') returns the set Γ of all FO2PE extensions γ' of (G', B'). In line 9, all FO2PE extensions of (G, B) can be obtained from $\Gamma \neq \emptyset$ according to the modifications stated in Lemmas 9, 10 and 11. Since **Algorithm EXTEND**(G', B') returns all FO2PE extensions when $n \leq 7$, we see by induction that **EXTEND**(G, B) correctly delivers all FO2PE extensions of (G, B).

We next show that **Algorithm EXTEND**(G, B) delivers a constant number of solutions. When $n \leq 7$, the graph G has at most $n - |B| \leq 4$ vertices to be arranged along the boundary of a possible FO2PE extension of (G, B), and at most 4! FO2PE extensions of (G, B) will be constructed. We construct exactly one FO2PE extension γ of (G, B) from an FO2PE extension γ' of (G', B'), except for the cases (ii) in Lemma 10 and (iii) in Lemma 11 wherein exactly two FO2PE extensions, say γ_1 and γ_2 of (G, B) will be generated from the same FO2PE extension γ' of (G', B'). Note that in this case, γ_1 is obtained from γ_2 by flipping a frill zw in the lemmas, and the frill in γ_i will be preserved in any extensions obtained from γ_i until it is output as a final solution. By Lemma 7, any FO2PE of a graph can contain at most two frills, which means that generating two FO2PE extensions in line 9 can occur at most twice. Therefore, **Algorithm EXTEND**(G, B) delivers a constant number of FO2PE extensions of (G, B).

As we have already observed, constructing a new instance (G', B') and modifying an FO2PE can be done in $O(1)$ time, **Algorithm EXTEND**(G, B) runs in $O(n)$ time.

This completes a proof of Lemma 8, thus proving Theorem 2.

References

1. Ackerman, E.: On the maximum number of edges in topological graphs with no four pairwise crossing edges. Discrete Comput. Geom. **41**(3), 365–375 (2009)
2. Agarwal, P.K., Aronov, B., Pach, J., Pollack, R., Sharir, M.: Quasi-planar graphs have a linear number of edges. Combinatorica **17**(1), 1–9 (1997)
3. Argyriou, E.N., Bekos, M.A., Symvonis, A.: The straight-line RAC drawing problem is NP-hard. J. Graph Algorithms Appl. **16**(2), 569–597 (2012)
4. Auer, C., Bachmaier, C., Brandenburg, F.J., Gleißner, A., Hanauer, K., Neuwirth, D., Reislhuber, J.: Recognizing outer 1-planar graphs in linear time. GD **107–118**, 2014 (2013)

5. Bekos, M.A., Cornelsen, S., Grilli, L., Hong, S.-H., Kaufmann, M.: On the recognition of fan-planar and maximal outer-fan-planar graphs. In: Duncan, C., Symvonis, A. (eds.) GD 2014. LNCS, vol. 8871, pp. 198–209. Springer, Heidelberg (2014)

6. Binucci, C., Di Giacomo, E., Didimo, W., Montecchiani, F., Patrignani, M., Tollis, I.G.: Fan-planar graphs: combinatorial properties and complexity results. In: Duncan, C., Symvonis, A. (eds.) GD 2014. LNCS, vol. 8871, pp. 186–197. Springer, Heidelberg (2014)

7. Cheong, O., Har-Peled, S., Kim, H., Kim, H.-S.: On the number of edges of fan-crossing free graphs. In: Cai, L., Cheng, S.-W., Lam, T.-W. (eds.) Algorithms and Computation. LNCS, vol. 8283, pp. 163–173. Springer, Heidelberg (2013)

8. Di Battista, G., Tamassia, R.: On-line planarity testing. SIAM J. Comput. **25**(5), 956–997 (1996)

9. Didimo, W., Eades, P., Liotta, G.: Drawing graphs with right angle crossings. Theor. Comput. Sci. **412**(39), 5156–5166 (2011)

10. Eades, P., Hong, S.-H., Katoh, N., Liotta, G., Schweitzer, P., Suzuki, Y.: A linear time algorithm for testing maximal 1-planarity of graphs with a rotation system. Theor. Comput. Sci. **513**, 65–76 (2013)

11. Fox, J., Pach, J., Suk, A.: The number of edges in k-quasi-planar graphs. SIAM J. Discrete Math. **27**(1), 550–561 (2013)

12. Fáry, I.: On straight line representations of planar graphs. Acta Sci. Math. Szeged **11**, 229–233 (1948)

13. Garey, M.R., Johnson, D.S.: Computers and Intractability: A Guide to the Theory of NP-Completeness. Freeman, San Francisco (1979)

14. Grigoriev, A., Bodlaender, H.L.: Algorithms for graphs embeddable with few crossings per edge. Algorithmica **49**(1), 1–11 (2007)

15. Hong, S., Eades, P., Katoh, N., Liotta, G., Schweitzer, P., Suzuki, Y.: A linear-time algorithm for testing outer-1-planarity. Algorithmica **49**(1), 1–11 (2014)

16. Hong, S.-H., Eades, P., Liotta, G., Poon, S.-H.: Fáry's theorem for 1-planar graphs. In: Gudmundsson, J., Mestre, J., Viglas, T. (eds.) COCOON 2012. LNCS, vol. 7434, pp. 335–346. Springer, Heidelberg (2012)

17. Hong, S., Nagamochi, H.: Beyond planarity: testing full outer-2-planarity in linear time. Technical report [2014-003], Department of Applied Mathematics and Physics, Kyoto University (2014)

18. Kaufmann, M., Ueckerdt, T.: The density of fan-planar graphs, CoRR abs/1403.6184 (2014)

19. Korzhik, V.P., Mohar, B.: Minimal obstructions for 1-immersions and hardness of 1-planarity testing. J. Graph Theor. **72**(1), 30–71 (2013)

20. Nagamochi, H.: Straight-line drawability of embedded graphs. Technical report [2013-005], Department of Applied Mathematics and Physics, Kyoto University (2013)

21. Pach, J., Toth, G.: Graphs drawn with few crossings per edge. Combinatorica **17**(3), 427–439 (1997)

22. Thomassen, C.: Rectilinear drawings of graphs. J. Graph Theor. **12**(3), 335–341 (1988)

Fixed Parameter Tractability

Triangulating Planar Graphs While Keeping the Pathwidth Small

Therese Biedl[✉]

David R. Cheriton School of Computer Science,
University of Waterloo, Waterloo, ON N2L 1A2, Canada
biedl@uwaterloo.ca

Abstract. Any simple planar graph can be triangulated, i.e., we can add edges to it, without adding multi-edges, such that the result is planar and all faces are triangles. In this paper, we study the problem of triangulating a planar graph without increasing the pathwidth by much. We show that if a planar graph has pathwidth k, then we can triangulate it so that the resulting graph has pathwidth $O(k)$ (where the factors are 1, 8 and 16 for 3-connected, 2-connected and arbitrary graphs). With similar techniques, we also show that any outer-planar graph of pathwidth k can be turned into a maximal outer-planar graph of pathwidth at most $4k + 4$. The previously best known result here was $16k + 15$.

1 Introduction

Let $G = (V, E)$ be an undirected simple graph that is *planar*, i.e., it has a crossing-free drawing in the plane. G is called *triangulated* if all maximal regions not intersecting the drawing are incident to three edges of G. (More detailed definitions will be given in Sect. 2.) Any planar simple graph with $n \geq 3$ vertices can be triangulated by adding edges without destroying planarity.

In this paper, we study the problem of triangulating a planar graph G such that the pathwidth of the resulting graph is proportional to the pathwidth of G. Here, the *pathwidth* $pw(G)$ of a graph G is a well-known graph parameter (defined formally in Sect. 2). Graphs of small pathwidth have many applications. Many graph problems can be solved in polynomial time if the pathwidth is constant. (See e.g. [6].) The pathwidth also serves as lower bound on the height of planar graph drawings [9]. Vice versa, some planar graphs G can be drawn with height $O(pw(G))$, notably trees [14] and 2-connected outer-planar graphs [2].

The latter paper raised the question whether any outer-planar graph can be made 2-connected by adding edges without increasing the pathwidth much. (For if so, then *all* outer-planar graphs can be drawn with height $O(pw(G))$.) This question was answered in the affirmative by Babu et al. [1], who showed that

Research was supported by NSERC and done while visiting Universität Salzburg.
Many thanks to Jasine Babu for sharing her manuscript of what later became [1], and the referees of an earlier version of this paper for helpful comments.

© Springer-Verlag Berlin Heidelberg 2016
E.W. Mayr (Ed.): WG 2015, LNCS 9224, pp. 425–439, 2016.
DOI: 10.1007/978-3-662-53174-7_30

any outer-planar graph G can be made into a 2-connected outer-planar graph G' with $pw(G') \leq 16pw(G) + 15$.

Our results: In this paper, we improve on the result by Babu et al. and show that we can add edges to any outer-planar graph G such that the result is a 2-connected outer-planar graph G' with $pw(G') \leq 4pw(G) + 4$. But our technique is much more general. Rather than working with outer-planar graphs, we prove that any planar 2-connected graph can be triangulated without increasing the pathwidth if we allow multi-edges. We can also remove multi-edges; this increases the pathwidth at most 8-fold. With much the same technique we can also handle graphs with cut-vertices and make them 2-connected while increasing the pathwidth (roughly) 16-fold. Outer-planar graphs can be handled as special cases and give an even smaller increase in the pathwidth.

Related results: Many papers have dealt with how to triangulate a planar graph under some additional constraint. For example, any 2-connected planar graph can be triangulated so that the result is 4-connected (except for wheel-graphs) [4]. Any k-outer-planar graph can be triangulated so that the result is $(k + 1)$-outer-planar [3]. Any planar graph G with treewidth $tw(G)$ can be triangulated so that the result has treewidth $\max\{3, tw(G)\}$ [5]. Triangulating planar graphs has also been studied while minimizing the maximum degree [12], and relates to planar graph connectivity-augmentation problems (see e.g. [11] and the references therein) since any triangulated graph is 3-connected.

2   Background

Let $G = (V, E)$ be a graph with at least 3 vertices. G is called *planar* if it can be drawn without crossing in the plane. A crossing-free drawing Γ of G defines a cyclic order of edges at a vertex v by enumerating them in clockwise order around v; we call such a set of orders a *planar embedding* of G. The maximal regions of $\mathbb{R}^2 - \Gamma$ are called *faces* of the drawing; they can be read from the planar embedding by computing the *facial circuit*, i.e., the order of vertices and edges encountered while walking around the face in clockwise order. A graph G is called *outer-planar* if $G \cup \{z^*\}$ is planar, where z^* is a newly-added *universal vertex* adjacent to all vertices of G.

A *loop* is an edge (v, v) for some vertex. A *multi-edge* is an edge (v, w) with *multiplicity* $\mu \geq 2$, i.e., there exist μ copies of (v, w). A graph is called *simple* if it has neither loops nor multi-edges. All input graphs in this paper are required to be simple, but we sometimes add multi-edges in intermediate steps. (We never add loops.) A *multi-graph* is a graph without loops (but possibly with multi-edges). The *underlying simple graph* of a multi-graph is obtained by deleting all but one copy of each multi-edge.

Connectivity: A multi-graph G is called *connected* if we can go from any vertex v to any vertex w while walking along edges of G. The *connected components* of a multi-graph are the maximal subgraphs that are connected. A multi-graph G is called k-*connected* if it remains connected even after deleting $k - 1$ arbitrary vertices. If G is connected but not 2-connected, then G has a *cut-vertex*, i.e., a vertex v such that $G - v$ is not connected. A graph that is 2-connected, but not 3-connected, has a *cutting pair*, i.e., a pair of vertices v, w such that $G - \{v, w\}$ is not connected.

If S is a set of vertices, then let $C'_1, \ldots, C'_L$ be the connected components of $G - S$ ($L = 1$ if S was not a cut-set). Define for $i = 1, \ldots, L$ the *cut-component* C_i of S to consist of C'_i, the edges from C'_i to S, and a complete graph added between the vertices of S. Define int $(C_i) := C'_i = C_i - S$ to be the *interior* of C_i.

Triangulating: A face (in a planar graph in some planar embedding) is called a *triangle* if its facial circuit contains three edges. A multi-graph G is called *multi-triangulated* if it has a planar embedding such that all faces of G are triangles. Such a graph may well have multi-edges, but duplicate copies of an edge must use different routes (no facial circuit may consist of two copies of the same edge). A graph G is called *triangulated* if it is multi-triangulated and simple. A triangulated graph is 3-connected (and hence has a unique planar embedding, up to reversal of all edge orders). A multi-triangulated graph G need not be 3-connected, but it is 2-connected since $n \geq 3$ and G has no loops. One can show (see a detailed proof in the appendix) that the cutting pairs of G correspond to multi-edges as follows: $\{u, v\}$ is a cutting pair that has L cut-components if and only if (u, v) is a multi-edge with multiplicity L. Further, G has at least one edge that is not a multi-edge.

The idea of *triangulating* is to add edges to a graph until it is triangulated. More formally, *multi-triangulating a planar multi-graph* G means adding edges to G so that the result is multi-triangulated. *Triangulating* a planar multi-graph G means to add edges to the underlying simple graph of G such that the result is triangulated. In particular, this operation is allowed to delete copies of a multi-edge from G.

Pathwidth: Let G be a multi-graph. Let $X_1, \ldots, X_N$ be sets of vertices of G; we call these *bags*. We say that $X_1, \ldots, X_N$ is a *path decomposition* $\mathcal{P}$ of G if

- every vertex appears in at least one bag,
- for every edge (u, v) in G, at least one bag X_i contains both u and v, and
- for every vertex v in G, the bags containing v form an interval. Put differently, if $v \in X_{i_1}$ and $v \in X_{i_2}$ then also $v \in X_i$ for all $i_1 < i < i_2$.

Bags naturally imply an order; we write $X_i \preceq X_j$ if $i \leq j$ and $X_i \prec X_j$ if $i < j$. The *bag-size* of such a path decomposition is $\max |X_i|$. The *width* of such a path decomposition is $\max |X_i| - 1$. A graph is said to have *pathwidth* at most k if it has a path decomposition of width k.

3   3-Connected Graphs

We first show how to multi-triangulate 2-connected graphs (which also triangulates 3-connected graphs).

Lemma 1. *Let G be a planar 2-connected multi-graph with a planar embedding for which any facial circuit has at least 3 edges. Then we can multi-triangulate G without increasing the pathwidth and without changing the planar embedding.*

Proof. [1] Fix a path decomposition $\mathcal{P}$ of G that has width $pw(G)$. Let G^+ be the graph induced by $\mathcal{P}$, i.e., G^+ has the same vertices as G, but an edge (v,w) for *any* pair of vertices that occur in a common bag. By properties of a path decomposition G^+ is an interval-graph, therefore chordal, therefore any simple cycle C of length ≥ 4 has a *chord* (an edge between two non-consecutive vertices of C). See Golumbic [10] for details of these concepts.

Let f be any facial circuit of G with 4 or more edges on it. By 2-connectivity f is a simple cycle, and hence G^+ contains a chord of C. Add this chord to G, routing it inside f. The resulting graph is still planar and 2-connected and all facial circuits have at least 3 edges, so repeat until G is multi-triangulated.   □

Our problem was motivated by planar graph drawing applications, where often one starts by triangulating the planar graph (or adding edges to the outer-planar graph to make it maximal outer-planar). For these applications, multi-edges are a problem. For example usually one triangulates so that one can use the canonical ordering [7] or a Schnyder wood [13], and these only exist for *simple* triangulated planar graphs. Hence one wonders whether the same lemma holds without allowing multi-edges. Thus, given a planar 2-connected graph, can we triangulate it without increasing the pathwidth? This turns out to be false. Consider a 4-cycle, which has pathwidth 2. The only way to triangulate a 4-cycle without multi-edges is to turn it in K_4, which has pathwidth 3.[2] However, if G was already 3-connected, then no multi-edges will happen.

Corollary 1. *Let G be a 3-connected simple planar graph with $n \geq 3$. Then we can triangulate G without increasing the pathwidth.*

Proof. Since G is simple, any face has at least 3 edges. Apply the previous lemma to get a multi-triangulated graph G'. Adding edges cannot decrease connectivity, so G' has no cutting pairs. Since multi-edges in multi-triangulated graphs correspond to cutting pairs, hence G' is simple.   □

4   2-Connected Graphs

We already know how to multi-triangulate 2-connected planar graphs with Lemma 1. The hard part, done in this section, is how to convert such a multi-triangulated graph into a triangulated one (i.e., remove the multi-edges and replace them with others) without increasing the pathwidth much. We state the required increase in terms of another parameter, c, because this will help to obtain a smaller bound for outer-planar graphs later.

---

[1] Babu et al. published a similar proof in an early version of [1], but omitted it in [1].
[2] More generally, it was shown by Haiko Müller (private communication) that for any $k \geq 2$ there exists a planar graph of pathwidth k that cannot be triangulated without increasing the pathwidth by at least 1.

Lemma 2. *Any multi-triangulated graph G can be triangulated, after possibly changing the planar embedding, such that the resulting graph G' has pathwidth $pw(G') \leq 2pw(G) + 1 + 2c$.*

Here c is the maximum number of cutting pairs that can exist in one bag, i.e., for any path decomposition $\mathcal{P}$ of width $pw(G)$ and any bag X_i of $\mathcal{P}$ there are at most c cutting pairs $\{u_1, v_1\}, \ldots, \{u_c, v_c\}$ such that $\{u_1, v_1, \ldots, u_c, v_c\} \subseteq X_i$.

The rest of this section is devoted to the proof of this lemma. We first give an outline of the proof. We add $|X_i| + 2c$ "tokens" to each bag X_i of $\mathcal{P}$; these are place-holders for vertices that need to be added to bags later when adding edges. These tokens are then redistributed so that in each bag X_i we have 2 tokens per cutting pair $\{u, v\} \subseteq X_i$, and one token for each cut-component of $\{u, v\}$ that "intersects" X_i in some sense. We then can read from the path-decomposition how to re-arrange the planar embedding such that we can replace a copy of a multi-edge by a new edge while using up only "few" tokens. In particular, the above invariant on what tokens exist in bags continues to hold. Repeating this until no multi-edges are left then gives the desired graph G'. Since we had $|X_i| + 2c$ tokens, the new bag-size is at most $2|X_i| + 2c$, and hence $pw(G') \leq (2(pw(G) + 1) + 2c) - 1 = 2pw(G) + 1 + 2c$.

For the detailed proof, fix one planar embedding of G such that all faces are triangles. (We later change this embedding, but all faces will continue to be triangles.) Fix one path decomposition $\mathcal{P}$ of G of width $pw(G)$.

Assigning tokens: We assign tokens to a bag X_i of $\mathcal{P}$ as follows: (1) Add one token to X_i for each vertex v in X_i; this is the *vertex-token* of v. (2) Add two tokens to X_i for every cutting pair $\{u, v\}$ with $\{u, v\} \subseteq X_i$; these are the *cutting-pair tokens*, or the *tokens of* $\{u, v\}$.

Peripheral pairs: Let $\{u, v\}$ be a cutting pair, and let $C_0, \ldots, C_L$ be its cut components. One can show (see a detailed proof in the appendix) that for $i \in \{0, \ldots, L\}$ the edges from v to $\text{int}(C_i)$ occur consecutively in the clockwise order of edges around v, surrounded by two copies of edge (u, v). See Fig. 1 for an illustration. Let b_i^ℓ and b_i^r be the first and last neighbor of v within this interval of edges to $\text{int}(C_i)$. We call $\{b_i^\ell, b_i^r\}$ the *peripheral pair of cut-component C_i*. Notice that $b_i^\ell = b_i^r$ if $\deg(b_i^\ell) = 2$ or (b_i^ℓ, v) is a multi-edge, but we use the term "pair" even then for ease of wording.

Observation 1. *Let G be a multi-triangulated graph that has a cutting pair $\{u, v\}$. Let C_i and C_j be two different cut-components of $\{u, v\}$. For any choice of $\alpha, \beta \in \{\ell, r\}$, deleting one copy of (u, v) and adding (b_i^α, b_j^β) results in a multi-triangulated graph (after possibly changing the planar embedding).*

Proof. This follows from the results in [8]. In a nutshell, we can reverse and swap cut-components until b_i^α and b_j^β both face one copy of (u, v). Deleting this copy gives a face with 4 edges; inserting edge (b_i^α, b_j^β) into this face gives a planar graph where all faces are triangles.   □

Bag-intervals: Let $\{b_i^\ell, b_i^r\}$ be the peripheral-pair of a cut-component C_i of a cutting pair $\{u, v\}$. Since G is multi-triangulated, $\{u, v, b_i^\alpha\}$ forms a triangle

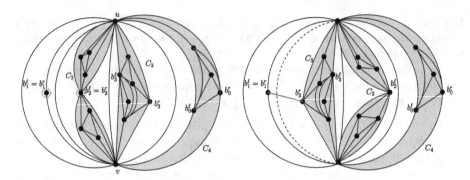

Fig. 1. A multi-triangulated graph with a cutting pair $\{u, v\}$ that has four cut-components. Dotted red lines are paths assigned to peripheral pairs as in Lemma 3. We can add edge (b_1^r, b_3^r) if we swap C_2 and C_3 and reverse C_3. (Color figure online)

for $\alpha \in \{\ell, r\}$. By the properties of the path decomposition there must exist at least one bag that contains all three vertices. Thus let $X(b_i^\alpha)$ be a bag containing $\{u, v, b_i^\alpha\}$; choose an arbitrary one if there is more than one. So far the superscripts ℓ and r for $\{b_i^\ell, b_i^r\}$ effectively meant "one" and "the other", since we can reverse the planar embedding of cut-component C_i. We now fix the superscripts such that $X(b_i^\ell) \preceq X(b_i^r)$, i.e., the bag of b_i^ℓ is left of the bag of b_i^r. The left-open set of bags $(X(b_i^\ell), X(b_i^r)] := \{X : X(b_i^\ell) \prec X \preceq X(b_i^r)\}$ is called the *bag-interval* of peripheral pair $\{b_i^\ell, b_i^r\}$. Notice that the bag-interval is empty if $X(b_i^\ell) = X(b_i^r)$; this will not pose problems. Occasionally we will also consider the *closed bag-interval* $\{X : X(b_i^\ell) \preceq X \preceq X(b_i^r)\}$.

Child-peripheral-pairs: So far all cut-components at a cutting pair have been treated equally. For token-accounting-purposes, we introduce a hierarchy among them. Fix one edge e of G that is not a multi-edge. For each cutting pair $\{u, v\}$ with cut-components $C_0, \ldots, C_L$, the *parent-component* of $\{u, v\}$ is the one that contains edge e, while all other cut-components are called *child-components*. Correspondingly we call a peripheral-pair of $\{u, v\}$ a *child-peripheral-pair* if it belongs to a child-component of $\{u, v\}$.

Redistributing tokens: Let $\mathcal{B}$ be the union, over all cutting pairs $\{u, v\}$, of all the child-peripheral-pairs of $\{u, v\}$. We want to redistribute vertex-tokens to child-peripheral-pairs, and for this we need an observation.

Lemma 3. *Let $\mathcal{B}$ be the set of all child-peripheral pairs in a multi-triangulated graph G. There exists a set of vertex-disjoint paths $P_1, \ldots, P_{|\mathcal{B}|}$ in G such that for any child-peripheral-pair $\{b^\ell, b^r\}$ in $\mathcal{B}$, one of the paths connects b^ℓ with b^r.*

Proof. Consider any child-peripheral-pair $\{b_i^\ell, b_i^r\}$, say at cut-component C_i of cutting pair $\{u, v\}$. Observe that there are three vertex-disjoint paths from b_i^ℓ to b_i^r: one via u, one via v, and one within $\text{int}(C_i) = C_i - \{u, v\}$ since the latter is connected by definition of cut-components. Since (u, v) is an edge, therefore $\{u, v, b_i^\ell, b_i^r\}$ with these paths form a subdivision of K_4. If follows that

$\{u, v, b_i^\ell, b_i^r\}$ all belong to one triconnected component, call it D. Since D is 3-connected, there must exist a path P from b_i^ℓ to b_i^r within $D - \{u, v\}$, and this is the path that we use for this child-peripheral pair.

It remains to argue that these paths are disjoint. Let $\{b', b''\}$ be some other child-peripheral-pair, say at cutting pair $\{u', v'\}$, such that $\{b', b'', u', v'\}$ belong to triconnected component D' and we assigned a path P' in $D' - \{u', v'\}$ to this child-peripheral pair.

Recall that cutting pair $\{u, v\}$ splits the graph into multiple cut-components. One of those is C_i, the child-component that contained b_i^ℓ and b_i^r and therefore also the triconnected component D and the path P. We have two cases:

- D' is part of a cut-component of $\{u, v\}$ other than C_i.
  We know that child-cut-components are vertex-disjoint except for $\{u, v\}$. Therefore D and D' are vertex-disjoint except for perhaps $\{u, v\}$. Hence P and P' are vertex-disjoint.
- D' is part of the child-component C_i of $\{u, v\}$.
  This implies that $\{u, v\} \neq \{u', v'\}$, since for each cutting pair, each cut-component gets only one peripheral pair. ($u = u'$ or $v = v'$ is possible, but not both.) Changing the point of view, now consider the cut-components of $\{u', v'\}$. Here D' belongs to a child-component (because $\{b', b''\}$ is a child-peripheral-pair), but D belongs to the parent-component (since D' belongs to a child-component of $\{u, v\}$). Exchanging the roles of the two cutting pairs hence shows as in the previous case that P and P' are vertex-disjoint.   □

We now redistribute vertex-tokens to child-peripheral pairs as follows. For every child-peripheral-pair $\{b^\ell, b^r\}$, find the path P connecting b^ℓ and b^r from Lemma 3. For every vertex $w \in P$, declare the vertex-token of w to belong to the child-peripheral-pair $\{b^\ell, b^r\}$; we now call it a *child-peripheral-pair token* and say that it *belongs to* $\{b^\ell, b^r\}$. Since the paths of child-peripheral-pairs are vertex-disjoint, every vertex-token is used at most once. By properties of a path decomposition, the set of bags $\mathcal{X}_P = \{X : X \text{ contains a vertex of } P\}$ forms an interval of bags since P is connected. Each bag in $\mathcal{X}_P$ obtains at least one token of $\{b^\ell, b^r\}$. Since $X(b^\ell), X(b^r) \in \mathcal{X}_P$, we therefore have:

Invariant 1. *(1) For every child-peripheral-pair $\{b^\ell, b^r\}$, every bag X in the bag-interval $(X(b^\ell), X(b^r)]$ contains at least one token of $\{b^\ell, b^r\}$. (2) For every cutting pair $\{u, v\}$, every bag containing both u and v contains two tokens of $\{u, v\}$.*

Adding edges: We now repeatedly delete one copy of a multi-edge (u, v) and replace it with some edge (b_i^α, b_j^β) between two different cut-components of $\{u, v\}$. Notice that no such edge can have existed before, so the sum of the multiplicities of multi-edges decreases. By Observation 1, adding these edges maintains a multi-triangulation. After repeated applications we hence end with a simple graph. Throughout these edge additions, we maintain a valid path decomposition for the graph by adding vertices to bags, if needed. This uses up some tokens, but we do it in such a way that Invariant 1 is maintained and hence the pathwidth is at most $2pw(G) + 1 + 2c$.

So let $\{u, v\}$ be a cutting pair. Let $C_0, \ldots, C_L$ be the cut components of $\{u, v\}$, with C_0 the parent-component. For each component C_i, let $\{b_i^\ell, b_i^r\}$ be the peripheral-pair of C_i. We distinguish cases.

1. There exists some $i \neq j$, $i > 0$, $j > 0$ such that $X(b_i^\ell) \prec X(b_j^\ell) \preceq X(b_i^r) \prec X(b_j^r)$. Put differently, there are two child components C_i and C_j whose closed bag-intervals intersect, but neither one contains the other. See also Fig. 2.

   Add an edge (b_j^ℓ, b_i^r). Since both C_i and C_j are child-components, by the invariant each bag X with $X(b_j^\ell) \prec X \preceq X(b_i^r)$ contains one token of $\{b_i^\ell, b_j^\ell\}$ and one token of $\{b_i^\ell, b_i^r\}$. We use one of them to add b_j^ℓ to all these bags; then b_j^ℓ and b_i^r share a bag, the bags containing b_j^ℓ continue to form an interval, and we hence have a valid path decomposition for the new graph.

   Adding the edge combines child-components C_i and C_j into one new child-component C' with peripheral-pair $\{b_i^\ell, b_j^r\}$. Since we used only one token in each bag, all bags X with $X(b_i^\ell) \prec X \preceq X(b_j^r)$ have a peripheral-pair-token left, which we now assign to C'. So the invariant holds.

2. There exists some $i \neq j$, $i > 0$, $j > 0$, such that $X(b_i^\ell) \preceq X(b_j^\ell) \preceq X(b_j^r) \preceq X(b_i^r)$. Put differently, there are two child components C_i and C_j whose closed bag-intervals intersect, and one is inside the other. See also Fig. 2.

   Add an edge (b_i^ℓ, b_j^ℓ). Each bag X with $X(b_i^\ell) \prec X \preceq X(b_j^\ell)$ contains a token of $\{b_i^\ell, b_i^r\}$. We use this to add b_i^ℓ to all these bags; then b_i^ℓ and b_j^ℓ share a bag and the bags containing b_i^ℓ are consecutive, hence we have a valid path decomposition of the new graph.

   Adding the edge combines components C_i and C_j into one new component C' with peripheral-pair $\{b_i^r, b_j^r\}$. Since we used only tokens in bags farther to the left, all bags X with $X(b_j^r) \prec X \preceq X(b_i^r)$ still have the token of $\{b_i^\ell, b_i^r\}$, and we assign these to the new peripheral-pair. So the invariant holds.

3. No two closed bag-intervals of two child-components intersect. After possible renaming of the child components $C_1, \ldots, C_L$, we may hence assume that $X(b_1^\ell) \preceq X(b_1^r) \prec X(b_2^\ell) \preceq X(b_2^r) \prec \cdots \prec X(b_L^\ell) \preceq X(b_L^r)$. (The bag-interval of the parent-component may be anywhere in this order.) See also Fig. 3. We will combine *all* cut components into one at once. Add edges (b_1^r, b_2^ℓ), $(b_2^r, b_3^\ell), \ldots, (b_{L-1}^r, b_L^\ell)$. To create a path decomposition for this, add b_i^r to all bags X with $X(b_i^r) \prec X \preceq X(b_{i+1}^\ell)$, for $i = 1, \ldots, L - 1$. Pay for these additions with the first token of (u, v). We know that each of these bag has such a token, since $X(b_1^\ell)$ and $X(b_L^r)$ contain $\{u, v\}$ by definition, and the bags between must contain $\{u, v\}$ by properties of a path decomposition. Finally add edge (b_1^ℓ, b_0^r). Create a path decomposition for this by adding b_1^ℓ to all bags from $X(b_1^\ell)$ to $X(b_0^r)$, and pay for it with the second cutting-pair-token of (u, v).

   Observation 1 applies to all added edges, since the ends of each edge are peripheral-vertices of two different cut-components, even after considering that previous edge-additions merged some them. Hence the resulting graph is a multi-triangulation after we deleted L copies of multi-edge (u, v).

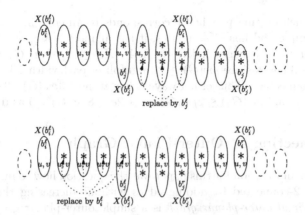

Fig. 2. Bag-intervals with peripheral-pair-tokens (shown with ∗). (Top) The closed bag-intervals intersect, but neither contains the other. (Bottom) One closed bag-interval contains the other.

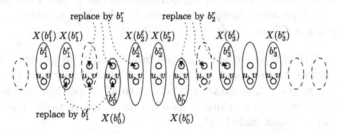

Fig. 3. Replacing cutting-pair-tokens (shown with ○) to combine all remaining cut-components of cutting pair $\{u, v\}$ if no closed bag-intervals intersect.

Since $\{u, v\}$ ceases to be a cutting pair after adding these edges, the invariant holds again since we only used tokens of $\{u, v\}$.

After repeatedly applying the above edge-additions to all cutting pairs, we hence end with a triangulated graph and a path decomposition of width at most $2pw(G) + 1 + 2c$ as desired. This proves Lemma 2.   □

Lemma 4. *Let G be a 2-connected planar graph with $n \geq 3$ vertices. Then we can triangulate G, after possibly changing the planar embedding, such that the result has pathwidth at most $8pw(G) - 5$.*

Proof. By Lemma 1 we ca multi-triangulate G without increasing the pathwidth. Call the result G_1. By Lemma 2 we can triangulate G_1 such that the resulting graph G_2 has $pw(G_2) \leq 2pw(G_1) + 1 + 2c = 2pw(G) + 1 + 2c$.

It remains to bound c. Recall that this is the maximum number of cutting pairs of G_1 for which all vertices occur in one bag X_i (of some path decomposition $\mathcal{P}$ of width $pw(G_1) = pw(G)$). Each such cutting pair corresponds to a multi-edge in G_1. Let $G[X_i]$ be the graph induced by X_i and G_s be its underlying simple graph. Each such cutting pair hence corresponds to an edge in G_s. Since G_s is planar and simple and has $|X_i|$ vertices, it has at most $3|X_i| - 6 \leq 3(pw(G) + 1) - 6 = 3pw(G) - 3$ edges if $|X_i| \geq 3$. If $|X_i| \leq 2$, then G_s has at most $1 \leq 3pw(G) - 3$ edges since $pw(G) \geq 2$ (a graph of pathwidth 1 is a forest and cannot be 2-connected). Thus either way G_s has at most $3pw(G) - 3$ edges, hence $c \leq 3pw(G) - 3$ and $pw(G_2) \leq 2pw(G) + 1 + 2c \leq 8pw(G) - 5$ as desired.   □

5   2-Connecting an Outer-Planar Graph

Recall that the motivation for this paper was the question how to make an outer-planar graph 2-connected by adding edges without increasing the pathwidth much. A *maximal outer-planar graph* is a simple outer-planar graph to which we cannot add edges without violating planarity, simplicity, or outer-planarity. Such a graph is 2-connected for $n \geq 3$.

Theorem 1. *Let G be a simple connected outer-planar graph. Then we can add edges to G, after possibly changing the planar embedding, to obtain a maximal outer-planar graph G' with $pw(G') \leq 4pw(G) + 4$.*

Proof. If $n = 1$ then G is already maximal outer-planar, so assume $n \geq 2$. Add a universal vertex z^* to G and call the result G_1; we know that G_1 is planar and $pw(G_1) = pw(G) + 1$ since we can add z^* to all bags. Observe that $G_1 - v$ is connected for any $v \neq z^*$ since z^* is adjacent to all vertices. Therefore any cutting pair of G_1 must include z^*.

Use Lemma 1 to multi-triangulate G_1 without increasing pathwidth, and call the result G_2; we have $pw(G_2) = pw(G) + 1$. Now use Lemma 2 to triangulate G_2, and call the result G_3. We have $pw(G_3) \leq 2pw(G_2) + 1 + 2c \leq 2pw(G) + 3 + 2c$.

Since any cutting pair includes z^*, we can get an improved bound for c as follows. Let $\mathcal{P}_2$ be any path decomposition of G_2 of width $pw(G_2)$ and let X_i be any bag of $\mathcal{P}_2$; we have $|X_i| \leq pw(G_2) + 1 = pw(G) + 2$. If X_i contains cutting pairs, then it must contain z^*. Each such cutting pair uses z^* and one other vertex in X_i, so there are at most $|X_i| - 1$ cutting pairs with both ends in X_i, and $c \leq |X_i| - 1 \leq pw(G) + 1$. Putting it all together, we have $pw(G_3) \leq 2pw(G) + 3 + 2(pw(G) + 1) = 4pw(G) + 5$.

Finally delete the added vertex z^* to obtain G_4, which has the same vertices as G. Since z^* was universal and G_3 was triangulated, G_4 is maximal outer-planar. Since z^* was universal, $pw(G_4) = pw(G_3) - 1 \leq 4pw(G) + 4$ and hence G_4 satisfies all conditions on G'.   □

We note here that the bound can be improved to $4pw(G) + 3$ by delving into the proofs of Lemmas 2 and 3 and observing that the vertex-token of z^* will never be used as child-peripheral-pair-token, since z^* is in all cutting pairs. We leave the details to the reader.

6 All Graphs

We now show how to handle cutvertices and disconnected graphs.

Lemma 5. *Any simple connected planar graph G with $n \geq 3$ can be triangulated, after possibly changing the planar embedding, so that the result has pathwidth at most $16pw(G) + 3$.*

Proof. Let v_1 be a cut-vertex of G. Add a new vertex z_1 as follows.

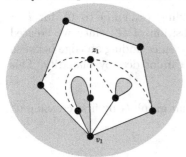

Let $C_0, \ldots, C_L$ be the cut-components of v_1. Rearrange the planar embedding at v_1 such that for each C_j the edges from v_1 to C_j are consecutive at v_1. In consequence, there now exists a face f_1 that is incident to all cut-components of v_1. Insert a new vertex z_1 in face f_1, and make it adjacent to v_1 and to all neighbors x of v_1 that are on f_1. Afterwards v_1 is no longer a cut-vertex, and z_1 is also not a cut-vertex.

We can obtain a path decomposition of $G \cup \{z_1\}$ by taking one of G and adding z_1 to all bags that contains v_1. This covers all new edges since all neighbors of z_1 are neighbors of v_1.

Repeat the process in the resulting graph until there are no cut-vertices left. Call the final graph G_1. Since none of the new vertices were cut-vertices, we added at most $|X_i|$ new vertices to each bag X_i of a path decomposition of G. Hence the bag-size at most doubles and $pw(G_1) \leq 2pw(G) + 1$.

Now multi-triangulate G_1 with Lemma 2 and call the result G_2. We have $pw(G_2) \leq 8pw(G_1) - 5 \leq 8(2pw(G) + 1) - 5 = 16pw(G) + 3$.

Now we must remove the added vertices while keeping a triangulated graph, and do this by contracting each into a suitable neighbor. Observe that the neighbors of z_1 form a simple cycle since G_2 is triangulated. Hence these neighbors induce a simple outer-planar 2-connected graph. It is well-known that every such graph has a vertex of degree 2. Therefore z_1 has a neighbor y_1 such that y_1 and z_1 have exactly two common neighbors (which are the third vertices on the faces incident to edge (z_1, y_1)). *Contract* edge (z_1, y_1), i.e., delete z_1 and re-route every incident edge of z_1 to end at y_1 instead. Delete resulting loops and multi-edges. Because z_1 and y_1 had exactly two neighbors in common, the resulting graph is again triangulated. Repeat the process for the other added vertices.

At the end the graph G_3 that results has the same vertices as G. It is well-known that contraction of an edge does not increase pathwidth, so $pw(G_3) \leq pw(G_2) \leq 16pw(G) + 3$ as desired.   □

As for disconnected graphs, one can easily show the following:

Lemma 6. *Let G be a planar graph. Then we can add edges to G so that the resulting graph G' is planar, connected, and $pw(G') = \max\{1, pw(G)\}$.*

Proof. Let $C_1, \ldots, C_L$ be the connected components of G. Each of them has pathwidth at most $pw(G)$ since they are subgraphs of G; let $\mathcal{P}_i$ be a path decomposition of C_i of width at most $pw(G)$. Start with path decomposition $\mathcal{P}_1$. Append one new bag, into which we insert one arbitrary vertex v_1 from the last bag of $\mathcal{P}_1$ and one arbitrary vertex u_2 from the first bag of $\mathcal{P}_2$. Then append $\mathcal{P}_2$. Repeat with the remaining components: insert a new bag after the last bag of $\mathcal{P}_i$, give it one vertex v_i from the last bag of $\mathcal{P}_i$ and one vertex u_{i+1} from the first bag of $\mathcal{P}_{i+1}$, and then append $\mathcal{P}_{i+1}$. Clearly we get a path decomposition $\mathcal{P}$ of G of width $\max\{1, pw(G)\}$.

Define G' to be the graph obtained by adding (u_i, v_{i+1}) to G, for $i = 1, \ldots, L-1$. Clearly $\mathcal{P}$ is also a path decomposition of G', since we created bags for each of these new edges. Also G' is planar since adding an edge between two vertices in different connected components cannot destroy planarity. This shows the result. $\qquad\square$

Hence we can triangulate a disconnected planar graph G by first creating G' and then triangulating G'.

7 Conclusion

In this paper, we studied how to add edges to a planar graph without increasing the pathwidth much. We summarize all our results with the following:

Theorem 2. *Let G be a simple planar graph with at least 3 vertices. Then we can triangulate G such that the result G' has*

- $pw(G') = pw(G)$ *if G is 3-connected,*
- $pw(G') \le 8pw(G) - 5$ *if G is 2-connected,*
- $pw(G') \le 16pw(G) + 3$ *otherwise.*

It may also be of interest to observe that our construction does not change a given path decomposition of the graph other than by adding more vertices to some bags. On the other hand, our construction often changes the planar embedding. Is it possible to triangulate a graph without increasing the pathwidth much and without changing the planar embedding?

Following the steps of the proof, one can see that the triangulation can be found in linear time, presuming that we are given a path decomposition of width $pw(G)$ in the form of the index of the first and last bag containing v for every vertex v. There is no need to compute triconnected components: One can find child-components via multi-edges, and the paths in Lemma 3 are only needed for accounting purposes and need not be computed.

The obvious open problem is to improve the factors, especially for 2-connected graphs. There are planar graphs that cannot be triangulated without increasing the pathwidth, but can every planar 2-connected graph G be triangulated so that the result has pathwidth at most $pw(G) + 1$?

It would also be of interest to study other width-parameters (such as the carving width, bandwidth, clique-width, etc.) and ask whether planar graphs can be triangulated while keeping the width-parameter asymptotically the same.

Finally, the concept of triangulating can be applied to graphs embedded in surfaces of higher genus. In such surfaces, multi-edges do not imply cutting pairs, and so already the equivalent of Corollary 1 fails. Can we triangulate graphs embedded in a surface without increasing the pathwidth much?

A   Properties of Multi-triangulated Graphs

In the main part of the paper, we used some properties of multi-triangulated planar graphs. These are quite easy to derive, but we include detailed proofs here for completeness' sake.

Lemma 7. *Let G be a multi-triangulated planar graph with $n \geq 3$ vertices. Fix an arbitrary planar embedding for which all faces are triangles. The following holds:*

1. *G is 2-connected.*
2. *Any cutting pair $\{u, v\}$ gives rise to a multi-edge (u, v).*
3. *For any multi-edge (u, v), $\{u, v\}$ is a cutting pair, and the number of its cut-components equals the multiplicity of the multi-edge.*
4. *Let $\{u, v\}$ be a cutting pair with a cut-component C, and consider the cyclic order of edges around u. Then the edges to $\mathrm{int}\,(C)$ appear consecutively, and are preceded and succeeded by copies of (u, v).*

Proof. Let S be a cut-set (i.e., either cut-vertex or cutting pair). Consider a vertex $v \in S$. Assume for contradiction that in the clockwise order around v there are two consecutive neighbors w_1, w_2 with $w_1 \in \mathrm{int}\,(C_1)$ and $w_2 \in \mathrm{int}\,(C_2)$ for two different cut-components C_1, C_2 of S. Consider the face f that is between edges $(v, w_1), (v, w_2)$ at v. Since w_1, w_2 are in the interior of different cut-components, we cannot have an edge (w_1, w_2). We must have $w_1 \neq v \neq w_2$, since otherwise there would be a loop. Therefore face f is incident to at least 4 edges. Contradiction.

Thus for any two cut-components of S, edges from v to the inside of the cut-component cannot be consecutive. Thus, there must an edge between any two cut-components (in the clockwise order around v) for which the other endpoint is also in S. If $|S| = 1$ then such an edge would be a loop, a contradiction. Therefore no cut-set can have size 1 and G is 2-connected; this proves (1).

If $|S| = 2$, say S is the cutting pair $\{u, v\}$, then the cut-components are separated by copies of edge (u, v). If there are L cut-components $C_1, \ldots, C_L$ for $L \geq 2$, then there are at least L places in the clockwise order around v where we switch from one cut-component to the next one, so we must have at least L copies of (u, v). This proves (2).

Let $e_0, \ldots, e_{\ell-1}$ be the copies of (u, v), enumerated in the clockwise order around v. We have just shown $\ell \geq L$. For $i = 1, \ldots, \ell$, edges e_{i-1} and e_i cannot be consecutive at v (where indices are modulo L), otherwise there would be a face of degree 2. So there must be vertices other than u between e_{i-1} and e_i. Further, the cycle formed by e_{i-1} and e_i separates everything on one side from everything on the other side. So the subgraph between e_{i-1} and e_i contains at least one cut-component of $\{u, v\}$. It follows that $\ell \leq L$, and so $\ell = L$. This proves (3).

Since $\ell = L$, the subgraph between e_{i-1} and e_i must contain exactly one cut-component of $\{u, v\}$. Therefore in the cyclic order around v we alternate between a copy of (u, v) and all edges to exactly one cut-component. This proves (4). □

Lemma 8. *Every multi-triangulated graph with $n \geq 3$ has at least one edge that is not a multiple edge.*

Proof. Fix one arbitrary planar drawing Γ of G for which all facial circuits have three edges. Nothing is to show if G is simple, so assume G has multi-edges. If e_1, e_2 are two copies of a multi-edge, then their drawing defines a closed curve C. This curve cannot be the boundary of a face since facial circuits have three edges. In consequence, at least one vertex must be inside any closed curve defined by two copies of a multi-edge.

Assume that e_1, e_2 has been chosen such that their closed curve encloses the minimum possible number of vertices among all such pairs. Let v be a vertex inside that curve, and let e be an edge incident to v. Then e must be simple by choice of e_1, e_2. □

References

1. Babu, J., Basavaraju, M., Chandran, L.S., Rajendraprasad, D.: 2-connecting outer-planar graphs without blowing up the pathwidth. Theor. Comput. Sci. **554**, 119–134 (2014)
2. Biedl, T.: A 4-approximation for the height of drawing 2-connected outer-planar graphs. In: Erlebach, T., Persiano, G. (eds.) WAOA 2012. LNCS, vol. 7846, pp. 272–285. Springer, Heidelberg (2013)
3. Biedl, T.: On triangulating k-outer-planar graphs. Discrete Appl. Math. **181**, 275–279 (2015)
4. Biedl, T., Kant, G., Kaufmann, M.: On triangulating planar graphs under the four-connectivity constraint. Algorithmica **19**(4), 427–446 (1997)
5. Biedl, T., Velázquez, L.E.R.: Drawing planar 3-trees with given face areas. Comput. Geom. Theor. Appl. **46**(3), 276–285 (2013)
6. Bodlaender, H.L.: Treewidth: algorithmic techniques and results. In: Privara, I., Ružička, P. (eds.) MFCS 1997. LNCS, vol. 1295, pp. 19–36. Springer, Heidelberg (1997)
7. de Frayysseix, H., Pach, J., Pollack, R.: How to draw a planar graph on a grid. Combinatorica **10**, 41–51 (1990)
8. Di Battista, G., Tamassia, R.: On-line planarity testing. SIAM J. Comput. **25**(5), 956–997 (1996)
9. Felsner, S., Liotta, G., Wismath, S.: Straight-line drawings on restricted integer grids in two and three dimensions. J. Graph Algorithms Appl. **7**(4), 335–362 (2003)

10. Golumbic, M.C.: Algorithmic Graph Theory and Perfect Graphs, 1st edn. Academic Press, New York (1980)
11. Gutwenger, C., Mutzel, P., Zey, B.: On the hardness and approximability of planar biconnectivity augmentation. In: Ngo, H.Q. (ed.) COCOON 2009. LNCS, vol. 5609, pp. 249–257. Springer, Heidelberg (2009)
12. Kant, G., Bodlaender, H.L.: Triangulating planar graphs while minimizing the maximum degree. In: Nurmi, O., Ukkonen, E. (eds.) SWAT 1992. LNCS, vol. 621, pp. 258–271. Springer, Heidelberg (1992)
13. Schnyder, W.: Embedding planar graphs on the grid. In: ACM-SIAM Symposium on Discrete Algorithms (SODA 1990), pp. 138–148 (1990)
14. Suderman, M.: Pathwidth and layered drawings of trees. Int. J. Comput. Geom. Appl. 14(3), 203–225 (2004)

Polynomial Kernelization
for Removing Induced Claws and Diamonds

Marek Cygan[1]([✉]), Marcin Pilipczuk[2], Michał Pilipczuk[1],
Erik Jan van Leeuwen[3], and Marcin Wrochna[1]

[1] Institute of Informatics, University of Warsaw, Warsaw, Poland
{cygan,michal.pilipczuk,m.wrochna}@mimuw.edu.pl
[2] Department of Computer Science, University of Warwick, Warwick, UK
m.pilipczuk@dcs.warwick.ac.uk
[3] Max-Planck Institut Für Informatik, Saarbrücken, Germany
erikjan@mpi-inf.mpg.de

Abstract. A graph is called {claw, diamond}-free if it contains neither a claw (a $K_{1,3}$) nor a diamond (a K_4 with an edge removed) as an induced subgraph, or, equivalently, it is a line graph of a triangle-free graph. We consider the parameterized complexity of the {CLAW, DIAMOND}-FREE EDGE DELETION problem, where given a graph G and a parameter k, the question is whether one can remove at most k edges from G to obtain a {claw, diamond}-free graph. Our main result is that this problem admits a polynomial kernel. We also show that, even on instances with maximum degree 6, the problem is NP-complete and cannot be solved in time $2^{o(k)} \cdot |V(G)|^{O(1)}$, assuming the Exponential Time Hypothesis.

1 Introduction

Graph modification problems form a wide class of problems, where one is asked to alter a given graph using a limited number of modifications in order to achieve a certain target property, for instance the non-existence of some forbidden induced structures. Depending on the allowed types of modification and the choice of the target property, one can consider a full variety of problems. Well-studied problems that can be expressed in the graph modification paradigm are VERTEX COVER, FEEDBACK VERTEX SET, and CLUSTER EDITING, among others.

It is natural to consider graph modification problems from the parameterized perspective, since they have an innate parameter: the number of allowed modifications, which is expected to be small in applications. As far as the set

The research was supported by Polish National Science Centre grants DEC-2013/11/D/ST6/03073 (Michał Pilipczuk and Marcin Wrochna) and DEC-2012/05/D/ST6/03214 (Marek Cygan and Marcin Pilipczuk). Michał Pilipczuk is currently holding a post-doc position at Warsaw Center of Mathematics and Computer Science. Moreover research of Marcin Pilipczuk was partially supported by the Centre for Discrete Mathematics and its Applications (DIMAP) at the University of Warwick and by Warwick-QMUL Alliance in Advances in Discrete Mathematics and its Applications.

© Springer-Verlag Berlin Heidelberg 2016
E.W. Mayr (Ed.): WG 2015, LNCS 9224, pp. 440–455, 2016.
DOI: 10.1007/978-3-662-53174-7_31

of allowed modifications is concerned, the most widely studied variants are vertex deletion problems (allowing only removing vertices), edge deletion problems (only removing edges), completion problems (only adding edges), and editing problems (both adding and removing edges). It is very easy to see that as long as the target property can be expressed as the non-existence of induced subgraphs from some finite, fixed list of forbidden subgraphs $\mathcal{F}$ (in other words, belonging to the class of $\mathcal{F}$-free graphs), then all the four variants can be solved in time $c^k \cdot |V(G)|^{\mathcal{O}(1)}$ via a straightforward branching strategy, where the constant c depends on $\mathcal{F}$ only. This observation was first pronounced by Cai [4].

From the perspective of kernelization, again whenever the property is characterized by a finite list of forbidden induced subgraphs, then a standard application of the sunflower lemma gives a polynomial kernel for the vertex deletion variant. The same observation, however, does not carry over to the edge modification problems. The reason is that altering one edge can create new obstacles from $\mathcal{F}$, which need to be dealt with despite not being contained in the original graph G. Indeed, Kratsch and Wahlström [23] have shown a simple graph H on 7 vertices such that the edge deletion problem for the property of being H-free does not admit a polynomial kernel unless NP $\subseteq$ coNP/poly. Later, the same conclusion was proved by Guillemot et al. [18] for H being a long enough path or cycle.

This line of study was continued by Cai and Cai [5] (see also the full version in the master's thesis of Cai [6]), who took up an ambitious project of obtaining a complete classification of graphs H on which edge modification problems for the property of being H-free admit polynomial kernels. The project was very successful: for instance, the situation for 3-connected graphs H is completely understood, and among trees there is only a finite number of remaining unresolved cases. In particular, the study of Cai and Cai revealed that the existence of a polynomial kernel for edge modification problems is actually a rare phenomenon that appears only for very simple graphs H.

One of the most tantalizing questions that is still unresolved is the case $H = K_{1,3}$, i.e., the CLAW-FREE EDGE DELETION problem (as well as the completion and editing variants). The study of this particular case is especially interesting in light of the recent powerful decomposition theorem for claw-free graphs, proved by Chudnovsky and Seymour [7]. For many related problems, having an equivalent structural view on the considered graph class played a crucial role in the design of a polynomial kernel, and hence there is hope for a positive result in this case as well. For this reason, determining the existence of a polynomial kernel for CLAW-FREE EDGE DELETION was posed as an open problem during Workshop on Kernels (WorKer) in 2013, along with the same question for the related LINE GRAPH EDGE DELETION problem [9].

Our results. As an intermediate step towards showing a polynomial kernel for CLAW-FREE EDGE DELETION, we study a related variant, where we forbid *diamonds* as well.[1] By a *diamond* we mean a K_4 with one edge removed, and

---

[1] A more detailed discussion of the relation between these two problems is provided in the conclusions section.

{claw, diamond}-free graphs are exactly graphs that do not contain claws or diamonds as induced subgraphs. This graph class is equal to the class of line graphs of triangle-free graphs, and to the class of *linear dominoes* (graphs in which every vertex is in at most two maximal cliques and every edge is in exactly one maximal clique) [21,24].

In this paper, we consider the {CLAW, DIAMOND}-FREE EDGE DELETION problem (CDF-ED for short) where, given a graph G and an integer k, one is asked to determine whether there exists a subset F of the edges of G with $|F| \leq k$ such that $G - F$ is {claw, diamond}-free; such a set F is also called an *HDS*.

Our main result is that CDF-ED admits a polynomial kernel.

Theorem 1. CDF-ED *admits a polynomial kernel.*

In order to prove Theorem 1, we give a *polynomial-time compression* of CDF-ED into a problem in NP. By a polynomial-time compression into an unparameterized problem R we mean a polynomial-time algorithm that, given an instance (G, k) of CDF-ED, outputs an equivalent instance y of R such that $|y| \leq f(k)$, for some computable function f called the *size* of the compression.

Theorem 2. CDF-ED *admits a polynomial-time compression algorithm into a problem in NP, where the size of the compression is* $\mathcal{O}(k^{24})$.

The problem in NP that Theorem 2 refers to actually is an annotated variant of CDF-ED. Unfortunately, we are unable to express the annotations in a clean manner using gadgets. Therefore, we compose the polynomial-time compression of Theorem 2 with the NP-hardness reduction that we present for CDF-ED (see Corollary 4 discussed below) in order to derive Theorem 1.

To prove Theorem 2, we apply the vertex modulator technique. We first greedily pack edge-disjoint claws and diamonds in the input graph. If more than k such obstacles can be packed, then we immediately infer that we are dealing with a no-instance. Otherwise, we obtain a set $X \subseteq V(G)$ with $|X| \leq 4k$ such that every induced claw and diamond in G has at least one edge with both endpoints in X; in particular, $G - X$ is {claw, diamond}-free. This means that we can start to examine the structure of $G - X$ understood as a line graph of a triangle-free graph: it consists of a number of maximal cliques (called henceforth *bags*) that can pairwise share at most a single vertex, and for two intersecting bags B_1, B_2 there is no edge between $B_1 \backslash B_2$ and $B_2 \backslash B_1$. Next, we prove that the neighborhood of every vertex $x \in X$ in $G - X$ is contained only in at most 2 bags, which gives us at most $8k$ bags that are important from the viewpoint of neighborhoods of vertices in X. The crux of the proof lies in observing that an optimum deletion set F consists only of edges that are close to these important bags. Intuitively, all the edges of F lie either in important bags or in bags adjacent to the important ones. A more precise combinatorial analysis leads to a set $S \subseteq V(G)$ of size polynomial in k such that every edge of F has both endpoints in S. After finding such a set S, a polynomial-time compression for the problem can be constructed using a generic argument that works for every edge modification problem with a finite list of forbidden induced subgraphs.

On a high level, our approach uses a vertex modulator technique that is similar to one used by Drange and Pilipczuk [13] for their recent polynomial kernel for TRIVIALLY PERFECT EDITING. However, since we are dealing with a graph class with fundamentally different structural properties, the whole combinatorial analysis of the instance with the modulator X (which forms the main part of the paper) is also fundamentally different. We also remark that Cai [6] obtained a kernel for the DIAMOND-FREE EDGE DELETION problem with $\mathcal{O}(k^4)$ vertices. However, the techniques used in that result seem unusable in our setting: their core observation is that a diamond can either be already present in the original graph G or be created by removing an edge of a K_4, and thus one can analyze an auxiliary 'propagation graph' with diamonds and K_4s of the original graph G as nodes. In our setting, we also forbid claws, and the core combinatorial properties of this propagation graph become much too complicated to handle.

Finally, we complement our positive result by proving that CDF-ED is NP-hard and does not admit a subexponential-time parameterized algorithm unless the Exponential Time Hypothesis of Impagliazzo et al. [20] fails.

Theorem 3. *There exists a polynomial-time reduction that, given an instance ϕ of 3SAT with n variables and m clauses, outputs an instance (G, k) of CDF-ED such that (a) (G, k) is a yes-instance if and only if ϕ is satisfiable, (b) $|V(G)|, k = \mathcal{O}(n + m)$, and (c) $\Delta(G) = 6$.*

Corollary 4. *Even on instances with maximum degree 6, CDF-ED is NP-complete and does not admit algorithms with running time $2^{o(k)} \cdot |V(G)|^{\mathcal{O}(1)}$ or $2^{o(|V(G)|)}$ unless the Exponential Time Hypothesis fails.*

Corollary 4 shows that, contrary to recent discoveries for a number of edge modification problems related to subclasses of chordal graphs [2,3,12,16,17], CDF-ED does not enjoy the existence of subexponential-time parameterized algorithms. The reduction of Theorem 3 resembles constructions for similar edge modification problems (see e.g. [12,13,22]): every variable is replaced by a cyclic variable gadget that has to be completely broken by the solution in one of two possible ways, and variable gadgets are wired together with constant-size clause gadgets that verify the satisfaction of the clauses.

In this extended abstract, we provide an almost complete proof of Theorem 2, with some simpler proofs expelled to the full version of the paper [10]. Due to lack of space the proof of Theorem 3 is deferred to the full version entirely.

2 Preliminaries

Graphs. We consider finite, undirected, simple graphs G with vertex set $V(G)$ and edge set $E(G)$. Edges $\{u, v\} \in E(G)$ will be written as uv for short. For a subset of vertices $S \subseteq V(G)$, the *subgraph of G induced by S*, denoted $G[S]$, is the graph with vertex set S and edge set $\{uv \in E(G) \mid u, v \in S\}$. We write $G - S$ for $G[V(G) \backslash S]$. For a subset of edges $F \subseteq E(G)$, we write $G - F$ for the subgraph of G obtained by deleting F, that is, $V(G - F) = V(G)$ and $E(G - F) = E(G) \backslash F$.

Two disjoint sets $X, Y \subseteq V(G)$ are *fully adjacent* if for every $x \in X$ and $y \in Y$, the vertices x and y are adjacent. If one of these sets is a singleton, say $X = \{v\}$, then we say that v and Y are fully adjacent.

For a vertex $v \in V(G)$, the (open) neighborhood $N_G(v)$ of v is the set $\{u \mid uv \in E(G)\}$. The *closed neighborhood* $N_G[v]$ of v is defined as $N_G(v) \cup \{v\}$. For a subset of vertices $S \subseteq V(G)$, we denote by $E_G(S)$ the set of edges of G with both endpoints in S. In this work N_G and E_G will always pertain to the graph named G, so we drop the subscript.

Cliques, claws and diamonds. A *clique* of G is a set of vertices that are pairwise adjacent in G; we often identify cliques with the complete subgraphs induced by them. A *maximal clique* is a clique that is not a proper subset of any other clique. A *claw* is a graph on four vertices $\{c, u, v, w\}$ with edge set $\{cu, cv, cw\}$, called *legs* of the claw; we call c the *center* of the claw, and u, v, w the *leaves* of the claw. When specifying the vertices of a claw we always give the center first. A *diamond* is a graph on four vertices $\{u, v, w, x\}$ with edge set $\{uv, uw, vw, vx, wx\}$.

Parameterized complexity. Parameterized complexity is a framework for refining the analysis of a problem's computational complexity by defining an additional "parameter" as part of a problem instance. Formally, a parameterized problem is a subset $\mathcal{Q}$ of $\Sigma^* \times \mathbb{N}$ for some finite alphabet Σ. The problem is fixed parameter tractable if there is an algorithm which solves an instance (x, k) of the problem in time $f(k) \cdot |x|^c$, where $f : \mathbb{N} \to \mathbb{N}$ is any computable function and c is any integer. If $f(k) = 2^{o(k)}$, we say the algorithm is a subexponential parameterized algorithm. A *kernelization algorithm* for $\mathcal{Q}$ is an algorithm that takes an instance (x, k) of $\mathcal{Q}$ and in time polynomial in $|x|+k$ outputs an equivalent instance (x', k') (i.e., (x, k) is in $\mathcal{Q}$ if and only if (x', k') is) such that $|x'| \leq g(k)$ and $k' \leq g(k)$ for some computable function g. If the *size* of the kernel g is polynomial, we say that $\mathcal{Q}$ admits a polynomial kernel. We can relax this definition to the notion of a *compression algorithm*, where the output is required to be an equivalent instance y of some unparameterized problem $\mathcal{Q}'$, i.e., $(x, k) \in \mathcal{Q}$ if and only if $y \in \mathcal{Q}'$. The upper bound $g(k)$ on $|y|$ will be then called the *size* of the compression. We refer the reader to the books of Downey and Fellows [11] and of Flum and Grohe [14] for a more rigorous introduction.

Forbidden induced subgraphs. Consider any finite family of graphs $\mathcal{H}$. A graph G is $\mathcal{H}$-free if for every $H \in \mathcal{H}$, G does not contain H as an induced subgraph. An *HDS* ($\mathcal{H}$-free deletion set) for G is a subset of edges $F \subseteq E(G)$ such that $G - F$ is $\mathcal{H}$-free. Whenever we talk about a *minimal* HDS, we mean inclusion-wise minimality. $\mathcal{H}$-FREE EDGE DELETION is the parameterized problem asking, for a graph G and a parameter k, whether G has an HDS of size at most k. In ANNOTATED $\mathcal{H}$-FREE EDGE DELETION we are additionally given a set $S \subseteq V(G)$ and the question is whether G has an HDS of size at most k that is contained in $E(S)$.

Let (G, k) be an instance of $\mathcal{H}$-FREE EDGE DELETION. Recall that we can easily find a subset X of the *vertices* of G of size polynomial in k such that (in particular) $G - X$ is $\mathcal{H}$-free. We refer to such a set as a *modulator* of G. The construction here is basically the same as in Lemma 3.3 of [13], and a slightly stronger construction based on the Sunflower Lemma can be found in [15].

Lemma 5. ($\spadesuit$)[2] *Let* $c = \max\{|V(H)| : H \in \mathcal{H}\}$. *Then one can in polynomial time either find a subset* $X \subseteq V(G)$ *of size at most* $c \cdot k$ *such that every induced* $H \in \mathcal{H}$ *in* G *has an edge in* $E(X)$, *or conclude that* (G, k) *is a no-instance.*

We finish this section by showing that it suffices to find a set S of vertices of size polynomial in k such that every minimal solution (every minimal HDS of size at most k) is contained in $E(S)$. Given such a set, we can compress the $\mathcal{H}$-FREE EDGE DELETION instance in polynomial time to an instance of the annotated version with $\mathcal{O}(|S|^{c-1})$ vertices, where $c = \max\{|V(H)| : H \in \mathcal{H}\}$ (we assume $c > 1$, as otherwise the problem is trivial). Since the annotated version is in NP (as an unparameterized problem), this compression, together with an algorithm to obtain S, concludes the proof of Theorem 2. Note that we do not require inclusion-wise minimal HDSs of size larger than k to be contained in $E(S)$.

Lemma 6. ($\spadesuit$) *There is an algorithm that, given an instance* (G, k) *of* $\mathcal{H}$-FREE EDGE DELETION *and a set* $S \subseteq V(G)$ *such that every inclusion-wise minimal HDS of size at most* k *is contained in* $E(S)$, *outputs in polynomial time a set* U, *where* $S \subseteq U \subseteq V(G)$ *and* $|U| \leq \mathcal{O}(|S|^{c-1})$, *such that* (G, k) *is a yes-instance if and only if* $(G[U], S, k)$ *is a yes-instance of* ANNOTATED $\mathcal{H}$-FREE EDGE DELETION.

3 Kernel

In this section, we prove Theorem 2. As discussed below the statement of Theorem 2, this yields the proof of Theorem 1 and thus the kernel. Throughout, let (G, k) to be an instance of {CLAW, DIAMOND}-FREE EDGE DELETION .

We first define a simple decomposition of {claw, diamond}-free graphs, which follows from the fact that they are precisely the line graphs of triangle-free graphs, as shown by Metelsky and Tyshkevich [24]. For a {claw, diamond}-free graph G', let $\mathcal{B}(G')$ be the family of vertex sets, called *bags*, containing:

- every maximal clique of G', and
- a singleton $\{v\}$ for each simplicial vertex v of G'
  (i.e., each vertex whose neighborhood is a clique).

Lemma 7. *Let* G' *be a* {*claw, diamond*}*-free graph. Consider the family* $\mathcal{B}(G')$ *of bags of* G'. *Then:*

(a) every non-isolated vertex of G' *is in exactly two bags;*
(b) for every edge $uv \in E(G')$ *there is exactly one bag containing both* u *and* v;

---

[2] The proofs of statements marked with ($\spadesuit$) are postponed to the full version of the paper [10].

(c) every two bags have at most one vertex in common;
(d) if two bags A, B have a common vertex v, then there is no edge between $A - v$
and $B - v$.

Moreover, $|\mathcal{B}(G')| \leq |V(G')| + |E(G')|$ and the family $\mathcal{B}(G')$ can be computed in polynomial time.

Proof. From the definitions of Sect. 3 and Theorem 5.2 of [24] it follows that {claw, diamond}-free graphs are precisely the *linear r-minoes* for $r = 2$, that is, graphs G' such that every vertex belongs to at most two maximal cliques and every edge belongs to exactly one maximal clique. In particular every edge of G' is contained in exactly one bag, which proves (b).

Let v be any non-isolated vertex of G'. If the neighborhood of v is a clique in G', then $N[v]$ is the only maximal clique containing v – hence v is in exactly two bags: the maximal clique and the singleton $\{v\}$, by definition. If the neighborhood of v is not a clique, then v has neighbors a, b that are not adjacent – hence v is contained in at least two bags: the maximal clique containing va and the (different) maximal clique containing vb. As G' is a linear 2-mino, v is not in any other maximal clique. Since v is not simplicial, by the definition of $\mathcal{B}(G')$ we conclude that also in this case v is in exactly two bags. This concludes the proof of (a).

Since all bags induce cliques in G', two different bags cannot have more than one vertex in common, as this would imply that an edge joining them is contained in both of them. This proves (c).

Finally, if two bags A, B had a common vertex v and there was an edge between $a \in A - v$ and $b \in B - v$, then since A is a maximal clique not containing b, there would be a vertex $a' \in A$ non-adjacent to b. But then the vertices a, a', b, v would induce a diamond subgraph in G', a contradiction. This proves (d).

To see that $|\mathcal{B}(G')| \leq |V(G')| + |E(G')|$, note that every bag of $\mathcal{B}(G')$ is either a singleton bag or it contains an edge. The number of singleton bags is bounded by $|V(G')|$, while the number of bags containing an edge is bounded by $|E(G')|$ due to (b). In order to compute $\mathcal{B}(G')$, it suffices to construct first singleton bags for all simplicial and isolated vertices, and then for every edge of G add the unique maximal clique containing it, constructed in a greedy manner. $\qquad \square$

Now run the algorithm of Lemma 5 on instance (G, k). In case the algorithm concludes it is a no-instance, we return a trivial no-instance of ANNOTATED {CLAW, DIAMOND}-FREE EDGE DELETION as the output of the compression. Otherwise, let X to be the obtained modulator; that is, X is a subset of $V(G)$ of size at most $4k$ such that every induced claw and diamond in G has an edge in $E(X)$. In particular, $G - X$ is a {claw, diamond}-free graph, so using Lemma 7 we compute in polynomial time the family of bags $\mathcal{B}(G - X)$. When referring to bags, we will refer to $\mathcal{B}(G - X)$ only, and implicitly use Lemma 7 to identify, for each non-isolated vertex v in $G - X$, *the two bags containing v*, and for each edge e of $G - X$, *the bag containing e*.

Knowing the structure of $G - X$, we proceed by describing the adjacencies between X and $G - X$. The following definition will play a central role. For $x \in X$, we call a bag B of $G - X$ *attached* to x if:

- B is fully adjacent to x, and
- if $B = \{v\}$ for some vertex v which is not isolated in $G - X$, then the other bag containing v is not fully adjacent to x.

We call a bag *attached* if it is attached to some $x \in X$. The next two propositions show that adjacencies between X and $G - X$ are fully determined by the attachment relation, see Fig. 1.

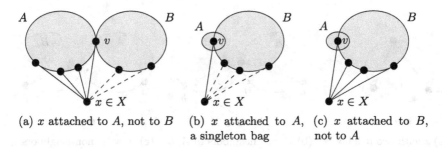

(a) x attached to A, not to B

(b) x attached to A, a singleton bag

(c) x attached to B, not to A

Fig. 1. Possible ways in which a vertex in X can neighbor a vertex v in $G - X$ and the two bags containing it.

Lemma 8. *Let $B \in \mathcal{B}(G - X)$ be a bag such that some vertex $x \in X$ has at least two neighbors in B. Then B is attached to x.*

Proof. Suppose x is adjacent to $u, v \in B$. If x was non-adjacent to some vertex $w \in B$, then since B induces a clique, the vertices x, u, v, w would induce a diamond subgraph in G (Fig. 2(a)). However, no edge of this induced diamond would be in $E(X)$, contradicting the properties of X as a modulator. Therefore, all vertices of B are adjacent to x (and $|B| > 1$), so B is attached to x.   □

Lemma 9. *Let v be a vertex in $G - X$ adjacent to a vertex $x \in X$. Then there is exactly one bag in $\mathcal{B}(G - X)$ that contains v and is attached to x.*

Proof. If v is an isolated vertex in $G - X$, then $\{v\}$ is the only bag containing v and is by definition attached to x.

Otherwise, let A, B be the two bags containing v. If one of the bags is a singleton, say $A = \{v\}$, then B, being unequal to A, contains some other vertices. If at least one vertex of $B\backslash\{v\}$ is adjacent to x, then it follows from Lemma 8 that B is attached to x and A is not. Otherwise, i.e. if no vertices of $B\backslash\{v\}$ are adjacent to x, then by definition A is attached to x and B is not.

It remains to consider the case when both $A - v$ and $B - v$ are not empty; see Fig. 2, (b) and (c). Suppose that x is adjacent to a vertex $a \in A - v$ and a vertex $b \in B - v$. Then a, b are non-adjacent by Lemma 7(d), so vertices v, a, b, x induce a diamond subgraph in G. However, no edge of this diamond is in $E(X)$, a contradiction.

Suppose x is non-adjacent to a vertex $a \in A - v$ and a vertex $b \in B - v$. Then a, b are non-adjacent by Lemma 7(d), so vertices x, a, b, v induce a claw subgraph in G. However, no edge of this claw is in $E(X)$, again a contradiction.

Therefore, if x is adjacent to a vertex in $A - v$, then A is attached to x (by Lemma 8) and x must be non-adjacent to all of $B - v$, implying B is not attached to x. Otherwise, if x is non-adjacent to all vertices in $A - v$, then x must be adjacent to every vertex of $B - v$. This means B is attached to x and A is not.   □

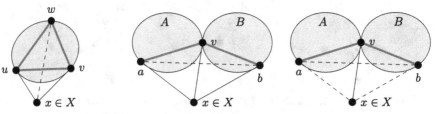

(a) x with two neighbors and a non-neighbor, giving a diamond

(b) x with neighbors in A, B, giving a diamond

(c) x with non-neighbors in A and B, giving a claw

Fig. 2. Adjacencies between X and $G - X$ that lead to a contradiction.

We can now limit the number of attached bags by $2|X|$, which is linear in k.

Lemma 10. *For any $x \in X$, there are at most two bags in $\mathcal{B}(G - X)$ attached to x.*

Proof. Let $x \in X$. We first show that bags attached to x must be pairwise disjoint and non-adjacent. If two bags attached to x contained a common vertex v, then v would be adjacent to x and, by Lemma 9, at most one of the bags would be attached to x, a contradiction.

If there was an edge uv between two different bags attached to x, then its endpoints u and v would be adjacent to x and, by Lemma 8, the bag containing the edge uv would be attached to x. But we have just shown that bags attached to x are disjoint, so no other bag attached to x could contain u or v, a contradiction. Therefore, every two bags attached to x are disjoint and non-adjacent.

Suppose that there are three or more bags adjacent to x. Let u, v, w be any vertices contained in three different bags. By the above observations, u, v, w are pairwise different and non-adjacent. Hence, vertices x, u, v, w induce a claw in G that has no edges in $E(X)$, a contradiction.   □

Having limited the number of attached bags, we want to show that unattached bags intersect solutions only in a simple way. The following technical proposition will help handle cases involving diamonds.

Lemma 11. *Let H be a subgraph (not necessarily induced) of G isomorphic to a diamond. Let $B \in \mathcal{B}(G - X)$ be an unattached bag containing at least two vertices of H. Then B contains all vertices of H.*

Proof. Let u, v be two vertices of H in B. Let w be a vertex of H adjacent to u and v in H (note that since H is a diamond, there always is such a vertex). Then w is also adjacent to u and v in G. Vertex w cannot be in X, as otherwise Lemma 8 would contradict the assumption that B is unattached. Hence, w is in $G - X$. Let A be the bag containing the edge uw. If w was not in B, then $B \neq A$ and vw would be an edge going between $v \in B - u$ and $w \in A - u$, contradicting Lemma 7(d). Therefore, $w \in B$.

Repeating this argument for the fourth vertex of the diamond H and an appropriate pair of vertices from $\{u, v, w\}$, all the vertices of H can be shown to be in B. □

It turns out that one may need to delete an edge of an unattached bag B, but in this case the intersection of any minimal HDS F with the edges of B has a very special structure: deleting the edges of F makes some of the vertices of B isolated, whereas the rest of B remains a smaller clique. This will later allow us to take only a limited number of unattached bags into account.

Lemma 12. *Let F be a minimal HDS of G and let $B \in \mathcal{B}(G - X)$ be an unattached bag. Then $G[B] - F$ consists of a clique and a number of isolated vertices.*

Proof. Let $B' \subseteq B$ be the set of vertices that are not isolated in $G[B] - F$. Consider the set $F' = F \backslash E(B')$. The graph $G - F'$ is obtained from $G - F$ by adding back all edges between vertices in B'. Thus the bag B induces in $G - F'$ a clique on B' plus isolated vertices $B \backslash B'$. We claim that F' is an HDS. By the minimality of F, this will imply that $F = F'$ and hence the claim.

Suppose to the contrary that $G - F'$ contains an induced claw or diamond H. Since $G - F$ contains neither an induced claw nor a diamond, H has an edge e in $F \cap E(B')$.

If H is a diamond in $G - F'$, then since e has both endpoints in B, by Lemma 11 we infer that all vertices of H are in B. But this contradicts that B induces a clique plus isolated vertices in $G - F'$.

If H is a claw in $G - F'$, then let c be its center and v, u_1, u_2 its leaves, so that $e = cv$. Since $e \in E(B')$, its endpoint c is in B', meaning c is not isolated in $G[B] - F$. Let w be a neighbor of c in $G[B] - F$. We show that vertices c, w, u_1, u_2 induce a claw in $G - F$. Consider where the leaves u_i may be. If $u_i \in B$ (for $i = 1$ or 2), then vertices c, v, u_i induce two legs of a claw (a P_3) in $G[B] - F'$, contradicting that $G[B] - F'$ is a clique plus isolated vertices. If $u_i \in X$, then since u_i is adjacent to $c \in B$ and B is not attached, by Lemma 8 we have that u_i cannot be adjacent to $w \in B$ in G. If $u_i \in G - (B \cup X)$, then it is in the bag A containing the edge cu_i and, by Lemma 7, $u_i \in A - c$ is not adjacent to $w \in B - c$ in G. In either case $u_1 w$ and $u_2 w$ are non-edges in G, thus also in $G - F$. By assumption, $u_1 u_2$ is a non-edge in $G - F'$, thus also in $G - F$. We showed that

$u_i \notin B$, so $cu_i \in E(G - F')$ are also edges in $G - F$. Finally, $cw \in E(G[B])\backslash F$, so indeed the vertices c, w, u_1, u_2 induce a claw in $G - F$, a contradiction.   □

Lemma 13. *If K is a clique in G with at least $2k + 2$ vertices, then every HDS F of G of size at most k satisfies $F \cap E(K) = \emptyset$.*

Proof. By contradiction, assume there exists $uv \in F$ with $u, v \in K$. However, then for every two distinct $w_1, w_2 \in K\backslash\{u, v\}$, the subgraph induced in $G - uv$ by u, v, w_1, w_2 is a diamond. As $|K| \geq 2k + 2$, we can find k edge-disjoint diamonds formed in this way in $G - uv$. Consequently, F needs to contains at least k edges apart from uv, a contradiction.   □

Corollary 14. *Let $B \in \mathcal{B}(G - X)$ be a bag with at least $2k + 2$ elements. Then for every HDS F of G of size at most k, $F \cap E(B) = \emptyset$. If furthermore B is attached to $x \in X$, then $F \cap E(B \cup \{x\}) = \emptyset$.*

Proof. Follows directly from Lemma 13, since every bag B is a clique, if B is attached to $x \in X$, then $B \cup \{x\}$ is a clique as well.   □

We are ready to present the main step of the compression procedure for CDF-ED.

Lemma 15. *One can in polynomial time find a set $S \subseteq V(G)$ of size $\mathcal{O}(k^4)$ such that every minimal HDS of size at most k is contained in $E(S)$.*

Proof. Call a bag *small* if it has less than $2k + 2$ vertices, *big* otherwise. We *mark* every small attached bag, every small unattached bag that shares a vertex with some small attached bag, and furthermore, for every vertex pair $x, y \in X$, we mark up to $k + 1$ small unattached bags of size at least two that have a vertex in $N(x) \cap N(y)$. We *mark* the following bags: every small attached bag, every small unattached bag that shares a vertex with some small attached bag, for every vertex pair $x, y \in X$, we mark up to $k + 1$ small unattached bags of size at least two that have a vertex in $N(x) \cap N(y)$ (if there are more such bags, we mark any $k + 1$ of them). Let S be the set of all vertices in marked bags and in X. Let us first show that $|S| = \mathcal{O}(k^4)$. By the construction of X in Lemma 5, we have that $|X| \leq 4k$. By Lemma 10, there are at most $2|X|$ attached bags. Hence, there are at most $2|X| \cdot (2k + 1)$ vertices in small attached bags. Since each vertex of $G - X$ is in at most two bags, there are at most $2|X| \cdot (2k + 1)$ small unattached bags that share a vertex with small attached bags. In the final point we mark at most $|X|^2 \cdot (k + 1)$ small bags. Therefore, we mark at most $2|X| + 2|X| \cdot (2k + 1) + |X|^2 \cdot (k + 1) = \mathcal{O}(k^3)$ small bags in total. The set $S\backslash X$ contains at most $(2k + 1)$ times as many vertices in total, which together with $|X| \leq 4k$ implies that $|S| = \mathcal{O}(k^4)$.

We want to show that a minimal HDS never deletes any edges in unmarked bags. Let Z be the set of edges that are either contained in a marked bag, or in $E(X)$, or connect a vertex of a marked bag with a vertex of X. Note that $Z \subseteq E(S)$, but the inclusion may be strict, due to an edge going between two vertices of some marked bags that belongs to an unmarked bag. Let F be a minimal HDS of size at most k. We will show that $F' = F \cap Z$ is also an HDS, concluding the proof of the lemma.

Claim 16. *If a bag does not induce a clique plus isolated vertices in $G - F'$, then it is a small attached bag.*

Proof. First consider $G - F$. By Lemma 12, every unattached bag induces a clique plus isolated vertices in $G - F$. By Corollary 14, every big bag induces a clique in $G - F$. Hence, if a bag does not induce a clique plus isolated vertices in $G - F$, then it is a small attached bag. Suppose now that a bag does not induce a clique plus isolated vertices in $G - F'$. Then it necessarily contains an edge of $F' \subseteq Z$ and thus must be marked. We infer that this bag induces the same subgraph in $G - F$ as in $G - F'$. Therefore, it must be small and attached.   □

Suppose to the contrary that $G - F'$ contains an induced claw or diamond H. Since $G - F$ contained none, H must have an edge $e \in F \setminus F' = F \setminus Z$. We consider the following cases depending on the location of e, each leading to a contradiction; see Fig. 3.

Case 1: edge e has an endpoint in the modulator X.
Then $e = vx$ for some $x \in X$ and $v \in V(G)$. If $v \in X$, then $e \in E(X) \subseteq Z$, contradicting $e \in F \setminus Z$. Otherwise, by Lemma 9, there is a bag B containing v that is attached to x. Since $e \in F$, by Corollary 14 we infer that B has less than $2k + 2$ elements. But then B is a small, attached, and hence marked bag, implying $e \in Z$, a contradiction.

Case 2: edge e has both endpoints in $G - X$ (and thus e is in $G - X$).
Let B be the bag containing e. Since $e \in F$, B is a small bag by Corollary 14. Since $e \notin Z$, B is not a marked bag. Since small attached bags are marked, B is unattached. By Claim 16, B induces a clique plus isolated vertices in $G - F'$.

Case 2a: H is a diamond (in $G - F'$).
Then the endpoints of e are in B, hence by Lemma 11 all vertices of H are in B. But B induces a clique plus isolated vertices in $G - F'$, a contradiction.

Case 2b: H is a claw (in $G - F'$).
Let c be the center of the claw H and let v, u_1, u_2 be its leaves, so that $e = cv$. Let A be the other bag containing c.

If u_i was in B (for $i = 1$ or 2), then B would not induce a clique plus isolated vertices in $G - F'$ because u_i, c, v induces a P_3, a contradiction.

If $u_i \notin X$, then u_i is in the bag containing cu_i but not in B, which means that u_i is in A. If both u_1, u_2 were not in X, then A would not induce a clique plus isolated vertices in $G - F'$ (because u_1, c, u_2 induces a P_3). By Claim 16, A would be a small attached bag that shares the vertex c with B, implying that B is marked, a contradiction.

If exactly one leaf of the claw is in X, e.g., $u_1 \in X$ and $u_2 \in G - X$, then u_2 is in A (as above). Because c is adjacent to $u_1 \in X$, by Lemma 9 we infer that one of A, B is attached to u_1. Since B is unattached, A is attached to u_1, so $u_1 u_2$ is an edge in G. Since $u_1 u_2$ is not an edge in $G - F'$, we have that $u_1 u_2 \in F' \subseteq F$. By Corollary 14 we infer that A is a small bag. It is also attached, and therefore B is marked, again a contradiction.

If both u_1, u_2 are in X, then note that B is an unattached bag of size at least two that has a vertex (namely c) in the common neighborhood of u_1 and u_2. By the definition of marked bags and as B was not marked in the third point, at least $k + 1$ different marked bags $B_1, \ldots, B_{k+1}$ are unattached, have size at least two, and have some vertex, respectively $c_1, c_2, \ldots, c_{k+1}$, in the common neighborhood of u_1 and u_2. If $c_i = c_j$ for some i, j with $1 \le i < j \le k + 1$, then B_i, B_j are the two bags that contain c_i. Since c_i is adjacent to u_1, one of those bags is attached to u_1 by Lemma 9, a contradiction. Hence, $c_i \ne c_j$ for all $1 \le i < j \le k + 1$. Let w_i be any vertex different from c_i in B_i. Since B_i is unattached, w_i is non-adjacent to u_1 and u_2 in G by Lemma 8. Clearly, c_i is adjacent to w_i, u_1, u_2 in G. Therefore, vertices c_i, w_i, u_1, u_2 induce $k + 1$ edge-disjoint claws in $G - u_1 u_2$. Since u_1, u_2 are leaves of the claw H in $G - F'$, they are non-adjacent in $G - F$. Hence, for each i with $1 \le i \le k + 1$, one of the edges $c_i w_i, c_i u_1, c_i u_2$ must be deleted by F. But $|F| \le k$, a contradiction. $\qquad \square$

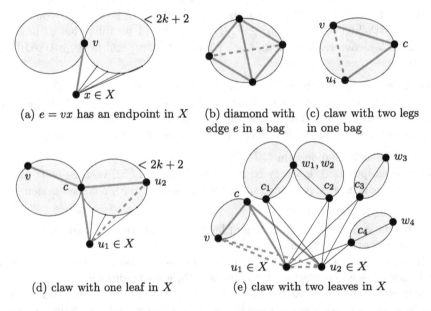

(a) $e = vx$ has an endpoint in X (b) diamond with edge e in a bag   (c) claw with two legs in one bag

(d) claw with one leaf in X (e) claw with two leaves in X

Fig. 3. The different situations where a claw or diamond (thick blue edges and dashed non-edges) might appear in $G - F'$, each leading to a contradiction. (Color figure online)

We can now conclude the proof of Theorem 2: given an instance (G, k), we first apply Lemma 15 to obtain a set $S \subseteq V(G)$, then apply Lemma 6 to the set S obtaining a set $S \subseteq U \subseteq V(G)$, and return an instance $(G[U], S, k)$ of ANNOTATED $\mathcal{H}$-FREE EDGE DELETION. The correctness of this procedure follows from Lemmata 6 and 15. Note that $|S| = \mathcal{O}(k^4)$, thus $|U| = \mathcal{O}(k^{12})$ and the adjacency matrix of the output instance can be encoded with $\mathcal{O}(k^{24})$ bits.

4   Conclusions

In this paper we have charted the parameterized and kernelization complexity of CDF-ED by proving that (i) the problem admits a polynomial kernel, and (ii) the simple $5^k \cdot n^{\mathcal{O}(1)}$ branching algorithm following from the observation of Cai [4] cannot be improved to a subexponential parameterized algorithm, unless the ETH fails.

It should not be a surprise for the reader that the results of this paper were obtained while working on kernelization for CLAW-FREE EDGE DELETION. In this problem, by applying the same vertex modulator principle we arrive at the situation where we have a modulator $X \subseteq V(G)$ with $|X| \leq 4k$, and $G - X$ is a claw-free graph. Then, one can use the structural theorem of Chudnovsky and Seymour [7,8] (see also variants suited for algorithmic applications, e.g., [19]) to understand the structure of $G - X$ and of the adjacencies between X and $G - X$. In essence, the structural theorem yields a decomposition of $G - X$ into *strips*, where each strip induces a graph from one of several basic graph classes; each strip has at most two distinguished cliques (possibly equal) called *ends*, and strips are joined together by creating full adjacencies between disjoint sets of ends. Thus, the whole decomposition looks like a line graph, where every vertex is replaced by a possibly larger strip; indeed, the degenerate case where all the strips are single vertices exactly corresponds to the case of line graphs. As far as base classes are concerned, probably the ones most important for understanding the whole decomposition are proper interval graphs and graphs with independent sets of size at most 2 or 3, in particular, co-bipartite graphs. Thus, we believe that for the sake of showing a polynomial kernel for CLAW-FREE EDGE DELETION, one needs to understand the three special cases when $G - X$ is (a) a line graph, (b) a proper interval graph, and (c) a co-bipartite graph.

We believe that the results of this paper present a progress towards this goal by providing a toolbox useful for tackling case (a). In our proof we have used in several places the fact that we exclude also diamonds. However, much of the structural analysis can translated also to the case when only claws are forbidden, so we hope that similar ideas can be also used for understanding case (a), and consequently how the whole decomposition structure should be dealt with in a polynomial kernel for CLAW-FREE EDGE DELETION. Unfortunately, we are currently unable to make any significant progress in cases (b) and (c), of which case (c) seems particularly difficult.

From another perspective, our positive result gives high hopes for the existence of a polynomial kernel for LINE GRAPH EDGE DELETION, which seems much closer to the topic of this work than CLAW-FREE EDGE DELETION. The problem is that {claw, diamond}-free graphs, or equivalently line graphs of triangle-free graphs, have much nicer structural properties than general line graphs. These properties, encapsulated in Lemma 7, were used several times to simplify the analysis, which would become much more complicated in the case of general line graphs. Also, note that in this paper the considered graph class can be characterized using only two relatively simple forbidden induced subgraphs. In the case of general line graphs, the classic characterization via forbidden induced subgraphs of Beineke [1] involves 9 different obstacles with up to 6 vertices.

References

1. Beineke, L.W.: Characterizations of derived graphs. J. Comb. Theor. **9**(2), 129–135 (1970)
2. Bliznets, I., Fomin, F.V., Pilipczuk, M., Pilipczuk, M.: A subexponential parameterized algorithm for Interval Completion (2014). CoRR, abs/1402.3473
3. Bliznets, I., Fomin, F.V., Pilipczuk, M., Pilipczuk, M.: A subexponential parameterized algorithm for proper interval completion. In: Schulz, A.S., Wagner, D. (eds.) ESA 2014. LNCS, vol. 8737, pp. 173–184. Springer, Heidelberg (2014)
4. Cai, L.: Fixed-parameter tractability of graph modification problems for hereditary properties. Inf. Process. Lett. **58**(4), 171–176 (1996)
5. Cai, L., Cai, Y.: Incompressibility of H-free edge modification. In: Gutin, G., Szeider, S. (eds.) IPEC 2013. LNCS, vol. 8246, pp. 84–96. Springer, Heidelberg (2013)
6. Cai, Y.: Polynomial kernelisation of H-free edge modification problems. Master's thesis. The Chinese University of Hong Kong, Hong Kong (2012)
7. Chudnovsky, M., Seymour, P.D.: Claw-free graphs. IV. Decomposition theorem. J. Comb. Theor. Ser. B **98**(5), 839–938 (2008)
8. Chudnovsky, M., Seymour, P.D.: Claw-free graphs. V. Global structure. J. Comb. Theor. Ser. B **98**(6), 1373–1410 (2008)
9. Cygan, M., Kowalik, L., Pilipczuk, M.: Open problems from workshop on kernels (2013). http://worker2013.mimuw.edu.pl/slides/worker-opl.pdf
10. Cygan, M., Pilipczuk, M., Pilipczuk, M., van Leeuwen, E.J., Wrochna, M.: Polynomial kernelization for removing induced claws and diamonds (2015). CoRR, abs/1503.00704
11. Downey, R.G., Fellows, M.R.: Parameterized Complexity. Monographs in Computer Science. Springer, New York (1999)
12. Drange, P.G., Fomin, F.V., Pilipczuk, M., Villanger, Y.: Exploring subexponential parameterized complexity of completion problems. In: STACS 2014, LIPIcs, vol. 25, pp. 288–299. Schloss Dagstuhl-Leibniz-Zentrum für Informatik (2014)
13. Drange, P.G., Pilipczuk, M.: A polynomial kernel for Trivially Perfect Editing. CoRR, abs/1412.7558 (2014)
14. Flum, J., Grohe, M.: Parameterized Complexity Theory. Texts in Theoretical Computer Science. An EATCS Series. Springer, Heidelberg (2006)
15. Fomin, F.V., Saurabh, S., Villanger, Y.: A polynomial kernel for proper interval vertex deletion. SIAM J. Discrete Math. **27**(4), 1964–1976 (2013)
16. Fomin, F.V., Villanger, Y.: Subexponential parameterized algorithm for minimum fill-in. SIAM J. Comput. **42**(6), 2197–2216 (2013)
17. Ghosh, E., Kolay, S., Kumar, M., Misra, P., Panolan, F., Rai, A., Ramanujan, M.S.: Faster parameterized algorithms for deletion to split graphs. In: Fomin, F.V., Kaski, P. (eds.) SWAT 2012. LNCS, vol. 7357, pp. 107–118. Springer, Heidelberg (2012)
18. Guillemot, S., Havet, F., Paul, C., Perez, A.: On the (non-)existence of polynomial kernels for P_l-free edge modification problems. Algorithmica **65**(4), 900–926 (2013)
19. Hermelin, D., Mnich, M., van Leeuwen, E.J.: Parameterized complexity of induced graph matching on claw-free graphs. Algorithmica **70**(3), 513–560 (2014)

20. Impagliazzo, R., Paturi, R., Zane, F.: Which problems have strongly exponential complexity? J. Comput. Syst. Sci. **63**(4), 512–530 (2001)
21. Kloks, T., Kratsch, D., Müller, H.: Dominoes. In: Mayr, E.W., Schmidt, G., Tinhofer, G. (eds.) WG 1994. LNCS, vol. 903, pp. 106–120. Springer, Heidelberg (1994)
22. Komusiewicz, C., Uhlmann, J.: Cluster editing with locally bounded modifications. Discrete Appl. Math. **160**(15), 2259–2270 (2012)
23. Kratsch, S., Wahlström, M.: Two edge modification problems without polynomial kernels. In: Chen, J., Fomin, F.V. (eds.) IWPEC 2009. LNCS, vol. 5917, pp. 264–275. Springer, Heidelberg (2009)
24. Metelsky, Y., Tyshkevich, R.: Line graphs of Helly hypergraphs. SIAM J. Discrete Math. **16**(3), 438–448 (2003)

Algorithms and Complexity
for Metric Dimension and Location-domination
on Interval and Permutation Graphs

Florent Foucaud[1], George B. Mertzios[2], Reza Naserasr[3], Aline Parreau[4(✉)],
and Petru Valicov[5]

[1] Université Blaise Pascal, LIMOS - CNRS UMR 6158, Clermont-Ferrand, France
florent.foucaud@gmail.com
[2] School of Engineering and Computing Sciences, Durham University, Durham, UK
george.mertzios@durham.ac.uk
[3] CNRS - IRIF, Université Paris Diderot, Paris, France
reza@lri.fr
[4] CNRS, LIRIS, UMR 5205, Université de Lyon, Villeurbanne, France
aline.parreau@univ-lyon1.fr
[5] CNRS, LIF, UMR 7279, Université d'Aix-Marseille, Marseille, France
petru.valicov@lif.univ-mrs.fr

Abstract. We study the problems LOCATING-DOMINATING SET and
METRIC DIMENSION, which consist of determining a minimum-size set
of vertices that distinguishes the vertices of a graph using either neighbourhoods or distances. We consider these problems when restricted to
interval graphs and permutation graphs. We prove that both decision
problems are NP-complete, even for graphs that are at the same time
interval graphs and permutation graphs and have diameter 2. While
LOCATING-DOMINATING SET parameterized by solution size is trivially
fixed-parameter-tractable, it is known that METRIC DIMENSION is $W[2]$-
hard. We show that for interval graphs, this parameterization of METRIC
DIMENSION is fixed-parameter-tractable.

1 Introduction

Combinatorial identification problems have been widely studied in various contexts. The common characteristic of these problems is that we are given a combinatorial structure, and we wish to distinguish (i.e. uniquely identify) its elements
by the means of a small set of selected elements. In this paper, we study two
such identification problems where the instances are graphs. In the LOCATING-
DOMINATING SET problem, we ask for a dominating set S such that the vertices
outside of S are distinguished by their neighbourhood within S. In METRIC
DIMENSION, we wish to select a set S of vertices of a graph G such that every
vertex of G is uniquely identified by its distances to the vertices of S.

This is a short version of the full paper [16] available on arXiv:1405.2424.

G. Mertzios—Partially supported by the EPSRC Grant EP/K022660/1.

© Springer-Verlag Berlin Heidelberg 2016
E.W. Mayr (Ed.): WG 2015, LNCS 9224, pp. 456–471, 2016.
DOI: 10.1007/978-3-662-53174-7_32

These problems have been extensively studied since their introduction in the 1970s and 1980s. They have been applied to various areas such as network verification [2], fault-detection in networks [36], graph isomorphism testing [1] or the logical definability of graphs [26].

Important Concepts and Definitions. All considered graphs are connected, finite and simple. We denote by $N[v]$, the *closed neighbourhood* of vertex v, and by $N(v)$ its *open neighbourhood*, i.e. $N[v] \setminus \{v\}$. A vertex is *universal* if it is adjacent to all the vertices of the graph. A set S of vertices of G is a *dominating set* if for every vertex v, there is a vertex x in $S \cap N[v]$. In the context of dominating sets we say that a vertex x *separates* two distinct vertices u, v if it dominates exactly one of them. Set S separates the vertices of a set X if all pairs of X are separated by a vertex of S. Given a partial set S, we say that two distinct vertices u, v *need to be separated* if S does not separate them. If the set S is clear from context, then we simply say x, y need to be separated. The distance between two vertices u, v is denoted $d(u, v)$. The following two definitions are the main concepts studied in this paper.

- (Slater [33,34]) A set L of vertices of a graph G is a *locating-dominating set* if it is a dominating set and it separates the vertices of $V(G) \setminus L$.

- (Harary and Melter [21], Slater [32]) A set R of vertices of a graph G is a *resolving set* if for each pair u, v of distinct vertices, there is a vertex x of R with $d(x, u) \neq d(x, v)$.

The smallest size of a locating-dominating set of G is the *location-domination number* of G, denoted $\gamma^{\mathrm{LD}}(G)$. The smallest size of a resolving set of G is the *metric dimension* of G, denoted $dim(G)$. The inequality $dim(G) \leq \gamma^{\mathrm{LD}}(G)$, relating these notions, holds for every graph G. If G has diameter 2, the two concepts are almost the same, as then, one can check that $\gamma^{\mathrm{LD}}(G) \leq dim(G) + 1$ holds. We consider the two associated decision problems:

LOCATING-DOMINATING SET   METRIC DIMENSION
Instance: A graph G, an integer k.   *Instance:* A graph G, an integer k.
Question: Is it true that $\gamma^{\mathrm{LD}}(G) \leq k$?   *Question:* Is it true that $dim(G) \leq k$?

We will study these problems on interval graphs and permutation graphs, which are classic graph classes that have many applications and are widely studied. They can be recognized efficiently, and many problems can be solved efficiently for graphs in these classes (see e.g. the book by Golumbic [19]). Given a set S of (geometric) objects, the *intersection graph* G of S is the graph whose vertices are associated to the elements of S and where two vertices are adjacent if and only if the corresponding elements of S intersect. Then, S is called an *intersection model* of G. An *interval graph* is the intersection graph of a set of (closed) intervals of the real line. Given two parallel lines B and T, a *permutation graph* is the intersection graph of segments of the plane which have one endpoint on B and the other endpoint on T.

Previous Work. The complexity of distinguishing problems has been studied by many authors. LOCATING-DOMINATING SET was first proved to be NP-complete in [7], a result extended to bipartite graphs in [5]. This was improved to planar bipartite unit disk graphs [29] and to planar bipartite subcubic graphs [14]. LOCATING-DOMINATING SET is hard to approximate within any $o(\log n)$ factor (n is the order of the graph), with no restriction on the input graph [35]. This result was extended to bipartite graphs, split graphs and co-bipartite graphs [14]. On the positive side, LOCATING-DOMINATING SET is constant-factor approximable for bounded degree graphs [20], line graphs [14,15], interval graphs [4] and is linear-time solvable for graphs of bounded clique-width (using Courcelle's theorem [8]). Furthermore, an explicit linear-time algorithm solving LOCATING-DOMINATING SET on trees is known [33].

METRIC DIMENSION, which has a non-local and more intricate flavour, was widely studied as well, and has (re)gained a lot of attention within the last few years. It was shown NP-complete in [18, ProblemGT61]. This result has recently been extended to bipartite graphs, co-bipartite graphs, split graphs and line graphs of bipartite graphs [11], to a special subclass of unit disk graphs [24], and to planar graphs [9]. Polynomial-time algorithms for the weighted version of METRIC DIMENSION for paths, cycles, trees, graphs of bounded cyclomatic number, cographs and partial wheels were given in [11]. A polynomial-time algorithm for outerplanar graphs was designed in [9] and one for chain graphs in [12]. It was shown in [2] that METRIC DIMENSION is hard to approximate within any $o(\log n)$ factor for graphs of order n. This is even true for bipartite subcubic graphs, as shown in [22,23].

In light of these results, the complexity of LOCATING-DOMINATING SET and METRIC DIMENSION for interval and permutation graphs is a natural open question (as posed in [11,28] for METRIC DIMENSION on interval graphs), since these classes are standard candidates for designing efficient algorithms.

Let us say a few words about the parameterized complexity of these problems. For standard definitions and concepts in parameterized complexity, we refer to the books [10,30]. It is known that for LOCATING-DOMINATING SET, any graph of order n and solution size k satisfies $n \le 2^k + k - 1$ [34]. Therefore, when parameterized by k, LOCATING-DOMINATING SET is trivially fixed-parameter-tractable (FPT): first check whether the above inequality holds (if not, return "no"), and if yes, use a brute-force algorithm checking all possible subsets of vertices. This is an FPT algorithm. However, METRIC DIMENSION (again parameterized by solution size k) is W[2]-hard even for bipartite subcubic graphs [22,23]. Remarquably, the bound $n \le D^k + k$ holds [6] (where n is the graph's order, D its diameter, and k is the size of a resolving set). Hence, for graphs of diameter bounded by a function of k, the same arguments as the previous ones yield an FPT algorithm for METRIC DIMENSION. This holds, for example, for the class of (connected) split graphs, which have diameter at most 3. Besides this, as remarked in [23], no standard class of graphs for which METRIC DIMENSION is FPT was previously known.

Our Results. We settle the complexity of LOCATING-DOMINATING SET and METRIC DIMENSION on interval and permutation graphs, showing that the two problems are NP-complete even for graphs that are at the same time interval graphs and permutation graphs and have diameter 2 (Sect. 2). Then, we present a dynamic programming algorithm (using path-decomposition) to solve METRIC DIMENSION in FPT time on interval graphs (Sect. 3). Up to our knowledge, this is the first nontrivial FPT algorithm for this problem. Due to space constraints, some proofs are deferred to the full version of the paper [16].

2 Hardness Results

We will now reduce 3-DIMENSIONAL MATCHING, which is a classic NP-complete problem [25], to LOCATING-DOMINATING SET on interval graphs.

3-DIMENSIONAL MATCHING
Instance: Three disjoint sets A, B and C each of size n, and a set T of m triples of $A \times B \times C$.
Question: Is there a perfect 3-dimensional matching $M \subseteq T$ of the hypergraph $(A \cup B \cup C, T)$, i.e. a set of disjoint triples of T such that each element of $A \cup B \cup C$ belongs to exactly one of the triples?

2.1 Preliminaries and Gadgets

We first define the following *dominating gadget* (a path on four vertices). The idea is to ensure that specific vertices are dominated locally, and therefore separated from the rest of the graph. We will use it extensively. The reduction is described as an interval graph, but we then show that it is also a permutation graph. In all that follows, we always consider interval graphs with an interval representation.

Definition 1 (Dominating gadget). *A dominating gadget D is a subgraph of an interval graph G inducing a path on four vertices, and such that each interval of $V(G) \setminus V(D)$ either contains all intervals of $V(D)$ or does not intersect any.*

In the following, a dominating gadget will be represented as in Fig. 1(a).

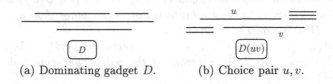

(a) Dominating gadget D.   (b) Choice pair u, v.

Fig. 1. Representations of dominating gadget and choice pair.

The following claim is easy to observe.

Claim 2. *If G is an interval graph containing a dominating gadget D and S is a locating-dominating set of G, then $|S \cap V(D)| \geq 2$.*

If x_1, x_2, x_3, x_4 denote the vertices of D, the set $S_D = \{x_1, x_4\}$ is called the *standard solution for D*. It is a locating-dominating set of D and if S is an optimal locating-dominating set of G, then replacing $S \cap V(D)$ by the standard solution S_D, one can obtain an optimal locating-dominating set S'.

Definition 3 (Choice pair). *A pair $\{u, v\}$ of intervals is called a* choice pair *if u, v both contain the intervals of a common dominating gadget (denoted $D(uv)$), and such that none of u, v contains the other.*

See Fig. 1(b) for an illustration of a choice pair. Intuitively, a choice pair gives us the choice of separating it from the left or from the right: since none of u, v is included in the other, the intervals intersecting u but not v can only be located at one side of u; the same holds for v. In our construction, we will make sure that, except for the choice pairs, all pairs of intervals will be easily separated using domination gadgets. Our aim will then be to separate the choice pairs. We have the following claim that follows directly from Claim 2:

Claim 4. *Let S be a locating-dominating set of an interval graph G and $\{u, v\}$ be a choice pair in G. If the solution $S \cap V(D(uv))$ for the dominating gadget $D(uv)$ is the standard solution, both vertices u and v are dominated, separated from all vertices in $D(uv)$ and from all vertices not intersecting $D(uv)$.*

We now define the central gadget of the reduction, the *transmitter gadget*. Roughly speaking, it allows to transmit information across an interval graph.

Definition 5 (Transmitter gadget). *Let P be a set of two or three choice pairs in an interval graph G. A* transmitter gadget *$Tr(P)$ is a subgraph of G consisting of a path on seven vertices $\{u, uv^1, uv^2, v, vw^1, vw^2, w\}$ and five dominating gadgets $D(u)$, $D(uv)$, $D(v)$, $D(vw)$, $D(w)$ such that the following properties are satisfied:*

- *u and w are the only vertices of $Tr(P)$ that separate the pairs of P.*
- *The intervals of the dominating gadget $D(u)$ (resp. $D(v)$, $D(w)$) are included in interval u (resp. v, w) and no interval of $Tr(P)$ other than u (resp. v, w) intersects $D(u)$ (resp. $D(v)$, $D(w)$).*
- *Pair $\{uv^1, uv^2\}$ is a choice pair and no interval of $Tr(P) \setminus (D(uv^1, uv^2) \cup \{uv^1, uv^2\})$ intersects both intervals of the pair. The same holds for pair $\{vw^1, vw^2\}$.*
- *The choice pairs $\{uv^1, uv^2\}$ and $\{vw^1, vw^2\}$ cannot be separated by intervals of G other than u, v and w.*

Figure 2 illustrates a transmitter gadget and shows the succinct graphical representation that we will use. As shown in the figure, we may use a "box" to denote $T_r(P)$. This box does not include the choice pairs of P but indicates where they are situated. Note that the middle pair $\{y_1, y_2\}$ could also be separated (from the left) by u instead of w, or it may not exist at all if P contains only two pairs.

The following claim shows how transmitter gadgets will be used in the main reduction.

Claim 6. *Let G be an interval graph with a transmitter gadget $Tr(P)$ and let S be a locating-dominating set of G. We have $|S \cap Tr(P)| \geq 11$ and if $|S \cap Tr(P)| = 11$, then no pair of P is separated by a vertex in $S \cap Tr(P)$. Moreover, there exist two sets of vertices of $Tr(P)$, $S^-_{Tr(P)}$ and $S^+_{Tr(P)}$ of size 11 and 12 respectively, such that the following holds:*

- *The set $S^-_{Tr(P)}$ dominates all the vertices of $Tr(P)$ and separates all the pairs of $Tr(P)$ but no pairs in P.*
- *The set $S^+_{Tr(P)}$ dominates all the vertices of $Tr(P)$, separates all the pairs of $Tr(P)$ and all the pairs in P.*

Proof. By Claim 2, we must have $|S \cap Tr(P)| \geq 10$ with 10 vertices of S belonging to the dominating gadgets. In order that uv^1, uv^2 are separated, at least one vertex of $\{u, uv^1, uv^2, v\}$ belongs to S (recall that by definition the intervals not in $Tr(P)$ cannot separate the choice pairs in $Tr(P)$), and similarly, for the choice pair $\{vw^1, vw^2\}$, at least one vertex of $\{v, vw^1, vw^2, w\}$ belongs to S. Hence $|S \cap Tr(P)| \geq 11$ and if $|S \cap Tr(P)| = 11$, vertex v must be in S and neither u nor w are in S. Therefore, no pair of P is separated by a vertex in $S \cap Tr(P)$.

We now prove the second part of the claim. Let S_{dom} be the union of the five standard solutions S_D of the dominating gadgets of $Tr(P)$. Let $S^-_{Tr(P)} = S_{dom} \cup \{v\}$ and $S^+_{Tr(P)} = S_{dom} \cup \{u, w\}$. The set S_{dom} has 10 vertices and so $S^-_{Tr(P)}$ and $S^+_{Tr(P)}$ have respectively 11 and 12 vertices. Each interval of $Tr(P)$ either contains a dominating gadget or is part of a dominating gadget and is therefore dominated by a vertex in S_{dom}. Hence, pairs of vertices that are not intersecting the same dominating gadget are clearly separated. By Claim 2, no vertex in a dominating gadget D is dominated by all vertices of S_D, hence a vertex adjacent to the whole of D is separated from all the vertices of D. Also, by Claim 2, all pairs of vertices inside a dominating gadget are separated by S_{dom}. Therefore, the only remaining pairs to consider are the choice pairs. Note that they are separated both at the same time either by v or by $\{u, w\}$. Hence the two sets $S^-_{Tr(P)}$ and $S^+_{Tr(P)}$ are both dominating and separating the vertices of $Tr(P)$. Moreover, since $S^+_{Tr(P)}$ contains u and w, it also separates the pairs of P. ☐

We will call the sets $S^-_{Tr(P)}$ and $S^+_{Tr(P)}$ the *tight* and *non-tight standard solutions* of $Tr(P)$.

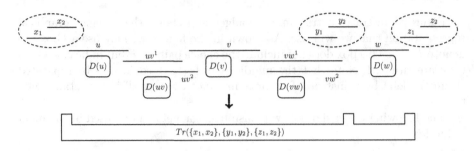

Fig. 2. Transmitter gadget $Tr(\{x_1, x_2\}, \{y_1, y_2\}, \{z_1, z_2\})$ and its "box" representation.

2.2   The Main Reduction

We now describe the reduction. Each element $x \in A \cup B \cup C$ is modelled by a choice pair $\{f_x, g_x\}$. Each triple of $\mathcal{T}$ is modelled by a triple gadget:

Definition 7 (Triple gadget). *Let* $T = \{a, b, c\}$ *be a triple of* $\mathcal{T}$. *The* triple gadget $G_t(T)$ *is an interval graph consisting of four choice pairs* $p = \{p_1, p_2\}$, $q = \{q_1, q_2\}$, $r = \{r_1, r_2\}$, $s = \{s_1, s_2\}$ *together with their associated dominating gadgets* $D(p)$, $D(q)$, $D(r)$, $D(s)$ *and five transmitter gadgets* $Tr(p, q)$, $Tr(r, s)$, $Tr(s, a)$, $Tr(p, r, b)$ *and* $Tr(q, r, c)$, *where:*

- $a = \{f_a, g_a\}$, $b = \{f_b, g_b\}$ *and* $c = \{f_c, g_c\}$;
- *Except for the choice pairs, for each pair of intervals of* $G_t(T)$, *its two intervals intersect different subsets of* $\{D(p), D(q), D(r), D(s)\}$;
- *In each transmitter gadget* $Tr(P)$ *and for each choice pair* $\pi \in P$, *the intervals of* π *intersect the same intervals except for the vertices* u, v, w *of* $Tr(P)$;
- *The intervals of* $V(G) \backslash V(G_t(T))$ *that are intersecting only a part of the gadget intersect according to the transmitter gadget definition and do not separate the choice pairs* p, q, r *and* s.

An illustration of a triple gadget is given in Fig. 3. We remark that p, q, r and s in $G_t(\{a, b, c\})$, are all functions of $\{a, b, c\}$ but to simplify the notations we simply write p, q, r and s.

The proof of the following claim is similar to the proof of Claim 6.

Claim 8. *Let* G *be a graph with a triple gadget* $G_t(T)$ *and* S *be a locating-dominating set of* G. *We have* $|S \cap G_t(T)| \geq 65$ *and if* $|S \cap G_t(T)| = 65$, *no choice pair corresponding to* a, b *or* c *is separated by a vertex in* $S \cap G_t(T)$. *Moreover, there exist two sets of vertices of* $G_t(T)$, $S^-_{G_t(T)}$ *and* $S^+_{G_t(T)}$ *of size* 65 *and* 66 *respectively, such that the following holds.*

- *The set* $S^-_{G_t(T)}$ *dominates all the vertices of* $G_t(T)$ *and separates all the pairs of* $G_t(T)$ *but does not separate any choice pairs corresponding to* $\{a, b, c\}$.
- *The set* $S^+_{G_t(T)}$ *dominates all the vertices of* $G_t(T)$, *separates all the pairs of* $G_t(T)$ *and separates the choice pairs corresponding to* $\{a, b, c\}$.

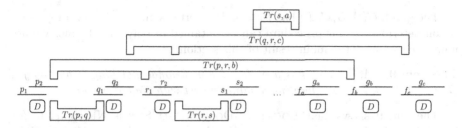

Fig. 3. Triple gadget $G_t(T)$ with $T = \{a, b, c\}$ together with the choice pairs of elements a, b and c. We recall that these choice pairs and their dominating gadgets are not part of $G_t(T)$.

Given an instance $(A, B, C, \mathcal{T})$ of 3-DIMENSIONAL MATCHING with $|A| = |B| = |C| = n$ and $|\mathcal{T}| = m$, we construct the interval graph $G = G(A, B, C, \mathcal{T})$ as follows.

- As mentioned previously, to each element x of $A \cup B \cup C$, we assign a distinct choice pair $\{f_x, g_x\}$ in G. The intervals of any two distinct choice pairs $\{f_x, g_x\}, \{f_y, g_y\}$ are disjoint and they are all in $\mathbb{R}^+$.
- For each triple $T = \{a, b, c\}$ of $\mathcal{T}$ we first associate an interval I_T in $\mathbb{R}^-$ such that for any two triples T_1 and T_2, I_{T_1} and I_{T_2} do not intersect[1]. Then inside I_T, we build the choice pairs $\{p_1, p_2\}, \{q_1, q_2\}, \{r_1, r_2\}, \{s_1, s_2\}$. Finally, using the choice pairs already associated to elements a, b and c we complete this to a triple gadget.
- When placing the remaining intervals of the triple gadgets, we must ensure that triple gadgets do not "interfere": for every dominating gadget D, no interval in $V(G) \setminus V(D)$ must have an endpoint inside D. Similarly, the choice pairs of each triple gadget or transmitter gadget must only be separated by intervals among u, v and w of its corresponding private transmitter gadget. For intervals of distinct triple gadgets, this is easily done by our placement of the triple gadgets. To ensure that the intervals of transmitter gadgets of the same triple gadget do not interfere, we proceed as follows. We place the whole gadget $Tr(p, q)$ inside interval u of $Tr(p, r, b)$. Similarly, the whole $Tr(r, s)$ is placed inside interval v of $Tr(p, r, b)$ and w of $Tr(q, r, c)$. One has to be more careful when placing the intervals of $Tr(p, r, b)$ and $Tr(q, r, c)$. In $Tr(p, r, b)$, we must have that interval u separates p from the right of p. We also place u so that it separates r from the left of r. Intervals uv^1, uv^2 both start in r_1, so that u also separates uv^1, uv^2 without these intervals interfering with the ones of r. Intervals uv^1, uv^2 continue until after pair s. In $Tr(q, r, c)$, we place u so that it separates q from the right, and we place w so that it separates r from the right; intervals uv^1, uv^2, v lie strictly between q and r; intervals vw^1, vw^2 intersect r_1, r_2 but stop before the end of r_2 (so that w can separate both pairs vw^1, vw^2 and r but without these pairs interfering). It is now easy to place $Tr(s, a)$ between s and a.

---

[1] Note that the intervals I_T are not part of the final construction.

The graph $G(A, B, C, \mathcal{T})$ has $159m + 18n$ vertices and the interval representation described by our procedure can be obtained in polynomial time. We are now ready to state the main result of this section.

Theorem 9. *$(A, B, C, \mathcal{T})$ has a perfect 3-dimensional matching if and only if $G = G(A, B, C, \mathcal{T})$ has a locating-dominating set with $65m + 7n$ vertices.*

Theorem 9 shows that LOCATING-DOMINATING SET is NP-complete for interval graphs. In fact, one can prove that the constructed graph $G(A, B, C, \mathcal{T})$ is also a permutation graph, see for example Fig. 4 for an illustration of the transmitter gadget as a permutation diagram intersection model.

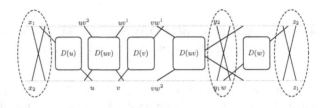

Fig. 4. Permutation diagram intersection model of a transmitter gadget.

Corollary 10. LOCATING-DOMINATING SET *is NP-complete for graphs that are both interval and permutation graphs.*

2.3   Diameter 2 and Consequence for METRIC DIMENSION

We now describe a self-reduction for LOCATING-DOMINATING SET for graphs with a universal vertex (hence, graphs of diameter 2), and a similar reduction from LOCATING-DOMINATING SET to METRIC DIMENSION.

Let G be a graph. Let $f_1(G)$ be the graph obtained from G by adding a universal vertex u and a neighbour v of u of degree 1. Let $f_2(G)$ be the graph obtained from G by adding two adjacent universal vertices u, u' and two non-adjacent vertices v and w that are only adjacent to u and u'. See Fig. 5 for an illustration. One can show that $\gamma^{\mathrm{LD}}(f_1(G)) = \gamma^{\mathrm{LD}}(G) + 1$, and $dim(f_2(G)) = \gamma^{\mathrm{LD}}(G) + 2$.

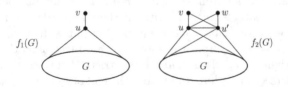

Fig. 5. Two reductions for diameter 2.

This implies the following two theorems.

Theorem 11. *Let $\mathcal{C}$ be a class of graphs that is closed under the graph transformation f_1. If* LOCATING-DOMINATING SET *is NP-complete for graphs in $\mathcal{C}$, then it is also NP-complete for graphs in $\mathcal{C}$ that have diameter 2.*

Theorem 12. *Let $\mathcal{C}$ be a class of graphs that is closed under the graph transformation f_2. If* LOCATING-DOMINATING SET *is NP-complete for graphs in $\mathcal{C}$, then* METRIC DIMENSION *is also NP-complete for graphs in $\mathcal{C}$ that have diameter 2.*

Since Theorems 11 and 12 can be applied to interval graphs and permutation graphs, Corollary 10 implies the following.

Corollary 13. LOCATING-DOMINATING SET *and* METRIC DIMENSION *are NP-complete for diameter 2-graphs that are both interval and permutation graphs.*

3   METRIC DIMENSION is FPT on Interval Graphs

We now prove that METRIC DIMENSION (parameterized by solution size) is FPT on interval graphs. The algorithm is based on dynamic programming over a path-decomposition.

Given an interval graph G, we can assume that in its interval model, all endpoints are distinct, and that the intervals are closed. We define two natural total orderings of $V(G)$ based on this model: $x <_L y$ if and only if the left endpoint of x is smaller then the left endpoint of y, and $x <_R y$ if and only if the right endpoint of x is smaller than the right endpoint of y. We will work with the fourth distance-power G^4 of the input graph G which is also an interval graph and has an interval model inducing the same orders $<_L$ and $<_R$ as G [17].

Our algorithm will use dynamic programming on a *nice path-decomposition* of G^4. The classic concepts of tree-decompositions and its "nice" variant, due to Kloks [27].

Definition 14. *A tree-decomposition of a graph G is a pair $(\mathcal{T}, \mathcal{X})$, where $\mathcal{T}$ is a tree and $\mathcal{X} := \{X_t : t \in V(\mathcal{T})\}$ is a collection of subsets of $V(G)$ (called bags), and they must satisfy the following conditions:*

(i) $\bigcup_{t \in V(\mathcal{T})} X_t = V(G)$;
(ii) for every edge $uv \in E(G)$, there is a bag of $\mathcal{X}$ that contains both u and v;
(iii) for every vertex $v \in V(G)$, the set of bags containing v induces a connected subtree of $\mathcal{T}$.

Given a tree-decomposition of $(\mathcal{T}, \mathcal{X})$, the maximum size of a bag X_t over all tree nodes t of $\mathcal{T}$ minus one is called the *width* of $(\mathcal{T}, \mathcal{X})$. The minimum width of a tree-decomposition of G is the *treewidth* of G. The notion of tree-decomposition has been used extensively in algorithm design, especially via dynamic programming over the tree-decomposition.

We consider a *rooted* tree-decomposition by fixing a root of $\mathcal{T}$ and orienting the tree edges from the root toward the leaves. A rooted tree-decomposition is *nice* (see Kloks [27]) if each node t of $\mathcal{T}$ has at most two children and falls into one of the four types:

(i) *Join* node: t has exactly two children t_1 and t_2, and $X_t = X_{t_1} = X_{t_2}$.
(ii) *Introduce* node: t has a unique child t', and $X_t = X_{t'} \cup \{v\}$.
(iii) *Forget* node: t has a unique child t', and $X_t = X_{t'} \setminus \{v\}$.
(iv) *Leaf* node: t is a leaf node in $\mathscr{T}$.

Given a tree-decomposition, a nice tree-decomposition of the same width always exists and can be computed in linear time [27].

If G is an interval graph, we can construct a tree-decomposition of G (in fact, a path-decomposition) with special properties.

Proposition 15. *Let G be an interval graph with clique number ω and an interval model inducing orders $<_L$ and $<_R$. Then, G has a nice tree-decomposition $(\mathscr{P}, \mathcal{X})$ of width $\omega - 1$ that can be computed in linear time, where moreover:*

(a) *$\mathscr{P}$ is a path (hence there are no join nodes);*
(b) *every bag is a clique;*
(c) *going through $\mathscr{P}$ from the leaf to the root, the order in which vertices are introduced in an introduce node corresponds to $<_L$;*
(d) *going through $\mathscr{P}$ from the leaf to the root, the order in which vertices are forgotten in a forget node corresponds to $<_R$;*
(e) *the root's bag is empty, and the leaf's bag contains only one vertex.*

Proof. Given a graph G, one can decide if it is an interval graph and, if so, compute a representation of it in linear time [3]. This also gives us the ordered set of endpoints of intervals of G.

To obtain $(\mathscr{P}, \mathcal{X})$, we first create the leaf node t, whose bag X_t contains the interval with smallest left endpoint. We then go through the set of all endpoints of intervals of G, from the second smallest to the largest. Let t be the last created node. If the new endpoint is a left endpoint $\ell(I)$, we create an introduce node t' with $X_{t'} = X_t \cup \{I\}$. If the new endpoint is a right endpoint $r(I)$, we create a forget node t' with $X_{t'} = X_t \setminus \{I\}$. In the end we create the root node as a forget node t with $X_t = \emptyset$ that forgets the last interval of G.

Observe that one can associate to every node t (except the root) a point p of the real line, such that the bag X_t contains precisely the set of intervals containing p: if t is an introduce node, p is the point $\ell(I)$ associated to the creation of t, and if t is a forget node, it is the point $r(I) + \epsilon$, where ϵ is sufficiently small and $r(I)$ is the endpoint associated to the creation of t. This set forms a clique, proving Property (b). Furthermore this implies that the maximum size of a bag is ω, hence the width is at most $\omega - 1$ (and at least $\omega - 1$ since every clique must be included in some bag).

Moreover it is clear that the procedure is linear-time, and by construction, Properties (a), (c), (d), (e) are fulfilled.

Let us now show that $(\mathscr{P}, \mathcal{X})$ is a tree-decomposition. It is clear that every vertex belongs to some bag, proving Property (i) of Definition 14. Moreover let u, v be two adjacent vertices of G, and assume $u <_L v$. Then, consider the introduce node of $\mathscr{P}$ where v is introduced. Since u has started before v but has not stopped before the start of v, both u, v belong to X_t, proving Property (ii).

Finally, note that a vertex v appears exactly in all bags starting from the bag where v is introduced, until the bag where v is forgotten. Hence Property (iii) is fulfilled, and the proof is complete.   □

Lemma 16. *Let G be an interval graph with an interval model inducing orders $<_L$ and $<_R$, let $d \geq 1$ be an integer and let $(\mathscr{P}, \mathcal{X})$ be a tree-decomposition of G^d obtained by Proposition 15 (recall that G^d is an interval graph, and it has an intersection model inducing the same orders $<_L$ and $<_R$ [17]). Then the following holds.*

(a) Let t be an introduce node of $(\mathscr{P}, \mathcal{X})$ with child t', with $X_t = X_{t'} \cup \{v\}$. Then, X_t contains every vertex w in G such that $d_G(v, w) \leq d$ and $w <_L v$.
(b) Let t' be the child of a forget node t of $(\mathscr{P}, \mathcal{X})$, with $X_t = X_{t'} \setminus \{v\}$. Then, $X_{t'}$ contains every vertex w in G such that $d_G(v, w) \leq d$ and $v <_R w$.

We now present the most crucial preliminary results necessary for our algorithm. We first start with a definition related to the linear structure of an interval graph that we will use extensively.

Definition 17. *Given a vertex u of an interval graph G, the rightmost path $P_R(u)$ of u is the path $u_0^R, \ldots, u_p^R$ where $u = u_0^R$, for every u_i^R ($i \in \{0, \ldots, p-1\}$) u_{i+1}^R is the neighbour of u_i^R with the largest right endpoint, and thus u_p^R is the interval in G with largest right endpoint. Similarly, we define the leftmost path $P_L(u) = u_0^L, \ldots, u_q^L$ where for every u_i^L ($i \in \{0, \ldots, q-1\}$) u_{i+1}^L is the neighbour of u_i^L with the smallest left endpoint.*

Note that $P_R(u)$ and $P_L(u)$ are two shortest paths from u to u_p^R and u_q^L, respectively. We say that a pair u, v of intervals in an interval graph G is separated by interval x *strictly from the right* (*strictly from the left*, respectively) if x starts after both right endpoints of u, v (ends before both left endpoints of u, v respectively). In other words, x is not a neighbour of any of u and v.

The next lemma is crucial for our algorithm.

Lemma 18. *Let u, v, x be three intervals in an interval graph G and let i be an integer such that x starts after both right endpoints of $u_i^R \in P_R(u)$ and $v_i^R \in P_R(v)$. Then the three following facts are equivalent:*

(1) x separates u_i^R, v_i^R;
(2) for every j with $0 \leq j \leq i$, x separates u_j^R, v_j^R;
(3) for some j with $0 \leq j \leq i$, x separates u_j^R, v_j^R.

A symmetric statement holds for $P_L(u)$.

We now define a *distance-2 resolving set* as a set S of vertices where for each pair u, v of vertices at distance at most 2, there is a vertex $x \in S$ such that $d(u, x) \neq d(v, x)$. Thanks to this local version of resolving sets, we will manage to "localize" the dynamic programming, as we will only need to distinguish pairs of vertices that will be present together in one bag, as claimed by the following lemma.

Lemma 19. *Any distance-2 resolving set of an interval graph is a resolving set.*

Proof. Assume that S is not a resolving set. It means that there is a pair of vertices u, v at distance at least 3 that are not separated by any vertex of S. Among all such pairs, we choose one, say $\{u, v\}$, such that $d(u, v)$ is minimized. Without loss of generality, we assume that u ends before v starts.

Consider u_1^R (v_1^L, respectively), the interval intersecting u (v, respectively) that has the largest right endpoint (smallest left endpoint, respectively). We have $u_1^R \neq v_1^L$ (since $d(u, v) \geq 3$) and $d(u_1^R, v_1^L) = d(u, v) - 2 < d(u, v)$. By minimality, u_1^R and v_1^L are separated by some vertex $s \in S$. But s does not separate u and v, thus $s \notin \{u_1^R, v_1^L\}$.

Without loss of generality, we can assume that $d(u_1^R, s) < d(v_1^L, s)$. In particular, $d(v_1^L, s) \geq 2$ and s ends before v_1^L starts. Thus there is a shortest path from s to v finishing by v_1^L and so $d(v, s) = d(v_1^L, s) + 1$. However, we also have $d(u, s) \leq d(u_1^R, s) + 1 \leq d(v_1^L, s) < d(v, s)$. Hence s is separating u and v, a contradiction. □

The next lemma enables us to bound the size of the bags in our path-decomposition, which will induce subgraphs of diameter 4 of G.

Lemma 20. *Let G be an interval graph with a resolving set of size k, and let $B \subseteq V(G)$ be a subset of vertices such that for each pair $u, v \in B$, $d_G(u, v) \leq d$. Then $|B| \leq 4dk^2 + (2d + 3)k + 1$.*

We are now ready to describe our algorithm.

Theorem 21. METRIC DIMENSION *can be solved in time $2^{O(k^4)}n$ on interval graphs, i.e. it is FPT on this class when parameterized by the solution size k.*

Proof. Let $(\mathscr{P}, \mathcal{X})$ be a path-decomposition of the interval graph G^4 obtained using Proposition 15. Our algorithm is a dynamic programming algorithm over $(\mathscr{P}, \mathcal{X})$.

Let t be a node of $\mathscr{P}$. We let $\mathcal{P}(X_t)$ be the set of pairs of X_t that are at distance at most 2 in G (by Lemma 19, these are the pairs that we need to separate). For each node t of $\mathscr{P}$, we compute a set of *configurations* using the configurations of the child of t in $\mathscr{P}$. A configuration contains full information about the local solution on X_t, but also stores necessary information about the vertex pairs that still need to be separated. More precisely, a configuration $C = (\mathsf{S}, \mathsf{sep}, \mathsf{toSepR}, \mathsf{cnt})$ of t is a tuple where:

- $\mathsf{S} \subseteq X_t$ contains the vertices of the sought solution belonging to X_t;
- $\mathsf{sep} : \mathcal{P}(X_t) \to \{0, 1, 2\}$ assigns, to every pair in $\mathcal{P}(X_t)$, value 0 if the pair has not yet been separated, value 2 if it has been separated strictly from the left, and value 1 otherwise;
- $\mathsf{toSepR} : \mathcal{P}(X_t) \to \{0, 1\}$ assigns, to every pair in $\mathcal{P}(X_t)$, value 1 if the pair needs to be separated strictly from the right (and it is not yet the case), and value 0 otherwise;

- cnt is an integer counting the total number of vertices in the partial solution that has led to C.

Starting with the leaf of $\mathscr{P}$, for each node our algorithm goes through all possibilities of choosing S; however, sep, toSepR and cnt are computed along the way. At each new visited node t of $\mathscr{P}$, a set of configurations is constructed from the configuration sets of the child of t. The algorithm makes sure that all the information is consistent, and that configurations that will not lead to a valid resolving set (or with cnt $> k$) are discarded.

The most crucial point of the algorithm is to use Lemma 18 to "localize" the problem by reducing it to separating pairs inside the current bag X_t. More precisely, for each pair $u, v \in \mathcal{P}(X_t)$, we can deduce from sep$(u, v)$ and S whether u, v are already separated from a previous step of the algorithm or by a solution vertex of S $\subseteq X_t$. If this is not the case and if t is a forget node that forgets u or v, then the pair u, v needs to be separated from the right in a future step of the algorithm corresponding to a bag that does not not contain the pair u, v. By Lemma 18, this will be done by separating some pair u_i^R, v_i^R (that will be present together in a bag considered later), and hence we can set toSepR$(u_1^R, v_1^R) = 1$. The algorithm will then make sure that u_i^R, v_i^R are separated from the right by carrying over this constraint until it is met, along $P_R(u)$ and $P_R(v)$.

The final step of the algorithm simply consists of checking whether, at the root node, we obtained a configuration with cnt $\leq k$. By Proposition 15(b), every bag of $(\mathscr{P}, \mathcal{X})$ is a clique of G^4 (i.e. a subgraph of diameter at most 4 in G) and hence by Lemma 20, it has $O(k^2)$ vertices. Since there are $2^{O(|X_t|^2)}$ configurations for a bag X_t and all computations are polynomial-time in terms of $|X_t|$, the running time is indeed $2^{O(k^4)}n$. $\square$

4   Conclusion

We proved that both LOCATING-DOMINATING SET and METRIC DIMENSION are NP-complete even for graphs of diameter 2 that are both interval and permutation graphs. This is in contrast to related problems such as DOMINATING SET, which is linear-time solvable on both classes. However, we do not know their complexity for unit interval graphs or bipartite permutation graphs (note that both problems are polynomial-time solvable on chain graphs, a subclass of bipartite permutation graphs [12]). We also note that our reduction can be adapted to related problems such as IDENTIFYING CODE (see the full version of this paper [16]).

Regarding our FPT algorithm for METRIC DIMENSION on interval graphs, we do not know whether the result holds for graph classes such as permutation graphs or chordal graphs. The main obstacles for adapting our algorithm to chordal graphs are (i) that Lemma 18, which is essential for our algorithm, heavily relies on the two orderings induced by intersection models of interval graphs, and (ii) that Lemma 19 is not true for chordal graphs.

Acknowledgments. We thank Adrian Kosowski for helpful discussions.

References

1. Babai, L.: On the complexity of canonical labelling of strongly regular graphs. SIAM J. Comput. **9**(1), 212–216 (1980)
2. Beerliova, Z., Eberhard, F., Erlebach, T., Hall, A., Hoffmann, M., Mihalák, M., Ram, L.S.: Network discovery and verification. IEEE J. Sel. Area Comm. **24**(12), 2168–2181 (2006)
3. Booth, K.S., Lueker, G.S.: Testing for the consecutive ones property, interval graphs, and graph planarity using PQ-tree algorithms. J. Comput. Syst. Sci. **13**(3), 335–379 (1976)
4. Bousquet, N., Lagoutte, A., Li, Z., Parreau, A., Thomassé, S.: Identifying codes in hereditary classes of graphs and VC-dimension. SIAM J. Discrete Math. **29**(4), 2047–2064 (2015)
5. Charon, I., Hudry, O., Lobstein, A.: Minimizing the size of an identifying or locating-dominating code in a graph is NP-hard. Theor. Comput. Sci. **290**(3), 2109–2120 (2003)
6. Chartrand, G., Eroh, L., Johnson, M., Oellermann, O.: Resolvability in graphs and the metric dimension of a graph. Disc. Appl. Math. **105**(1–3), 99–113 (2000)
7. Colbourn, C., Slater, P.J., Stewart, L.K.: Locating-dominating sets in series-parallel networks. Congr. Numer. **56**, 135–162 (1987)
8. Courcelle, B.: The monadic second-order logic of graphs. I. Recognizable sets of finite graphs. Inf. Comput. **85**(1), 12–75 (1990)
9. Díaz, J., Pottonen, O., Serna, M., van Leeuwen, E.J.: On the complexity of metric dimension. In: Epstein, L., Ferragina, P. (eds.) ESA 2012. LNCS, vol. 7501, pp. 419–430. Springer, Heidelberg (2012)
10. Downey, R.G., Fellows, M.R.: Fundamentals of Parameterized Complexity. Texts in Computer Science. Springer, Heidelberg (2013)
11. Epstein, L., Levin, A., Woeginger, G.J.: The (weighted) metric dimension of graphs: hard and easy cases. Algorithmica **72**(4), 1130–1171 (2015)
12. Fernau, H., Heggernes, P., van't Hof, P., Meister, D., Saei, R.: Computing the metric dimension for chain graphs. Inform. Process. Lett. **115**, 671–676 (2015)
13. Flotow, C.: On powers of m-trapezoid graphs. Disc. Appl. Math. **63**(2), 187–192 (1995)
14. Foucaud, F.: Decision and approximation complexity for identifying codes and locating-dominating sets in restricted graph classes. J. Discrete Alg. **31**, 48–68 (2015)
15. Foucaud, F., Gravier, S., Naserasr, R., Parreau, A., Valicov, P.: Identifying codes in line graphs. J. Graph Theor. **73**(4), 425–448 (2013)
16. Foucaud, F., Mertzios, G., Naserasr, R., Parreau, A., Valico, P.: Identification, location-domination and metric dimension on interval and permutation graphs. II. Algorithms and complexity. Algorithmica, to appear (2016). arXiv:1405.2424
17. Foucaud, F., Naserasr, R., Parreau, A., Valicov, P.: On powers of interval graphs and their orders. arXiv:1505.03459
18. Garey, M.R., Johnson, D.S.: Computers and Intractability: A Guide to the Theory of NP-Completeness. W. H. Freeman, San Francisco (1979)
19. Golumbic, M.C.: Algorithmic Graph Theory and Perfect Graphs. Elsevier, Amsterdam (2004)
20. Gravier, S., Klasing, R., Moncel, J.: Hardness results and approximation algorithms for identifying codes and locating-dominating codes in graphs. Algorithmic Oper. Res. **3**(1), 43–50 (2008)

21. Harary, F., Melter, R.A.: On the metric dimension of a graph. Ars Comb. **2**, 191–195 (1976)
22. Hartung, S.: Exploring parameter spaces in coping with computational intractability. Ph.D. Thesis, TU Berlin, Germany (2014)
23. Hartung, S., Nichterlein, A.: On the parameterized and approximation hardness of metric dimension. In: Proceedings of the CCC 2013, pp. 266–276 (2013)
24. Hoffmann, S., Wanke, E.: METRIC DIMENSION for gabriel unit disk graphs Is NP-complete. In: Bar-Noy, A., Halldórsson, M.M. (eds.) ALGOSENSORS 2012. LNCS, vol. 7718, pp. 90–92. Springer, Heidelberg (2013)
25. Karp, R.M.: Reducibility among combinatorial problems. In: Complexity of Computer Computations, pp. 85–103. Plenum Press, New York (1972)
26. Kim, J.H., Pikhurko, O., Spencer, J., Verbitsky, O.: How complex are random graphs in first order logic? Random Struct. Alg. **26**(1–2), 119–145 (2005)
27. Kloks, T.: Treewidth: Computations and Approximations. LNCS, vol. 842. Springer, Heidelberg (1994)
28. Manuel, P., Rajan, B., Rajasingh, I., Chris-Monica, M.: On minimum metric dimension of honeycomb networks. J. Discrete Alg. **6**(1), 20–27 (2008)
29. Müller, T., Sereni, J.-S.: Identifying and locating-dominating codes in (random) geometric networks. Comb. Probab. Comput. **18**(6), 925–952 (2009)
30. Niedermeier, R.: Invitation to Fixed-Parameter Algorithms. Oxford University Press, Oxford (2006)
31. Raychaudhuri, A.: On powers of interval graphs and unit interval graphs. Congr. Numer. **59**, 235–242 (1987)
32. Slater, P.J.: Leaves of trees. Congr. Numer. **14**, 549–559 (1975)
33. Slater, P.J.: Domination and location in acyclic graphs. Networks **17**(1), 55–64 (1987)
34. Slater, P.J.: Dominating and reference sets in a graph. J. Math. Phys. Sci. **22**(4), 445–455 (1988)
35. Suomela, J.: Approximability of identifying codes and locating-dominating codes. Inform. Process. Lett. **103**(1), 28–33 (2007)
36. Ungrangsi, R., Trachtenberg, A., Starobinski, D.: An implementation of indoor location detection systems based on identifying codes. In: Aagesen, F.A., Anutariya, C., Wuwongse, V. (eds.) INTELLCOMM 2004. LNCS, vol. 3283, pp. 175–189. Springer, Heidelberg (2004)

On Structural Parameterizations of Hitting Set: Hitting Paths in Graphs Using 2-SAT

Bart M.P. Jansen$^{(\boxtimes)}$

Eindhoven University of Technology, Eindhoven, The Netherlands
b.m.p.jansen@tue.nl

Abstract. HITTING SET is a classic problem in combinatorial optimization. Its input consists of a set system $\mathcal{F}$ over a finite universe U and an integer t; the question is whether there is a set of t elements that intersects every set in $\mathcal{F}$. The HITTING SET problem parameterized by the size of the solution is a well-known W[2]-complete problem in parameterized complexity theory. In this paper we investigate the complexity of HITTING SET under various structural parameterizations of the input. Our starting point is the folklore result that HITTING SET is polynomial-time solvable if there is a tree T on vertex set U such that the sets in $\mathcal{F}$ induce connected subtrees of T. We consider the case that there is a tree-like graph with vertex set U such that the sets in $\mathcal{F}$ induce connected subgraphs; the parameter of the problem is a measure of how treelike the graph is. Our main positive result is an algorithm that, given a graph G with cyclomatic number k, a collection $\mathcal{P}$ of simple paths in G, and an integer t, determines in time $2^{5k}(|G| + |\mathcal{P}|)^{\mathcal{O}(1)}$ whether there is a vertex set of size t that hits all paths in $\mathcal{P}$. It is based on a connection to the 2-SAT problem in multiple valued logic. For other parameterizations we derive W[1]-hardness and para-NP-completeness results.

1 Introduction

HITTING SET is a classic problem in combinatorial optimization that asks, given a set system $\mathcal{F}$ over a finite universe U, and an integer t, whether there is a set of t elements that intersects every set in $\mathcal{F}$. It was one of the first problems to be identified as NP-complete [15]. Parameterized complexity theory is a refined view of computational complexity that aims to attack NP-hard problems by algorithms whose running time is exponential in a problem-specific *parameter value*, but polynomial in terms of the overall input size. The standard parameterization of HITTING SET by the size of the desired solution is unlikely to admit such a fixed-parameter tractable algorithm, as it is W[2]-complete [8]. The goal of this paper is to consider other parameterizations of HITTING SET, with the aim of obtaining FPT algorithms. Our starting point is the folklore result that HITTING SET is polynomial-time solvable when there is a tree T on vertex set U

Supported by NWO Veni grant "Frontiers in Parameterized Preprocessing" and NWO Gravity grant "Networks".

© Springer-Verlag Berlin Heidelberg 2016
E.W. Mayr (Ed.): WG 2015, LNCS 9224, pp. 472–486, 2016.
DOI: 10.1007/978-3-662-53174-7_33

Table 1. Parameterized complexity overview for hitting subgraphs by the minimum number of vertices, parameterized by measures of structure of the host graph.

Parameter	Complexity for type of subgraphs to be hit			
	Path		3-leaf subtree	
Cyclomatic number	FPT, no $k^{\mathcal{O}(1)}$ kernel	Theorem 2	W[1]-hard	Theorem 4
Feedback vertex number	Para-NP-complete	Theorem 5	Para-NP-complete	Theorem 5

such that all sets $S \in \mathcal{F}$ induce connected subtrees of T. The HITTING SET problem on such an instance can be solved by a greedy strategy (Sect. 2). Motivated by this result, we consider whether HITTING SET can be solved efficiently if there is a graph G that is close to being a tree, such that all $S \in \mathcal{F}$ induce connected subgraphs of G. We therefore parameterize the problem by measures of closeness of G to a tree, which forms an example of parameterizing by distance from triviality [18].

Our Results. One way to measure how close a connected graph is to a tree is to consider its *cyclomatic number* $k := m - (n - 1)$. This is the size of a minimum feedback edge set of the graph, i.e., of a minimum set of edges whose removal breaks all cycles in the graph. As a tree has cyclomatic number zero, it is natural to ask if HITTING SET can be solved efficiently if the set system $\mathcal{F}$ can be represented by a graph G on vertex set U having small cyclomatic number, such that every set $S \in \mathcal{F}$ induces a connected subgraph of G. To decouple the difficulty of finding a representation of $\mathcal{F}$ in this form from the problem of exploiting this representation to solve HITTING SET, we consider the situation when such a representation is given. In this setting, the problem can be phrased more naturally in graph-theoretical terms: given a graph G of cyclomatic number k, a collection $\mathcal{S}$ of connected subgraphs of G, and an integer t, is there a vertex set of size t that hits all subgraphs in $\mathcal{S}$?

Our first result for the parameterization by cyclomatic number is a hardness proof showing this problem to be W[1]-hard. In fact, we prove W[1]-hardness even when all subgraphs in $\mathcal{S}$ are trees with at most three leaves. To establish this hardness result we prove that a variation of 3-SAT in multiple valued logic (see Sect. 2) is W[1]-hard, which may be of independent interest. Concretely, we show the following. Given a set of n variables $x_1, \ldots, x_n$ that can take values from 1 to N, and a formula that is a conjunction of clauses of size at most three, where each literal is of the form $x_i \geq c$ or $x_i \leq c$ for $c \in [N]$, it is W[1]-hard parameterized by n to determine whether there is an assignment to the variables satisfying all clauses. This parameterized logic problem reduces to the discussed structural parameterization of HITTING SET in a natural way.

The hardness result motivates us to place further restrictions on the problem in search of fixed-parameter tractable cases. We consider the situation of hitting a set $\mathcal{P}$ of *simple paths* in a graph G of cyclomatic number k. This corresponds to HITTING SET instances where there is a graph G on U such that for all sets S in $\mathcal{F}$, there is a *simple path* in G on vertex set S. We prove that this problem is fixed-parameter tractable and can be solved in time $2^{5k}(|G| + |\mathcal{P}|)^{\mathcal{O}(1)}$, which is the main algorithmic result in this paper. The algorithm is based on a reduction to 2^{5k} instances of the 2-SAT problem in multiple valued logic, which is known to be polynomial-time solvable [3, 17]. The reduction exploits the fact that in tree-like parts of the graph, the local structure of minimum hitting sets can be determined by greedily computed optimal hitting sets for subtrees of a tree. After branching in 2^{5k} directions to determine the form of a solution, the interaction between such canonical subsolutions is then encoded in a 2-SAT formula in multiple valued logic, which can be evaluated efficiently.

There are several other parameters that measure the closeness of a graph to a tree, such as the *feedback vertex number* and *treewidth* (cf. [9]). As these parameters have smaller values than the cyclomatic number, one might hope to extend the FPT result mentioned above to these parameters. However, we show that this is impossible, unless $P = NP$. In particular, we prove that the problem of hitting simple paths in a graph of feedback vertex number 2 is NP-complete, showing the parameterizations by feedback vertex number and treewidth to be para-NP-complete. Table 1 gives an overview of the results in this paper.

Related Work. Several authors [5, 10, 20] have considered problems parameterized by cyclomatic number; this is also known as parameterizing by feedback edge set. In parameterized complexity, HITTING SET is often studied when the sets to be hit have constant size. In this setting, several FPT algorithms and kernelizations bounds are known [1, 6, 21]. The weighted SET COVER problem, which is dual to HITTING SET, has been analyzed for tree-like set systems by Guo and Niedermeier [12]. Recently, Lu et al. [16] considered SET COVER and HITTING SET for set systems representable as subtrees of a (restricted type of) tree, distinguishing polynomial-time and NP-complete cases.

Organization. Preliminaries are given in Sect. 2. The FPT algorithm for hitting paths is developed in Sect. 3. Section 4 contains the hardness proofs. Due to space restrictions, the proofs of statements marked (★) have been deferred to the full version [14].

2   Preliminaries

Parameterized Complexity. A parameterized problem is a set $Q \subseteq \Sigma^* \times \mathbb{N}$, where Σ is a fixed finite alphabet. The second component of a tuple $(x, k) \in \Sigma^* \times \mathbb{N}$ is the *parameter*. A parameterized problem is (strongly uniformly) *fixed-parameter tractable* if there is an algorithm that decides every input (x, k) in time $f(k)|x|^{\mathcal{O}(1)}$. Evidence that a problem is not fixed-parameter tractable is given by proving that it is W[1]-hard. We refer to one of the textbooks [8, 11] for more background.

Graphs. All graphs we consider are simple, undirected and finite. A graph G consists of a set of vertices $V(G)$ and edges $E(G)$. Notation not defined here is standard. For a set of vertices S we denote by $N_G(S)$ the set $\bigcup_{v \in S} N_G(v) \setminus S$.

A path in a graph G is a sequence of distinct vertices such that successive vertices are connected by an edge. The first and last vertices on the path are its endpoints, the remaining vertices are its interior vertices. Given a graph G and a vertex subset $S \subseteq V(G)$, the operation of *identifying* the vertices of S into a new vertex z is performed as follows: delete the vertices in S and their incident edges, and insert a new vertex z that is adjacent to $N_G(S)$, i.e., to all remaining vertices of G that were adjacent to at least one member of S.

Proposition 1 (★). *Let G be a connected graph of minimum degree at least two with cyclomatic number k. The number of vertices in G with degree at least three is bounded by $2k - 2$.*

Proposition 2 (★). *Let G be a connected graph of minimum degree at least two with cyclomatic number k and let S be the set of vertices of degree at least three. If $S \neq \emptyset$ then the number of connected components of $G - S$ is at most $k + |S| - 1$.*

Hitting Set. A set system $\mathcal{F} \subseteq 2^U$ can be viewed as a hypergraph whose vertices are U and whose hyperedges are formed by the sets in $\mathcal{F}$. A set system $\mathcal{F}$ is a *hypertree* if there is a tree T on vertex set U such that every set in $\mathcal{F}$ induces a subtree of T. Testing whether a set system is a hypertree, and constructing a tree representation if this is the case, can be done in polynomial time [19].

We frequently use the fact that a minimum hitting set for a hypertree can be found in polynomial time (cf. [12, Sect. 2] for a view from a dual perspective). When a tree representation is known, a greedy algorithm can be used to find a minimum hitting set. If we root the tree at a leaf and find a vertex v of maximum depth for which there is a set $S \in \mathcal{F}$ whose members all belong to the subtree rooted at v, then it is easy to show there is a minimum hitting set containing v. Consequently, we may add v to the solution under construction, remove all sets hit by v, and remove all elements in the subtree rooted at v from the universe.

This idea can be extended for the following setting. Suppose we have a graph G that is isomorphic to a simple cycle and a set $\mathcal{P}$ of paths in G. To find a minimum vertex set that hits all the paths in $\mathcal{P}$, we try for each vertex v of G whether there is a minimum solution containing it. After removing v and the paths hit by v, the remaining structure is a hypertree since the cycle breaks open when removing v. The minimum over all choices of v gives an optimal hitting set. We will use this in our FPT algorithm to deal with a corner case.

Multiple Valued Logic. The hitting set problems we are interested in turn out to be related to variations of the SATISFIABILITY problem that have been studied in the field of multiple valued logic. In a multiple valued logic, variables can take on more values than just 0 and 1: there is a *truth value set* containing the possible values. For our application, the truth value set is totally ordered; it is a range of integers $[N] = \{1, \ldots, N\}$. A *regular sign* is a constraint of the form $\geq j$ or $\leq j$ for $j \in [N]$. By constraining variables with regular signs, resulting in (generalized) literals of the form $x_i \geq j$ or $x_i \leq j$, and combining such literals with the usual logical connectives, one creates totally ordered regular signed formulas. As expected, the satisfiability problem for such formulas is to determine whether every variable can be assigned a value in the range $[N]$ such that the formula is satisfied. We shall be interested in the case of CNF formulas with clauses having at most two (2-SAT) or at most three (3-SAT) literals.

---

n-TOTALLY ORDERED REGULAR SIGNED 3-SAT   **Parameter:** n.
Input: A totally ordered regular signed 3-CNF formula with n variables and truth value set $[N]$.
Question: Is the formula satisfiable?

For brevity we sometimes refer to this problem as n-TORS 3-SAT. We also consider TORS 2-SAT, where clauses have at most two literals, which is polynomial-time solvable [17]. In particular, TORS 2-SAT can be reduced to the 2-SAT problem in classical logic [3, Sect. 3], which is well-known to be solvable in linear time [2].

3   Algorithms

The goal of this section is to develop an FPT algorithm for the following parameterized problem.

HITTING PATHS IN A GRAPH   **Parameter:** k.
Input: An undirected simple graph G with cyclomatic number k, an integer t, and a set $\mathcal{P}$ of simple paths in G.
Question: Is there a set $X \subseteq V(G)$ of size at most t that hits all paths in $\mathcal{P}$?

The algorithm consists of two reductions. An instance of HITTING PATHS IN A GRAPH is reduced to a hitting set problem on a more structured graph, called a flower. An instance with such a flower structure can be reduced to a polynomial-time solvable 2-SAT problem in multiple valued logic. This section is structured as follows. We first describe the flower structure and the reduction to 2-SAT in Sect. 3.1. Afterward we show how to build an FPT algorithm from this ingredient, in Sect. 3.2.

3.1   Hitting Paths in Flowers

The key notion in this section is that of a *flower graph*, which is a graph G with a distinguished vertex z called the *core* such that all connected components of $G - \{z\}$ are paths $R_1, \ldots, R_n$ of which no interior vertex is adjacent to z. These paths are called *petals* of the flower. When working with flower graphs we will assume an arbitrary but fixed ordering of the petals as $R_1, \ldots, R_n$, as well as an orientation of each petal R_i as consisting of vertices $r_{i,1}, \ldots, r_{i,|V(R_i)|}$. For ease of discussion we will interpret each petal to be laid out from left to right in order of increasing indices. We will give an FPT branching algorithm that reduces HITTING PATHS IN A GRAPH to solving several instances of the following more restricted problem.

HITTING PATHS IN A FLOWER WITH BUDGETS
Input: A flower graph G with core z and petals $R_1, \ldots, R_n$, a set of simple paths $\mathcal{P} = \{P_1, \ldots, P_m\}$ in G, and a budget function $b \colon [n] \to \mathbb{N}_{\geq 1}$.
Question: Is there a set $X \subseteq V(G) \backslash \{z\}$ that hits all paths in $\mathcal{P}$ such that $|X \cap V(R_i)| = b(i)$ for all $i \in [n]$?

We show that HITTING PATHS IN A FLOWER WITH BUDGETS can be solved in polynomial time. The following notion will be instrumental to analyze the structure of solutions to this problem.

Definition 1. *Let R_i be a petal of an instance $(G, z, \mathcal{P}, b)$ of* HITTING PATHS IN A FLOWER WITH BUDGETS *and let $1 \leq \ell \leq |V(R_i)|$. The* canonical ℓ-th solution *for petal R_i is defined by the following process.*

1. *If there is a path in $\mathcal{P}$ that is contained entirely within $\{r_{i,1}, \ldots, r_{i,\ell-1}\}$, then define the canonical ℓ-th solution to be NIL.*
2. *Otherwise, initialize $X_{i,\ell}$ as the singleton set containing $r_{i,\ell}$.*
   (a) *While there is a path in $\mathcal{P}$ that is contained entirely within R_i and is not intersected by $X_{i,\ell}$, consider a path among this set that minimizes the index j' of its right endpoint and add $r_{i,j'}$ to $X_{i,\ell}$.*
   (b) *While $|X_{i,\ell}| < b(i)$ and $Y := \{r_{i,\ell}, \ldots, r_{i,|V(R_i)|}\} \backslash X_{i,\ell} \neq \emptyset$, add the highest-indexed vertex from Y to $X_{i,\ell}$. (Recall that $b(i)$ is the budget for petal R_i.)*
   (c) *If $|X_{i,\ell}| = b(i)$, the canonical ℓ-th solution is $X_{i,\ell}$. If $|X_{i,\ell}| \neq b(i)$, define the canonical ℓ-th solution to be NIL.*

A set $X_i \subseteq V(R_i)$ is a *canonical solution* for petal R_i if there is an integer ℓ for which X_i is the canonical ℓ-th solution for R_i. A canonical solution is *well defined* if it is not NIL. A solution X to the instance $(G, z, \mathcal{P}, b)$ is *globally canonical* if $X \cap R_i$ is a well-defined canonical solution for all i.

Figure 1 illustrates these concepts. For a set $X_i \subseteq V(R_i)$ we will denote by $\max(X_i)$ the highest index of any vertex in X_i, i.e., the index of the rightmost vertex of X_i. Similarly, we denote by $\min(X_i)$ the index of the leftmost vertex of X_i. The following observations about the procedure will be useful.

Observation 1. *If $X_{i,\ell}$ is a well-defined canonical solution, then $\min(X_{i,\ell}) = \ell$.*

Observation 2. *Let $X_{i,\ell}$ result from Definition 1, and assume that Step 1 does not apply and that Step 2b is never triggered during the procedure. Partition the interval $\{r_{i,\ell}, \ldots, r_{i,\max(X_{i,\ell})}\}$ into $|X_{i,\ell}|$ maximal subpaths that each end at a vertex of $X_{i,\ell}$ and contain no other vertices of $X_{i,\ell}$. Then, for every such subpath R' except the singleton subpath $\{r_{i,\ell}\}$, there is a path in $\mathcal{P}$ contained entirely within R'.*

The main strategy behind our reduction of HITTING PATHS IN A FLOWER WITH BUDGETS to TORS 2-SAT will be as follows. We will show that, if a solution to the hitting set problem exists, then there is a globally canonical solution.

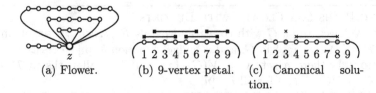

(a) Flower.   (b) 9-vertex petal.   (c) Canonical solution.

Fig. 1. (a) A flower graph with 4 petals and core z. (b) A 9-vertex petal whose endpoints are adjacent to z. The target paths within the petal that must be hit by a solution are drawn stacked on top of each other. (c) The set $\{3, 8\}$ is the 3-rd canonical solution of size 2 for the petal, with respect to the target paths drawn in (b). The corresponding partition of $\{3, \ldots, 8\}$ into two subpaths described in Observation 2 is shown above the petal. It includes the singleton path $\{3\}$. The canonical 1-st solution of size 2 is NIL, since the procedure of Definition 1 produces the set $X_{i,\ell} = \{1, 6, 9\}$, which is too large and is rejected in Step 2c.

Such a solution can be fully characterized by indicating, for each petal, the index of the canonical solution on the petal (i.e., the leftmost vertex of the petal that is in the solution). Hence finding a solution reduces to finding a choice of canonical solutions on the petals. It turns out that for every path $P \in \mathcal{P}$, one can create a signed 2-clause on the variables controlling the choices on two petals, such that the path is hit by the selected solution if and only if the indices of the canonical subsolutions satisfy the 2-clause. This allows the hitting set problem to be modeled by TORS 2-SAT. We now formalize these ideas.

Lemma 1 (★). *Let $(G, z, \mathcal{P}, b)$ be an instance of* HITTING PATHS IN A FLOWER WITH BUDGETS *and let R_i be a petal. The indices for which R_i has a well-defined canonical solution form a contiguous set of integers.*

As the procedure of Definition 1 can be implemented in polynomial time, the set of indices for which a petal has a canonical solution can be computed in polynomial time. We continue describing the structure of canonical solutions.

Lemma 2 (★). *Let $(G, z, \mathcal{P}, b)$ be an instance of* HITTING PATHS IN A FLOWER WITH BUDGETS *and let R_i be a petal. If $\ell_1 < \ell_2$, and the ℓ_1-th and the ℓ_2-th canonical solutions are well-defined as X_{i,ℓ_1} and X_{i,ℓ_2}, then $\max(X_{i,\ell_1}) \leq \max(X_{i,\ell_2})$.*

We now establish that the hitting set problem has a globally canonical solution, if it has a solution at all. The proof exploits the fact that, after selecting the leftmost vertex of a petal to be used in the hitting set, removing it from the graph, and removing the paths hit by this vertex from the graph, the remainder of the petal turns into a pendant path that connects to the rest of the graph at vertex z. The hitting set problem has a greedy solution within this resulting path, which reflects the structure of the canonical solution. Formalizing this line of reasoning is tedious but straight-forward.

Lemma 3 (★). *Let* $(G, z, \mathcal{P}, b)$ *be an instance of* HITTING PATHS IN A FLOWER WITH BUDGETS *having petals* $R_1, \ldots, R_n$. *If the instance has a solution* X', *then it has a globally canonical solution* X.

Lemma 4 (★). *Let* $(G, z, \mathcal{P}, b)$ *be an instance of* HITTING PATHS IN A FLOWER WITH BUDGETS. *There is a polynomial-time algorithm that, given a path* P *(not necessarily contained in* $\mathcal{P}$*) which is a suffix or a prefix of a petal* R_i, *either correctly determines that no well-defined canonical solution for* R_i *hits* P, *or produces a literal of the form* $x_i \geq c$ *or* $x_i \leq c$ *for* $c \in \mathbb{N}_{\geq 1}$, *such that the following holds.*

1. *If* X *is a globally canonical solution for the instance that hits* P *and contains the* ℓ-*th canonical solution for petal* R_i, *then the literal is satisfied by setting* $x_i = \ell$.
2. *If* $x_i = \ell$ *satisfies the literal and the* ℓ-*th canonical solution* $X_{i,\ell}$ *is well-defined, then* P *is hit by* $X_{i,\ell}$.

Using the lemmata developed so far, we can present a polynomial-time algorithm for the problem in flower graphs.

Theorem 1. HITTING PATHS IN A FLOWER WITH BUDGETS *can be solved in polynomial time.*

Proof. We show how to reduce an instance $(G, z, \mathcal{P}, b)$ with petals $(R_1, \ldots, R_n)$ to an equivalent instance of the polynomial-time solvable TORS 2-SAT problem. The main work will be done by Lemma 4 to create the literals of the formula. Let $N := \max_{i \in [n]} |V(R_i)|$ be the maximum size of a petal. The truth value set for our multiple valued logic formula will be $[N]$. We create a variable x_i for every petal i. The clauses in the formula are produced as follows.

1. For every petal index $i \in [n]$, we compute the values of $1 \leq \ell \leq |V(R_i)|$ for which the ℓ-th canonical solution for petal R_i is well-defined, using the procedure of Definition 1. By Lemma 1 these values form a contiguous interval, say $\ell_1, \ldots, \ell_2$. We add the singleton clause $x_i \geq \ell_1$ to the formula, as well as the singleton clause $x_i \leq \ell_2$. If there is no well-defined canonical solution for R_i then, by Lemma 3, the hitting set instance has no solution. In this case we simply output the answer NO.
2. For every path $P \in \mathcal{P}$ that is not contained entirely within a single petal (i.e., for every path that contains the core vertex z of the flower) we do the following. If $P = \{z\}$ is the singleton path containing only vertex z, then we output NO as a solution is not allowed to contain vertex z; this path can never be hit. Otherwise, let P_1, P_2 be the two connected components of $P - \{z\}$. (In the exceptional case that $P - \{z\}$ has only a single component because P has z as an endpoint, take $P_1 = P_2$ to be equal to $P - \{z\}$.) For $k \in \{1, 2\}$ let R_{i_k} be the petal containing P_k and invoke Lemma 4 on P_k with R_{i_k}. If the invocations for both values of k produce a literal, say ϕ_1 and ϕ_2, then add the disjunction $\phi_1 \vee \phi_2$ as a 2-clause to the formula. If one invocation concludes that no well-defined canonical solution hits the path, but the other invocation

produces a literal, then add a singleton clause with the latter literal. Finally, if neither P_1 nor P_2 produces a literal, then neither of the subpaths of $P - \{z\}$ is hit by any well-defined canonical solution, and therefore the path P is not hit by any canonical solution. (Recall that solutions are forbidden to contain z.) Since, by Lemma 3, a canonical solution exists if a solution exists at all, it follows that we can safely output NO and halt.

The process above results in a totally ordered regular signed 2-SAT formula Φ on n variables with $\mathcal{O}(n + |\mathcal{P}|)$ clauses, which is polynomial in the size of the total input. All numbers involved are in the range $[N]$ which is bounded by the order of the input graph G. The reduction can therefore be performed in polynomial time, and produces an instance of TORS 2-SAT of polynomial size, even when encoding the numbers in unary. It remains to prove correctness of the reduction.

Claim ($\bigstar$). Formula Φ is satisfiable if and only if $(G, z, \mathcal{P}, b)$ has a solution.

The claim shows that to solve the hitting set problem, it suffices to check the satisfiability of the polynomial-sized TORS 2-SAT instance. As the latter can be done in polynomial time, this proves Theorem 1. $\square$

3.2   Hitting Paths in Graphs

In this section we will show that an instance of HITTING PATHS IN A GRAPH can be reduced to 2^{5k} instances of HITTING PATHS IN A FLOWER WITH BUDGETS. By the results of the previous section, this leads to an FPT algorithm.

We will frequently use the following observation. It formalizes that if v is a degree-one vertex in G and we are looking for a set that hits all paths in $\mathcal{P}$, then either there is a single-vertex path $P = \{v\} \in \mathcal{P}$, forcing v to be in any solution, or there is an optimal solution that does not contain v.

Observation 3. *Let $(G, k, t, \mathcal{P})$ be an instance of* HITTING PATHS IN A GRAPH *and let $v \in V(G)$ have degree at most one.*

1. *If the singleton path $P = \{v\}$ is contained in $\mathcal{P}$, then $(G, k, t, \mathcal{P})$ is equivalent to the instance obtained by decreasing t by one, removing v from the graph, and removing all paths containing v from $\mathcal{P}$.*
2. *Otherwise, $(G, k, t, \mathcal{P})$ is equivalent to the instance obtained by removing v from the graph and replacing every path $P \in \mathcal{P}$ by $P \backslash \{v\}$.*

The cyclomatic number is not affected by these operations.

For an instance $(G, \mathcal{P}, k, t)$ of HITTING PATHS IN A GRAPH and a vertex subset $S \subseteq V(G)$, the *cost of the subgraph induced by S*, denoted OPT(S), is defined as the minimum cardinality of a set that hits all paths $P \in \mathcal{P}$ for which $V(P) \subseteq S$. Equivalently, OPT(S) is the minimum cardinality of a set that hits all paths $\{P \in \mathcal{P} \mid V(P) \subseteq S\}$ in the graph $G[S]$. Observe that if S induces an acyclic subgraph of G, then this value is computable in polynomial time as discussed in Sect. 2. To reduce the general HITTING PATHS IN A GRAPH problem to the version with budget constraints discussed in the previous section, the following lemma is useful for determining relevant values for the budgets.

Lemma 5 (★). *Let* $(G, \mathcal{P}, k, t)$ *be an instance of* HITTING PATHS IN A GRAPH. *Let* S *be the vertices of degree unequal to two in* G. *There is a minimum-size hitting set* X *for* $\mathcal{P}$ *such that, for every connected component* C *of* $G - S$, *we have* $\text{OPT}(C) \leq |X \cap V(C)| \leq \text{OPT}(C) + 1$.

Using these ingredients we give an algorithm for HITTING PATHS IN A GRAPH.

Theorem 2. HITTING PATHS IN A GRAPH *parameterized by cyclomatic number can be solved in time* $2^{5k} \cdot (|G| + |\mathcal{P}|)^{\mathcal{O}(1)}$.

Proof. When presented with an input $(G, \mathcal{P}, k, t)$, the algorithm proceeds as follows. First, as a preprocessing step, the algorithm repeatedly removes vertices of degree at most one from the graph using Observation 3. If the resulting graph is empty, then we can simply decide the problem: the answer is YES if and only if the value of t was not decreased below zero by these operations. Otherwise we obtain a graph with minimum degree at least two. While this graph is disconnected, add an arbitrary edge between two distinct connected components. This does not change the answer to the instance (the paths $\mathcal{P}$ to be hit are unchanged) and leaves the cyclomatic number unchanged. From now on we therefore assume that the instance we work with has minimum degree at least two and consists of a connected graph. For ease of notation, we refer to instance resulting from these steps simply as $(G, \mathcal{P}, k, t)$. If G consists of just a simple cycle (i.e., G is 2-regular) then we can decide the problem in polynomial time as discussed in Sect. 2, so we focus on the case that the set S of vertices of degree at least three is nonempty. By Proposition 1, the size of S is bounded by $2k$. The main idea of the algorithm is to use branching make two successive guesses.

- First, we guess which vertices from S are used in a solution. Concretely, we try all subsets $S' \subseteq S$ and test whether there is a solution X for which $X \cap S = S'$.
- For every such set S', we do the following. By Lemma 5, there is a minimum-size hitting set that intersects every component C of $G - S$ (which is a path) in either $\text{OPT}(C)$ or $\text{OPT}(C) + 1$ vertices. Let $\mathcal{C}$ denote the set of these components. By Proposition 2, we have $|\mathcal{C}| < k + |S| \leq 3k$. We now guess the collection $\mathcal{C}' \subseteq \mathcal{C}$ of components C for which the solution uses $\text{OPT}(C)$ vertices. Every guess $\mathcal{C}'$ defines a budget $b(C)$ for each component $C \in \mathcal{C}$ as follows: $b(C) = \text{OPT}(C)$ if $C \in \mathcal{C}'$, and $b(C) = \text{OPT}(C) + 1$ otherwise.

Having guessed both S' and $\mathcal{C}'$, we create an instance of HITTING PATHS IN A FLOWER WITH BUDGETS to verify whether there is a hitting set X for the paths $\mathcal{P}$ such that $X \cap S = S'$ and for all components C of $G - S$ we have $|X \cap C| = b(C)$. Observe that these constraints on X completely determine its size, which must be $|S'| + \sum_{C \in \mathcal{C}} b(C)$. Hence if the size exceeds t, then these guessed sets will not lead to a hitting set of the desired size, and can therefore be skipped. When we have a guess that leads to a hitting set size of at most t, we aim to produce an instance of HITTING PATHS IN A FLOWER WITH BUDGETS to check whether there is a solution consistent with the guesses. To this end, initialize G' as a copy of G, and $\mathcal{P}'$ as a copy of $\mathcal{P}$. We modify these structures to create an input on a flower graph. Throughout these modifications there will be a clear correspondence between components of $G' - S$ and those of $\mathcal{C}$, so that we may refer to the budgets of components C of $G' - S$. For each guess S' and $\mathcal{C}'$, we proceed as follows.

1. Remove all the vertices of S' from the graph G' and remove all paths hit by S' from $\mathcal{P}'$.
2. For all paths $P \in \mathcal{P}'$ for which there is a component C of $G' - S$ such that all vertices of C belong to P and $b(C) > 0$, remove P from the set $\mathcal{P}'$. All hitting sets that contain $b(C)$ vertices from C must hit P, so we can drop the constraint P because we will introduce a budget constraint on C.
3. For all components C of $G' - S$ such that $b(C) = 0$, do the following. Remove the vertices of C from the graph G'. For every $P \in \mathcal{P}'$, replace P by the subgraph $P - V(C)$. This may cause the elements of $\mathcal{P}'$ to become disconnected subgraphs, rather than paths, but this will be resolved in the next step.
4. The final step identifies several vertices in the graph into a single core vertex, to obtain a flower structure. Concretely, update the graph G' by identifying all vertices of $S \backslash S'$ into a single vertex z. Similarly, update every subgraph $P \in \mathcal{P}'$ by identifying all vertices of $(S \backslash S') \cap V(P)$ into a single vertex z.

Let $G^*, \mathcal{P}^*$ denote the resulting graph and system of subgraphs. Refer to Fig. 2 for an illustration of these steps.

Claim 1 (★). G^* is a flower with core z; all subgraphs in $\mathcal{P}^*$ are simple paths.

The claim shows that we can use the structures G^* and $\mathcal{P}^*$ resulting from the process above to formulate an instance of HITTING PATHS IN A FLOWER WITH BUDGETS. To that end, we use G^* as the flower graph, z as the core, and $\mathcal{P}^*$ as the set of paths to be hit. We number the connected components of $G^* - \{z\}$, which are the petals of the flower, as $R_1, \ldots, R_n$. Each such petal corresponds to a connected component of $G - S$ for which we assigned a budget when guessing $\mathcal{C}' \subseteq \mathcal{C}$; we define the budget function b^* for the instance by letting $b^*(i)$ be $b(C_i)$ where C_i is the component of $G - S$ corresponding to R_i. This results in a valid instance $(G^*, z, \mathcal{P}^*, b^*)$ of HITTING PATHS IN A FLOWER WITH BUDGETS. For the correctness of the algorithm, the following claim is crucial.

Claim 2 (★). For every guess of $S' \subseteq S$ and $\mathcal{C}' \subseteq \mathcal{C}$, the following are equivalent.

1. There is a hitting set X for the paths $\mathcal{P}$ in graph G such that $X \cap S = S'$ and all components C of $G - S$ satisfy $|X \cap C| = b(C)$.
2. The produced instance $(G^*, z, \mathcal{P}^*, b^*)$ of HITTING PATHS IN A FLOWER WITH BUDGETS has a solution.

Using the claim, the final part of the algorithm becomes clear. For every guess $S' \subseteq S$ and $\mathcal{C}' \subseteq \mathcal{C}$ that leads to a solution of size at most t, we construct

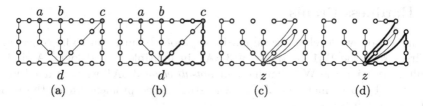

Fig. 2. (a) A graph with cyclomatic number 5, whose vertices of degree ≥ 3 are $S :=$ $\{a, b, c, d\}$. (b) A simple path P in the graph. (c) Illustration of reduction Steps 1 and 4 in the algorithm for the guess $S' = \{a, b\}$. Vertices a and b are deleted, while c and d are identified into a single vertex z to obtain a flower structure. (d) Merging c and d into z turns P into a cyclic subgraph P'. The bottom right petal is contained entirely within P'. If its budget is positive, any solution hits P' in that petal, causing P to be removed in Step 2. If its budget is zero, the vertices of the petal are removed from P' instead in Step 3, to eliminate the cycle.

the corresponding instance of HITTING PATHS IN A FLOWER WITH BUDGETS and solve it using Theorem 1. Since the flower instances are not larger than the input instance, this can be done in time $(|G| + |\mathcal{P}|)^{\mathcal{O}(1)}$ for every guess. As there are $2^{|S|} \cdot 2^{|C|} \leq 2^{2k} \cdot 2^{3k}$ options for S' and C' to check, the total running time is bounded by $2^{5k}(|G| + |\mathcal{P}|)^{\mathcal{O}(1)}$. If one of the HITTING PATHS IN A FLOWER WITH BUDGETS instances has answer YES, then we output YES; otherwise we output NO. In one direction, the correctness of this approach follows from the previous claim together with the facts that flower instances are only produced when the size of the resulting hitting sets is at most t. For the other direction, if $(G, \mathcal{P}, t, k)$ has a hitting set of size at most t, then by Lemma 5 there is a minimum-cardinality hitting set X (whose size is at most t) whose intersection with every component C of $G - S$ is either OPT(C) or OPT(C) + 1. In the branch where $S' = X \cap S$ and C' consists of the components where we use OPT(C) vertices, this leads to a YES-instance of HITTING PATHS IN A FLOWER WITH BUDGETS. This concludes the proof of Theorem 2. $\square$

We remark that, while the previous theorem shows that HITTING PATHS IN A GRAPH is fixed-parameter tractable parameterized by the cyclomatic number, this problem is unlikely to admit a polynomial kernel. The general HITTING SET problem parameterized by the number of universe elements n can be reduced to an instance of HITTING PATHS IN A GRAPH with cyclomatic number $\mathcal{O}(n^2)$: if we let G be a complete n-vertex graph, which has cyclomatic number $\mathcal{O}(n^2)$, then we can model any subset of the universe as a simple path in G. Hence there is a polynomial-parameter transformation from HITTING SET parameterized by the universe size to HITTING PATHS IN A GRAPH parameterized by cyclomatic number. Since HITTING SET parameterized by universe size has no polynomial kernel unless NP $\subseteq$ coNP/poly [7, Theorem 5.3], the same holds for HITTING PATHS IN A GRAPH parameterized by cyclomatic number.

4   Hardness Proofs

In this section we develop several hardness proofs. It turns out to be convenient to first prove the W[1]-hardness of 3-SAT in multiple valued logic. A similar result concerning the W[1]-hardness of *not-all-equal* 3-SAT was obtained independently by Bringmann et al. [4], who studied the problem under the name NAE-INTEGER-3-SAT.

Theorem 3 (★). *The problem n-TOTALLY ORDERED REGULAR SIGNED 3-SAT is W[1]-hard.*

Theorem 3 is used as the starting point for the next hardness proof.

Theorem 4. *It is W[1]-hard to determine, given a graph G with cyclomatic number k, a set S of subgraphs of G, each isomorphic to a tree with at most three leaves, and an integer t, whether there is a set of t vertices in G that intersects all subgraphs in S.*

Proof. We give an FPT-reduction from n-TORS 3-SAT. Consider an instance of that problem, consisting of a signed 3-CNF formula over variables $x_1, \ldots, x_n$ whose truth value set is $[N]$. We assume that there are no clauses that are trivially satisfied (that contain literals $x_i \leq c_1$ and $x_i \geq c_2$ for $c_2 \leq c_1 + 1$), as they can be efficiently recognized and removed without changing the answer.

We construct a hitting set problem on a flower graph G that has a core z and n petals $R_1, \ldots, R_n$. Each petal is a path on N vertices whose endpoints are adjacent to z. It is easy to see that this gives a cyclomatic number of at most $k = n$ for the graph G, as removing the n edges from z to the last vertex of each petal gives an acyclic graph. We seek a hitting set of size at most $t := n$.

Signed literals of the formula have the form $x_i \leq c$ or $x_i \geq c$ for $c \in [N]$. We associate every literal to a prefix or suffix of a petal: a literal $x_i \leq c$ corresponds to the prefix $\{r_{i,1}, \ldots, r_{i,c}\}$ of petal R_i, while a literal $x_i \geq c$ corresponds to the suffix $\{r_{i,c}, \ldots, r_{i,N}\}$. For every clause C of the formula, we consider the pre/suffixes associated to its literals. We add the subgraph S_C that is induced by their vertices, together with z, to the set S of subgraphs to be hit. Observe that, since there are no clauses that are trivially satisfied, each such subgraph S_C induces a tree in G with at most three leaves. In addition, for every petal R_i we add the path R_i as a subgraph to S. This concludes the description of the hitting set instance.

Claim 3 (★). *There is a hitting set of size at most t if and only if the formula is satisfiable.*

The claim shows the correctness of the reduction. It is a valid FPT-reduction since it can be executed in polynomial time and the new parameter k equals the old parameter n. Since n-TOTALLY ORDERED REGULAR SIGNED 3-SAT is W[1]-hard by Theorem 3, this concludes the proof.   □

By slightly modifying the construction, we can also obtain the following result which shows that hitting paths in graphs is para-NP-complete [11] parameterized by the feedback vertex number of the graph.

Theorem 5 (★). *It is NP-complete to determine, given a graph G with a feedback vertex set of size two, a set $\mathcal{P}$ of simple paths in G, and an integer t, whether there is a set of t vertices in G that intersects all paths in $\mathcal{P}$.*

We close this section on hardness by a discussion of subexponential-time algorithms. The construction in Theorem 5 can be used to reduce an n-variable instance of the classical 3-SAT problem (with binary variables) to the problem of hitting simple paths in a graph of cyclomatic number $\mathcal{O}(n)$. This implies that, assuming the exponential-time hypothesis [13], the dependence on k in Theorem 2 cannot be improved to $2^{o(k)}$.

5 Conclusion

We have analyzed the problem of hitting subgraphs of a restricted form within a larger host graph, parameterized by structural measures of the host graph. There are several research directions related to this work that remain unexplored. For example, we have not touched upon the issue of computing, given a generic hitting set instance consisting of a set system $\mathcal{F}$ over a universe U, how complex graphs on vertex set U must be in which every set in $\mathcal{F}$ induces a connected subgraph. What is the complexity of finding, given $\mathcal{F}$ and U, a graph of minimum cyclomatic number that embeds $\mathcal{F}$ in this way? Alternatively, what is the complexity of finding the minimum cyclomatic number of a graph G such that for every set $S \in \mathcal{F}$, there is a simple path in G on vertex set S? Efficient algorithms for this task could be used to transform generic hitting set instances into inputs of HITTING PATHS IN A GRAPH, on which Theorem 2 can be applied.

One can also consider aggregate parameterizations of the hitting set problem using the measure of structure introduced here. We have shown that HITTING PATHS IN A GRAPH is FPT parameterized by the cyclomatic number. It is well known that the general HITTING SET problem is FPT parameterized by the number of sets, as it can be solved by dynamic programming. Suppose we have a HITTING SET instance where there are k_1 arbitrary sets, and there is a graph G of cyclomatic number k_2 such that the remaining sets correspond to paths in G. Is HITTING SET parameterized by $k_1 + k_2$ FPT, when this structure is given?

Acknowledgments. We are grateful to Mark de Berg and Kevin Buchin for interesting discussions that triggered this research.

References

1. Abu-Khzam, F.N.: A kernelization algorithm for d-Hitting set. J. Comput. Syst. Sci. **76**(7), 524–531 (2010)
2. Aspvall, B., Plass, M.F., Tarjan, R.E.: A linear-time algorithm for testing the truth of certain quantified Boolean formulas. Inf. Process. Lett. **8**(3), 121–123 (1979)
3. Béjar, R., Hähnle, R., Manyà, F.: A modular reduction of regular logic to classical logic. In: Proceedings of 31st International Symposium on Multiple-Valued Logic, pp. 221–226 (2001)

4. Bringmann, K., Hermelin, D., Mnich, M., van Leeuwen, E.J.: Parameterized complexity dichotomy for Steiner multicut. In: Proceedings of 32nd STACS, pp. 157–170 (2015)
5. Coppersmith, D., Vishkin, U.: Solving NP-hard problems in 'almost trees': vertex cover. Discrete App. Math. **10**(1), 27–45 (1985)
6. Dell, H., van Melkebeek, D.: Satisfiability allows no nontrivial sparsification unless the polynomial-time hierarchy collapses. J. ACM **61**(4), 23:1–23:27 (2014)
7. Dom, M., Lokshtanov, D., Saurabh, S.: Kernelization lower bounds through colors and IDs. ACM Trans. Algorithms **11**(2), 13 (2014)
8. Downey, R.G., Fellows, M.R.: Fundamentals of Parameterized Complexity. Texts in Computer Science. Springer, London (2013)
9. Fellows, M.R., Jansen, B.M.P., Rosamond, F.: Towards fully multivariate algorithmics: parameter ecology and the deconstruction of computational complexity. Eur. J. Combin. **34**(3), 541–566 (2013)
10. Fiala, J., Kloks, T., Kratochvíl, J.J.: Fixed-parameter complexity of λ-labelings. Discrete Appl. Math. **113**(1), 59–72 (2001)
11. Flum, J., Grohe, M.: Parameterized Complexity Theory. Springer, New York (2006)
12. Guo, J., Niedermeier, R.: Exact algorithms and applications for tree-like weighted set cover. J. Discrete Algorithms **4**(4), 608–622 (2006)
13. Impagliazzo, R., Paturi, R., Zane, F.: Which problems have strongly exponential complexity? J. Comput. Syst. Sci. **63**(4), 512–530 (2001)
14. Jansen, B.M.P.: On structural parameterizations of hitting set: hitting paths in graphs using 2-SAT. CoRR, abs/1507.05890 (2015)
15. Karp, R.M.: Reducibility among combinatorial problems. In: Complexity of Computer Computations, pp. 85–103. Plenum Press (1972)
16. Lu, M., Liu, T., Tong, W., Lin, G., Xu, K.: Set cover, set packing and hitting set for tree convex and tree-like set systems. In: Gopal, T.V., Agrawal, M., Li, A., Cooper, S.B. (eds.) TAMC 2014. LNCS, vol. 8402, pp. 248–258. Springer, Heidelberg (2014)
17. Manyà, F.: The 2-SAT problem in signed CNF-formulas. In: Multiple-Valued Logic (2000)
18. Niedermeier, R.: Reflections on multivariate algorithmics and problem parameterization. In: Proceedings of 27th STACS, pp. 17–32 (2010)
19. Trick, M.A.: Induced subtrees of a tree and the set packing problem. Technical Report 377, Institute for Mathematics and Its Applications (1987)
20. Uhlmann, J., Weller, M.: Two-layer planarization parameterized by feedback edge set. Theor. Comput. Sci. **494**, 99–111 (2013)
21. Wahlström, M.: Algorithms, measures and upper bounds for satisfiability and related problems. Ph.D. thesis, Linköpings universitet, Sweden (2007)

Recognizing k-equistable Graphs in FPT Time

Eun Jung Kim[1], Martin Milanič[2], and Oliver Schaudt[3(✉)]

[1] CNRS-Université Paris-Dauphine, Place du Maréchal de Lattre de Tassigny,
75775 Paris Cedex 16, France
eun-jung.kim@dauphine.fr
[2] UP IAM and UP FAMNIT, University of Primorska, Muzejski trg 2,
6000 Koper, Slovenia
martin.milanic@upr.si
[3] Universität zu Köln, Institut Für Informatik, Weyertal 80, 50931 Köln, Germany
schaudto@uni-koeln.de

Abstract. A graph $G = (V, E)$ is called *equistable* if there exist a positive integer t and a weight function $w : V \to \mathbb{N}$ such that $S \subseteq V$ is a maximal stable set of G if and only if $w(S) = t$. Such a function w is called an *equistable function* of G. For a positive integer k, a graph $G = (V, E)$ is said to be k-*equistable* if it admits an equistable function which is bounded by k.

We prove that the problem of recognizing k-equistable graphs is fixed parameter tractable when parameterized by k, affirmatively answering a question of Levit et al. In fact, the problem admits an $O(k^5)$-vertex kernel that can be computed in linear time.

Keywords: Equistable graphs · Recognition algorithm · Fixed parameter tractability

1 Introduction

The main notion studied in this paper is the class of equistable graphs, introduced by Payan in 1980 [17] as a generalization of the well known and well studied class of threshold graphs [1,9]. A *stable* (or *independent*) *set* in a (finite, simple, undirected) graph G is a set of pairwise non-adjacent vertices. A *maximal stable set* is a stable set not contained in any other stable set. A graph $G = (V, E)$ is said to be *equistable* if there exists a function $\varphi : V \to \mathbb{R}_+$ such that for every $S \subseteq V$, set S is a maximal stable set of G if and only if $\varphi(S) := \sum_{x \in S} \varphi(x) = 1$. Equivalently, G is equistable if and only if there exist a positive integer t and a weight function $w : V \to \mathbb{N} := \{1, 2, 3, \ldots\}$ such that $S \subseteq V$ is a maximal stable set of G if and only if $w(S) = t$. Such a function w is called an *equistable function* of G, while the pair (w, t) is called an *equistable structure*. Equistable

This work is supported in part by the Slovenian Research Agency (I0-0035, research program P1-0285 and research projects N1-0032, J1-5433, J1-6720, and J1-6743) and by the bilateral project BI-FR/15–16–PROTEUS-003.

© Springer-Verlag Berlin Heidelberg 2016
E.W. Mayr (Ed.): WG 2015, LNCS 9224, pp. 487–498, 2016.
DOI: 10.1007/978-3-662-53174-7_34

graphs were studied in a series of papers [5–8,10,13–18]; besides threshold graphs and cographs, they also generalize the class of general partition graphs [4,11,13]. The complexity status of recognizing equistable graphs is open, even the membership in NP of this problem is not known. No combinatorial characterization of equistable graphs is known. Given a graph G and a function $w : V(G) \to \mathbb{N}$, it is co-NP-complete to determine if w is an equistable function of G [14].

Levit et al. introduced in [8] the notion of k-equistable graphs. For a positive integer k, a graph $G = (V, E)$ is said to be k-equistable if it admits an equistable function $w : V \to [k] := \{1, \ldots, k\}$. Such a weight function is called a k-equistable function, and the corresponding structure (w, t) is a k-equistable structure. We remark that there exist equistable graphs such that the smallest k for which the graph is k-equistable is exponential in the number of vertices of G [14].

For a positive integer t, an equistable graph $G = (V, E)$ is said to be target-t equistable if it admits an equistable function $w : V \to \mathbb{N}$ with equistable structure (w, t). Clearly, every target-t equistable graph is also t-equistable (but not vice versa).

As mentioned above, the complexity of recognizing equistable graphs is open, but it seems plausible that the problem could be NP-hard. It thus makes sense to search ways to simplify the recognition problem. To this end, we consider the following two parameterized problems related to equistability.

---

k-EQUISTABILITY

    Input: A graph $G = (V, E)$, a positive integer k.
Parameter: k.
Question: Is G k-equistable?

TARGET-t EQUISTABILITY

    Input: A graph $G = (V, E)$, a positive integer t.
Parameter: t.
Question: Is G target-t equistable?

Similarly as for the recognition of equistable graphs, it is not known whether the above two problems are NP-hard, and whether they belong to NP.

Apart from being natural parameterizations of the equistability problem, the first problem has been tackled before (in a non-parameterized variant) in a paper by Levit et al. [8]. There they prove the following.

Theorem 1 (Levit et al. [8]). *For every fixed k, there is an $O(n^{2k})$ algorithm to decide whether a given n-vertex graph is k-equistable. In case of a positive instance, the algorithm also produces a k-equistable structure of G.*

Also, the authors ask whether Theorem 1 can be strengthened in the sense that there is an FPT algorithm for recognizing k-equistable graphs. We answer this question affirmatively.

More precisely, we prove the following results:

- There is an $O(k^5)$-vertex kernel for the k-EQUISTABILITY problem that can be computed in linear time. This yields an FPT algorithm for the k-EQUISTABILITY problem of running time $O(k^{9k+1} + m + n)$, given a graph with n vertices and m edges. This affirmatively answers the question posed by Levit et al. [8].
- The TARGET-t EQUISTABILITY problem admits an $O(t^2)$-vertex kernel, computable in linear time. Moreover, there is an $O(t^{3t+1} + m + n)$ time algorithm to solve the TARGET-t EQUISTABILITY problem.

The first result we prove in Sect. 5, and the second in Sect. 4.

In order to achieve the above mentioned running times of our FPT algorithms, we present a refinement of the algorithm proposed by Levit et al. in [8] in their proof of Theorem 1. This is done in Sect. 3.

2 Preliminaries

2.1 Twin Classes

Following [8], we say that vertices u and v of a graph G are *twins* if they have exactly the same set of neighbors other than u and v. It is easy to verify that the twin relation is an equivalence relation. We recall some basic properties of the twin relation (see [8]):

Lemma 1. *Let $G = (V, E)$ be a graph. The twin relation is an equivalence relation, and every equivalence class is either a clique or a stable set.*

An equivalence class of the twin relation will be referred to as a *twin class*. Twin classes that are cliques will be referred to *clique classes*, and the remaining classes will be referred to as *stable set classes*. We say that two disjoint sets of vertices X and Y in a graph G *see* each other if every vertex of X is adjacent to every vertex of Y, and they *miss* each other if every vertex of X is non-adjacent to every vertex of Y. A vertex x *sees* a set $Y \subseteq V(G) \setminus \{x\}$ if the singleton $\{x\}$ sees Y, and similarly x *misses* Y if $\{x\}$ misses Y. The set of all twin classes will be denoted by $\Pi(G)$ and referred to as the *twin partition* of G. The number of twin classes of G will be denoted by $\pi(G) = |\Pi(G)|$. The following observation is an immediate consequence of the fact that the twin classes are equivalence classes under the twin relation.

Observation 2. *Every two distinct twin classes either see each other or miss each other.*

By Observation 2, the *quotient graph* of G, denoted $\mathcal{Q}(G)$, is thus well defined: Its vertex set is $\Pi(G)$, and two twin classes are adjacent if and only if they see each other in G. Given a graph G, it is possible to find in linear time the twin partition $\Pi(G)$, the quotient graph $\mathcal{Q}(G)$ and $\pi(G)$, using any of the linear time algorithms for modular decomposition [2,12,19] and observing that one can derive the twin classes from the modular decomposition tree.

The following two lemmas due to Levit et al. [8] show why twin partitions are important in the study of equistable graphs.

Lemma 2. *For every equistable function w of G and for every i, every set of the form $V_i^w = \{x \in V : w(x) = i\}$ is a subset of a twin class of G. In particular, if G is a k-equistable graph, then $\pi(G) \leq k$.*

Corollary 1. *If G is a target-t equistable graph, then $\pi(G) \leq t$.*

Lemma 3. *For every equistable function w of an equistable graph G and for every clique class C there exists an i such that $V_i^w = C$.*

2.2   Parameterized Complexity

A decision problem parameterized by a problem-specific parameter k is called *fixed-parameter tractable* if there exists an algorithm that solves it in time $f(k) \cdot n^{O(1)}$, where n is the instance size. Such an algorithm is called an *FPT algorithm*. The function f is typically super-polynomial and depends only on k. One of the main tools to design such algorithms is the kernelization technique. A *kernelization* is a polynomial-time algorithm which transforms an instance (I, k) of a parameterized problem into an equivalent instance (I', k') of the same problem such that the size of I' is bounded by $g(k)$ for some computable function g and k' is bounded by a function of k. The instance I' is said to be a *kernel* of size $g(k)$. It is a folklore that a parameterized problem is fixed-parameter tractable if and only if it admits a kernelization. In the remainder of this paper, the kernel size is expressed in terms of the number of vertices. For more background on parameterized complexity the reader is referred to Downey and Fellows [3].

3   A Refined XP-algorithm for k-EQUISTABILITY

In this section we propose a revised version of the algorithm of Levit et al. [8] for checking whether a given graph is k-equistable. We implement some speed-ups and give a more careful analysis of the running time. Let us remark that this improvement does not speed up the running time when k is fixed, and it is thus not relevant for the main result of Levit et al. [8]. However, the improved running time is essential when the algorithm is applied to a kernelized instance, for the k-EQUISTABILITY resp. TARGET-t EQUISTABILITY problem. We refrain from formally restating the whole algorithm from [8] in order not to create redundancy.

Theorem 3. *Let G be a graph on n vertices and m edges, and let $k \in \mathbb{N}$. Then there is an algorithm of running time $O(n + m + \max\{n^{2k} k^{1-k}, k^{3k+1}\})$ to check whether G is k-equistable. This algorithm computes a k-equistable structure, if one exists, and the same holds if a target t is prescribed.*

We emphasize that unlike in the statement of Theorem 1, the constant hidden in the O notation in Theorem 3 does not depend on k (in Theorem 3, k is not restricted to be a constant).

Before we prove Theorem 3, we state the following observation.

Lemma 4. *Let $k, n \in \mathbb{N}$ and let $a \in \mathbb{N}_0^k$ with $\sum_{i=1}^k a_i = n$. Then*

$$\prod_{i=1}^k (a_i + 1) \leq (n/k + 1)^k .$$

Proof. If $k = 1$, the statement is immediate. So, let $k > 1$, and assume the statement is true for $k - 1$. Let $a \in \mathbb{N}_0^k$ with $\sum_{i=1}^k a_i = n$. We know that $\prod_{i=1}^{k-1}(a_i + 1) \leq ((n - a_k)/(k - 1) + 1)^{k-1}$, and thus $\prod_{i=1}^k (a_i + 1) \leq (a_k + 1) \cdot ((n - a_k)/(k - 1) + 1)^{k-1}$. A straightforward calculation shows that the right hand side is maximized (over $a_k \geq 0$) for $a_k = n/k$. Thus,

$$\prod_{i=1}^k (a_i + 1) \leq \left(\frac{n}{k} + 1\right) \cdot \left(\frac{n - \frac{n}{k}}{k - 1} + 1\right)^{k-1} = \left(\frac{n}{k} + 1\right)^k ,$$

which completes the proof. $\qquad\square$

We can now prove Theorem 3.

Proof (of Theorem 3). Recall that by Lemma 2, any equistable weight function for G assigns the same weight only to vertices of the same twin class. Following the algorithm of Levit et al. [8], we proceed as follows. First, we compute in time $O(n+m)$ the twin partition of G and the quotient graph $\mathcal{Q}(G)$ (cf. Section 2.1). Fix any ordering $V(G) = \{v_1, \ldots, v_n\}$ such that vertices in each twin class appear consecutively in this ordering. Clearly, the permutation of the weights within a twin class produces an equivalent weight function, i.e., a weight function is an equistable function of G if and only if after any permutation of the weights within a twin class we still have an equistable function. We aim to produce a family $\mathcal{F}$ which contains all equistable functions up to permutations of the weights within a twin class. It suffices to produce all mappings $w : V(G) \to [k]$ such that the vertices in the set $w^{-1}(i)$, $i \in [k]$, appear consecutively in the ordering of $V(G)$. Let $K(n, k)$ be the number of partitions of $[n]$ into k labeled intervals, where some of the intervals may be empty. It is straightforward to verify that $|\mathcal{F}|$ is bounded by $K(n, k)$. A standard counting argument yields $K(n, k) \leq k! \cdot \binom{n+k-1}{n}$.

The set $\mathcal{F}$ can be computed in time $O(k^k(n+k-1)^{k-1})$ as follows. Generate all one-to-one mappings from the set $[k - 1]$ to an $(n + k - 1)$-element set. Using the above ordering of $V(G)$, each such mapping determines a partition of $V(G)$. If the partition refines the twin partition of G, then compute all the $O(k^k)$ one-to-one mappings from the resulting set of (at most k) non-empty intervals to the set $[k]$. Each of these mappings specifies, in a natural way, a function in $\mathcal{F}$.

Let us now estimate more carefully the size of $\mathcal{F}$. Let $\hat{n} := \max\{n, k^2\}$. We have

$$\frac{k \cdot (\hat{n} + k)^{k-1}}{\hat{n}^k} = \frac{k}{\hat{n}} \cdot \left(1 + \frac{k}{\hat{n}}\right)^{k-1} \leq \frac{1}{k} \cdot \left(1 + \frac{1}{k}\right)^{k-1} \leq \frac{e}{k} ,$$

implying $k \cdot (\hat{n} + k)^{k-1} \leq e\hat{n}^k/k$. We thus obtain

$$K(n, k) \leq K(\hat{n}, k) \leq k! \cdot \binom{\hat{n} + k - 1}{\hat{n}} = k \cdot \frac{(\hat{n} + k - 1)!}{\hat{n}!} < k \cdot (\hat{n} + k)^{k-1} \leq \frac{e\hat{n}^k}{k}.$$

Thus, we have to consider only $|\mathcal{F}| = O(\hat{n}^k/k)$ many weight functions, which can be computed in time $O(k^k(n + k - 1)^{k-1}) = O((k\hat{n})^k)$.

It remains to check if any of these $O(\hat{n}^k/k)$ weight functions in $\mathcal{F}$ is an equistable function. For every weight function $w \in \mathcal{F}$, the algorithm from [8] first computes the target value t by evaluating the w-weight of an arbitrary (fixed) maximal stable set of G (see [8] for details); in our setting, this computation can be implemented in time $O(k)$. The algorithm then computes the set X_w of all k-dimensional vectors x with integer coordinates such that $0 \leq x_i \leq |w^{-1}(i)|$ for all $i \in [k]$. A vector $x \in X_w$ represents the set of all subsets of $V(G)$ such that the number of vertices of w-weight i in the set equals x_i.

Note that the number of vectors in X_w is bounded by $\prod_{i=1}^{k}(|w^{-1}(i)| + 1)$, which, by Lemma 4, is in turn bounded by $(\hat{n}/k + 1)^k = O((\hat{n}/k)^k)$, for each function w. For each vector $x \in X_w$, the algorithm then checks whether the corresponding sets are of the right weight, that is, whether $\sum_{i=1}^{k} ix_i = t$ if and only if the vector encodes a set of maximal stable sets. This latter condition can be verified in time $O(k^2)$ using the quotient graph $\mathcal{Q}(G)$ (see [8] for details).

The running time of this algorithm is thus

$$O\left(n + m + (k\hat{n})^k + \frac{\hat{n}^k}{k}\left(k + \left(\frac{\hat{n}}{k}\right)^k k^2\right)\right).$$

This expression simplifies to

$$O(n + m + \hat{n}^{2k}k^{1-k}) = O(n + m + \max\{n^{2k}k^{1-k}, k^{3k+1}\}),$$

as desired.

We remark that, in case of a prescribed target value t, the above algorithm can be modified in an obvious way to accept only those equistable functions under which all maximal stable sets have total weight t. This completes the proof. □

4   An $O(t^2)$-vertex Kernel for the TARGET-t EQUISTABILITY Problem

Given a graph G, the following reduction rule is specified by a positive integer r as a parameter.

r-Clique Reduction. If a clique class C contains more than r vertices, delete from C all but r vertices.

The following lemma shows why r-Clique Reduction rule is safe for both problems, TARGET-t EQUISTABILITY and k-EQUISTABILITY.

Lemma 5. *Let G be a graph, $T \subseteq \mathbb{N}$ a finite set, C a clique class of G with $|C| > r$ where $r := \max T$, and k a positive integer. Then, for every $t \in T$, graph G is target-t k-equistable if and only if G' is target-t k-equistable, where G' is a graph obtained after the r-Clique Reduction rule has been applied to G with respect to the clique class C.*

Proof. Let $t \in T$. First assume that G is target-t k-equistable, say with a k-equistable structure (w, t). It is immediate that the restriction w' of w to $V(G')$ yields a k-equistable structure (w', t) of G'. Therefore G' is target-t k-equistable.

Now assume that G' is target-t k-equistable, with a k-equistable structure (w', t). We define a function $w : V(G) \to \{1, \ldots, k\}$ by extending w' to the set $V(G)$. Indeed, we simply put $w(u) := w'(u)$ for all $u \in V(G')$, and $w(u) := w(v)$ for all $u \in C \setminus V(G')$ where $v \in C \cap V(G')$. The choice of $v \in C \cap V(G')$ is arbitrary, since w' is constant on $C \cap V(G')$ by Lemma 3.

We claim that (w, t) is an equistable structure of G. To show this, pick an arbitrary maximal stable set X of G. Then $|X \cap C| \leq 1$, and so we may assume that $X \subseteq V(G')$. Clearly X is a maximal stable set of G', and so $w'(X) = t$. Therefore $w(X) = t$.

Conversely, let $X \subseteq V(G)$ be a set with $w(X) = t$. Since $w(X \cap C) \leq w(X) = t$, we have $|X \cap C| \leq t$. As w is constant on C and $|C| > \max T \geq t$, we may w.l.o.g. assume that $X \subseteq V(G')$. Hence, $w'(X) = w(X) = t$, and so X is a maximal stable set of G'. Thus, X is a maximal stable set of G which completes the proof. □

In particular, r-Clique Reduction rule is safe for the TARGET-t EQUISTABILITY problem. This is seen by putting $k := t$ and $T = \{t\}$ in the statement of Lemma 5.

Theorem 4. *The TARGET-t EQUISTABILITY problem admits a kernel of at most t^2 vertices, computable in linear time. Moreover, there is an $O(t^{3t+1} + m + n)$ time algorithm to solve the TARGET-t EQUISTABILITY problem, given a graph with n vertices and m edges.*

Proof. Let G be a graph on n vertices and m edges. Using one of the linear time algorithms for modular decomposition [2,12,19], we can compute $\Pi(G)$ and $\pi(G)$ in linear time. If $\pi(G) > t$, then we conclude that G is not target-t equistable, by Corollary 1. Similarly, if there exists a stable set class S with $|S| > t$, then we conclude that G is not target-t equistable. Also, we can apply r-Clique Reduction rule with parameter t, to every clique class, in linear time. Afterward, the graph has at most t^2 vertices, which proves the first statement of the theorem.

Our FPT algorithm works as follows. First we compute in time $O(m + n)$ a kernel G' with $n' \leq t^2$ many vertices. Then we apply Theorem 3 to check whether G' is target-t equistable. For this, we can put $k := t$ and decide whether G' is k-equistable with target t. We thus obtain a running time of $O(|V(G')| + |E(G'| + \max\{n'^{2k} k^{1-k}, k^{3k+1}\}) = O(t^{3t+1})$. □

5   An $O(k^5)$-vertex Kernel for the k-Equistability Problem

This section is devoted to the proof of the following result.

Theorem 5. *The k-Equistability problem admits an $O(k^5)$-vertex kernel, computable in linear time. Moreover, there is an $O(k^{9k+1} + m + n)$ time algorithm to solve the k-Equistability problem, given a graph with n vertices and m edges.*

Proof. Let us first prove that the second statement follows from the first one. Assume that we can compute an $O(k^5)$-vertex kernel for the k-Equistability problem in linear time. By Theorem 3, we can then decide whether this kernel is k-equistable in time $O(k^{9k+1})$.

We now turn to the construction of the $O(k^5)$-vertex kernel. In case of a no-instance, our algorithm simply returns a non-equistable graph, say the 4-vertex path P_4. In what follows, we will assume that the input graph G satisfies $\pi(G) \leq k$, since otherwise G is not k-equistable, by Lemma 2. The following claim is the main step of our kernelization.

Claim 1. *If there exist two distinct twin classes X and Y such that one of them is a stable set and $\min\{|X|, |Y|\} \geq k(k+1)$, then G is not k-equistable.*

Proof. Suppose for a contradiction that G is k-equistable, with an equistable weight function $w : V(G) \to [k]$, and that there exist two distinct twin classes X and Y with $\min\{|X|, |Y|\} \geq k(k+1)$ such that X is a stable set. If the set $X \cup Y$ is contained in every maximal stable set of G, then $X \cup Y$ forms a twin class, a contradiction. Thus, we may assume without loss of generality that there exists a maximal stable set S of G such that $X \subseteq S$ and $Y \not\subseteq S$.

Recall that Y is either a clique class or a stable set class. Since every clique intersects every stable set in at most one vertex and every stable set class is either entirely contained in S or disjoint from it, the fact that $Y \not\subseteq S$ implies $|Y \cap S| \leq 1$. Let $i, j \in [k]$ be weights such that $|\{x \in X : w(x) = i\}| \geq k + 1$, and $|\{y \in Y : w(y) = j\}| \geq k + 1$. Since $j \leq k$, there exists a set X' of j vertices in X of weight i. Since $i \leq k$ and $|Y \cap S| \leq 1$, there exists a set Y' of i vertices in Y of weight j such that $Y' \cap S = \emptyset$. Then, the set $S' = (S \setminus X') \cup Y'$ is not a stable set, since otherwise by Observation 2 the set $S \cup Y$ would be a stable set properly containing S, contrary to the maximality of S. Note that $w(S') = w(S)$, contradicting the assumption that w is an equistable weight function of G.   □

We consider the following two cases.

Case 1. *Every twin class X with $|X| \geq k(k+1)$ is a clique class.*

In this case, every stable set class has less than $k(k+1)$ vertices, which implies that every maximal stable set of G contains at most $k(k+1)$ vertices from each twin class and is thus of total size at most $k^2(k+1)$. This implies that in every k-equistable structure (w, t) of G, we have $t \leq k^3(k+1)$.

We now perform r-Clique Reduction rule from Sect. 4 with $r := k^3(k+1)$. By Lemma 5 applied with $T = [r]$ and k, the application of r-Clique Reduction rule is safe. When the rule can no more be applied, we have a graph G' with at most k twin classes, each of size at most $k^3(k+1)$. We are done since $|V(G')| = O(k^5)$.

Case 2. *There exists a stable set twin class X with $|X| \geq k(k+1)$.*

By Claim 1, we may assume that X is the unique twin class of size at least $k(k+1)$ (since otherwise G is not k-equistable).

Note that $V(G) \setminus X$ contains at most $k-1$ twin classes, each containing less than $k(k+1)$ vertices, hence $|V(G) \setminus X| \leq (k-1)k(k+1) \leq k^3$.

Suppose first that X corresponds to an isolated vertex in the quotient graph $Q(G)$. If $|X| < k^5$, then $|V(G)| < k^5 + k^3 = O(k^5)$ and we are done.

So suppose that $|X| \geq k^5$.

Claim 2. *G is k-equistable if and only if it admits a k-equistable function that is constant on X.*

Proof. The if part being trivial, assume that G is k-equistable, and let (w, t) be a k-equistable structure of G. Let $i \in \{1, \ldots, k\}$ be such that $|X_i^w| \geq k^4$, where $X_i^w = \{v \in X : w(v) = i\}$. Now we define a weight function w' that equals w outside X, and is constantly i on X. We claim that w' is a k-equistable function of G. Clearly, w' is bounded by k. Under w', all maximal stable sets of G have weight $t' := t - w(X) + w'(X)$.

The only possible problem is that $w'(S) = t'$ for some vertex set S that is not a maximal stable set of G. In this case, we claim that $r := |X \setminus S| \leq k^4$. To see this, suppose $r > k^4$. Since $|S \setminus X| \leq |V(G) \setminus X| \leq k^3$, we get $w'(S \setminus X) \leq k^4$. Therefore $w'(S) = w'(X) - w'(X \setminus S) + w'(S \setminus X) \leq i(|X| - r) + k^4$. But $k^4 < ir$, since $i \geq 1$ and $r > k^4$. Thus $i(|X| - r) + k^4 < i(|X| - r) + ir = i|X| \leq t'$, a contradiction.

So, $r \leq k^4$, and since $k^4 \leq |X_i^w|$ and w' is constant on X we may assume that $X \setminus S \subseteq X_i^w$. But this yields

$$
\begin{aligned}
w(S) &= w'(S) - w'(X \cap S) + w(X \cap S) \\
&= t' - i(|X| - r) + w(X \cap S) \\
&= t' - i(|X| - r) + w(X \cap S) - ir + ir \\
&= t' - i|X| + (w(X \cap S) + ir) \\
&= t' - w'(X) + (w(X \cap S) + w(X \setminus S)) \\
&= t' - w'(X) + w(X) = t.
\end{aligned}
$$

A contradiction.   □

According to Claim 2, it suffices to test if G is k-equistable by considering all possible functions $w : V(G) \to [k]$ that are constant on X, and test for each of them whether it is a k-equistable function.

Before that, we reduce size of X. For this, we compute a graph G' from G by deleting all but k^4 many vertices from X. Note that, since X is a twin class, G' is unique up to isomorphism.

Claim 3. *G is k-equistable if and only if G' is k-equistable.*

Proof. Let $X' := X \cap V(G')$ and $Y' := V(G') \setminus X'$.

First we assume that G is k-equistable, say with an equistable structure (w, t). By Claim 2, we may assume that w is constant on X, say $w|_X \equiv i$. We now consider the weight function $w' := w|_{V(G')}$ with target value $t' := t - i|X \setminus X'|$, and claim that (w', t') is a k-equistable structure of G'. Since every maximal stable set of G (resp., G') contains X (resp., X') as a subset, it is straightforward that every maximal stable set of G' has weight t'. Suppose that there is a set $S \subseteq V(G')$ with $w(S) = t'$ that is not a maximal stable set of G'. Then the set $S \cup (X \setminus X')$ has total weight t, but is not a maximal stable set of G, a contradiction. This proves that G' is k-equistable.

Now we assume that G' is k-equistable, say with an equistable structure (w', t'). By Claim 2 applied to G', we may assume that w' is constant on X', say $w'|_{X'} \equiv i$. Consider the weight function $w : V(G) \to [k]$ defined as $w(x) = w'(x)$ for all $x \in V(G')$ and $w(x) = i$ for all $x \in X \setminus X'$ with target value $t := t' + i|X \setminus X'|$. We claim that (w, t) is a k-equistable structure of G. Again it is straightforward that any maximal stable set of G has weight t. Suppose that there is a set $S \subseteq V(G)$ with $w(S) = t$ that is not a maximal stable set of G.

Recall that $|Y'| \le (k-1)k(k+1) \le k^3$ and consequently $w(Y') \le k^4$. If $|X \setminus S| > k^4$, we thus obtain

$$
\begin{aligned}
w(S) &\le w(Y') + i|X| - i(k^4 + 1) \\
&\le k^4 + i|X| - (k^4 + 1) \\
&= i|X| - 1 \\
&< w(X) \le t,
\end{aligned}
$$

a contradiction. Thus, $|X \setminus S| \le k^4$, and so we may assume that $X \setminus S \subseteq X'$. Let $S' := S \cap V(G')$. Then $w'(S') = w(S) - i|X \setminus X'| = t'$, but S' is not a maximal stable set of G'. This is contradictory, and so G is k-equistable.   □

Proof. By Claim 3, it suffices to check whether G' is k-equistable. Since $|V(G')| \le k^4 + k^3 = O(k^4)$, we are done.

Now, suppose that X corresponds to a non-isolated vertex in the quotient graph $Q(G)$. Then, there exists a twin class Y that sees X. Let S be a maximal stable set of G containing a vertex of Y. Then, $S \subseteq V(G) \setminus X$. Since $|V(G) \setminus X| \le k^3$, we have in particular that $|S| \le k^3$.

If $|X| > k|S|$, then for every k-equistable function w of G and every maximal stable set, say S', such that $X \subseteq S'$, we have $w(S') \ge |X| > k|S| \ge w(S)$, hence G is not k-equistable.

If $|X| \le k|S|$, then $|V(G)| \le (k+1)k^3 = O(k^4)$.

Since it is clear that the above algorithm runs in time $O(n + m)$, the proof is complete.   □

6 Future Work

Several open problems surrounding our work remain, some of which we want to mention here in order to stimulate research on this topic.

Firstly, we believe it is NP-hard to determine, given a graph G and an integer k, whether G is k-equistable. It would be satisfying to see this proven, especially for the purpose of this paper. As mentioned in the introduction, the smallest such k (if existing) might have to be exponential in the number of vertices of G [14], which might serve as a hint for the hardness of this problem.

The analogous question is open also for the problem of TARGET-t EQUISTA-BILITY: what is the computational complexity of determining, given a graph G and an integer t, whether G is target-t equistable? Again, the smallest such t (if existing) might have to be exponential in the number of vertices of the input graph [14].

A different computational problem in this context would be the following: given a graph G and a number k, does it admit an equistable weight function using at most k different weights? Here, both the parameterized and classical complexity are unknown. Although we did not study this problem in depth, our impression is that it should be NP-hard, but FPT when parameterized by k. In view of the results of the present paper, there might very well be a polynomial kernel for this problem. Another problem that seems similar at first sight is whether equistability is FPT when parameterized by $\pi(G)$, the number of twin-classes of G.

Apart from these recognition problems, it is apparently open whether the maximum stable set problem is FPT in the class of equistable graphs. Here we do at least know that this problem is APX-hard in this class [14].

References

1. Chvátal, V., Hammer, P.L.: Aggregation of inequalities in integer programming. In: Proceedings of the Workshop on Studies in Integer Programming, Bonn (1975). Ann. of Discrete Math., **1**, 145–162. North-Holland, Amsterdam (1977)
2. Cournier, A., Habib, M.: A new linear algorithm for modular decomposition. In: Tison, S. (ed.) CAAP 1994. LNCS, vol. 787, pp. 68–84. Springer, Heidelberg (1994)
3. Downey, R.G., Fellows, M.R.: Fundamentals of Parameterized Complexity. Texts in Computer Science. Springer, London (2013)
4. Kloks, T., Lee, C.-M., Liu, J., Müller, H.: On the recognition of general partition graphs. In: Bodlaender, H.L. (ed.) WG 2003. LNCS, vol. 2880, pp. 273–283. Springer, Heidelberg (2003)
5. Korach, E., Peled, U.N.: Equistable series-parallel graphs. Discrete Appl. Math. **132**(1–3), 149–162 (2003). Stability in graphs and related topics
6. Korach, E., Peled, U.N., Rotics, U.: Equistable distance-hereditary graphs. Discrete Appl. Math. **156**(4), 462–477 (2008)
7. Levit, V.E., Milanič, M.: Equistable simplicial, very well-covered, and line graphs. Discrete Appl. Math. **165**, 205–212 (2014)
8. Levit, V.E., Milanič, M., Tankus, D.: On the recognition of k-equistable graphs. In: Golumbic, M.C., Stern, M., Levy, A., Morgenstern, G. (eds.) WG 2012. LNCS, vol. 7551, pp. 286–296. Springer, Heidelberg (2012)

9. Mahadev, N.V.R., Peled, U.N.: Threshold graphs and related topics. Ann. Discrete Math. **56**. North-Holland Publishing Co., Amsterdam (1995)

10. Mahadev, N.V.R., Peled, U.N., Sun, F.: Equistable graphs. J. Graph Theor. **18**(3), 281–299 (1994)

11. McAvaney, K., Robertson, J., DeTemple, D.: A characterization and hereditary properties for partition graphs. Discrete Math. **113**(13), 131–142 (1993)

12. McConnell, R.M., Spinrad, J.P.: Modular decomposition and transitive orientation. Discrete Math. **201**(1–3), 189–241 (1999)

13. Miklavič, Š., Milanič, M.: Equistable graphs, general partition graphs, triangle graphs, and graph products. Discrete Appl. Math. **159**(11), 1148–1159 (2011)

14. Milanič, M., Orlin, J., Rudolf, G.: Complexity results for equistable graphs and related classes. Ann. Oper. Res. **188**, 359–370 (2011)

15. Milanič, M., Rudolf, G.: Structural results for equistable graphs and related classes. RUTCOR Research Report 25-2009 (2009)

16. Milanič, M., Trotignon, N.: Equistarable graphs and counterexamples to three conjectures onequistable graphs (2014). arXiv:1407.1670 [math.CO]

17. Payan, C.: A class of threshold and domishold graphs: equistable and equidominating graphs. Discrete Math. **29**(1), 47–52 (1980)

18. Peled, U.N., Rotics, U.: Equistable chordal graphs. Discrete Appl. Math. **132**(1–3), 203–210 (2003). Stability in graphs and related topics

19. Tedder, M., Corneil, D.G., Habib, M., Paul, C.: Simpler linear-time modular decomposition via recursive factorizing permutations. In: Aceto, L., Damgård, I., Goldberg, L.A., Halldórsson, M.M., Ingólfsdóttir, A., Walukiewicz, I. (eds.) ICALP 2008, Part I. LNCS, vol. 5125, pp. 634–645. Springer, Heidelberg (2008)

Beyond Classes of Graphs with "Few" Minimal Separators: FPT Results Through Potential Maximal Cliques

Mathieu Liedloff, Pedro Montealegre, and Ioan Todinca[✉]

Univ. Orléans, INSA Centre Val de Loire, LIFO EA 4022, Orléans, France
{mathieu.liedloff,pedro.montealegre,ioan.todinca}@univ-orleans.fr

Abstract. In many graph problems, like LONGEST INDUCED PATH, MAXIMUM INDUCED FOREST, etc., we are given as input a graph G and the goal is to compute a largest induced subgraph $G[F]$, of treewidth at most a constant t, and satisfying some property $\mathcal{P}$. Fomin et al. [12] proved that this generic problem is polynomial on the class of graphs $\mathcal{G}_{\text{poly}}$, i.e., the graphs having at most poly(n) minimal separators for some polynomial poly, when property $\mathcal{P}$ is expressible in counting monadic second order logic (CMSO).

Here we consider the class $\mathcal{G}_{\text{poly}} + kv$, formed by graphs of $\mathcal{G}_{\text{poly}}$ to which we may add a set of at most k vertices with arbitrary adjacencies, called *modulator*. We prove that the generic optimization problem is fixed parameter tractable on $\mathcal{G}_{\text{poly}} + kv$, with parameter k, if the modulator is also part of the input. The running time is of type $\mathcal{O}\left(f(k+t,\mathcal{P}) \cdot n^{t+5} \cdot (\text{poly}(n)^2)\right)$, for some function f.

1 Introduction

Many classical optimization problems on graphs, e.g., MAXIMUM INDEPENDENT SET, MAXIMUM INDUCED FOREST (whose optimal solution is the complement of a MINIMUM FEEDBACK VERTEX SET), LONGEST INDUCED PATH and MAXIMUM INDUCED MATCHING consist in finding a maximum induced subgraph $G[F]$ of the input graph G such that $G[F]$ has a tree-like structure (i.e., the treewidth is bounded by a constant) and satisfies some particular property $\mathcal{P}$ (like being a path, a matching, etc.). All these properties are expressible in Counting Monadic Second Order Logic (CMSO). We do not need in this paper the technical definition of CMSO formulae, for which the reader may refer to [10] or [12]. We only need to keep in mind that many natural properties (connectivity, excluding a fixed minor, etc.) are expressible in CMSO, and the fact that CMSO properties are *regular*, in a sense to be defined in the next section.

Fomin et al. [12] introduced the following generic optimization problem called OPTIMAL INDUCED SUBGRAPH FOR $\mathcal{P}$ AND t, which encompasses those cited above and many others. In this generic problem, t is an integer constant and $\mathcal{P}$ is a property on graphs and vertex sets, expressible in CMSO.

© Springer-Verlag Berlin Heidelberg 2016
E.W. Mayr (Ed.): WG 2015, LNCS 9224, pp. 499–512, 2016.
DOI: 10.1007/978-3-662-53174-7_35

OPTIMAL INDUCED SUBGRAPH FOR $\mathcal{P}$ AND t

Input: A graph $G = (V, E)$
Output: A pair (F, X) of vertex subsets $X \subseteq F \subseteq V$ such that
  – $\mathrm{tw}(G[F]) \leq t$,
  – $\mathcal{P}(G[F], X)$ is true, and
  – X is of maximum size under these constraints.

In the problems that we have mentioned, the vertex set X is equal to F. Nevertheless, set X allows to optimize other criteria than the size of the induced subgraph. E.g., in the INDEPENDENT $\mathcal{H}$-PACKING PROBLEM [8], we are given a fixed family of graphs $\mathcal{H}$, and the goal is to find an induced subgraph $G[F]$ with a maximum number of connected components such that each of its components is isomorphic to an element of $\mathcal{H}$. For this problem, property $\mathcal{P}$ expresses the constraints on the components of $G[F]$ and the fact that X intersects each such component in exactly one vertex.

Consider a polynomial poly, and let $\mathcal{G}_{\mathrm{poly}}$ be the class of graphs such that, for any $G \in \mathcal{G}_{\mathrm{poly}}$, graph G has at most $\mathrm{poly}(n)$ minimal separators. (As usually, we denote by n and m the number of vertices, respectively of edges of graph G.) Fomin et al. [12] proved that, for any constant t and any CMSO property $\mathcal{P}$, problem OPTIMAL INDUCED SUBGRAPH FOR $\mathcal{P}$ AND t is polynomial-time solvable on class $\mathcal{G}_{\mathrm{poly}}$.

The approach is based on the notion of *potential maximal clique*. Given an arbitrary graph $G = (V, E)$, a *minimal triangulation* $H = (V, F)$ is a minimal chordal supergraph of G (recall that a graph is chordal if it contains no induced cycle with four or more vertices). A *potential maximal clique* of G is a vertex subset Ω inducing a maximal clique in some minimal triangulation of G. Potential maximal cliques are strongly related to minimal separators. Graphs in $\mathcal{G}_{\mathrm{poly}}$ have $\mathcal{O}(n \cdot (\mathrm{poly}(n))^2)$ potential maximal cliques [7], and the set of all these objects can be enumerated in polynomial time. The algorithm of [12] takes as input all the potential maximal cliques of the input graph. Then it proceeds by dynamic programming on potential maximal cliques, constructs the induced subgraph $G[F]$ and in the meantime applies Courcelle's theorem [9], in the version proposed by Borie, Parker and Tovey [5], for testing CMSO properties on graphs on bounded treewidth. Altogether, this solves the generic optimization problem. Many graph classes, e.g., *weakly chordal*, *circle*, *polygon-circle* or *circular-arc* graphs are known to be in $\mathcal{G}_{\mathrm{poly}}$ for some particular polynomial poly. Therefore, the generic problem and all its particular instances are polynomial on all these classes. We refer to [12] for further discussions on graph classes and applications of the problem.

Our results. Our goal is to study the problem from a parameterized perspective, for classes of graphs with "few" minimal separators, to which we are allowed to add k vertices with arbitrary adjacencies. Let $\mathcal{G}_{\mathrm{poly}} + kv$ denote the class of graphs $G = (V, E)$ containing a vertex subset $M \subseteq V$ of size at most k, such that $G - M \in \mathcal{G}_{\mathrm{poly}}$. The set M is called the *modulator* of G.

Let OPTIMAL INDUCED SUBGRAPH FOR $\mathcal{P}$ AND t ON $\mathcal{G}_{\text{poly}} + kv$ be the problem OPTIMAL INDUCED SUBGRAPH FOR $\mathcal{P}$ AND t on the graph class $\mathcal{G}_{\text{poly}} + kv$, with parameter k. Moreover, we assume that the input graph G is given together with a modulator M (see also Sect. 4 for discussions on this point). Our main result is that problem OPTIMAL INDUCED SUBGRAPH FOR $\mathcal{P}$ AND t ON $\mathcal{G}_{\text{poly}} + kv$ is fixed-parameter tractable (FPT), i.e., there is an algorithm solving the problem in time $f(k) \cdot n^{O(1)}$ for some function f. More specifically, the running time is of type $\mathcal{O}\left(f(k+t,\mathcal{P}) \cdot n^{t+5} \cdot (\text{poly}(n)^2)\right)$, where function f depends on property $\mathcal{P}$ and on $k+t$ (see also the Conclusion section for further discussions).

Theorem 1. *Problem* OPTIMAL INDUCED SUBGRAPH FOR $\mathcal{P}$ AND t ON $\mathcal{G}_{\text{poly}} + kv$ *with parameter k is fixed-parameter tractable, when a modulator is also part of the input.*

A similar result was obtained by Fomin and the authors of this article [11], on graphs with vertex cover at most k, i.e., formed by an independent set plus at most k vertices. In this class, the number of potential maximal cliques is $4^k \cdot n^{O(1)}$, hence the algorithm of [12] can be used as is. But, as shown in [11], even for the class of graphs formed by a tree plus one vertex, or an induced matching plus two vertices, the number of potential maximal cliques may be exponential in n, therefore we need another approach.

A natural idea is to "guess" how the optimal solution intersects with the modulator M. But we still need to carefully express how the rest of the solution intersects with the graph $G - M$. Think, e.g., of the longest induced path problem: we need to make sure that the solution restricted to $G - M$ forms indeed a connected path with the selected vertices of the modulator. Our algorithm extends, in a non-trivial way, the one of [12] in order to handle this situation.

We also point out that several authors considered classes of graphs of similar flavor, e.g. *Chordal* + kv [19] and *Split* + kv [18], providing both FPT and hardness results for various problems, parameterized by k.

2   Preliminaries

Treewidth, minimal triangulations and potential maximal cliques. Given a graph $G = (V, E)$, we denote by n the number of its vertices and by m the number of its edges. $G[C]$ denotes the subgraph of G induced by a vertex subset C, and $N(C)$ is the neighborhood of C in G. We say that a set of vertices C is a connected component of G if $G[C]$ is connected and C is inclusion-maximal for this property. Given a set $S \subseteq V$, let $G - S$ denote the graph $G[V \setminus S]$. If there are two distinct connected components C and D of $G - S$ such that $N(C) = N(D) = S$, we say that S is a *minimal separator* of G. The set of all minimal separators of G is denoted Δ_G.

A *tree decomposition* of graph $G = (V, E)$ is a pair $(\mathcal{T}, \mathcal{X})$, where $\mathcal{T}$ is a tree and $\mathcal{X}$ are vertex subsets of G, called *bags*. Moreover, each node i of the tree corresponds to a bag $X_i \in \mathcal{X}$, the bags cover all vertices and all edges of G, and for each vertex x of G, the set of nodes $\{i \mid x \in X_i\}$ form a connected subtree of $\mathcal{T}$. The *width* of the decomposition is $\max\{|X_i| - 1 \mid X_i \in \mathcal{X}\}$. Finally, the *treewidth* of G, denoted $tw(G)$, is the minimum width among all mobile tree decompositions of G.

A graph H is *chordal* if it has no induced cycle with four or more vertices. Let $G = (V, E)$ be an arbitrary graph. We say that $H = (V, F)$ is a *triangulation* of G if H is a chordal super-graph of G (i.e., $E \subseteq F$). If, moreover, H is inclusion-minimal for this property, then H is a *minimal triangulation* of G. It is well-known that chordal graphs have tree decompositions whose bags correspond to maximal cliques (see, e.g., [15]). The treewidth of G is also equal to the minimum integer w such that G has a (minimal) triangulation H, and each clique of H has at most $w + 1$ vertices.

Proposition 1 (respecting triangulations [13]). *Consider an arbitrary graph $G = (V, E)$ and let $F \subseteq V$ be a set of vertices. For any minimal triangulation T_F of $G[F]$, there is a minimal triangulation T_G of G such that T_F is an induced subgraph of T_G.*

A *potential maximal clique* of G is a vertex subset Ω such that Ω induces a maximal clique in some minimal triangulation H of G. The set of all potential maximal cliques of G is denoted Π_G. E.g., if G is a cycle, then its potential maximal cliques are exactly the triples of vertices. See also [6] for a characterization of potential maximal cliques.

If Ω is a potential maximal clique, then the neighborhoods of the components of $G - \Omega$ are exactly the minimal separators of G, contained in Ω.

Given a polynomial poly, let $\mathcal{G}_{\text{poly}}$ denote the class of graphs having at most $poly(n)$ minimal separators. The minimal separators Δ_G and the potential maximal cliques Π_G of these graphs can be listed in polynomial time, by [1] and [7] respectively.

A pair (S, C) such that S is a minimal separator of G and C is a component such that $N(C) = S$ is called a *full block* associated to S. For convenience, we will also consider the empty set as being a minimal separator of $G = (V, E)$, and the pair $(\emptyset, V)$ is considered as a full block. Let (S, C) be a full block and Ω be a potential maximal clique with $S \subset \Omega \subseteq S \cup C$. The triple (S, C, Ω) is called a *good triple*. Our dynamic programming is based on full blocks and good triples, which is again polynomially bounded on class $\mathcal{G}_{\text{poly}}$.

Terminal recursive graphs and regular properties. Graphs of bounded treewidth can be defined recursively, based on a graph grammar. Let w be a non-negative integer. A *w-terminal graph* is a triple (V, T, E), where (V, E) is a graph and T is a *totally ordered* subset of V, of size at most w. The vertices of T are called the *terminals* of the graph. Since T is totally ordered, we can speak of the ith terminal, for $i \le |T|$.

The class of *w-terminal recursive graphs* is defined by the following operations. A *base graph* is a w-terminal recursive graph of the form (V, T, E) with $T = V$. Hence it has at most w vertices, all of them being terminals.

The *gluing* operation takes two disjoint w-terminal recursive graphs $G_1 = (V_1, T_1, E_1)$ and $G_2 = (V_2, T_2, E_2)$ and creates a new graph $G = glue_m(G_1, G_2)$, depending on a matrix m. The matrix m has two rows and at most w columns, with elements in $\{0, 1 \ldots, w\}$. The gluing operation takes the disjoint union of G_1 and G_2, and then identifies the terminal number i in G_1 (resp. in G_2) to terminal m_{1i} (resp. m_{2i}) in G. Each terminal of G_1 (resp. G_2) is mapped on at most one terminal of G. We take $m_{ji} = 0$, for $j \in \{1, 2\}$, if terminal number i in G_j does not exist or it is not mapped on any terminal of G.

The *forget* operation takes a w-terminal recursive graph $G_1 = (V, T, E)$ and creates the graph $G = forget_m(G_1)$ with $G = (V, T', E)$ such that T' is a subset of T. The matrix m has only one row and $|T|$ columns, and m_{1i} specifies as before that the ith terminal of G_1 is mapped on terminal m_{1i} of G. The mapping is injective, and if $m_{1i} = 0$ then the ith terminal of G_1 is removed, in G, from the set of terminals.

We point out that the number of possible different matrices and hence of different operations is bounded by a function on w.

Proposition 2 (see [2,12]). *Graph $H = (V, T, E)$ is $(w+1)$-terminal recursive if and only if there exists a tree decomposition of $G = (V, E)$, of width at most w, having T as one of its bags. Hence the grammar of $(w+1)$-terminal recursive graphs constructs exactly the graphs of treewidth at most w (see [2,12]).*

Let $\mathcal{P}(G, X)$ be a property assigning to each graph G and vertex subset X of G a boolean value. We extend the gluing and forget operations to pairs (G, X) in the natural way (see, e.g., [5,12]). In particular, when we perform a gluing on (G_1, X_1) and (G_2, X_2), the result is a pair (G, X) where X is obtained by the the gluing of X_1 and X_2. Therefore the intersections of sets X_1 and X_2 with the terminals of G_1 and respectively G_2 must be coherent with the gluing, in the sense that if two terminals x_1 of G_1 and x_2 of G_2 are identified in G, then we either have $x_1 \in X_1$ and $x_2 \in X_2$, or we have $x_1 \notin X_1$ and $x_2 \notin X_2$.

Definition 1 (regular property). *Property $\mathcal{P}$ is called* regular *if, for any value w, we can associate a finite set $\mathcal{C}$ of classes and a homomorphism function h, assigning to each w-terminal recursive graph G and to each vertex subset X a class $h(G, X) \in \mathcal{C}$ such that:*

1. *If $h(G_1, X_1) = h(G_2, X_2)$ then $\mathcal{P}(G_1, X_1) = \mathcal{P}(G_2, X_2)$.*
2. *For each gluing operation $glue_m$ there exists a function $\odot_{glue_m} : \mathcal{C} \times \mathcal{C} \to \mathcal{C}$ such that, for any two pairs (G_1, X_1) and (G_2, X_2),*

$$h(glue_m((G_1, X_1), (G_2, X_2))) = \odot_{glue_m}(h(G_1, X_1), h(G_2, X_2))$$

and for each operation $forget_m$ there is a function $\odot_{forget_m} : \mathcal{C} \to \mathcal{C}$ such that, for any pair (G, X),

$$h(forget_m(G, X)) = \odot_{forget_m}(h(G, X)).$$

The first condition separates the classes into *accepting* ones (i.e., classes $c \in \mathcal{C}$ such that $h(G, X) = c$ implies that $\mathcal{P}(G, X)$ is true) and *rejecting* ones (s.t. $h(G, X) = c$ implies that $\mathcal{P}(G, X)$ is false). In full words, the second condition states that, if we perform a *glue* (resp. *forget*) operation on two graphs (resp. one graph) and corresponding vertex subsets, the homomorphism class of the result can be obtained from the homomorphism classes of the graphs on which these operations are applied. Therefore, if a w-terminal recursive graph is given together with its expression in this grammar, and if moreover we know how to compute the classes of the base graph, then the homomorphism class of the whole graph, for a regular property $\mathcal{P}$, can be obtained by dynamic programming. We simply need to parse the expression from bottom to top and, at each node, we compute the class of the corresponding sub-expression thanks to the second condition of regularity. At the root, the property is true if and only if we are in an accepting class.

Proposition 3 (Borie, Parker, and Tovey [5], Courcelle [9]). *Any property $\mathcal{P}(G, X)$ expressible by a CMSO formula is regular.*

Moreover, the result of Borie, Parker, and Tovey shows how to compute explicitly the set of classes, the homomorphism function for base graphs as well as the composition functions $\odot_{glue_m}$ and $\odot_{forget_m}$. Altogether, this provides an effective algorithm for checking the property in $O(n)$ time.

The reader may try to express the homomorphism classes and function for his/her favorite CMSO property. Let us consider the property "$G[X]$ is connected". We can choose, as homomorphism $h((V, T, E), X)$, the set of subsets of T, which correspond to the intersections of T with components of $G[X]$. Observe that each such subset T_i of T is encoded by the indices of its elements in the ordered set T. Hence each homomorphism class will be a set of (disjoint) subsets of $\{1, \ldots, w\}$.

We may assume w.l.o.g. that the homomorphism class $c = h(G, X)$, for $G = (V, T, E)$, encodes the intersection of X with the set of terminals. This is not explicitly required by the definition of regular properties, but it can be done since it only costs w bits to encode the number of the terminals contained in X. Therefore we assume there is a function $trm(c, T)$ that, given a homomorphism class c and an ordered set of terminals T returns the unique possible set $X \cap T$, over all pairs $(G = (V, T, E), X)$ mapped to c. Thanks to this function, when we will glue two terminal recursive graphs with their corresponding vertex subsets, we will be able to check that the gluing is coherent.

3   The Algorithm

Our goal is to provide an FPT algorithm for the problem OPTIMAL INDUCED SUBGRAPH FOR $\mathcal{P}$ AND t ON $\mathcal{G}_{\text{poly}} + kv$, thus proving our Main Theorem 1. Recall that in this problem, the input is a graph $G \in \mathcal{G}_{\text{poly}} + kv$, together with a modulator M of size at most k.

We may assume w.l.o.g. that we also have as input the set $\Pi_{G'}$ of potential maximal cliques of graph $G' = G - M$. Indeed these objects can be computed in polynomial time by [7].

The following easy observation is crucial for the correctness of our algorithm.

Lemma 1 (compatibility lemma). *Let G be the input graph, M be a vertex subset, and (F, X) be an optimal solution for $\mathcal{P}$ and t. Let $G' = G - M$ and $F' = F \setminus M$. There is a minimal triangulation $T_{F'}$ of $G'[F']$ of width at most t, and a minimal triangulation $T_{G'}$ of G' respecting $T_{F'}$, i.e., such that $T_{F'}$ is the subgraph induced by F' in $T_{G'}$.*

Proof. Since $G[F]$ is of treewidth at most t, so is its subgraph $G'[F']$. Therefore it exists a minimal triangulation $T_{F'}$ of $G'[F']$ of width at most t. By Proposition 1 applied to graph G' and to $T_{F'}$, there is a minimal triangulation $T_{G'}$ of G', respecting the minimal triangulation $T_{F'}$.   □

We will "guess", by brute force, the intersections of F and X with the modulator M. Let us fix $F^M = F \cap M$ and $X^M = X \cap M$, with $X^M \subseteq F^M$. For each such pair (F^M, X^M), we need to construct $F' = F \setminus M$ and $X' = X \setminus M$ such that the pair $(F, X) = (F' \cup F^M, X' \cup X^M)$ satisfies the constraints and $X' \cup X^M$ is of maximum size under these conditions. (Eventually, the global solution is obtained by trying all the 3^k possible combinations for subsets $X^M \subseteq F^M \subseteq M$.)

Our algorithm will construct X' and F' by dynamic programming on minimal separators and potential maximal cliques. The graph $G[F' \cup F^M]$ has to be of treewidth at most t, and while we construct this graph we also need to check property $\mathcal{P}$ on it. Unfortunately we will not be able to handle $G[F' \cup F^M]$ as a $(t+1)$-terminal recursive graph. Instead, we will see it as a $(t + k^M + 1)$-terminal recursive graph, where $k^M = |F^M|$ (and we will explicitly check that $\mathrm{tw}(G[F]) \leq t$). Informally, while we construct F', we also maintain some information on a tree-decomposition of $G'[F']$, of width at most t. To this decomposition, we simply add the whole set F^M in each bag, hence obtaining a tree-decomposition of width at most $t + k^M$ of $G[F \cup F^M]$.

Now, on the $(t + k^M + 1)$-terminal recursive graph $G[F' \cup F^M]$ having $W \cup F^M$ as set of terminals for some W (W will be memorized during the dynamic programming), we need to check the property $\mathcal{P}$ but also the fact that the graph is of treewidth at most t. Therefore, let $\mathcal{Q}(H, Y)$ be the property $\mathcal{P}(H, Y) \wedge (\mathrm{tw}(H) \leq t)$. Property $\mathcal{Q}$ is also regular:

Lemma 2. *Let $\mathcal{P}(G, X)$ be a regular property on graphs and vertex sets, and let $\mathcal{Q}(G, X)$ be the property $\mathcal{P}(G, X) \wedge (\mathrm{tw}(G) \leq t)$. Property $\mathcal{Q}$ is regular.*

Proof. As proven by Borie, Parker, and Tovey [5], if two properties $\mathcal{P}(G, X)$ and $\mathcal{P}'(G)$ are regular, then property $\mathcal{Q}(G, X) = \mathcal{P}(G, X) \wedge \mathcal{P}'(G)$ is also regular. In order to observe this, one can simply notice that the couple $(h_{\mathcal{P}}(G, X), h_{\mathcal{P}'}(G))$ formed by the respective classes of $\mathcal{P}$ and $\mathcal{P}'$ directly provides the homomorphism class $h_{\mathcal{Q}}(G, X)$ for property $\mathcal{Q}$.

It remains to argue that property $\mathcal{P}'(G)$ defined by "$\mathrm{tw}(G) \leq t$" is regular. One classical argument is that the class of graphs of treewidth at most t is minor-closed (see, e.g., [3] for a similar discussion). Hence, by the Graph Minors theorem [21], the class is defined by a finite set of forbidden minors, denoted $Obs(t)$. Therefore, $\mathrm{tw}(G) \leq t$ if and only if G has no minor among the graphs of $Obs(t)$. The property that a fixed graph is a minor of G is expressible in CMSO (see, e.g., [5]), hence the property "$\text{tw}(G) \leq t$" is regular by Proposition 3. In order to turn this argument into a completely constructive one, we must build the obstruction set $Obs(t)$ for graphs of treewidth at most t. This can be done by brute force, thanks to the result of Lagergren [17] showing that such obstructions are of size (number of vertices) at most $f(t)$, for some function doubly exponential in t^5. Hence, one could enumerate all the graphs with number of vertices bounded by this function, test the ones of treewidth strictly larger than t and extract the minimal ones w.r.t. the minor relation. The output is precisely $Obs(t)$.

For our purpose, a better alternative is provided by the celebrated algorithm of Bodlaender and Kloks [4]. This algorithm takes as input a graph of treewidth at most w, for some constant w, and decides if this graph has treewidth at most t. Its running time is $O(n)$, and the hidden constant is single exponential in a polynomial in w. Moreover, the algorithm precisely provides an effective way for constructing the homomorphism class of property $\mathcal{P}'(G) = (\text{tw}(G) \leq t)$ for $(w+1)$-terminal recursive graphs. Such a class is addressed in [4] as a *full set of characteristics*. Recall that, in our case, we need to construct these class for $(w+1)$-terminal recursive graphs where $w \leq t + k$, k being the size of the modulator M. Therefore, for computing the homomorphism classes for property $\mathcal{Q}(G, X) = \mathcal{P}(G, X) \wedge (\text{tw}(G) \leq t)$ one can combine the Borie, Parker, and Tovey approach (to obtain $h_{\mathcal{P}}(G, X)$) and the Bodlaender and Kloks approach (to obtain $h_{\mathcal{P}'}(G)$ for $\mathcal{P}'(G) = (\text{tw}(G) \leq t)$), and the couple $(h_{\mathcal{P}}(G, X), h_{\mathcal{P}'}(G))$ is the homomorphism class $h_{\mathcal{Q}}(G, X)$.   □

Our algorithm is an extension of the dynamic programming scheme proposed by Fomin et al. [12], which considers the same optimization problem, without the modulator M (and thus checking property $\mathcal{P}$ instead of $\mathcal{Q}$). For a better understanding we completely describe the new algorithm, trying to follow the same notations as in [12], and we emphasize the points that differ from the result of Fomin et al.

Recall that sets X^M and F^M are fixed, and $X^M \subseteq F^M \subseteq M$. We consider a total order $(v_1, \ldots, v_n)$ on the vertices of $G = (V, E)$. When we speak of a subset T of vertices as a set of terminals, T is considered as the ordered set, with the ordered induced by $(v_1, \ldots, v_n)$ on its vertices.

Definition 2 (partial compatible solution). *Consider a full block (S, C) and a good triple (S, C, Ω) of graph G'. Let $W \subseteq S$ (resp. $W \subseteq \Omega$) a vertex set of size at most $t+1$. Let c be a homomorphism class for property $\mathcal{Q}$ on $(t+k^M+1)$-terminal recursive graphs. We say that a pair (F, X) is a partial solution compatible with (S, C, W, c) (resp. with (S, C, Ω, W, c)) if the following conditions hold:*

1. $X \cap M = X^M$, $F \cap M = F^M$, and $X \subseteq F$.
2. $F \setminus M \subseteq S \cup C$ and $F \cap S = W$ (resp. $F \cap \Omega = W$).
3. $H = (F, W \cup F^M, E(G[F]))$ is a $(t + k^M + 1)$-terminal recursive graph, and the homomorphism class $h_{\mathcal{Q}}(H, X)$ for property $\mathcal{Q}$ is exactly c.

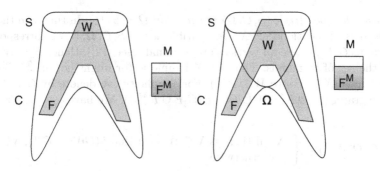

Fig. 1. (a) Partial solutions compatible with (S, C, W, c) (left), and (b) with (S, C, Ω, W, c) (right). Set F is depicted in grey. Note that set W corresponds to $F \cap S$ in the first case, and to $F \cap \Omega$ in the second case.

4. There is a minimal triangulation $T_{F'}$ of $G'[F']$ (here $F' = F \setminus M$) and a minimal triangulation $T_{G'}$ of G' respecting $T_{F'}$, such that S is a minimal separator (resp. Ω is a maximal clique) of $T_{G'}$.

With the same notations as above, let $\alpha(S, C, W, c)$ (resp. $\beta(S, C, \Omega, W, c)$) be the maximum size of X over all partial solutions (F, X) compatible with (S, C, W, c) (resp. (S, C, Ω, W, c)). The situation is depicted in Fig. 1. For simplicity, we did not represent set X. The algorithm orders the full blocks (S, C) by the size of $S \cup C$. It proceeds by dynamic programming on full blocks (S, C) in this order, and on good triples (S, C, Ω), computing all possible values $\alpha(S, C, W, c)$ and $\beta(S, C, \Omega, W, c)$. The outline of Algorithm 1 is the same as in [12], the differences appear in the details of the computations of α and β values.

Algorithm 1. Optimal Induced Subgraph for $\mathcal{P}$ and t on $\mathcal{G}_{\text{poly}} + kv$

Input: graph $G = (V, E)$ and a modulator M of size at most k s.t. $G' = G - M$ is in $\mathcal{G}_{\text{poly}}$; the potential maximal cliques of G'; sets $X^M \subseteq F^M \subseteq M$
Output: sets $X \subseteq F \subseteq V(G)$ such that $G[F]$ has treewidth at most t, $\mathcal{P}(G[F], X)$ is true, $X \cap M = X^M$, $F \cap M = F^M$ and, subject to these constrains, X is of maximum size

1 Order all full blocks (S, C) of G' by inclusion on $S \cup C$;
2 **for** *all full blocks (S, C) in this order* **do**
3   **for** *all good triples (S, C, Ω) of G', all $W \subseteq \Omega$ of size $\leq t + 1$ and all $c \in C$* **do**
4   **if** $\Omega = S \cup C$ **then** Compute $\beta(S, C, \Omega, W, c)$ using Eq. 1;
5   ;
6   **else** Compute $\beta(S, C, \Omega, W, c)$ using Eqs. 3, 4, 5, and 6;
7   ;
8   **for** *all $W \subseteq S$ of size $\leq t + 1$ and all $c \in C$* **do**
9   Compute $\alpha(S, C, W, c)$ using Eq. 2;
10 Compute an optimal solution using Eq. 7;

Base case: the good triple (S, C, Ω) *is such that* $\Omega = S \cup C$. In this case the only possible partial solutions (F, X) compatible with (S, C, Ω, W, c) correspond to base graphs $G[F]$, where all vertices are terminals (see [12]). Hence $F = W \cup F^M$ is also the set of terminals. Thus set X is unique (or might not exist), because we must have $X = trm(c, W \cup F^M)$. For simplicity, we denote by $G[W \cup F^M]$ the base graph $(W \cup F^M, W \cup F^M, E(G[F \cup F^M]))$. We have:

$$\beta(S, C, \Omega, W, c) = \begin{cases} |X| & \text{if there is } X \subseteq W \text{ such that } h(G[W \cup F^M], X) = c \\ -\infty & \text{otherwise} \end{cases}$$

(1)

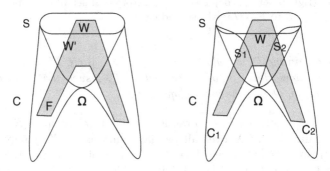

Fig. 2. Computing α form β (left), and β from α (right). Set M is not depicted for simplicity.

Computing α from β. We aim to compute $\alpha(S, C, W, c)$ from β values on good triples of type (S, C, Ω) (see also Fig. 2(a)).

Let (F, X) be an optimal solution compatible with (S, C, W, c) and $F' = F \setminus M$. We denote by H the $(t + k^M + 1)$-terminal recursive graph $G[F]$ with $W \cup F^M$ as set of terminals. By Definition 2, there is a minimal triangulation $T_{F'}$ of F' and a minimal triangulation $T_{G'}$ of G' respecting $T_{F'}$, such that S is a minimal separator of $T_{G'}$. By [6], there is a potential maximal clique Ω of G', inducing a maximal clique in $T_{G'}$, and such that (S, C, Ω) form a good triple of G'. Let $W' = F' \cap \Omega$. The graph H', corresponding to $G[F]$ with set of terminals $W' \cup F^M$, is also a $(t + k^M + 1)$-terminal recursive graph (see Proposition 2, or [12] for full details). Let $c' = h(H', X)$. Note that $H = forget_{W' \cup F^M \to W \cup F^M}(H)$, where the *forget* operation corresponds to the fact that the set of terminals $W' \cup F^M$ is reduced to $W \cup F^M$. Therefore, we have (see [12] for full details):

$$\alpha(S, C, W, c) = \max \beta(S, C, \Omega, W', c'),$$

(2)

where the maximum is taken over potential maximal cliques Ω such that (S, C, Ω) is a good triple, all subsets $W' \subseteq \Omega$ of size at most $t + 1$ such that $W' \cap S = W$ and all classes $c' \in \mathcal{C}$ such that $\odot_{forget_{W' \cup F^M \to W \cup F^M}}(c') = c$.

Computing β from α. Let (S, C, Ω) be a good triple of G. Denote by $C_1, \ldots, C_p$ the components of $G' - \Omega$ contained in C, and let S_i be the neighborhood of C_i in G'. The pairs (S_i, C_i) are also full blocks (see [6] for more details) and they have been processed by the algorithm before (S, C). Our goal is to compute $\beta(S, C, \Omega, W, c)$. Let $W_i = W \cap S_i$, for all $i \in \{1, \ldots p\}$. We will use, as in [12], two intermediate functions γ_i and δ_i.

Let $\delta_i(S, C, \Omega, W, c_i^+)$ denote $\max |X_i|$ over the partial solutions (F_i^+, X_i) compatible with (S, C, Ω, W, c_i^+) and such that $F_i^+ \setminus F^M \subseteq \Omega \cap C_i$. Let H_i^+ be the graph $G[F_i^+]$ with set of terminals $W \cup F^M$. Let also H_i be the graph $H_i^+[S_i \cup C_i \cup M]$, with set of terminals $W_i \cup F^M$. Note that H_i^+ is obtained by gluing H_i with the base graph $G[W \cup F^M]$, with the "canonical" gluing, obtained by identifying the vertices of S_i from both sides. Let $glue_{W_i \cup F^M ; W \cup F^M}$ denote this gluing operation.

$$\delta_i(S, C, \Omega, W, c_i^+) = \max \alpha(S_i, C_i, W_i, c_i) + |trm(c_W, W \cup F^M) \setminus trm(c_i, W_i \cup F^M)|, \quad (3)$$

where the maximum is taken over all $c_i, c_W \in \mathcal{C}$ s.t. $\odot glue_{W_i \cup F^M ; W \cup F^M}(c_i, c_W) = c_i^+$ and $c_W = h(G[W \cup F^M], X_W \cup X^M)$ for some $X_W \subseteq W$. Here $G[W \cup F^M]$ denotes the base graph with terminals $W \cup F^M$.

Notice the part $|trm(c_W, W \cup F^M) \setminus trm(c_i, W_i \cup F^M)|$ in the formula, which avoids the overcounting of the vertices of $X_i \cap S_i$.

These partial solutions (F_i^+, X_i^+) corresponding to $\delta_i(S, C, \Omega, W, c_i^+)$ cannot be glued together in one step, since we are only allowed to glue two graphs at a time. Hence the need of the γ function which allows to add, one by one, the partial solutions to the gluing. Now let $\gamma_i(S, C, \Omega, W, c)$ denote the size of the optimal partial solution compatible with (S, C, Ω, W, c) and contained in $M \cup \Omega \cup C_1 \cdots \cup C_i$. So we only consider the first i components, the partial solution is the union of (F_1^+, X_1^+) to (F_i^+, X_i^+). By definition,

$$\gamma_1(S, C, \Omega, W, c) = \delta_1(S, \Omega, C, W, c) \quad (4)$$

We then compute γ_i, for i from 2 to p as follows.

$$\gamma_i(S, C, \Omega, W, c) = \max \gamma_{i-1}(S, C, \Omega, W, c') + \delta_i(S, \Omega, C, W, c'') - |trm(c', W \cup F^M)|, \quad (5)$$

where the maximum is taken over all $c', c'' \in \mathcal{C}$ s. t. $\odot glue_{W \cup F^M ; W \cup F^M}(c', c'') = c$, where the gluing operation is the canonical gluing, the set of terminals for both arguments being $W \cup F^M$.

By definition of γ_p, we have

$$\beta(S, C, \Omega, W, c) = \gamma_p(S, C, \Omega, W, c). \quad (6)$$

It remains to retrieve the optimal solution for the algorithm. The maximum is taken over all accepting classes c, i.e., classes such that $h(G, X) = c$ implies that $\mathcal{P}(G, X)$:

$$\max \alpha(\emptyset, V, \emptyset, c), \quad (7)$$

We refer to [12] for detailed proofs of correctness and for complexity issues. Altogether, the algorithm takes time $\mathcal{O}(f(t+k^M,\mathcal{P}) \cdot n^{t+4} \cdot |\Pi_{G'}|)$. The function $f(t+k^M,\mathcal{P})$ comes from the application of Proposition 3 on $t+k^M+1$-recursive graphs. By applying Algorithm 1 on all possible subsets $X^M \subseteq F^M \subseteq F$, we have proved Theorem 1.

We point out that our algorithm really needs to keep track of the homomorphism classes of partial solutions (F,X) for property $\mathcal{Q}(G[F],X) = \mathcal{P}(G[F],X) \wedge (\mathrm{tw}(G[F]) \leq t)$. A naïve approach would be to only keep the class $h_\mathcal{P}(G[F],X)$ (for property $\mathcal{P}$) and to reject partial solutions that do not satisfy $\mathrm{tw}(G[F]) \leq t$. We could have two different partial solutions (F,X) and (F',X) for the same part of the graph (e.g., corresponding to the same parameters for function α), such that $h_\mathcal{P}(G[F],X) = h_\mathcal{P}(G[F'],X')$, and both $\mathrm{tw}(G[F])$ and $\mathrm{tw}(G[F'])$ are at most t. Or, it may happen that one of the solution, say (F,X), can be extended into a better one of type $(F \cup F'', X \cup X'')$, while the other cannot because such an extension$(F' \cup F'', X' \cup X'')$ would violate the condition $\mathrm{tw}(G[F' \cup F'']) \leq t$.

Of course our approach, keeping the class $h_\mathcal{Q}(G[F],X)$ for property $\mathcal{Q}$, ensures that if two partial solutions are of the same class, they are equivalent w.r.t. extensions.

4   Conclusion and Discussion

We gave an FPT algorithm for the problem OPTIMAL INDUCED SUBGRAPH FOR $\mathcal{P}$ AND t ON $\mathcal{G}_{\mathrm{poly}} + kv$. The problem encompasses many classical ones [12]. As it will be shown in the full version of the paper, the result can be extended to the classes of graphs $\mathcal{G}_{\mathrm{poly}} - ke$ and $\mathcal{G}_{\mathrm{poly}} + ke$ (i.e., graphs of $\mathcal{G}_{\mathrm{poly}}$ minus or plus ar most k edges), and the generic problem is polynomial on $\mathcal{G}_{\mathrm{poly}} - kv$, if k is small. One of the limits of our algorithm is that we explicitly need the modulator of the input graph. Let us consider the following problem:

DELETION TO $\mathcal{G}_{\mathrm{poly}}$

Input: A graph $G = (V,E)$ and a polynomial poly
Parameter: k
Output: A vertex subset M of size at most k, such that $G - M$ is in $\mathcal{G}_{\mathrm{poly}}$

Our main open question is the existence of an FPT algorithm for problem DELETION TO $\mathcal{G}_{\mathrm{poly}}$. We recall that the problem CHORDAL DELETION is FPT [20], but on the other hand the problem WEAKLY CHORDAL DELETION is $W[2]$-hard [16]. The latter does not rule out the possibility that DELETION TO $\mathcal{G}_{\mathrm{poly}}$ could be FPT. Moreover, even an FPT approximation for DELETION TO $\mathcal{G}_{\mathrm{poly}}$ would allow us to conclude that the problem OPTIMAL INDUCED SUBGRAPH FOR $\mathcal{P}$ AND t is FPT on the class $\mathcal{G}_{\mathrm{poly}} + kv$, without needing to require to have the modulator M as part of the input.

Another direction for improvement concerns the complexity of our algorithm, which is $\mathcal{O}\left(f(k+t,\mathcal{P}) \cdot n^{t+5} \cdot (\mathrm{poly}(n)^2)\right)$. The dependency on $\mathcal{P}$ and $t+k$ comes from Courcelle's theorem (Proposition 3), applied for deciding property $\mathcal{P}$ on graphs of treewidth $t + k$. As shown by Frick and Grohe [14], function f can be very huge, typically a tower of exponentials in $t + k$, the height of the tower depending on the property to be checked. Our algorithm actually constructs an induced subgraph of treewidth t, although we were only able to build a decomposition of width $t + k$. In particular, if we do not need to check a particular property $\mathcal{P}$ on the induced graph, but we only ask this graph to be of treewidth at most t, then function f becomes $(k+t)^{O((k+t)^3)}$ — coming from the algorithm of Bodlaender and Kloks [4] that, given a graph and tree decomposition of width $k + t$, checks whether the treewidth of the graph is at most t. For easier cases, when $t = 0$ to $t = 1$, the function f becomes 2^k and $(k + t)^{O(k+t)}$ respectively, as we shall discuss in the full version. Also, for natural properties $\mathcal{P}$, like "being connected" or "being a path" one can perform the property checking using standard (ad-hoc) dynamic programming tools which avoid the heavy machinery of Proposition 3. Again, the extra-cost becomes much more reasonable.

A natural and challenging question would be to separate the dependency on t and k, typically to obtain a complexity of type $f(t, \mathcal{P}) \cdot g(k) \cdot n^{t+\mathcal{O}(1)}$, where g would be a "more reasonable" function. For that purpose we would need to construct the partial solutions as a $(t + 1)$-terminal recursive graph, maybe by a more clever way of dealing with the intersection between this solution and the modulator.

Acknowledgements. We would like to thank Fedor Fomin and Nicolas Nisse for fruitful discussions on this subject.

References

1. Berry, A., Bordat, J.P., Cogis, O.: Generating all the minimal separators of a graph. Int. J. Found. Comput. Sci. **11**(3), 397–403 (2000)
2. Bodlaender, H.L.: A partial k-arboretum of graphs with bounded treewidth. Theor. Comput. Sci. **209**(1–2), 1–45 (1998)
3. Bodlaender, H.L.: Fixed-parameter tractability of treewidth and pathwidth. In: Bodlaender, H.L., Downey, R., Fomin, F.V., Marx, D. (eds.) The Multivariate Algorithmic Revolution and Beyond. LNCS, vol. 7370, pp. 196–227. Springer, Heidelberg (2012)
4. Bodlaender, H.L., Kloks, T.: Efficient and constructive algorithms for the path-width and treewidth of graphs. J. Algorithms **21**(2), 358–402 (1996)
5. Borie, R.B., Gary Parker, R., Tovey, C.A.: Automatic generation of linear-time algorithms from predicate calculus descriptions of problems on recursively constructed graph families. Algorithmica **7**(5–6), 555–581 (1992)
6. Bouchitté, V., Todinca, I.: Treewidth and minimum fill-in: grouping the minimal separators. SIAM J. Comput. **31**(1), 212–232 (2001)
7. Bouchitté, V., Todinca, I.: Listing all potential maximal cliques of a graph. Theor. Comput. Sci. **276**(1–2), 17–32 (2002)
8. Cameron, K., Hell, P.: Independent packings in structured graphs. Math. Program. **105**(2–3), 201–213 (2006)
9. Courcelle, B.: The monadic second-order logic of graphs. I. Recognizable sets of finite graphs. Inf. Comput. **85**(1), 12–75 (1990)

10. Courcelle, B., Engelfriet, J.: Graph Structure and Monadic Second-Order Logic. Cambridge University Press, Cambridge (2012)
11. Fomin, F.V., Liedloff, M., Montealegre, P., Todinca, I.: Algorithms parameterized by vertex cover and modular width, through potential maximal cliques. In: Ravi, R., Gørtz, I.L. (eds.) SWAT 2014. LNCS, vol. 8503, pp. 182–193. Springer, Heidelberg (2014)
12. Fomin, F.V., Todinca, I., Villanger, Y.: Large induced subgraphs via triangulations and CMSO. SIAM J. Comput. **44**(1), 54–87 (2015)
13. Fomin, F.V., Villanger, Y.: Finding induced subgraphs via minimal triangulations. In: STACS 2010, LIPIcs, pp. 383–394. Schloss Dagstuhl - Leibniz-Zentrum fuer Informatik (2010)
14. Frick, M., Grohe, M.: The complexity of first-order and monadic second-order logic revisited. Ann. Pure Appl. Logic **130**(1–3), 3–31 (2004)
15. Golumbic, M.C.: Algorithmic Graph Theory and Perfect Graphs. Academic Press, New York (1980)
16. Heggernes, P., van't Hof, P., Jansen, B.M.P., Kratsch, S., Villanger, Y.: Parameterized complexity of vertex deletion into perfect graph classes. Theor. Comput. Sci. **511**, 172–180 (2013)
17. Lagergren, J.: Upper bounds on the size of obstructions and intertwines. J. Comb. Theor. Ser. B **73**(1), 7–40 (1998)
18. Mancini, F.: Minimum fill-in and treewidth of split+ke and split+kv graphs. Discrete Appl. Math. **158**(7), 747–754 (2010)
19. Marx, D.: Parameterized coloring problems on chordal graphs. In: Downey, R.G., Fellows, M.R., Dehne, F. (eds.) IWPEC 2004. LNCS, vol. 3162, pp. 83–95. Springer, Heidelberg (2004)
20. Marx, D.: Chordal deletion is fixed-parameter tractable. Algorithmica **57**(4), 747–768 (2010)
21. Robertson, N., Seymour, P.D.: Graph minors. XX. Wagner's conjecture. J. Comb. Theor. Ser. B **92**(2), 325–357 (2004). Special Issue Dedicated to Professor W.T. Tutte

Author Index